# ANALOG VLSI
## Signal and Information Processing

# McGraw-Hill Series in Electrical and Computer Engineering

**Senior Consulting Editor**
*Stephen W. Director, Carnegie Mellon University*

**Circuits and Systems**
**Communications and Signal Processing**
**Computer Engineering**
**Control Theory**
**Electromagnetics**
**Electronics and VLSI Circuits**
**Introductory**
**Power and Energy**
**Radar and Antennas**

**Previous Consulting Editors**
Ronald N. Bracewell, Colin Cherry, James F. Gibbons, Willis W. Harman, Hubert Heffner, Edward W. Herold, John G. Linvill, Simon Ramo, Ronald A. Rohrer, Anthony E. Siegman, Charles Susskind, Frederick E. Terman, John G. Truxal, Ernst Weber, and John R. Whinnery

# ELECTRONICS AND VLSI CIRCUITS

**Senior Consulting Editor**
*Stephen W. Director, Carnegie Mellon University*

**Consulting Editor**
*Richard C. Jaeger, Auburn University*

# *ANALOG VLSI*
## Signal and Information Processing

## Mohammed Ismail
*Ohio State University*

## Terri Fiez
*Washington State University*

**McGraw-Hill, Inc.**

New York   St. Louis   San Francisco   Aukland   Bogotá   Caracas
Lisbon   London   Madrid   Mexico City   Milan   Montreal   New Delhi
San Juan   Singapore   Sydney   Tokyo   Toronto

The editor was George T. Hoffman;
the production supervisor was Diane Ficarra.
The cover was designed by Initial Graphic Systems, Inc.
R. R. Donnelley & Sons Company was printer and binder.

**ANALOG VLSI**
**Signal and Information Processing**

 This book is printed on recycled, acid-free paper containing a minimum
of 50% total recycled fiber with 10% postconsumer de-inked fiber.

1 2 3 4 5 6 7 8 9 0 DOC DOC 9 0 9 8 7 6 5 4

ISBN 0-07-032386-0

Library of Congress Catalog Card Number: 93-80913

# CONTENTS

## 3  Continuous-Time Signal Processing      57
*Mohammed Ismail, Shu-Chuan Huang, and Satoshi Sakurai*

## 4  Low-Voltage Signal Processing      135
*Johan Huijsing, Ron Hogervorst, Jeroen Fonderie,*
*Bernard Van den Dool, Klaas - Jan de Langen, and Gert Groenevold*

## 5  Basic BiCMOS Circuit Techniques      184
*Seyed Zarabadi, Mohammed Ismail, and Frode Larsen*

## 6  Current-Mode Signal Processing      248
*Terri Fiez, Russell Croman, Edmund Schneider, and Marius Goldenberg*

## PREFACE

This textbook presents the first comprehensive treatment of analog VLSI. It blends circuit design concepts used in traditional analog signal processing with design techniques used in emerging contemporary information processing applications. It is designed to serve as a textbook in senior- or graduate-level courses and as a reference book for practicing engineers. It is presumed that the reader is familiar with basic electronics, including biasing, modeling and analysis of electronic circuits, and has some familiarity with device physics and IC fabrication.

Chapter 1 discusses the renewed role of analog signal processing in the VLSI environment. Chapters 2-8 are primarily devoted to circuit design techniques at the transistor level while Chapters 9-13 are devoted to design issues at the system level. Chapters 14, 15, and 16 respectively, discuss statistical modeling and simulation, computer-aided design, and analog VLSI layout. The organization of this text is illustrated in Fig. 1.5 which draws the "big picture" of the analog VLSI design hierarchy and shows the correlation between the various book chapters.

Chapter 2 gives background material on modeling of the MOS transistor for modern circuit analysis. It introduces a graphical MOS model valid for both strong inversion and subthreshold operations and discusses the concept of current division in CMOS circuits. Furthermore, it presents the concept of the "super MOS" transistor and discusses its applications in the design of high-gain CMOS fully-differential operational amplifiers.

Chapter 3 covers the basic concepts of modern CMOS circuit design for analog VLSI continuous-time signal and information processing. Basic analog circuit techniques used in the design of voltage-to-current converters, analog multipliers, MOS resistors, and winner-take-all circuits are presented. The chapter concludes with simple amplifier-based analog building blocks using op amps, OTAs, and DDAs as basic circuit elements.

Chapter 4 moves to the emerging topic of low-voltage signal processing and discusses basic circuit techniques leading to the design of low-voltage op amps in bipolar, CMOS, and BiCMOS technologies. It concludes with applications in low-voltage, continuous-time filters with optimum dynamic range.

Chapter 5 is devoted almost entirely to BiCMOS analog circuits and signal processing. It discusses the advantages gained in using BiCMOS over bipolar and MOS technologies to design high performance analog circuits and offers many examples in the design of BiCMOS voltage-to-current converters, amplifiers, computational and telecommunication circuits. It concludes with an introduction to the translinear principle.

Chapter 6 is devoted to current-mode signal processing where currents rather than the conventional voltages are the input and output signal medium. Basic circuit techniques, both continuous-time and sampled-data, are discussed. Sampled-data current-mode circuits are called "switched-current" circuits to distinguish them from sampled-data switched-capacitor (voltage-mode) circuits which are covered in Chapter 9. Chapter 9 also discusses a systematic methodology to convert switched-capacitor circuits to switched-current.

Chapters 7 and 8 cover basic continuous-time CMOS analog building blocks for neural-based computational and information processing applications. The implementation of artificial neurons and neural networks for engineering applications are discussed. The chapters clearly demonstrate that analog VLSI has a tremendous potential to address real-world problems.

Chapter 10 discusses data converters where the emphasis is entirely devoted to the design aspects of one of the most important types of converters, namely oversampled data converters. It also demonstrates their use in multi-channel voice-band coders.

Chapter 11 covers silicon sensors which find many applications in modern instrumentation and control applications. Several types of sensors and their properties are discussed together with simple analog signal processing circuits for their interfacing to electronics.

Chapter 12 provides comprehensive coverage of basic techniques used in analog and mixed-signal design-for-test. It begins with a discussion on fault modeling and fault simulation and covers several techniques such as built-in self test, analog test busses, and design for electron-beam testability.

Chapter 13 covers the physics of interconnects in VLSI and describes a model for the estimation of circuit wiring density. It then presents a configurable architecture for the fast prototyping of analog VLSI circuits.

Chapter 14 covers the basic techniques used in the statistical modeling and simulation of analog VLSI circuits. It describes a statistical model to account for random device mismatches and discusses its implementation in a circuit simulator.

Chapter 15 gives a comprehensive treatment of CAD tools used in both analog circuit and automated layout design while Chapter 16 covers basic layout design techniques used in analog and mixed-signal VLSI chips.

Appendix A provides SPICE model parameters to assist with computer simulation which is used throughout the book, particularly with end-of-chapter problem sets.

All the chapters, except the last two, include a comprehensive set of problems which emphasize analysis, design, and applications. An instructor's manual containing solutions to these problems is available.

The entire book may be covered in a two-course senior/graduate level sequence on analog VLSI. In schools where the subject of analog VLSI is taught in one senior/graduate course, the teacher will find the coverage of material in the book flexible enough to assemble the course based on his/her school's needs.

# CONTRIBUTORS

**Christopher J. Abel** Solid - State Microelectronics Laboratory, The Ohio State University, Columbus, Ohio, USA

**Andreas G. Andreou** Department of Electrical and Computer Engineering, Johns Hopkins University Baltimore, Maryland, USA

**Henry Baltes** Institute of Quantum Electronics, Physical Electronics Laboratory, ETH, Zurich, Switzerland

**Kwabena A. Boahen** Computation and Neural Systems, California Institute of Technology, Pasadena, California, USA

**Klaas Bult** Philips Research Laboratories, Eindhoven, The Netherlands

**L. Richard Carley** Department of Electrical and Computer Engineering, Carnegie Mellon University, Pittsburgh, Pennsylvania, USA

**Joongho Choi** Department of Electrical Engineering, University of Southern California, Los Angeles, California, USA

**Russ Croman** Crystal Semiconductor, Austin, Texas, USA

**Terri Fiez** School of Electrical Engineering and Computer Science, Washington State University, Pullman, Washington, USA

**Igor Filanovsky** Department of Electrical Engineering, University of Alberta, Edmonton, Canada

**Jeroen Fonderie** Signetics Company, Sunnyvale, California, USA

**Umberto Gatti** Department of Electronics, University of Pavia, Pavia, Italy

**Marius Goldenberg** Crystal Semiconductor, Austin, Texas, USA

**Gert Groenewold** Delft University of Technology, Delft, The Netherlands

**Ron Hogervorst** Delft University of Technology, Delft, The Netherlands

**Shu - Chuan Huang** Solid - State Microelectronics Laboratory, The Ohio State University, Columbus, Ohio, USA

**Johan H. Huijsing** Delft University of Technology, Delft, The Netherlands

**Mohammed Ismail** Solid - State Microelectronics Laboratory, The Ohio State University, Columbus, Ohio, USA

**Klaas - Jan de Langen** Delft University of Technology, Delft, The Netherlands

**Hans G. Kerkhoff** Department of Electrical Engineering, University of Twente, Enschede, The Netherlands

**Frode Larsen** Solid - State Microelectronics Laboratory, The Ohio State University, Columbus, Ohio, USA

**Bosco Leung** Department of Electrical Engineering, University of Waterloo, Waterloo, Ontario, Canada

**Franco Maloberti** Department of Electronics, University of Pavia, Pavia, Italy

**Christopher Michael** National Semiconductor, Inc., Santa Clara, California, USA

**Gordon W. Roberts** Department of Electrical Engineering, McGill University, Montreal, Canada

**Satoshi Sakurai** Solid - State Microelectronics Laboratory, The Ohio State University, Columbus, Ohio, USA

**Edmund Schneider** School of Electrical Engineering and Computer Science, Washington State University, Pullman, Washington, USA

**Adel S. Sedra** Department of Electrical Engineering, University of Toronto, Toronto, Canada

**Bing J. Sheu** Department of Electrical Engineering, University of Southern California, Los Angeles, California, USA

**Massimo Sivilotti** Tanner Research, Inc., Pasadena, California, USA

**Bernard J. Van den Dool** Delft University of Technology, Delft, The Netherlands

**Seyed R. Zarabadi** Delco Electronics Corporation, Kokomo, Indiana, USA

## ACKNOWLEDGMENTS

To the many who have encouraged and assisted us, we offer our sincere thanks, especially Christopher Abel, Robert Brannen, Rajesh John, Frode Larsen, Satoshi Sakurai, and Thomas To of the Ohio State University, and Marius Goldenberg, Subba Somanchi, Aria Eshraghi, and Farbod Aram of Washington State University. We owe special thanks to Edmund Schneider, at Washington State University, for formatting the entire textbook and to Shu-Chuan Huang, at Ohio State University, for assembling the instructor's manual. We would also like to thank Ronny Khan of the Norwegian Institute of Technology for his help in assembling the chapters in the early stages. Oddvar Aaserud of the Norwegian Institute of Technology and Tor Lande of the University of Oslo were also very helpful while M. Ismail was in Norway on leave during 1992.

The book has benefited immesurably from careful reviews by Eric Donkor of the University of Conneticut, William Eisenstadt and Robert Fox of the University of Florida, Godi Fischer of the University of Rhode Island, Ron Gyurcsik of North Carolina State University, Ramesh Harjani of the University of Minnesota, Khalil Najafi of the University of Michigan, and Andy Robinson formerly of the University of Michigan now with Advanced Technology Laboratories, Jaime Ramirez-Angulo and Jay Steelman of New Mexico State University, and Yosef Shacham of Cornell University. The encouragement and support of Stephen Director of Carnegie Mellon University and Richard Jaeger of Auburn University have been very much appreciated.

Financial support from many organizations is greatly appreciated, especially the National Science Foundation, the Semiconductor Research Corporation, and the Norwegian NTNF.

The authors of Chapter 5 would like to thank David W. Stringfellow, Gregory J. Manlove, Roukoz M. Atallah and Abhi Chavan for proof-reading the manuscript and their helpful comments. They also thank Paul. J. Ainslie and Dave W. Osburn for the support and encouragement. The great work of Roger Poisson and John Frazee for processing the circuits is acknowledged.

The authors of Chapter 8 thank Professor Carver Mead for making them believe in the power of low energy analog computation and for his encouragement over the last 6 years. Robert Jenkins, Kim Strohbehn, Aleksandra Pavasovic, Marc Cohen, Philippe Pouliquen, Weimin Liu, Kewei Yang, Paul Furth, Richard Meitzler, Mark Martin at Jonhs Hopkins have influenced the design of the integrated circuits and systems. At Caltech, they thank Tobi Delbruck for providing them with the scanner circuitry and for helpful discussions on the "Bump" circuits. Thanks also goes to Ron Benson for fruitful discussions on current–mode signal representations and neural network architectures. The circuit model of diffusion using only one device type was suggested to KAB by Carver Mead.

The author of Chapter 10 would like to acknowledge the contributions of Feng Chen, Seyed Moussavi, and Charlie Ma toward the problem sets and solutions.

The author of Chapter 12 would like to thank Mr. A. Haggenburg of Philips

Components in Hijmegen, the Netherlands, for providing an example of a complex industrial mixed-signal chip incorporating analog DFT, and Mr. K. Lutz of Advantest System Engineering GmbH (ASEG) in Munich, Germany, for general support.

Christopher Abel, one of the co-authors of Chapter 13, would like to acknowledge The Ohio Aerospace Institute for their encouragement, and for their financial support.

Finally, all the contributors would like to thank our families for their understanding and support.


M. Ismail, Columbus, Ohio
T. Fiez, Pullman, Washington
November 8, 1993

# CHAPTER

# 1

# INTRODUCTION
# TO ANALOG VLSI

Analog signal processing is showing signs of dramatic change. This is motivated by advances in technology, increased market demands, and sophisticated and innovative signal and information processing applications. This chapter will discuss the renewed role of analog signal processing in the VLSI environment. An interdisciplinary view of VLSI design in general, and of analog VLSI design in particular, is presented. Two examples of VLSI chips which combine analog and digital circuits will be given together with a general description of the topics covered in the book.

## 1.1  VLSI MICROELECTRONICS

Prior to the 1970's, electronic circuits and systems were designed almost exclusively using analog design techniques implemented with discrete components. With the advent of integrated circuits, digital circuits have become the basis of many systems of today. As a result, the development of analog integrated circuits (ICs) has been rather slow compared to digital ICs. In addition to the knowledge-intensive nature of analog design, very-large-scaled-integrated (VLSI) technologies have primarily been developed to meet the needs of digital computation and storage. Early attempts to integrate analog designs have been in hybrid, bipolar, and single-channel Metal-oxide semiconductor (MOS) technologies. In the past decade, rapid advances in silicon VLSI technology, the maturing of digital IC design, and increased market demands have created the need for more

1

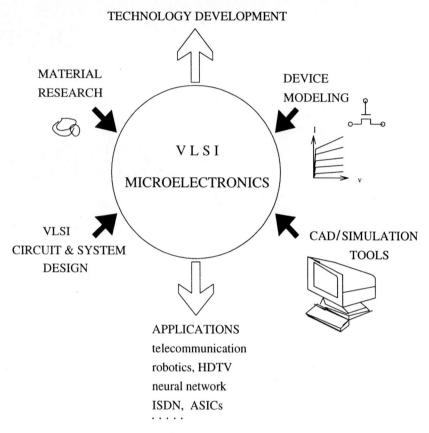

**FIGURE 1.1**
Components of VLSI microelectronics.

integration. The ultimate goal is to realize a system-on-chip with mixed analog and digital circuitry. More recently, attention has been focused on scaled complementary MOS (CMOS), and Bipolar-CMOS (BiCMOS) technologies, where system integration is showing unprecedented success. In CMOS technologies, both n- and p-channel MOS devices are fabricated on the same substrate. BiCMOS technologies combine both bipolar and MOS devices on the same substrate.

VLSI microelectronics, in the context of modern technologies, is indeed an interdisciplinary field, as illustrated in Fig. 1.1. Microelectronic devices are developed and modeled as a result of research studying the electrical, thermal, magnetic, and other properties of materials. Devices are then used in very large numbers to build VLSI circuits and systems. The only meaningful way to complete a design of a complex VLSI system in a reasonable time is to use computers in simulating, optimizing, and actually designing both the circuit and the layout.

The four inputs in Fig. 1.1, namely materials, devices, circuits, and computer-aided design (CAD), constitute the components of VLSI microelectronics. New technologies and innovative applications continue to emerge as a result of successful interdisciplinary efforts combining these four areas. If one claims to be a specialist in any one of these four areas, it is becoming crucial these days to also acquire minimum knowledge of the other three areas. For instance, useful CAD tools can only be developed if needs of circuit designers are successfully identified. High speed circuits, which are expected to be a major trend in the 1990's and beyond, can perform well only if successful device design and modeling is achieved, and so on. The VLSI designer should have at least a minimum knowledge of device characteristics and CAD tools performance. The designer would then be able to optimize the use of the device in building a circuit, to efficiently use CAD tools, and to identify the needs for new CAD tools. The synergies between circuits, CAD, and device design are becoming very important particularly for VLSI technologies that will soon reach their feature size limits.

## 1.2   MIXED-SIGNAL VLSI CHIPS

The role of analog VLSI signal processing has changed dramatically in recent years. In fact, we can now recognize two general categories of signal processing systems where both analog and digital circuits are integrated on the same chip, and we classify them here with respect to the roles analog and digital play on the chip.

### 1.2.1   Mixed Analog/Digital

These are mixed-signal chips where the analog part provides the I/O interface to the core of the chip which is digital. In addition to data converters, the interface circuits may include amplifiers, filters, compressors, power drivers, sensor interfaces, etc. These circuits are mainly located at the periphery of the chip, but their performance is critical to the overall system function. As an example, Fig. 1.2 shows the analog front end (AFE) of the U-Interface transceiver in an Integrated Service Digital Network or ISDN [1].

This interface is the final link between the digital world of ISDN, and the end user. In addition to data converters, the AFE contains a pulse density modulation (PDM) signal generator, a line driver, an adaptive echo precancellor or hybrid balance filter, an anti-aliasing filter, a low pass receive filter, and an adaptive gain control (AGC). The role of the AFE is crucial in that it is tailored to improve the far end signal-to-echo ratio, achieve higher immunity to noise, and allow baud-rate sampling. It must also handle nonlinear echo cancellation, and share with the digital the burden of overcoming the jitter-induced echo problem.

### 1.2.2   Mixed Digital/Analog

These are mixed-signal chips where the digital part provides the I/O interface to the core of the chip which is analog. In other words, analog and digital interchange their traditional roles. This is an emerging category in VLSI design, and it

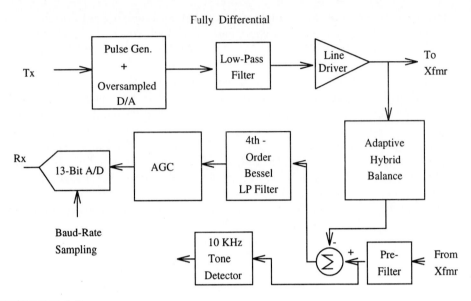

**FIGURE 1.2**
U-Interface analog front end for ISDN [1].

is used in the implementation of low-precision, massively parallel interconnected systems based on cognitive neural paradigms [2,3]. These systems or networks can also be implemented with purely analog VLSI circuits [3,4]. The digital part is primarily used to facilitate handling of the increased complexity of the system I/O. It is interesting to note that although technology scaling provides an increasing advantage to digital circuitry over analog circuitry to implement signal processing circuits, analog will remain the best candidate [5] for the VLSI implementation of these low-precision artificial neural networks. In these networks, the analog components are often different from those used in mixed analog/digital systems discussed earlier. They include four-quadrant multipliers, vector/matrix multipliers, programmable synaptic elements, winner-take-all (WTA) circuits, analog memories, comparators, etc. MOS transistors used in these circuits may be biased in strong or weak (subthreshold) inversion.

As an example, Fig. 1.3 shows the block diagram of a mixed digital/analog VLSI neural processor for high speed image compression [2] which can be used in many applications such as HDTV, teleconferencing, facsimile, data-base management, etc. The analog part, which contains input neurons, programmable synapses, summing neurons, and winner-take-all circuits, performs massively parallel neural computation while the digital part processes multiple-bit address information. The chip [2], see Fig. 1.4, was built in a 2.0 $\mu$m scalable CMOS process, and provides a computing capability as high as 3.2 billion connections.

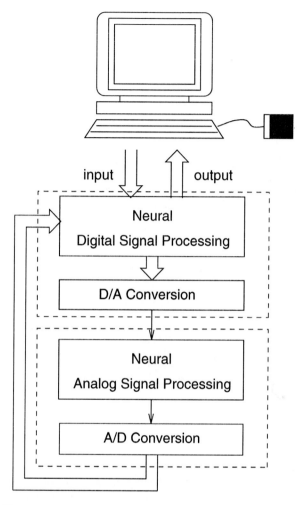

**FIGURE 1.3**
A mixed-signal neural processor for image data compression [2]. The digital part is a VQ
codebook generator and consists of a host interface and timing control, a vector address handler,
an input vector, a codebook, and a self-organizing trainer. The analog part is a neural-based
vector quantizer and consists of a synapse matrix, current summing neurons, sample-and-hold
circuits, WTA cells, and a digital encoder.

## 1.3 POTENTIAL OF ANALOG VLSI

The interdisciplinary view of VLSI discussed earlier is particularly important
for analog design since VLSI technologies and CAD tools have, for the most
part, been developed primarily for digital VLSI [6]. One should be able to think
"integrated" rather than "discrete." It is important to think of analog design
techniques or methodologies which make the use of a CAD environment much
easier for analog. For instance, the more configurable and versatile an analog

**FIGURE 1.4**
A die photo of the neural-based vector quantizer chip (©1992 IEEE).

cell is, the more it becomes like a digital cell. Design of primitive or well-defined reconfigurable, programmable, input-output-compatible analog cells must be encouraged. These cells can be interconnected to achieve different linear or nonlinear functions. Examples of "CAD-compatible" [7,8] analog circuit design start to emerge in the literature. A class of Field Programmable Analog Arrays (FPAAs) has recently been reported [9] which attempts to reap the advantages of methods used in Field Programmable Digital Arrays. These efforts should bridge the gap in analog work which exists between CAD- and circuit-design-oriented efforts and should eventually bring analog design a step closer toward automation.

While digital technology is still very much in its heyday, there is tremendous potential in analog VLSI for addressing real-world problems. As discussed earlier, analog VLSI is well suited for construction of artificial neurons, the functional units of the brain. This means that with analog VLSI, a chip can follow the brain's lead with exciting new information processing applications in a variety of areas such as image processing, speech recognition, associative recall, and hand writing recognition.

Mead [3] has built a family of silicon retinas. Each silicon retina is a VLSI

· **TABLE 1.1**
Comparison of analog VLSI with digital VLSI in terms of cost, power, and computational density. (©1992 BYTE)

| ANALOG VLSI VS. DIGITAL VLSI | | | | | |
|---|---|---|---|---|---|
| *Analog VLSI is strikingly superior to digital technology in terms of cost, power, and computation density. ( Estimated by Federico Faggin.)* | | | | | |
| | Cost (MCS*/$) | | Power (MCS/watt) | | Computation density (MCS/ft.$^3$) |
| | 1991 | 2000 | 1991 | 2000 | 1991 | 2000 |
| Conventional digital | 0.002 | 0.1 | 0.1 | 10 | 0,2 | 10 |
| Special-purpose digital | 0.1 | 4 | 10 | 10,000 | 10 | 1,000 |
| Dedicated digital | 5 | 200 | 500 | 50,000 | 40 | 3,000 |
| Dedicated analog | 500 | 20,000 | 50,000 | 5,000,000 | 4,000 | 4,000,000 |
| Human brain | $10^{9**}$ | | $10^{10}$ | | $10^{11}$ | |

\* MCS=A million connection updates per second.
\*\* This calculation assumes that the cost of a human brain is $10,000,000.

chip that is a square centimeter in area, weighs about a gram, and consumes about a milliwatt of power. Between arrays of photo-transistors etched in silicon, dedicated circuits execute smoothing, contrast 0, and motion processing. The transistors within the chip operate in the subthreshold (analog mode) region.

Compared with a typical CCD (charged-couple device) camera and standard digital image processor, the Mead chip is a paragon of efficiency in performance, power consumption, and compactness. A special-purpose digital equivalent would be about the size of a standard washing machine. Unlike cameras that must time sample, typically at 60 frames per second, the analog retina works continuously without needing to sample until the information leaves the chip already preprocessed.

Operations performed with Mead's chip capture some of the functions that real retinas perform; however, real retinas contain many more circuits than Mead's synthetic one. While it makes sense to build chips to maximize efficiency in the three critical elements (i.e., power, cost, and density), we must still push analog VLSI techniques a long way to approximate neural efficiency [10]. The incentive to go forward with this technology will depend on whether the payoff looks promising in the long term (see Table 1.1).

## 1.4 THE BOOK FLOOR PLAN

The objective of this book is to address many of the important areas of analog VLSI for signal and information processing. A "floor plan" for the book is shown in Fig. 1.5. Topics covered in the book range from circuit design (Chapters 2-8) to system-level design (Chapters 9-13) and include statistical modeling and

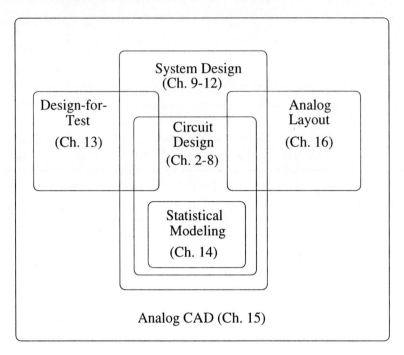

**FIGURE 1.5**
A "floor plan" of the book chapters. The large box represents the entire analog IC where CAD is used to complete the different tasks involved in the design process.

simulation of analog VLSI circuits (Chapter 14), analog CAD (Chapter 15) and analog IC layout techniques (Chapter 16). Below is a brief description of some of the pressing issues in analog VLSI in general and the topics which are addressed in this text.

- Analog circuit design in emerging technologies such as fine-line CMOS, Bi-CMOS, GaAs/silicon, heterojunction, floating-gate, and superconducting technologies. The goal is primarily to extend the performance of analog circuits to meet higher system demands in the marketplace. In this regard, fine-line CMOS and BiCMOS will emerge as the dominant technologies. The floating-gate technology will provide a means for the implementation of extremely high-resolution analog ICs. GaAs and $Si_xGe_{1-x}$ devices and perhaps the two-dimensional electron GaAs FET devices, such as the MODFET, will have specific applications particularly in optoelectronic systems. In this text emphasis is placed on the design of fine-line CMOS and BiCMOS analog ICs (Chapters 2-8).

- Current-mode and low-voltage signal processing is another important area of activity which attempts to circumvent the problem of limited dynamic range of analog circuits in scaled low-voltage (logic) technologies. The development of integrated circuit design techniques which process currents rather than voltages will constitute a major activity in the future (Chapter 6). The design of low-voltage low-power analog circuits in bipolar, CMOS, and BiCMOS technologies will also receive considerable attention (Chapter 4).

- Analog artificial VLSI neural networks, and mixed digital/analog information systems are other areas that have witnessed significant growth. In the future, more emphasis will be given to application-driven analog VLSI neural networks which take advantage of new circuit design concepts, CAD design tools, and emerging VLSI technologies. Chapters 7 and 8 present a practical view of Neuromorphic computing and implementation with Chapter 13 describing analog interconnect and system-level chip prototyping strategies.

- Analog interface systems for high performance mixed analog/digital VLSI. High performance digital radio/TV and other advanced communication system designs will be made possible with the development of high performance analog interface systems. Further advancement in time-interweaved, parallel-series, and high order sigma-delta converters would certainly help to clear the path. Chapter 9 and 10 deal with sampled-data systems and oversampled data converters, respectively. These may be used to interface to silicon sensors which are presented in Chapter 11.

- With the increased complexity of analog ICs and the introduction of analog VLSI systems, analog design-for-test or analog testability will receive more attention. Analog testable design methodologies adapted from the digital VLSI realm, such as functional and structural test, are now being developed for analog. Chapter 12 carefully reviews the analog design-for-test methods.

- Modeling and simulation which combine device/circuit/process issues so that non-ideal effects and VLSI technology limitations are incorporated into the design process. Due to high demands for continuous performance improvement, advanced devices are often used despite their problems. Therefore, it becomes absolutely essential to carefully model devices and simulate circuits, particularly using statistical techniques, to assure high yields. In addition to device mismatches, advanced small-features-size CMOS processes suffer from hot electron, and impact ionization caused by the channel high vertical field. These problems are detrimental to the yield and must be predicted and monitored. Chapter 14 covers in detail how statistical modeling can be used for analog circuit design that takes these characteristics into account.

- CAD tools for analog and mixed-signal VLSI, mixed-mode simulators, analog layout tools, automatic synthesis tools, and the development of analog hardware description languages are all important for analog design. Industrial competition and rapid changes in system needs create a desperate need for smart and dependable CAD tools, especially for mixed-signal ICs. A general description of analog CAD tools is given in Chapter 15, and layout design of analog and mixed-signal VLSI is the subject of Chapter 16. Appendix A provides CMOS and Bipolar transistor SPICE models. These models are used in conjunction with the end-of-chapter problems. The reader may also use them to reinforce understanding of many circuits discussed in the book.

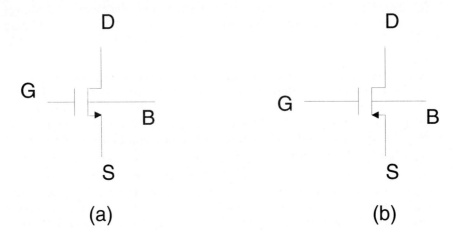

**FIGURE 1.6**
(a) n-channel MOS (NMOS) and (b) p-channel MOS (PMOS) symbols used when the bulk connection (B) is nonstandard. The source arrow may be omitted if the device operates in the linear region and is biased at $V_{DS} = 0$. In this case, drain and source terminals interchange. Arrows may also be used at the bulk (B), instead of the source (S). In this case, an arrow pointing inward is used for n-channel devices and output for p-channel devices.

## 1.5   NOTATION AND SYMBOLOGY

The conventions used in this book are consistent with those used in undergraduate electronic texts and with standards proposed by technical societies. The International System of Units has been used throughout. Every effort has been made to maintain consistent use of notations and symbols.

Unless otherwise stated, bias or dc quantities, such as supply voltage $V_{DD}$, are represented by uppercase symbols with uppercase subscripts. Small-signal quantities and elements of a small-signal equivalent circuit, such as the incremental change in a transistor drain current, $i_d$ and small-signal transconductance, $g_m$, are represented by lowercase symbols with lowercase subscripts. Finally quantities such as total drain current $I_d$, which represents the sum of the bias and the signal, are represented by an uppercase symbol with lowercase subscript.

Transistor symbols will already be familiar to the reader. In CMOS technology, devices of one type or another are fabricated in individual, separate isolation regions, which may or may not be connected to the appropriate power supply. If these isolation regions (wells) are connected to the appropriate power supply, the substrate connection will not usually be shown on the circuit diagram. If the individual isolation regions are connected elsewhere, the devices will be represented by the symbols of Fig. 1.6. Occasionally MOS symbols which distinguish the PMOS device with a bubble at its gate will be used in conjunction with circuits operating in subthreshold.

# REFERENCES

[1] R. Batroni, P. Lemaitre, and T. Fensch, "Mixed Digital/Analog Signal Processing for Single-chip 2B1Q U-Interface Transceiver," *IEEE J. Solid-State Circuits*, vol. 25, pp. 1414-1425, December 1990.

[2] W.C. Fang, B.J. Sheu, O. Chen, and J. Choi, "A VLSI Neural Processor for Image Data Compression Using Self-Organizing Networks," *IEEE Trans. Neural Networks*, vol. 3, pp. 506-518, May 1992.

[3] C. Mead, *Analog VLSI and Neural Systems*, Addison-Wesley, 1989.

[4] C. Mead and M. Ismail, *Analog VLSI Implementation of Neural Systems*, Kluwer Academic Publishers, Norwood, Mass., 1989.

[5] E. A. Vittoz, "Future of Analog in the VLSI Environment," Proc. ISCAS, pp. 1372-1375, May 1990.

[6] P.E. Allen, "Future of Analog Integrated Circuits," Ch-18 in *Analog IC Design: The Current-mode Approach*, Toumazou, Lidgey and Haigh, Eds. Peter Peregriuns Ltd., London, 1990.

[7] M. Ismail and S. Bibyk, "CAD Latches onto New Techniques for Analog ICs," IEEE Circuits and Devices Magazine, pp. 12-17, September 1991.

[8] M. Ismail and D. Evans, "Recent Advances in Analog VLSI," Proc. 15th Nordic Semiconductor Conference, pp. 305-312, Helsinki, June 1992.

[9] E.K.F. Lee and P.G. Gulak, "Field Programmable Analog Array Based on MOSFET Transconductors," Elect. Letters, vol. 28, pp. 28-29, January 1992.

[10] T.J. Sejnowski and P.S. Churchland, "Silicon Brains" in BYTE Magazine, pp. 137-146, October 1992.

# CHAPTER
# 2

# BASIC CMOS
# CIRCUIT
# TECHNIQUES

## 2.1 INTRODUCTION

The number of analog circuits designed in standard CMOS processes has grown enormously in the past two decades with one of the major driving forces being the need for one-chip solutions for systems with combined analog and digital circuitry. As most analog circuits have historically been designed in bipolar processes, many analog designers have strong bipolar circuit backgrounds. Because of this background, they tend to try and fit CMOS transistors into a bipolar model, which may produce non-optimal MOS designs. As a result, the MOS transistor has been viewed as a slow and noisy device with bad current drive capabilities. Good analog MOS design takes "MOS thinking" and a proper understanding of the MOS device operation is of utmost importance. This requires a different view of the MOS transistor. Characteristics such as the continuously changing threshold voltage along the channel must be incorporated into a fully symmetrical expression for the current-voltage relationship with respect to the source and drain terminals. In this chapter, rather than giving an overview of the latest CMOS circuit tricks, the fundamentals of MOS transistor operation and the basic gain-stage design are presented. Recently developed design techniques are discussed as an example of applying these basics to practical applications.

## 2.2   MOS MODELS

In order for the MOS transistor model to be useful to the circuit designer, simplified expressions are necessary. In this section, we present a general expression
for the current flowing through the MOS device.

This expression is graphically represented to provide insight into circuit
operation and it is useful for designing and understanding MOS circuits [1]. The
model expression describes the drain current of an MOS transistor as a function
of the four terminal voltages. All potentials are referenced to the substrate potential instead of the source potential, which emphasizes the symmetry between
source and drain terminals, both in the model expressions and in the graphical
representations.

### 2.2.1   General MOS Transistor Models

Figure 2.1 shows a cross section of an MOS transistor. For purposes of clarity, we
will confine the explanation to n-channel MOS transistors, but the model applies
equally well for p-channel MOS transistors. We will start by defining the channel
voltage $V_C(x)$ at a certain position $x$ in the channel. This channel voltage $V_C(x)$
is an imaginary voltage which varies along the channel from the source voltage $V_S$
$(x = 0)$ to the drain voltage $V_D$ $(x = L)$. The voltage in the channel at position
$x$ can be measured externally at the source and drain diffused regions via a
silicon probe with a similar doping profile and metal contacts. As all internal
contact potentials are included in the probe construction, we can describe the
voltage-current relations in the MOS transistor with external terminal voltages.
The current at an arbitrary location $x$ in the channel can be caused by both drift
and diffusion. If the inversion layer current at $x$ is denoted by $I(x)$, we will have
[12]:

$$I(x) = I_{drift}(x) + I_{diff}(x) \tag{2.1}$$

The current due to drift is proportional to the local channel charge density $Q_C$,
the electron mobility $\mu$, the local electric field along the channel $dV_C/dx$, and
the channel width $W$:

$$I_{drift}(x) = -W\mu Q_C \cdot (dV_C/dx) \tag{2.2}$$

The current due to diffusion is proportional to the electron mobility, $\mu$, the
channel width, $W$, the thermal voltage, $kT/q$, and the derivative of the charge
density, $Q_C$, with respect to the position $x$:

$$I_{diff}(x) = W\mu(kT/q)(dQ_C/dx) \tag{2.3}$$

Substitution of (2.2) and (2.3) into (2.1) yields:

$$I(x) = W\mu[-Q_C(dV_C/dx) + (kT/q)(dQ_C/dx)] \tag{2.4}$$

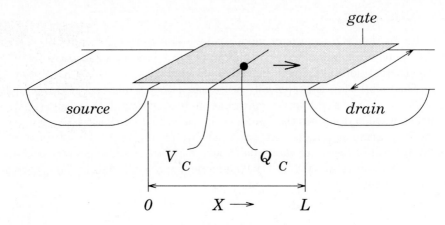

**FIGURE 2.1**
A cross-view of an MOS transistor.

As the current is constant along the channel (quasi steady state), integration from source to drain leads to the following equation:

$$IL = W \int_{x=0}^{x=L} -\mu \left[ Q_C (dV_C/dx) - (kT/q)(dQ_C/dx) \right] dx \qquad (2.5)$$

Changing the running variable to the channel voltage, $V_C$, and dividing by $L$ yields the following expression for the drain current, $I_D \ (= -I)$:

$$I_D = (W/L) \int_{V_C=V_S}^{V_C=V_D} f(V_G, V_C) dV_C \qquad (2.6)$$

with

$$f(V_G, V_C) = -\mu Q_C + \mu(kT/q)(dQ_C/dV_C) \qquad (2.7)$$

The expression clearly shows the symmetry of the MOS device; interchanging the source and the drain results in a drain current of equal magnitude but of opposite direction to the source current. This expression can be scaled by adjusting the device dimensions $W$ and $L$. By first-order approximation, this expression is valid in all operation regions: in strong as well as weak inversion and in the saturated as well as the non-saturated regions! The integrand can be quite a complicated function of $V_G$ and $V_C$ when including characteristics such as mobility reduction and body-effect. In general though, the accuracy of this expression can be very good, especially in the linear region also called the ohmic or the triode region. The influence of the drain on the device current in saturation however, due to channel-length shortening and drain-induced barrier lowering, are poorly modeled.

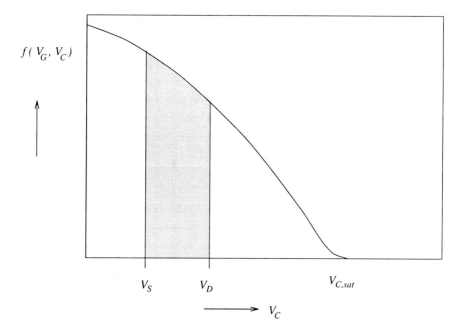

**FIGURE 2.2**
The function $f(V_G, V_C)$ as a function of the channel voltage $V_C$. The MOS transistor current is proportional to the shaded area, with W/L being the proportionality constant. The MOS transistor operates in the linear region.

### 2.2.2   Graphical Representation of the Model

A graphical representation of (2.6) is shown in Fig. 2.2. The curve shows the function $f(V_G, V_C)$, for a certain value of $V_G$, as a function of the channel voltage $V_C$. According to (2.6), integration of this function, by the channel voltage $V_C$ running from the source voltage, $V_S$, to drain voltage, $V_D$, and multiplying the result with the $W/L$ ratio, yields the drain current, $I_D$. Therefore, apart from the factor $W/L$, the shaded area represents the current through the MOS device. This graphical representation gives us a clear view of what happens to the drain current if the terminal voltages change. Clearly the transistor becomes saturated as the drain voltage becomes larger than $V_{Csat}$.

In the first order approximation, assuming strong inversion operation (i.e. neglecting the second term in (2.7)), the function $f(V_G, V_C)$ in Fig. 2.2 would be a straight line, as shown in Fig. 2.3. Integration as done in (2.6) leads to an ideal quadratic voltage-current relationship. This approach yields a simple MOS model for hand calculation. The bending of the $f(V_G, V_C)$ curve toward the lower values of $V_C$ is caused primarily by mobility reduction.

The deviation from the straight line in the vicinity of $V_C = V_{Csat}$ is caused by weak inversion. In this regime, the function $f(V_G, V_C)$ is an exponential function, leading to an exponential voltage-current relationship. Figure 2.4 shows an enlargement of Fig. 2.2 around $V_C = V_{Csat}$ and again, apart from the factor

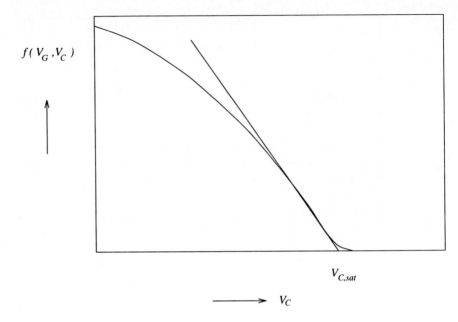

$f(V_G, V_C)$

$V_C$

$V_{C,sat}$

**FIGURE 2.3**
A first-order approximation of the function $f(V_G, V_C)$. The deviation from the straight line for the smaller values of $V_C$ is caused by mobility reduction mainly, whereas the deviation from this line around $V_C = V_{C\,sat}$ is caused by weak inversion.

$W/L$, the shaded area represents the current through the MOS transistor, now operating in weak inversion.

Although it is easy to see when the transistor becomes saturated, the dependence of the drain current in saturation on the drain voltage cannot be modeled in this way. No channel-shortening, static feedback, drain-induced barrier lowering, weak avalanche, or other drain dependent effects on the saturated drain current are modeled.

### 2.2.3   A Simplified Graphical Model

In this section we begin with (2.6) and derive a simplified expression for the drain current in strong inversion. For this purpose, we will use the first term of (2.7) only and fill in expressions for $\mu$ and $Q_C$ as a function of $V_G$ and $V_C$. As in most hand formulas, we will neglect mobility dependence on $V_G$ and $V_C$ and thus:

$$\mu = \mu_o \tag{2.8}$$

Conforming to the classical derivation of the MOS transistor current-voltage relationship [2], the inversion charge density is expressed as:

$$Q_C(V_G, V_C) = -[V_G - V_{TB}(V_C)]C_{ox} \tag{2.9}$$

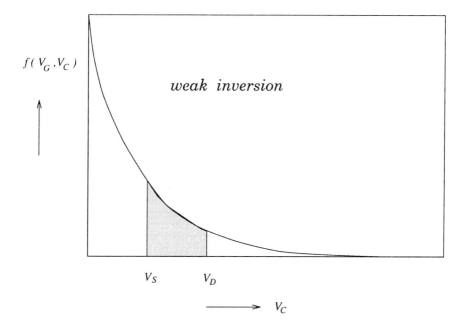

$f(V_G,V_C)$

*weak inversion*

$V_S$    $V_D$

$V_C$

**FIGURE 2.4**
An enlargement of Fig. 2.2 around $V_C = V_{Csat}$, showing weak inversion operation. In this region, the function $f(V_G, V_C)$ is an exponential function of $V_C$.

in which $V_{TB}(V_C)$ is the threshold voltage as a function of the channel voltage, $V_C$, referenced to the substrate potential. The threshold voltage, $V_{TB}$, defines the gate voltage, referred to the substrate potential, at the onset of strong inversion and is a function of the channel voltage, $V_C$. The first-order MOS model used in many classical MOS textbooks [2] neglects the increase in depletion region charge with increasing source-substrate voltage, i.e. the body effect. Here we will approximate the body effect by a linear function:

$$V_{TB}(V_C) = V_{TO} + \alpha V_C \qquad (2.10)$$

with $V_{TO} = V_{TB}(0)$. This yields a much better approximation without adding too much computational complexity. In modern MOS processes $\alpha$ is between 1.05 and 1.35 obtained from process measurements. If we do not want to include the body effect, we have to take $\alpha = 1$ (and not $\alpha = 0$, as all voltages are referred to the substrate potential). Substituting (2.8), (2.9), and (2.10) into (2.7) and using this in (2.6) yields:

$$I_D = \mu_o C_{ox}(W/L) \int_{V_C=V_S}^{V_C=V_D} (V_G - V_{TO} - \alpha V_C)dV_C \qquad (2.11)$$

More explicitly, we can construct a graphical representation of (2.11). In Fig. 2.5, the threshold voltage, $V_{TB}$, according to (2.10) is shown as a function of the channel voltage, $V_C$. It is a straight line starting at $V_C = 0$ with $V_{TB} =$

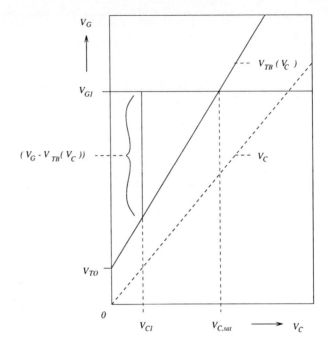

**FIGURE 2.5**
The gate voltage $V_G$ and the threshold voltage $V_{TB}$ as a function of the channel voltage $V_C$. The difference between the two curves, as indicated by the arrow, equals the amount of overdrive voltage, i.e., the effective voltage across the oxide capacitance which determines the charge density in the channel.

$V_{TO}$ and a slope of $\alpha$. Also shown in this figure is the gate voltage $V_G$. It is a straight horizontal line, intersecting the y-axis at $V_{G1}$. At a given channel voltage $V_{C1}$, the difference between these two functions equals the integrand function $(V_{G1} - V_{TO} - \alpha V_{C1})$ of (2.11). In Figs. 2.6 and 2.7, the terminal voltages $V_S$ and $V_D$ are indicated along the horizontal axis. Hence, the shaded area equals the integral of $(V_G - V_{TO} - \alpha V_C)$ from $V_S$ to $V_D$. Therefore, apart from the factor $\mu_o C_{ox}(W/L)$, the shaded area equals the total amount of drain current. In the linear region ($V_D < V_{Csat}$), the shaded area is trapezoidal (Fig. 2.6), whereas in the saturated region ($V_D > V_{Csat}$), $V_{Csat}$ should be taken as the upper bound and the area becomes triangular (Fig. 2.7). The drain saturation voltage, $V_{Csat}$, is the channel voltage, $V_C$, for which the channel pinches and $(V_G - V_{TO} - \alpha V_C)$ becomes equal to zero:

$$V_{Csat} = (V_G - V_{TO})/\alpha \tag{2.12}$$

We can also evaluate (2.11) analytically. In the linear region, this leads to:

$$I_D = (K/2\alpha)[(V_G - V_{TO} - \alpha V_S)^2 - (V_G - V_{TO} - \alpha V_D)^2] \tag{2.13}$$

with

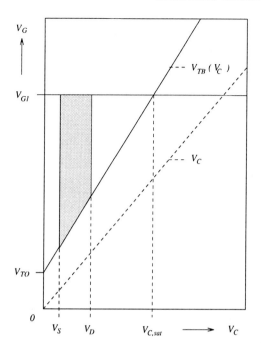

**FIGURE 2.6**
The MOS transistor current is proportional to the shaded area, with $K = \mu C_{ox}(W/L)$ being the proportionality constant. The MOS transistor operates in the linear region.

$$K = \mu_o C_{ox}(W/L) \tag{2.14}$$

Equation (2.13) clearly shows the symmetry of the device between the source and drain terminals. The drain current is written as a difference of two quadratic expressions. For $V_D > V_{Csat}$, the channel pinches off and $V_D = V_{Csat}$ should be substituted into (2.13), reducing it to:

$$I_D = (K/2\alpha)(V_G - V_{TO} - \alpha V_S)^2 \tag{2.15}$$

If all the voltages are referred to the source, the drain current can be rewritten as:

$$I_D = (K/2\alpha)(V_{GS} - V_T)^2 \tag{2.16}$$

where $V_T$ is the threshold voltage at the source node referred to the source potential, and is given by:

$$V_T = V_{TB}(V_S) - V_S$$
$$= V_{TO} + (\alpha - 1)V_S \tag{2.17}$$

If we substitute $\alpha = 1$ into (2.13) and (2.15), the expressions reduce to the classical first-order MOS transistor model [2].

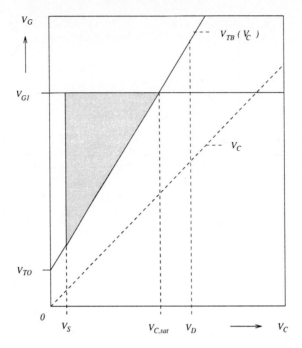

**FIGURE 2.7**
The MOS transistor current is proportional to the shaded area, with $K = \mu C_{ox}(W/L)$ being the proportionality constant. The MOS transistor operates in the saturated region.

### 2.2.4 A Simplified Model for the Weak-Inversion Region

As in the previous section, we begin with (2.6) and derive a simplified expression for the drain current, now for the weak-inversion region. In this region the second term of (2.7) dominates. Using this term only yields together with (2.6):

$$I_D = -\frac{W}{L}\int_{Q_C=Q_S}^{Q_C=Q_D} \frac{\mu kT}{q}dQ_C \tag{2.18}$$

With $Q_S$ and $Q_D$ being the charge-density at the source and drain end of the channel, respectively. In [1] we find the following expression for the channel-charge density:

$$Q_C = Q_{C0}\exp^{q(V_G-V_C)kT} \tag{2.19}$$

with $Q_{C0} = Q_C$ at $V_G = \alpha V_C$. Using this in (2.18) results in:

$$I_D = \frac{W}{L}I_{D0}(\exp^{q(V_G-V_S)kT} - \exp^{q(V_G-V_D)kT}) \tag{2.20}$$

with

$$I_{D0} = \mu\frac{kT}{q}Q_{C0} \tag{2.21}$$

Again we observe the symmetry of the device between the source and the drain terminals. Figure 2.4 gives a graphical representation of this expression.

### 2.2.5 Merits and Shortcomings of the Models

The graphical representations include the dependence of the drain current on all the MOS transistor terminal voltages. The effect of a change in one of the terminal voltages is clearly visualized. The symmetry of the device between the source and drain is also strikingly clear. Interchanging the source and drain terminals has no effect on the model behavior (but keep in mind how we define the drain current: if the source and drain are interchanged, the shaded area represents the negative value of the drain current). As the body effect is also included, we can study its influence on the behavior of our circuits. The graphical representation clearly visualizes the changing threshold voltage, going from source to drain. But moreover, note that without linearizing the $V_{TB}(V_C)$ characteristic as is done in (2.10), the same graphical representation may still be used.

If however, we want to also include the mobility reduction effect, we have to take the approach given in Sec. 2.2.2. Analytically, we would have to solve (2.6) and (2.7) in which $f(V_G, V_C)$ may be a complicated function. This would (if solvable) lead to very complicated device expressions which do not contribute to our understanding of the circuits. Using the graphical representation as shown in Fig. 2.2, we are able to gain insight into this nonlinear effect as well as when the transistor becomes saturated. However, the dependence of the drain current in saturation on the drain voltage cannot be modeled in this way. So no channel-shortening, static feedback, drain induced barrier lowering, weak avalanche, or other drain dependent effects on the saturated drain current are included.

### 2.3   THE CURRENT DIVISION TECHNIQUE

A technique for dividing currents (or voltages) very accurately and linearly is useful for various kinds of analog signal processing such as controllable attenuation, analog-to-digital and digital-to-analog conversion. A common technique is to use resistors or capacitors for the linear and accurate division of current (or voltage) while using MOS transistors as switches or amplifying elements [3, 4]. The same MOS transistors for both the signal division as well as the switching function may be used, thus eliminating the need for resistors or capacitors. Although an MOS transistor exhibits a non-linear relationship between current and voltage (even in the linear region), it is shown here that the current division function is inherently linear. To show the feasibility of this technique, a volume control circuit has been realized with a measured THD better than -85dB and a dynamic range better than 100dB [10, 11].

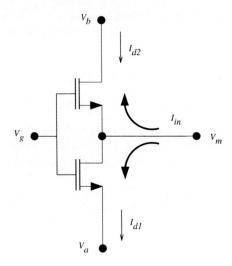

**FIGURE 2.8**
The basic principle of current division.

## 2.3.1   The Basic Principle

The basic principle of this current division technique is shown in Fig. 2.8. Both transistors have the same gate voltage (with respect to the substrate potential). Voltages $V_a$ and $V_b$ are dc voltages and may have any value as long as the transistors are in the on-state. A current $I_{in}$ flowing into or out of the circuit will be divided into two parts; one part flowing into the node connected to $V_a$ and the other flowing into the node connected to $V_b$. We will now show that the fraction of this input current flowing to one side is:

- constant and independent of $I_{in}$ (implying low distortion),
- independent of the values of $V_a$ and $V_b$,
- independent of whether one or both devices are operating in saturated or non-saturated regions, and
- also independent of whether one or both devices operate in strong or in weak inversion.

Figure 2.9 shows some measured results [11] for the circuit in Fig. 2.8. A gate-substrate voltage of 3.0 V was applied and an input current $I_{in}$ was varied from $-100\,\mu$A to $+100\,\mu$A. $V_a$ was kept constant at 0.0V whereas $V_b$ was varied from 0 to 5.0V. Depending on the value, $V_b$, a dc current flows in the circuit even if no input current is applied ($I_{in} = 0$). This dc current acts as an offset current and, as the principle states, the division of the input current to either side of the circuit is not affected by the value of $V_b$.

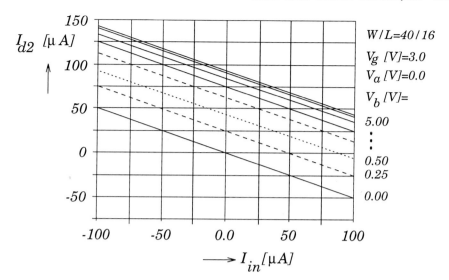

**FIGURE 2.9**
The current $I_b$ as a function of the input current $I_{in}$ for different values of $V_b$.

## 2.3.2 Proof of the Principle

To prove the current division principle we need an MOS model valid in all regions of operation, therefore we will use (2.6). In this expression, $f(V_G, V_C)$ can be quite a complicated function including all kinds of secondary effects such as the mobility reduction and the body-effect. We can now prove the validity of the current division principle without any mathematical effort by using the graphical representation presented in Sec. 2.2.1. Figure 2.3 shows the graphical representation of the current through an MOS transistor. In the same way, Fig. 2.10 shows the current through the transistors of Fig. 2.8. As both transistors share the same gate voltage (with respect to the substrate potential), we can use the same curve for the function $f(V_g, V_c)$ as a function of the channel voltage, $V_c$. Moreover, as the drain voltage of the lower transistor equals the source voltage of the upper transistor, the areas representing the current through each transistor are always adjacent.

If an input current is applied (Fig. 2.8), the voltage $V_m$ at the input node will be a (non-linear) function of the input current $I_{in}$, where $V_m = V_m(I_{in})$. Let $V_{m1}$ be the initial input voltage and let $V_{m2}$ be the input voltage when an input current has been applied. From Fig. 2.11 we can see that, apart from a factor $(W_1/L_1)$, the increase of the current through the lower transistor, $\Delta I_{d_1}$, is represented in the figure by the double shaded area. In the same way, apart from a factor $(W_2/L_2)$, the decrease in the current through the upper transistor, $-\Delta I_{d2}$, is represented by the same shaded area. This means that, although $V_m(I_{in})$ is a non-linear function of $I_{in}$, the ratio:

$$\Delta I_{d1}/\Delta I_{d2} = -(W_1 L_2)/(W_2 L_1) \tag{2.22}$$

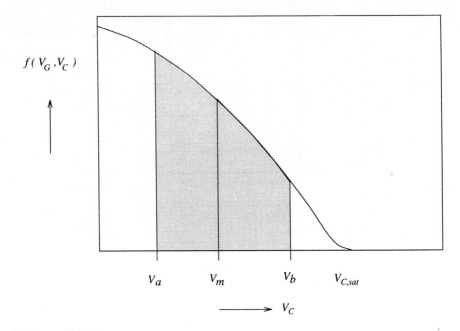

**FIGURE 2.10**
The drain currents of the two MOS transistors of Fig. 2.9.

is independent of $I_{in}$ and dependent on the geometry of the devices only. As:

$$I_{in} = \Delta I_{d1} - \Delta I_{d2} \tag{2.23}$$

the same holds for $\Delta I_{d1}/I_{in}$ and $\Delta I_{d2}/I_{in}$. This implies that the current division is inherently linear and insensitive to second-order effects like mobility reduction and body effect and valid in all operating regions of an MOS transistor. In fact, the current division technique is based on the symmetry of an MOS device expressed by (2.6) only.

### 2.3.3   Secondary Effects

The principle presented above is based on the validity of (2.6) and on ideal matching of both devices. Therefore, secondary effects that may influence the accurate current division are channel-length shortening, drain-induced barrier lowering, static feedback, and mismatch. The first three are only of interest if the devices are biased in the saturation region and, in that case, can give rise to (still very low) distortion. Mismatches in device geometry or oxide thickness influence the accuracy of the division only and do not affect distortion. Mismatch in threshold voltage however, will give rise to distortion. According to this conclusion, we expect the best results when the devices are biased in the linear region, with a large effective gate-source voltage and a large gate area. The volume control circuit presented in [10, 11] uses this technique and reports a measured THD

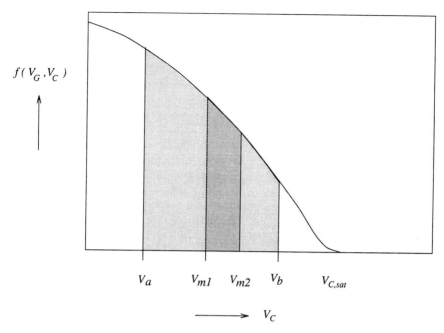

**FIGURE 2.11**
The drain currents of the two MOS transistors of Fig. 2.9 before and after a current $I_{in}$ is injected at the input. The double-shaded area represents as well the increase of the drain current through the lower transistor as the decrease of the drain current through the upper transistor.

better than -85 dB and a dynamic range better than 100 dB. The total chip area of the current division part was only 0.22 mm$^2$.

## 2.4  THE BASIC GAIN STAGE

Proper understanding of the functionality of MOS transistor basic gain stage is important in almost all analog designs. In this section, we will discuss the operation and the properties of the MOS basic gain stage and the cascoded versions.

### 2.4.1  DC Gain

Consider the basic gain stage shown in Fig. 2.12. In this configuration we will assume the MOS transistors operate in the saturated region. To obtain the small signal parameters we start with (2.15). Substituting $V_S = 0$ into (2.15) results in:

$$I_D = (K/2\alpha)V_{GT}^2 \qquad (2.24)$$

with

$$K = \mu_o C_{ox}(W/L) \qquad (2.25)$$

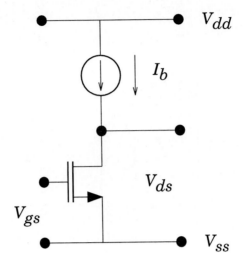

**FIGURE 2.12**

The basic gain stage.

and

$$V_{GT} = (V_G - V_{TO} - \alpha V_S) \qquad (2.26)$$

being the effective gate-source voltage. The transconductance, $g_m$, is defined as the derivative of the drain current with respect to the gate voltage:

$$g_m = (\delta I_D / \delta V_G) = (K/\alpha) V_{GT} \qquad (2.27)$$

where $V_{GT} = V_G - V_T$, the excess gate voltage referenced to the substrate potential.

Two independent design parameters (in this case $(W/L)$ and $V_G$) out of a set of three design parameters $((W/L), V_G$ and $I_D)$, may be chosen for determining the small signal parameters. Hence, $g_m$ can also be written as:

$$g_m = \sqrt{2(K/\alpha)I_d} \qquad (2.28)$$

or:

$$g_m = 2I_d / V_{GT} \qquad (2.29)$$

The output conductance of an MOS transistor is very difficult to model and only very rough estimates are suitable for hand calculations. Here we will use the following expression:

$$g_o = \lambda' I_d / L \qquad (2.30)$$

with $\lambda'$ being a process dependent parameter. Note that the constant *lambda'* is given approximately by $\sqrt{\frac{2}{qN_{SUB}}}$, and is related to the channel length modulation parameter, $\lambda$ by $\lambda = \lambda'/L$. Now we can analyze the circuit of Fig. 2.12 and determine the small signal dc-gain. For any MOS transistor we may write:

$$i_d = g_m v_{gs} + g_o v_{ds} \qquad (2.31)$$

In the circuit a current source is connected to the drain of the MOS transistor which results in $i_d = 0$ and thus:

$$0 = g_m v_{gs} + g_o v_{ds} \tag{2.32}$$

Substituting of $v_i = v_{gs}$ and $v_o = v_{ds}$ yields:

$$\frac{v_o}{v_i} = A_o = -\frac{g_m}{g_o} = -g_m r_{out} \tag{2.33}$$

with:

$$r_{out} = \frac{1}{g_o} \tag{2.34}$$

and $A_o$ being the dc-gain of this stage. The dc-gain can be written as the product of the transconductance with the output impedance of the circuit. This is further explained by realizing that the transfer of the signal happens in two steps. First, the input voltage is transferred via the transconductance, $g_m$, into a signal current and next the current is transferred again into an output voltage via the output impedance, $r_{out}$, of the stage. Again we can express this dc-gain with different sets of independent parameters. Using (2.25), (2.28), (2.30), (2.33), and (2.34) we find:

$$A_o = \left(\frac{1}{\lambda'}\right)\sqrt{\frac{2\mu C_{ox} W L}{\alpha I_d}} \tag{2.35}$$

and using (2.29) instead of (2.28) yields:

$$A_o = \frac{2L}{\lambda' V_{GT}} \tag{2.36}$$

In (2.35) if the amplifier is biased at a constant current level, to the first order, the dc-gain is determined by the square root of the gate area. This expression is particularly useful when simultaneously optimizing for $1/f$ noise or matching and dc-gain. Equation (2.36), however, shows us that if biased at a constant gate voltage, the dc-gain only depends on the channel length, $L$, and not on the channel width, $W$.

## 2.4.2 The AC Behavior

Consider again the circuit of Fig. 2.12. As we have seen in the previous section, for low frequencies, the transconductance transfers the input signal into a signal current and next, the signal current is transferred into an output voltage by the output impedance. For higher frequencies, we have to take into account the effect of the various capacitances present in the circuit. A small signal equivalent circuit of the basic gain stage of Fig. 2.12 is shown in Fig. 2.13. Three capacitances have been added to the model. First we have $C_{gs}$, the total capacitance between gate and source, consisting of $C_{gsch}$ and $C_{ovs}$:

$$C_{gs} = C_{gsch} + C_{ovs} \tag{2.37}$$

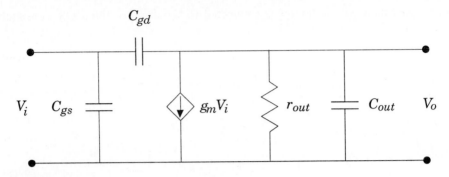

**FIGURE 2.13**
The small signal equivalent circuit of the basic gain stage.

with $C_{gsch}$ being the source-associated channel capacitance and $C_{ovs}$ being the overlap capacitance between gate and source. There is also $C_{gd}$, which represents the overlap capacitance between gate and drain. Finally, we have $C_{out}$ representing the capacitance connected to the output node:

$$C_{out} = C_{ds} + C_{load} \qquad (2.38)$$

with $C_{ds}$ being the junction capacitance between the drain and substrate and $C_{load}$, the total load-capacitance connected to the output node. As we drive our circuit with an ideal voltage source, this yields 1 pole and 1 zero in the transfer function:

$$\frac{V_o}{V_i} = -g_m r_{out} \frac{1 - sC_{gd}/g_m}{1 + s(C_{out} + C_{gd})r_{out}} \qquad (2.39)$$

The Bode plot of this transfer function is shown in Fig. 2.14. From this plot, the bandwidth of our circuit is:

$$BW = \frac{1}{r_{out}C_{tot}} \qquad (2.40)$$

with $C_{tot} = C_{out} + C_{gd}$. The gain-bandwidth product becomes:

$$GBW = A_o BW = \frac{g_m}{C_{tot}} \qquad (2.41)$$

Using (2.27) for the transconductance, $g_m$, yields:

$$GBW = \frac{K V_{GT}}{\alpha C_{tot}} \qquad (2.42)$$

whereas using (2.28) results in:

$$GBW = \sqrt{2(K/\alpha)I_d}/C_{tot} \qquad (2.43)$$

Equation (2.42) illustrates that, biased at a constant effective gate-source voltage $V_{GT}$, the GBW is proportional to a process dependent factor $(\mu C_{ox}/\alpha)$, the $W/L$

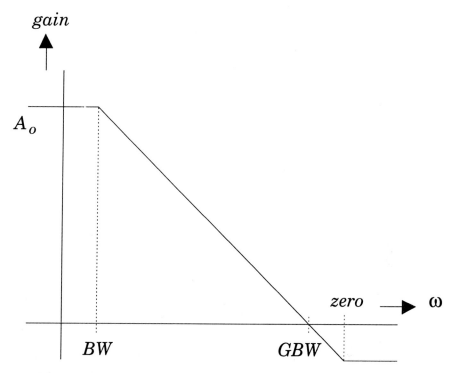

**FIGURE 2.14**
A Bode diagram of the transfer function of the basic gain stage.

**TABLE 2.1**
Small signal parameters as a function of design parameters.

|  | $W, L, V_{GT}$ | $W, L, I_d$ | $L, V_{GT}, I_d$ |
|---|---|---|---|
| $g_m$ | $\frac{\mu C_{ox} W V_{GT}}{\alpha L}$ | $\sqrt{2\mu C_{ox}(W/L)I_d/\alpha}$ | $\frac{2I_d}{V_{GT}}$ |
| $r_o$ | $\frac{2L^2\alpha}{\lambda'\mu C_{ox} W V_{GT}^2}$ | $\frac{L}{\lambda' I_d}$ | $\frac{L}{\lambda' I_d}$ |
| $A_o$ | $\frac{2L}{\lambda' V_{GT}}$ | $\frac{1}{\lambda'}\sqrt{\frac{2\mu C_{ox} W L}{\alpha I_d}}$ | $\frac{2L}{\lambda' V_{GT}}$ |
| $GBW$ | $\frac{\mu C_{ox} W V_{GT}}{\alpha C_{tot} L}$ | $\frac{\sqrt{2\mu C_{ox}(W/L)I_d/\alpha}}{C_{tot}}$ | $\frac{2I_d}{V_{GT}C_{tot}}$ |

ratio, and inversely proportional to the total load capacitance, $C_{tot}$. Biased at a constant current however, GBW is proportional to the square-root of $(\mu C_{ox}/\alpha)$, the square-root of $(W/L)$, and again inversely proportional to $C_{tot}$. Table 2.1 shows an overview of the small signal parameters as a function of different sets of design parameters.

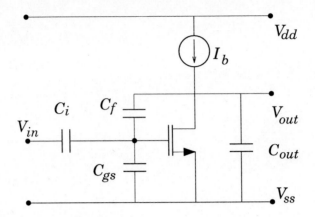

**FIGURE 2.15**

The basic gain stage embedded in a capacitive feedback circuit.

### 2.4.3   The Settling Behavior

Consider the circuit of Fig. 2.15. This is the basic gain stage with capacitive feedback and it is frequently encountered in CMOS design (for instance in switched-capacitor circuits). Using this circuit model, we will try to optimize the settling behavior of this stage.

The capacitors $C_i$ and $C_f$ are normally determined by the rest of the system (for instance a switched-capacitor filter) and $C_{out}$ and $C_{gs}$ are parasitic capacitors of the MOS transistor (for simplicity we will neglect the overlap capacitance between gate and drain). If we choose a very large transistor width, W, we obtain a large transconductance, $g_m$. However, at the same time, the combination of both a small feedback factor and a large load capacitance results in a very small bandwidth. Choosing a small width, on the other hand, yields a small transconductance and hence also yields a small bandwidth. Thus, there must be an optimum and the designer is faced with the problem of finding the optimal bandwidth of the gain stage.

To determine this, we will first have to find the width dependence of the parasitic capacitances $C_{out}$ and $C_{gs}$. The parasitic capacitance at the input can be written as [2]:

$$C_{gs} = (2/3)C_{ox}WL \qquad (2.44)$$

We will use a simple expression for the parasitic junction capacitance, $C_{ds}$, at the output node:

$$C_{out} = \delta W \qquad (2.45)$$

where $\delta$ is a fitting parameter which shows the dependence of $C_{out}$ on the width $W$.

Settling speed can be calculated as follows. In Fig. 2.15 the feedback factor

$\beta$ is given by:

$$\beta = \frac{C_f}{C_i + C_{gs} + C_f} \qquad (2.46)$$

whereas the total load capacitance seen at the output equals:

$$C_{load} = C_{out} + C_f \frac{C_i + C_{gs}}{C_i + C_{gs} + C_f} \qquad (2.47)$$

i.e. $C_{out}$ in parallel with the series connection of $C_f$ and $(C_i + C_{gs})$. The unity-gain frequency, $\omega_u$, given by $g_m/C_{load}$, now becomes:

$$\omega_u = g_m \frac{C_i + C_{gs} + C_f}{C_{out}(C_i + C_{gs} + C_f) + C_f(C_i + C_{gs})} \qquad (2.48)$$

The theoretical settling time-constant, $\tau = \frac{1}{u}$, can now be calculated:

$$\tau = \frac{C_{gs} + C_i + C_{out} + (C_i + C_{gs})C_{out}/C_f}{g_m} \qquad (2.49)$$

To optimize the stage for minimal settling time, we have to substitute (2.44) and (2.45) for the parasitic capacitances, and one of the expressions of Sec. 2.4.1 for the transconductance, $g_m$. In this case we choose (2.27) which gives the $g_m$ as a function of $W$, $L$ and the effective gate-source voltage, $V_{GT}$, and we obtain:

$$\tau = \frac{L\alpha\{(2/3)C_{ox}L + (C_i/W) + \delta + (C_i/C_f)\delta + \delta W L[(2/3)(C_{ox}/C_f)]\}}{\mu C_{ox} V_{GT}} \qquad (2.50)$$

$V_{GT}$ is kept at a fixed value as it is usually determined by characteristics such as noise and output swing. We will take the shortest possible channel length $L_{min}$ and we can now find the optimum value for the width, W, by taking the derivative of (2.50) with respect to $W$ and equating the result to zero. This yields:

$$W_{opt} = \sqrt{\frac{C_i C_f}{\delta(2/3)C_{ox}L_{min}}} \qquad (2.51)$$

There is a (finite) optimum value for W and the optimum value of W is independent of the effective gate-source voltage. Note, that in the optimum situation

$$C_i C_f = C_{out} C_{gs} \qquad (2.52)$$

which means that the geometric mean of input and output parasitic capacitances equals the geometric mean value of the feedback network capacitors. The effective gate-source voltage can be optimized independently according to the requirements regarding noise, output swing, and settling speed.

## 2.5   LIMITATIONS OF THE BASIC GAIN STAGE

The important fundamental limitations of the basic gain stage are bandwidth and dc-gain. Using cascoding techniques, it is possible to improve the dc-gain without degrading the high frequency behavior. In this section we will discuss these fundamental limits and improved circuit techniques to maximize the performance of MOS analog ICs.

### 2.5.1   A Fundamental Gain-Bandwidth Limitation

In Sec. 2.4.1 we found expressions for the dc-gain of the simple gain stage of Fig. 2.12, and in Sec. 2.4.2 we found similar expressions for the gain-bandwidth product GBW. Very often it is desirable to design for both high dc-gain, $A_o$, as well as high gain-bandwidth product, GBW. In this section, we present the fundamental limitation to obtaining both of these conditions. Comparing (2.36) and (2.42) in combination with (2.25), we see that increasing the effective gate-source voltage, $V_{GT}$, and decreasing the channel length, $L$, increases the gain-bandwidth product, GBW, but decreases at the same time the dc-gain, $A_o$, by the same factor. This means that, in the first-order approximation, the product of gain and gain-bandwidth product, $A_o GBW$, is independent of both the effective gate-source voltage, $V_{GT}$, and the channel-length, $L$:

$$A_o GBW = \frac{2\mu C_{ox} W}{\alpha \lambda' C_{out}} \tag{2.53}$$

In fact, (2.53) shows us that this product $A_o GBW$ is, apart from some process dependent parameters, dependent on the ratio $W/C_{out}$ only. From this result, we might conclude that increasing the channel width, $W$, results in an increasing $A_o GBW$. Hence it enables us to simultaneously achieve both a high dc-gain and a high gain-bandwidth product, but that is not true. If we increase the channel length, $W$, we also increase all parasitic capacitors associated with the width, i.e. $C_{gsch}$, $C_{ovs}$, $C_{ds}$, and $C_{ovd}$ and thereby both $C_{gs}$ and $C_{out}$. This means that increasing $W$ only increases the $W/C_{out}$ ratio up to a certain limit. In cases with feedback (as, for instance, discussed in Sec. 2.4.3), there is always an optimum value of $W$ with respect to the total bandwidth of the circuit. In any case, we may write:

$$C_{out} = \chi W \tag{2.54}$$

with $\chi$ being a feedback and process dependent parameter. Applying this to (2.53) yields:

$$A_o GBW = \frac{2\mu C_{ox}}{\alpha \lambda' \chi} = constant \tag{2.55}$$

Here we clearly see that we can only increase $A_o$ at the cost of a lower GBW or vice versa. Figure 2.16 shows the result of this with a Bode diagram. Suppose that in a certain process and given a certain amount of feedback, curve

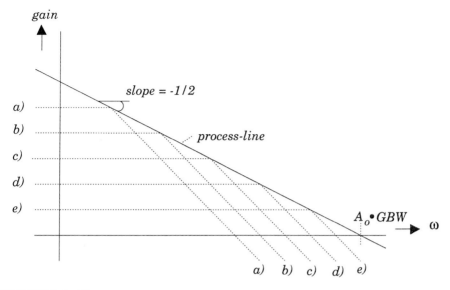

**FIGURE 2.16**
The "process-line" of a CMOS process.

(c) is an achievable transfer function of a simple gain stage. Increasing the ratio $L/V_{GT}$ increases the gain and decreases the GBW resulting in curves (b) and (a). Whereas decreasing the same ratio results in a higher GBW and a lower dc-gain as shown in curves (d) and (e). According to (2.55) the -3 dB points of the different curves lie on a straight line with a slope of -1/2 (-3 dB/octave). We call this line the process-line. It also shows the maximum possible unity-gain bandwidth frequency. Anything under this line is achievable. Anything above this line is not achievable using a simple gain stage. In the next section, we will see that increasing $A_oGBW$ is possible, to a certain extent, using cascoding, and in Sec. 2.6 we present a technique that truly decouples dc-gain and gain-bandwidth.

### 2.5.2   The Cascode Gain Stage

Consider the simple cascode stage shown in Fig. 2.17. In this section we will show that the dc-gain can be expressed as the product of an effective transconductance, $g_{meff}$, and the output impedance, $R_{out}$. Moreover, we will show that for low frequencies, cascoded transistors increase the output impedance with only a small effect on the transconductance and, for high frequencies, the effect of cascoding transistors is almost negligible. The transfer function of the cascode stage can be found as follows. Suppose an ideal voltage source is connected to the output of the stage. If a voltage $\Delta V_i$ is applied to the input, a current $\Delta I_o$ will flow into the voltage source. For low frequencies:

$$\frac{\Delta I_o}{\Delta V_i} = g_{m1} \frac{g_{m2}r_{o1} + r_{o1}/r_{o2}}{g_{m2}r_{o1} + r_{o1}/r_{o2} + 1} = g_{meff} \qquad (2.56)$$

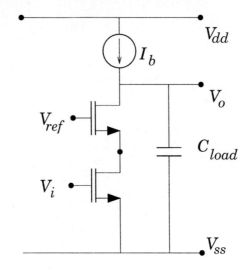

**FIGURE 2.17**
The cascode stage.

This effective transconductance is approximately equal to $g_{m1}$, the reduction is due to the feedback via the drain of the input transistor. Removing the voltage source causes a change $\Delta V_o = \Delta I_o R_{out}$, where $\Delta I_o$ is calculated using (2.56) and $R_{out}$ is the output resistance of the total circuit. The voltage-gain, $A_o$, is:

$$A_o = g_{meff} R_{out} \qquad (2.57)$$

The output impedance of the total circuit can easily be calculated to be:

$$R_{out} = (g_{m2} r_{o2} + 1) r_{o1} + r_{o2} \qquad (2.58)$$

This is roughly the output resistance of the input transistor, $r_{o1}$, multiplied with the gain of the cascode transistor, $g_{m2} r_{o2}$. Note the effect of the cascoded transistors with respect to the non-cascoded situation. The cascoded transistors shield the drain of the input transistor from the effect of the signal swing at the output, and reduce the swing at the drain of the input transistor by a factor $g_{m2} r_{o2}$, i.e. the gain of the cascoded transistors. This leads to the following expression for the dc-gain:

$$A_o = g_{m1} r_{o1} (g_{m2} r_{o2} + 1) \qquad (2.59)$$

This expression shows us that the dc-gain of the cascode stage approximately equals the square of the gain of the basic gain stage:

$$A_{ocascode} \approx g_{m1} r_{o1} g_{m2} r_{o2} \approx A_{osimple}^2 \qquad (2.60)$$

The behavior of the circuit, in the vicinity of the unity-gain frequency, is similar to the situation with the voltage source connected to the output. Now the load capacitor, $C_{load}$, forms the short circuit to ground. At the unity-gain frequency,

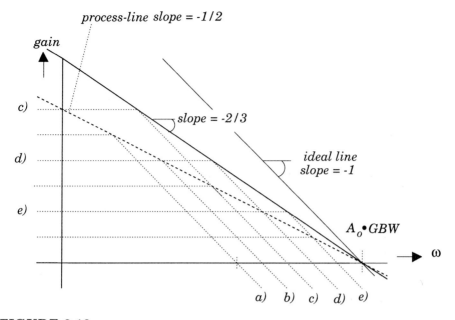

**FIGURE 2.18**
The process-line and the equivalent line for the cascode stage.

we use the effective transconductance as given by (2.56), and this results in the following expression for the gain-bandwidth product, (GBW):

$$GBW = \frac{g_{meff}}{C_{load}} \tag{2.61}$$

As (2.56) shows us that the effective transconductance, $g_{meff}$, is almost equal to the transconductance of the simple stage, we see that:

$$GBW_{cascode} \approx GBW_{simple} \tag{2.62}$$

As seen in (2.59) and (2.61), cascoding increases the dc-gain of the simple gain stage without sacrificing speed, i.e. at the same GBW. In Sec. 2.4.1 we have shown that there is a fundamental limitation to achieving simultaneously high dc-gain and a high GBW product, (2.55). Substituting (2.60) and (2.62) into (2.56) yields a similar limitation for the cascode stage:

$$\sqrt{A_{ocascode}}GBW_{cascode} = constant \tag{2.63}$$

The equivalent process-line for the cascode stage is shown in Fig. 2.18. This line has a slope of $-2/3$ and intersects the frequency axis at the same frequency as the basic gain stage process-line. The ideal situation for an opamp would of course be to be able to achieve any dc-gain at a given (possible) GBW product. This would be a line with slope of -1 intersecting the frequency axis at the same

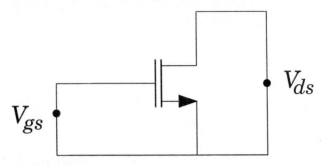

**FIGURE 2.19**
"Measurement set-up" for the MOS $f_T$.

frequency again. It is clear that this cannot be achieved by simple cascoding. In the next section we will present a technique that truly decouples the dc-gain and GBW product.

### 2.5.3 The MOS Transistor $f_T$

For bipolar transistors the $f_T$ is the ubiquitous measure of speed. A comparable definition for the speed of an MOS transistor will be presented.

Consider the circuit of Fig. 2.19. The input voltage at the gate terminal is transformed into a current flowing into the voltage source at the drain of the transistor. If we increase the input signal frequency, the capacitive current flowing into the gate will also increase and, at some point in frequency, equal the drain current. Here we will call that frequency the $f_T$ of an MOS transistor. Using (2.25) and (2.42) yields:

$$f_T = \frac{g_m}{2\pi C_{gs}} = \frac{3\mu}{4\pi\alpha} \frac{V_{GT}}{L^2} \tag{2.64}$$

Note that $f_T$ approximately equals the unity-gain frequency of a simple gain stage loaded with an identical stage. In fact, in that case, the unity-gain frequency is somewhat lower as the parasitics at the drain have to be included. Equation (2.64) furthermore shows us that the $f_T$ of an MOS transistor depends on the effective gate-source voltage and on the channel length. Taking the minimum channel length, $L_{min}$, yields the highest possible $f_T$ at a certain effective gate-source voltage.

Searching for very high frequency circuits, (2.64) might give the impression that increasing the effective gate-source voltage results in an ever increasing $f_T$. Unfortunately this is not true due to the mobility reduction effect which in modern processes is already quite severe at moderate effective gate-source voltages. Mobility reduction can, to the first order be modeled as [2]:

$$\mu = \frac{\mu_o}{1 + V_{GT}(\theta + (\mu_o/\alpha v_{max}L)} \tag{2.65}$$

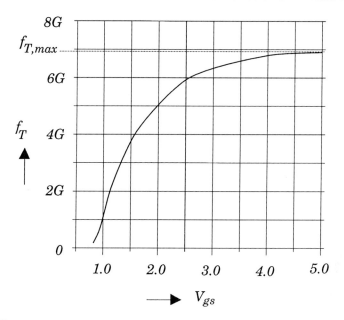

**FIGURE 2.20**
The MOS transistor $f_{Tmax}$ as a function of the gate source voltage $V_{gs}$.

where $v_{max}$ is the maximum carrier velocity, $\mu_o$ is the mobility at $V_{GT} = 0$, and $\theta$ is the mobility reduction parameter. Substituting this in (2.24) and recalculating the transconductance, $g_m$, yields:

$$g_m = \frac{\delta I_D}{\delta V_G} = \frac{\mu C_{ox} W V_{GT}\{2 + V_{GT}[\theta + (\mu_o/\alpha v_{max}L)]\}}{2L\alpha\{1 + V_{GT}[\theta + (\mu_o/\alpha v_{max}L)]\}^2} \qquad (2.66)$$

Using this expression in (2.64) yields a more accurate expression for $f_T$. The maximum $f_T$ is then found with minimum channel length and infinite gate-source voltage, resulting in:

$$f_{Tmax} = \lim_{V_{GT} \to \infty} f_T = \frac{3\mu_o}{8\pi\alpha L_{min}[\theta L_{min} + (\mu_o/\alpha v_{max})]} \qquad (2.67)$$

Figure 2.20 shows $f_T$ as a function of the gate-source voltage. Increasing the effective gate-source voltage beyond 2.5 V does not bring a much further increase in $f_T$. Modern processes ($L_{min}$=0.5 $\mu$m) yield $f_{Tmax}$'s of about 5-10 GHz. Depending on biasing and circuit techniques, a bandwidth of up to 1 GHz is achievable.

## 2.6   THE GAIN-BOOSTING TECHNIQUE

Speed and accuracy are two of the most important properties of analog circuits; as we have seen in the previous sections, optimizing circuits for both aspects

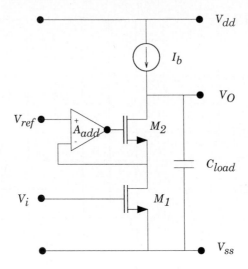

**FIGURE 2.21**
Cascoded gain stage with gain enhancement.

leads to contradictory demands. This section gives an overview of a gain-boosting technique, which improves the accuracy of cascoded CMOS circuits without any speed penalty. This is achieved by increasing the effect of the cascode transistor by means of an additional gain stage, thus increasing the output impedance of the sub-circuit. Used in op amp design, this technique allows the combination of the high-frequency behavior of a single-stage opamp with a dc-gain compatible to a multi-stage amplifier design.

### 2.6.1    The Gain-Boosting Principle

We have shown that the dc-gain of Fig. 2.16 can be expressed as the product of an effective transconductance, $g_{meff}$, and the output impedance. From (2.57) and (2.61) we may conclude that the only way to improve $A_o$ without reducing the GBW is to increase the output impedance. Note that this is the effect of the cascode transistor itself with respect to the non-cascoded situation. The cascode transistor shields the drain of the input transistor from the effect of the signal swing at the output.

The technique presented here is based on increasing the cascoding effect of M2 by adding an additional gain stage as shown in Fig. 2.21. This stage further reduces the feedback from the output to the drain of the input transistor. Thus, the output impedance of the circuit is further increased by the gain of the additional gain stage, $A_{add}$:

$$R_{out} = [g_{m2}r_{o2}(A_{add} + 1) + 1]r_{o1} + r_{o2} \qquad (2.68)$$

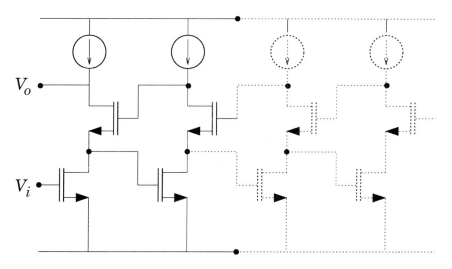

**FIGURE 2.22**
Repetitive implementation of gain enhancement.

Moreover, the effective transconductance is increased slightly:

$$\frac{\Delta I_o}{\Delta V_i} = g_{m1} \frac{g_{m2} r_{o1} (A_{add} + 1) + r_{o1}/r_{o2}}{g_{m2} r_{o1} (A_{add} + 1) + r_{o1}/r_{o2} + 1} = g_{meff} \tag{2.69}$$

Hence, using the same reasoning as in (2.57), the total dc-gain now becomes:

$$A_{otot} = g_{m1} r_{o1} [g_{m2} r_{o2} (A_{add} + 1) + 1] \tag{2.70}$$

If the additional stage is implemented as a cascode stage, the gain-enhancement technique can also be applied to this additional stage. In this way, a repetitive implementation of the gain-enhancement technique can be obtained as shown in Fig. 2.22. The limitation on the maximum voltage gain is then set by factors such as leakage currents, weak avalanche, and thermal feedback. From the above discussion, we may conclude that the repetitive usage of the gain-enhancement technique yields a decoupling of the op amp gain and unity-gain frequency.

## 2.6.2   High Frequency Behavior

In Fig. 2.23, the magnitude vs. frequency plot is shown for the original cascoded gain stage of Fig. 2.17 ($A_{orig}$), the additional gain stage ($A_{add}$), and the improved cascoded gain stage of Fig. 2.21 ($A_{tot}$). At DC, the gain enhancement $A_{tot}/A_{orig} \approx [1 + A_{add(0)}]$, according to (2.59) and (2.70). For $\omega > \omega_1$, the output impedance is mainly determined by $C_{load}$. In fact, we have to substitute ($R_{out}//C_{load}$) into (2.57) for $R_{out}$. This results in a first-order roll-off of $A_{tot()}$. Moreover, this implies that $A_{add()}$ may have a first-order roll-off for $\omega > \omega_2$ as long as $\omega_2 > \omega_1$. This is equivalent to the condition that the unity-gain frequency ($\omega_4$) of the additional

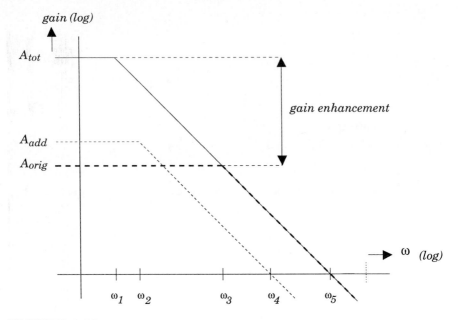

**FIGURE 2.23**
Gain Bode-plots of the original cascoded gain stage ($A_{orig}$), the additional gain stage ($A_{add}$), and the improved cascoded gain stage ($A_{tot}$).

gain stage has to be larger than the -3 dB bandwidth ($\omega_3$) of the original stage, but it can be much lower than the unity-gain frequency ($\omega_5$) of the original stage. The unity-gain frequencies of the improved gain-stage and the original gain stage are the same.

To obtain a first-order roll-off for the total transfer function, the additional gain stage does not have to be a fast stage. In fact, this stage can be a cascoded gain stage as shown in Fig. 2.17, with smaller width and non-minimal length transistors biased at low current levels. Moreover, as the additional stage forms a closed loop with M2, stability problems may occur if this stage is too fast. There are two important poles in this loop. One is the dominant pole of the additional stage and the other is the pole at the source of M2. The latter is equal to the second pole, $\omega_6$, of the main amplifier. For stability reasons, we set the unity-gain frequency of the additional stage lower than the second pole frequency of the main amplifier. A safe range for the location of the unity-gain frequency $\omega_4$ of the additional stage is given by:

$$\omega_3 < \omega_4 < \omega_6 \qquad (2.71)$$

This can easily be implemented. However, it can be shown [6, 7] that a single-pole settling behavior demands a higher unity-gain frequency for the additional stage than a simple first-order roll-off in the frequency domain requires. Fast settling is guaranteed when the unity-gain frequency of the additional stage is higher than the -3 dB bandwidth of the closed-loop circuit [7]. On the other hand, for

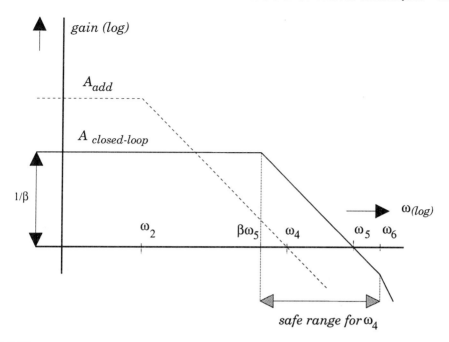

**FIGURE 2.24**
The "safe" range for the unity-gain frequency of the additional stage.

reasons concerning stability, the unity-gain frequency must be lower than the second-pole frequency of the main amplifier as indicated by (2.71). This results in the "safe" area for the unity-gain frequency of the additional stage:

$$\beta \omega_5 < \omega_4 < \omega_6 \qquad (2.72)$$

as shown in Fig. 2.24. Note that this "safe" area is smaller than given by (2.71). A satisfactory implementation however is still no problem even if $\beta = 1$, because the load capacitor of the additional stage (which determines $\omega_4$), is much smaller than the load capacitor of the op amp (which determines $\omega_5$). In most cases, the $W/L$ ratio of the additional stage can be as small as 5% of the $W/L$ ratio of the main stage.

### 2.6.3 Circuit Implementation

The circuit diagram of a folded cascode with gain boosting is shown in Fig. 2.25. The main stage is a folded-cascode amplifier [8]. The simplest implementation of the additional stage is one MOS transistor [9]. Here, a cascode version is chosen because of its high gain and the possibility of repetitive usage of the gain-boosting technique. The input stage design of the additional amplifier is determined by the common-mode range requirement, which is close to $V_{SS}$ if a large signal swing is required. As a consequence, a folded-cascode structure with

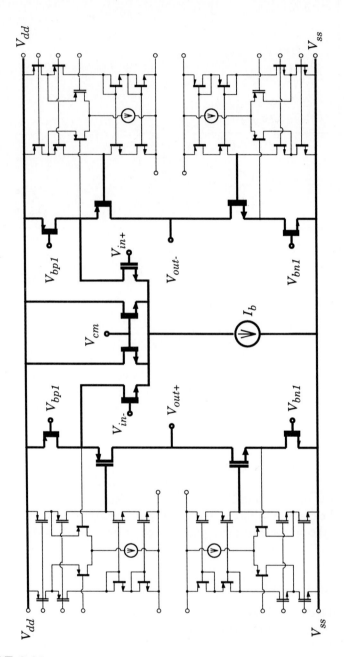

**FIGURE 2.25**
Complete circuit diagram of the op amp.

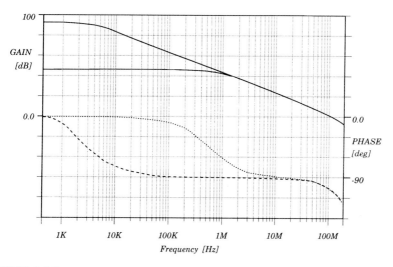

**FIGURE 2.26**
Results of gain and phase measurements both with and without gain enhancement.

PMOS input transistors has been chosen for the additional amplifier. In [5, 7] a maximum output swing of 4.2 V is achieved with a single 5.0 V supply. To realize very high output impedance, the current source in Fig. 2.21 is also realized as a cascoded structure with an additional stage. The two input transistors connected to common-mode voltage, $V_{cm}$, have been added to control the common-mode bias voltage at the output. Results of gain measurements both with and without the gain-boosting technique are shown in Fig. 2.26. A gain enhancement of 45 dB was measured without affecting the gain or phase for higher frequencies, resulting in a total dc gain of 90 dB combined with a unity-gain frequency of 116 MHz.

## 2.7 THE SUPER MOS TRANSISTOR

A disadvantage of the implementation of the gain-boosting technique discussed in the previous section is the complexity of the design and layout. A disadvantage of cascoded amplifiers in general is the number of required biasing voltages resulting in long wires across the chip. These long wires consume a considerable amount of space and, what is worse, are susceptible to cross-talk and therefore also instability. A more general use of this technique is in the form of a transistor-like building block [7]: the Super MOS. This compound circuit behaves as a regular MOS transistor but has an intrinsic gain $g_m r_o$ of more than 90 dB. The building block is self biasing and therefore very easy to use in design. An opamp consisting of only 8 Super MOS and 4 regular MOS transistors [7] shows results equivalent to the design discussed above.

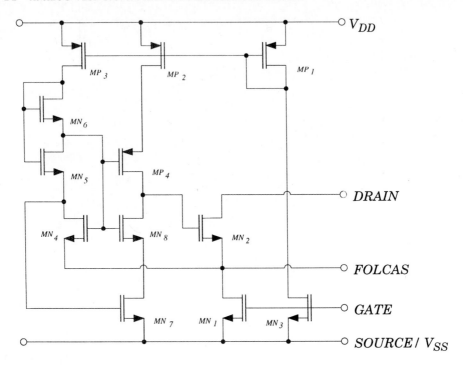

**FIGURE 2.27**
N-type Super-MOS.

## 2.7.1   Circuit Description

The Super MOS is a compound circuit which behaves like a cascoded MOS transistor and has, like a regular MOS transistor, a source, a gate, and a drain terminal. The Super MOS however has an extremely high output impedance due to implementation of the gain-boosting technique. Moreover, it does not require any biasing voltage or current other than one single power supply. The circuit of an N-type Super MOS transistor is shown in Fig. 2.27. The circuit consists of three parts:

- Transistors MN1 and MN2 are the main transistor and its cascode transistor, respectively. They form the core of the circuit. Their size determines the current-voltage relationships and the high frequency behavior of the "device."

- Transistors MN7, MN8, MP2, and MP4 form the additional stage for the gain-boosting effect.

- Transistors MN3, MP1, MP3, MN4, MN5, and MN6 are for biasing purposes and ensure that MN2 is always biased in such a way that MN1 is just 50 - 100 mV above the edge of saturation, independent of the applied gate voltage. This ensures a low saturation voltage of the Super MOS device.

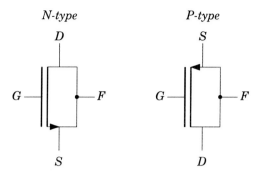

**FIGURE 2.28**
Super-MOS symbols.

The size of the transistors of the additional stage and of the biasing branch is as small as a few percent of the main transistors. Figure 2.28 shows the symbol for the Super MOS transistor; S is the source terminal, G is the gate terminal, and D is the drain terminal. Furthermore an extra low impedance current input terminal F is available which can be used for folded-cascode structures.

### 2.7.2   Measured "Device" Characteristics

Figure 2.29 shows the measured characteristics of a single N-type MOS transistor and a Super MOS transistor for $V_{gs}$ ranging from $0.7V(a)$ to 1.0 V(g). The single MOS transistor exhibits an early voltage of approximately 5 V whereas the Super MOS has an early voltage which is several orders of magnitude higher. Note that the "device" is already saturated at a voltage only slightly above the saturation voltage of one single MOS transistor, indicating a large possible output swing. An increase in the output impedance of 350 times compared to a single MOS has been measured [7].

### 2.7.3   Applications of the Super MOS Transistor

The Super MOS can be used as a regular MOS transistor. As an example, a current mirror has been realized as shown in Fig. 2.30. Figure 2.31 shows the measured results for a P-type current mirror at several input currents and output voltages. The mirror accuracy is independent of output voltage and is only determined by matching.

The F-terminal of the Super MOS provides a very low ohmic current input which can be used for folded-cascode structures [8]. As an example, Fig. 2.32 shows a straightforward circuit topology of a folded-cascode op amp but now realized with Super MOS transistors. Note that the tail current of the differential input pair also consists of a Super MOS. In this way, the common-mode rejection ratio, CMRR, is increased several orders of magnitude for low frequencies. For the input pair, regular transistors have been used. Measured results of this straightforward design are equivalent to the results discussed in Section 2.6.3.

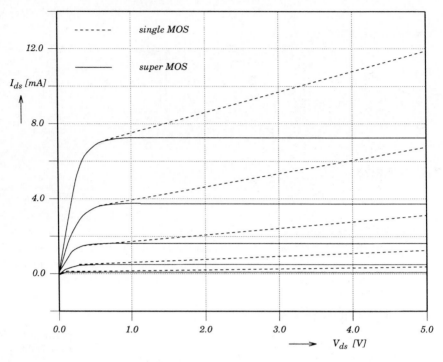

**FIGURE 2.29**
Measured $I_{ds} - V_{ds}$ characteristics of single N-type MOS transistor and Super-MOS transistor for $V_{gs}$ ranging from 0.7 V (a) to 1.0 V (g).

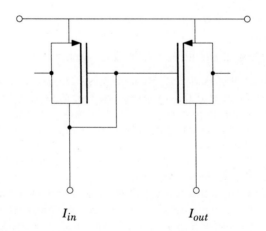

**FIGURE 2.30**
Current mirror with Super-MOS transistors.

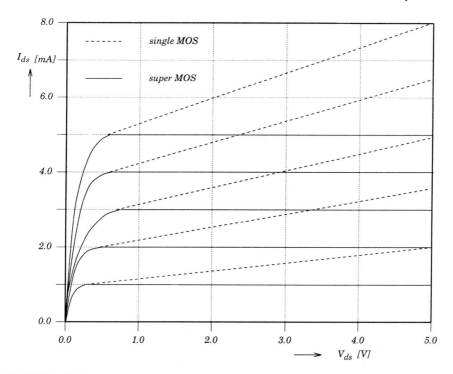

**FIGURE 2.31**
Measured output current of a P-type current mirror with $I_{in}$ ranging from 1.0mA (a) to 5.0mA (e).

There are several advantages in using a Super MOS as a building block in circuit design. The design of high-quality op amps, OTAs, current sources, etc., can be split into two parts. First the design of the Super MOS circuit itself, in which the knowledge of proper biasing, gain-boosting, optimal settling, and stability is used. Much attention can be paid to an optimal layout also, as this sub-circuit is going to be used as a basic block in many designs of larger circuits or systems. Secondly, the design of the larger circuit now becomes rather straightforward. In this higher level design, no knowledge of the gain-boosting principle or of optimal biasing of a cascode transistor is required. This eases the design of such circuits and shortens the design time considerably. Moreover, Super MOS transistors are very suitable for the automatic generation of analog parameterized cells. As each Super MOS is self biasing, no long biasing wires are required on chip.This leads to a design which is much less susceptible to crosstalk and therefore it is less likely to suffer from high frequency instability problems. And finally, as wiring usually consumes a relatively large part of the chip area, the approach presented here consumes considerably less chip area.

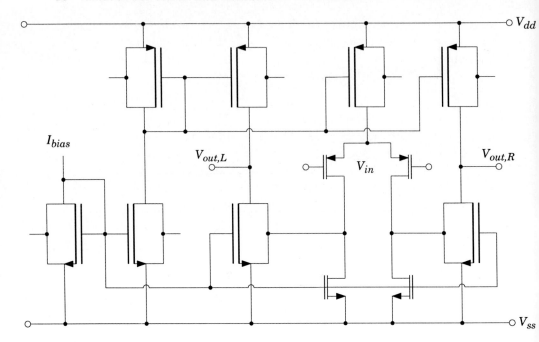

**FIGURE 2.32**
Folded-cascode op amp realized with Super-MOS transistors.

## 2.8   CONCLUSION

A short deduction of the relationship between the current through an MOS transistor and its terminal voltages has been presented. Without making approximations or simplifications, this leads to a general expression for the device current in the form of the W/L ratio times an integral of local channel conductivity from the source to drain voltage. The usefulness of this expression, especially in combination with the graphical representation, is shown in the current division technique. By neglecting the mobility reduction effect and approximating the body effect with a linear expression, this general expression is then translated into a simple, but fully symmetrical expression for the device current as a function of the terminal voltages. Again, the graphical representation of this model is shown to be helpful in gaining insight into device and circuit operation. Using the current division technique enables us to define a very linear signal relationship in the current domain. This technique is an example of a typical MOS design and its validity is proven using the graphical representation of the general current voltage relationship.

For the basic gain stage, the dc and ac behavior is discussed and an overview of the small signal parameters as a function of different sets of design parameters is given. The settling behavior is shown to have optimal dimensions for a gain stage for a given capacitive feedback situation. The limitations with respect to gain and bandwidth are shown. The product of gain and gain-bandwidth (or the gain bandwidth product) is constant and determined by process parameters only. Furthermore, it is shown that cascoding improves the gain without damaging the high frequency behavior, but also is subject to a similar limitation concerning gain and bandwidth. Finally the definition of an $f_T$ for an MOS transistor is discussed.

The gain-boosting technique is a typical CMOS technique which truly decouples the dc-gain and unity-gain frequency of a gain stage. A very high dc-gain can be achieved in combination with any unity-gain frequency achievable by a (folded-) cascode design and an example is given.

The Super MOS transistor is a transistor-like building block incorporating the gain-boosting technique in a completely self-biased implementation, to achieve high intrinsic gain without any speed penalty. The Super MOS eases the design of high gain amplifiers considerably.

## PROBLEMS

**2.1.** Series- and parallel-combinations of MOS transistors as shown in Fig. P2.1 find many useful applications and are often seen as equivalent to a single MOS transistor.

  (*a*) Include channel-length modulation effect ($\lambda$) in the drain equation (2.16) and find the equivalent values of $K$ and $\lambda$ ($K_{eq}$ and $\lambda_{eq}$) for the equivalent transistor of the parallel combination.

  (*b*) Neglect the channel-length modulation effect ($\lambda$) in the drain equation and find the equivalent value of $K$ ($K_{eq}$) for the series combination. (Hint: M1 is always operating in the triode region, why?)

  (*c*) Repeat (*b*) using the simplified graphical method discussed in Sec. 2.2.3.

**2.2.** Use the graphical representation technique of Sec. 2.2.1 to compare the drain current of M1 in Fig. P2.2 with the difference of the drain currents of M2 and M3.

**2.3.** Figure P2.3 shows a programmable V-I converter. [13].

  (*a*) Derive an expression for $I_{out_1} - I_{out_2}$ using the classical first-order MOS transistor model. Given that $\frac{W}{L}_1 = \frac{W}{L}_2$, $\frac{W}{L}_3 = \frac{W}{L}_4$ and $\frac{W}{L}_5 = \frac{W}{L}_6 = \frac{W}{L}_7 = \frac{W}{L}_8$.

  (*b*) Design the transconductor to obtain a transconductance of 75 $\mu$ A/V and then simulate the design using SPICE. Refer to Appendix A for the transistor parameters.

**2.4.** Figure P2.4 shows an attenuator which is the basic building block of the volume control circuit mentioned in Sec 2.3.3. If voltages $V_A$ and $V_B$ are fixed and are equal, find the relationship between $I_b$ and $I_{in}$. Also find the relationship between $I_{b1}$ and $I_{in}$. Assume that all transistors are identical and operating in the triode region.

**2.5.** Figure P2.5 shows a two-bit D/A converter based on the current attenuator of Fig. P2.4.

  (*a*) How does the circuit work? Derive an expression for $V_{out}$ in terms of $V_{ref}$, $b_1$ and $b_2$, where $b_1$ and $b_2$ are the two input bits and $V_{ref}$ the reference

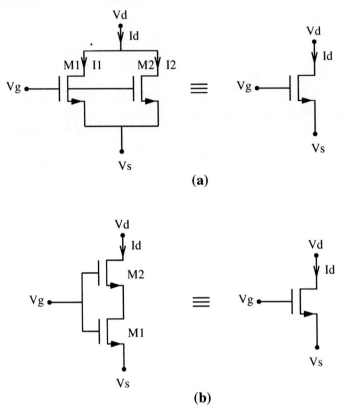

**(a)**

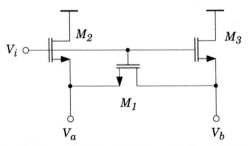

**(b)**

**FIGURE P2.1** (a) Parallel- (b) Series- combinations of MOS transistors.

**FIGURE P2.2** The voltages $V_g$, $V_a$ and $V_b$ are chosen in such a way that M1 is in the linear region and M2 and M3 are in the saturation region. Use the graphical representation of Sec. 2.2.2 to compare the drain current of transistor M1 with the difference of the drain currents of M2 and M3.

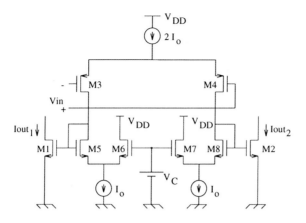

**FIGURE P2.3** A programmable transconductor circuit

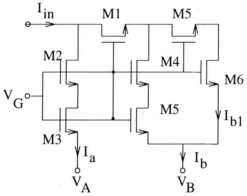

**FIGURE P2.4** A current attenuator based on the current division technique.

voltage. Assume the operational amplifiers to be ideal.

(b) Draw the circuit of a five-bit D/A converter based on the same technique.

**2.6.** Calculate the dc-gain from the input to the source of the cascode transistor in the 3 different configurations shown in Fig. P2.6.

**2.7.** Calculate the optimum feedback capacitor value, with respect to settling speed for the OTA in the circuit given in Fig. P2.7. The input and output parasitic capacitance of the op amp are equal and amount to $C_{ip} = C_{op} = 10$ pF and the ratio of $\frac{C_{fb}}{C_{fi}} = 4$.

**2.8.** Calculate the dc-gain of a source follower with back-gate connected to source and with back-gate connected to ground as shown in Fig. P2.8.

**2.9.** Calculate the gain margin of the amplifier of Fig. 2.12.

**2.10.** Consider the small signal equivalent circuit of Fig. P2.10 and neglect the overlap capacitance, $C_{gd}$.

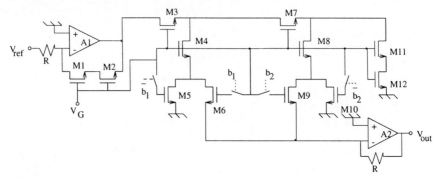

**FIGURE P2.5** 2-bit D/A converter based on the current division technique.

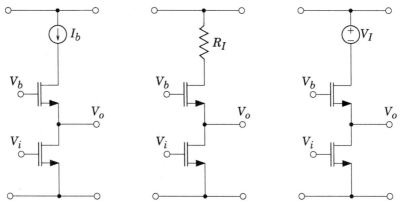

**FIGURE P2.6** The cascode stage with 3 different load conditions. Calculate the DC-gain from input to the source of the cascode transistor in the 3 different situations.

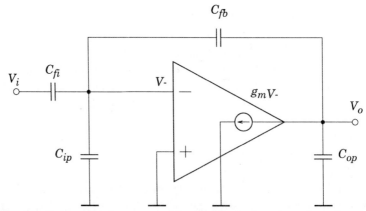

**FIGURE P2.7** An OTA in a capacitive feedback situation often encountered in switched capacitor circuits. If $C_{ip} = C_{op} = 10$ pF and $C_{fb} = 25C_{fi}$, calculate the optimum values for $C_{fi}$ and $C_{fb}$ with respect to bandwidth.

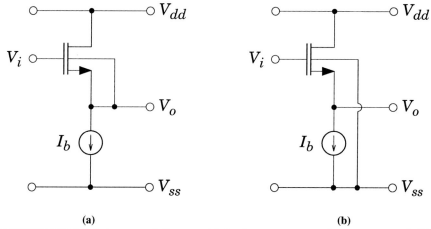

(a)                                          (b)

**FIGURE P2.8** (a) A source-follower with back-gate connected to source, and (b) a source-follower with back-gate connected to ground.

(a) Show that the small-signal short circuit ($V_o = 0$) current gain of an MOS transistor is given by

$$A_{i_{s.c.}} = \frac{i_d}{i_g} = \frac{g_m}{j\omega C_{gs}} = \frac{\omega_T}{j\omega}$$

Note that $A_{i_{s.c.}}$ is analogous to the familiar short circuit current gain $\beta$ of a bipolar junction transistor.

(b) The circuit in Fig. P2.10 is used in the design of very high frequency (VHF) transistor-only filters [14]. Find an expression for the small-signal input impedance $Z_{in} = \frac{v_{in}}{i_{in}}$ assuming matched transistors, and show that $Z_{in}$ simulates a passive RLC and approximates an inductance for $\omega < \omega_T$.

**2.11.** Figure P2.11 shows the P-type super MOS transistor corresponding to the N-type super MOS of Fig. 2.26.

(a) Determine the regions of operation of the various transistors in the P-type super MOS transistor.

(b) Now using the transistor parameters in Appendix A calculate the $\frac{W}{L}$ ratio for the various transistors so that they operate in their respective regions of operation while delivering a drain current of 150 $\mu$A for a gate voltage of -3.8 V. $V_{SS}$=-5 V ; $V_{DD}$=5 V.

(c) Simulate the above design and then adjust the aspect ratios of the transistors so that $V_{SD}$ of $M_{P1}$ is just 50 to 100 mV above its saturation voltage and the output resistance is greater than 150 M$\Omega$. Refer to Appendix A for the transistor parameters.

**2.12.** Consider the single-ended folded-cascode op amp of Fig. P2.12. Instead of the normal cascode stage the super MOS transistors are used to increase the output resistance and thereby the dc-gain of the op amp.

(a) Carry out the design for the N-type super MOS transistor of Fig 2.26 using the same conditions as in Prob. 2.11.

(b) Now design the folded-cascode op amp for a dc-gain of 90 dB and gain bandwidth of 50 MHz. Assume a load capacitance of 16 pF.

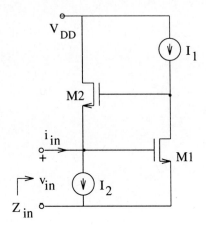

**FIGURE P2.10**

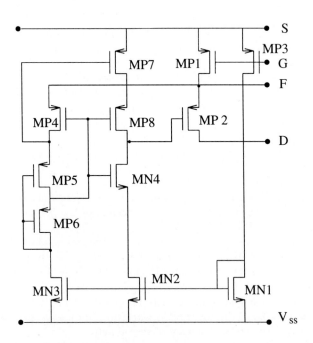

**FIGURE P2.11** A P-type super MOS transistor.

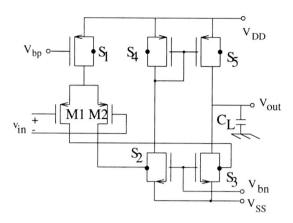

**FIGURE P2.12** A single-ended folded-cascode op amp using super MOS transistors.

## REFERENCES

[1] H. C. de Graaff and F. M. Klaassen, "Compact Transistor Modelling for Circuit Design," Springer-Verlag, New York, 1990.

[2] H. Wallinga and K. Bult, "Design and Analysis of CMOS Analog Signal Processing Circuits by Means of a Graphical MOST Model," *IEEE J. of Solid-State Circuits*, vol. 24, no. 3, pp. 672-680, June 1989.

[3] R. S. Muller and T. I. Kamins, "Device Electronics for Integrated Circuits," Wiley, New York, 1977.

[4] J. L. McCreary and P. R. Gray, "All-MOS charge redistribution analog-to-digital conversion techniques Part I," *IEEE J. of Solid-State Circuits*, vol. SC-10, no. 6, pp. 371-379, December 1975.

[5] H. Lee, D. A. Hodges and P. R. Gray, "A self-calibrating 15 bit CMOS A/D Converter," *IEEE J. of Solid-State Circuits*, vol. SC-19, no. 6, pp. 813-819, Dec. 1984.

[6] K. Bult and G. J. G. M. Geelen, "A Fast-Settling CMOS Opamp with 90-dB DC gain and 116 MHz Unity-Gain Frequency," *ISSCC Dig. Tech. Papers*, pp. 108-109, February 1990.

[7] K. Bult and G. J. G. M. Geelen, "A Fast-Settling CMOS Opamp for SC Circuits with 90-dB DC Gain," *IEEE J. of Solid-State Circuits*, vol. 25, no. 6, pp. 1379-1384, December 1990.

[8] K. Bult and G. J. G. M. Geelen, "The CMOS Gain-Boosting Technique," *Analog Integrated Circuits and Signal Processing*, vol. 1, no. 2, pp. 119-135, April 1991.

[9] T. C. Choi et al., "High-Frequency CMOS Switched-Capacitor Filters for Communications Application," *IEEE J. of Solid-State Circuits*, vol. SC-18, no. 6, pp. 652-664, December 1983.

[10] E. Sackinger and W. Guggenbuhl, "High-Swing, High-Impedance MOS Cascode Circuit," *IEEE J. of Solid-State Circuits*, vol. 25, no. 1, pp. 289-298, February 1990.

[11] K. Bult and G. J. G. M. Geelen, "An Inherently Linear and Compact MOST-Only Current-Division Technique," *ISSCC Dig. Tech. Papers*, pp. 198-199, February 1992.

[12] K. Bult and G. J. G. M. Geelen, "An Inherently Linear and Compact MOST-Only Current-Division Technique," *IEEE J. of Solid-State Circuits*, vol. 27, no. 6, pp. 1730-1735, December 1992.

[13] Y. Tsividis, "Operation and modeling of the MOS transistor," McGraw-Hill, 1987.

[14] E. Klumpernik, E. v.d. Zwan and E. Seevinck, "CMOS variable transconductance circuit with constant bandwidth," *Electron. Lett.*, vol. 25, no. 10, pp. 657-676, May 1989.

[15] M. Ismail, R. Wassenaar and W. Morrison, "A high-speed continuous-time bandpass VHF filter in MOS technology," *Proc. IEEE Int. Symp. on Circuits and Systems*, pp. 1761-1764, April 1991.

# CHAPTER
# 3

## CONTINUOUS-TIME SIGNAL PROCESSING

## 3.1  INTRODUCTION

Continuous-time analog signal processing circuits have long been designed in technologies other than CMOS. Modern analog and mixed-signal VLSI applications in areas such as artificial neural computation, telecommunications, smart sensors and battery-operated consumer electronics require subtle CMOS analog design solutions.

In recent years, CMOS continuous-time signal processing has shown signs of dramatic change. Field programmable analog arrays (FPAAs) [1], configurable and modular analog circuits [2, 3], are representative of emerging analog design philosophies leading to a whole new generation of analog circuit and layout design methodologies.

This chapter presents the fundamental concepts of modern CMOS circuit design for analog VLSI continuous-time signal and information processing. It attempts to blend design concepts used in traditional applications, such as monolithic amplifiers and filters, with design techniques used in emerging contemporary analog IC's such as winner-take-all circuits, vector multipliers and MOS resistive networks. We adopt a bottom-up approach and start with discussion of basic two-transistor "primitive" analog cells that are used in the design of more complex analog circuits and systems.

## 3.2   PRIMITIVE ANALOG CELLS

Figure 3.1 shows several two-transistor MOS analog cells which include the common-source and common-gate differential pairs, the composite CMOS transistor [9], the composite NMOS transistor or COMFET [5], a voltage follower and the CMOS inverter. Studying the operation of these "primitive" cells carefully will help us to easily understand the operation of many analog MOS circuits described in this chapter as well as in several other chapters in the book.

The current equations for these primitive cells are now derived with transistors operating in the saturation region, where the simple square-law current equation, i.e. Eq. (2.15) with $\alpha = 1$, is used.

For the matched common-source differential pair shown in Fig. 3.1(a), where $K_p = \mu_p C_{ox} W/L$ and $V_{Tp}$ is the threshold voltage of the PMOS transistors, we have

$$I_1 = \frac{K_p}{2}(V_{x1} - V_C - V_{Tp})^2 \tag{3.1}$$

$I_2$ will have the same form with $V_{x1}$ replaced by $V_{x2}$. Recognizing that $I_1 - I_2$ constitutes the difference between squares, one can use $a^2 - b^2 = (a - b)(a + b)$ and hence the differential output current is given by

$$I_1 - I_2 = \frac{K_p}{2}(V_{x1} - V_{x2})(V_{x1} + V_{x2} - 2V_C - 2V_{Tp}) \tag{3.2}$$

Similarly, for the common-gate differential pair shown in Fig. 3.1(b), the differential output current is given by

$$I_1 - I_2 = \frac{K_p}{2}(V_{x1} - V_{x2})(V_{x1} + V_{x2} - 2V_C + 2V_{Tp}) \tag{3.3}$$

Fig. 3.1(c) shows the composite CMOS transistor which can be viewed as equivalent to a single transistor with a floating gate-source voltage denoted by nodes $V_{Gn}$ and $V_{Gp}$. The equivalent transistor has the parameters $K_{eq}$ and $V_{Teq}$. Assuming that the equivalent transistor behaves as an n-channel transistor, $K_{eq}$ and $V_{Teq}$ are developed as follows.

$$I_D = \frac{K_n}{2}(V_{GS_n} - V_{Tn})^2 \tag{3.4}$$

$$= \frac{K_p}{2}(V_{GS_p} - V_{Tp})^2 \tag{3.5}$$

$$= \frac{K_{eq}}{2}(V_{Gn} - V_{Gp} - V_{Teq})^2 \tag{3.6}$$

which result in

$$V_{GS_n} = \sqrt{\frac{2I_D}{K_n}} + V_{Tn} \tag{3.7}$$

$$V_{GS_p} = -\sqrt{\frac{2I_D}{K_p}} + V_{Tp} \tag{3.8}$$

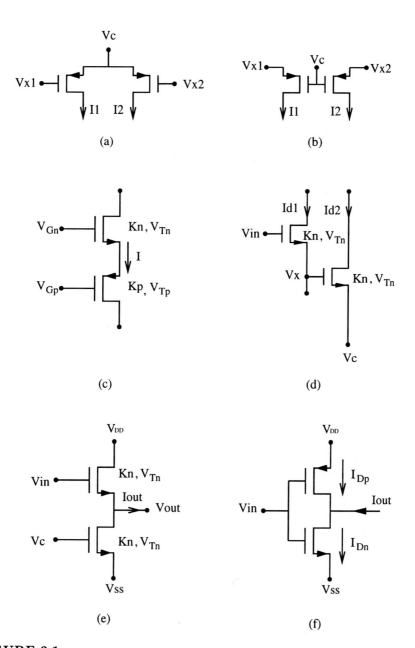

**FIGURE 3.1**
Primitive analog cells: (a) Common-source differential pair, (b) Common-gate differential pair, (c) Composite CMOS transistor, (d) Composite NMOS transistor, (e) Voltage follower, (f) CMOS Inverter.

$$V_{Gn} - V_{Gp} = \sqrt{\frac{2I_D}{K_{eq}}} + V_{Teq} \qquad (3.9)$$

That is

$$V_{Gn} - V_{Gp} = \sqrt{2I_D}(\frac{1}{\sqrt{K_n}} + \frac{1}{\sqrt{K_p}}) + (V_{Tn} - V_{Tp}) \qquad (3.10)$$

Comparing (3.9) with (3.10), we can obtain the equivalent transconductance parameter and threshold voltage of the resulting MOS transistor as below.

$$\frac{1}{\sqrt{K_{eq}}} = \frac{1}{\sqrt{K_n}} + \frac{1}{\sqrt{K_p}} \qquad (3.11)$$

$$V_{Teq} = V_{Tn} - V_{Tp} \qquad (3.12)$$

The differential current of the composite NMOS transistor shown in Fig. 3.1(d) is given by

$$I_{d1} - I_{d2} = \frac{K_n}{2}(V_{in} - 2V_x + V_C)(V_{in} - V_C - 2V_{Tn}) \qquad (3.13)$$

In the voltage follower shown in Fig. 3.1(e), since the two NMOS transistors are matched, the gate-source voltages are equal. Assuming $I_{out} = 0$ we have

$$V_{in} - V_{out} = V_C - V_{SS} \qquad (3.14)$$

where $V_C$ is a constant bias voltage. Therefore,

$$V_{out} = V_{in} - V_C + V_{SS} \qquad (3.15)$$

The drain currents of the CMOS inverter shown in Fig. 3.1(f) with transistors operating in the saturation region are given as

$$I_{Dn} = \frac{K_n}{2}(V_{GSn} - V_{Tn})^2 \qquad (3.16)$$

$$I_{Dp} = \frac{K_p}{2}(V_{GSp} - V_{Tp})^2 \qquad (3.17)$$

The output current of the inverter is

$$I_{out} = I_{Dn} - I_{Dp} = a(V_{in} - V_{Tn})^2 + bV_{in} + c \qquad (3.18)$$

where

$$a = \frac{1}{2}(K_n - K_p)$$

$$b = -K_n V_{SS} + K_p(V_{DD} - V_{Tn} + V_{Tp})$$

$$c = \frac{K_n}{2}(2V_{SS}V_{Tn} + V_{SS}^2) + \frac{K_p}{2}[V_{Tn}^2 - (V_{DD} + V_{Tp})^2]$$

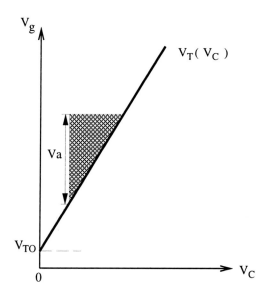

**FIGURE 3.2**

Representation of a constant current by the graphical method.

In addition to the hand analysis performed above, the graphical method, discussed in Chap. 2, can be used to evaluate the desired output current. Although the method is convenient, it is often used in a circuit where the drain, source, gate, and substrate voltages are known. In many MOS circuits, terminal voltages are not usually specified. An example of such circuits is the simple differential pair shown in Fig. 3.3(a) and biased by a constant current source. To circumvent this problem, one may recall the drain current equation given in Eq. (2.1). If a constant current source, $I$, is implemented by an MOS transistor operating in the strong inversion and biased by a constant voltage, where channel-length modulation effect is neglected, the current source $I$ can be treated as a triangle as shown in Fig. 3.2. The area of the triangle is given by

$$\triangle = \frac{I}{K} = \frac{V_a^2}{2\alpha} \tag{3.19}$$

Therefore,

$$V_a = \sqrt{\frac{2\alpha I}{K}} \tag{3.20}$$

Now, one may derive the I-V relation of the NMOS differential pair shown in Fig. 3.3(a) as follows. First, construct a graph as shown in Fig. 3.3(b). In the figure, $\triangle ABC$, $\triangle ADE$, and $\triangle AFG$ represent $I_2/K$, $I_1/K$ and $I/K$, respectively. Since

$$I_1 + I_2 = I \tag{3.21}$$

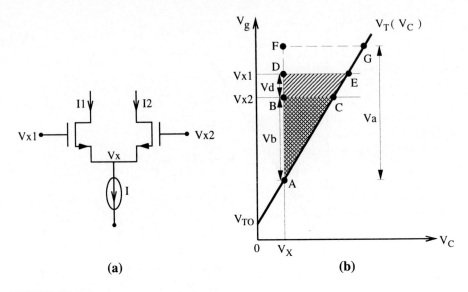

**(a)**                                                                           **(b)**

## FIGURE 3.3
(a) NMOS differential pair, drain voltages are not specified, (b) Graphical representation. The dependence of $V_T$ on the channel voltage, $V_C$, is approximated by a straight line having a slope $\alpha$.

the area of $\triangle ABC$ must equal that of $\Box DEGF$, and the following relation is obtained

$$V_b^2 = V_a^2 - (V_b + V_d)^2 \tag{3.22}$$

and

$$V_b = \frac{-V_d + \sqrt{2V_a^2 - V_d^2}}{2} \tag{3.23}$$

The difference in area between $\triangle ADE$ and $\triangle ABC$ is given by

$$\Box DBCE = V_d \times \frac{V_b + \frac{V_d}{2}}{\alpha}$$

$$= \frac{1}{2\alpha} V_d \sqrt{2V_a^2 - V_d^2}$$

$$= \frac{1}{2\alpha} V_d \sqrt{\frac{4\alpha I}{K} - V_d^2}$$

and

$$I_1 - I_2 = \frac{K}{2\alpha} V_d \sqrt{\frac{4\alpha I}{K} - V_d^2} \tag{3.24}$$

where $V_d$ is the differential input current.

## 3.3   LINEAR VOLTAGE-CURRENT CONVERTERS

Voltage-to-current (V-I) converters or VICs, also called transconductance amplifiers or simply transconductors, find many applications in both traditional analog ICs, e.g. amplifiers and filters, and contemporary analog VLSI, e.g. field programmable analog arrays [1] and cellular neural networks [6]. The simplest and most widely used transconductor is the source-coupled differential pair. While offering excellent high frequency performance (fundamentally limited by transmission line effects in the gate) and low noise, its large signal characteristics are nonlinear [7].

This section discusses simple techniques used in the design of linear V-I converters with relatively large signal handling. First we present a technique exploiting the body effect to improve the linearity of differential pairs.

### 3.3.1   A Simple Linear V-I Converter

If the body effect of the transistors in the PMOS common-source differential pair is included, (3.2) can be rewritten as

$$I_1 - I_2 = -\frac{K_p}{2\alpha_p} V_x \left(V_{x1} - V_{x2}\right) \tag{3.25}$$

where $V_x = -V_{x1} - V_{x2} + 2V_C + 2V_{Tp}$ and $V_{Tp} = V_{Tpo} + (\alpha_p - 1)V_{SB}$ is the threshold voltage. According to Eq. (3.25), linear relation between $(V_{x1} - V_{x2})$ and $(I_1 - I_2)$ can be achieved if $V_x$ is constant. This may be achieved by using a simple NMOS differential attenuator as shown in Fig. 3.4; composed of transistors M1–M5 so that its differential outputs ($V_{x1}$ and $V_{x2}$) are used as inputs to the PMOS differential pair, where M1-M2 ($K_d$) and M3-M4 ($K_u$) are matched respectively and M5 is a simple current sink biased by $V_{B1}$. The threshold voltage variations of M3 and M4 due to the body effect are characterized by the following equation:

$$V_{Tn(V_{SB})} = V_{Tno} + (\alpha_n - 1)V_{SB} \tag{3.26}$$

Writing down KCL equations at nodes $V_{x1}$ and $V_{x2}$, one gets

$$\frac{K_u}{2\alpha_n}(V_{DD} - V_{x1} - V_{Tn(V_{x1}-V_{ss})})^2 = \frac{K_d}{2\alpha_n}(V_{GS1} - V_{Tno})^2 \tag{3.27}$$

$$\frac{K_u}{2\alpha_n}(V_{DD} - V_{x2} - V_{Tn(V_{x2}-V_{ss})})^2 = \frac{K_d}{2\alpha_n}(V_{GS2} - V_{Tno})^2 \tag{3.28}$$

where $V_{GS1}$ and $V_{GS2}$ are the gate-source voltages of M1 and M2, respectively. By substituting (3.26) into the above equations, $V_{x1}$ and $V_{x2}$ are given as:

$$V_{x1} = \frac{1}{\alpha_n}\left[V_{DD} + (\alpha_n - 1)V_{SS} - V_{Tno} - \sqrt{\frac{K_d}{K_u}}(V_{GS1} - V_{Tno})\right] \tag{3.29}$$

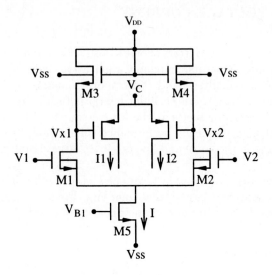

**FIGURE 3.4**
Simple linear V-I converter.

$$V_{x2} = \frac{1}{\alpha_n}\left[V_{DD} + (\alpha_n - 1)V_{SS} - V_{Tno} - \sqrt{\frac{K_d}{K_u}}(V_{GS2} - V_{Tno})\right] \quad (3.30)$$

Since $V_d \equiv V_1 - V_2 = V_{GS1} - V_{GS2}$ and $V_{GS1} + V_{GS2} = 2V_{Tno} + \sqrt{-V_d^2 + \frac{4_n I}{K_d}}$
(using (3.21) and (3.24)), then for the simple differential pair composed by M1
and M2, we get

$$V_{x1} - V_{x2} = -\frac{1}{\alpha_n}\sqrt{\frac{K_d}{K_u}}V_d \quad (3.31)$$

$$V_{x1} + V_{x2} = \frac{2}{\alpha_n}[V_{DD} + (\alpha_n - 1)V_{SS} - V_{Tno}] -$$

$$\frac{1}{\alpha_n}\sqrt{\frac{K_d}{K_u}}\sqrt{-V_d^2 + \frac{4\alpha_n I}{K_d}} \quad (3.32)$$

According to (3.31) and (3.32), the differential current expressed in (3.25)
can be rewritten as

$$I_d = \frac{K_p}{2\alpha_p\alpha_n}\sqrt{\frac{K_d}{K_u}}\left\{2V_C + 2V_{Tp} - \frac{2}{\alpha_n}[V_{DD} + (\alpha_n - 1)V_{SS} - V_{Tno}]\right.$$

$$\left.+\frac{1}{\alpha_n}\sqrt{\frac{K_d}{K_u}}\sqrt{-V_d^2 + \frac{4\alpha_n I}{K_d}}\right\}V_d \quad (3.33)$$

$$= \frac{K_p}{2\alpha_p\alpha_n}\sqrt{\frac{K_d}{K_u}}(V_K + \Delta V)V_d \quad (3.34)$$

where

$$V_K = 2V_C + 2V_{Tp} - \frac{2}{\alpha_n}[V_{DD} + (\alpha_n - 1)V_{SS} - V_{Tno}] \tag{3.35}$$

and

$$\Delta V = \frac{1}{\alpha_n}\sqrt{\frac{K_d}{K_u}}\sqrt{-V_d^2 + \frac{4\alpha_n I}{K_d}} \tag{3.36}$$

Note that $\Delta V$ represents the nonlinearity in a simple differential pair biased by a constant current (see (3.24)).

Since the device parameters and bias voltages are fixed, the nonlinearity of the differential current is due to $\Delta V$, which is practically small and can also be minimized by selecting small $K_d/K_u$ ratios as illustrated in the following example.

**Example 3.1.** In the circuit of Fig. 3.4, if $V_C=V_{DD}=$ 5 V, $V_{SS}=$ -5 V, $V_{Tp}=$ -0.73 V, $V_{Tno}=$ 1.068 V, $(W/L)_u=$ 100 $\mu$m/3$\mu$ m, $(W/L)_d=$ 50 $\mu$m/30$\mu$m, $K_d$ =30 $\mu$A/V$^2$ and I = 125 $\mu$A, the value of $\sqrt{-V_d^2 + \frac{4I}{K_d}}$ varies from 4.08 V to 3.23 V with $V_d$ increasing from 0 to 2.5 V. With no body effect ($\alpha_n = 1$), $V_K=$ 0.676 V and the variation of the nonlinear term with $\sqrt{\frac{K_d}{K_u}}=$ 0.134 is $\Delta V =$ 0.134 $\times$ (4.08 $-$ 3.23)= 0.114 V, which is not negligible compared to $V_K$. The total harmonic distortion (THD) obtained from SPICE [8], when the body terminals of M3 and M4 are connected to their sources, is 5.4% with $V_d$ equal to 1 V. If the body effect is taken into account, i.e. $\alpha_n > 1$, $V_K=$ 4.92 V and $\Delta V = \frac{0.134}{1.3117}$(4.68 $-$ 3.95)= 0.075 V with $\alpha_n=$ 1.3117. Since $\Delta V \ll V_K$, the nonlinearity is negligible and the V-I converter is practically linear. The THD obtained from SPICE (with the body terminals of M3 and M4 connected to $V_{SS}$) is 0.022% and 0.325% for $V_d$ equal to 1 V and 2.5 V, respectively.

We can then conclude that the body effect improves the linearity of the circuit. Further SPICE simulations using the parameters mentioned above are performed. The resultant I-V curves at different $V_C$ are illustrated in Fig. 3.5(a), where $V_2=$ 0 and $V_d = V_1$. This demonstrates that the transconductance can be electronically programmed by changing $V_C$. Different attenuation ratios ($\sqrt{\frac{K_d}{K_u}}$) are also used and the THD at 1KHz with $V_d = \pm$ 2 V is 0.167%, 0.211% and 0.311% for $\sqrt{\frac{K_d}{K_u}}$ equal to 0.109, 0.134 and 0.190 respectively, and the corresponding I-V curves are shown in Fig. 3.5(b).

### 3.3.2  A COMFET Transconductor

Here we will discuss linear V-I converters (VICs) composed of COMFETs [5, 9]. According to Eq. (3.13), a linear relation is achieved if

$$2V_x = V_{in} + V_B \tag{3.37}$$

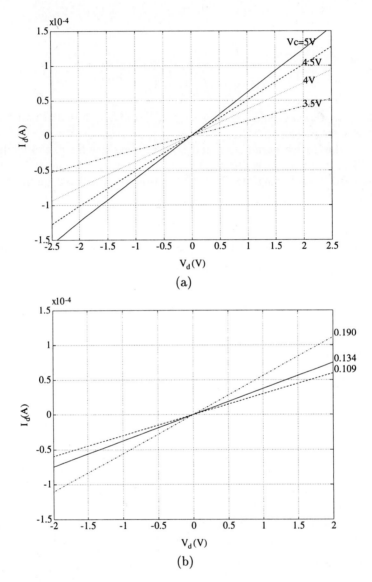

**FIGURE 3.5**
I-V curves with (a) different $V_C$, and (b) different attenuation ratios.

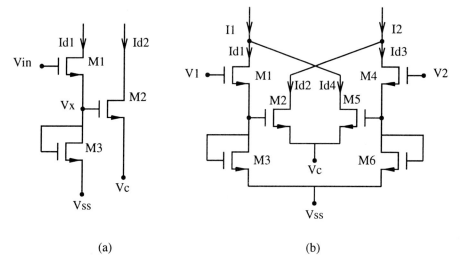

(a)                                    (b)

**FIGURE 3.6**
(a) Linear composite cell, (b) Linear NMOS V-I converter.

where $V_B$ is a constant voltage. It can be realized as shown in Fig. 3.6(a) [10], where $V_B = V_{SS}$ and M1, M2, and M3 are matched. The two-transistor circuit, $M_1, M_3$ resembles the voltage follower in Fig. 3.1(e). The difference in the drain currents with $\alpha = 1$ is then given by

$$I_{d1} - I_{d2} = \frac{K_n}{2}(V_C - V_{SS})(V_{in} - V_C - 2V_{Tn}) \qquad (3.38)$$

The resultant equation shows that the differential current and the input voltage, $V_{in}$, have a linear relation but with a dc-voltage offset. Therefore, a linear transconductance can be implemented by cross-coupling two basic cells as shown in Fig. 3.6(b) [10]. The differential current is then given by

$$
\begin{aligned}
I_1 - I_2 &= (I_{d1} - I_{d2}) - (I_{d3} - I_{d4}) \\
&= \frac{K_n}{2}(V_C - V_{SS})(V_1 - V_C - 2V_{Tn}) - \frac{K_n}{2}(V_C - V_{SS})(V_2 - V_C - 2V_{Tn}) \\
&= \frac{K_n}{2}(V_C - V_{SS})(V_1 - V_2) \qquad (3.39)
\end{aligned}
$$

This linear V-I converter has a constant $g_m$ given by $g_m = K_n(V_C - V_{SS})/2$, which can be electronically tuned by $V_C$. A disadvantage of this circuit is that the linear input range is limited by the condition needed to turn on transistors M1, M3, M4, and M6, which is $V_{SS} + 2V_{Tn} \leq V_{12}$.

In order to increase the linear range, the circuit is modified by replacing transistors M2 and M5 with composite CMOSFETs as shown in Fig. 3.7. If the mobilities of the NMOS and PMOS transistors are known, one can attempt to

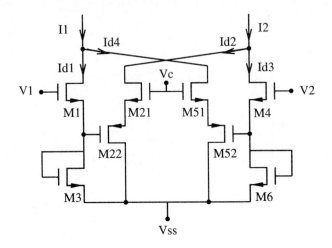

**FIGURE 3.7**
Modified linear CMOS V-I converter.

closely match $K_{eq}$ of the composite device and $K_n$. In fact, this is usually not the case, and a more practical solution is to choose the transconductance parameter of the PMOS transistor to be much larger than that of the NMOS transistor. As a result, $K_{eq} \approx K_n$. The differential current is obtained as performed before and is

$$I_1 - I_2 = \frac{K_n}{2}(V_C - V_{SS} - V_{Tn} - V_{Teq})(V_1 - V_2) \qquad (3.40)$$

and

$$g_m = \frac{K_n}{2}(V_C - V_{SS} - V_{Tn} - V_{Teq}) \qquad (3.41)$$

The previous analyses have assumed that $K_{eq} = K_n$, which is true if $K_p$ is relatively large. To avoid using large device sizes, only a reasonably large value of $K_p$ is chosen, and therefore results in distortion. Besides, nonlinearity occurs when transistors M1 and M3 (M4 and M6) leave the strong inversion region and enter subthreshold (weak inversion), i.e. $V_{12} - V_{SS} < 2(V_{Tn} + \alpha kT/q)$, where $kT/q$ is the thermal voltage. Since transistors M1 and M3 (M4 and M6) are assumed exactly matched, (3.37) is valid for transistors operating in both the strong and weak inversion regions with $V_{DS} \gg 3kT/q$ and $I_d \gg I_{cs}$, where $I_{cs}$ is the leakage current of the parasitic bipolar transistor associated with the MOS transistor. Although $I_{d1}$ is almost zero below $V_{SS} + 2V_{Tn}$ due to weak inversion, the resultant distortion is not significant. Since $I_{d2}$ ($I_{d4}$) is increasing and $I_{d1}$ ($I_{d3}$) is decreasing as $V_1$ ($V_2$) is decreasing, $I_{d2}$ ($I_{d4}$) is always much larger than $I_{d1}$ ($I_{d3}$) for small $V_1$ ($V_2$). The linear input range is approximately given as $(V_{SS} + MAX(3kT/q, V_{GScr})) \leq V_{12} \leq (V_{DD} + V_{Tn})$, where $V_{GScr}$ is the voltage $V_{GS}$ at which $I_d = I_{cs}$. Unlike Fig. 3.7, $I_{d1}$ ($I_{d3}$) is not negligible compared to $I_{d2}$ ($I_{d4}$) in Fig. 3.6(b).

The circuit shown in Fig. 3.7 is simulated by SPICE with MOSIS [11] p-

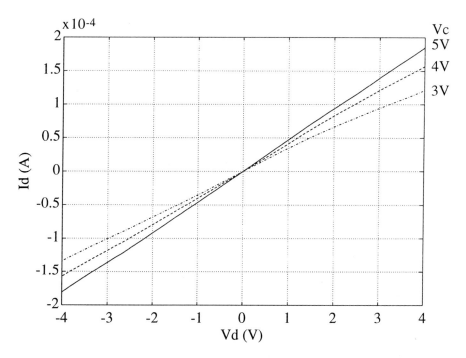

**FIGURE 3.8**
I-V curves of the COMFET transconductor.

well level 2 parameters. The aspect ratios of the NMOS and PMOS transistors are $10\mu m/20\mu m$ and $80\mu m/2\mu m$, respectively. The I-V curves at different values of $V_C$ are shown in Fig. 3.8 with $V_2 = 0$, $V_{SS} = -5V$, $V_{Tn} = 0.975V$ and $V_{Tp} = -0.769V$.

### 3.3.3  An Inverter-Based V-I Converter

Now we will discuss a linear CMOS V-I converter for high frequency applications [12]. It is based on the well-known CMOS inverter (Fig. 3.1(f)). According to (3.18), linear V-I conversion is achieved, if $a = 0$, i.e. $K_n = K_p$. However, mismatching causes nonlinearities, and a balanced structure as shown in Fig. 3.9(a) is used to cancel nonlinearity. The two matched inverters are driven by a differential input voltage $V_{id}$, balanced around the common-mode voltage $V_C$. $V_C$ is generated by the circuit shown in Fig. 3.9(b). With zero inverter output current, $V_C$ in Fig. 3.9(b) is obtained by (3.18) with $V_{in} = V_C$ and is given by

$$V_C = \frac{V_{DD} - V_{SS} - V_{Tn} + V_{Tp}}{1 + \sqrt{\frac{K_n}{K_p}}} + (V_{SS} + V_{Tn}) \qquad (3.42)$$

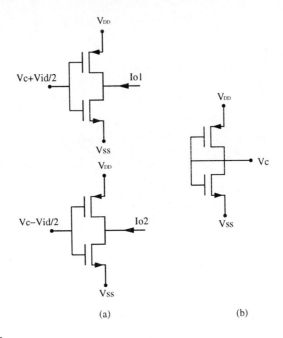

**FIGURE 3.9**

(a) Two balanced inverters driven by balanced signals to perform linear conversion, (b) Circuit to generate common-mode voltage $V_C$.

For $K_n = K_p$ and $V_{Tn} = -V_{Tp}$, $V_C$ is simply $(V_{DD} + V_{SS})/2$. The differential output current of Fig. 3.9(a) can therefore be obtained as follows.

$$
\begin{aligned}
I_{od} &= I_{o1} - I_{o2} = [a(V_C + V_{id}/2 - V_{Tn})^2 + b(V_C + V_{id}/2) + c] - \\
&\quad [a(V_C - V_{id}/2 - V_{Tn})^2 + b(V_C - V_{id}/2) + c] \\
&= V_{id}[b + 2a(V_C - V_{Tn})] \\
&= V_{id}[K_p(V_{DD} - V_C + V_{Tp}) + K_n(-V_{SS} + V_C - V_{Tn})]
\end{aligned}
$$
(3.43)

Substituting $V_C$ by (3.42), one obtains

$$I_{od} = V_{id}(V_{DD} - V_{SS} - V_{Tn} + V_{Tp})\sqrt{K_n K_p}$$
(3.44)

Therefore, the differential output current is linear with respect to the input differential voltage even when $K_n \neq K_p$. $K_n$ and $K_p$ should be matched however, in order to reduce the common-mode output current.

Figure 3.10 is the complete V-I converter, where inverters Inv3-Inv6 control the common-mode level of the output voltages $V_{o1}$ and $V_{o2}$. The output currents of inverters Inv3-Inv6, driven by the output voltages, are given by

$$I'_{o1} = I_{out5} + I_{out6}$$
(3.45)

$$I'_{o2} = I_{out4} + I_{out3}$$
(3.46)

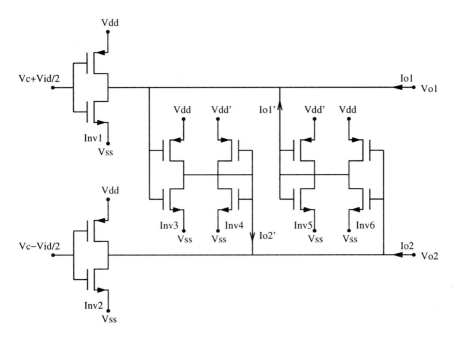

**FIGURE 3.10**
Complete linear V-I converter.

where $I_{out i}$ is the output current of $Inv i$. Expressing the output voltages in terms of the common-mode output voltage $V_{oc}$ and the differential-mode output voltage $V_{od}$, we get $V_{o1} = V_{oc} + V_{od}/2$ and $V_{o2} = V_{oc} - V_{od}/2$. To simplify the explanation, the transconductances of these inverters are assumed linear, i.e. $K_n = K_p = K$ and Eq. (3.18) is reduced to $I_{out i} = g_{mi}(V_{in} - V_C)$, where $g_{mi} = K(V_{DD} - V_{SS} - V_{Tn} + V_{Tp})$ and $V_C = (V_{DD} + V_{SS} + V_{Tn} + V_{Tp})/2$. The currents are

$$I'_{o1} = g_{m5}(V_{o1} - V_C) + g_{m6}(V_{o2} - V_C)$$

$$= (g_{m5} + g_{m6})(V_{oc} - V_C) + (g_{m5} - g_{m6})\frac{V_{od}}{2} \qquad (3.47)$$

$$I'_{o2} = (g_{m4} + g_{m3})(V_{oc} - V_C) - (g_{m4} - g_{m3})\frac{V_{od}}{2} \qquad (3.48)$$

As a result, nodes $V_{o1}$ and $V_{o2}$ are virtually loaded with resistances $1/(g_{m5} + g_{m6})$ and $1/(g_{m3} + g_{m4})$, respectively, for common-mode output signals. For differential-mode output signals, nodes $V_{o1}$ and $V_{o2}$ are virtually loaded with resistances $1/(g_{m5} - g_{m6})$ and $1/(g_{m4} - g_{m3})$, respectively. If all $g_m$'s of the four inverters are matched, the common-mode resistance is low and the differential-mode resistance is infinite. This results in a controlled common-mode voltage of the outputs, and the quiescent voltage is $V_C$.

This V-I converter is suitable for high frequency applications due to the absence of internal nodes, i.e. nodes which have no direct connection to either an input or an output terminal or a bias or supply terminal. Therefore the circuit has no internal parasitic capacitances and, in turn, has no parasitic poles or zeroes. All parasitic capacitances in the circuit are present at the input and output nodes. In a transconductance-C filter which uses the circuit as a basic block [12], the parasitic capacitances are lumped with the filter capacitors and incorporated into the design of the filter. It was shown [12, 13] that such filters operate successfully up to 100 MHz in a 3-micron CMOS process. Furthermore, the circuit has recently been used to implement the synapses of a high-speed application-specific, neural network [14].

## 3.4 MOS MULTIPLIERS

Analog four-quadrant multipliers find many applications in modern analog VLSI signal and information processing [15, 16]. One of the design techniques for a CMOS analog multiplier circuit is based on the square-law characteristics of an MOS transistor [4, 17, 18, 19]. A multiplier circuit with a 3-dB bandwidth of 30MHz has been described [18, 19]; however, the input range for a linear output swing did not exceed 50% of the power supply($\pm 5V$). In a bipolar technology, a Gilbert multiplier cell [20] is used, and a modified CMOS version of it with $\pm 4V$ input range is described [21]; however, its bandwidth is less than 1MHz. In this section, we will discuss a general circuit design technique for CMOS analog multipliers. As we will see, both in this section and in Chapter 5, this technique could lead to high-speed multipliers that are also suited for implementation in emerging low-voltage VLSI technologies. These low-voltage multipliers have many useful applications particularly in portable battery-operated consumer products.

The design technique is illustrated by the multiplier block diagram shown in Fig. 3.11(a), where each building block, $K_{BB}$, has an output current that is $K_{BB}(A/V^2)$ times the square of the difference of two input voltages [22]. That is $I_1 = K_{BB}(V_1 - V_2)^2, I_2 = K_{BB}(V_3 - V_4)^2....etc$. One can show that

$$I_1 + I_2 - I_3 - I_4 = 2K_{BB}(V_1 - V_3)(V_4 - V_2) \qquad (3.49)$$

Hence an analog multiplication of two differential input voltages is obtained. A voltage output is easily obtained when the above current is sourced into a resistor.

### 3.4.1 Square-Law Operation

Each building block in Fig. 3.11(a) can be implemented by a single MOS transistor, Fig. 3.11(b), operating in the saturation region assuming the ideal square-law drain current model. The resulting four-transistor circuit was first discussed by Zarabadi et al. [23], and more recently by Raut [24], and Ramirez-Angulo [25]. It was used in the implementation of a BiCMOS multiplier/mixer [23] which will be discussed further in Chapter 5. In this case, inputs will be applied at the gate($V_a$) and the source($V_b$) as shown in Fig. 3.11(b). Assuming that all four transistors

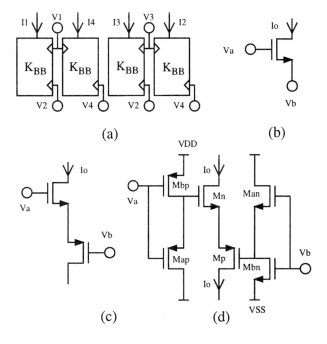

**FIGURE 3.11**
(a) Block diagram of the multiplier. (b) A single transistor implementation of the basic block.
(c) A composite transistor implementation. (d) CMOS implementation.

are matched, all nonlinearities including those due to the body effect will be canceled (see Example 3.3). However, the source terminal needs to be preceded by a buffer circuit because of its low input impedance [23, 22]. Additionally, the requirement $V_a - V_b - V_{Tn} > 0$, to keep the transistor on, limits the input range significantly. A possible alternative is to use the composite CMOSFET shown in Fig. 3.11(c). This provides high input impedance for both inputs, but the input range is still limited since the condition $V_a - V_b - V_{Teq} > 0$ must be satisfied. These limitations are overcome by the CMOS circuit shown in Fig. 3.11(d). The output current $I_O$, which flows through the composite $M_n - M_p$ pair, is given by

$$I_O = \frac{K_{eq}}{2}(V_{Gn} - V_{Gp} - V_{Teq})^2 \tag{3.50}$$

The PMOS transistors $M_{bp}$ and $M_{ap}$ and the NMOS transistors $M_{an}$ and $M_{bn}$ act as active attenuators [21] for inputs $V_a$ and $V_b$, respectively. $V_{Gn}$ and $V_{Gp}$ are given by

$$V_{Gn} = (1 - A_p)(V_a - V_{Tp}) + A_p V_{DD} \tag{3.51}$$

$$V_{Gp} = (1 - A_n)(V_b - V_{Tn}) + A_n V_{SS} \tag{3.52}$$

where $A_n^2 = K_{bn}/(K_{bn} + K_{an})$ and $A_p^2 = K_{bp}/(K_{bp} + K_{ap})$. The input range of the attenuators is limited by the threshold voltage of $M_{bp}$ or $M_{bn}$, whichever is larger.

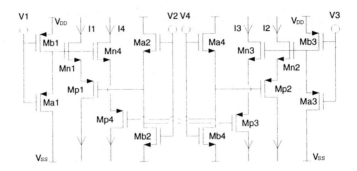

**FIGURE 3.12**
A complete multiplier circuit diagram: $K_{n1} = K_{n2} = K_{n3} = K_{n4} = K_n, K_{p1} = K_{p2} = K_{p3} = K_{P4} = K_p, K_{b1} = K_{b3} = K_{bp}, K_{b2} = K_{b4} = K_{bn}, K_{a1} = K_{a3} = K_{ap}$, and $K_{a2} = K_{a4} = K_{an}$.

Thus, $|V_{ab}| < min[V_{DD} - |V_{Tp}|, -V_{Tn} - V_{SS}]$. Furthermore, it can be shown that the requirement, $V_{Gn} - V_{Gp} > V_{Teq}$ is satisfied for the above input range as well if $A_n + A_p \geq 1$ (Prob. 3.6). It should be noted however that a larger $A_{np}$ results in more signal attenuation and hence the signal-to-noise ratio may be degraded. Fig. 3.12 shows the entire circuit; note that only one attenuator is needed for each multiplier input. Bulks are connected either to $V_{DD}$ or $V_{SS}$. Neglecting all the second order effects of MOSFETs, one can write

$$I_1 = \frac{K_{eq}}{2}[(1 - A_p)V_1 - (1 - A_n)V_2 + A_p(V_{DD} + V_{Tp}) - A_n(V_{SS} + V_{Tn})]^2 \quad (3.53)$$

$I_2$, $I_3$ and $I_4$ can be obtained similarly, and their combination becomes

$$I_1 + I_2 - I_3 - I_4 = K_{eq}(1 - A_p)(1 - A_n)(V_1 - V_3)(V_4 - V_2). \quad (3.54)$$

In Eq. (3.54), each of the inputs can take any value greater than (or less than) $V_{DD} - |V_{Tp}|$ (or $V_{SS} + V_{Tn}$). The circuit can also operate as a programmable VIC with either of the differential input voltages $V_1 - V_3$ or $V_4 - V_2$ operating as a dc-control voltage to program the transconductance. Example 3.2 discusses a design of the circuit in a $2\mu$ CMOS technology which shows that the circuit may successfully handle input voltages that cover 80% of the supply voltage and has a completely flat magnitude and phase response up to several mega-hertz.

**Example 3.2.** In this design example, the circuit of Fig. 3.12 is designed and simulated using level 2 SPICE parameters from the MOSIS [11] $2\mu m$ n-well CMOS process. The output $R(I_1 + I_2 - I_3 - I_4)$ can be produced in several ways using resistors, current mirrors, or combinations of both [22]. Current addition is performed by simply shorting the respective nodes while the subtraction is achieved by one or two current mirrors. To ensure better output signal swing and to isolate the accuracy of the current mirror from the overall performance, two ideal current mirrors can be used to source $I_1 + I_2$ into and sink $I_3 + I_4$ from a load resistor R. The aspect ratio of transistors used are: $(W/L)_n = 10/5$,

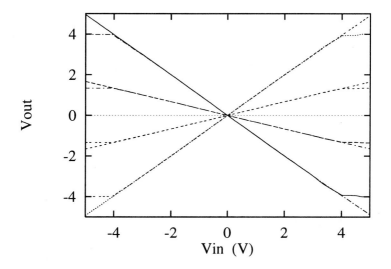

**FIGURE 3.13**
DC-transfer characteristics of the new multiplier.

$(W/L)_p = 55/5$, $(W/L)_{bp} = 10/5$, $(W/L)_{bn} = 20/5$, $(W/L)_{ap} = 5/5$, and $(W/L)_{an} = 23/5$. Inputs $V_3$ and $V_4$ were grounded to demonstrate the circuit operation with single-ended inputs. Fig. 3.13 shows the dc-characteristics of the circuit for $V_{DD} = -V_{SS} = 5$ V. Curves that saturate for $V_{IN} > 4$ V have $V_1$ as the input voltage and $V_2$ as the dc-voltage ($\pm 3$ V and $\pm 1$ V). When $V_1$ and $V_2$ are interchanged, we get the curves that saturate for $V_{IN} < -4$ V. For $|V_1|, |V_2| < |3\ V|$, the linearity error of the multiplier, defined as the percent deviation from the ideal output value, is less than 1%. The percent nonlinearity increases to about 2% for $|V_1|$ and $|V_2|$ equal to 4 V; this covers 80% of the supply voltage. Total harmonic distortion (THD) with 3 $V_{PEAK}$ input signal at either input terminal ($V_1$ or $V_2$) with $\pm 3$ V dc-voltage at the other terminal is less than 0.25% for frequencies up to 1 MHz and increases to .8% at 10 MHz. An ac small signal frequency response shows a very flat response up to several mega-hertz. The frequency at which the gain changes by 0.086 dB(1%) from the low frequency gain for $V_1 = V_{IN}(V_2 = \pm 3$ V) is 30 MHz with 5° phase shift; for the case $V_2 = V_{IN}(V_1 = \pm 3$ V) it is at 55 MHz with 1° phase shift.

The body effect causes the threshold voltage to shift and introduces nonlinearity in the output current. However, for the $M_n - M_p$ pairs, as the source voltage changes, $V_{Tn}$ and $V_{Tp}$ shift to the opposite direction from one another and their sum $V_{Teq}$ remains almost constant. Also, degradations caused by the body effect on the attenuators are small because each attenuator output supplies the signal to two $M_n - M_p$ pairs whose currents are subtracted at the output. $\lambda_n, \lambda_p, \theta_n$, and $\theta_p$ are the constants that model (in a usual way) the channel length

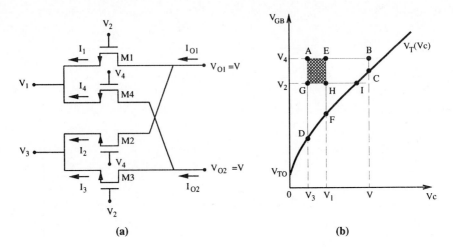

**FIGURE 3.14**

(a) The four-transistor multiplier implementation, (b) Graphical representation. $V_T(V_C)$ represents the nonlinear dependence of $V_T$ on the channel voltage, $V_C$.

modulation and the mobility degradation effects for the NMOS and PMOS transistors. Theoretical analysis with some approximations (*e.g.* assuming $\theta^2 = 0$) shows that in order to minimize the distortion behavior of the circuit due to $\lambda-$ and $\theta-$effects, $K_n \lambda_p^2 - K_p \lambda_n^2$ and $(\theta_n/K_n - \theta_p/K_p) K_{eq}$ must be respectively minimized.

The multiplier circuits discussed so far use MOS transistors operating in the saturation region. We will now discuss circuit examples which use transistors operating in the triode region.

### 3.4.2    Saturation/Triode Operation

Let us go back to Fig. 3.11 and use a single MOS transistor implementation for each building block, $K_{BB}$. The resulting four-transistor multiplier is shown in Fig. 3.14(a), where $I_{O1} = I_1 + I_2$, $I_{O2} = I_3 + I_4$ and, according to (3.49), the output current is simply $I_{O1} - I_{O2}$. The graphical representation of currents in the circuit is shown in Fig. 3.14(b). We will now show that under certain conditions all nonlinearities in the multiplier, including those due to the body effect, are also cancelled when all four matched transistors operate in the triode region or when the cross-coupled pair in Fig. 3.14(a) operates in the triode region while the other pair operates in saturation or vice versa [26]. The operation of the circuit as such allows the multiplier to handle large input signals while operating from a lower supply voltage. First, let us assume that $V_{O1} = V_{O2} = V$ and study the following two cases:

**1.** Both input nodes, $V_1$ and $V_3$, act as drains and the two $V$ nodes act as sources which is the case when the circuit is biased so that $V_1$ and $V_3$ are greater than $V$. The gate-source voltages of $M_1$ and $M_3$ are equal, and the same is true for $M_4$ and $M_2$. Recall that transistor currents in saturation are ideally dependent only on $V_{GS}$ while in triode both $V_{GS}$ and $V_{DS}$ change the current. Therefore, if all transistors operate in the saturation region, they will carry pairwise identical currents resulting in a zero differential output current.

2. Both input nodes $V_1$ and $V_3$ act as sources, Fig. 3.14(a), which is the case if $V_1$ and $V_3$ are smaller than $V$. Then either one of the transistor pairs $M_1 - M_3$, $M_4 - M_2$, or both pairs may operate in the saturation or the triode region.

This shows that in case 1, transistors may only operate in the triode region while in case 2, they may operate in saturation, triode, or both. Let us now consider case 2 and use (2.13) to write the following current equations in the triode region:

$$I_1 = \frac{K}{2\alpha}[(V_2 - V_{To} - \alpha V_1)^2 - (V_2 - V_{To} - \alpha V)^2] \tag{3.55}$$

$$I_3 = \frac{K}{2\alpha}[(V_2 - V_{To} - \alpha V_3)^2 - (V_2 - V_{To} - \alpha V)^2] \tag{3.56}$$

$$I_4 = \frac{K}{2\alpha}[(V_4 - V_{To} - \alpha V_1)^2 - (V_4 - V_{To} - \alpha V)^2] \tag{3.57}$$

$$I_2 = \frac{K}{2\alpha}[(V_4 - V_{To} - \alpha V_3)^2 - (V_4 - V_{To} - \alpha V)^2] \tag{3.58}$$

By subtracting $I_1$ and $I_3$, we find:

$$I_1 - I_3 = -\frac{K}{2}(V_1 - V_3)[2V_2 - 2V_{To} - \alpha(V_1 + V_3)] \tag{3.59}$$

Note that the differential current $(I_1 - I_3)$ will remain the same, (3.59), if the right quadratic term in both (3.55) and (3.56) disappears. This is actually the case if $M_1$ and $M_3$ operate in saturation (see (2.15)). In a similar way, $I_4 - I_2$ can be derived and is given by

$$I_4 - I_2 = -\frac{K}{2}(V_1 - V_3)[2V_4 - 2V_{To} - \alpha(V_1 + V_3)] \tag{3.60}$$

and hence,

$$I_{OUT} = I_{O1} - I_{O2} = (I_1 - I_3) - (I_4 - I_2)$$
$$= K(V_1 - V_3)(V_4 - V_2) \tag{3.61}$$

which shows that the output current is not only independent of the value of $V$ but also insensitive to both the body-effect factor ($\alpha$) and the threshold voltage ($V_{To}$). Therefore, both enhancement and depletion type transistors can be used.

It is also important to note from (3.55)–(3.58) that we must have $V_{O1} = V_{O2} = V$ to cancel nonlinearities when the transistors operate in the triode region and that this condition does not have to exist when they operate in the saturation region. Note that the above analysis neglects both $\lambda$- and $\theta$-effects.

The four-quadrant multiplier circuit with the four transistors operating in triode was first used by Song [16] in the implementation of high frequency data communication circuits. An operational amplifier-based version of the circuit was also developed [27] for low frequency applications and was shown to operate as both a multiplier and divider [28] (Prob. 3.7). It was also shown [29] that these triode-region MOS multipliers are practically insensitive to the distributed effects in the channel, and that the four-transistor circuit itself possesses inherent compensation to such effects [30] (Prob. 3.9). High speed low-voltage BiCMOS multipliers which exploit the operation of the circuit in triode and saturation will be discussed further in Chapter 5.

The different modes of operation of this simple MOS circuit indicate that while MOS circuits may look topologically identical, they may exhibit different behaviors under different operating conditions. In fact, an earlier version of the four-transistor circuit was used by Hsieh et al. [31] as a chopper multiplier to reduce the effect of flicker noise in low-noise CMOS operational amplifiers. The circuit operates quite differently in this case where the gate voltages are used to switch on the cross-coupled pair while the other pair is switched off and vice versa. This multiplies the differential input signal by a chopping square wave having an amplitude of $\pm 1$. When the two transistors in each pair are switched on, they operate simultaneously in saturation/triode [32] and the circuit performs in almost the same way as described earlier assuming that the chopper frequency is sufficiently high. Thus, the circuit operation as a chopper multiplier provides yet another circuit example which exploits operation in both triode and saturation.

**Example 3.3.** The result obtained in Eq. (3.61) can also be made visible by using the graphical method discussed in Chapter 2. $I_1$ corresponds to the area enclosed by IHF in Fig. 3.14(b) while $I_3$ corresponds to the area surrounded by IGD. The difference in current $(I_3 - I_1)$ is consequently K times the area enclosed by DGHF. It becomes visible by means of this graphical representation of currents that the difference in the drain-currents $(I_3 - I_1)$ is independent of the value of V. Therefore, transistor pair (M1-M3) may operate in either the linear or the saturation region.

In a similar way it can be shown that $K$ times the area enclosed by DAEF equals $(I_2 - I_4)$ and is also independent of the value of V. Thus, transistor pair (M2-M4) may also operate in the linear or the saturation region. It also becomes clear that, in the case that both pairs are operating in the saturation mode, their "V voltages" may even differ, i.e. $V_{O1} \neq V_{O2}$.

The output current $(I_{OUT})$ corresponds to the area indicated by AEHG. Since this area is always a rectangle, all nonlinearities due to the body effect represented by the $V_T(V_c)$ curve have been canceled. It shows at the same time that the circuit can be used as a continuous-time-four-quadrant multiplier.

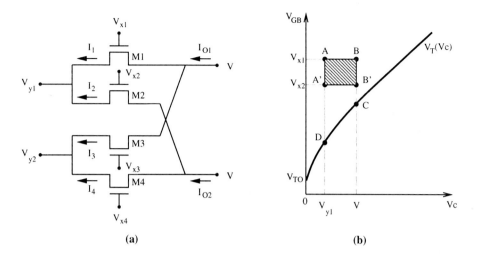

**FIGURE 3.15**

(a) The four transistors in triode, (b) Graphical representation of $I_1$ and $I_2$.

### 3.4.3  Triode-Only Operation

Let us further examine the multiplier circuit when we restrict its operation to the triode region. As we discussed earlier, the voltages $V_{O1}$ and $V_{O2}$ must be equal in this case, i.e. $V_{O1} = V_{O2} = V$, to achieve nonlinearity cancellation. However, it is interesting to note that complete nonlinearity cancellation is still achieved when each of the four transistors operates with a different gate voltage [33]. Furthermore, the four transistors do not even have to be all matched. To demonstrate this let us consider the circuit shown in Fig. 3.15(a) which is topologically identical to the circuit in Fig. 3.14(a). Note, however, that the arrow representing the source is omitted and that currents can flow in either direction, e.g. when the dc-bias voltage $V_{DS}$ is zero, unlike the case when devices are in saturation where inputs $(V_y)$ must be driving transistor sources. Now consider the graphical representation of $I_1$ and $I_2$ shown in Fig. 3.15(b). It is easy to see that

$$I_1 = K_1 \times \text{Area enclosed by ABCD} \qquad (3.62)$$

and

$$I_2 = K_2 \times \text{Area enclosed by A'B'CD} \qquad (3.63)$$

Now if M1 and M2 are matched, i.e. $K_1 = K_2 = K_A$, then $I_1 - I_2$ is linear as represented by the rectangle ABB'A' and is given by

$$I_1 - I_2 = K_A(V_{x1} - V_{x2})(V - V_{y1}) \qquad (3.64)$$

This simple two-transistor circuit [34], M1-M2, multiplies the differential gate voltage $(V_{x1} - V_{x2})$ by $(V - V_{y1})$. An op amp-based multiplier/divider [35] which uses the circuit is discussed in Probs. 3.11 and 3.12. Similarly, we can show that the two-transistor circuit, M3-M4, has

$$I_3 - I_4 = K_B(V_{r3} - V_{r4})(V - V_{y2}) \tag{3.65}$$

where $K_3 = K_4 = K_B$. Now it is clear that the four transistors in Fig. 3.15(a) do not have to be all matched; rather the pair M1-M2 is matched and so is M3-M4. This is unlike the cases discussed earlier when all four operate in saturation or when the cross-coupled pair operates in saturation (triode) while the other pair operates in triode (saturation). In those cases, all four transistors must be matched with pairwise, equal gate voltages as shown in Fig. 3.14(a). When the two matched pairs are interconnected (Fig. 3.15(a)), the resulting differential output current is given by

$$
\begin{aligned}
I_{O1} - I_{O2} &= (I_1 + I_3) - (I_2 + I_4) \\
&= K_A(V_{r2} - V_{r1})V_{y1} + K_B(V_{r4} - V_{r3})V_{y2} \\
&\quad + V[K_A(V_{r1} - V_{r2}) - K_B(V_{r4} - V_{r3})]
\end{aligned}
\tag{3.66}
$$

Equation (3.66) provides useful insight to the operation of the circuit and several special cases can easily be deduced from it [33]. First, if we are to eliminate the dependence of the differential output current on the voltage V, we must have

$$K_A(V_{r1} - V_{r2}) = K_B(V_{r4} - V_{r3}) \tag{3.67}$$

The circuit in Fig. 3.14(a), with all devices operating in the triode region, is obtained as a special case when $K_A = K_B$, $V_{r1} = V_{r4}$, and $V_{r2} = V_{r3}$.

Another circuit technique to eliminate $V$ from (3.66) is to force $I_{O1}$ to equal $I_{O2}$. In this case if $V_{y1}$ and $V_{y2}$ are grounded the differential signals $V_{r1} - V_{r2}$ and $V_{r4} - V_{r3}$ will follow (3.67). This technique is very useful in the implementation of fully-differential multiplier/divider circuits [36] and finite gain amplifiers with very high input impedance (Prob. 3.29). This is the case where M1 and M2, with gate voltages $V_{r1}$ and $V_{r2}$ are placed in the feedforward path of the amplifier while M3 and M4, with $V_{r3}$ and $V_{r4}$, are connected in the feedback path. A third circuit technique is to simply force $V$ to zero [34, 35].

Useful applications are also possible if the voltage $V$ is instead used as an input signal. Specifically, we will describe a simple fully-differential, four-quadrant multiplier based on the two transistor circuit, M1-M2 or M3-M4 in Fig. 3.15(a). This simple two–transistor circuit was used in an op amp multiplier/divider circuit [35] which forces the voltage $V$ to zero (see Prob. 3.11). The multiplier we are about to describe, though, uses the circuit but with $V_y = 0$ instead and uses $V$ as an input signal. The circuit is shown in Fig. 3.16. It is a bipolar-MOS, or BiCMOS, circuit where the bipolar transistors $Q_1$, $Q_2$, $Q_1'$ and $Q_2'$ are biased in the active region and operate as voltage followers. Assuming $v_{BE1} \approx v_{BE2} = v_{BE}$, the drain voltages of $M_1$ and $M_2$ will follow the input voltage $V_1$. That is $V = V_1 - v_{BE}$. Similarly with $v_{BE1}' \approx v_{BE2}' = v_{BE}$, the drain voltages of $M_1'$ and $M_2'$ will follow $V_2$. Neglecting base currents, the differential output current, $I_{out}$, is given by

$$I_{out} = (I_1 + I_2') - (I_2 + I_1')$$
$$= (I_1 - I_2) - (I_1' - I_2') \tag{3.68}$$

Using Eq. (3.64) with $V_{y1} = 0$, we have

$$I_1 - I_2 = K_A(V_{x1} - V_{x2})(V_1 - v_{BE}) \tag{3.69}$$

Similarly for $I_1' - I_2'$, we have

$$I_1' - I_2' = K_A'(V_{x1} - V_{x2})(V_2 - v_{BE}) \tag{3.70}$$

Hence, for $K_A = K_A' = K$, we get

$$I_{out} = K(V_{x1} - V_{x2})(V_1 - V_2) \tag{3.71}$$

It is important to note that the drain voltages in each MOS pair are made approximately equal by using bipolar devices. The exponential nature of these devices permits the assumption that the base-emitter voltages of devices having different emitter currents are approximately equal. Proper biasing of this BiCMOS multiplier should allow operation of the circuit from a relatively low supply voltage, e.g. 3.3V. The circuit can also operate successfully at higher frequencies, e.g. 50 MHz or more in a $2\mu$ BiCMOS process. This is because its internal nodes, $V$ and $V'$ have low impedance (outputs of emitter followers). It is also because the two-transistor circuit is inherently insensitive to distributed parasitic capacitances [37]. In fact, that is why the interconnection of two, or more, of these circuits, e.g. the four-transistor circuit of Fig. 3.15, is also insensitive to distributed parasitic effects (Prob. 3.9). More discussion on the circuit will follow in Chapter 5 (Prob. 5.21).

### 3.4.4 Multipliers and Programmable Transconductors

As indicated earlier in this section, if one of the differential voltage inputs is used as a dc-control voltage, the circuit operates as a programmable voltage-to-current converter (transconductor). An important application of linear transconductors is in the implementation of the function of a linear resistor. We will now show that the linear output current of any differential two-port linear transconductor can be seen as equivalent to a differential current flowing in a pair of floating resistors. To demonstrate this, let us consider the four-transistor circuit of Fig. 3.14(a). Its differential output current is given by (3.61). If the differential input, $(V_4 - V_2)$, is used as a dc-control voltage $(V_{C2} - V_{C1})$ then the differential transconductance is given by

$$g_{m1} = \frac{I_{O1} - I_{O2}}{V_1 - V_3} = K(V_{C2} - V_{C1}) \tag{3.72}$$

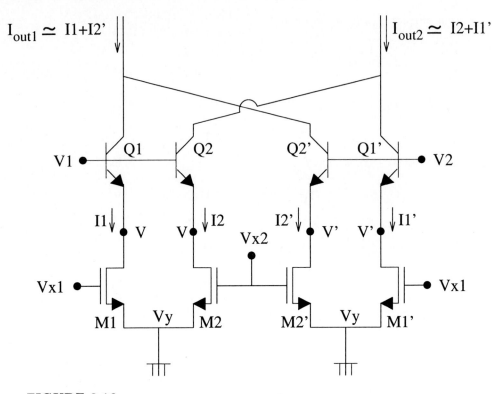

**FIGURE 3.16**
Fully-differential BiCMOS four-quadrant multiplier.

Alternatively, if $V_1 - V_3$ is viewed as a dc-control voltage, $V_{C2} - V_{C1}$, the differential transconductor becomes

$$g_{m2} = \frac{I_{O1} - I_{O2}}{V_4 - V_2} = K(V_{C2} - V_{C1}) \tag{3.73}$$

In both transconductors, transistors may operate in the saturation region ($V_{O1}$ may or may not equal $V_{O2}$), the triode region [38, 39] ($V_{O1}$ must equal $V_{O2}$) or in both regions as discussed earlier.

Now let us consider the resistor pair shown in Fig. 3.17(b). Its differential current is given by

$$I_{O1} - I_{O2} = \frac{1}{R}(V_3 - V_1) \tag{3.74}$$

which shows that the four-transistor circuit is equivalent to the voltage-controlled resistor pair, R, and that the equivalent resistance, $R_{eq}$, is given by

$$R_{eq} = R = \frac{1}{K(V_{C1} - V_{C2})} \tag{3.75}$$

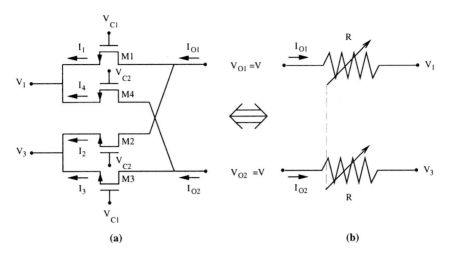

**FIGURE 3.17**
The double-MOSFET method.

which, assuming operation in the triode region , is rewritten as

$$\frac{1}{R} = K(V_{C1} - V_T) - K(V_{C2} - V_T) = \frac{1}{R_1} - \frac{1}{R_2} \tag{3.76}$$

where $R_1$ and $R_2$ are the small-signal resistance, also called the on-resistance $R_{on}$ [32], of the MOSFETs having $V_{C1}$ and $V_{C2}$, respectively, at their gates, $R_{on}$ is defined as $1/(dI_d/dV_{ds})$ when $V_{ds}$ approaches zero.

Fig. 3.17 shows this equivalence when $V_2 = V_{C1}$ and $V_4 = V_{C2}$ are used as dc-control voltages. The linear equivalence as depicted in Fig. 3.17 is referred to as the double-MOSFET method [40] and is used to systematically convert a classical analog circuit, e.g. active-RC circuit, to an all-MOS implementation, e.g. MOSFET-C filter. Care must be exercised in making use of the equivalence, since it is unidirectional in the sense that it is valid only when we look into the MOS circuit from the output port. The reader is encouraged to verify that the difference in currents leaving nodes $V_1$ and $V_3$ is nonlinear (find $I_1 + I_4 - I_2 - I_3$ from (3.55)-(3.58), or use the graphical representation). Circuit techniques which overcome this problem and provide bidirectional "MOS resistors" are discussed in the next section.

## 3.5  MOS RESISTORS

Resistors implemented by MOS transistors find many useful applications in modern analog VLSI signal and information processing. This is because linear passive resistors require a large area and cannot be electronically programmed to compensate for spread in their absolute values caused by random process variations. In these analog applications, the MOS resistor implementation takes a variety

of forms. In high-precision continuous-time MOS integrated filters, where implementation of accurate RC time-constants is essential, the function of a resistor is
often implemented by a linearized *two-port* V-I converter [40, 42, 43]. Accurate
time constants are achieved using an on-chip automatic tuning scheme which
controls the transconductance of the V-I converter in order to compensate for
random process variations [42].

In emerging low-precision analog VLSI information processing applications [44, 45, 46], *e.g.* image processing, large-scale resistive grids are used to
process signals in parallel and obtain results in real time. In such applications,
a resistor, which can be linear or nonlinear(resistive fuse), is implemented by
a *one-port* MOS circuit. Both two-port V-I converters and one-port "MOS resistors," or "active resistors," use circuit techniques similar to those discussed
earlier to cancel the nonlinearities in the MOS transistor current.

A one-port active MOS resistor architecture directly simulates the function
of a one-port real passive resistor, allowing the direct replacement of a resistor in
an analog circuit by an all-MOS implementation. This of course makes an active
MOS resistor an attractive basic circuit element in analog VLSI applications.

All-MOS resistors could be implemented by MOSFETs operating in either
the linear [47] or the saturation region [48, 49]. The lower MOS transconductance
and distributed channel capacitance in the triode region may limit the frequency
response. Dependence of the resistor on the threshold voltage of MOSFETs,
particularly when body effects cannot be ignored, hinders resistor tunability
(programmability) and limits its range. It also injects substrate noise into the
signal path and makes the design more sensitive to process variations. We will
describe several implementations of MOS programmable resistors which utilize
the strong or the weak inversion (subthreshold) mode of an MOS transistor. We
will also show that resistive circuits, where currents are the only signals of interest, can be implemented solely by transistors. Several system-level information
processing applications of MOS resistors will be discussed in Chapters 7 and 8.

### 3.5.1   A CMOS Floating Resistor

In this subsection, we will discuss a CMOS architecture for a floating one-port
linear resistor based on transistors operating in the saturation region [50]. It will
be shown that the architecture is threshold-voltage independent.

The design of the MOS resistor is based on two matched transistors operating in the saturation region as shown in Fig. 3.18, where nodes $V_x$ and $V_y$ are
the two terminals of the resistor, and $I_{in} = I_{out}$ is the current that flows through
the resistor. Note that the current, $I_1$, flowing into M1 is equal to the source
current of M2 and vice versa for $I_2$. Currents $I_1$ and $I_2$ are obtained with the
use of the square-law drain current equations as follows:

$$I_2 = \frac{K}{2}(V_x - V_1 - V_T)^2 \tag{3.77}$$

$$I_1 = \frac{K}{2}(V_y - V_2 - V_T)^2 \tag{3.78}$$

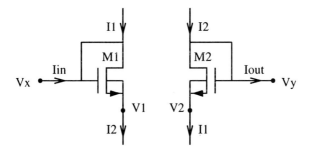

**FIGURE 3.18**
Basic cell of the MOS resistor.

The current of the MOS resistor is given by

$$I_{in} = I_{out} = I_2 - I_1$$
$$= \frac{K}{2}[(V_x - V_y) - (V_1 - V_2)] \times$$
$$[(V_x + V_y) - (V_1 + V_2) - 2V_T] \qquad (3.79)$$

The equivalent resistance, $R$, is given by

$$R = \frac{V_x - V_y}{I_{in}} = \frac{V_x - V_y}{I_{out}} \qquad (3.80)$$

In order to achieve linear resistance, we let

$$V_1 = V_y + V_B \qquad (3.81)$$
$$V_2 = V_x + V_B \qquad (3.82)$$

where $V_B$ is a constant voltage so that

$$V_1 - V_2 = -(V_x - V_y) \qquad (3.83)$$
$$V_1 + V_2 = V_x + V_y + 2V_B \qquad (3.84)$$

Obviously, this makes $I_{in} = I_{out}$ linear in $V_x - V_y$ and is expressed as follows:

$$I_{in} = I_{out} = 2K(-V_B - V_T)(V_x - V_y) \qquad (3.85)$$

The above relations can easily be implemented using two simple voltage followers of the type described earlier in Fig. 3.1(e). The complete circuit of the MOS resistor is shown in Fig. 3.19(a), where transistor pairs M3-M5 and M4-M6 are two voltage followers connected symmetrically between M1 and M2 and have the same $K$ value, $K_2$. $M_7$ through $M_{20}$ are current mirrors or constitute a part of a current mirror. These current mirrors force the condition $I_a = I_b = 0$ (needed for the voltage followers to operate as described in Sec. 3.2) and $I_{in} = I_{out}$

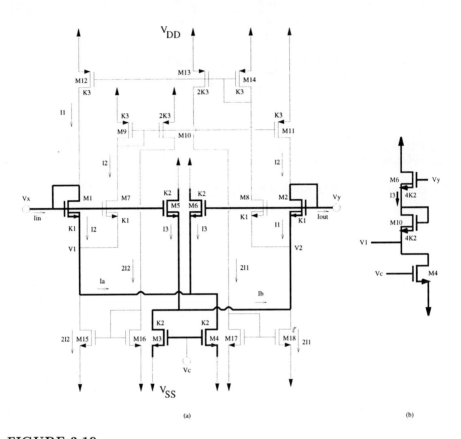

**FIGURE 3.19**
(a) MOSFET resistor circuit, (b) Modified voltage follower.

which is required in order for this circuit to behave as a floating resistor. The reader is encouraged to verify that these mirrors do indeed force these conditions. According to (3.15), $V_1$ and $V_2$ are given by

$$V_1 = V_y - V_C + V_{SS} \tag{3.86}$$

$$V_2 = V_x - V_C + V_{SS} \tag{3.87}$$

Substituting $V_1$ and $V_2$ in (3.79) by the above expressions, we have

$$I_2 - I_1 = 2K(V_x - V_y)(V_C - V_{SS} - V_T) \tag{3.88}$$

The equivalent resistance $R$ is expressed in terms of dc-biasing voltages and the transistor parameters as follows

$$R = \frac{1}{2K(V_C - V_{SS} - V_T)} \tag{3.89}$$

and a MOSFET linear resistor which is tuned by the dc-voltage $V_C$ is obtained. Clearly there is a tradeoff between the tuning range and the maximum voltage that can be applied across the resistor without losing the linearity. Let us study this tradeoff between the resistance value and the linearity by grounding $V_y$ to simplify the analysis. The resistor becomes nonlinear when the $R$ value chosen is too large ($V_C$ too small), or too small ($V_C$ too large) and when $V_r$ is positive (too close to $V_{DD}$), or negative (too close to $V_{SS}$). When the resistance value is too large, $V_C - V_{SS}$ in (3.89) becomes too small and $M_3$ ($M_4$) may turn off. Thus, this will give the following restriction which has to be met:

$$V_C - V_{SS} > V_T \tag{3.90}$$

In order to obtain a small resistance, $V_C - V_{SS}$ must be set large. This implies that the gate-to-source voltage of $M_4$ (and $M_3$) have to be increased, and this results in an increase in $I_3$, and thus lowers the voltage $V_1$ toward $V_{SS}$. This will cause $M_4$ to go into the linear region, and in order to avoid this situation, $V_C$ must satisfy the following condition.

$$V_C < \frac{V_{SS} + V_T}{2} \tag{3.91}$$

This is obtained from the saturation condition, $V_{DS4} > V_{GS4} - V_T$ and (3.86). The combination of (3.90) and (3.91) provides a design range for the control voltage $V_C$ given by

$$V_T + V_{SS} < V_C < \frac{V_{SS} + V_T}{2} \tag{3.92}$$

Note that the condition of (3.92) is still independent of $V_r$, however, the effect of $V_r$ is revealed next.

When $V_r$ is too large compared to $V_y$ (0V in this analysis), $V_2$ will follow and will increase by the same amount as $V_r$. Then transistors $M_8$ and $M_2$ will turn off. This can be avoided by satisfying the following condition.

$$V_{GS26} - V_T > 0 \tag{3.93}$$

which yields

$$V_C - V_{SS} - V_T > V_r \qquad (V_r > 0) \tag{3.94}$$

For the operation of this circuit with large negative $V_r$, the gate voltage of $M_1$ will come too close to $V_1$, which is set by $I_3$ and $V_y$, and $M_1$ and $M_7$ will turn off. To avoid this, the following condition must be satisfied.

$$V_C - V_{SS} - V_T > -V_r \qquad (V_r < 0) \tag{3.95}$$

Combining the above equations, an overall condition for the proper operation of the circuit is given by

$$V_T + V_{SS} + |V_r| < V_C < \frac{V_{SS} + V_T}{2} \tag{3.96}$$

**A Modified Architecture Independent of $V_T$**   It was shown previously that transistors $M_1$ and $M_2$ will turn off when $V_x$ is too close to $V_{DD}$ or $V_{SS}$. This is because, for a given value of the current $I_3$, the voltage $V_2$ is fixed by $V_x$ and $V_1$ is fixed by $V_y$. Thus there is not enough range between $V_x$ and $V_1$ (also $V_y$ and $V2$) for $V_1$ (and $V_2$) to increase or decrease before turning off the transistors. In order to increase the range of $V_x$ or $V_y$ and still keep $M_1 - M_2$ on, the two voltage followers are modified by adding a diode-connected transistor as shown in Fig. 3.19(b) with $M_6$ and $M_{10}$ having a $K$ value of $4K_2$. The resistance $R$ of the circuit between node $V_x$ and $V_y$ is determined in a similar fashion. This time we can write three equations for $I_3$; one for each of $M_6, M_{10}$, and $M_4$ as follows

$$I_3 = 2K_2(V_{gs6} - V_T)^2 \tag{3.97}$$

$$I_3 = 2K_2(V_{gs10} - V_T)^2 \tag{3.98}$$

$$I_3 = \frac{K_2}{2}(V_C - V_{SS} - V_T)^2 \tag{3.99}$$

Rewriting (3.97) and (3.98) for $V_{gs6}$ and $V_{gs10}$ and summing them to obtain $V_y - V_1$, we have

$$V_{gs6} + V_{gs10} = V_y - V_1 = 2V_T + 2\sqrt{\frac{I_3}{2K_2}} \tag{3.100}$$

and substituting (3.99) into (3.100) for $I_3$, we obtain

$$V_y - V_1 = 2V_T + V_C - V_{SS} - V_T$$

$$V_1 = V_y - V_C + V_{SS} - V_T \tag{3.101}$$

Comparing (3.101) to (3.86), we find that the difference, $V_y - V_1$, is increased by $V_T$. Similarly, for $V_2$ of this modified circuit, we obtain

$$V_2 = V_x - V_C + V_{SS} - V_T \tag{3.102}$$

Equations (3.101) and (3.102) are substituted into (3.79) and we obtain

$$I_2 - I_1 = 2K(V_x - V_y)(V_C - V_{SS}) \tag{3.103}$$

The resistance of the modified circuit is then given by

$$R = \frac{V_x - V_y}{I_{in}} = \frac{V_x - V_y}{I_2 - I_1}$$

$$= \frac{1}{2K(V_C - V_{SS})} \tag{3.104}$$

Thus, the resistance becomes independent of the threshold voltage.

Again we study the conditions that have to be met in order for the circuit to act as a linear resistor. By going through the same four instances when

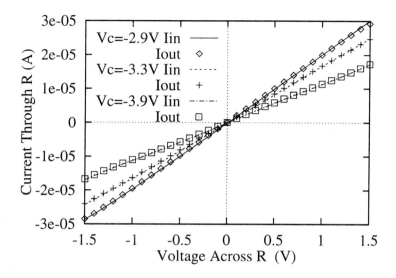

**FIGURE 3.20**
I–V curves of the modified MOS resistor circuit.

the resistor becomes nonlinear, as for the circuit in Fig. 3.19(a), the following
equations provide the linear ranges of operation of the modified circuit.

$$V_C - V_{SS} > V_T \qquad (3.105)$$

$$V_C < V_{SS}/2 \qquad (3.106)$$

$$V_C - V_{SS} > |V_x| \qquad (3.107)$$

Comparing (3.94) and (3.107), the voltage that can be applied across the resistor
without affecting the linearity should increase by $V_T$.

Figure 3.20 shows the SPICE simulation results of the modified circuit using
models obtained from a $2\mu$ CMOS process. Table 3.1 shows the resistance as a
function of the control voltage $|V_C|$. In summary, the linear range of operation of
the MOS resistor circuit is increased while $V_T$ is eliminated from the equivalent
resistance expression.

**Circuit Modeling** In order to perform a quick hand analysis of an analog
system that utilizes the MOS resistor circuit, a small signal circuit model of the
resistor circuit is needed. When all the MOSFETs in the modified circuit are
replaced by a simple MOS equivalent circuit model, a small signal model of the
resistor is obtained, and it is simplified and approximated (*e.g.* $\frac{1}{g_m} \ll r_O$) down
to a very compact model as shown in Fig. 3.21(a) in which $g_m$ is the small signal
transconductance and $C_{gs}$ is the gate-to-source capacitance.

**TABLE 3.1**
Resistance of the Modified MOS Resistor Circuit.

| $-V_C(V)$ | 2.5 | 2.7 | 2.9 | 3.1 | 3.3 | 3.5 | 3.7 | 3.9 |
|-----------|-----|-----|-----|-----|-----|-----|-----|-----|
| $R(K\Omega)$ | 44 | 47 | 50 | 54 | 60 | 66 | 74 | 85 |

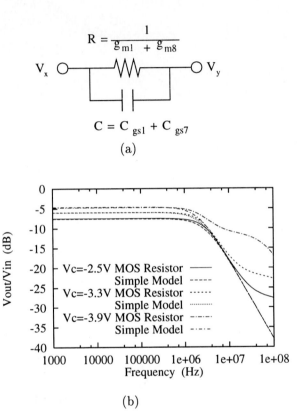

**FIGURE 3.21**
Small signal modeling of the MOS resistor circuit. (a) A simple model. (b) Frequency response of the voltage divider with the MOS resistor and with the resistor replaced by the model.

Using $g_m = K(V_{gs}-V_T)$ ((2.24) with $\alpha = 1$), we can show that the resistance value derived for the model is actually equivalent to (3.104) derived earlier. Observing that

$$g_{m1} = g_{m7} = K(V_X - V_1 - V_T) \tag{3.108}$$

$$g_{m2} = g_{m8} = K(V_Y - V_2 - V_T) \tag{3.109}$$

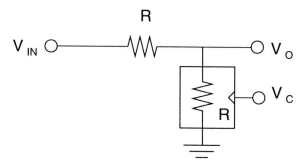

**FIGURE 3.22**
A voltage divider circuit used for the frequency response.

and substituting Eq. (3.101) for $V_1$ in Eq. (3.108), we obtain

$$g_{m1} = g_{m7} = K(V_X - V_Y + V_C - V_{SS}) \qquad (3.110)$$

and similarly,

$$g_{m2} = g_{m8} = K(V_Y - V_X + V_C - V_{SS}) \qquad (3.111)$$

Thus,

$$R = \frac{1}{g_{m1} + g_{m8}} = \frac{1}{g_{m2} + g_{m7}} = \frac{1}{2K(V_C - V_{SS})} \qquad (3.112)$$

Figure 3.21(b) shows the frequency response of the resistor model used in the voltage divider circuit of Fig. 3.22. The ideal resistor is $60K\Omega$ and the values of $C_{gs}$(2.1PF) and $g_m$ at different values of $V_C$ were obtained from SPICE. The simple model predicts the frequency response of the MOS resistor circuit well up to 10MHz.

### 3.5.2   Low-Precision MOS Resistor Circuits

Resistive networks [51, 52] can be used to smooth signals and images by filtering out noise. Furthermore, resistive fuses [44, 45] achieve noise filtering while preserving the edges in an image. Resistors in these networks need not be high-precision, but their I-V characteristics must resemble that of a resistor. Figure 3.23(a) shows a typical I-V curve of a low-precision resistor; for a large voltage drop across the resistor, the current slowly saturates to a constant value. On the other hand, when a resistive fuse detects a large voltage drop, it opens the circuit itself in order to maintain the original contrast of that segment. The I-V curve of a resistive-fuse is shown in Fig. 3.23(b).

In this subsection we will discuss low-precision resistor and resistive fuse circuits which utilize the weak or the strong inversion region.

**3.5.2.1   RESISTOR CIRCUITS.**  Since the VLSI applications of low-precision resistors require them to be placed in large numbers in the form of an array or

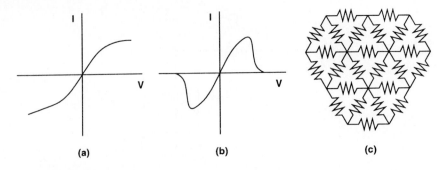

**FIGURE 3.23**
I-V curve of a low precision resistor (a) and a resistive fuse (b). (c) A resistive network.

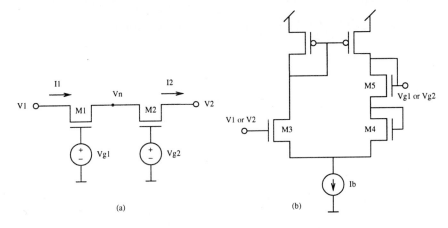

**FIGURE 3.24**
(a) An MOS low-precision resistor in the weak inversion mode. (b) A bias circuit.

a grid (Fig. 3.23(c)), the area of each resistor circuit, and hence the number and size of transistors used in the circuit, must be minimized. The circuit shown in Fig. 3.24(a) [51] operates in the weak inversion region and behaves as a resistor.

Equation (2.20) is used to analyze the circuit. For simplicity, we will drop the term $kT/q$ and keep in mind that all voltages should be divided by $kT/q$ in order for an expression to be correct. Equation (2.20) can be written as:

$$I_D = I_O e^{V_g}(e^{-V_s} - e^{-V_d}) \tag{3.113}$$

where $I_O = (W/L)I_{DO}$. Nodes $V_1$ and $V_2$ are the two terminals of the resistor and $V_{g1}$ and $V_{g2}$ are the bias voltages provided by the CMOS circuit shown in Fig. 3.24(b). The transistors with bubbles at their gates are p-channel devices. This symbology is often used with analog MOS circuits operating in subthreshold (weak inversion) [51]. Assuming that $V_1 > V_2$ and using (3.113), we can write expressions for $I_1$ and $I_2$ flowing from left to right.

$$I_1 = I_O e^{(V_{g1}-V_1)}(e^{(V_1-V_n)} - 1) \tag{3.114}$$

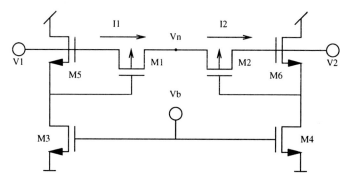

**FIGURE 3.25**
A CMOS low precision resistor in strong inversion mode.

$$I_2 = I_0 e^{(V_{g2} - V_2)}(1 - e^{(V_2 - V_n)}) \qquad (3.115)$$

Note that $I_1$ and $I_2$ are the same current which is the current flowing through the resistor. The bias current $I_b$ forces $V_{g1}/\alpha - V_1$ and $V_{g2}/\alpha - V_2$ to a constant $V_q$ ($I_0 e^{V_q} = I_b/2$; Prob. 3.17). With this in mind, we can find $V_n$ in terms of $V_1$ and $V_2$ by equating (3.114) and (3.115). Substituting $V_n$ back into (3.114) or (3.115) results in an expression for the resistor current in terms of $V_1$ and $V_2$.

$$I = I_0 e^{V_q} \frac{e^{V_1} - e^{V_2}}{e^{V_1} + e^{V_2}} \qquad (3.116)$$

$$I = I_0 e^{V_q} \tanh(\Delta V/2) \qquad (3.117)$$

where $\Delta V = V_1 - V_2$ is the voltage across the resistor. Now, since the *tanh* function resembles Fig. 3.23(a), the circuit shown in Fig. 3.24(a) indeed behaves as a low-precision resistor. We can compute the approximate equivalent resistance of the resistor by differentiating (3.117) with respect to $\Delta V$ and evaluating the result at $\Delta V = 0$ which gives

$$R_{eq} = \frac{2kT/q}{I_0 e^{V_q}} \qquad (3.118)$$

The next resistor circuit shown in Fig. 3.25 [52] has transistors operating in the strong inversion. Note that the basic subcircuit formed by $M_1$ and $M_2$ is the same as in Fig. 3.24(a). However, $M_1 - M_2$ are biased differently and operate in the triode region; the conditions for such an operation is explored in Prob. 3.18. The bias circuit consisting of $M_3 - M_6$ operates in the saturation region and has the gate-to-source voltage of each transistor equal to $V_b$. Thus, assuming the current flow as indicated in Fig. 3.25, we immediately see that

$$V_{sg1} = V_1 - V_{g1} = V_b \qquad (3.119)$$

$$V_{dg2} = V_2 - V_{g2} = V_b \qquad (3.120)$$

Since $M_1$ and $M_2$ are assumed to be in the triode region and have small $V_{ds}$, we can write approximate expressions for drain currents $I_1$ and $I_2$ by neglecting their $V_{ds}^2$ terms.

$$I_1 \approx K(V_{sg1} - |V_T|)(V_1 - V_n) \tag{3.121}$$

$$I_2 \approx K(V_{sg2} - |V_T|)(V_n - V_2)$$

$$= K(V_n - V_2 - V_b - |V_T|)(V_n - V_2) \tag{3.122}$$

where $K$ is the transconductance parameter and $V_T$ is the threshold voltage. Equating (3.121) and (3.122) and using (3.119) and (3.120), we find $V_n$ to be given by $(V_1 + V_2)/2$; note that the $(V_n - V_2)^2$ term was again neglected. Now, substituting $V_n$ back into (3.121), we find the current flowing through the resistor to be

$$I \approx \frac{K}{2}(V_b - V_T)(V_1 - V_2) \tag{3.123}$$

and hence the equivalent resistance between terminals $V_1$ and $V_2$ is given approximately by

$$R_{eq} = \frac{2}{K(V_b - V_T)} \tag{3.124}$$

**3.5.2.2  RESISTIVE FUSE CIRCUITS.**  As mentioned earlier, a resistive fuse is nothing but a resistor that becomes an open circuit when a large voltage drop is detected. The resistor circuit of Fig. 3.24 can be modified such that it works as a fuse [44]. First, note that since $I_o e^{V_q} = I_b/2$, we can rewrite (3.117) as follows

$$I = \frac{I_b}{2} \tanh(\Delta V/2) \tag{3.125}$$

Thus, what we need is a circuit which provides $I_b$ such that it reduces to zero for a large $\Delta V$. In Fig. 3.26(a), $I_{abs}$ is a current that is a function of $\Delta V$ and $I_c$ is a constant bias current. $I_b$ is given by $I_c - I_{abs}$ and becomes zero as soon as $I_{abs}$ reaches the value of $I_c$; the diode-connected transistor on the right side does not allow $I_b$ to be negative. $I_{abs}$ is produced by the circuit shown in Fig. 3.26(b), and its expression is given by (Prob. 3.19):

$$I_{abs} = I_a \tanh(\frac{|\Delta V|}{2\alpha}) \tag{3.126}$$

Thus, as the voltage drop across the resistor increases, $I_{abs}$ increases toward $I_c$ and inhibits the current flow through the resistor. Hence, this circuit works as a resistive-fuse operating in the weak inversion region.

The next resistive fuse circuit shown in Fig. 3.27 [45] operates in the strong inversion region. First, note that the resistor is formed by $M_3, M_4$, and $M_5$ in series, all operating in the triode region. We let the length of $M_4$ be much longer than that of $M_3$ and $M_5$ so that the equivalent resistance is provided mainly by $M_4$. This also implies that $V_a \approx V_1$ and $V_b \approx V_2$, which will be assumed in the rest of the analysis. The current source at the bottom of the circuit is set

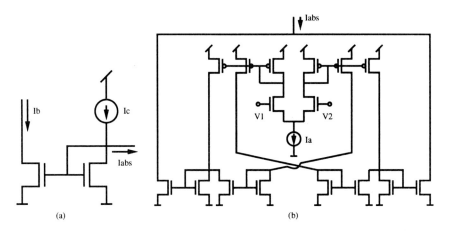

**FIGURE 3.26**
Bias circuits used to obtain a resistive fuse. (a) A circuit for $I_b$. (b) A circuit for $I_{abs}$.

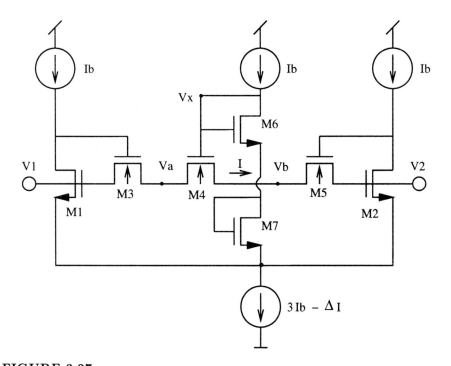

**FIGURE 3.27**
A resistive fuse utilizing the on resistance of an MOS operating in the triode region.

to be slightly less than $3I_b$ (by the amount $\Delta I$). When $\Delta V = V_1 - V_2$ is small, currents through $M_1$ and $M_2$ are slightly less than $I_b$. This will cause the drain of $M_1$ and $M_2$ to be pulled up toward the upper rail. Hence $M_3$ and $M_5$ will be in the triode region. Transistors $M_1, M_2, M_6,$ and $M_7$ have the same aspect ratios $(K_1 = K_2 = K_6 = K_7 = K_r)$ and have $I_b$ flowing through them. They are in the saturation region and their gate-to-source voltage is given by

$$V_{gs1267} = V_T + \sqrt{2I_b/K_r} \tag{3.127}$$

Then the gate voltage of $M_4, V_x$, is given by

$$V_x \approx V_1 + V_T + \sqrt{2I_b/K_r} \tag{3.128}$$

$$\approx V_2 + V_T + \sqrt{2I_b/K_r} \tag{3.129}$$

Using the above information, we find the gate-to-drain voltage of $M_4, V_{gd4}$, to be given by $V_T + \sqrt{2I_b/K_r}$, and hence $M_4$ is operating in the triode region. The current flowing through the resistive-fuse is given by

$$I \approx K_4(V_x - V_2 - V_T)(V_a - V_b) \tag{3.130}$$

Substituting Eq.(3.129) into Eq.(3.130), we find the equivalent resistance of the circuit,

$$R_{eq} = \frac{1}{K_4\sqrt{2I_b/K_r}} \tag{3.131}$$

When $V_1$ becomes large compared to $V_2$, more of the bottom current source tries to sink current from $M_1$, and hence the drain of $M_1$ is pulled down toward ground. This will cause $M_3$ to turn off and the path between $V_1$ and $V_2$ becomes open circuit.

All the circuits discussed above are symmetrical and the results obtained by assuming the current flow in the opposite direction will be the same as above.

### 3.5.3 Pseudo Resistive Networks

In Chapter 2, an interesting graphical model of a current $I$ in an MOS transistor was developed. Using this model for the MOS transistor shown in Fig. 3.28, where all voltages are referenced to the bulks (either the substrate of the chip, or a local well), the current is expressed as

$$I_D = I_S[F(V_G, V_S) - F(V_G, V_D)] \tag{3.132}$$

where, according to (2.6), $I_S$ is a constant current proportional to the aspect ratio $W/L$ and $I_S F(V_G, V_S) = W/L \int f(V_G, V_C)dV_C$. The above expression is general, i.e. it is valid for both the strong and weak inversion modes of operation. It is also symmetrical with respect to $V_D$ and $V_S$ and has been exploited in Chapter 2 in the development of a linear current division technique.

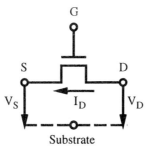

**FIGURE 3.28**
MOS transistor.

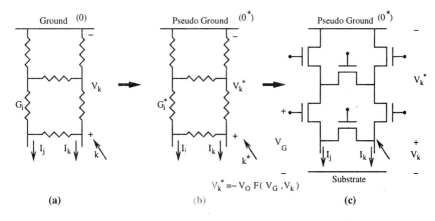

$$V_k^* = -V_O F(V_G, V_k)$$

(a)                    (b)                    (c)

**FIGURE 3.29**
Resistive network (a) in the $V$-domain, (b) in the $V^*$-domain, (c) MOS implementation of (b).

Vittoz and Arreguit [53] have rewritten the above equation in the form of a *pseudo Ohm's Law* as follows

$$I_D = G^*(V_D^* - V_S^*) \tag{3.133}$$

where $V^*$ is a *pseudo voltage* defined as

$$V^* = -V_O F(V_G, V) \tag{3.134}$$

and $V_O$ is an arbitrary scaling voltage. $G^*$ is a *pseudo conductance* equal to $I_S/V_O$. $V^*$ is always negative for an n-channel transistor, positive for a p-channel, and zero at a *pseudo ground*. Based on this *pseudo Ohm's Law*, any arbitrary network of linear resistors (conductances $G_i = 1/R_i$) can be transformed into the $V^*$-domain by replacing each resistor by an MOS transistor having a *pseudo conductance* $G_i^*$. Fig. 3.29 demonstrates such transformation. If all gate voltages are the same, the MOS implementation (Fig. 3.29(c)) behaves linearly with respect to current. It behaves linearly also with respect to voltages in the $V^*$-domain. The value of each $G_i^*$ can be made equal to the ratio $W/L$ of transistor $M_i$. In the weak inversion, $F(V_G, V)$ can be approximated by the product of two separate functions of $V_G$ and $V$, respectively [53]. That is

$$F(V_G, V) = e^{\frac{q(V_G - V_T)}{kT}} e^{\frac{-qV}{kT}} \tag{3.135}$$

The linear *pseudo Ohm's Law* is still valid but with the following new definitions:

$$V^* = -V_O \exp{\frac{-qV}{kT}} \tag{3.136}$$

and

$$G^* = \frac{I_S}{V_O} e^{\frac{q(V_G - V_T)}{kT}} \tag{3.137}$$

In this case $G^*$ of each MOS transistor is electronically programmable by the value of its gate voltage $V_G$. The *pseudo resistive* MOS network could find many useful applications in modern analog VLSI information processing systems. One such application, which is based on the "silicon retina" [51], is the MOS implementation of a large resistive array [53] intended for edge enhancement in a two-dimensional image. It could also be used in traditional analog applications. Prob. 2.5 discusses an MOS implementation of the familiar R-2R ladder used in data converters.

## 3.6 WINNER-TAKE-ALL CIRCUITS

Winner-take-all (WTA) circuits constitute yet another important class of circuits in analog VLSI information processing applications. They have recently been used in a number of applications including image feature extraction [54], nonlinear inhibition [55], subthreshold-region signal processing, etc. These applications take advantage of artificial neural systems which emulate certain aspects of the powerful organizing principles found in the nervous or neurobiological systems [51].

Neural systems contain a large number of massively-interconnected neurons with a large number of inputs. A neuron, like any other logic device, makes a decision based on the values of its inputs. This decision-making process is analog and involves processing of continuous-time signals. An important part of the function of the neural system is to be able to learn from "training" samples drawn from the environment. Among architectures suggested for the electronic implementation of artificial neural systems is the self-organizing network [56] using competitive learning [57]. It has the desirable property of effectively producing spatially-organized representation of various features of the input signals. Competitive learning depends upon competition between the output nodes of the neural network. The WTA function is executed to select a single winner. In this operation, the largest output value emerges as the winning node and it inhibits all other nodes.

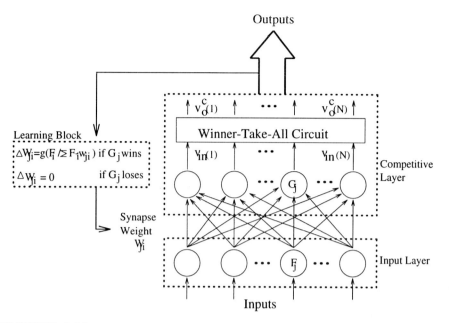

**FIGURE 3.30**
Winner-take-all (WTA) circuit used in self-organizing neural networks for competitive learning.

A typical self-organizing neural network consists of two layers as shown in Fig. 3.30. The first layer, which is also called the input layer, includes an array of input neurons. The second layer, which is also called the output layer, or the competitive layer, includes an array of output neurons and the WTA circuit. The input and output neurons are fully connected through a matrix of programmable synapses. An inner-product, or vector multiplication operation is typically performed in the synapse matrix with the operands from the synapse weights and the input signals. The output neuron functions as a current summer and a current-to-voltage converter to produce an analog voltage proportional to the inner-product result. It is usually realized by an operational amplifier with controllable voltage gains (see Section 3.7). The WTA circuit searches for the largest analog voltage from the output neurons and produces a sufficiently large output voltage level for the winning unit against the others.

A transistor-level schematic diagram for the WTA circuit is shown in Fig. 3.31 [58]. All transistors operate in the saturation region. Each WTA cell consists of two branches. The first branch ($M_1$, $M_2$, and $M_5$) converts an input voltage into the cell current as

$$I_j = \frac{K_1}{2}(V_{in}(j) - V_{CM} - V_T)^2, \quad j = 1...N \tag{3.138}$$

where $K_1$ and $V_T$ are the transconductance parameter and the effective threshold voltage of transistor $M_1$, respectively. These currents are compared and redistributed along the common signal node $V_{CM}$. All source terminals of the input

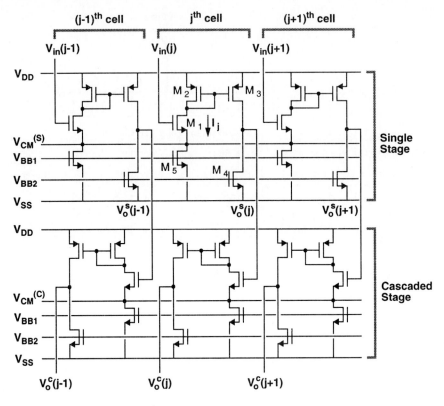

**FIGURE 3.31**
Schematic diagram of the basic WTA circuit.

transistors $M_1$'s are tied to this common signal node. In the second branch ($M_3$ and $M_4$), the current in each cell is converted into the output voltage as

$$V_{out}(j) = \frac{1}{\lambda_n} \left[ \frac{2mI_j}{K_4(V_{BB2} - V_{SS} - V_{T4})^2} - 1 \right] + V_{SS} \qquad (3.139)$$

where $\lambda_n$ is the channel length modulation of an NMOS transistor, $m$ is the current gain between transistors $M_2$ and $M_3$, and $V_{T4}$ is the threshold voltage of transistor $M_4$. Since the source terminals of transistor $M_1$ in each cell are at the same potential for all the cells, the current flowing through each cell is related to the square of the individual input voltage. Thus, the largest input voltage can secure the largest amount of current from the total bias current. This largest current is converted and amplified to produce the output voltage of the winning cell. If the input voltage differences are sufficiently large, the output voltage of the winning cell is saturated at the positive supply value, while the output voltage of the other cells is saturated at the negative power supply value. In fact, the output voltage behaves in a manner similar to the output of a digital CMOS inverter. This is no surprise since the output side, $M_3$-$M_4$, resembles a

CMOS digital inverter. Through the use of a common signal node, the total bias current, $I_B$, is provided by the transistor $M_5$ in every cell.

For the simplified case of only two competitive cells, this WTA circuit degenerates to a simple differential amplifier with differential voltage outputs. In Fig. 3.32(a), the simulated dc characteristics of a two-cell WTA circuit are shown [58]. A single power supply voltage of 5 V is used. One input voltage is set to 2.5 V, and the other input voltage is increased linearly from 2 to 3 V. To obtain the full binary output values for the winning and losing cells, the required input voltage difference is found to be at least 100 mV for the single-stage configuration. The simulated transient response is shown in Fig. 3.32(b). The response time of the single-stage design is about 100 ns at the load capacitance of 0.5 pF. The circuit performance can be enhanced by cascading the identical stages as shown in Fig. 3.31.

Performance improvement is apparent as shown in Fig. 3.32 for the two-input network. The cascading configuration drastically increases the entire voltage gain of the cell so that the transition region between the winner and the losers is greatly reduced. The operation speed of the WTA circuit is also improved because the load capacitance can be quickly driven by the stronger output signal.

## 3.7 AMPLIFIER-BASED SIGNAL PROCESSING

In previous sections, we discussed signal processing circuits designed at the transistor level. The primitive cells described in Section 3.2 were used as basic building blocks in the design of these circuits.

In this section, we go one step higher in the hierarchy of analog VLSI towards the implementation of large-scale systems and discuss signal processing circuits which use amplifiers as basic circuit elements. The design of the amplifier itself is discussed in this and other chapters in the book. For example, the operational amplifier (op amp) whose symbol is shown in Fig. 3.33(a) is discussed in Chapters 2, 4, and 5. The operational transconductance amplifier (OTA) (Fig. 3.33(c)), or V-I converter, is discussed in earlier sections of this chapter and in Chapters 4 and 5.

The differential difference amplifier (DDA), Fig. 3.33(b), is an emerging analog building block [59, 60, 2]. It has recently been used in a limited number of applications including the implementation of fully-differential MOS switched-capacitor filters [59], common-mode detection [61, 62], telephone line adaptation [63], and continuous-time filters [61, 34, 64, 65]. The DDA is an extension of the concept of the op amp. The main difference is that instead of two single-ended inputs, as is the case in op amps, it has two differential input ports, $(V_{pp} - V_{pm})$ and $(V_{np} - V_{nn})$. The output of the DDA can be written as

$$V_o = A_o[(V_{pp} - V_{pm}) - (V_{np} - V_{nn})] \qquad (3.140)$$

where $A_o$ is the open-loop gain of the DDA. When a negative feedback is introduced, i.e. to $V_{pm}$ or $V_{np}$ or both, which appear in (3.140) with a negative sign,

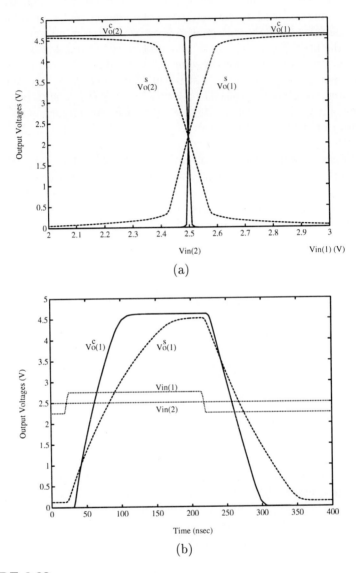

**FIGURE 3.32**
Simulation results for the two-input WTA circuit. (a) DC-transfer curve, (b) Transient behavior of the winning output with $C_L$ of 0.5pF. $V_o^s$ and $V_o^c$ are the single-stage and cascaded-stage output voltages, respectively.

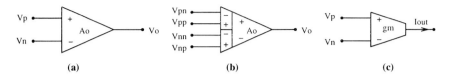

**FIGURE 3.33**
Amplifier symbols (a) Op amp, (b) DDA and (c) OTA, $I_{out} = g_m(V_p - V_n)$.

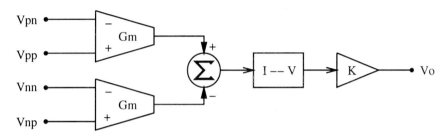

**FIGURE 3.34**
Block diagram of the DDA.

the following expression is obtained:

$$V_{pp} - V_{pm} = V_{np} - V_{nn} \quad with \quad A_o \to \infty \tag{3.141}$$

For a finite open-loop gain $A_o$, the difference between the two differential voltages increases as $A_o$ decreases. Therefore, the open-loop gain should be as large as possible in order to achieve high performance. According to (3.141), some interesting circuits can be realized with the DDA. These will be discussed later in the section.

In practice, the DDA may be implemented as shown in Fig. 3.34, where two V–I converters (the two input ports of the DDA) convert the two differential voltages into two currents which are then subtracted, converted into a voltage, and amplified. The simplest way to implement the DDA is to use two differential pairs to realize the V–I converters. One can recall that the differential current $I_d$ of the differential pair shown in Fig. 3.3 is given by (3.24). For linear conversion, $I_d$ should be proportional to $V_d$ only; however, the existence of the square-root term results in nonlinearity. Nonlinearity of $I_d$ is defined as the percent deviation from the ideal value of $g_m V_d$, where $g_m$ is the transconductance at $V_d = 0$

$$g_m = \left.\frac{dI_d}{dV_d}\right|_{V_d=0} = \sqrt{\frac{KI}{\alpha}} \tag{3.142}$$

When used as an input stage of an op amp, the differential pair is always operating at $V_d \approx 0$ due to the virtual short property in op amps and therefore its transconductance is approximately $g_m$. Unlike the op amp, the two inputs of each V–I converter in the DDA are not virtually shorted, and the differential pairs are not operating at a fixed $V_d$. This makes the input side of the DDA similar to that of the OTA. As $V_d$ increases, the transconductance of the differential pair decreases, and becomes zero for $|V_d| \geq \sqrt{\frac{2I}{K}}$, which occurs when $I_d = I$. The reduction of the transconductance due to nonlinearities in the V–I converter may degrade the open-loop gain of the DDA, which makes (3.141) no longer valid. Therefore, CMOS, Bipolar, or BiCMOS V–I converters with wide linear input range, such as those discussed in earlier sections and in Chapters 4 and 5, are essential for the design of a wide input range DDA [66]. DDA-based circuits operate in the frequency range of op amp-based circuits. However, they combine the low component count and high input impedance properties of OTA circuits with the low output impedance and ease of design of op amp circuits.

### 3.7.1   System-Level Building Blocks

Here we study some basic amplifier-based building blocks for system-level applications. Difference amplifiers, active resistors, multiplier/divider circuits, and integrators will be discussed. These basic circuits find important system-level applications in both traditional analog signal processing and modern information processing in analog VLSI [15, 44]. For simplicity, we will describe single-ended output implementations of these basic blocks. Their conversion to fully-differential architectures [31], which improves the system signal-to-noise ratio, is straightforward (see Chapters 4 and 9).

**Adder/Subtractor**   Using the DDA with $V_{pp} = V_{1+}$, $V_{pn} = V_-$, $V_{nn} = V_{2+}$, and $V_{np} = V_o$, (3.141), (negative feedback) as shown in Fig. 3.35(a), the output of the DDA circuit is given by

$$V_o = V_{1+} + V_{2+} - V_- \tag{3.143}$$

Note that not only are no additional components external to the DDA needed but also high input impedances are achieved. Some special functions can therefore be realized by this circuit. For example, with $V_{1+}$ and $V_{2+}$ grounded, the circuit becomes a voltage inverter. A voltage doubler, a basic block in a pipeline implementation of an algorithmic A/D converter [32], can also be realized by connecting $V_{1+}$ and $V_{2+}$ as the input and grounding the $V_-$ node. In addition, a level shifter and a voltage subtractor (differential to single-ended converter) can be implemented by the same architecture. A simple DDA instrumentation amplifier is shown in Fig. 3.35(b) which has an output given by

$$V_o = (1 + \frac{R_1}{R_2})(V_{in2} - V_{in1}) \tag{3.144}$$

**MOS Grounded Resistor**   An equivalent grounded resistor can be realized by an MOS transistor operating in the triode region with its drain and source

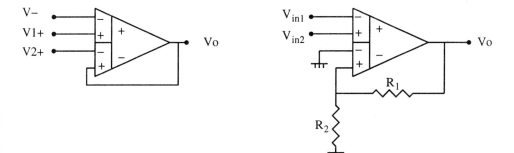

**FIGURE 3.35**
(a) DDA adder/subtractor and (b) instrumentation amplifier.

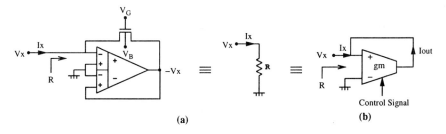

**(a)**                                          **(b)**

**FIGURE 3.36**
Grounded resistor implementations.

voltages out of phase [4] as shown in Fig. 3.36(a). Using the simple drain current equation for an NMOS enhancement transistor in the triode region, one may obtain the current $I_x$, which is linearly dependent on $V_x$ and is given by

$$I_x = 2K(V_G - V_T)V_x \quad for \ V_G \geq |V_x| + V_T \qquad (3.145)$$

where $K$ is the transconductance parameter and $V_T$ is the threshold voltage of the transistor. Note that a DDA inverter is connected between the drain and the source nodes of the transistor. The equivalent resistance $R$ is given by

$$R = \frac{V_x}{I_x} = \frac{1}{2K(V_G - V_T)} \qquad (3.146)$$

A simple implementation which uses an OTA is shown in Fig. 3.36(b). A linear resistance $R = 1/g_m$ over a wide voltage range is obtained when we use a linearized V-I converter in which $I_{out}$ is linear (see earlier sections and Chapter 5). The resistance value is electronically programmed using the V-I converter's control signal.

**Multiplication/Division**   A four-quadrant multiplication/modulation cell is realized using a circuit concept similar to that of the linear grounded resistor

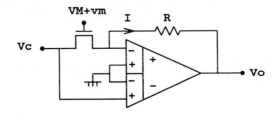

**FIGURE 3.37**
DDA-based multiplication/modulation cell.

and is shown in Fig. 3.37, where $V_M$ is a dc-bias and $v_m$ is an ac-signal. Due to the virtual short property between the two differential input ports of the DDA with negative feedback, $V_{pn} = -V_c$. The current $I$ is therefore linear and given by $2K(V_M + v_m - V_T)V_c$ with the transistor operating in the triode region. As a result, the output of the cell is given by

$$V_o = -V_c - 2KR(V_M + v_m - V_T)V_c$$
$$= -[1 + 2KR(V_M - V_T)] \times$$
$$[1 + \frac{2KR}{1 + 2KR(V_M - V_T)}v_m]V_c \qquad (3.147)$$

where $V_M + v_m \geq |V_c| + V_T$ for an NMOS transistor. Recall that an amplitude modulated (AM) waveform can be expressed by

$$\phi_{AM}(t) = A(1 + mf_m(t))\cos\omega_c t \qquad (3.148)$$

where $A$ is a dc-signal and $f_m(t)$ and $\cos\omega_c t$ are the modulating and carrier signals, respectively. $m$ is the modulation index, where $m \leq 1$ is required for demodulation using an envelope detector. Comparing (3.147) and (3.148), one can see that $v_m$ operates as the modulating signal and $V_c$ as the carrier signal, and thus an AM signal can be generated by the circuit.

A programmable all-MOS implementation of the circuit in Fig. 3.37 is easily achieved using voltage-controlled floating MOS resistors to replace $R$, e.g. the MOS square-law floating resistor discussed in Sec. 3.5. In this case the output $V_o$ is programmed by the dc-voltage controlling $R$.

Analog multipliers and dividers which use op amps have a wide range of applications and have long been designed in technologies other than CMOS. Modern VLSI applications such as analog neural computation [44], sensor linearization [29], and other important mixed-signal applications require CMOS design solutions. In this regard, CMOS multiplier design has received most of the attention. Dividers, on the other hand, require more complex circuitry and, in their simplest form, employ a multiplier in the feedback path of an op amp-based inverting amplifier. A simple design solution which simultaneously implements multiplication and division is shown in Fig. 3.38(a) [15, 28]. It uses the four-transistor MOS multiplier circuit, described earlier in Sec. 3.4, at both the

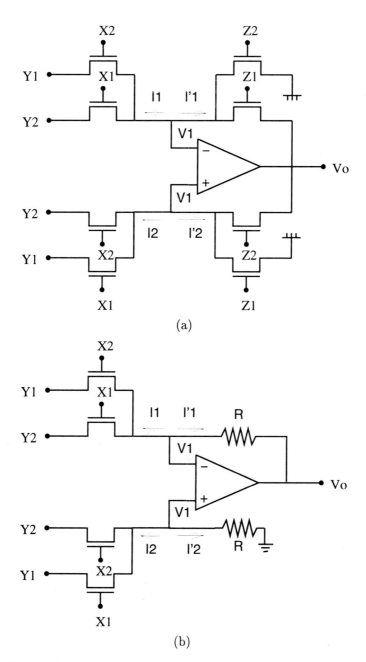

**FIGURE 3.38**
Opamp-based multiplier/divider (a) and multiplier (b) circuits.

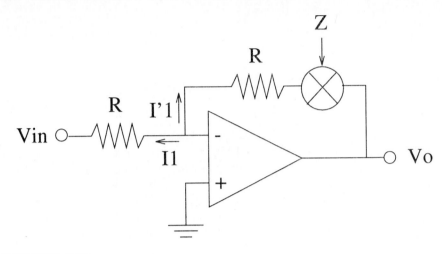

**FIGURE 3.39**
Conventional divider implementation.

feedforward and feedback paths. In order to understand the operation of the circuit and appreciate its simplicity, let us first consider the multiplier circuit shown in Fig. 3.38(b). Using (3.61), we have

$$I_1 - I_2 = K_i(Y_1 - Y_2)(X_1 - X_2) \tag{3.149}$$

Also writing KVL at the input nodes of the op amp (assumed ideal) we have

$$V_o = I_2'R - I_1'R = R(I_2' - I_1') \tag{3.150}$$

Since $I_1 + I_1' = 0$ and $I_2 + I_2' = 0$ then $I_2' - I_1' = I_1 - I_2$ and

$$V_o = K_iR(Y_1 - Y_2)(X_1 - X_2) \tag{3.151}$$

which provides the multiplication of the differential inputs $\Delta Y = Y_1 - Y_2$ and $\Delta X = X_1 - X_2$. Traditionally a divider is implemented using a multiplier connected in the feedback path of an inverting amplifier as shown in Fig. 3.39. Writing KCL at the inverting input node, we get

$$V_o = -\frac{V_{in}}{Z} \quad Z > 0 \tag{3.152}$$

where the polarity of the divisor is restricted to maintain negative feedback and hence ensures stability. An all-MOS implementation of this divider circuit is possible. However, the resulting MOS circuit will use a large number of MOS transistors and two op amps.

Now going back to the multiplier circuit of Fig. 3.38(b), the resistor pair in the circuit can easily be replaced by MOS resistors according to the double-MOSFET method discussed earlier in Section 3.4. The resulting circuit is shown in Fig. 3.38(a), and it has

$$I_1' - I_2' = K_oV_o(Z_2 - Z_1) \tag{3.153}$$

Again, since $I_2' - I_1' = I_1 - I_2$, we get

$$K_o V_o (Z_1 - Z_2) = K_i (Y_1 - Y_2)(X_1 - X_2) \qquad (3.154)$$

which can be written as

$$V_o = \frac{(\frac{W}{L})_i}{(\frac{W}{L})_o} \frac{\Delta Y \Delta X}{\Delta Z} \qquad \Delta Z > 0 \qquad (3.155)$$

where $\Delta Z = Z_1 - Z_2$, and $(\frac{W}{L})_{io}$ are the width-to-length ratios of the four MOS devices at the input and output side of the circuit, respectively. Again, the polarity of the divisor is restricted to $\Delta Z > 0$ or $Z_1 > Z_2$ in order to ensure stability. While the reason for this is easy to see in Fig. 3.39, it is not very obvious here. In order to clarify this, let us assume that the MOS transistors are operating in the triode region and that the open-loop gain of the op amp is modeled by $GB/s$ where $GB$ is the finite op amp gain-bandwidth. A small signal analysis of the circuit when all 8 MOS transistors are replaced by their triode-region small-signal equivalent resistances $R_{r1}$, $R_{r2}$, $R_{z1}$, and $R_{z2}$ yields [29]

$$V_o = \frac{(\frac{W}{L})_i}{(\frac{W}{L})_o} \frac{\Delta Y \Delta X}{\Delta Z} \frac{1}{1 + \frac{R_f}{R_{eq}} \frac{s}{GB}} \qquad (3.156)$$

where $R_f$ is the effective feedback resistance and is given by

$$\frac{1}{R_f} = \frac{1}{R_{z1}} - \frac{1}{R_{z2}} = K_o (Z_1 - Z_2) \qquad (3.157)$$

and $R_{eq}$ is the equivalent grounded resistance seen at either the inverting or noninverting input node of the op amp and is given by

$$\frac{1}{R_{eq}} = \frac{1}{R_{r1}} + \frac{1}{R_{r2}} + \frac{1}{R_{z1}} + \frac{1}{R_{z2}} \qquad (3.158)$$

It is now obvious to see that we must have $Z_1 > Z_2$ so that the parasitic op amp pole is in the left half of the plane.

In addition to multiplication and division, the circuit has many useful signal-processing applications such as signal squaring, amplitude modulation, signal inversion, rms-to-dc conversion, and neural computation. A scalar product of two n-tuple vector inputs can easily be achieved by a straightforward extension of the two input multiplier to realize an output $V_o$ that is proportional to $\sum_{i=1}^{n} \Delta X_i \Delta Y_i$. In this case, $n$ four-transistor circuits are used at the input. The resulting vector multiplier lends itself naturally to the VLSI implementation of feedback/feedforward adaptive neural networks where adaptive weights are represented by positive or negative voltage levels $\Delta Y_i$ and outputs of other identical neurons, used as inputs, are represented by $\Delta X_i$.

A generalized multiplier/divider structure with m four-transistor circuits connected in the feedback is shown in Fig. 3.40. The circuit has an output given

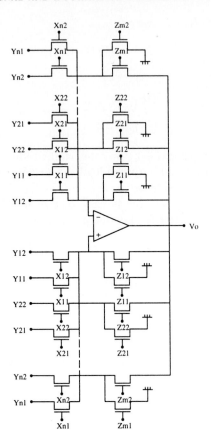

**FIGURE 3.40**
A generalized multi-input multiplier/divider.

by [15]

$$V_O = \frac{\sum_{i=1}^{n} (\frac{W}{L})_{iin} \Delta X_i \Delta Y_i}{\sum_{i=1}^{m} (\frac{W}{L})_{iout} \Delta Z_i} \qquad (3.159)$$

where $\Delta X_i = X_{i1} - X_{i2}$, $\Delta Y_i = Y_{i1} - Y_{i2}$ and $\Delta Z_i = Z_{i1} - Z_{i2}$.
Certain conditions must be met for the transistors to operate in a certain region.
For example, in the triode region, we have

$$max(Y_{11}, Y_{12}...Y_{n1}Y_{n2}, V_O) \leq min(\overline{X}_{11}, \overline{X}_{12}...\overline{X}_{n1}, \overline{X}_{n2}, \overline{Z}_{11}, \overline{Z}_{12}...\overline{Z}_{m1}, \overline{Z}_{m2})$$
$$(3.160)$$

where $\overline{X}_{ij} = X_{ij} - V_T$ and $\overline{Z}_{ij} = Z_{ij} - V_T$.

**Integrators**   Differential integrators are key components in many frequency-
selective analog systems such as filters and oscillators. There are several ways to
realize continuous-time differential integrators by using op amps as basic blocks
[67, 68, 40] (see Fig. 3.41(c), also see end-of-chapter problems). However, these

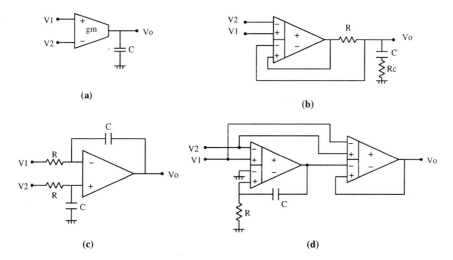

**FIGURE 3.41**
Differential integrators (a) Transconductance-C integrator, (b) DDA integrators with phase lag compensation, (c) Balanced time-constant integrator (d) A DDA integrator with low output impedance.

differential integrators, whether implemented using discrete RC components or in monolithic form, have low input impedance. Also, component-matching (e.g. $R$ or $\frac{W}{L}$ and $C$ matching) is required to achieve differential integration. Component mismatch leads to phase and magnitude errors [67] and increases harmonic distortion, particularly in monolithic implementations where matching is essential for nonlinearity cancellation [40, 69, 68, 70, 71]. In contrast, the DDA compares two differential voltages instead of single-ended voltages, and therefore it would lend itself naturally to the implementation of CMOS differential integrators. Figures 3.41(b) and (d) show two possible ways to realize lossless differential integrators with DDAs [66]. The DDA together with $R$ in Fig. 3.41(b) implements an OTA (Fig. 3.41(a)) with $g_m = 1/R$ [60]. A lossless integrator is realized when a resistor is connected in parallel with the integrator capacitor, $C$. In Fig. 3.41(b), $R_c$ provides passive compensation [72] to mitigate the effect of the finite op amp gain-bandwidth product (see below). Since the inputs of these integrators are directly the inputs of the DDAs (gates of MOS transistors) the DDA-based differential integrators have high input impedances. In addition, only a single resistor and a single capacitor are used; hence, there are no passive component-matching requirements. The output of the two DDA circuits found by using (3.141) is identical and is given by (assuming ideal DDAs)

$$V_o = \frac{V_1 - V_2}{sRC} \tag{3.161}$$

We see that the two DDA integrator circuits perform the same function, but the integrator in Fig. 3.41(d), or simply integrator (d), employs an extra DDA but uses a grounded resistor in its lossless implementation. Grounded resistors are much easier and simpler to be implemented by MOS circuit techniques than floating resistors. This second DDA also functions as a buffer stage, where the output impedance of the DDA used is small. The integrator in Fig. 3.41(b), integrator (b), on the other hand, must drive a high input impedance node in order to maintain proper operation when cascaded. Fortunately, if the output is directly connected to the input of another DDA, no buffer stage is needed. However, the parasitic capacitances associated with the input of the latter DDA may result in variations in the integrator time constant, and should be incorporated with the integrator capacitor, $C$, in the design process.

Similar to op amp-based integrators, finite gain-bandwidth (GB) also results in excess phase lag in DDA-based integrators, which causes Q-enhancement in the response of filters built by these integrators. If the DDAs frequency-dependent open-loop gain is approximated by $A(s) = GB/s$, the nonideal transfer function of integrator (b) can be found as follows, where $\omega_o = 1/RC$.

$$\frac{V_o}{V_1 - V_2} = \frac{\omega_o}{s} \times \frac{1}{(1 + \frac{\omega_o}{GB}) + \frac{s}{GB}} \tag{3.162}$$

Hence, the error caused by the finite $GB$ of the DDA's is:

$$\epsilon(s) = \frac{1}{(1 + \frac{\omega_o}{GB}) + \frac{s}{GB}} \tag{3.163}$$

According to $\epsilon(s)$, a parasitic pole in the s plane (phase lag) is introduced by the nonideal open-loop gain. To circumvent this problem, the same compensation techniques used in op amp-based integrators, i.e. passive, active [40] or the recently introduced adaptive compensation technique [41], apply to those of the DDA as well. One possible passive compensation approach is shown in Fig. 3.41(b) which uses $R_c$ to compensate for the finite GB effect of the DDA. Let $\omega_c = 1/R_c C$. The resultant output of the nonideal integrators is

$$\frac{V_o}{V_1 - V_2} = \frac{1}{sRC} \times \frac{GB}{\omega_o + \omega_c} \frac{\frac{\omega_c}{GB} + \frac{s}{GB}}{\frac{\omega_c(\omega_o + GB)}{GB(\omega_o + \omega_c)} + \frac{s}{GB}} \tag{3.164}$$

and the phase error is given by

$$\phi_r(\omega) = \tan^{-1}\frac{\omega}{\omega_c} - \tan^{-1}\frac{\omega(\omega_o + \omega_c)}{\omega_c(\omega_o + GB)} \tag{3.165}$$

Therefore, the phase error $\phi_r(\omega)$ could be minimized to zero by tuning $R_c$ to achieve

$$\omega_c = GB \tag{3.166}$$

Note that $R_c$ can be implemented by a single MOSFET operating in the triode region. Its small-signal channel resistance, which is controlled by the gate voltage, is linear since the ac-voltage across the drain-source terminals is practically very small and is not going to affect the linearity of the integrator.

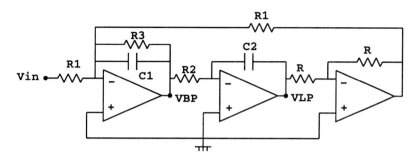

**FIGURE 3.42**
Op amp-based resonator filter.

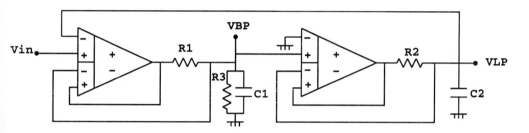

**FIGURE 3.43**
DDA-based resonator filter.

### 3.7.2   Continuous-Time Filters

In system-level applications, filters can be built by using the basic cells discussed in previous subsections. Two second-order filter examples, a state-variable filter and a resonator filter, and one high-order low-pass filter will be discussed.

**Resonator filter**   An op amp-based configuration of a resonator filter is composed by a lossy integrator, a lossless integrator and an inverter connected in a loop as shown in Fig. 3.42 [73]. A DDA-based resonator is developed as shown in Fig. 3.43. Note that the DDA circuit needs only two DDAs to build the filter, a lossy integrator and a lossless integrator. The function of the inverter in Fig. 3.42 is implemented by exploiting the differential feature of the DDA. If the DDAs are assumed to be ideal, the transfer function of the bandpass and lowpass outputs are given by

$$\frac{V_{BP}}{V_{in}} = \frac{s\left(\frac{1}{R_1 C_1}\right)}{s^2 + \frac{1}{R_3 C_1}s + \frac{1}{R_1 R_2 C_1 C_2}} \tag{3.167}$$

$$\frac{V_{LP}}{V_{in}} = \frac{\frac{1}{R_1 R_2 C_1 C_2}}{s^2 + \frac{1}{R_3 C_1}s + \frac{1}{R_1 R_2 C_1 C_2}} \tag{3.168}$$

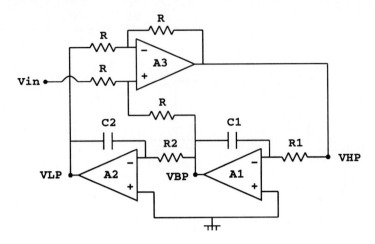

**FIGURE 3.44**
Op amp-based state-variable filter.

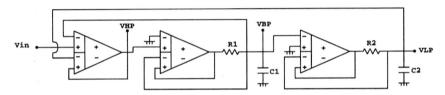

**FIGURE 3.45**
DDA state-variable filter.

where $\omega_n = 1/\sqrt{R_1 R_2 C_1 C_2}$ and $Q = R_3\sqrt{\frac{C_1}{R_1 R_2 C_2}}$. To simplify the design, let $R_1 = R_2 = R$ and $C_1 = C_2 = C$. As a result,

$$\omega_n = \frac{1}{RC} \tag{3.169}$$

$$Q = R_3/R \tag{3.170}$$

Therefore, high $Q$ can be achieved by increasing the $R_3/R$ ratio. Infinite $Q$ ($j\omega$-axis poles for oscillation) is achieved by disconnecting $R_3$.

**State-variable filter** A state-variable filter, also called the universal filter [40, 73], realized by op amps is shown in Fig. 3.44, where $V_{HP}$, $V_{BP}$, and $V_{LP}$ are the outputs of the highpass, the bandpass, and the lowpass functions, respectively.

The DDA counterpart shown in Fig. 3.45 is achieved by converting the op amp building blocks into the corresponding DDA-based circuits. The ideal transfer functions of the filter are

$$\frac{V_{HP}}{V_{in}} = \frac{s^2}{s^2 + \frac{1}{R_1 C_1}s + \frac{1}{R_1 R_2 C_1 C_2}} \tag{3.171}$$

$$\frac{V_{BP}}{V_{in}} = \frac{\frac{s}{R_1 C_1}}{s^2 + \frac{1}{R_1 C_1} s + \frac{1}{R_1 R_2 C_1 C_2}} \qquad (3.172)$$

$$\frac{V_{LP}}{V_{in}} = \frac{-\frac{1}{R_1 R_2 C_1 C_2}}{s^2 + \frac{1}{R_1 C_1} s + \frac{1}{R_1 R_2 C_1 C_2}} \qquad (3.173)$$

where $\omega_n = 1/\sqrt{R_1 R_2 C_1 C_2}$ and $Q = \sqrt{\frac{R_1 C_1}{R_2 C_2}}$.

According to the above two examples, one may observe that the DDA-based filters are simpler than those of the op amp. High input impedances and a minimum number of resistors are achieved in DDA-based filters. However, a drawback of these filters is the fact that the filter outputs (except for $V_{HP}$ in Fig. 3.45) are not taken at an amplifier output which may require the use of a buffer stage, or alternatively the differential integrator of Fig. 3.41(b) as a basic block. In addition to resonator and state-variable second-order filters, leap-frog ladder, high-order filters can also be built by DDA integrators in a straightforward manner [64, 65] (Prob. 3.25). All-MOS monolithic implementations of these filters will make use of MOS voltage-controlled resistors [4, 50] and must use proper automatic tuning to compensate for random process variations [74, 69, 75] (Prob. 3.28). MOSFET-C integrators [76, 40] and op amp-based integrated amplifiers are used to implement opamp-based filters. Usually the basic filter blocks have to be converted to balanced structures. For example, the integrators in Fig. 3.42 are first converted to balanced integrators, e.g. the integrator in Fig. 3.41(c). The double-MOSFET method discussed in Section 3.4 can then be used to replace resistor pairs in the filter by MOS transistors and thus convert the balanced RC circuit to an all-MOS implementation (see Prob. 3.32).

For high-frequency applications, Transconductance-C or OTA-C filters [77, 12, 78, 79, 80] are often used. This is because the transconductor operates in an open-loop topology, i.e. without local feedback around the OTA itself (see Fig. 3.41(a)). The non-dominant poles in this case are at much higher frequencies compared to non-dominant poles in op amp- or DDA-based filters, which are limited by the amplifier $GB$. Furthermore, parasitic capacitances are often present at input or output nodes and can be contained (lumped) with the filter's main capacitors [12].

Using the basic OTA-C integrator block discussed previously in Fig. 3.41, an OTA-C implementation of a fifth-order all-pole low-pass filter (see Prob. 3.24) is shown in Fig. 3.46 [81].

Further discussions on continuous-time filters can be found in Chapters 4 (low-voltage filters) and 6 (current-mode filters). Their discrete-time counterparts are discussed in Chapter 6 (switched-current filters) and Chapter 9 (switched-capacitor filters.)

## 3.8 SUMMARY

In this chapter, and in Chapter 2, we discussed basic circuit techniques for analog VLSI continuous-time signal processing. The main focus has been on design prin-

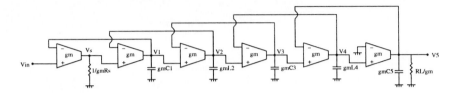

**FIGURE 3.46**
OTA-C fifth-order low-pass filter.

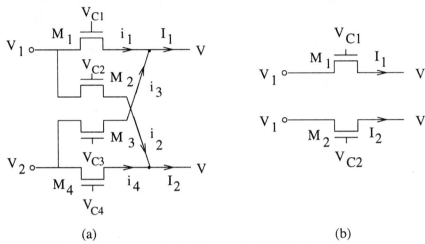

(a)                               (b)

**FIGURE P3.1** V-I converters for implementating MOS resistor.

ciples which use CMOS technologies. In the following five chapters, other types of continuous-time signal processing which use CMOS, bipolar, and BiCMOS technologies will be discussed. The literature covering continuous-time signal processing is broad, encompassing the research efforts of many individuals. The topics were chosen to provide an introduction to the most useful and well-known techniques in traditional applications and to techniques that are recognized to have the most potential in emerging analog VLSI applications.

## PROBLEMS

**3.1.** Figure P3.1 shows two V-I converters [33, 34] which can be used to implement MOS resistors. Assume that all the transistors are operating in the linear region, $M_1 - M_2$ are matched and $M_3 - M_4$ are also matched.

(a) Derive an expression for $I_1 - I_2$ for each circuit using the simplified graphical MOS model.

(b) Find the relationship between the gate voltages in circuit (a) such that $I_1 - I_2$ is independent of $V$, then design the circuit so that it has a transconductance $g_{ma} = (V_1 - V_2)/(I_1 - I_2)$ equal to $g_{mb}$ of circuit (b) where $g_{mb}$ is given by $V_1/(I_1 - I_2)$ when $V = 0$.

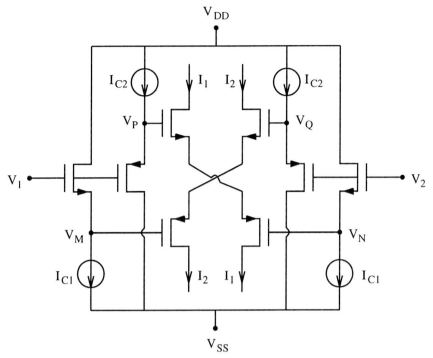

**FIGURE P3.2** Linear V–I converter.

**3.2.** Verify that the circuit of Fig. P3.2 [82] acts as a linear V-I converter, and find an expression for $I_1 - I_2$ in terms of the differential input $V_1 - V_2$ and the bias currents $I_{C1}$ and $I_{C2}$.

**3.3.** The V-I converter shown in Fig. 3.7 can be modified to be biased by a constant current source as shown in Fig. P3.3, where transistor MD $(K_d)$ is a level shifter. Derive the differential output current $I_1 - I_2$.

**3.4.** In Fig. P3.4, the four PMOS transistors are matched. Also M3, M4 are matched and M6, M8 are matched. All transistors are operating in saturation.
  (a) Find an expression for $i_{out}$ in terms of the four input voltages $V_1$, $V_2$, $V_{c1}$, and $V_{c2}$. Use the simple square-law model in saturation and ignore $\lambda$ effects. What function or functions can the circuit perform?
  (b) The circuit has two current mirrors. Is it possible to use one current mirror and still get the same output current? If yes, how?

**3.5.** Consider the multiplier circuit concept of Fig. 3.11 when a single MOS implementation of the building block $K_{BB}$ is used:
  (a) Using the graphical MOS model, show that nonlinearity cancellation in $I_O$ is achieved.
  (b) Verify part (a) analytically.
  (c) Repeat (a) and (b) above when transistors are operating in the triode region. What is the condition for the cancellation to take place?
  (d) Show that nonlinearity cancellation is also achieved if a pair of MOSFETs is operating in triode while the other is operating in saturation. Identify each

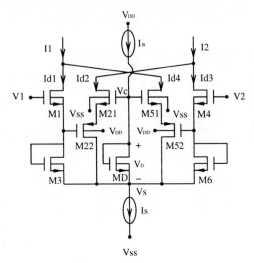

**FIGURE P3.3** Linear V–I converter biased by a constant current source.

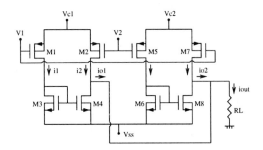

**FIGURE P3.4**

pair.

**3.6.** Using (3.51) and (3.52), prove that $A_n + A_p \geq 1$ is necessary to satisfy the condition $V_{Gn} - V_{Gp} > V_{Teq}$. For the simplicity of calculation, assume $V_{DD} = -V_{SS}$ and $V_{Tn} = -V_{Tp}$. (Hint: the worst case happens when $V_a$ is minimum and $V_b$ is maximum of the allowed input.)

**3.7.** Consider the circuit shown in Fig. P3.7.
  (a) Let $Z_1 - Z_2 = \Delta Z$, $Y_1 - Y_2 = \Delta Y$ and $X_1 - X_2 = \Delta X$. Find an expression for $V_O$.
  (b) Show that the circuit operates as a programmable four-quadrant multiplier. Let $Y_1 = -Y_2 = Y$, $X_1 = Q_O + X$, $X_2 = Q_O - X$, $Z_1 = V_{C1}$, and $Z_2 = V_{C2}$. Determine the necessary operating conditions.
  (c) Develop a vector multiplier circuit of two n-tuple vector inputs based on the circuit in (b).
  (d) Using balanced input signals, discuss the operation of the circuit as a divider and its operating conditions.

**3.8.** This problem explores the application of multiplier/divider circuits in the linearization of second-order sensor models [29]. A typical sensor is normally char-

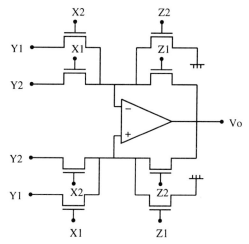

**FIGURE P3.7** Multiplier/divider circuit.

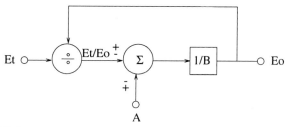

**FIGURE P3.8(a)**

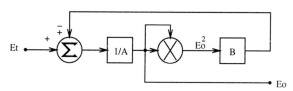

**FIGURE P3.8(b)**

acterized by

$$E_t = at \pm bt^2 \tag{3.174}$$

where $E_t$ is the quadratic signal to be linearized.

(a) Show that the block diagram in Fig. P3.8(a) produces $E_O = \gamma t$, a linear function of $t$. Determine $A$ and $B$ in terms of $a$, $b$, and $\gamma$.

(b) Using the multiplier/divider circuit of Fig. P3.7, implement the linearization circuit in MOS technology.

(c) An alternative implementation which uses a vector multiplier is based on Fig. P3.8(b). Sketch the MOS implementation.

**3.9.** In this problem we will investigate the nonideal effects on the performance of the multiplier/divider circuit of Fig. P3.7 [29], assume that all transistors are

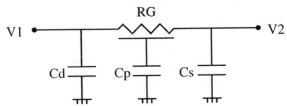

**FIGURE P3.9** Small-signal high-frequency model of an MOS transistor in triode region.

operating in the triode region.

(a) The first nonideal effect to be considered is the op amp finite gain-bandwidth product, $GB$. Let the open-loop gain of the op amp be given by $A(s) = GB/s$ and the output voltage be given by $v'_O = v_O \epsilon_{MD}(s)$, where $v_O$ is the ideal output voltage and $\epsilon_{MD}(s)$ is the error term due to the nonideal op amp. Determine $\epsilon_{MD}(s)$. (Hint: replace each MOSFET with a small-signal resistance given by $R_{Gi} = \dfrac{1}{\mu C_{OX} \frac{W}{L}(V_{Gi} - V_{TB})}$ where $V_{Gi}$ is the gate voltage of the MOSFET.)

(b) Determine the location of the parasitic pole in $\epsilon_{MD}(s)$.

(c) What is the condition of $Z_1$ and $Z_2$ in order for this circuit to be stable? We will now investigate the high-frequency distributed effects of the MOS-FET. A high-frequency small signal MOS model is shown in Fig. P3.9, and is characterized by the following first-order y-parameters.

$$y_{11} = y_{22} = \frac{\frac{s\tau}{2} + 1}{R_{Gi}\left(\frac{s\tau}{6} + 1\right)}$$

$$y_{12} = y_{21} = \frac{-1}{R_{Gi}\left(\frac{s\tau}{6} + 1\right)}$$

where $\tau = R_{Gi}C_P$ and $C_P = C_{OX}WL$.

For the following parts assume that the distributed effects are the only source of nonideality.

(d) The effect of $C_d$ and $C_s$ are neglected. Why?

(e) Let the differential transconductance of the four-transistor circuit at the input side be given by $G'_{mi} = G_{mi}\epsilon_i(s)$ where $G_{mi}$ is the ideal transconductance and $\epsilon_i(s)$ is the error term. Determine $\epsilon_i(s)$.

(f) Repeat (e) for the output side.

(g) Determine the overall error function caused by the distributed effects and comment on its influence on the circuit performance at very high frequencies.

**3.10.** Consider the circuit shown in Fig. P3.10.

(a) Find an expression for $V_O$ where the transistors are operating in the triode region.

(b) What is the function of the circuit?

(c) What are the operating conditions?

**3.11.** Consider the two identical multiplier/divider circuits of Fig. P3.11 [35]. All MOS-FETs are biased in the triode region. Assume that all MOSFETs are matched and that the op amps are ideal.

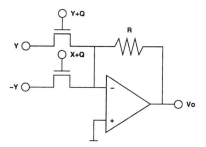

**FIGURE P3.10**

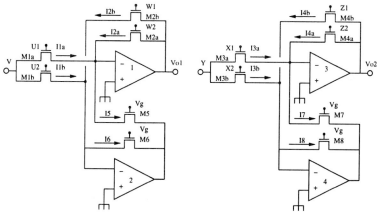

**FIGURE P3.11** Identical multiplier/divider circuits.

(a) Explain why $I_5 = I_6$ and $I_7 = I_8$.

(b) Find expressions relating $(I_{1a}-I_{1b})$ to $(I_{2a}-I_{2b})$, and $(I_{3a}-I_{3b})$ to $(I_{4a}-I_{4b})$.

(c) Let $U_1 - U_2 = \Delta U$, $W_1 - W_2 = \Delta W$, $X_1 - X_2 = \Delta X$, and $Z_1 - Z_2 = \Delta Z$. Find expressions relating $V_{01}$ to $V$, and $V_{02}$ to $Y$.

(d) How could these circuits be used as amplifiers? as multipliers? as dividers?

**3.12.** Consider once again the two multiplier/divider circuits of Fig. P3.11. Op amps 2 and 4 are primarily used to supply virtual grounds to some of the transistors. It is possible that both circuits can share one of these op amps and still perform in the desired manner. This will reduce the number of op amps in an analog VLSI system which uses a large number of multipliers (*e.g.*, neural networks). The resulting circuit is shown in Fig. P3.12. Note that two current sources have been added to insure complete cancellation of nonlinearities. Assume that all MOSFETs are matched and that the op amps are ideal.

(a) Use KCL at the negative input of op amp 2 to find the current $I_5$.

(b) Explain why $I_5 = I_6 = I_7$.

(c) Use KCL at the negative inputs of op amps 1 and 3 to find expressions for $(I_{2a} - I_{2b})$, and $(I_{4a} - I_{4b})$.

(d) What must $I_9$ and $I_{10}$ be in order for the circuit to operate in the same manner as the two circuits of Fig. P3.11?

(e) Draw a circuit replacing the current sources $I_9$ and $I_{10}$ with MOSFET transistors.

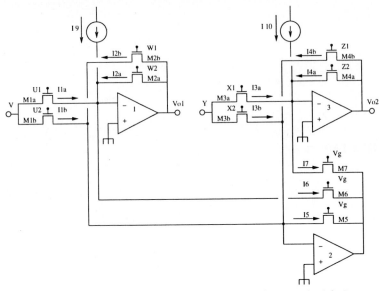

**FIGURE P3.12** The two multipliers of Fig. P3.11 implemented with 3 op amps.

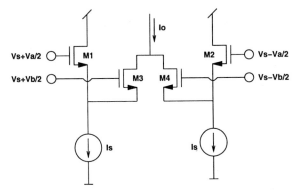

**FIGURE P3.13**

(*f*) Find expressions relating $V_{01}$ to $V$, and $V_{02}$ to $Y$.

**3.13.** In this problem a design of a multiplier based on the quarter-square technique is considered [17]. First, note that $(V_1 + V_2)^2 - (V_1 - V_2)^2 = 4V_1V_2$, thus, if we can develop a circuit that squares the sum and the difference of two signals, we can design a multiplier. Verify that the circuit shown in Fig. P3.13 can be used as a building block, and complete the design. Assume that all the transistors are in saturation. (Hint: let $(W/L)_{1,2} \gg (W/L)_{3,4}$. The circuit as shown produces the square of the difference; therefore you must adjust the inputs to obtain the square of the sum.)

**3.14.** Consider the circuit shown in Fig. P3.14.

(*a*) Show that the circuit shown in Fig. P3.14(a) behaves as a floating resistor. Let $K_1 = K_2 = K_3 = K_4 = K$ and $V_{Tn} = V_{Tp}$. (Hint: assume that

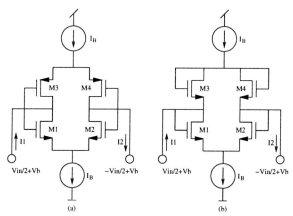

FIGURE P3.14

$KV_{IN}^2/4I_B$ is much less than 1.)

(b) What is the resistance obtained by this circuit?

(c) Discuss any practical considerations that may be associated with the design of the circuit.

(d) Repeat (a) for Fig. P3.14(b). What is the advantage of this circuit over the circuit of Fig. P3.14(a)?

**3.15.** The circuit shown in Fig. P3.15 simulates the function of a floating tunable resistance [83] seen between nodes A and B. The two PMOS transistors, M1A and M1B, are operating in the triode region while the other four transistors are operating in saturation. Find an expression of the equivalent resistance and determine its value if $\mu C_{ox} = 20\ \mu A/V^2$, $V_T = 1$ V, and $V_c = 2$ V. Use the following simple model for a PMOS operating in triode:

$$i_D = \mu C_{ox} \frac{W}{L}[(V_{SG} - V_T)V_{SD} - \frac{V_{SD}^2}{2}]$$

and ignore body and other nonideal effects. Assume identically-sized transistors with $\frac{W}{L} = \frac{10\mu m}{10\mu m}$. In the square-law resistor of Fig. 3.19, two diode-connected transistors were used to eliminate the dependence on $V_T$. A similar technique can be used with the triode resistor of Fig. P3.15. What type of devices (NMOS or PMOS) should be used for the diode-connected transistors? Where should they be connected in the circuit diagram and how are they sized? Discuss the matching requirements in this case.

**3.16.** Using the conditions derived in Sec. 3.5.1, we can calculate the ratio, $M$, of the maximum to the minimum resistance obtainable for the resistor circuit shown in Fig. 3.19(a).

(a) Derive the expression for the ratio and calculate $M$, for $V_{SS} = -5$ V, $V_T = 1$ V, and the maximum desired input voltage $|V_X| = 1V (V_Y = 0)$.

(b) Repeat (a) for the resistor circuit with the modified voltage followers.

**3.17.** For the circuit in Fig. 3.24(b), derive $V_{G1}$ in terms of $V_1$ and $I_B$. Assume that the current mirror is perfect and all the transistors are operating in the weak inversion region. You may drop the $\frac{kT}{q}$ term for simplicity.

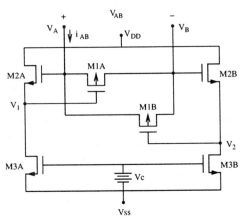

**FIGURE P3.15** Floating resistor.

**3.18.** For the circuit in Fig. 3.25, find the conditions that must be satisfied to keep $M_1$ and $M_2$ operating in the triode region and $M_3$ and $M_4$ in the saturation region. For $V_B = 2$ V and $V_T = 1$ V, what is the maximum voltage drop that this circuit can take while still operating properly? $M_5$ and $M_6$ can be assumed to be operating in the saturation region at all times.

**3.19.** For the circuit in Fig. 3.26(b), derive $I_{abs}$ in terms of $V_1 - V_2$ and $I_a$. Assume that the current mirrors are perfect and all the transistors are operating in the weak inversion region. You may drop the $\frac{kT}{q}$ term for simplicity.

**3.20.** (a) Given that $V_{j-1} = 2.475$V, $V_j = 2.5$V, and $V_{j+1} = 2.525$V in the single-stage WTA circuit of Fig. 3.31, find
   i.   $V_{CM}$, common signal node voltage.
   ii.  $I_j$'s, cell currents flowing through the first branch ($M_1$, $M_2$, and $M_5$).
   iii. $I_{oj}$'s, cell output currents in $M_3 - M_4$.
   iv.  $V_{oj}$'s, output voltages.
   Assume that $K_n = 2\,K_p = 42\ \mu\text{A/V}^2$, $V_{DD} = 5$ V, $V_{SS} = 0$ V, $V_{BB1} = V_{BB2} = 1$ V, $V_{Tn} = -V_{Tp} = 0.8$ V, $\lambda = 0.01V^{-1}$, $M_3 = M_5 = 2M_4 = 16\mu\text{m}/4\mu\text{m}$, and $M_1 = 8\mu\text{m}/2\mu\text{m}$. All transistors are in the saturation region. Verify your results using SPICE (use the model parameters in Appendix A).

   (b) Did the circuit provide a single winner in part (a)? Why? Assume $V_{j-1}$ and $V_j$ are kept the same as in (a) while $V_{j+1} = 3$ V and run SPICE to find the output voltages $V_{oj}$'s. Determine the winner, if any, and verify the results analytically.

**3.21.** In the single-stage WTA circuit of Fig. 3.31, assume that there are only two competitve cells. Provided that the input of the first cell, $V_{in}(1)$, is 2.5 V, find the minimum $V_{in}(2)$- input to the second cell, which forces the first cell to cut-off and $M_1$ of the second cell to operate in the saturation region. Also find the output voltages of each cell. Use parameters given in Prob. 3.20.

**3.22.** Figure P3.22 shows a two-input version of an alternative winner-take-all architecture. The basic operation of the circuit is as follows. If the input currents, $I_{i1}$ and $I_{i2}$ are equal, the bias current $I_B$ is split evenly between transistors $M1_A$ and $M1_B$; that is $I_{o1} = I_{o2} = I_B/2$. If one of the two currents is sufficiently larger than the other, the circuit steers all of $I_B$ to the output corresponding to the larger input.

(a) Verify this operation by deriving an expression for $I_{o1} - I_{o2}$ as a function of $I_{i1} - I_{i2}$. Use the simple PMOS $I_D$ model (including the channel length modulation term) and neglect the body effect for all transistors. What is the minimum value of input current difference, $I_{i1} - I_{i2}$, for which $I_{o1} = I_B$, that is, for which the circuit completely identifies input 1 as the "winner."

(b) Simulate the circuit in SPICE to verify the expression derived in part (a).

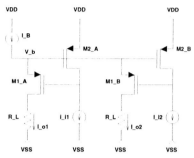

**FIGURE P3.22** Alternative WTA.

**3.23.** Implement a threshold-voltage-independent, DDA-based, four-quadrant multiplier using the circuit of Fig. 3.37 as a basic building block.

**3.24.** Consider the passive RLC circuit shown in Fig. P3.24,

(a) Express the branch currents $I_s$, $I_2$, $I_4$ in terms of the node voltages $V_1$, $V_3$, and $V_5$.

(b) Express the node voltages in terms of the branch currents.

(c) Scale the above six expressions by $1\Omega$ so that all currents are seen as voltages, $V_s$, $V_2$, and $V_4$.

(d) Verify that the expressions in (c) are identical to those obtained by analysis of the OTA-C filter in Fig. 3.46.

**3.25.** Using the DDA integrator of Fig. 3.41(b) as a basic block, design a DDA-based fifth-order, low-pass leapfrog filter using the passive ladder shown in Fig. P3.24 with the terminating resistors $R_s$ and $R_L$ normalized to $1\Omega$. Follow design steps similar to those outlined in the previous problem.

**3.26.** Consider the lossless integrator cell of Fig. P3.26 [84]. All MOSFETs are biased in the triode region. The positive and negative inputs of both op amps are always tied directly to ground or are at virtual ground. Transistors M3 and M4 are identically, matched transistors. They both have the same drain, gate, source, and bulk voltages; therefore, the currents I3 and I4 must be equal, i.e.

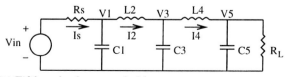

**FIGURE P3.24** Fifth-order low-pass ladder.

$I_3 = I_4$. In this way, the current $I_2$ gets copied by transistors M3 and M4. The current difference, $I_1 - I_2$, of matched input transistors, M1 and M2, is thus forced through the capacitor. Using the accurate strong inversion model of an NMOS transistor, this current difference can be shown to be

$$I_C = I_1 - I_2 = \mu C_{ox} \frac{W}{L}(V_{g1} - V_{g2})(V_{IN})$$

(a) Use the graphical method to show that the current difference is linear.
(b) Derive an expression for $V_O$.
(c) Four other variations of the lossless integrator are obtained by reconfiguring it in such a way that the above current equation remains the same while changing the way the virtual grounds are provided to the op amp inputs. Draw the circuit schematic for two of these variations.

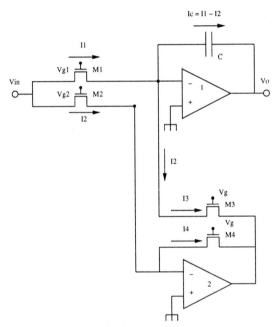

**FIGURE P3.26** Lossless integrator.

**3.27.** Consider the lossless integrator cell of Fig. P3.26.
(a) Find an expression for the RC time-constant. Your expression should include the mobility, $\mu$, and the gate oxide, $C_{ox}$.
(b) What is the value of the RC time-constant. Assume NMOS transistors, $V_{g1} = 5$ V, $V_{g2} = 4.5$ V, C=20 pF, W=5$\mu$m, L=100$\mu$m, and nominal SPICE parameters as listed in Appendix A for a typical CMOS process.
(c) Assume that during manufacturing, random process variations may cause a 30% change in the mobility from its nominal value; and a 20% change in value of the capacitor, C. What are the minimum and maximum expected values for the RC time-constant of the integrator after manufacturing?

**3.28.** A crucial design issue in many MOSFET-C continuous-time analog circuits is the accurate on-chip implementation of RC time-constants. Problem 3.27 illustrated that random process variations may make this goal difficult to acheive. An *automatic tuning circuit* is often employed with MOSFET-C circuits to insure accurate time-constants regardless of unknown process variations. The job of the tuning circuit is to tune other circuit's time-constants by automatically adjusting their dc-control voltages. As an example, consider the automatic tuning circuit shown in Fig. P3.28.

(a) An amplifier circuit using switched capacitor resistors (see Chapter 9) and MOS transistors is embedded in the tuning circuit. $\Phi_1$ and $\Phi_2$ are non-overlapping clocks with a frequency of $f_s$. Two identical switched capacitor resistors are shown, each with an equivalent resistance of $R_{SC} = 1/f_s C$. Find an expression for $V_{out}/V_{in}$.

(b) The amplifier circuit has been placed in a feedback circuit to insure it maintains unity gain. Set $V_{out}/V_{in} = 1$ and find an expression for the difference between the output voltages $(V_{g1} - V_{g2})$.

(c) The output voltages, $V_{g1}$ and $V_{g2}$, of the automatic tuning circuit can be used to control the voltages, $V_{g1}$ and $V_{g2}$, of the integrator shown in Fig. P.28. Find a new expression for the transfer function of the integrator by replacing $(V_{g1} - V_{g2})$ with the solution of part (b). Label the integrator capacitor as $C_i$, and the tuning circuit capacitors as $C_t$. (Hint: If the tuning circuit and the integrator are placed in close proximity on chip, then $\mu$ and $C_{ox}$ for both circuits will be the same.)

Explain the dependence of the new integrator transfer function on the process paramaters $\mu$ and $C_{ox}$.

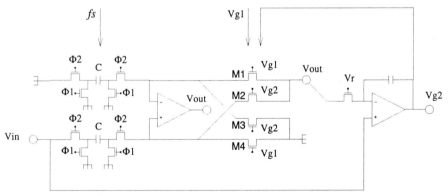

**FIGURE P3.28** Automatic tuning circuit.

**3.29.** In the fully-differential amplifier circuit shown in Fig. P3.29 [85], the $M_1 - M_2$ matched pair has $K = K_1$, $M_3 - M_4$ matched pair has $K = K_2$. All transistors are operating in the triode region. Assume the op amp is ideal.

(a) Find the differential gain of the amplifier.

(b) Determine the necessary signal conditions which keep the transistors in the triode region.

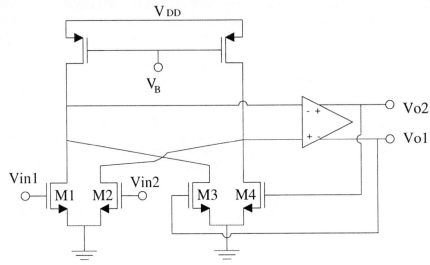

**FIGURE P3.29**

(c) Disconnect $M_3$ and $M_4$ from the circuit and connect two matched capacitors in the feedback path; one capacitor is between $V_{O2}$ and the inverting input node of the op amp while the other is connected between $V_{O1}$ and the positive op amp input node. Show that the resulting circuit is a fully-differential lossless integrator [85, 86] and find its time constant.

**3.30.** The fully-differential lossy integrator of Fig. P3.30 [87, 3] can be reconfigured to design filters with different time-constants, $\tau$, and different quality factors, $Q$, without altering device sizes. Assume an ideal op amp. The integrator transfer function can be expressed as

$$\frac{V_o}{V_{in}} = \frac{1}{s\tau + \frac{1}{Q_o}}$$

Assume all transistors operating in the triode region with $C^+ = C^- = C$ and find $\tau$ and $Q_o$ if

(a) $C^-$ is connected to node 2 and $C^+$ connected to node 3.
(b) $C^-$ is connected to node 2 and $C^+$ connected to node 4.
(c) $C^-$ is connected to node 1 and $C^+$ connected to node 4.

In your analysis, use the following second-order transistor model:

$$I_D = K[(V_g - V_T)(V_d - V_s) + a_2(v_d^2 - V_s^2)]$$

where $a_2$ is a constant.

   Two of these configurable integrators, one lossy and one lossless, are connected in a loop to design a second-order configurable resonator filter. Since each integrator is configurable according to (a), (b), (c) above, the resulting filter is also configurable. How many possible filter topologies are there? Find the pole-frequency and pole-Q of each filter topology.

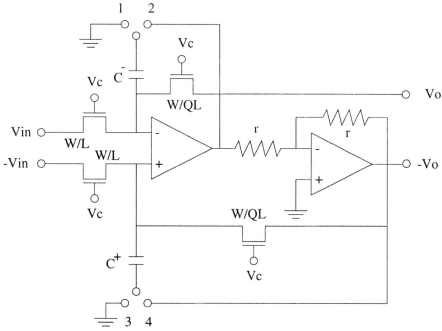

**FIGURE P3.30** Configurable fully-differential MOSFET-C integrator.

**3.31.** The lossless MOSFET-C integrator of Fig. P3.31(b) is obtained by configuring the integrator circuit in Fig. P3.31(a). Note that the two circuits have the same number of components with the respective components having the same size. The only difference is that, in Fig. P3.31(b), two different gate voltages, $V_{C1}$ and $V_{C2}$, are used. Assume transistors in both circuits are operating in the triode region.

In one of the circuits, all-MOS nonlinearities are cancelled and the integrator time-constant is independent of $V_T$. Identify the circuit. Use the following model for MOS transistor operation in the triode region:

$$I_D = K[(V_g - V_T)(V_d - V_s) + a_2(v_d^2 - V_s^2) + a_3(v_d^3 - V_s^3) + ...]$$

where $a_i(i = 2, 3, ...)$ are constants.

**3.32.** (a) Use the concept of the double-MOSFET design methodology and convert the active-RC-state variable filter of Fig. 3.44 to a MOSFET-C implementation [40]. Find expressions for the pole-frequency and pole-Q of the MOSFET-C filter. Assume ideal op amps.

(b) Repeat (a) for the resonator circuit of Fig. 3.42.

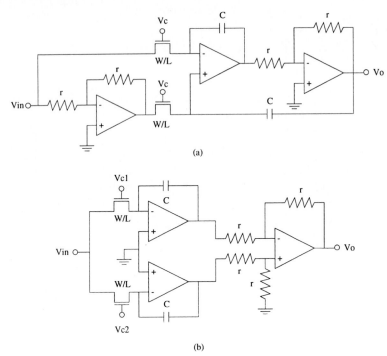

(a)

(b)

**FIGURE P3.31**

## REFERENCES

[1] E. K. F. Lee and P. G. Gulak, "Field programmable analogue array based on MOSFET transconductance," *Electron. Lett.*, vol. 28, pp. 28–29, Jan. 1992.

[2] S. R. Zarabadi, F. Larsen, and M. Ismail, "A configurable op amp/DDA CMOS amplifier architecture," *IEEE Trans. Circuits Syst. I*, vol. 39, pp. 484–487, June 1992.

[3] M. Ismail and S. Bibyk, "CAD latches onto new techniques for analog ICs," *IEEE Circuits and Devices Magazine*, pp. 11–17, Sept. 1991.

[4] H. Wallinga and K. Bult, "Design and analysis of CMOS analog signal processing circuits by means of a graphical MOST model," *IEEE J. Solid-State Circuits*, vol. SC-24, pp. 672–680, June 1989.

[5] M. C. H. Cheng and C. Toumazou, "Linear composite MOSFETs (COMFETs)," *Electron. Lett.*, pp. 1802–1804, Sept. 1991.

[6] M. A. Tan, "Synthesis of artificial neural networks by transconductors only," *J. Analog Integrated Circuits and Signal Processing*, vol. 1, pp. 339–351, Dec. 1991.

[7] S. T. Dupuie and M. Ismail, "High frequency CMOS transconductors," in *Analog IC design: the current-mode approach* (C. Toumazou, F.J. Lidgey, and D. G. Haigh, eds.), ch. 5, London: Peter Peregrinus Ltd., 1990.

[8] L. W. Nagel, *SPICE2: a computer program to simulate semiconductor circuits*. Electronics Research Laboratory, Univ. of California, Berkeley, May 1975.

[9] E. Seevinck and R. F. Wassenaar, "A versatile CMOS linear transconductor/square-law function circuit," *IEEE J. Solid-State Circuits*, vol. SC-22, pp. 366–377, June 1987.

[10] S. C. Huang and M. Ismail, "Linear tunable COMFET transconductors," *Electron. Lett.*, vol. 29, pp. 459–461, March 1993.

[11] C. Tomovich, "MOSIS–a gateway to silicon," *IEEE Circuits and Devices Magazine*, vol. 4, pp. 22–23, March 1988.

[12] B. Nauta, "A CMOS transconductance-C filter technique for very high frequencies," *IEEE J. Solid-State Circuits*, vol. 27, pp. 142–153, Feb. 1992.

[13] B. Nauta, *Analog CMOS filters for very high frequencies,* Kluwer Academic Publishers, 1993.

[14] P. Masa, K. Hoen and H. Wallinga, "20 million patterns per second analog CMOS neural network pattern classifier," *Proc. 11th European Conference on Circuit Theory and Design, Part I*, pp. 497-502, Elsevier, September 1993.

[15] N. I. Khachab and M. Ismail, "A nonlinear CMOS analog cell for VLSI signal and information processing," *IEEE J. Solid-State Circuits*, vol. 26, pp. 1689–1699, Nov. 1991.

[16] B. Song, "CMOS RF circuits for data communications applications," *IEEE J. Solid-State Circuits*, vol. 25, pp. 310–317, April 1986.

[17] J. Pena-Finol and J. Connelly, "An MOS four-quadrant analog multiplier using the quarter-square technique," *IEEE J. Solid-State Circuits*, vol. SC-22, pp. 1064–1073, Dec. 1987.

[18] H. Song and C. Kim, "An MOS four-quadrant analog multiplier using simple two-input squaring circuits with source followers," *IEEE J. Solid-State Circuits*, vol. 25, pp. 841–848, June 1990.

[19] C. Kim and S. Park, "Design and implementation of a new four-quadrant MOS analog multiplier," *J. Analog Integrated Circuits and Signal Processing*, vol. 2, pp. 96–103, April 1992.

[20] B. Gilbert, "A high-performance monolithic multiplier using active feedback," *IEEE J. Solid-State Circuits*, vol. SC-9, pp. 267–276, Dec. 1974.

[21] S. Qin and R. Geiger, "A ±5-V CMOS analog multiplier," *IEEE J. Solid-State Circuits*, vol. SC-22, pp. 1143–1146, Dec. 1987.

[22] S. Sakurai and M. Ismail, "A high-frequency wide range CMOS analog multiplier," *Electron. Lett.*, vol. 28, pp. 2228–2229, Nov. 1992.

[23] S. R. Zarabadi, M. Ismail, and N. I. Khachab, "High frequency BiCMOS linear V-I converter, voltage multiplier, mixer." U.S. Patent No. 5,151,625, filed Nov. 8, 1990, issued Sept. 29, 1992.

[24] R. Raut, "A novel VCT for analog IC applications," *IEEE Trans. Circuits Syst., II*, vol. 39, pp. 882–883, Dec. 1992.

[25] J. Ramirez-Angulo, "Highly linear four-quadrant analog BiCMOS multiplier for +-1.5v supply operation," *Proc. IEEE Int. Symp. Circuits Syst.*, pp. 1467–1470, May 1993.

[26] R. F. Wassenaar and M. Ismail, "A four-transistor continuous-time MOS transconductor operating in both triode and saturation," Tech. Rep. C93139, Semiconductor Research Corp., 1993.

[27] N. I. Khachab and M. Ismail, "Novel continuous-time all-MOS four-quadrant multipliers," *Proc. IEEE Int. Symp. Circuits Syst.*, pp. 762–765, May 1987.

[28] N. I. Khachab and M. Ismail, "MOS multiplier/divider cell for analog VLSI," *Electron. Lett.*, vol. 25, pp. 1550–1552, Nov. 1989.

[29] N. I. Khachab and M. Ismail, "Linearization techniques for nth-order sensor models in MOS VLSI technology," *IEEE Trans. Circuits Syst.*, vol. 38, pp. 1439–1450, Dec. 1991.

[30] M. Ismail and D. Rubin, "Improved circuits for the realization of MOSFET-capacitor filters," *Proc. IEEE Int. Symp. Circuits Syst.*, pp. 1186–1189, May 1986.

[31] K. C. Hsieh, P. R. Gray, D. Senderowicz, and D. G. Messerschmidt, "A low-noise chopper-stabilized differential switched-capacitor filtering technique," *IEEE J. Solid-State Circuits*, vol. 16, pp. 708–715, Dec. 1981.

[32] P. E. Allen and D. R. Holberg, *CMOS analog circuit design.* Holt, Rinehart and Winston, 1987.

[33] M. Ismail, "Four-transistor continuous-time MOS transconductor," *Electron. Lett.*, vol. 23, pp. 1099–1100, Sept. 1987.

[34] M. Ismail and J. Prigeon, "A novel technique for designing continuous-time filters in MOS technology," *Proc. IEEE ISCAS, Heloinki, Finland*, pp. 1665–1668, June 1988.

[35] M. Ismail, R. Brannen, S. Takagi, R. Khan, O. Aaserud, and N. Khachab, "A configurable CMOS multiplier/divider for analog VLSI," *Proc. IEEE Int. Symp. Circuits Syst.*, pp. 1085–1089, May 1993.

[36] M. Ismail, "A novel fully-differential MOS multiplier/divider circuit," *Internal Technical Report*, Solid-State Microelectronics laboratory, The Ohio State University, April 1993.

[37] M. Ismail, D. Kim, and H. Shin, "Finite GB and MOS parasitic capacitance effects in a class of MOSFET-C filters," *Proc. 33rd Midwest Symp. on Circuits and Systems*, Aug. 1990.

[38] Z. Czarnul, "Modification of the Banu-Tsividis continuous-time integrator structure," *IEEE Trans. Circuits Syst.*, vol. 33, pp. 714–716, July 1986.

[39] P. Ryan and D. G. Haigh, "Novel fully-differential MOS transconductor for integrated continuous–time filters," *Electron. Lett.*, vol. 23, pp. 742–743, July 1987.

[40] M. Ismail, S. V. Smith, and R. G. Beale, "A new MOSFET-C universal filter structure for VLSI," *IEEE J. Solid-State Circuits*, vol. SC-23, pp. 183–194, Feb. 1988.

[41] R. John and M. Ismail, "Adaptive compensation of MOSFET-C filters," *Proceedings IEEE Midwest Symposium on Circuits and Systems,* pp. 1402-1405, August 1993.

[42] S. Sakurai, M. Ismail, J.-Y. Michel, E. Sanchez-Sinencio, and R. Brannen, "A MOSFET-C variable equalizer circuit with simple on–chip automatic tuning," *IEEE J. Solid-State Circuits*, vol. 27, pp. 927–934, June 1992.

[43] R. Schaumann, M. S. Ghausi, and K. R. Laker, *Design of analog filters, passive, active RC and switched–capacitor*, ch. 7. Prentice-Hall, Englewood Cliffs, NJ, 1990.

[44] C. Mead and M. Ismail, *Analog VLSI implementation of neural systems*. Kluwer, 1989.

[45] P. C. Yu, S. J. Decker, H.-S. Lee, C. G. Sodini, and J. L. Wyatt, Jr., "CMOS resistive fuses for image smoothing and segmentation," *IEEE J. Solid-State Circuits*, vol. 27, pp. 545–553, April 1992.

[46] H. Kobayashi, J. L. White, and A. A. Abidi, "An active resistor network for Gaussian filtering of images," *IEEE J. Solid-State Circuits*, vol. 26, pp. 738–748, May 1991.

[47] K. Nagaraj, "New CMOS floating voltage-controlled resistor," *Electron. Lett.*, vol. 22, pp. 667–668, 1986.

[48] S. P. Singh, J. V. Hanson, and J. Vlach, "A new floating resistor for CMOS technology," *IEEE Trans. Circuits Syst.*, vol. 36, pp. 1217–1220, Sept. 1989.

[49] M. Steyaert, J. Silva-Martinez, and W. Sansen, "High-frequency saturated CMOS floating resistor for fully-differential analog signal processors," *Electron. Lett.*, vol. 27, pp. 1609–1611, 1991.

[50] S. Sakurai and M. Ismail, "A CMOS square-law programmable floating resistor independent of the threshold voltage," *IEEE Trans. Circuits Syst. II*, vol. 39, pp. 565–574, Aug. 1992.

[51] C. Mead, *Analog VLSI and neural systems*, ch. 7. Addison-Weslley, 1989.

[52] T. Shimmi, H. Kobayashi, T. Yagi, T. Sawaji, T. Matsumoto, and A. A. Abidi, "A parallel analog CMOS signal processor for image contrast enhancement," *Proc. European Solid-State Circuits Conference*, pp. 163–166, Sept. 1992.

[53] E. A. Vittoz and X. Arreguit, "Linear networks based on transistors," *Electron. Lett.*, vol. 29, pp. 297–299, Feb. 1993.

[54] M. A. Mahowald and T. Delbruck, "Coorperative stereo matching using static and dynamic image features," in *Analog VLSI implementation of neural systems* (C. Mead and M. Ismail, eds.), Norwell, MA: Kluwer Academic, 1989.

[55]  J. Lazzaro, R. Ryckebush, M. A. Mahowald, and C. Mead, "Winner-take-all networks of 0(n) complexity," in *Advances in neural information processing systems* (D. S. Touretzky, ed.), vol. 1, pp. 703–711, Los Altos, CA: Morgan Kaufmann, 1989.

[56]  T. Kohonen, "The self-organizing map," *Proc. IEEE*, pp. 1464–1480, Sept. 1990.

[57]  S. Grossberg, "Competitive learning: From interactive activation to adaptive resonance," *Cognitive Sci.*, vol. 11, pp. 23–63, 1987.

[58]  J. Choi and B. J. Sheu, "A high-precision VLSI winner-take-all circuit for self-organizing neural networks," *IEEE J. Solid-State Circuits*, vol. 28, pp. 576–583, May 1993.

[59]  A. D. L. Plaza and P. Morlon, "Power-supply rejection in differential switched-capacitor filters," *IEEE J. Solid-State Circuits*, vol. SC-19, pp. 912–918, Dec. 1984.

[60]  E. Sackinger and W. Guggenbuhl, "A versatile building block: the CMOS differential difference amplifier," *IEEE J. Solid-State Circuits*, vol. SC-22, pp. 287–294, April 1987.

[61]  S. T. Dupuie, S. Bibyk, and M. Ismail, "A novel all-MOS high-speed continuous-time filter," *Proc. IEEE Int. Symp. Circuits and Systems (ISCAS), Portland OR.*, pp. 675–680, May 1989.

[62]  T. Kwan and K. Martin, "An adaptive analog continuous-time CMOS biquadratic filter," *IEEE J. Solid-State Circuits*, vol. 26, pp. 859–867, June 1991.

[63]  O. Alminde, U. Gatti, V. Liberali, F. Maloberti, and P. O'Leary, "An integrated CMOS telephone line adaptor," *J. Analog Integrated Circuits and Signal Processing*, vol. 2, pp. 71–78, April 1992.

[64]  S. C. Huang and M. Ismail, "Novel fully-integrated active filters using the CMOS differential difference amplifier," *32nd Midwest Symposium on Circuits and Systems*, pp. 173–176, Aug. 1989.

[65]  S.-C. Huang, "A wide dynamic range CMOS differential difference amplifier with applications to continuous-time filters," Master's thesis, The Ohio State University, Columbus, 1990.

[66]  S.-C. Huang, M. Ismail, and S. R. Zarabadi, "A wide range differential difference amplifier: a basic block for analog signal processing in MOS technology," *IEEE Trans. Circuits Syst., Part II*, vol. 40, no. 5, pp. 289-301, May 1993.

[67]  L. C. Thomas, "The biquad: part I–some practical design considerations, part II–a multipurpose active filtering system," *IEEE Trans. Circuit Theory*, vol. CT-18, pp. 350–361, May 1971.

[68]  M. Banu and Y. Tsividis, "Fully-integrated active RC filters in MOS technology," *IEEE J. Solid-State Circuits*, vol. SC-18, pp. 644–651, Dec. 1983.

[69]  R. Schaumman, K. Laker, and M. S. Ghausi, *Design of analog filters, passive, active RC and switched-capacitor*, ch. 7. Prentice-Hall, Englewood Cliffs, NJ, 1990.

[70]  S. Smith, F. Liu, and M. Ismail, "Active-RC building blocks for MOSFET-C integrated filters," *Proceeding of ISCAS*, pp. 342–346, May 1987.

[71]  R. Unbehauen and A. Cichocki, *MOS switched-capacitor and continuous-time integrated circuits and systems*. Springer-Verlag, Berlin, 1989.

[72]  A. Soliman and M. Ismail, "Phase correction in two-integrator loop filters using a single compensating resistor," *Electron. Lett.*, vol. 14, pp. 375–376, June 1978.

[73]  L. P. Huelsman and P. E. Allen, *Introduction to the theory and design of active filters*. McGraw-Hill, 1980.

[74]  K. S. Tan and P. R. Gray, "Fully integrated analog filters using bipolar-JFET technology," *IEEE J. Solid-State Circuits*, vol. 13, pp. 814–821, Dec. 1978.

[75]  S. Sakurai, M. Ismail, J.-Y. Michel, E. Sanchez-Sinencio, and R. Brannen, "A MOSFET-C variable equalizer circuit with simple on-chip automatic tuning," *IEEE J. Solid-State Circuits*, vol. 27, pp. 927–934, June 1992.

[76] Y. Tsividis, M. Banu, and J. Khoury, "Continuous-time MOSFET-C filters in VLSI," *IEEE Trans. Circuits Syst.*, vol. 33, pp. 125–139, Feb. 1986.

[77] H. Khorramabadi and P. R. Gray, "High-frequency CMOS continuous-time filters," *IEEE J. Solid-State Circuits*, vol. 19, pp. 939–948, Dec. 1984.

[78] R. Alini, A. Baschirotto, and R. Castello, "8-32 MHz tunable BiCMOS continuous-time filters," *Digest European Solid-State Circuits Conference*, pp. 9–12, Sept. 1991.

[79] Y.-T. Wang and A. A. Abidi, "CMOS active filters at very high frequencies," *IEEE J. Solid-State Circuits*, vol. 25, pp. 1562–1574, Dec. 1990.

[80] J. Pennock, P. Frith, and R. G. Barker, "CMOS triode transconductor continuous-time filters," *Proc. IEEE Custom IC Conference*, pp. 378–381, 1986.

[81] R. Nawrock, "Electronically tunable all-pole low-pass leapfrog ladder filter with operational transconductance amplifier," *Int. J. Electronics*, vol. 62, no. 5, pp. 667–672, 1987.

[82] S. Szczepanski, R. Schaumann, and P. Wu, "Linear transconductor based on crosscoupled CMOS pairs," *Electron. Lett.*, vol. 27, pp. 783–785, April 1991.

[83] M. Banu and Y. Tsividis, "Floating voltage-controlled resistors in CMOS technology," *Electron. Lett.*, vol. 18, pp. 678–679, July 1982.

[84] S. Takagi, M. Ismail, N. Fujii, and R. Brannen, "Novel MOSFET-C filter realization using a canonical number of capacitors," *Proc. IEEE Int. Symp. Circuits Syst.*, pp. 1339–1342, June 1991.

[85] S. Wong, "Novel drain-biased transconductance building blocks for continuous–time filter applications," *Electron. Lett.*, vol. 25, pp. 100–101, Jan. 1989.

[86] C. A. Laber and P. Gray, "A 20-MHz sixth-order BiCMOS parasitic-insensitive continuous-time filter and second-order equalizer optimized for disk-drive read channels," *IEEE J. Solid-State Circuits*, vol. 28, pp. 462–470, April 1993.

[87] M. Ismail and D. Ganow, "MOSFET-Capacitor continuous-time filter structures for VLSI," *Proc. IEEE Int. Symp. Circuits Syst.*, pp. 1196–1200, May 1986.

# CHAPTER
# 4

# LOW-VOLTAGE
# SIGNAL
# PROCESSING

## 4.1 INTRODUCTION

The push toward higher component densities per unit chip area requires smaller isolation barriers and consequently lower voltages in VLSI chips. The voltages go from 5 V to 3.3 V, to levels of 2 V, and ultimately to 1 V. It is inevitable that this will be true for both digital as well as analog circuits, particularly in the context of mixed analog/digital VLSI signal processing systems.

The design techniques of low-voltage digital cells are not very different from the design techniques of digital cells that operate with a relatively high voltage. However, low-voltage analog design techniques do differ considerably from the design techniques of analog cells that operate with a relatively high power supply voltage. Therefore, analog circuits must be completely reconsidered and redesigned in the low-voltage environment.

To keep the signal-to-offset, or signal-to-noise ratio as large as possible, circuits must be designed which have rail-to-rail input and output voltage swing. Many conventional circuit topologies must be abandoned and replaced by circuits adopting innovative design solutions.

This chapter gives an overview of low-voltage design techniques for basic analog building blocks. In Sec. 4.2, 4.3, and 4.4, the design of high-quality low-voltage CMOS, bipolar, and BiCMOS operational amplifiers is treated. The operational amplifiers are capable of operating from rail-to-rail at the input as

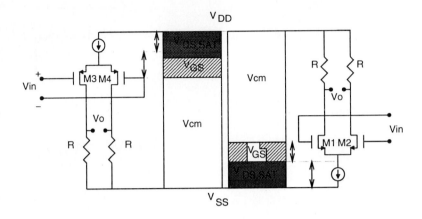

**FIGURE 4.1**
P-channel or n-channel input stage.

well as at the output. Sec. 4.5 describes a low-voltage bipolar instrumentation amplifier of which the input common-mode voltage range lies from below the negative supply rail voltage up to one diode voltage below the positive supply rail. Finally, Sec. 4.6 describes the design of low-voltage continuous-time filters with an optimal signal-to-noise ratio.

## 4.2  CMOS OPERATIONAL AMPLIFIER DESIGN

Mixed analog/digital electronics are becoming increasingly important. One example is the growing demand for signal processing circuits for smart sensors with digital outputs. Digital electronics are mostly designed in CMOS technology. To be able to integrate the digital and analog parts on to one chip, high performance analog CMOS circuits are required.

This section describes some design techniques of CMOS operational amplifiers. These techniques are primarily suitable for low-voltage operational amplifiers.

### 4.2.1  Input Stages

The input stage of conventional operational amplifiers consists of a single differential pair. This input pair can either be an n-channel, $M_1$-$M_2$, or a p-channel, $M_3$-$M_4$, input pair, as shown in Fig. 4.1. This differential pair amplifies the differential input signals and rejects the common-mode input voltages.

The common-mode voltage input range is a key parameter of an input stage. It is defined as the range of common-mode voltages at which the input stage properly responds to differential input signals. To be able to deal with a wide range of common-mode input voltages, the common-mode input range should also be wide.

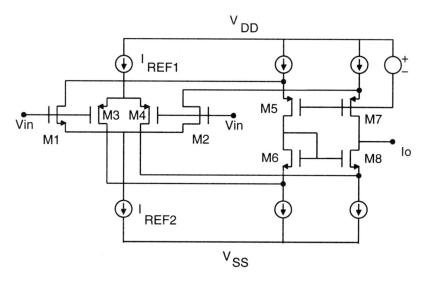

**FIGURE 4.2**
Complementary input stage with rail-to-rail common-mode input voltage range.

From Fig. 4.1 it can be seen that the common-mode input voltage range of the p-channel input pair is given by:

$$V_{SS} < V_{common} < V_{DD} - V_{DS(sat)} - V_{SGp} \qquad (4.1)$$

where $V_{SGp}$ is the source-gate voltage of a p-channel input transistor. $V_{DS(sat)}$ is the voltage across the current source which is necessary to ensure a current flowing out of it. $V_{DD}$ and $V_{SS}$ are the positive and the negative supply rail, respectively. The common-mode input voltage range of the n-channel input pair, as shown in Fig. 4.1, is given by:

$$V_{SS} + V_{DS(sat)} + V_{GSn} < V_{common} < V_{DD} \qquad (4.2)$$

where $V_{GSn}$ is the gate-source voltage of an n-channel input transistor.

In low-voltage design, the common-mode input voltage range of a single differential pair becomes very small, and therefore, the input stage can only deal with a small range of common-mode input voltages. To increase the common-mode input range the n-channel, $M_1$-$M_2$, and p-channel, $M_3$-$M_4$, input pairs have to be placed in parallel [1], as is shown in Fig. 4.2. The common-mode input range of this complementary input stage can be divided into three parts:

- Low common-mode input voltages; only the p-channel pair operates.
- Intermediate common-mode input voltages; both the input pairs operate.
- High common-mode input voltages; only the n-channel input pair operates.

The circuit, as shown in Fig. 4.2, has a rail-to-rail common-mode input range. In practice, the common-mode input range can even extend beyond the supply

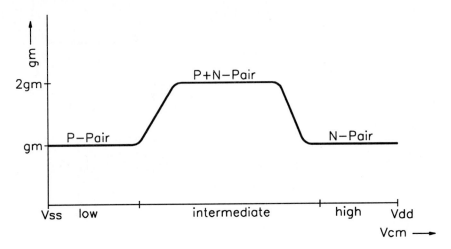

**FIGURE 4.3**
$g_m$ versus the common-mode input voltage for the complementary input stage in Fig.4.2.

rails by approximately 0.3 V. The cascodes, $M_5$-$M_8$, add the signal currents of the complementary input pairs.

A drawback of the complementary input stage is that the transconductance ($g_m$) changes by a factor of two within the common-mode input range, as shown in Fig. 4.3, where it is assumed that the $g_m$ of the p-channel and the $g_m$ of the n-channel input pairs are equal. This can easily be achieved by properly choosing the $W$ over $L$ ratio of the input transistors.

The variation in the $g_m$ impedes optimal frequency compensation because the bandwidth of an operational amplifier is proportional to the $g_m$ of its input stage. In order to obtain an optimal frequency compensation, the $g_m$ of the input stage has to be constant.

If the input transistors operate in weak inversion, the $g_m$ is proportional to the tail-current. To obtain a constant $g_m$, the sum of the tail-currents has to be constant. The complementary input stage with $g_m$ control is shown in Fig. 4.4. The current-switch, $M_5$, together with the one-to-one current mirror, $M_6$-$M_7$, keeps the sum of the tail-currents constant. Since the tail-current of an input pair operating in weak inversion is proportional to its $g_m$, the $g_m$ of the complementary input is constant, as shown in Fig. 4.5. A bipolar version of this input stage can be found in Sec. 4.3.2.

A drawback of an input stage operating in weak inversion is that the $g_m$ is rather small. The $g_m$ of an input stage can easily be increased by biasing the input transistors in strong inversion rather than in weak inversion. But this means that the $g_m$ is proportional to the square-root of the tail-current instead of proportional to the tail-current. If the $g_m$ control with one-to-one current mirror is used, as shown in Fig. 4.4, the $g_m$ of the complementary input stage still varies 41%. Therefore another method to control the $g_m$ has to be developed.

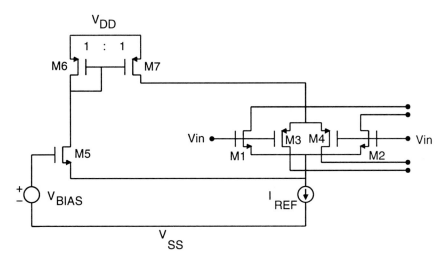

**FIGURE 4.4**
Complementary input stage with one-to-one current mirror.

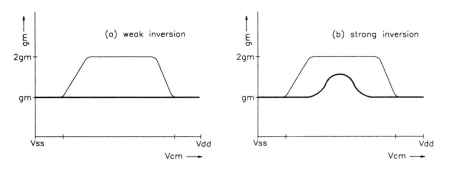

**FIGURE 4.5**
$g_m$ versus common-mode input voltage for the complementary input stage of Fig. 4.4; (a) weak inversion and (b) strong inversion.

From Fig. 4.3, it can be concluded that if the $g_m$ is increased by a factor of two at the lower and the upper part of the common-mode input range, the $g_m$ is constant. In order to increase the $g_m$ when only one input pair is operating, the tail-current of the actual active input pair has to be increased. Since the $g_m$ of an input pair is proportional to the square-root of its tail-current, the tail-current of the actual active input pair has to be increased by a factor of four [2].

This principle has been realized in circuit form, as shown in Fig. 4.6. If low common-mode voltages are applied to the input, the voltages between $V_{SS}$ and a few hundred millivolts less than $V_{SS}+V_{REF2}$, the current-switch, $M_8$, is off while the current-switch, $M_5$, is on. The current $I_{REF2}$ flows through $M_5$ to the current multiplier, $M_6$-$M_7$, where it is multiplied by a factor of 3. The output current of the current multiplier is added to $I_{REF1}$. Since $I_{REF1}$ and $I_{REF2}$ are

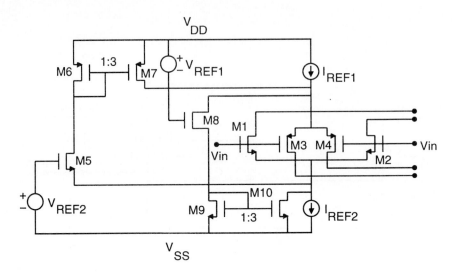

**FIGURE 4.6**
Complementary input stage with three-times current mirror.

equal to $I_{REF}$, the tail-current of the p-channel input transistors equals $4I_{REF}$.

If intermediate common-mode voltages are applied to the input, the voltages between a few hundred millivolts more than $V_{SS}+V_{REF2}$ and a few hundred millivolts less than $V_{DD}-V_{REF1}$, both the current-switches, $M_5$ and $M_8$, are off. Now, the tail-current of the n-channel as well as the tail-current of the p-channel input pairs equal $I_{REF}$.

If high common-mode voltages are applied to the input, voltages between a few hundred millivolts more than $V_{DD}-V_{REF1}$ and $V_{DD}$, the current-switch, $M_5$, is off while the current-switch, $M_8$, is on. The current $I_{REF1}$ flows through $M_8$ to the current multiplier, $M_9$-$M_{10}$, where it is multiplied by a factor of three. The output current of the current multiplier, $M_9$-$M_{10}$, is added to the current $I_{REF2}$. This results in a tail-current of the n-channel input pair of $4I_{REF}$. The result is a constant $g_m$ which is given by

$$g_m = 2\sqrt{KI_{REF}} \qquad (4.3)$$

where

$$K = \mu_n \left[\frac{W}{L}\right]_n C_{ox} = \frac{1}{2}\mu_p \left[\frac{W}{L}\right]_p C_{ox} \qquad (4.4)$$

In the turn-over ranges of the current-switches the $g_m$ changes only 15.5%, as is shown in Fig. 4.7.

### 4.2.2   Output Stages

The function of an output stage is to drive a relatively large current into a load. Also the maximum output voltage swing has to be reasonably large. In CMOS

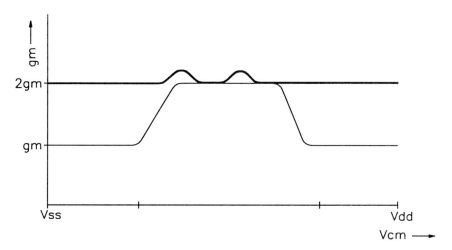

**FIGURE 4.7**
$g_m$ versus the common-mode input voltage for the complementary input stage of Fig 4.6.

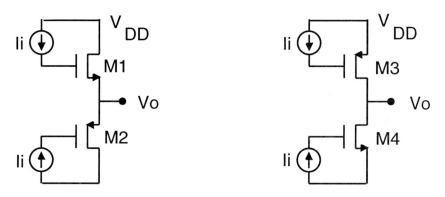

**FIGURE 4.8**
Common-drain and common-source output stage.

operational amplifiers, we can distinguish two main types of output stages. The first type, shown in Fig. 4.8, consists of two transistors connected in a common-drain configuration. The second type consists of two transistors connected in a common-source configuration, as is shown in Fig. 4.8. The top transistor of both the output stages pushes a current into the load. The bottom transistor pulls a current from the load. Not surprisingly, the output stages are called push-pull output stages.

The maximum output voltage swing of the first output stage is limited by:

$$V_{SS} + V_{GSp} < V_o < V_{DD} - V_{GSn} \tag{4.5}$$

where $V_{GSn}$ and $V_{GSp}$ are the gate-source voltages of the n-channel and p-channel output transistors, respectively. $V_o$ is the output voltage. In low-voltage applica-

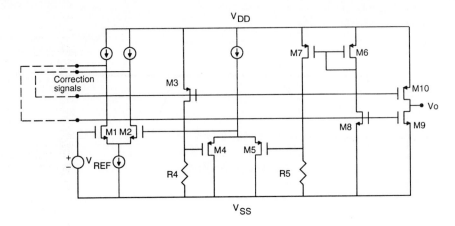

**FIGURE 4.9**
Output stage with feedback-biased class-AB control.

tions in particular, this output voltage swing can become quite small.

The maximum output voltage swing of the second stage is limited by:

$$V_{SS} + V_{DS(sat)n} < V_o < V_{DD} - V_{DS(sat)p} \tag{4.6}$$

This output swing is, even in low-voltage applications, relatively large. Therefore, this output stage is used in low-voltage design. Since the output voltage can swing to within one saturation voltage of each supply rail, it is called a rail-to-rail output stage.

An output stage which is able to deliver a relatively high current to a load and also has small quiescent current, requires a class-AB biasing. A feedback-biased class-AB control, which can be used in CMOS operational amplifier design, and which is also used in bipolar low-voltage design, is shown in Fig. 4.9 [3].

Transistors $M_3$ and $M_8$ measure the currents through the output transistors, $M_9$-$M_{10}$. The resistors, $R_4$ and $R_5$, convert these currents into voltages. These voltages are compared by the differential pair, $M_4$-$M_5$. Suppose that $M_{10}$ is delivering the output current. This means that the voltage across $R_4$ is much larger than the voltage across $R_5$. The result is that the tail-current of $M_4$-$M_5$ flows completely through $M_5$. The source voltage of $M_4$-$M_5$ reflects the minimum current which is the current of the output transistor that is not delivering the output current. If a difference occurs between the source voltage of $M_4$-$M_5$ and that of $V_{REF}$, the differential amplifier, $M_1$-$M_2$, feeds a correction signal to the output transistors. In this way, the minimum current of the output transistors is controlled.

Suppose that the output stage is next in rest. The voltages across $R_4$ and $R_5$ are equal and therefore, the tail-current of $M_4$ and $M_5$ is equally divided between both transistors. The source voltage of $M_4$ and $M_5$ reflects the quiescent current of the output transistors. Again, if a difference occurs between the source voltage

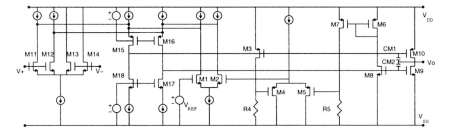

**FIGURE 4.10**
Intermediate stage and class-AB output stage.

of $M_4$ and $M_5$ and $V_{REF}$, the differential amplifier feeds a correction signal to the output transistors. In this way, the quiescent current can be controlled.

The signals, which have to be delivered to the output transistors, have to be in-phase currents. For example, these currents can be realized by using two differential pairs which are placed in parallel, $M_{11}$-$M_{14}$, as is shown in Fig 4.10. These differential pairs can be used as an intermediate stage of a three-stage operational amplifier.

The cascode transistors, $M_{15}$-$M_{18}$, are used to increase the gain of the operational amplifier. Further, the capacitors, $C_{M1}$ and $C_{M2}$, are used to frequency-compensate the stage, as will be explained in Sec. 4.3.4.

### 4.2.3 Conclusion

The input and output stage, which are described in this section, have rail-to-rail capability. Therefore, they are very well suited to low-voltage applications. The input stage and the output stage can easily be combined to obtain a rail-to-rail operational amplifier. In a 2 $\mu$m process, a unity-gain frequency between 1 MHz and 10 MHz can be obtained. The input is able to extend both supply rails by approximately 0.3 V. The output voltage can swing within 100 mV of the supply rail. The complete amplifier is able to operate on supply voltages as low as 3 V, assuming that the threshold voltages of both the n-channel and p-channel transistors are about 0.7 V.

### 4.3 BIPOLAR OPERATIONAL AMPLIFIER DESIGN

### 4.3.1 Introduction

Bipolar operational amplifiers that run on a supply voltage of 1 V have been in existence for over a decade [4], but until now, the properties of these operational amplifiers have not allowed accurate signal processing. In this section, techniques to design high-quality low-voltage bipolar operational amplifiers are highlighted [5].

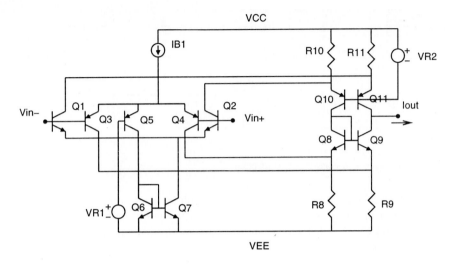

**FIGURE 4.11**
Bipolar differential input stage with rail-to-rail common-mode input voltage range and a constant transconductance over the full common-mode range.

## 4.3.2 Input Stages

A commonly used input stage of an operational amplifier is a differential input stage. The differential input stage that is shown in Fig. 4.11 has a rail-to-rail common-mode input voltage range. In this way, a reasonable ratio between the signal and additive interferences, such as noise, is obtained. Further, a rail-to-rail common-mode input voltage range also enhances the general-purpose nature of the operational amplifier.

If the common-mode input voltage is near the negative supply rail, current source $I_{B1}$ activates the PNP pair, $Q_3$-$Q_4$, provided that the supply voltage is high enough. This PNP pair is now able to handle the input signal. If the common-mode input voltage is raised above the reference voltage, $V_{R_1}$, transistor $Q_5$ takes away the current from current source $I_{B1}$, and through $Q_6$ and $Q_7$, supplies the NPN pair, $Q_1$-$Q_2$. The PNP pair, $Q_3$-$Q_4$, is now switched off and the signal operation is performed by the NPN pair, $Q_1$-$Q_2$. Even in the turnover range of $Q_5$, the sum of the tail-currents of the input pairs, and therefore also the total transconductance of the input stage, is kept constant. The value of the reference voltage source, $V_{R_1}$, is 0.8 V, high enough to enable the NPN pair to function properly when the common-mode input voltage exceeds this value. The transistors $Q_8$-$Q_{11}$ and resistors $R_8$-$R_{11}$ sum the collector currents of the input pairs.

Although not examined here, an input stage with a fully rail-to-rail common-mode input range that operates at a minimum supply voltage of 1 V can also be used [6].

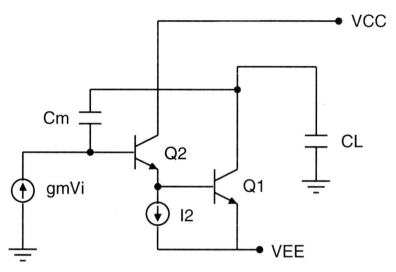

**FIGURE 4.12**
Darlington output stage (NPN side).

### 4.3.3   Output Stages

As mentioned in Sec. 4.1, the output stage of a low-voltage operational amplifier should be able to deliver an output voltage that extends from rail-to-rail. After all, the supply voltage range is by definition not very large, so we want to be able to make full use of it. This rules out the use of emitter followers as output transistors, because such a use would result in the loss of one diode voltage.

The maximum bandwidth of the operational amplifier is determined by the current flowing through the output transistors, which determines their transconductance, and by the capacitive load it has to drive. Any transistors preceding the actual common emitter (CE) output transistors more or less reduce the obtainable bandwidth. This reduction should be kept as small as possible.

The output stage should be able to push and pull output currents on the order of 10 mA, which means that its current gain should be on the order of 1000. One output transistor in a CE configuration does not meet this condition and therefore, emitter followers are normally placed in front of it. This contradicts, however, the previous demand. To illustrate the consequences of the demands stated above, two examples of low-voltage NPN output stages are examined. The theory is, however, also valid for PNP output stages.

Figure 4.12 shows the Darlington output stage. Transistors $Q_1$ and $Q_2$ are the CE stage and emitter follower, respectively, and pole splitting is realized with $C_m$. Current source $I_2$ ensures a minimum current through $Q_2$ at low output currents. The current source $g_m V_i$ represents the output of the preceding intermediate stage. The minimum supply voltage for this output stage is about 1.7 V.

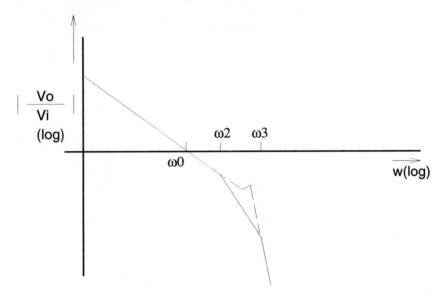

**FIGURE 4.13**
Frequency response of the Darlington output stage. The dashed line shows the "output bump,"
at large output currents, as a result of complex poles.

Figure 4.13 shows the frequency response of the Darlington output stage.
Because the bandwidth, $\omega_0$, always has to be chosen smaller than $\omega_2$ in order to
keep a positive phase margin, $\omega_2$ ultimately sets the bandwidth of the output
stage and, consequently, of the complete operational amplifier. If the base-emitter
capacitance of $Q_2$ is small compared to $C_m$, then $\omega_2$ can be calculated as:

$$\omega_2 = \frac{g_o}{C_L} \tag{4.7}$$

where $g_o$ is the transconductance of the output stage and $C_L$ is the load capacitor. This leads to the important conclusion that the bandwidth of the operational
amplifier with a current-driven output stage is determined by the quiescent current through that output stage and by the maximum load capacitor that has to
be driven.

If the output current of the output stage increases, $\omega_2$ and the high-
frequency pole, $\omega_3$, which is situated at the base of $Q_1$, move toward each other
and finally collide. The complex poles that then arise cause the famous "output
bump" [7], as is shown by the dashed line in Fig. 4.13. These complex poles may
have a negative influence on the overall performance of the operational amplifier. The presence of complex poles causes damped oscillations in the transient
response.

In the circuit shown in Fig. 4.14, the multi-path-driven (MPD) output stage
[8, 9] combines the advantages of the simple CE stage with a higher current gain,
and therefore bypasses the trade-off between bandwidth and gain. Further, the

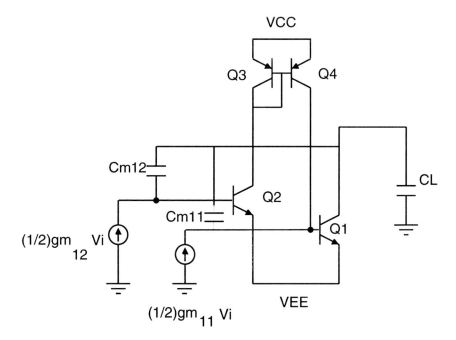

**FIGURE 4.14**
The NPN multi-path-driven output stage.

MPD output stage is also able to operate at a supply voltage as low as 1 V.

The transistors $Q_2$, $Q_3$, and $Q_4$ drive the CE output transistor $Q_1$ and parallel to this path there is a feedforward path directly from the intermediate stage to transistor $Q_1$. The path through $Q_2$-$Q_4$ supplies the necessary current gain, while the direct path to $Q_1$ guarantees good high-frequency behavior. The intermediate stage has to supply the output stage with two identical, but uncoupled, input signals. This can, for instance, be realized with two differential stages in parallel, both driven by the same input voltage [8]. The two outputs of the intermediate stage are symbolized in Fig. 4.14 by the two input current sources, $(1/2)g_{m11}V_i$ and $(1/2)g_{m12}V_i$. The poles at the output and at both inputs are split with Miller capacitors, $C_{m11}$ and $C_{m12}$. The current gain from the bases of $Q_1$ and $Q_2$ to the output is $1/2\,\beta_1(\,\beta_2 + 1) \approx 1/2\,\beta_1\,\beta_2$, half the value of the Darlington output stage and sufficient to drive large output currents.

The frequency response of the MPD output stage is shown in Fig. 4.15. The dashed and dotted lines represent the response of the gain and feedforward paths, respectively. The gain path has a large gain but a smaller bandwidth, and the feedforward path has a small gain but larger bandwidth. The slopes of the two responses coincide if the Miller capacitors $C_{m11}$ and $C_{m12}$ as well as the transconductances of the intermediate transistors, and therefore, the current

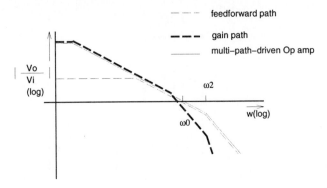

**FIGURE 4.15**
The principle of the multi-path-driven strategy. The overall transfer characteristic has the gain of the gain path, and the bandwidth of the feedforward path.

through these transistors, are matched:

$$\frac{g_{m11}}{C_{m11}} = \frac{g_{m12}}{C_{m12}} \tag{4.8}$$

A mismatch would result in a pole-zero doublet but, fortunately, the matching can be very accurate. Figure 4.15 shows that complex poles do not occur and the bandwidth is only limited by the current flowing through $Q_1$ and by the load capacitor, $C_L$, as given by (4.8).

## 4.3.4   Overall Design and Frequency Compensation

**Intermediate stages and class-AB current control**   A simple operational amplifier can consist of an input stage, directly coupled to an output stage. Although the simplicity of this topology is attractive, a major disadvantage is that the dc open-loop gain is relatively low, on the order of 80 dB. If a larger gain is desired, an additional differential amplifier stage should be inserted between the input and output stage. An example of such an intermediate stage, that also effectively compensates the Early effect acting on the transistors, is shown in Fig. 4.16

Transistors $Q_{11}$ and $Q_{14}$ supply the PNP output stage, and have a current mirror $Q_{15}$ and $Q_{18}$ that is connected to the positive supply rail, just as the PNP output transistor is. The other differential stage, $Q_{12}$ and $Q_{13}$, is connected to the NPN output stage and has an NPN current mirror, $Q_{16}$ and $Q_{17}$, that is connected to the negative supply rail, just as the NPN output transistor is. In order to be able to fully drive the differential stage, $Q_{12}$ and $Q_{13}$, the current sources $I_{12}$ and $I_{13}$ should be equal to the tail-current sources $I_{10}$ and $I_{15}$. Although this

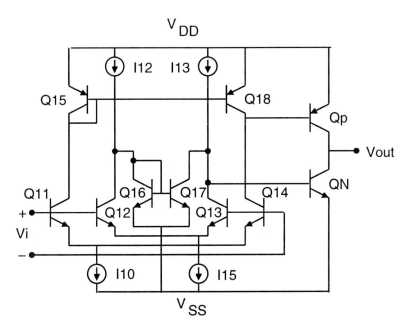

**FIGURE 4.16**
Intermediate stage with separate current mirrors and tail-current sources, to compensate the
Early effect on the intermediate-stage transistors.

may seem a wasteful dissipation of supply power, the consumption is actually
less than when the intermediate stage is equipped with folded cascodes.

The complementary CE output stage needs a class-AB biasing to assure
that neither one of the output transistors is ever completely turned off, and
the quiescent power consumption is still limited. An approach to achieve this is
shown in Fig. 4.17 [3, 6].

An image of the current through PNP output transistor, $Q_{11}$, is transferred
by $Q_{33}$ into a voltage over $R_{35}$, and, similarly, the voltage over $R_{34}$ is proportional
to the current through the NPN output transistor $Q_{21}$. The feedback amplifier
$Q_{31}$, $Q_{32}$ equates the voltage at the emitters of $Q_{35}$ and $Q_{36}$ to the reference voltage, $V_R$, and thus controls the quiescent current through the output transistors.
If one output transistor is driven hard, the current of $I_{35}$ flows through either
$Q_{35}$ or $Q_{36}$ and the voltage over the resistors $R_{35}$ or $R_{34}$, respectively, is 18 mV
lower than in the quiescent situation. This results in a residual current through
the non-active output transistor that is about half the quiescent value. If $V_R$ is
composed of one diode matched to $Q_{35}$ and $Q_{36}$, and a resistor matched to $R_{35}$
and $R_{34}$, it is independent of variations in temperature and process parameters
such as the current gain, $\beta$. Because the current through the output transistors
is measured directly at their bases, the influence of supply-voltage variations on
the performance of the output stage is also largely cancelled. For the sake of
simplicity, the output stage shown in Fig. 4.17 is composed of just two single CE

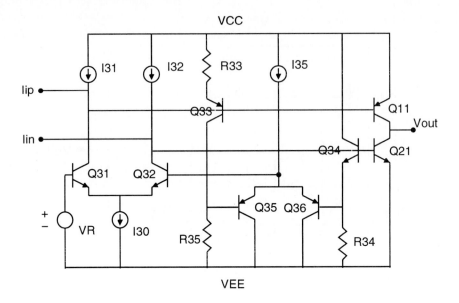

**FIGURE 4.17**
Output stage with feedback-biased class-AB control; the residual current is half the quiescent current.

output transistors, $Q_{11}$ and $Q_{21}$, but the use of this class-AB current control can also be extended to all kinds of compound output stages.

The class-AB feedback-control loop operates independently of the feedforward driving by the transistors of the intermediate stage, or in other words, the functions of biasing and signal driving are separated. The signal driving benefits from this method, because the inputs of the two output pairs are directly accessible, and the signal does not need to pass through circuit elements which are for biasing purposes only. The feedback class-AB control has an additional advantage. Suppose the intermediate stage drives the nodes $I_{ip}$ and $I_{in}$ in Fig. 4.17 with two small in-phase signals, $+i$. The current at node $I_{ip}$ cannot flow into $Q_{11}$ and it must therefore go through $Q_{31}$. The opposite signal, $-i$, then flows through the other half of the differential pair, $Q_{32}$. This means that $Q_{21}$ is not driven with $+i$, but with $+2i$; the feedback amplifier thus doubles the input signal that is supplied to the output transistors.

**Frequency Compensation.** The input, output, and intermediate stages of low-voltage operational amplifiers, as described in the previous sections, all have collector outputs and hence contribute dominating poles to the transfer from input to output. If there are two or more dominating poles, the phase margin of the frequency response is negative, and overall feedback of the operational amplifier is not possible. To change this, frequency compensation techniques must be applied that give the operational amplifier one dominating pole frequency with

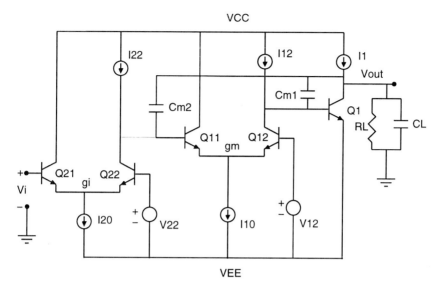

**FIGURE 4.18**
Three-stage operational amplifier with three poles which are split by two nested Miller capacitors.

a straight 6-dB/octave frequency roll-off. The most effective way to realize this is by applying the technique of nested pole splitting [1].

Figure 4.18 shows a simplified representation of a three-stage operational amplifier, comprised of an input stage, an intermediate stage, and an NPN output transistor. The intermediate stage and the output transistor are compensated with a Miller capacitor, $C_{m1}$, and the combination of these two, together with the input stage, are in turn compensated with $C_{m2}$. The open-loop bandwidth of the operational amplifier, $\omega_0$, should satisfy:

$$\omega_o = \frac{g_i}{C_{m2}} = \frac{1}{2}\frac{g_m}{C_{m1}} = \frac{1}{4}\frac{g_o}{C_L} \qquad (4.9)$$

The ratio of the transconductance of the input stage, $g_i$, and the Miller capacitor, $C_{m2}$, should be chosen to be half the ratio of the intermediate-stage transconductance, $g_m$, and $C_{m1}$. This ratio of $g_m$ and $C_{m1}$, which is in fact the bandwidth of the intermediate stage combined with the output stage should, in turn, be chosen to be half the value of the output pole: the ratio of the output transconductance, $g_o$, and the load capacitor, $C_L$. The most difficult condition, as far as frequency compensation is concerned, is if overall unity-gain feedback is applied to the operational amplifier. The condition given in (4.9) then results in a flat frequency response, and the poles move into the third-order Butterworth positions.

**TABLE 4.1**
Differences between MOS and bipolar transistors.

|  | *MOS* | *BJT* |
|---|---|---|
| advantages | high input impedance<br>high current gain<br>low input bias current | high transconductance<br>high voltage gain<br>low offset and noise voltage<br>wide range of collector currents |
| disadvantages | low transconductance<br>low voltage gain<br>high offset and noise voltage<br>small range of drain currents | low input impedance<br>low current gain<br>high input bias current |

## 4.4   LOW-VOLTAGE BICMOS OP AMP DESIGN

There is currently great interest in the BiCMOS technology, because it is easy to implement modern systems comprising both analog and digital signal processing in BiCMOS. The bipolar part is very well suited for use in analog circuits and the MOS part for use in digital circuits. Further, the availability of both MOS and bipolar devices offers additional freedom to the designer. However, in spite of these advantages, BiCMOS is not yet extensively used. The reason for this is that BiCMOS is an expensive technology. It is much more expensive than CMOS and therefore it is very attractive to try to implement the analog part also in CMOS.

### 4.4.1   BiCMOS Technology

MOS devices and bipolar devices are fundamentally very different, and therefore, each type of device is best suited to different tasks. This means that in order to get the best possible performance, we have to select the best type of device for each task. The combination of the high voltage gain of the bipolar junction transistor (BJT) and the very high input impedance of the MOS transistor is especially interesting and should be exploited in BiCMOS designs. Several important advantages and disadvantages of MOS and bipolar transistors are listed in Table 4.1. The reader is referred to Chapter 5 for more details on BiCMOS technologies and analog BiCMOS IC design.

For low-voltage operation, the minimum output voltage (the voltage across the collector and the emitter for the BJT, and the voltage across the drain and the source for an MOS transistor), is important because this determines the minimum voltage across a current source. The output voltage of the BJT is limited by the transistor saturation voltage. For low output voltages, the base-collector junction conducts. It can be time consuming to get the transistor out of saturation, which results in considerable delay. This delay causes distortion and is therefore not allowable. Because of the above mentioned factors, the voltage across the collector and emitter must not become much smaller than approximately 100 mV. The only junctions of the MOS transistor, the source to substrate and the drain to the substrate are normally reverse biased, so that the voltage across the drain and the source can become very small. However, for these voltages, the MOS transistor operates in the linear region and has a low output impedance. Therefore, an MOS current source needs a much higher voltage, usually more than 0.5 V.

## 4.4.2   Input Stages

When designing an input stage in BiCMOS, the first thing to decide is which type of device to use. We can use either a simple input differential pair or a more complicated input stage. For low-voltage operation sometimes it is necessary to have a rail-to-rail voltage range input stage, such as those described in Sec. 4.3.2 (bipolar) and Sec. 4.2.1 (CMOS).

For a long time, JFET or MOSFET devices have been used as the input stage of bipolar operational amplifiers [7, 10, 11] because of their high input impedance, negligible bias current and because a JFET or MOSFET input stage has a better slew rate. This is particularly the case in applications where the operational amplifier is loaded by a capacitor, such as in switched-capacitor circuits and in read-out of capacitive sensors. Bipolar transistors can not be used and therefore most BiCMOS designs employ an MOS input stage. However, when the bias current is not important, the bipolar transistor is an attractive and often better alternative, because of its low offset voltage and low voltage noise.

An interesting input stage demonstrating a nice cooperation between bipolar and FET devices is used in the OP-275 [12]. The input stage of the OP-275 consists of a bipolar input stage and a JFET input stage in parallel. In Fig. 4.19, a simplified schematic of an operational amplifier is shown with a BiCMOS version of the input stage of the OP-275.

The input stage consists of input transistors $Q_1$, $Q_2$, $M_1$, $M_2$ and current mirror transistors $Q_3$, $Q_4$. Transistor $Q_5$ represents the second stage of the operational amplifier. This input stage combines the high transconductance, low offset voltage, and the noise of a bipolar stage with the better slew rate of an MOS input stage. For small signals, the bipolar stage dominates, while for large voltage steps, the tail-current of the MOS input stage, $I_2$, is additionally available to charge the compensation capacitor $C_M$. The following relation can be found for the slew rate, S  [7]:

$$S = \frac{I_t}{C_M} \tag{4.10}$$

in which $I_t$ is the tail-current of the input pair and $C_M$ the compensation capacitor. We can therefore conclude that the extra current $I_2$ increases the slew rate. In fact, a compromise is made between the advantages of a full bipolar and a full MOS input stage. The ratio of the currents $I_1$ and $I_2$ determines how dominant the bipolar stage is for small signals and how much the slew rate is enhanced.

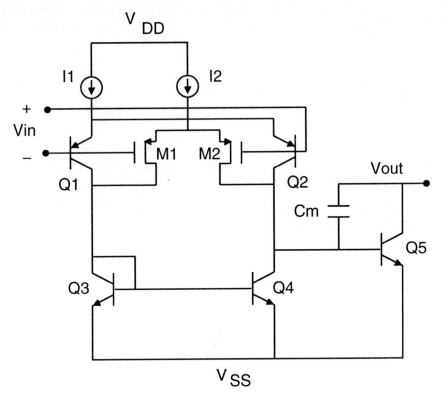

**FIGURE 4.19**
Operational amplifier with BiCMOS input stage.

Because MOS input stages are often needed, we will now consider MOS input stages. An example of an operational amplifier with MOS input transistors is shown in Fig. 4.20. The MOS input transistors provide high input impedance and very low bias currents, but because of their low transconductance, we need to add some extra gain by adding a gain stage or, preferably by means of cascodes. PMOS transistors are mainly used because the 1/f noise of PMOS transistors is better than the 1/f noise of NMOS transistors. The current mirror $Q_1, Q_2$, $M_3$ is an improved BiCMOS version of the mirror used in Fig. 4.19. The MOS transistor $M_3$ cancels the influence of the base current of the bipolar transistors in such a way that a very accurate current mirror is realized. It is possible to cascode the output of the current mirror, but the output impedance of the input stage will be limited by the output impedance of the differential input pair. Transistors $Q_3$ and $M_4$ form the second stage, which loads the input stage in such a way that the voltage on both sides of the differential input is equal, so that influence of supply variations is minimized.

The disadvantage of this kind of input stage is that the common-mode input voltage cannot reach the negative supply rail. The only way to achieve this is by

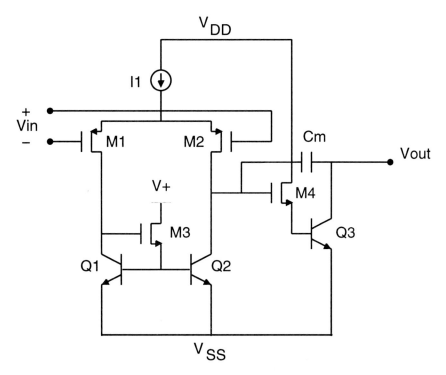

**FIGURE 4.20**
BiCMOS operational amplifier with MOS input stage.

using a folded cascode as shown in Fig. 4.21. The bipolar current sources, $Q_1$-$Q_4$, need only 100 mV and therefore the common-mode voltage of this input stage can go 0.5 V or more (depending on the gate-source voltage of the input transistors) below the supply rail. The MOS cascodes provide a high output impedance, so that a reasonable voltage gain can be obtained when this stage is followed by an MOS transistor. When a single-ended output is necessary, the BiCMOS current mirror that was used in Fig. 4.20 can be applied. The resulting circuit is shown in Fig. 4.22. In-phase outputs (for example for driving a rail-to-rail output stage) can be made by placing transistors parallel to $M_4$ and $M_6$.

### 4.4.3   Output Stages

In low-power operational amplifiers, the output stage plays a very important role, because most of the power is dissipated in the output stage, while the power that is dissipated in the other stages is determined by the current gain of the output stage. Thus, the construction of the output stage determines the power dissipation of the whole operational amplifier. In order to keep the power dissipation of the output stage low, the quiescent current must be small. And, as the output stage must still be able to deliver high currents, this means that

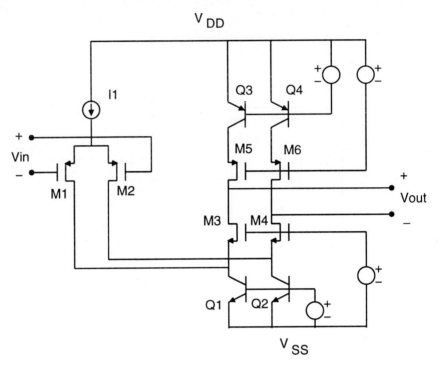

**FIGURE 4.21**
Folded cascode input stage with differential output.

the output transistor must handle a wide range of currents. Therefore, a bipolar transistor is best-suited as an output transistor. To keep the power dissipation of the other stages low, they must be biased at low currents and therefore the current gain of the output stage must be high. For this job, the MOS transistor is very useful. The basic structure of such a low power BiCMOS output stage is shown in Fig. 4.23(a). This configuration is not suited for low-voltage applications, because the base-emitter voltage of a BJT and the gate-source voltage of an MOS transistor are stacked. Using a mirror, a low-voltage configuration can be constructed as shown in Fig. 4.23(b).

Frequency compensation of this output stage can be done by means of pole splitting, but this will cause complex poles for high output currents. Therefore the same compensation method should be used as in the NPN multi-path-driven output stage [8, 9], which has been discussed in Sec. 4.3.3. The resulting BiC-MOS multi-path-driven output stage is shown in Fig. 4.24. The current sources symbolize two in-phase outputs of a preceding stage. For example, the input stage shown in Fig. 4.22 can be used and the two in-phase signals can be made by doubling the size of the cascode transistors $M_4$ and $M_5$.

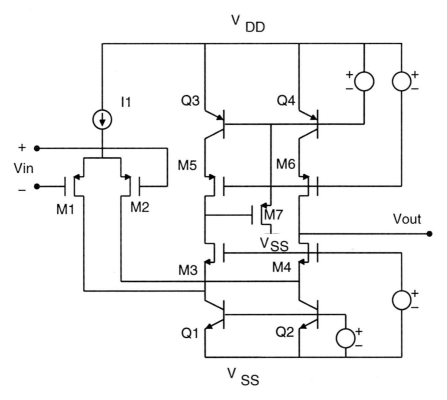

**FIGURE 4.22**
Folded cascode input stage with single ended output.

### 4.4.4 Conclusion

In the preceding sections some examples have been shown that utilize the possibilities of the BiCMOS technology. Based upon the discussed input and output stages, simple two-stage operational amplifiers can be built with sufficient gain, while the simple two-stage topology guarantees high bandwidth. Using techniques that increase the gain of these operational amplifiers, it is possible to realize low-voltage operational amplifiers with 100 dB dc-gain and 30 MHz unity-gain frequency.

As the availability of reasonably priced BiCMOS processes to designers is quite new, a lot of research is needed to find new clever circuit topologies. In the future, these new BiCMOS circuit topologies will result in more compact and more power efficient circuits. Further discussion of BiCMOS circuits is presented in Chapter 5.

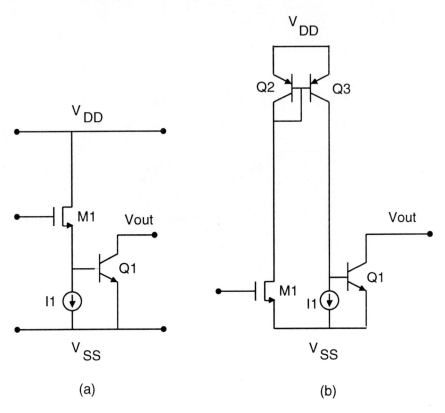

**FIGURE 4.23**
BiCMOS output stage (a) principle and (b) low-voltage version.

## 4.5 INSTRUMENTATION AMPLIFIER DESIGN

An instrumentation amplifier is used in applications where a small differential output voltage of a sensor must be accurately amplified in the presence of strong common-mode input voltages. Figure 4.25 shows a typical environment in which an instrumentation amplifier can be used. A sensor generates the desired signal $V_s$ and has a source resistance $R_s$. Both the sensor and the amplifier are connected to a common ground terminal, e.g., the metal frame of a car. If the sensor and the amplifier are separated some distance, a voltage $V_{cm}$ will be present between the two ground connections. The amplifier must be insensitive to this common-mode voltage, thus it must have a high common-mode rejection. Moreover, the amplifier must be able to operate around the negative rail, i.e., the common-mode input range must include ground.

In the following section, two alternative approaches to the design of an instrumentation amplifier are compared; resistive feedback and current feedback. Both methods are well suited to IC implementations, although current feedback

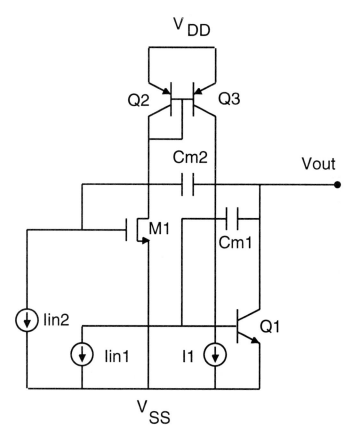

**FIGURE 4.24**
BiCMOS multi-path-driven output stage.

more fully exploits the benefits of monolithic techniques, as will become clear
later. After the comparison, a design based on current feedback is dealt with in
more detail. The emphasis is on three instrumentation amplifier properties: good
accuracy, high common-mode rejection, and negative rail capability.

### 4.5.1   Comparison of Two Alternative Approaches

**Resistive feedback**   The simplest type of an instrumentation amplifier with
resistive feedback uses one operational amplifier and four resistors to set the gain
(Fig. 4.26). One major disadvantage of this one-operational amplifier instrumentation amplifier is the absence of input buffering, which results in a small input
resistance. An improved circuit is the three-operational amplifier instrumentation amplifier, shown in Fig. 4.27. The input resistance is very high now and the
gain can be set by $R_2$ without altering the resistor bridge. However, a common-

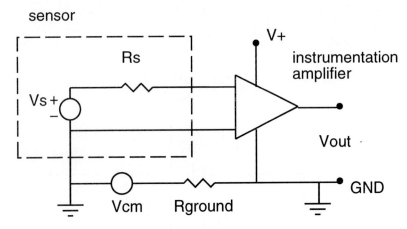

**FIGURE 4.25**
Measurement system with interfering signal $V_{cm}$.

mode input voltage is passed on to nodes (a) and (b) in Fig. 4.27, and therefore, relatively large common-mode currents will flow through $R_4 + R_6$ and $R_5 + R_7$ toward the output and ground nodes. If a small imbalance exists between the two branches, this common-mode input current will cause a differential mode (DM) output voltage. Therefore, expensive on-chip trimming is required to obtain a good CMRR [13]. Another disadvantage of this circuit is that the outputs of operational amplifier 1 and operational amplifier 2 cannot become zero or negative and therefore sensors which are grounded at one side, as shown in Fig. 4.25, cannot be connected to this amplifier.

**Current feedback**   To improve the common-mode rejection, the current feedback method isolates the common-mode voltage at the input of the amplifier from the rest of the circuit, thus preventing large common-mode currents flowing through poorly balanced circuit parts. Because common-mode currents can never be completely avoided, additional balancing is used to further improve the CMRR. In the following section, an example is presented of an instrumentation amplifier design based on this isolation principle.

### 4.5.2   Instrumentation Amplifier with Current Feedback

In the design that follows, emphasis is on achieving high common-mode rejection, good accuracy, and a common-mode input range that includes ground.

**Input stage**   An input stage that can perform the isolation action is shown in Fig. 4.28. For large values of $R_1$, the transfer function for the DM input voltages is $I_{out}/V_{in} \approx 1/R_1$. If $Q_1$, $Q_2$ and the current sources have infinite

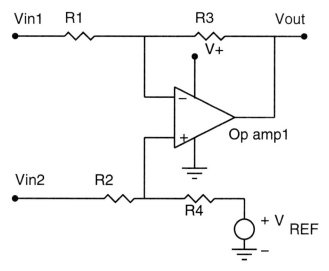

**FIGURE 4.26**
One op amp instrumentation amplifier.

output resistance, common-mode input voltages are confined to the base emitter circuit of $Q_1$ and $Q_2$ and the common-mode input currents will be zero. However, output resistances are not infinite and small input common-mode currents flow through $r_{o1}$, $r_{o2}$ and $R_{I1}$, $R_{I2}$. Small resistor value mismatches result in undesired differential output currents. $R_{I1}$ and $R_{I2}$ can be made very large in comparison with $r_{o1}$ and $r_{o2}$, e.g., by the use of cascoded current sources. In the case of Fig. 4.28, the CMRR due to $r_{o1}$ and $r_{o2}$ can be written as:

$$CMRR \approx \mu \frac{\mu}{\Delta \mu} \qquad (4.11)$$

where $\mu$ is the intrinsic voltage gain of the input transistors and $\Delta \mu$ is the difference in voltage gain between $Q_1$ and $Q_2 (\mu = g_m r_o = V_{early}/(kT/q))$. The 'isolation factor' $\mu$ is approximately 2000 and the 'balancing factor' $\Delta \mu / \mu$ is approximately 2% which gives an estimated CMRR of 100 dB, without on-chip trimming. The common-mode rejection can be increased by the use of composite input transistors with high output resistance [14].

**Amplifier configuration** There are various ways of making an instrumentation amplifier with overall accurate voltage transfer function using the input stage of Fig. 4.28 [15, 16, 17]. One configuration is shown in Fig. 4.29. The first transconductance stage $Q1$ converts the input voltage into a current $i_1$ and the second transconductance stage converts the attenuated output voltage into a current $i_2$ which is fed back to the input stage. Loop amplifier A makes $i_1 - i_2 = 0$ and therefore the overall transfer is:

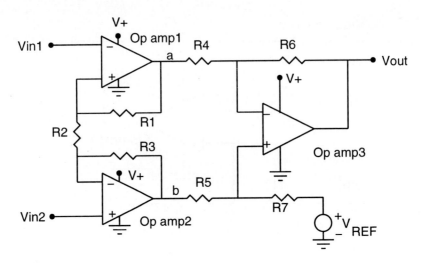

**FIGURE 4.27**
Three op amp instrumentation amplifier.

$$\frac{V_{out}}{V_{in}} = \frac{T1}{T2}\frac{R_3 + R_4}{R_4} \qquad (4.12)$$

The gain can be set in two ways. The first method takes $(R_3 + R_4)/R_4 = 1$ and sets the gain with resistors $R_1$ and $R_2$. Both transconductance stages should be accurate and linear so that $T1/T2 = R_2/R_1$ [15, 18]. The second method takes $R_1 = R_2$ and sets the gain with $R_3$ and $R_4$. The advantage of this approach is that the absolute error of $T1$ and $T2$ is not important. As long as $T1$ and $T2$ are well matched, their ratio is still exactly one. The second method is used here.

**Common-mode Input Range**  A demand on the present design is that the common-mode input range should include ground. This can simply be achieved by using PNP differential pairs at the input, as shown in Fig. 4.30. The voltage across $R_{31}$ and $R_{32}$ is 0.3 V and therefore, the input common-mode voltage can reach GND-0.2 V without forward biasing the collector-base junctions of $Q_{11}$ and $Q_{12}$.

**Accuracy**  To obtain good accuracy in closed loop amplification, two conditions have to be met. The first condition is that the two transconductance stages must have equal transfer functions. Their transfer functions are approximately:

$$Q1 \approx \frac{1}{R_1 + r_{c11} + r_{c12}} \qquad Q2 \approx \frac{1}{R_2 + r_{c21} + r_{c22}} \qquad (4.13)$$

As can be seen from these equations, $Q1$ and $Q2$ depend on the value of $r_c$ which is a function of bias current and temperature. In the present design, the ratio

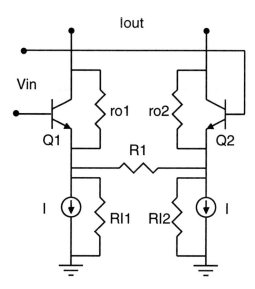

**FIGURE 4.28**
Instrumentation amplifier input stage.

$Q1/Q2$ must be exactly one, hence all resistor values, bias currents, and component temperatures must have well matched values. To obtain good matching, the resistors, transistors and current sources of the two input stages must be placed close together on the chip in a symmetrical layout. (See Chapter 16 for more description of analog layout techniques.)

The second requirement for good accuracy is that the gain of the loop amplifier, A, (Fig. 4.30) is large. For an accuracy of 0.1% at a closed loop gain of 1000, the open loop gain must be at least $1\times10^6$. One restriction that has to be taken into account is that the complete loop transfer function should not contain more than two dominant poles, to simplify frequency compensation. The loop gain of the circuit shown in Fig. 4.31 can be found as follows. Starting at the output, $V_{out}$ is attenuated by $R_3$ and $R_4$ and converted into a current by the PNP differential pair. Transistors $Q_{31}$ and $Q_{32}$ form the input stage of the operational amplifier. They provide the necessary level shift which is required to drive the NPN gain stage $Q_{41}$-$Q_{44}$. The gain stage amplifies the collector currents of $Q_{31}$ and $Q_{32}$ by a factor of $\beta^2$. The current mirror $Q_{45}$ and $Q_{46}$ doubles the gain.

The output current of the gain stage causes a change in the voltage at node X which is determined by the total resistance from node X to ground. This resistance is made up of the output resistance of $Q_{44}$ in parallel with the output resistance of $Q_{46}$, which gives a total value of $\frac{1}{2} r_o$. Transistors $Q_{51}$ and $Q_{52}$ have unity voltage gain. Their function is to achieve a low output resistance of the amplifier. With this output configuration, the output voltage can reach GND

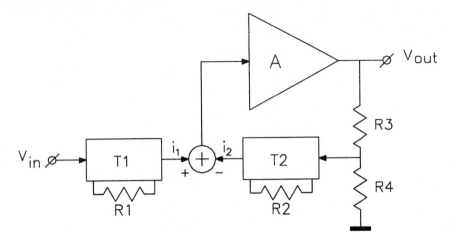

**FIGURE 4.29**
Current feedback instrumentation amplifier.

level. The total loop gain is:

$$A_{loop} \approx \frac{R_4}{R_3 + R_4} \frac{1}{R_2} 2\beta^2 \frac{1}{2} r_o \qquad (4.14)$$

Inserting $R_2$=10 k$\Omega$, $R_3 = 0$, $\beta = 75$, $r_o$=2 M$\Omega$ gives $A_{loop} \approx 1.1 \times 10^6$. However, in practice the gain will not exceed $10^5$. This is mainly due to the fact that signal current escapes from the loop through $R_{31}$, $R_{32}$, and through the current sources in the collectors of $Q_{31}, Q_{32}$, $Q_{45}$ and $Q_{46}$. The gain can be boosted by a gain boosting technique as described in [19].

The complete loop in the circuit shown in Fig. 4.31 has two dominant poles, one at the input of the gain stage and one at the output of the gain stage. Frequency compensation can simply be achieved by placing a Miller capacitor, $C_{m1}$, across the gain stage, with an extra capacitor, $C_{m2}$, for reasons of symmetry.

### 4.5.3 Conclusion

In the design of an instrumentation amplifier two approaches can be used to obtain good accuracy and common-mode rejection. Resistive feedback relies on balancing techniques and requires three operational amplifiers and on-chip laser trimming to obtain good performance. Current feedback, on the other hand, uses a combination of isolation and balancing techniques to obtain good common-mode rejection and good accuracy. Without on-chip trimming, CMRR and accuracy figures of 100 dB and 0.1%, respectively, can be achieved.

## 4.6  LOW-VOLTAGE FILTERS

When designing integrated filters, especially filters for low supply voltages, one of the most fundamental problems that is encountered is the dynamic range. It

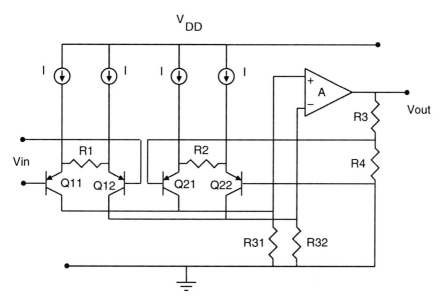

**FIGURE 4.30**
Instrumentation amplifier with PNP input stages.

is defined as the ratio of the largest to the smallest signal voltage which the filter can handle. The maximum signal voltages are limited by the power-supply voltage or by distortion, and the minimum signal is limited by noise.

In a number of applications, the dynamic range required is very high. Typical examples of such applications are intermediate-frequency filtering and high-quality audio. In these applications, everything possible should be done to optimize the dynamic range of the filters in order to make the filter usable at all.

There are also applications where the dynamic range is less critical, such as video and low-quality audio applications. The latter is found, for instance, in hearing aids. In these applications, dynamic-range optimization is not necessary to make a filter realizable, but it can be used to minimize the chip area and hence, the production cost.

In this section, the design of optimal dynamic-range integrated filters is discussed. We confine ourselves to continuous-time filters, which inherently have a larger dynamic range than switched-capacitor filters. Design methods for integrated continuous-time filters are discussed. It is shown that such a filter is designed as a network of integrators. Both the network and the integrators can be designed optimally for the dynamic range. We restrict ourselves to discussing the design of dynamic-range optimal integrators. Finally, a design example of a low-voltage bandpass filter is given [20].

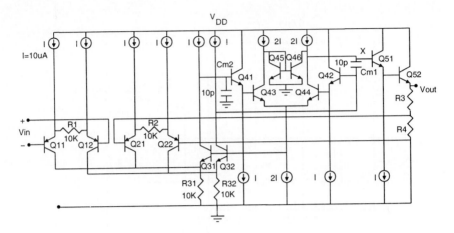

**FIGURE 4.31**
Complete instrumentation amplifier circuit.

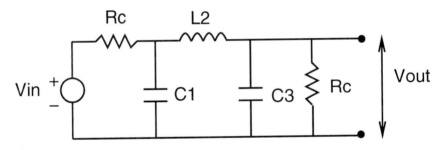

**FIGURE 4.32**
An LC prototype ladder filter.

### 4.6.1 Filter Design

A very popular synthesis method for integrated filters is the active simulation of a passive LC prototype filter. A third-order LC ladder filter is shown in Fig. 4.32. An active integrated filter can be derived from this prototype by simulating the inductor with a gyrator and a capacitor [21, 22], or by simulating the network equations of this filter with a network of three active integrators [23]. These integrators can be realized as a capacitively-loaded transconductor, as shown in Fig. 4.33. The two methods then give rise to the same active filter, shown in Fig. 4.34. In this way, an integrated filter is realized as a network of active integrators. It must be noted that there is an infinite number of possible network configurations. For a good filter design, a good filter network has to be selected. This problem is discussed at length elsewhere [20, 24], and is not addressed here.

A practical problem that arises when this approach is chosen, is the fact that usually the time constants of the integrators cannot be realized accurately

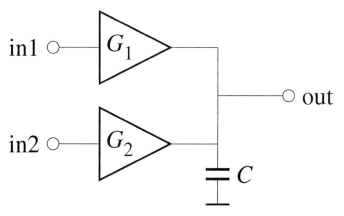

**FIGURE 4.33**
A two-input transadmittance-capacitance integrator.

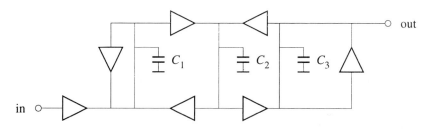

**FIGURE 4.34**
A network of active integrators that realizes the same transfer function as the filter of Fig. 4.32.

enough, because of the inaccuracies in the transconductor and capacitor values. This problem is overcome by making the time constants of the integrators tunable, and furnishing the filter with an automatic tuning system [25,26]. Therefore, either the capacitances or the transconductances must be tunable. Tunable capacitances are very non-linear, and thus they limit the dynamic range which is why the transconductors should be made tunable.

Figure 4.35 shows a different way of realizing an integrator for on-chip filters [27]. Two MOSFETs are used as resistors that are tunable via their gate voltage, and two capacitances are used in the feedback path of a balanced integrator. The circuit is balanced to cancel even-order nonlinearities in the MOSFETs. If the filter shown in Fig. 4.32 is simulated using these integrators, the circuit shown in Fig. 4.36 is obtained.

### 4.6.2  Integrators

To optimize the dynamic range of low-voltage filters, the integrators must be optimized. In this section, we consider the different possibilities and reach a conclusion as to which type can best be used.

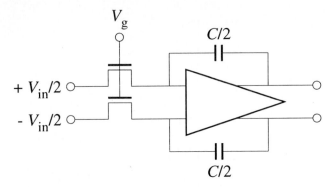

**FIGURE 4.35**
An op amp-MOSFET-C integrator.

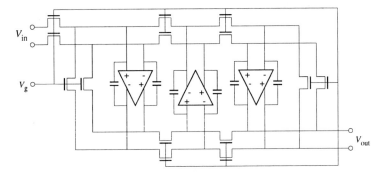

**FIGURE 4.36**
Active realization of the filter shown in Fig. 4.32 with op amp integrators.

**Current-Mode Compared to Voltage-Mode**   Since the maximum signal voltages imposed by the supply voltages limit the dynamic range, it seems logical to use currents as signal-carrying quantities. To do this, current-mode integrators are necessary. Figure 4.37 shows such an integrator with two inputs and two outputs. If, with this integrator as a building block, the prototype shown in Fig. 4.32 is simulated, the circuit shown in Fig. 4.38 is found.

When this circuit is compared to the one of Fig. 4.34, it appears that the only difference between them is that the input transconductor has been moved to the output. It can easily be proven that the noise contribution of this transconductor can be made negligible in both realizations. This means that in both cases, the dynamic range is limited on one side by the noise contributions of the other transconductors, and on the other side by the maximum signal voltages over the capacitors. Therefore the dynamic range of both circuits is the same. It is true that in the current-mode version, the (current) output signal level can be increased to any level by increasing the transfer value of the output transconductor. However, this also amplifies the noise level, so that the dynamic range is not changed by this action.

This statement does not apply only to transconductance-capacitance filters.

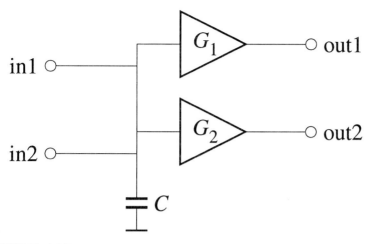

**FIGURE 4.37**
A two-input-two-output current-mode integrator.

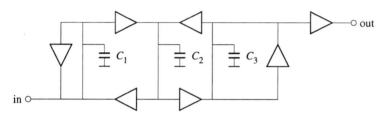

**FIGURE 4.38**
A simulation of the filter shown in Fig. 4.32 with the integrator shown in Fig. 4.37.

It is also valid for any filter structure that contains resistors, transistors, and capacitors; the currents that flow through the capacitors in the circuit cause signal voltages over these capacitors. As these voltages are limited by the supply voltages, there is a fundamental limit to the dynamic range. This limit is the same for both current-mode and voltage-mode filters.

There is a situation in which current-mode can be advantageous to the effective dynamic range, namely when signal compression and expansion is used to limit the capacitor voltages to a safe region [28]. If this is done, the output noise level increases in proportion to the output signal level, so strictly speaking, the dynamic range is not enlarged. In situations where noise appreciation is dependent on the signal level, however, the dynamic range effectively increases. This is typically the case in audio applications. In applications where large and small signals must be processed at the same time, as in an intermediate-frequency filter, this scheme does not work.

**Optimal Conditions**   To maximize the dynamic range of an integrator, its output signal level should be maximized and its noise generation should be minimized. The noisy components of a transconductor-type integrator are its transconductors. The input-referred noise voltage spectrum of a transconductor can be expressed as:

$$S_{ni} = \frac{2kT\xi}{G} \tag{4.15}$$

where $k$ is Boltzmann's constant, $T$ is the absolute temperature, and $G$ is the transconductance value. This equation defines $\xi$ as the noise factor of the transconductor. If $\xi = 1$, the transconductor generates the same amount of noise as a passive conductance $G$. Because of the active circuitry inside the transconductor, $\xi$ can be larger than 1. $\xi = 1$ corresponds to an optimal situation. If the unity-gain frequency $B = G/C$, where $C$ is the integration capacitance, is used as an estimate for the noise bandwidth of the integrator, the total input noise voltage is obtained:

$$\overline{V_{ni}^2} = \frac{1}{2\pi} \int_{-B}^{B} S_{ni}(\omega)d\omega = \frac{2kT\xi}{\pi C} \tag{4.16}$$

If the maximum effective signal voltage at the input or output of the integrator is V$_{maxeff}$, the dynamic range is

$$DR = \frac{V_{maxeff}^2}{\overline{V_{ni}^2}} = \frac{\pi V_{max\,eff}^2 C}{2kT\xi} \tag{4.17}$$

From (4.16), we see that the input noise voltage of an integrator is proportional to $kT\xi/C$. This is also true for the operational amplifier integrator. Therefore, the total noise voltage in any frequency band over the integration capacitor, $C$, is proportional to $kT\xi/C$. Thus, if the capacitor value is fixed, the dynamic range of the integrator is maximized by minimizing its noise factor, and by maximizing the maximum signal voltage over the capacitor.

**Transconductance Integrators**   An advantage of transconductor integrators over operational amplifier integrators is the fact that the capacitor can be used differentially, so that the maximum signal voltage over it can be doubled, as shown in Fig. 4.39. If the transconductors have rail-to-rail output stages, and are driven as antiphase signals, the maximum peak-to-peak signal voltage over the capacitor is twice the supply voltage. This is the largest value that can ever be reached in any configuration. Therefore, a fundamental maximum dynamic-range can be reached if two conditions are met:

- The noise factor of the transconductors is 1, and
- The transconductors have a rail-to-rail output stage.

These two conditions, however, are conflicting.

The problem is that, since the transconductors have a current output, internally the output current must be measured. This measurement can take place

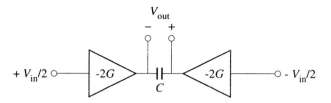

**FIGURE 4.39**
If the capacitance is used floating between the outputs of two transadmittances, the signal voltage across it can be doubled.

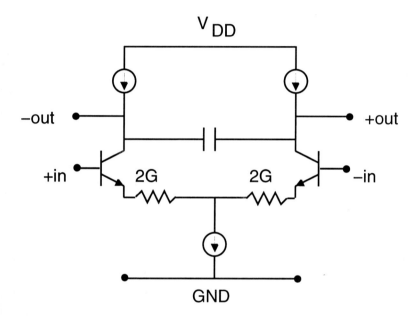

**FIGURE 4.40**
An example of a transconductor integrator, where the two conductances, 2$G$, serve to measure the capacitor current.

inside a transistor but usually this happens outside the transistors, for instance when source degeneration is applied. In Fig. 4.40 the measuring components are the two conductors, 2$G$.

In any case, *something* internally must be put in series with the output to perform this measurement. It is also this *something* that forms the most principal noise source in the transconductor. To clarify this, Fig. 4.41 shows a model of a transconductor with transconductance $-2G$.

The series component for output current measurement is a conductor with value 2$aG$, where $a$ is a dimensionless scaling factor. If this component is the only noisy component, and if this noise is thermal, it can readily be proven that the noise factor, $\xi$, of this stage equals $a$.

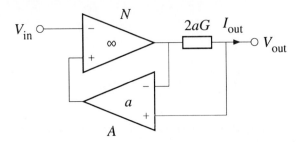

**FIGURE 4.41**
A model for a transconductor with tranconductance factor -2G. It consists of a nullor N, a voltage amplifier with amplification factor a, and a measuring conductance, 2aG.

A large output swing is incompatible with a low noise factor. If the output swing must be large, the voltage drop over the series conductance must be low, so that $a$ must be large. This implies a large noise factor. For an optimal dynamic range, an optimal value for $a$ can be found. It is not difficult to prove that this optimum is $a = 1$. In this case, the maximum peak-to-peak voltage over the capacitance is equal to the supply voltage $V_{SUP}$. This maximum is only reached if the active stages shown in Fig. 4.41 have rail-to-rail properties. For sinusoidal signals, this maximum corresponds to a maximum effective output voltage of $V_{SUP}/2\sqrt{2}$, and a noise factor $\xi=1$. It can be proven [20] that under these circumstances there is no transconductor that has a larger dynamic range than the one shown in Fig. 4.41. Therefore, with (4.17) we have found a fundamental maximum for the dynamic range of transconductor integrators given by

$$DR_{max} = \frac{\pi V_{SUP}^2 C}{16kT} \qquad (4.18)$$

**Operational Amplifier Integrators**  A typical operational amplifier integrator is shown in Fig. 4.35. A disadvantage of this integrator is the fact that the capacitors cannot be driven differentially as in Fig. 4.39. An advantage, however, is that no current-measuring component needs to be connected in series with the output of the integrator. Therefore, it is possible to realize this integrator with a rail-to-rail output. The maximum input voltage range is determined by the MOSFETs at the input. The maximum input voltage is $V_G$-$V_T$, where $V_G$ is the gate-voltage, and $V_T$ is the threshold voltage of the MOSFETs. This means that for a rail-to-rail input voltage swing to be possible, either the gate voltage must be higher than the supply voltage, or the threshold voltage must be negative. The latter requires depletion MOSFETs to be available, the former requires an on-chip supply voltage multiplier.

The noise of this integrator originates from the operational amplifier and the MOSFETs. If the noise production of the operational amplifier is negligible when compared to the noise production of the MOSFET's, the integrator has a noise factor close to 1. Therefore, if the above conditions are met, a fundamental

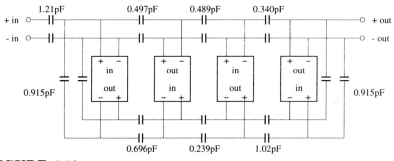

**FIGURE 4.42**
The filter network. The building blocks are biquads, which are realized as shown in Fig. 4.43.

upper limit for the dynamic range of any active integrator is given by (4.18).

**Conclusion** We may conclude that there is an upper limit to the dynamic range of active integrators. This limit is given by (4.18) and can be reached using transconductors as well as by operational amplifier integrators. With the use of transconductor integrators it will be difficult to reach this limit, because for this purpose a circuit to implement the model in Fig. 4.41 must be realized [20, 29]. Therefore, operational amplifier integrators are best suited to realize on-chip filters for low supply voltages and with a high dynamic range. In the following section we discuss design examples of such filters.

### 4.6.3   Design Examples

To demonstrate the theory of the previous sections, two design examples are given below. Both examples are eighth-order Butterworth bandpass filters with a central frequency of 100 kHz and a bandwidth of 7 kHz. The filter networks in the two examples are the same, but the active circuits used are different. One example is designed for a 3 V power supply, the other for a 1 V power supply.

**The Network** A suitable filter network is shown in Fig. 4.42. The building blocks are biquads, the realization of which is shown in Fig. 4.43. Each biquad contains two integrators, each consisting of one operational amplifier with balanced outputs, two capacitances of 10 pF, and two MOSFETs that are used as tunable resistors. As explained in the previous section, in order to realize rail-to-rail capabilities, either the MOSFETs should be of the depletion type, or the gate voltages should be operated above the supply voltage. We will assume that one of these conditions is met. The filter network has been discussed in detail in [24]; a filter realized with this network has a dynamic range that is only 1 dB below a network-related optimum, which is satisfactory. The design of the operational amplifiers with balanced output follows. A 3 V operational amplifier is given first and then a 1 V version of it is given.

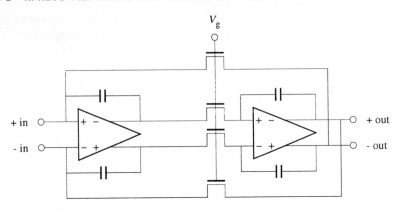

**FIGURE 4.43**
The biquads shown in Fig. 4.42 are realized with op amp integrators. The MOSFET's realize
tunable resistors with a nominal value of 159 k$\Omega$. The capacitances are 10 pF each.

**3-Volt Amplifier and Filter Design** A 3 V operational amplifier with balanced output is shown in Fig. 4.44. The input stage has been built around $Q_1$
and $Q_2$, the output stage around $Q_3$ and $Q_4$. For the output transistors, vertical
PNPs should be used. $Q_{12}$ and $Q_{13}$ together with four 170 k$\Omega$ resistors form
a common-mode feedback circuit. This circuit is necessary to ensure common-
mode stability and accurate output balancing. The latter is necessary to cancel
even-order nonlinearities in the integrators that are connected to the output of
the integrator in question.

The input stage is a dominant factor in the noise generated in the integrator. A bipolar input stage has been chosen so that a very low noise factor
can be realized with a low bias current. The output stage has also been realized
with bipolar transistors so as to combine a large gain with a small chip area.
Rail-to-rail properties have been obtained by using the output transistors $Q_3$
and $Q_4$ in a common-emitter configuration. Thus, the maximum output voltage
is only a saturation voltage of about 200 mV lower than the supply voltage. A
similar situation applies to the current sinks $Q_{10}$ and $Q_{12}$.

The operational amplifier with balanced output is used in the biquads
according to Fig. 4.43, and these are intercoupled to form the filter shown in
Fig. 4.42. Simulations show a maximum differential output signal amplitude of
2.8 V. This corresponds to nearly rail-to-rail operation, which would be 3 V
total amplitude. The difference is only -0.6 dB. The total output noise voltage
of the filter is 331 $\mu$V. This corresponds to a noise factor of only 1.1, or -0.4 dB.
Therefore, the integrator is 1 dB below optimum. We have found that the filter
network 4.42 is also 1 dB below optimum, so the complete filter, when real op
amps are used, is 2 dB below the optimum, which is quite acceptable.

The dynamic range of the filter is the ratio of the maximum effective output
voltage and the output noise voltage, or 75.6 dB. The optimum is 2 dB more,
which means that it is impossible to design an integrated eighth-order Butterworth bandpass filter with a $Q$ of 14.3, a 3 V supply voltage and a total of 160 pF

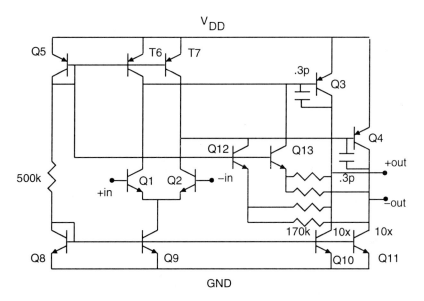

$V_{DD}$

**FIGURE 4.44**
A 3 V operational amplifier with balanced output.

integration capacitance with a larger dynamic range than 77.6 dB.

**1-Volt Amplifier and Filter Design**   When the filter is to be designed for a
1 V supply voltage, the quiescent input and output voltage level must be about
0.5 V for a maximum swing. Therefore a floating differential input stage can no
longer be used, and the operational amplifier with balanced output, shown in
Fig. 4.45, can be employed.

The quiescent input voltage is the junction voltage of the input transistor.
This voltage can be kept small by choosing a large transistor, and has a value of
about 0.65 V. The voltage swing is maximum if the supply voltage is twice this
value, so the operational amplifier with balanced output is optimum at 1.3 V
supply voltage, although it works at 0.8 V. We assume a supply voltage of 1.3 V.

As the input stage no longer has common-mode rejection, the common-
mode output level must be fed back to the input of this stage to ensure common-
mode stability. In Fig. 4.45, we show that this has been done via four resistors.
The filter is common-mode stable if the common-mode gain of the integrators is
less than unity. For this circuit, this means that the resistance of the common-
mode feedback resistors must be smaller than twice the integration resistance
(i.e., the resistance of the MOSFETs shown in Fig. 4.43). This means that the
common-mode feedback resistances contribute to the noise of the integrator,
and the noise factor has at least a value of two. In practice, the common-mode
resistors must have a value that is sufficiently smaller than that stated above
to overcome component tolerances, so the noise factor is somewhat higher. In a

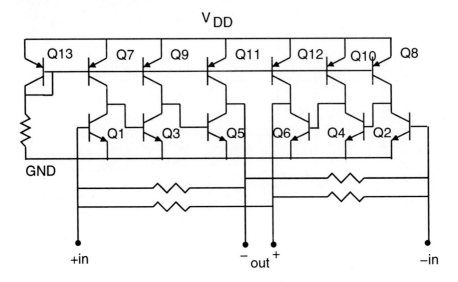

**FIGURE 4.45**
A 1 V operational amplifier with balanced output.

practical implementation a noise factor of 3.15 (-5 dB) was obtained.

The maximum peak differential output voltage is 1.2 V. This is 0.7 dB below rail-to-rail operation. Thus, in total, the filter is 6.7 dB below optimum. The dynamic range of the filter is 63.5 dB.

Since the common-mode rejection is as small as possible to obtain a minimum noise factor, a noninverting two-stage amplifier would be unstable. Therefore, an inverting three-stage amplifier has been used. This makes compensation difficult. The compensation scheme is not discussed here. The simulated transfer function is shown in Fig. 4.46.

### 4.6.4  Conclusion

There is a fundamental upper limit to the dynamic range of integrated filters. As this limit decreases rapidly if the supply voltage is decreased, dynamic range is a considerable problem when working with low supply voltages. Therefore, it is important to optimize the dynamic range in a low supply voltage situation. It is possible to approach the fundamental limit closely by optimizing the filter network and the integrators that are the building blocks in the network. The integrator type that can be optimized most easily is the operational amplifier MOSFET-C integrator. It has been shown that with a 3 V supply voltage the dynamic range can be optimized to within 2 dB from the fundamental limit. When the supply voltage is reduced down to 1 V, the limit can be approached to within 7 dB.

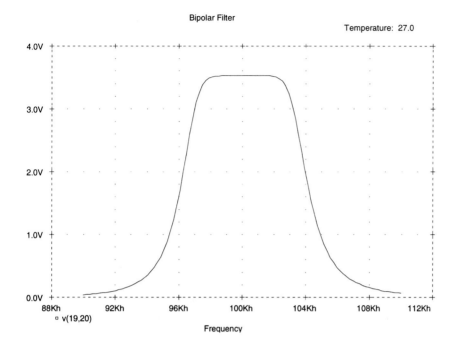

**FIGURE 4.46**
The transfer function of the 1 V filter by simulation.

## PROBLEMS

**4.1.** Consider the input stages, as shown in Fig. 4.1. Given that $V_{SS}=0$ V, $V_{DD}=3$ V, $V_{SGp}=V_{GSn}=1$ V, $V_{DS(satp)}=V_{DS(satn)}=250$ mV and the saturation voltage of the current-sources is equal to $V_{DS(sat)}$, determine the following.
  (a) Calculate the common-mode input range of the p-channel input pair.
  (b) Calculate the common-mode input range of the n-channel input pair.
  (c) To obtain the complementary input stage, as shown in Fig. 4.2, the n-channel and p-channel pair are placed in parallel. Calculate the common-mode input range of this complementary input stage.

**4.2.** Given $V_{be}=600$ mV and $V_{CE(sat)}=100$ mV for the bipolar transistors, $V_{GSp}=V_{GSn}=1$ V, $V_{DS(sat)}=100$ mV, $V_{DS(satp)}=V_{DS(satn)}=250$ mV for the MOS transistors, and the saturation voltage of the current-sources is equal to $V_{CE(sat)}$ or $V_{DS(sat)}$, depending on the technology.
  (a) Calculate the minimum supply voltage which allows a fully rail-to-rail operation of the bipolar input stage, as shown in Fig. 4.11.
  (b) Do the same for the CMOS input stage, as shown in Fig. 4.2.
  (c) What could you conclude from the results of 4.2(a) and 4.2(b)?

**4.3.** Consider the op amp circuit shown in Fig. P4.3. Use the CMOS model parameters given in Appendix A:
 (a) For $V_{DD} = -V_{SS} = 1.5$ V, size $M_{R_1}$ and $M_{R_2}$ such that $I_{R_1} = I_{R_2} = 10$ $\mu$A.
 (b) Determine the transistor sizes of $M_1 - M_4$ such that the maximum value of the total $g_m$ is 100 $\mu$ A/V and the minimum value is 50 $\mu$A/V.
 (c) Design the gain stage such that the op amp has an open-loop dc-gain of 40 dB and a unity gain frequency of 1 MHz when $g_m$ of the input stage is $100\mu$A/V.

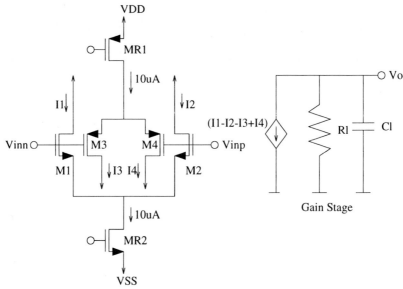

**FIGURE P4.3** A simple CMOS op amp with a rail-to-rail input stage.

**4.4.** For the op amp circuit in Problem 4.3, perform the following SPICE simulations. Use the CMOS model parameters given in Appendix A.
 (a) The total $g_m$ of the input stage as a function of the common-mode input voltage $V_{CM}$.
 (b) The open-loop frequency response with $V_{CM}=0$, 1.3, and -1.3 V.
 (c) Run a Fourier analysis to determine the THD of a unity-gain feedback connected op amp. Use 1 kHz sinusoidal signal with amplitude of 0.1 and 1.3 V.

**4.5.** Consider the Op-Amp circuit shown in Fig. P4.5. Use the CMOS model parameters given in Appendix A:
 (a) For $V_{DD}=-V_{SS}=1.5$ V, size $M_{R_1}$ and $M_{R_2}$ such that $I_{R_1}=I_{R_2}=10$ $\mu$A.
 (b) Determine the transistor sizes such that the value of the total $g_m$ is approximately 100 $\mu$A/V over the entire common mode range.
 (c) Design the gain stage such that the op amp has an open-loop dc-gain of 40 dB and the a unity gain frequency of 1 MHz.

**4.6.** Consider the circuit shown in Fig. P4.6[31]. $M_1$ through $M_4$ make up the differential pair input stage and are biased by $I_p$ and $I_n$. For all the transistors

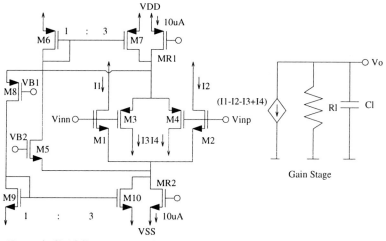

**Figure P4.5** A CMOS op amp with rail-to-rail constant $g_m$ input stage.

$K=K_n = K_p$. Show that the total transconductance of the input stage is constant for any value of $V_{cm}$ by considering the following cases.

(a) For $V_{cm}$ near $V_{DD}$, only the n-channel differential pair contributes to the total $g_m$. In this case the entire $4I_o$ flows into $M_5$.

(b) For $V_{cm}$ near mid-rail, both pairs contribute to the total $g_m$. $M_5$ and $M_6$ each receives $2I_o$.

(c) For $V_{cm}$ near $V_{SS}$, only the p-channel differential pair contributes to the total $g_m$. In this case the entire $4I_o$ flows into $M_6$.

(Hint: write a loop equation around $M_7 - M_9$ in terms of $I_o, I_p$, and $I_n$ and compare it to the expression for the total $g_m$.)

**4.7.** Run SPICE simulations as in Prob. 4.4. and compare the THD of the two op amp circuits in Fig. P4.4 and P4.5. Comment on their differences.

(Hint: $V_{B1}$ and $V_{B2}$ must be chosen such that $M_8$ and $M_5$ are turned off when $V_{CM}$ is near mid-rail.)

**4.8.** Consider the output stages, as shown in Fig. 4.8 with $V_{SS} = 0$ V, $V_{DD} = 3$ V, $V_{GSp} = V_{GSn} = 1$ V, and $V_{DS(satn)} = V_{DS(satp)}$.

(a) Calculate the maximum output voltage swing of the common-drain output stage and the common-source output stage.

(b) What could you conclude from these results?

Given: $V_{SGp} = V_{GSn}=1$ V (For all transistors and assuming that the output stage is in rest); $V_{Tp} = V_{Tn} =.6$ V; $V_{ref}=1.3$ V; $R_4 = R_5=20$ kΩ; $10(W/L_{M9}) = W/L_{M8}$; $10(W/L_{M3}) = W/L_{10}$

**4.9.** Consider the output stage, as shown in Fig. 4.9 and $V_{SGp} = V_{GSn}=1$ V (For all transistors and assuming that the output stage is in rest); $V_{Tp} = V_{Tn} =.6$ V; $V_{ref}=1.3$ V; $R_4 = R_5=20$ kΩ; $10(W/L_{M9}) = W/L_{M8}$; $10(W/L_{M3}) = W/L_{10}$.

(a) Calculate the quiescent currents through the output transistors $M_9$ and $M_{10}$.

(b) Suppose $M_9$ is delivering the total output current, calculate the bias current of $M_{10}$.

**4.10.** Consider the three-stage operational amplifier, as shown in Fig. 4.18: *Given:* $C_L=100$ pF; $C_{m2}= 12$ pF; $g_o=5$ mmho; $g_m = .1$ mmho.

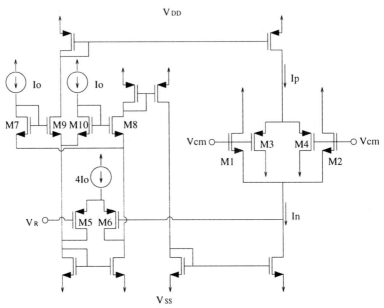

**FIGURE P4.6** A CMOS input stage with rail-to-rail constant $g_m$.

    (a) Calculate the open-loop bandwidth ($f_o$) of the operational amplifier.
    (b) Calculate the value of the capacitor $C_{m1}$.
    (c) Calculate the value of the transconductance of the input stage.

**4.11.** Consider the op amp with BiCMOS input stage as shown in Fig. 4.19 and assume that the tail-current of the MOS input stage $I_2$ is the same as the tail-current $I_1$ of the bipolar input stage. *Given:* $I_1 = 20\ \mu A$; $K = 20\ \mu A/V^2$; $C_m = 5$ pF; $W/L = 9$.
    (a) Calculate the transconductance $g_m$ of the bipolar input transistors.
    (b) Calculate the transconductance of the MOS transistors and compare it to the transconductance of the bipolar input stage.
    (c) Suppose that the CMOS stage is not present and calculate the slew rate $S$ of this bipolar op amp.
    (d) Calculate the slew rate of the op amp with BiCMOS input stage.
    (e) Assume $I_2 = 9I_1$ and again calculate the transconductance of the MOS transistors and the slew rate of the op amp with a BiCMOS input stage. Compare these results with the results of (b) and (d).

**4.12.** Show that the common-mode rejection of the input stage shown in Fig. 4.28 is given by Eq. (4.11) where $\mu$ is the intrinsic voltage gain of the input transistors and $\Delta\mu$ is the difference in voltage gain between $Q_1$ and $Q_2$ ($\mu = g_m r_o = V_{early}$ (kT/q)). Neglect the influence of $R_{I1}$ and $R_{I2}$.
    (Hint: find the common-mode currents through $r_{o1}$ and $r_{o2}$ due to a common-mode input voltage, and then calculate the equivalent base emitter voltages.)

**4.13.** Consider the filter network, as shown in Fig. 4.42, and the operational amplifier with balanced output, as shown in Fig. 4.44: *Given:* T=300 K, $V_{sup}$ =3 V, $V_{dsat}$ =100 mV, C=10 pF, $\xi = 1.2$.

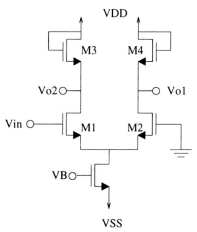

**FIGURE P4.16** A single-ended-to-differential signal converter/level shifter.

(*a*) Calculate the dynamic range of the filter, assuming that the filter configuration is optimal.

(*b*) Calculate the fundamental maximum of the dynamic range of the filter

**4.14.** Repeat Prob. 4.13 for $V_{sup} = 1\ V$.

**4.15.** Compare the results of Prob. 4.14 and 4.13. What is your conclusion?

**4.16.** In low-voltage circuit design, it is often required to level shift the incoming signal $V_{in}$ to maximize the range of $V_{o1} - V_{o2}$ which is used as input to drive other low-voltage circuits. It is also desired to convert a single ended signal to a differential signal for this purpose. The circuit shown in Fig. P4.16 [30] achieves both tasks.

(*a*) Find $V_{o1}$ and $V_{o2}$ and verify that the output voltage $V_{o1} - V_{o2}$ equals $V_{in}$. Assume that $M_1 - M_4$ have the same size.

(*b*) Find the valid range for $V_{in}$.

(Hint: $M_1$ and $M_2$ must stay in saturation region. The circuit does not operate properly when the current through $M_1$ or $M_2$ becomes zero.)

# REFERENCES

[1] J.H. Huijsing and D. Linebarger, "Low-voltage operational amplifier with rail-to-rail input and output ranges," *IEEE J. Solid-State Circuits*, vol. SC-20, pp. 1144-1150, Dec. 1985.

[2] R. Hogervorst, R.J. Wiegerink, P.A.L. de Jong, J. Fonderie, R.F. Wassenaar, and J.H. Huijsing, "CMOS low-voltage operational amplifiers with constant-$g_m$ rail-to-rail input stage," *Proc. IEEE Intl. Symp. Circ. Syst.* (San Diego), pp. 2876-2879, May 10-13, 1992.

[3] E. Seevinck, W. de Jager, and P. Buitendijk, "A low-distortion output stage with improved stability for monolithic power amplifiers," *IEEE J. Solid-State Circuits*, vol. SC-23, pp. 794-801, June 1988.

[4] R.J. Widlar, "Low-voltage techniques," *IEEE J. Solid-State Circuits*, vol. SC-13, pp. 838-846, Dec. 1978.

[5] J. Fonderie, *Design of Low-Voltage Bipolar Operational Amplifiers*. Ph.D. thesis, Delft Univ. Techn., Delft, The Netherlands, 1991.

[6] J.Fonderie, M.M. Maris, E.J. Schnitger, and J.H. Huijsing, "1-V operational amplifier with rail-to-rail input and output ranges," *IEEE J. Solid State Circuits*, vol. SC-24, pp. 1551-1559, Dec. 1989.

[7] J.E. Solomon, "The monolithic op amp: a tutorial study," *IEEE J. Solid-State Circuits*, vol. SC-9, no. 6, pp. 314-332, Dec. 1974.

[8] J. Fonderie and J.H. Huijsing, "Operational amplifier with 1-V rail-to-rail multipath-driven output stage," *IEEE J. Solid-State Circuits*, vol. 26, no. 12, pp. 1817-1824, Dec. 1991.

[9] J.H. Huijsing and J. Fonderie,"Multi-Stage Amplifier with Capacitive Nesting and Multi-Path-Driven Forward Feeding for Frequency compensation," U.S. Pat. Appl. Ser. No. 654.855, filed February 11, 1991.

[10] O.H. Schade, Jr. and E.J. Kramer, "A low-voltage BiMOS op amp," *IEEE J. Solid-State Circuits*, vol. SC-16, no. 6, pp. 661-668, December 1981.

[11] A.N. Karanicolas, K.K. O, J.Y.A. Wang, H.-S. Lee, and R.L. Reif, "A high-frequency fully-differential BiCMOS operational amplifier," *IEEE J. Solid-State Circuits*, vol. SC-26, no. 3, pp. 203-208, March 1991.

[12] D.F. Bowers, "The impact of new architectures on the ubiquitous operational amplifier," *Proc. of the Workshop Advances in Analog Circuit Design*, 1992.

[13] M.A. Smither, D.R. Pugh, and L.M. Woolard, "CMRR analysis of the 3-operational amplifier instrumentation amplifier," *Electron. Lett.*, vol. 13, pp. 594., 1977.

[14] R.J. van de Plassche, "A wide-band monolithic instrumentation amplifier," *IEEE J. Solid State Circuits*, vol. SC-10, no. 6, pp. 424-431, Dec. 1975.

[15] J.H. Huijsing, "Instrumentation amplifiers; A comparative study on behalf of monolithic integration," *IEEE Trans. on Instrumentation and Measurement*, vol. IM-25, no. 3, pp. 227-231. September 1976.

[16] A.P. Brokaw and P.M. Timko, "An improved monolithic instrumentation amplifier," *IEEE J. Solid-State Circuits*, vol.SC-10, no. 6, pp. 417-423, December 1975.

[17] G.H. Hamstra, A. Peper, and C.A. Grimbergen, "Low-power, low-noise instrumentation amplifier for physiological signals," *Medical and Biological Engineering and Computing*, pp. 272-274, May 1984.

[18] E. Säckinger and W. Guggenbühl, "A versatile building block; the CMOS differential difference amplifier," *IEEE J. Solid-State Circuits*, vol. SC-22, no. 2, pp. 287-294, April 1987.

[19] E. Seevinck, "*Analysis and Synthesis of Translinear Integrated Circuits*," Elsevier, Amsterdam, pp. 131-133, 1988.

[20] G. Groenewold, *Optimal Dynamic Range Integrated Continuous-Time Filters*. Ph.D. thesis, Delft University of Technology, 1992.

[21] F. Krummenacher and N. Joehl, "A 4-Mhz CMOS continuous-time filter with on-chip automatic tuning," *IEEE J. Solid-State Circuits*, vol. SC-23, pp. 750-758, June 1988.

[22] B. Nauta, "CMOS VHF transconductance-C lowpass filter," *Electron. Lett.*, vol. 26, pp. 421-422, March 1990.

[23] J.M. Khoury, "Design of high dynamic range continuous-time filter with on-chip tuning," *IEEE J. Solid-State Circuits*, vol. SC-26, pp. 1988-1997, December 1991.

[24] G. Groenewold, "The design of high dynamic range continuous-time integratable bandpass filters," *IEEE Trans. Circuits and Syst.*, vol. 38, pp. 838-852, August 1991.

[25] Khen-Sang Tan and P.R. Gray, "Fully integrated analog filters using bipolar JFET technology," *IEEE J. Solid-State Circuits*, vol. SC-13, pp. 814-821, December 1978.

[26] H. Khorramabadi and P.R. Gray, "High-frequency CMOS continuous-time filters," *IEEE J. Solid-State Circuits*, vol. SC-19, pp. 939-948, December 1984.

[27] M. Banu and Y. Tsividis, "Fully integrated active RC filters in MOS technology," *IEEE J. Solid-State Circuits*, vol. SC-18, pp. 644-651, December 1983.

[28]  E. Seevinck, "Companding current-mode integrator: A new principle for continuous-time monolithic filters," *Electron. Lett.*, vol. 26, pp. 2046-2047, November 1990.

[29]  G. Groenewold, "Optimal dynamic range integrators," To be published in *IEEE Trans. Circuits Syst.*

[30]  J. Ramerez-Angulo, "Highly linear four-quadrant analog BiCMOS multiplier for ±1.5V operation," *IEEE Proc. ISCAS*, pp. 1467-1470, May 1993.

[31]  J. H. Botma, R. F. Wassenaar, and R. J. Wiegerink, "A low-voltage CMOS op amp with rail-to-rail constant-gm input stage and a class AB rail-to-rail output stage," *IEEE Proc. ISCAS*, pp. 1314-1317, May 1993.

# CHAPTER
# 5

# BASIC BICMOS CIRCUIT TECHNIQUES

## 5.1 INTRODUCTION

In the past decade, CMOS technology has played a major role in the rapid advancement and the increased integration of digital VLSI systems. CMOS devices feature high input impedance, extremely low offset switches, high packing density, low switching power consumption, and most importantly, they are easily scaled. Analog circuits have also greatly benefited from CMOS technology [1], however, the relatively low transconductance and high $\frac{1}{f}$ noise in CMOS devices have led most designers to favor the use of bipolar transistors in many analog applications.

Recently, there has been an increasing interest in BiCMOS technologies for implementing high-performance digital systems [4]-[7]. This is mainly due to the high speed capability of BiCMOS digital gates compared with their CMOS counterparts. Also the emergence of BiCMOS technologies as a viable approach to VLSI offers new opportunities for improving system performance by combining the complementary devices in both bipolar and CMOS technologies. Recent reports also indicate a significant speed improvement in BiCMOS static and dynamic RAMs, relative to the CMOS memories [2]. Although BiCMOS digital circuits have received considerable attention for improving their speed performance, the progress in the area of analog BiCMOS has been rather slow. The translinear property of bipolar transistors in conjunction with the square-law

184

**TABLE 5.1**
Performance comparison of bipolar, CMOS, and BiCMOS devices.

| Available features | Bipolar | CMOS | BiCMOS |
|---|---|---|---|
| Minimum current drain | | | X |
| High input impedance | | X | X |
| Automatically scalable for low and medium speed digital control logic | | X | X |
| Greater than 1GHz toggle frequency | X | | X |
| Low 1/F noise | X | | X |
| High dc gain | X | | X |
| Low input offset for differential pair | X | | X |
| Zero offset analog switches | | X | X |
| High gain-bandwidth product | | | X |
| Good voltage reference | X | | X |

behavior of MOS devices should open up numerous possibilities for new analog circuits [8]-[12].

The analog section of many VLSI communication systems is constructed from building blocks such as amplifiers, multipliers, oscillators, computational circuits and filters (just to mention a few) which are required to operate with high signal-to-noise ratio at high speed. Simultaneous achievement of these requirements is very difficult to obtain with CMOS devices. High-performance bipolar junction transistors are usually used to implement the analog section of these systems. On the other hand, the digital section becomes much cheaper when implemented with CMOS. Using two separate technologies increases the cost of packaging and testing, and results in yield losses. BiCMOS processes offer both high-performance bipolar and CMOS devices on a single substrate. Of course, the complexity (cost), the packing density, the performance, and reproducibility of the devices and many other practical issues should be carefully considered before substituting CMOS and bipolar processes with a BiCMOS process. Table 5.1 compares the available features in bipolar, CMOS, and BiCMOS technologies.

The primary goal of this chapter is to study the capabilities of BiCMOS processes for analog circuit design. A comparison between CMOS and BiCMOS analog building blocks will be provided. Speed, noise, and large-signal performance will be the key parameters in the comparison. Practical issues such as the matching properties and silicon area will be discussed. The chapter begins with a brief discussion of a BiCMOS process. Formulation of the key properties of MOS and bipolar devices will be provided. Finally, we will introduce several analog BiCMOS circuit building blocks and compare their performance with that of their CMOS counterparts. Lengthy algebra has been avoided where possible in our derivations, and we encourage the reader to prove the validity of the presented results. Numerical problems at the end of this chapter are intended to

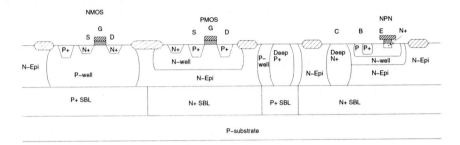

**FIGURE 5.1**
Simplified cross section of a BiCMOS process.

clarify the presented material and further enhance the reader's understanding of the subject. The device parameters in Table 5.2 will be used in the numerical computations of these problems.

## 5.2   DEVICES AND TECHNOLOGY

A BiCMOS technology combines bipolar and CMOS transistors on a single substrate. A high performance BiCMOS technology requires advanced process techniques, which require small critical dimensions, thin gate oxides, epitaxial layers, advanced isolation (e.g., trench isolated devices), silicided gates, silicided emitters and diffusions having low sheet resistance, and multiple metal layers. These result in small geometry, and wide-band bipolar transistors and high quality CMOS devices [2, 3].

Figure 5.1 shows a simplified cross-section of a BiCMOS process which offers quality CMOS and NPN transistors. The substrate is p-type and acts as an isolator for the bipolar transistors. The P+ and N+ selective buried layers are obtained through boron and phosphorus ion-implantation and diffusion, respectively. Observe that the N+ selective buried layers (SBLs) provide the foundation for both the PMOS and NPN transistors. A low-doped n-type epitaxial layer is grown or deposited on the p-substrate, then N-wells and P-wells are formed in the N-epi. The NMOS, PMOS, and bipolar devices are then constructed. The emitter is formed using poly-silicon and ion-implantation. This allows a good base-width control and reduces parasitic capacitances (for improved high-frequency performance). The deep N+ and P+ regions are introduced to minimize the collector resistance of the NPN transistors and to isolate devices, respectively. PNP bipolar transistors can also be formed by introducing deep trench-isolations [2] or oxide-isolations, which isolate each device from the others. Siliciding the gates and the emitters will further improve the speed.

## 5.2.1   Performance Comparison of Bipolar, CMOS and BiCMOS Devices

Sophisticated bipolar processes offer small devices, which are optimized to achieve a large gain-bandwidth product and low noise. Fast emitter-coupled logic (ECL) and high-performance analog circuits can be designed using these devices. Unfortunately, bipolar devices consume significant static power which makes the design of a large bipolar power-hungry integrated circuit (IC) formidable. On the other hand, CMOS devices consume negligible static power and offer high circuit densities. Integration of analog CMOS circuits, which have comparable signal-to-noise ratios at high speeds, usually require more IC "real estate." These circuits will also consume more power than the bipolar counterparts.

Virtually all communication systems, from radios to the radar systems, incorporate: an RF section, a synthesizer section (can be simple or elaborate), and an IF section. Due to many practical considerations such as performance requirements, power consumption, packing densities, and required external components, these sections are designed as several ICs. It should be clear that more ICs imply a greater reliability concern, higher IC testing and packaging costs, more performance uncertainty due to different ICs, etc. It is very difficult to design a high-performance, low-cost synthesizer IC in either a bipolar or CMOS process. For instance, very low-noise VCOs, low-power and low-noise RF amplifiers, and very high-speed, low-power prescaler logic circuits can be only achieved with bipolar devices. However, the intermediate and low-speed logic sections, the loop filter and many programmability features are best implemented in CMOS. Since BiCMOS processes offer both bipolar and CMOS devices, it becomes possible to integrate high-performance mixed analog/digital ICs while improving the reliability, performance, and testability and reducing the packaging costs.

This section presents the characteristics of both bipolar and MOS devices, including their matching properties, noise characteristics and ac-performance. Details of the key claims in Table 5.1 will be provided in the following sections of this chapter.

Advanced BiCMOS processes offer both NPN and PNP bipolar transistors. Poly-emitters are used to reduce the emitter area and the junction depth in order to reduce device capacitance and to improve device speed. As the base width is decreased, the emitter depth has to follow as well. Use of conventional metal contacts cause a severe degradation of current gain, and the reduced back-injected base current associated with poly emitter contacts is used to circumvent this phenomenon [2]. Also, small feature sizes of MOS and bipolar transistors should result in high digital packing density and faster CMOS and bipolar devices. There is usually some trade-off between the process complexity (number of masks and process steps) and the achievable device performance. Table 5.2 presents typical measured device parameters of a BiCMOS process which offers only NPN transistors. These parameter values are used in the simulations of the circuits presented in this chapter.

Two important bipolar transistor parameters in a BiCMOS process are

**TABLE 5.2**
Typical measured parameters of BiCMOS devices.

| Parameter | Symbol | NMOS | PMOS |
|-----------|--------|------|------|
| Substrate Doping $(cm^3)$ | $N_{sub}$ | $6 \cdot 10^{16}$ | $7 \cdot 10^{15}$ |
| Gate Oxide Thickness $(nm)$ | $t_{ox}$ | 40 | 40 |
| Channel Mobility $(cm^2/V \cdot sec)$ | $\mu_n, \mu_p$ | 577 | 200 |
| Threshold Voltage $(V)$ | $V_T$ | .85 | -.85 |
| Minimum Gate Length $(\mu m)$ | $L$ | 2 | 2 |
| S/D Junction Depth $(\mu m)$ | $X_j$ | .18 | .26 |
| S/D Lateral Diffusion $(\mu m)$ | $L_D$ | .15 | .2 |
| S/D Overlap Capacitance $(fF/\mu m^2)$ | $C_{ovl}$ | .166 | .222 |
| S/D Junction Capacitance $(fF/\mu m^2)$ | $C_{j0}$ | .11 | .3 |
| S/D Sidewall Capacitance $(fF/\mu m)$ | $C_{jsw0}$ | .50 | .95 |

| Parameter | Symbol | NPN Bipolar | PNP Bipolar |
|-----------|--------|-------------|-------------|
| Forward Current Gain | $\beta$ | 160 | 55 |
| Early Voltage $(V)$ | $V_A$ | 90 | 30 |
| Collector Resistance $(\Omega)$ | $r_c$ | 150 | 300 |
| Emitter Resistance $(\Omega)$ | $r_e$ | 11 | 15 |
| Base Resistance $(\Omega)$ | $r_b$ | 300 | 400 |
| B-E Junction Capacitance $(fF)$ | $C_{je0}$ | 35 | 50 |
| B-C Junction Capacitance $(fF)$ | $C_{jc0}$ | 25 | 45 |
| C-S Junction Capacitance $(fF)$ | $C_{jcs0}$ | 200 | 200 |
| Forward Transit Time $(p - sec)$ | $\tau_f$ | 30 | 60 |
| Forward Sat. Current $(fA)$ | $I_S$ | .138 | .138 |
| B-E Leakage Sat. Current $(fA)$ | $I_{SE}$ | .2 | .2 |
| B-C Leakage Sat. Current $(fA)$ | $I_{SC}$ | .2 | .2 |

the base transit time, $\tau_f$, and the transition frequency, $f_T$. The base transit time is determined by the effective base charging capacitance, while the transition frequency is a convenient encapsulation of all device parasitics affecting its maximum operating speed. These parameters determine the speed capability of a bipolar transistor. The base transit time is the ratio of the minority-carrier charge of the base in transit to the device collector current flow. The plot in Fig. 5.2 is an extracted NPN bipolar transistor base transit time (from the actual measurements of the transition frequency assuming a one-pole model) which is a function of the inverse of the collector bias current for two different collector-to-emitter bias voltages, $V_{CE}$. Note that at a collector current 300 $\mu$A, the base transit time is 170 and 185 psec for $V_{CE}$ of 4 V and 2 V, respectively. Meanwhile, the maximum measured transition frequencies (at $I_c = 1$ mA) were 1.51 GHz and 1.6 GHz for $V_{CE}$ of 2 V and 4 V, respectively.

The analyses in this chapter assume the following expressions for the drain

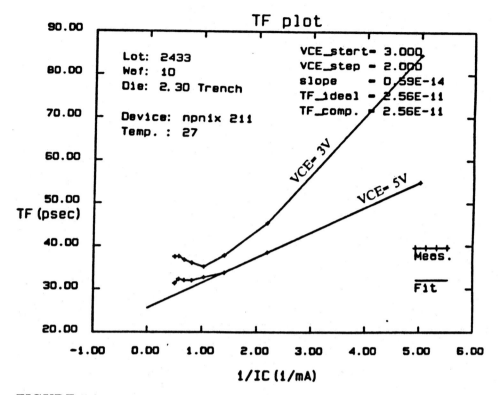

**FIGURE 5.2**

Measured base-transit time of a 1X NPN bipolar device.

current, $I_d$, of an MOS transistor operating in the saturation region and its transconductance, $g_{mM}$, and the collector current, $I_c$, of a bipolar transistor operating in the active region and its transconductance, $g_{mQ}$:

$$I_d = \frac{K'}{2}\frac{W}{L}(V_{GS} - V_T)^2 \qquad \text{where} \qquad K' = \mu_{eff}C_{ox} \qquad (5.1)$$

$$g_{mM} = \frac{\partial I_D}{\partial V_{GS}} = K'\frac{W}{L}(V_{GS} - V_T) = \sqrt{2K'WI_D/L} \propto \sqrt{I_D} \qquad (5.2)$$

In Eqs. (5.1) and (5.2), $\mu_{eff}$ is the effective carrier mobility, $C_{ox}$ is the per unit area gate oxide capacitance, $W$ is the channel width, $L$ is the channel length, $V_{GS}$ is the gate-to-source voltage, and $V_T$ is the threshold voltage of an MOS transistor. Similarly,

$$I_c = I_S e^{(\frac{qV_{BE}}{kT})} \qquad (5.3)$$

and

$$g_{mQ} = \frac{\partial I_C}{\partial V_{BE}} = \frac{qI_C}{kT} \propto I_C \qquad (5.4)$$

where $q$ is the electron charge, $I_c$ is the collector, $k$ is the Boltzmann constant, $T$ is the absolute temperature, $I_s$ is the base-emitter junction saturation current, $V_{BE}$ is the base-emitter voltage of a bipolar transistor. Although the simplified current equations are adequate for a first order analysis, more precise analysis should be based on improved transistor models [26, 27].

## 5.2.2    AC Characteristics

The maximum useful frequency of bipolar and MOS transistors is the transition frequency, $f_T$, which is defined as the frequency where the magnitude of the short-circuit, common-emitter/common-source current gain falls to unity. The simplified transition frequency expressions for the bipolar and MOS transistors are [14]:

$$f_{Tbipolar} = \frac{g_{mQ}}{2\pi(C + C)} = \frac{qI_c}{kT\left(2\pi\frac{qI_{cf}}{kT} + C_{je} + C\right)} \tag{5.5}$$

where $\tau_f$ is the base transit time in the forward direction, $C$ (in this chapter $C_p$ and $C$ are used interchangeably) $= g_m t_f + C_{je}$, $C_{je}$ is the emitter-base junction capacitance, $C$ (in this chapter $C_m$ and $C$ are used interchangeably) is the base-collector junction Miller capacitance of the bipolar transistor.

$$f_{TMOS} = \frac{g_{mM}}{2\pi C_{gs}\sqrt{1 + \frac{2C_{gd}}{C_{gs}}}} \approx \frac{g_{mM}}{2\pi C_{gs}} = \frac{\sqrt{\frac{2I_{deff}}{W_{eff}L_{eff}C_{ox}}}}{L_d + 2\frac{L_{eff}}{3}} \approx \frac{\mu_{eff}V_{on}}{2\pi L_{eff}^2} \tag{5.6}$$

where $W_{eff}$ is the effective channel width, $L_{eff}$ is the effective channel length, $L_d$ is the lateral diffusion, $C_{gs}$ is the gate-to-source capacitance, $C_{gd}$ is the gate-to-drain capacitance, and $V_{on} = (V_{GS} - V_T)$ which is the excess gate voltage, $V_{DS(sat)}$ (also called the "on voltage," or $V_{GT}$).

Advanced BiCMOS processes offer small size with shallow diffusion bipolar transistors, resulting in a small base-width, base transit time, and large transition frequency for the bipolar transistor. This is because the base transit time is proportional to the square of the base-width. From Eq. (5.5) one observes that the transition frequency is a direct function of the collector current. By choosing an appropriate collector current, the speed of a bipolar transistor can be maximized. On the other hand, the transition frequency of an MOS transistor is a function of the square-root of the drain current and is inversely proportional to the square of the effective channel length of the device.

The optimum design of a BiCMOS circuit demands a good understanding of the capabilities of bipolar and CMOS transistors. We consider the gain-bandwidth product as a design parameter to be maximized. A common-emitter bipolar transistor amplifier and a common-source MOS amplifier are shown in Fig. 5.3. Analyzing their small-signal equivalent circuits (these small-signal models will be used throughout this chapter) results in the following dc-gain ($A_{dc}$) and pole-zero locations ($P_1$, $P_2$, and $Z$), where $I_b$ is the base current, $V_A$ is

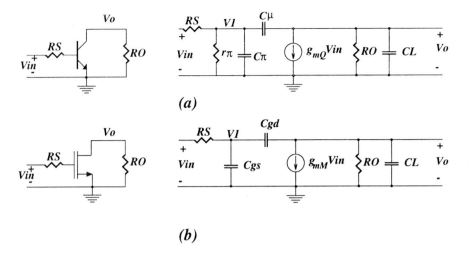

**FIGURE 5.3**
Basic amplifiers and their small signal models (a) bipolar and (b) MOS.

the Early voltage, and $V_t$ is the thermal voltage. We will also assume that the poles are far apart and that $R_o = R_L \| r_o$, where $r_o$ is the transistor output resistance, $R_L$ is the external load resistance, $g_{mQ}$ is the transconductance of the bipolar transistor, and $g_{mM}$ is the transconductance of the MOS transistor. The subscripts (a) and (b) refer to the circuits in Fig. 5.3.

$$A_{dc(a)} = -\frac{g_{mQ} R_o r}{r + R_s} = -\frac{V_A}{V_t + I_b R_s}\bigg|_{R_L \gg r_o} \tag{5.7}$$

$$Z_{(a)} = \frac{g_{mQ}}{C} \tag{5.8}$$

$$P_{1(a)} = -\frac{(C + C_L)(\frac{1}{R_s} + \frac{1}{r}) + \frac{C+C}{R_o} + g_{mQ}C}{CC + C_L(C + C)} \tag{5.9}$$

$$P_{2(a)} = -\left[\frac{\frac{1}{R_o}(\frac{1}{r} + \frac{1}{R_s})}{(C + C_L)(\frac{1}{R_s} + \frac{1}{r})} + \frac{(C_m + C)}{R_o} + g_{mQ}C_m\right] \tag{5.10}$$

$$GBW_{(a)} = \frac{g_{mQ} R_o}{R_o(C_m + C_L)(1 + \frac{R_s}{r}) + R_s(C_m + C) + g_{mQ} R_o R_s C_m} \tag{5.11}$$

where $r = \frac{o}{g_{mQ}}$ (in this chapter $r_p$ and $r$ are used interchangeably), $R_s$ is the input source resistance, $C_L$ is the total load capacitance. Similar expressions for the MOS amplifier in Fig. 5.3 are:

$$A_{dc(b)} = -g_{mM} R_o \tag{5.12}$$

$$Z_{(b)} = \frac{g_{mM}}{C_{gd}} \tag{5.13}$$

$$P_{1(b)} = -\frac{\frac{C_{gs}+C_{gd}}{R_o} + \frac{C_L+C_{gd}}{R_s} + g_{mM}C_{gd}}{C_{gs}(C_{gd} + C_L) + C_{gd}C_L} \tag{5.14}$$

$$P_{2(b)} = -\frac{1}{R_s(C_{gs} + C_{gd}) + R_o(C_{gd} + C_L) + g_{mM}R_oR_sC_{gd}} \tag{5.15}$$

$$GBW_{(b)} = \frac{g_{mM}R_o}{R_s(C_{gs} + C_{gd}) + R_o(C_{gs} + C_L) + g_{mM}R_oR_sC_{gd}} \tag{5.16}$$

where $C_{gd}$ is the gate-to-drain capacitance. Note that the zero, $Z$, resides in the right-half plane, and it has a phase characteristic similar to that of a pole in the left-half plane. Using this basic amplifier configuration in the design of a wide-band amplifier can degrade the phase margin. Since, in general, the capacitance $C_p$ of a bipolar transistor is large compared to the capacitance $C_{gs}$ of an MOS transistor, from (5.11) and (5.16), one notices that the source resistance, $R_s$, influences the gain-bandwidth product of a bipolar amplifier more than it does in an MOS amplifier. When the source resistance approaches zero, the gain-bandwidth product of both amplifiers is maximized,

$$GBW_{(a)}\big|_{R_s \to 0} = \frac{g_{mQ}}{C + C_L} \tag{5.17}$$

$$GBW_{(b)}\big|_{R_s \to 0} = \frac{g_{mM}}{C_{gd} + C_L} \tag{5.18}$$

Since the bipolar transistor has a significantly higher transconductance than the MOS transistor (for the same bias current, $g_{mM} = \frac{2I_D}{V_{gs}-V_T}$ and $g_{mQ} = \frac{I_C}{V_t}$), bipolar devices are more commonly used in the design of high-speed analog circuits. Circuits that incorporate an emitter-coupled input pair (as opposed to a source-coupled pair) exhibit superior dc-gain, lower input referred dc-offset voltage and lower input referred noise (assuming a relatively low source impedance). However, their inputs require appreciable input bias current.

Another basic analog building block commonly used in the design of analog circuits is the current mirror/amplifier. The current mirror/amplifier is widely used in many analog circuits such as operational transconductance amplifiers, translinear circuits, digital-to-analog converters, and in current-mode signal processing circuits. Therefore, its ac-response is of prime importance. The ac-response of the simple bipolar and MOS current mirrors shown in Fig. 5.4 will now be considered.

By referring to the circuits in Fig. 5.4 (disregard M3, Q3, and Q4, and assume that the base and the collector terminals of Q1 are shorted together), and assuming equal geometry for similar devices, the -3 dB bandwidths, $\omega_{-3dB}$, of these simple current mirrors are found to be:

$$\omega_{-3dBMOS} \approx \frac{g_{mM}}{\sqrt{C_{1M}^2 - C_{gd2}^2 + 2C_{1M}C_{gd2}}} \approx \frac{g_{mM}}{C_{1M}} \tag{5.19}$$

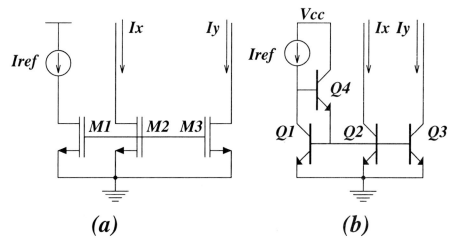

**FIGURE 5.4**
Current mirrors/current amplifiers (a) MOS (b) bipolar.

$$\omega_{-3dB bipolar} \approx \frac{g_{mQ}\sqrt{1 - \frac{4}{} - \frac{4}{2}}}{C_{1Q} - 2C_2} \approx \frac{g_{mQ}}{C_{1Q}} \tag{5.20}$$

where

$$C_{1M} = C_{gs1} + C_{db1} + C_{gs2} \tag{5.21}$$

and

$$C_{1Q} = C_1 + C_2 + C_{cs1} + C_2 \tag{5.22}$$

$C_{cs1}$ is the collector-to-substrate junction capacitance of Q1, $\beta$ is the current gain of Q1, $C_{db1}$ is the drain-to-bulk capacitance of M1, and the other parameters as previously defined.

Most analog building blocks incorporate a differential input pair which is desirable to have a large dc-input impedance. This greatly reduces the static current to the input transistors. Extremely large dc-input impedance is inherent with gates of MOS devices. In choosing MOS devices for inputs, one should consider the effect of noise, dc-offset voltage, operating speed, and the silicon real estate. Sample-hold amplifiers, charge sharing digital-to-analog converters, and switched capacitor filters require amplifiers with high input impedance to avoid the practical problems associated with charge losses. In such circuits MOS devices should be used for charged nodes and bipolar devices, or subtle combinations of bipolar and MOS, in the intermediate amplifying stage(s) to improve precision and maximize the gain bandwidth product.

As an example, consider the folded-cascode amplifier, an essential block in the aforementioned circuits, whose CMOS version and a possible BiCMOS version are shown in Fig. 5.5.

Several different combinations of bipolar and CMOS devices could be used in the design of this amplifier. Here, we assume that PNP bipolar transistors are

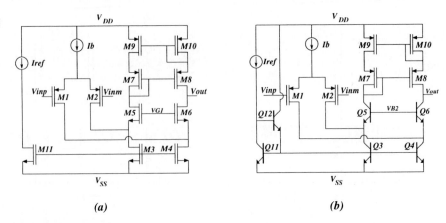

**FIGURE 5.5**
Folded-cascode amplifiers, (a) common-source, common-gate and (b) common-source, common-base.

not available, which implies a BiCMOS process having less complexity. However, the addition of high-quality PNP bipolar transistors offers extra flexibility in the design of many analog subcircuits such as high-speed, sample-and-hold amplifiers and high-speed data converters.

In situations where the input voltage source has a relatively large output impedance, or if we are operating in the charge domain where high impedance is required to store charge, one needs to use MOS devices for the input stage. Depending upon the desired characteristics of the amplifier, different combinations of CMOS and bipolar devices can be used in the rest of the amplifier. Here, we simplify the analysis and only compare the equivalent half-circuit small-signal response of the folded-cascode amplifiers in Fig. 5.5. The simplified half-circuits of the amplifiers are shown in Fig. 5.6. Note that we are neglecting the excess phase lag associated with the PMOS transistors (M7-M10), which comprise the differential to single-ended circuit.

The small-signal equivalent circuit of amplifiers in Fig. 5.6 is shown in Fig. 5.7. The following analysis assumes that the bipolar transistors have a reasonably high transition frequency ($f_T \approx 1GHz$).

The transfer function of the circuit in Fig. 5.7 is developed as it follows,

$$H(s) = \frac{V_o(s)}{V_{in}(s)} = \frac{N(s)}{D(s)} \tag{5.23}$$

where

$$N(s) = \frac{(sC_{gd1} - G_{m1})(1 + G_{m6}r_{o6})}{r_{o6}C_L(C_{gd1} + C_1)} \tag{5.24}$$

$$D(s) = s^2 + s\left(\frac{G_{m6} + \frac{1}{R_L} + \frac{1}{r_{o6}}}{C_{gd1} + C_1} + \frac{\frac{1}{r_{o6}} + \frac{1}{R_L}}{C_L}\right)$$

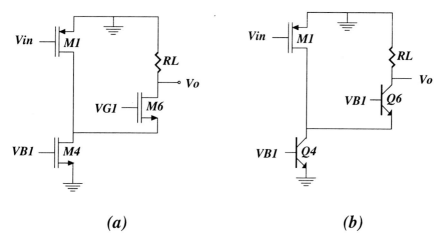

**(a)**                             **(b)**

**FIGURE 5.6**
Simplified half-circuit of the folded-cascode amplifiers, (a) common-source, common-gate and (b) common-source, common-base.

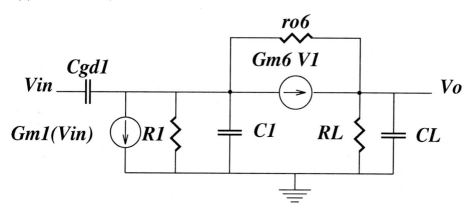

**FIGURE 5.7**
Small-signal equivalent circuit of amplifiers in Fig. 5.6.

$$+ \quad \frac{\frac{G_{m6}}{R_L} + \frac{1}{r_{o6} R_L} + \frac{1}{r_{o6} R_1} + \frac{1}{R_1 R_L}}{C_L (C_{gd1} + C_1)} \tag{5.25}$$

The parameters in $N(s)$ and $D(s)$ for the circuit in Fig. 5.6(a) are defined as:

$$C_1 = C_{gd1} + C_{db1} + C_{gb4} + C_{gd4} + C_{gs6} + C_{sb4} \tag{5.26}$$

$$R_1 = r_{o1} \| r_{o4} \tag{5.27}$$

$$G_{m6} = g_{m6} + g_{mb6} \tag{5.28}$$

$$r_{o6} = \frac{1}{g_{ds6}} \tag{5.29}$$

$$C_L = C_{load} + C_{db6} + C_{gd6} \tag{5.30}$$

and similarly the parameters in $N(s)$ and $D(s)$ for the circuit in Fig. 5.6(b) are:

$$C_1 = C_{gd1} + C_{db1} + C_{cs4} + C_4 + C_{Q6} \tag{5.31}$$

$$R1 = r_{o1} \| r_{oQ4} \| r_{Q6} \tag{5.32}$$

$$r_{Q6} = \frac{\beta_0}{G_{m6}} \tag{5.33}$$

$$G_{m6} = g_{mQ6} \tag{5.34}$$

$$r_{o6} = \frac{V_{A6}}{I_{C6}} \tag{5.35}$$

$$C_L = C_{load} + C_{cs6} + C_6 \tag{5.36}$$

where $C_{sb4}$ is the source-to-bulk capacitance of M4, $C_{load}$ is the external load capacitance, $g_{ds6}$ is the drain-to-source conductance of M6, and $g_{mb1}$ is the body transconductance of the MOS transistor.

The dc-gain and the poles/zero of the amplifiers (assuming poles are far apart) are:

$$A_{dc} \approx -\frac{g_{m1} g_{m6} R_L}{g_{m6} + \frac{R_L}{r_{o6} R_1} + \frac{1}{R_1} + \frac{1}{r_{o6}}} \tag{5.37}$$

$$Z = \frac{g_{m1}}{C_{gd1}} \tag{5.38}$$

$$P_1 = -\frac{\frac{g_{mG}}{R_L} + \frac{1}{r_{o6} R_1} + \frac{1}{r_{o6} R_L} + \frac{1}{R_1 R_L}}{(g_{m6} + \frac{1}{R_L} + \frac{1}{r_{o6}})C_L + (\frac{1}{r_{o6}} + \frac{1}{R_L})(C_{gd1} + C_1)} \approx -\frac{1}{R_L C_L} \tag{5.39}$$

$$P_2 = -\frac{g_{m6} + \frac{1}{R_L} + \frac{1}{r_{o6}}}{C_{gd1} + C_1} + \frac{\frac{1}{r_{o6}} + \frac{1}{R_L}}{C_L} \approx -\frac{g_{m6}}{C_{gd1} + C_1} \tag{5.40}$$

Since the $G_{m6} r_{o6}$ of Fig. 5.6(b) is larger than the $G_{m6} r_{o6}$ of Fig. 5.6(a), it is evident from Eq. (5.37), the BiCMOS op amp provides larger dc-gain than the CMOS op amp. In these amplifiers when $G_{m6} r_{o6} R_1 \gg R_L$, the unity gain frequency (at which $A_{dc} P_1$ has a unity value) is simply $\frac{G_{m1}}{C_L}$, which can be very high for the BiCMOS op amp. It is also observed from Eq. (5.40) that the second pole value in the BiCMOS op amp is higher than that of the CMOS op amp. This is mainly due to the high transconductance of the bipolar transistor, Q6, in Fig. 5.6(b). Both op amps should have comparable values for the parasitic capacitance $C_1$. Therefore, one can conclude that a BiCMOS op amp should have a higher gain-bandwidth product than the CMOS counterpart.

The parameters in Table 5.2 were used to simulate the small-signal response of the amplifiers in Fig. 5.5. The simulation result (magnitude and phase) is shown in Fig. 5.8.

The design assumes equal power consumption and equal CMOS device sizes for both amplifiers. These plots clearly indicate that the BiCMOS op amp has

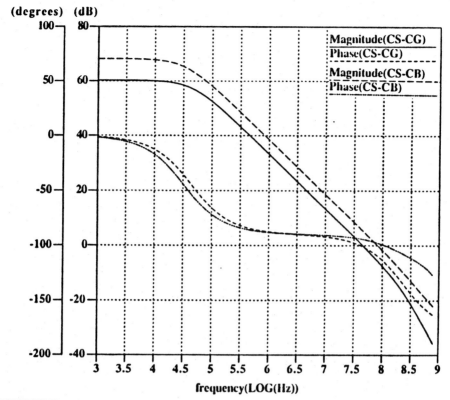

**FIGURE 5.8**

Magnitude and phase response of the amplifiers in Fig. 5.5.

both a higher gain-bandwidth product and a better phase response than that of the CMOS amplifier. In addition, the silicon real estate consumed by the BiCMOS amplifier was estimated to be approximately 10% less than that of the CMOS amplifier.

## 5.2.3   Matching Properties

Device mismatch limits the precision of circuits. In particular, the performance of operational amplifiers, current-mode, digital-to-analog converters and current-mode signal processors is strongly influenced by device matching. Random device mismatches show up as dc-offset voltages in differential pairs and as current ratio errors in the mismatched transistor current sources and mirrors [13]. As an example, consider the current sources in Fig. 5.4. If the devices in the current mirror have equally drawn sizes, any difference between the currents $I_r$ and $I_y$ caused by device mismatches will limit the precision of these current mirrors. Here, we assume that the mismatch in MOS transistors is due to mismatch in their gain factors, $K_n = \frac{eff C_{ox} W}{2L}$, and threshold voltages $(V_T)$. The mismatch in bipolar transistors is caused by emitter-area $(A_E)$ mismatches and by mismatches in total base impurity doping per unit area ($Q_B$). The current mismatch expressions for these current mirrors are found to be [14],

$$\frac{\Delta I_D}{I_D}, MOS \approx \frac{-\Delta V_T}{V_{GS} - V_T} + \frac{\Delta K_n}{K_n} \qquad (5.41)$$

where $I_D = I_x + I_y$, $\Delta I_D = I_x - I_y$, $K_n = K_{n2} + K_{n3}$, $\Delta K_n = K_{n2} - K_{n3}$, $\Delta V_T = V_{Tn2} - V_{Tn3}$, and $V_{GS} - V_T$ is the average excess gate voltage of M2 and M3 in Fig. 5.4(a).

$$\frac{\Delta I_C}{I_C}, BJT \approx \frac{-\Delta A_E}{A_E} + \frac{\Delta Q_B}{Q_B} \qquad (5.42)$$

where $I_C = I_x + I_y$, $\Delta I_C = I_x - I_y$, $A_E = A_{E2} + A_{E3}$ ($A_{E2}$ is the emitter area of Q2), $\Delta A_E = A_{E2} - A_{E3}$, $Q_B = Q_{B2} + Q_{B3}$, and $\Delta Q_B = Q_{B2} - Q_{B3}$.

Equation (5.41) indicates that any threshold mismatch results in current mismatch whose magnitude can be reduced by an increase in the on-voltages of MOS devices. It is observed from Eq. (5.42) that the emitter area mismatch causes current mismatch which can be minimized by choosing large emitter-area. In general, bipolar transistors match better compared with MOS transistors.

In practice, random errors from process variations, which affect $K_n$, and finite photolithography resolution limit the aspect ratio matching between two adjacent devices. Layout techniques such as cross-coupled, common-centroid, and the inclusion of dummy devices to reduce the etching error can improve device matching (refer to Chapter 16). Improved matching is also possible through the use of good design practices. For instance, the inclusion of emitter/source re-sistor degeneration may also improve device matching at the expense of some additional noise. In Fig. 5.4, if a resistor $R_1$ is inserted between the emitter (source) terminal of Q2 (M2) and the negative supply voltage, and a resistor $R_2$ between the emitter (source) terminals of Q3 (M3) and the negative supply volt-age, the following current mismatch expressions, Eqs. (5.43) and (5.44), indicate that matching improves when larger degeneration resistors are used.

$$\frac{\Delta I_D}{I_D}(degen.\ MOS) \approx \frac{\frac{\Delta K_n}{K_n}(V_{GS} - V_T) - \Delta V_T - I_D \frac{\Delta R}{4}}{V_{GS} - V_T + \frac{I_D R}{2}} \qquad (5.43)$$

where $\Delta R = R_1 - R_2$ and $R = R_1 + R_2$.

$$\frac{\Delta I_C}{I_C}(degen.\ BJT) \approx \frac{-1}{2 + \eta}\left(\left[\frac{\Delta A_E}{A_E} - \frac{\Delta Q_B}{Q_B}\right] + \frac{\eta}{2}\left[\frac{\Delta R}{R} - \frac{\Delta \alpha_F}{\alpha_F}\right]\right) \qquad (5.44)$$

where $\alpha_F = \frac{\beta}{1+\beta}$, $\beta$ is the average dc-current gain, $\eta = \frac{g_m R}{F}$ and $g_m$ is the average transconductance of Q2 and Q3.

The above derivations can be used to find an expression for the input referred dc-offset voltage/current mismatch caused by device mismatches in a more complex circuit. The input referred dc-offset voltage of amplifiers such as those shown in Fig. 5.5 is usually dominated by device mismatches of the input gain stage. However, this assumption is no longer valid when the average current density of the cascode stage is large, or when the devices in the cascode section are very small. A simplified schematic of the BiCMOS amplifier of Fig. 5.5(b) is shown in Fig. 5.9. Due to the extremely low input dc-current of M1 and M2 gate terminals, there is a very small mismatch in the input bias currents and its contribution to the dc-offset is negligible. We also assume that the base resistance, $r_b$, mismatches in Q3-Q4 and Q5-Q6 transistor pairs are insignificant. We can now apply the above derivation results to the circuit of Fig. 5.5(b), and derive an expression for its input referred dc-offset voltage caused by any mismatch between transistors M1 and M2, Q3 and Q4, Q5 and Q6. That is,

$$V_{os}^{\left(\text{total input referred}\right)} = V_{os}^{(M1M2)} + V_{os}^{(Q3Q4)} + V_{os}^{(Q5Q6)} \tag{5.45}$$

$$V_{os}^{(M1M2)} \approx \Delta V_{T12} - (V_{GS_{12}} - V_{T12})\frac{\Delta K_{p12}}{K_{p12}} \tag{5.46}$$

$$V_{os}^{(Q3Q4)} \approx \frac{(I_{C3} + I_{C4})\left(\dfrac{\Delta A_E^{(34)}}{A_E^{(34)}} - \dfrac{\Delta Q_B^{(34)}}{Q_B^{(34)}}\right)}{g_{m12}} \tag{5.47}$$

$$V_{os}^{(Q5Q6)} \approx \frac{(I_{E5} + I_{E6})\left(\dfrac{\Delta A_E^{(56)}}{A_E^{(56)}} - \dfrac{\Delta Q_B^{(56)}}{Q_B^{(56)}} - \dfrac{\Delta_F^{(56)}}{F^{(56)}}\right)}{g_{m12}} \tag{5.48}$$

where $I_{Ci}$ is the collector current of the ith transistor, $I_{Ej}$ is the emitter current of the jth transistor, $\Delta A_E^{(ij)}$ is the difference between the areas of the ith and the jth transistors, $A_E^{(ij)}$ is the sum of the areas of the ith and the jth transistors, $g_{m12}$ is the average transconductance of the PMOS transistors M1 and M2, and the rest of the symbols used follow previous definitions. Typically, the offset of an MOS differential pair will be 10-20 mV compared to a bipolar differential pair offset of 1-2 mV or less.

From Eq. (5.45) one realizes that the total input referred dc-offset can be minimized by using large sizes for the input devices, employing emitter degeneration resistors at Q3 and Q4 (presuming that the resistors do not appreciably degrade the noise performance), and by keeping the collector current, $I_c$, small in Q3, Q4, Q5, and Q6.

In general, the matching properties of bipolar transistors are superior to that of MOS transistors. This is because the threshold voltage mismatch in MOS transistors is the dominant contributor to the matching error, which is absent in bipolar devices. Choosing small $V_{GT} = (V_{GS} - V_T)$ for MOS transistors, which implies small quiescent current, should improve its dc-offset voltage. This is at the expense of lower operating speed and higher noise level. Reduced over drive, $(V_{GS} - V_T)$ will improve the input offset voltage, but the current matching will, however, deteriorate due to the larger contribution from the threshold voltage mismatch. Due to the larger transconductance of bipolar transistors, the

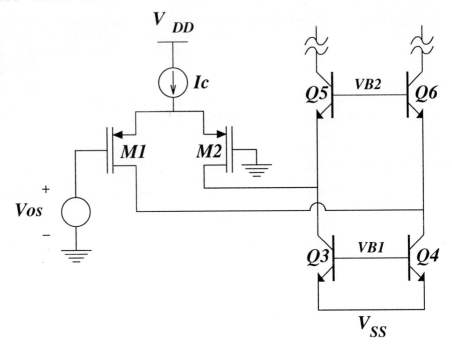

**FIGURE 5.9**
Simplified circuit of the common-source, common-base amplifier.

resistor degeneration technique benefits bipolar circuits more than their MOS counterparts. Low dc offset BiCMOS circuits may be obtained by employing dc-offset reduction circuit techniques and by special layout practices. The popular double-correlated sampling dc-offset cancellation technique is routinely used in the sample-data circuits having MOS input transistors, and trimming techniques such as laser, fuse, and current calibration are used in circuits with bipolar transistors, where extremely low dc-offset is essential.

### 5.2.4 Noise Properties

Three major noise sources exist in BiCMOS devices: flicker ($\frac{1}{f}$) noise, thermal noise, and shot noise. The flicker noise in an MOS transistor is caused by the presence of carrier energy states at $Si/SiO_2$ interface of the gate. Due to the averaging effect of the surface states, the magnitude of this noise is inversely proportional to the gate-oxide capacitance. Therefore, an MOS device having thin gate-oxide and a large gate area, assuming a clean process, will exhibit small flicker noise. The flicker noise in a bipolar transistor originates from charge traps caused by process contamination and crystal defects in the emitter-base depletion layer. The magnitude of this flicker noise is a function of the direct current and is, in general, negligible.

Thermal noise in a resistor is due to random thermal motion of carriers.

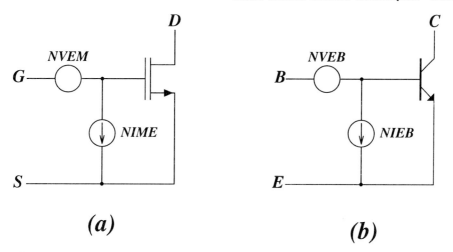

**FIGURE 5.10**
Equivalent input noise representation of (a) MOS and (b) bipolar transistors.

The thermal noise in an MOS transistor is generated from the channel resistance, and its magnitude is inversely proportional to the device transconductance. The thermal noise in a bipolar transistor is caused by the various resistive components of the device. The base resistance, $r_b$, and the emitter resistance, $r_e$, (in poly-emitter bipolar transistors) are usually the dominant contributors to the thermal noise in a bipolar transistor.

Shot noise is associated with a direct-current flow across a junction. In reality, carriers cross a junction in a random fashion, and the device static current which appears to be a steady current is actually a conglomeration of a large number of random independent current components. This shows up as a fluctuation in the static current and is called shot noise. Therefore, shot noise is more pronounced in a bipolar transistor (a junction device) than in an MOS transistor (a surface device). Any gate-leakage current in an MOS transistor has a random nature and will contribute to its shot noise. There is another noise generator in a bipolar transistor known as a burst or popcorn noise generator. It is a low frequency type noise source, and it has been related to the presence of heavy-metal ion contamination in the base region [1, 14].

Noise in a transistor can be represented by two input-referred equivalent noise generators, as shown in Fig. 5.10 for MOS and bipolar transistors.

The mean-square input-referred noise values of the noise sources in an MOS and a bipolar transistor are [10]:

$$NVEM = \frac{8KT\Delta f}{3g_m} + \frac{K_f \Delta f}{WLC_{ox}f} \tag{5.49}$$

$$NIEM = 2qI_G\Delta f \tag{5.50}$$

$$NVEB = 4KTr_b\Delta f + \frac{2qI_C\Delta f}{g_m^2} \tag{5.51}$$

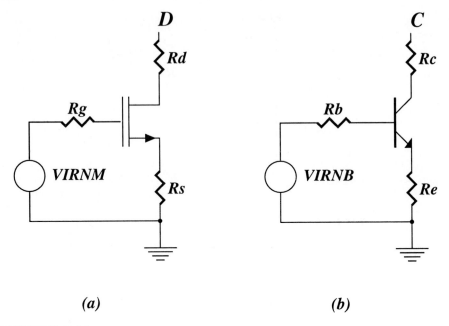

*(a)*                                        *(b)*

**FIGURE 5.11**
Input-referred noise from active and terminal impedance components of (a) MOS and (b)
bipolar amplifiers.

$$NIEB = \left[ 2qI_B + \frac{K_1 I_B^a}{f} + \frac{I_B^c}{1 + (\frac{f}{f_T})^2} \right] \Delta f \qquad (5.52)$$

where $K_f$, $K_1$, $a$, $c$ are process-related constants, $f$ is the frequency (Hz), $T$ is
the absolute temperature, $q$ is the electron charge, $I_G$ is the MOS gate-leakage
current, and $\Delta f$ is the noise bandwidth.

The noise expressions, Eqs. (5.49)-(5.52), can be used to derive the equivalent input-referred noise of any BiCMOS circuit. The exact noise derivation
becomes rather tedious when the number of devices increases. In general, the
noise computation of an amplifier is straightforward, because the noise sources
generated after the input gain stage are scaled down by its gain and each has
a negligible contribution to the overall noise power. In Fig. 5.11 we have shown
two basic bipolar and MOS amplifiers.

In these amplifiers the noise from their active noise generators and terminal
impedances are lumped into an equivalent input-referred noise source, (VIRNM
for the MOS amplifier and VIRNB for the bipolar amplifier).

Derivation of the noise power expressions for these amplifiers reveals the effect of the terminal impedances on the magnitude of the noise. These expressions
are provided below. We have assumed that the collector-to-base junction capacitance and gate-to-drain Miller capacitance have negligible effect on the total
noise power of these amplifiers. One can easily show that these parasitic capaci-

tances have lesser effect on the magnitude of the noise when the load impedance is large.

$$(VIRNM)^2 = 4KTR_g\Delta f + \left[(\omega C_g(R_g + R_s))^2 + (1 + g_m R_s)^2\right]$$

$$\times \left[NVEM + \frac{4KT\Delta f}{g_m^2 R_d}\right] \tag{5.53}$$

$$(VIRNB)^2 = \left[4KT(R_b + r_b) + 2qI_B(R_b + r_b)^2 \right.$$

$$+ \frac{1 + g_m R_c + \left(\frac{R_b + r_b + R_e}{r}\right)^2 + \omega^2 C^2 (R_b + r_b + R_e)^2}{gm^2}$$

$$\left. \times \left(\frac{4KT}{R_c} + 2qI_C\right)\right]\Delta f \tag{5.54}$$

We note from Eqs. (5.53) and (5.54) that the total noise power increases with an increase in the values of $R_s$, $R_g$, and $C_g$ in the MOS amplifier, and $R_b$, $r_b$, $R_c$, and $C_p$ in the bipolar amplifier. The total noise power decreases as $g_m$, $W$, and $L$ increase in the MOS amplifier, and as $g_m$ and $r_p$ increase in the bipolar amplifier. Advanced BiCMOS processes offer small devices, hence $C_g$ and $C_p$ are small, and their noise contributions are also small. From Eqs. (5.53) and (5.54) one also notices that as the input source impedance, $R_g$ or $R_b$, increases, the noise performance of the bipolar transistor circuit deteriorates much faster than in the MOS transistor circuit. This is because the second term in $(VIRNB)^2$ is missing in $(VIRNM)^2$. However, for a low source impedance, the opposite is true. We can thus conclude that in a situation when the source impedance is small, a bipolar transistor should be used. This will result in superior noise performance compared to its MOS counterpart. This optimization option does not exist for circuits consisting of only CMOS or only bipolar transistors.

The above noise results can be used to derive the input-referred noise expressions for the transconductance amplifiers in Fig. 5.5. To simplify the analysis, we lump the noise from the active cascode differential to single-ended circuit into an equivalent noise generator RL, as shown in Fig. 5.12. In this figure, each transistor noise is referred to its input and represented by an equivalent noise source. An expression for the input-referred noise of the circuit in Fig. 5.12 may be easily obtained by finding an expression for the equivalent impedance appearing at nodes N1 and N2.

The expression for this equivalent impedance is provided below. We have assumed that the Miller capacitances of the transistors have a negligible effect on the total noise magnitude.

$$Z_{eq} = \frac{1 + \frac{r_{o5(6)}}{R_L}}{\frac{1}{R_L} + \frac{1}{Z_1} + \frac{r_{o5(6)}}{Z_1 R_L} + \frac{g_{m5(6)} r_{o5(6)}}{R_L}} \tag{5.55}$$

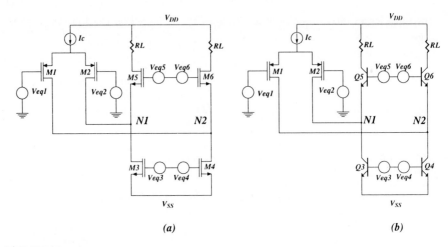

**FIGURE 5.12**
Input-referred noise representations of (a) CMOS and (b) BiCMOS amplifiers.

where the parameters in (5.55) for the circuit in Fig. 5.12(a) are:

$$Z_1 = r_{o1(2)} \| r_{o3(4)} \| \frac{1}{jwC_{ptot}} \tag{5.56}$$

$$C_{ptot} = C_{gd1(2)} + C_{db1(2)} + C_{gd3(4)} + C_{db3(4)} + C_{sb5(6)} + C_{gs5(6)} \tag{5.57}$$

and for the circuit in Fig. 5.12(b), $C_{ptot}$ is defined by:

$$C_{ptot} = C_{gd1(2)} + C_{db1(2)} + C_{3(4)} + C_{cs3(4)} + C_{5(6)} \tag{5.58}$$

The product of Eq. (5.54) and the transconductance of M1(M2) will be used to find the input-referred noise power expression due to M3-M6 and Q3-Q6. The equivalent input-referred noise power expressions of the amplifiers in Fig. 5.12 are found to be:

$$V_{eqtotal}^2 = V_{eq1}^2 + V_{eq2}^2 + (\frac{g_{m3(4)}}{g_{m1(2)}})^2(V_{eq3}^2 + V_{eq4}^2) + \kappa^2(V_{eq5}^2 + V_{eq6}^2) \tag{5.59}$$

where $\kappa = \frac{|1+g_{m5(6)}Z_{eq}|}{g_{m1(2)}g_{m5(6)}|Z_{eq}^2|}$, $V_{eqi}^2$ (i = 1-6) in Fig. 5.12(a) and $V_{eqi}^2$ (i = 1-2) in Fig. 5.12(b) are the same as that of Eq. (5.49), and $V_{eqi}^2$ (i = 3-6) in Fig. 5.12(b) is defined by Eq. (5.54) where $R_b$, $R_c$, and $R_e$ set to zero. Due to the higher transconductance of a bipolar transistor, the BiCMOS amplifier should provide better noise performance than the CMOS amplifier. In designing a high performance BiCMOS amplifier, using the basic amplifier in Fig. 5.12, one should consider an optimum transconductance for the bipolar transistors to minimize the amplification of noise sources in Q3 through Q6, when referred to the input of the amplifier. When a bipolar transistor exhibits small $r_b$, $R_b$, and $R_c$, then the input-referred noise expression of a bipolar transistor can be approximated by:

$$V_{eq}^2 \approx \frac{(\frac{4KT}{Z_{eq}} + 2qI_C)\Delta f}{g_m^2} \tag{5.60}$$

When Eq. (5.60) is substituted into the $V_{eqtotal}^2$ expression for each bipolar transistor, the amplified components of the bipolar transistor noise become negligible (because of $g_m^2$ in the denominator of Eq. (5.60)), and an improved amplifier noise characteristic results. In the case when the differential input stage provides a reasonable amount of gain, the noise of the cascode stage will not be important when referred to the input.

## 5.2.5  Second-Order MOS Model Effects

The MOS transistors used in most of the circuits discussed in this chapter operate in the saturation region. The simplified square-law drain current equation in Eq. (5.1) is conveniently used for first-order derivations. A more accurate analysis accounts for second-order effects such as channel length modulation, body effect, source terminal series resistance, and mobility degradation. These effects are discussed in Chapter 2, and are briefly reviewed here. Assuming that the expression in Eq. (5.1) is denoted by $I_{d(ideal)}$, an approximate drain current expression that includes the channel length modulation effect may be written as follows [1, 14];

$$I_d = I_{d(ideal)}(1 + \lambda V_{DS}) \tag{5.61}$$

$$\lambda = \frac{\sqrt{\frac{\varrho_{si}}{qN_{eff}(V_{DS}-V_{Dsat})}}}{L} \approx \frac{\sqrt{\frac{\varrho_{si}}{(2qN_{eff}V_{DS})}}}{L}(1 + \frac{V_{DS}}{2V_{Dsat}}) \tag{5.62}$$

$$V_{Dsat} = V_{GS} - V_T \tag{5.63}$$

where $\epsilon_o$ is the permitivity of free space, $\epsilon_{si}$ is the dielectric constant of silicon, and $N_{eff}$ is the substrate doping density. Note that the effect of $\lambda$ is small when $L$, the channel length, is large. The mechanisms that contribute to the drain-voltage dependence of the current are: (1) the widening of the drain depletion region due to an increase in the drain voltage, which modulates the effective channel length, (2) the space-charge limited current caused by the extension of the drain depletion into the source region (this is referred to as punch-through effect), and (3) the electrostatic feedback of the drain field into the channel causing the drain terminal to act as a second gate affecting the channel conductance.

A large electric field is applied across the gate insulator when the transistor gate oxide is thin and a large gate voltage is applied. This electric field induces large perpendicular forces on the charge carriers and decreases the carrier mobility along the interface. An approximate model of the mobility reduction is,

$$\mu = \frac{\mu_o}{1 + \theta(V_{GS} - V_T)} \tag{5.64}$$

where $\mu_o$ is the zero field mobility, $\theta = \frac{1}{t_{oz}E_{cr}}$ which is process dependent and could have values ranging from 0.01 to 0.25 $V^{-1}$, $t_{or}$ is the oxide thickness, and $E_{cr}$ is the critical field.

Any variation in the source-to-substrate voltage causes the source-to-substrate depletion region charge and the threshold voltage to vary accordingly. This is modeled as,

$$V_T = V_{TO} + \gamma(\sqrt{2\phi_f + V_{SB}} - \sqrt{2\phi_f}) \tag{5.65}$$

where $V_{TO}$ is the threshold voltage when the source-to-substrate voltage is zero, $\gamma = \frac{t_{oz}\sqrt{2q_{on}N_{eff}}}{oz}$ and is called the body effect (BE), $\phi_f$ is the Fermi potential, $\epsilon_{or}$ is the gate oxide dielectric constant, and $V_{SB}$ is the source-to-substrate voltage.

The metal contacts to source and drain regions and the resistance of wire interconnects may cause (for large drain currents) some voltage drop between the source and drain terminals and the intrinsic source and drain. The source resistance modifies the dc-operating points of the transistor when biased in the saturation region. This is modeled as,

$$V_{GS} = V_{GS}^{intrinsic} + I_d R_s \tag{5.66}$$

where $R_s$ is the source series resistance, $V_{GS}^{intrinsic}$ is the gate-to-source voltage when $R_s = 0$.

Temperature variations affect the characteristic of an MOS transistor. Carrier mobility and the threshold voltage are predominantly temperature dependent. The effect of temperature variation on these parameters should be modeled and accounted for in the design of an MOS circuit required to operate over a wide range of temperatures.

## 5.3 BASIC ANALOG SUBCIRCUITS

Differential amplifiers, transconductance and transresistance amplifiers, linear voltage-to-current and current-to-voltage converters, current mirrors, current sources, voltage and current references, translinear computational circuits, output stages, and buffers are some of the fundamental building blocks used in the design of analog subsystems. In this section, we will introduce some of these building blocks in BiCMOS technology and make performance comparisons with those in CMOS.

### 5.3.1 Current Mirrors/Current Sources

The implementation of BiCMOS current sources follows directly from the bipolar-only and MOS-only current sources. A high-performance current mirror features an extremely high output impedance, a very low input impedance, and a large output voltage swing [18]. An extremely high output impedance can be obtained by stacking MOS and bipolar devices in a double or triple cascade configuration as shown in Fig. 5.13. The low frequency output impedances of these current mirrors are:

$$r_{out}^{(a)} = \frac{r_{o3}[(g_{M3} + g_{M3b})(1 + \frac{r_{o2}}{r_{o1}} + r_{o2}g_{M2}) + g_{M2}(\frac{r_{o2}}{r_{o3}} + \frac{1}{1}) + G_X]}{\frac{1}{r_{o1}} + \frac{1}{r_{o2}} + \frac{g_{M2}+g_{M2b}}{1} + \frac{1}{r_{o1}}}$$

$$\approx \frac{r_{o2}g_{M2}r_{o3}(g_{M3} + g_{M3b})}{\frac{1}{r_{o1}} + \frac{g_{M2}}{1}} \tag{5.67}$$

$$r_{out}^{(b)} = \beta r_{o2}r_{o3}(g_{M3} + g_{M3b})r_{o4}(g_{M4} + g_{M4b}) \tag{5.68}$$

where $G_X = \frac{1}{r_{o2}} + \frac{1}{r_o} + \frac{1}{r_{o1}} + \frac{r_{o2}}{r_{o1}r_{o3}} + \frac{1}{r_{o3}}$, $r_{oi}$ is the output resistance, $g_{Mi}$ is the transconductance of the ith transistor, $b$ denotes the back bias effect ($g_{mb} = \frac{dI_D}{dV_{SB}}$ is the bulk-channel transconductance), and $\beta$ is the current gain of the bipolar transistors.

Although these current mirrors feature very high output impedance, they have a limited output voltage swing and a large input impedance. The input impedances and the minimum allowable voltages at the outputs of these current mirrors are expressed as follows:

$$R_{in}^{(a)} \approx \frac{1}{g_{M4}} + \frac{1}{g_{M5}} + \frac{1}{(g_{M6} + g_{M6b})} \tag{5.69}$$

$$R_{in}^{(b)} \approx \frac{1}{g_{M5}} + \frac{1}{g_{M6}} + \frac{1}{g_{M7} + g_{M7b}} + \frac{1}{g_{M8} + g_{M8b}} \tag{5.70}$$

$$V_{(outmin)}^{(a)} = V_{ss} + V_{BE1} + V_{BE2} + V_{GS3} - V_{T3} \tag{5.71}$$

$$V_{(outmin)}^{(b)} = V_{ss} + V_{BE1} + V_{BE2} + V_{GS3} + V_{GS4} - V_{T4} \tag{5.72}$$

The BiCMOS current mirrors in Fig. 5.14 alleviate the shortcomings described above using a cascode active feedback approach, which increases the output impedance while achieving lower input impedance, and significantly improves the output voltage swing capability [15, 19]. These current mirrors require a reference current, $I_{ref}$, for the active feedback circuit.

The expressions for the low frequency output impedance (assuming that the output impedances of the current reference circuits are large), minimum output voltage swing, and input impedance are:

$$r_{out}^{(a)} = \frac{r_{o1}r_{o2}r_3(g_{M2} + g_{M2}g_{M3}r_{o3} + g_{M2b} + \frac{1}{r_3} + \frac{1}{r_{o1}} + \frac{1}{r_{o2}})}{r_{o1} + r_3}$$

$$\approx \beta r_{o2}g_{M2}V_A I_{ref} \tag{5.73}$$

$$r_{out}^{(b)} = r_{o1} + r_{o2} + r_{o1}r_{o2}(g_{M2} + g_{M2b}) +$$

$$r_{o1}r_{o2}r_{o3}r_{o4}g_{M2}g_{M3}(g_{M4} + g_{Mb4} + \frac{1}{r_{o4}})$$

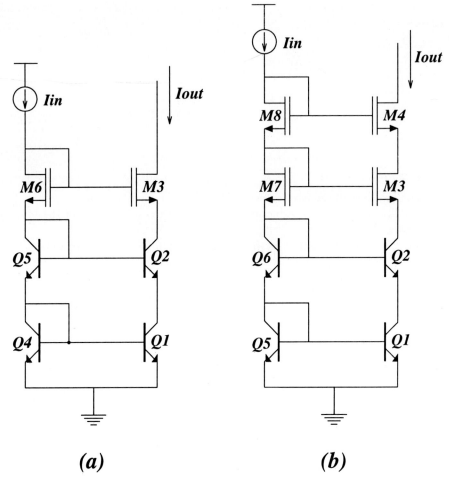

**(a)**                                   **(b)**

**FIGURE 5.13**
Current sources, (a) double cascode and (b) triple cascode.

$$\approx r_{o1}r_{o2}r_{o3}r_{o4}g_{M2}g_{M3}\left(g_{M4} + g_{Mb4} + \frac{1}{r_{o4}}\right) \tag{5.74}$$

$$R_{in}^{(a)} = \frac{r_{o4}(r_{b4} + r_4)}{r_{o4} + r_{b4} + r_4 + r_{o4}r_4 g_{M4}} \approx \frac{1}{g_{M4}} \tag{5.75}$$

$$R_{in}^{(b)} \approx \frac{1}{g_{M5}} + \frac{1}{g_{M6} + g_{M6b}} \tag{5.76}$$

$$V_{(outmin)}^{(a)} = V_{ss} + V_{BE3} + V_{GS2} - V_{T2} \tag{5.77}$$

$$V_{(outmin)}^{(b)} = V_{ss} + V_{GS3} + V_{GS2} - V_{T2} \tag{5.78}$$

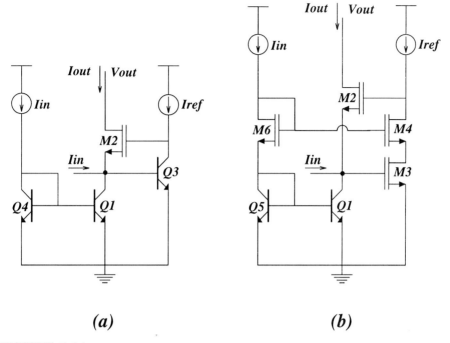

**FIGURE 5.14**
BiCMOS cascode current sources with active feedback.

### 5.3.2 Amplifier Stages

Figure 5.15 shows three basic single-ended, high-performance BiCMOS amplifiers. High gain-bandwidth product and extremely high dc-input impedance are
the main performance objectives. The differential configuration of each of these
basic amplifiers in conjunction with an output stage can be used to implement a general purpose high-performance amplifier. Amplifiers using these basic
blocks (with no output stage) are referred to as transconductance amplifiers,
and are suitable for use in sampled-data, switched-capacitor, and continuous-
time transconductance-capacitor analog integrated circuits.

An amplifier based on the circuit shown in Fig. 5.15(a) was introduced in
Fig. 5.5. As was shown earlier, the amplifier out performs its CMOS counterpart
in terms of its ac-response. The ac-equations are given in Eqs. (5.23)-(5.40).

An amplifier with a single-ended output which uses the basic circuit in
Fig. 5.15(b) is developed in Fig. 5.16 [19]. This amplifier is composed of an
input common-source gain stage and an output transresistance circuit. The input
stage transistors M1 and M2 convert the differential input voltages to differential
output currents. These currents are fed to the transresistance circuits comprised
of transistors Q3, Q5, M4 and M6, M7, Q8, Q9, Q10. The diode-connected
transistor Q11 is used to provide a single-ended output. Note that the 3-transistor
subcircuits used at the output side exploit the active feedback concept shown in

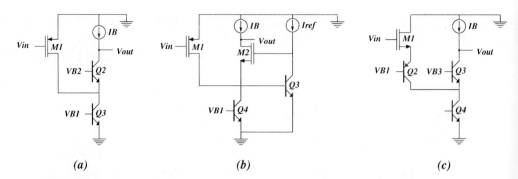

**(a)** **(b)** **(c)**

**FIGURE 5.15**

BiCMOS basic amplifier circuits (a) common-source, common-base (b) common-source, common-gate with active feedback, (c) common-drain, common-base common-base.

the current source circuit of Fig. 5.14(a). A similar amplifier architecture can also be developed using the concept demonstrated in Fig. 5.14(b) [18] which would result in a very high dc-gain.

The small-signal behavior of the amplifier in Fig. 5.16 can be determined by studying the ac-response of the basic amplifier circuit in Fig. 5.15(b). If we assume that the current source $I_B$ has a parallel output impedance, $R_L (R_L \gg r_{o1}$, $r_{o2}$, $r_{o3}$ and $r_{o4})$ and that $g_{M2} + g_{M2b} \gg (\frac{1}{r_{o1}||r_{o4}||r_3} + \frac{1}{r_{o2}})$, then one can find that the basic amplifier has two zeros (one in the right-half plane and the other in the left-half plane) and three poles in the left-half plane. The left-half plane zero approximately cancels one of the poles. The expressions for the dc-gain and the poles/zero are:

$$A_{dc} = \frac{-g_{m1}[g_{m2}g_{m3} + \frac{g_{m2}+g_{mb2}}{R_2}]}{\frac{g_{m2}g_{m3}}{R_L} + \frac{g_{m2}+g_{mb2}+1R_1+1r_{o2}}{R_2 R_L} + \frac{1}{R_1 R_2 r_{o2}} + \frac{1}{R_1 R_2 r_{o2}}}$$

$$\approx -\frac{g_{m1}g_{m2}g_{m3}R_1 R_2 r_{o2} R_L}{g_{m2}g_{m3}R_1 R_2 r_{o2} + R_L} \tag{5.79}$$

$$Z = \frac{g_{m1}}{C_{gd1}} \tag{5.80}$$

$$P_1 \approx -\frac{g_{m2} + g_{mb2}}{C_1 + C_{gd1}} \tag{5.81}$$

$$P_2 \approx -\frac{1}{R_L C_L} - \frac{1}{R_1 R_2 r_{o2} g_{m2} g_{m3} C_L} \tag{5.82}$$

where $R_1 = r_{o1}||r_{o4}||r_3$, $R_2 = r_{o3}||r_{oIrcf}$, $C_1 = C_{ds1} + 2C_{gd1} + C_3 + C_3 + C_{cs4} + C_4 + C_{sb2} + C_{gs2}$ and $C_2 = C_{gs2} + C_{gd2} + C_{cs3} + C_3$. Comparison of Eqs. (5.79)-(5.82) with Eqs. (5.37)-(5.40) reveals that the basic amplifier in Fig. 5.15(b) is capable of providing significantly higher gain-bandwidth product than the amplifier in Fig. 5.15(a). The amplifier of Fig. 5.15(b) has a pole-zero pair which nearly cancels each other. This could affect its high speed settling behavior.

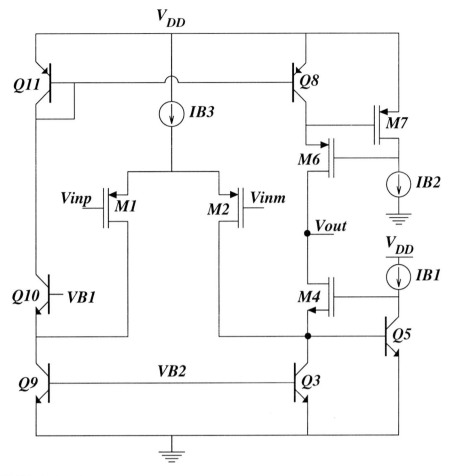

**FIGURE 5.16**
BiCMOS amplifier based on the basic amplifier in Fig.5.15(b).

Equations (5.38) and (5.80) indicate that amplifiers based on the circuits in Figs. 5.15(a) and (b) have a right-half plane zero which tends to degrade their phase-margins. The phase lag due to the right-half plane zero (caused by the input device Miller parasitic capacitance) limits the maximum useful speed. The amplifier in Fig. 5.15(c) has been introduced here as a basic building block for the design of high-performance amplifiers. This circuit does not suffer from the device Miller capacitance problems.

An amplifier using the circuit in Fig. 5.15(c) is shown in Fig. 5.17 [19]. This implementation assumes high-performance PNP transistors are available. Otherwise, PMOS devices can be used to substitute the PNPs. The static currents flowing through Q3 and Q6 are determined by the value of $I_B$ and the ratio of their emitter area to that of Q4 and Q5. Also, the transconductance of M1 is proportional to the square-root of the sum of the currents in Q3 and Q4.

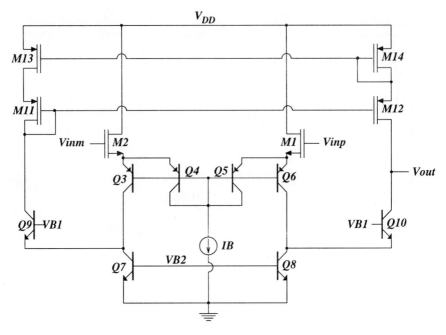

**FIGURE 5.17**
BiCMOS amplifier based on the basic amplifier in Fig.5.15(c).

Therefore, wide range of transconductance values for M1 (M2), with respect to Q3 (Q6), can be chosen through appropriate ratioing of the emitter areas of Q3 (Q6) and Q4 (Q5).

There are numerous ways to design amplifier circuits using this basic building block. All-CMOS or all-bipolar implementations are possible and will have competitive speed performance over that of conventional amplifier circuits. The small-signal behavior of the basic amplifier in Fig. 5.15(c) is now investigated which will help us gain some insight about the ac-response of the amplifier in Fig. 5.17. From the small-signal equivalent circuit of this basic amplifier we find that there is a left-half plane zero and three left-half plane poles. If the NMOS transistor M1 is designed to have a large transconductance (large $\frac{W}{L}$), then the amplifier can be broad-banded further by using a feedforward capacitor between the emitter and collector of transistor Q2. The expressions for the dc-gain and the zero/poles (assuming that the poles are widely apart) of the basic amplifier in Fig. 5.15(c) are:

$$A_{dc} \approx \frac{-g_{m1}g_{m3}r_{o2}}{(g_{m1} + g_{mb1} + g_{m2})(1 + \frac{r_{o2}}{R_L})} \tag{5.83}$$

$$Z = \frac{-g_{m1}}{C_{gs1}} \tag{5.84}$$

$$P_1 \approx -\frac{g_{m1} + g_{mb1} + g_{m2}}{C_{gs1} + C_2} - \frac{g_{m2}}{C_3} \tag{5.85}$$

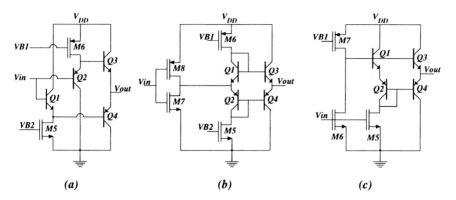

**FIGURE 5.18**
Output stages with unity gain (a) and with dc-amplification (b) and (c).

$$P_2 \approx \frac{-g_{m2}(g_{m1} + g_{mb1} + g_{m2})}{C_3(g_{m1} + g_{mb1} + g_{m2}) + g_{m2}(C_{gs1} + C_2)} \tag{5.86}$$

$$P_3 \approx -\frac{1}{R_L C_L} - \frac{1}{r_{o2} C_L} \tag{5.87}$$

where $C_2 = C_{gs1} + C_{sb1} + C_2$, $C_3 = C_2 + C_{cs2} + C_4 + C_{cs4} + C_3$, and $R_L$ is the output impedance of the current source $I_B$. The pole-zero pair $P_1$ and Z can easily be forced to cancel each other, and the amplifier will then have a dominant pole, $P_3$, and a non-dominant pole, $P_2$.

### 5.3.3   Output Stages

There are occasions when an amplifier is required to drive a large capacitive or a small resistive load (or a combination of the two). In such cases, the amplifier incorporates an appropriate output stage. The circuits in Fig. 5.18 are examples of BiCMOS output stages [19]. They assume that good vertical, isolated, PNP bipolar transistors are available. If not, a "pseudo-PNP" [25, 30] consisting of a PMOS transistor with an NPN emitter-follower for added speed and current gain might be used (Prob. 5.13).

The circuit in Fig. 5.18(a) functions as a unity gain buffer. Its low-frequency input/output impedance and output voltage expressions are:

$$R_{in} \approx [r_4(1 + \beta_1) + \beta_1\beta_4 R_L]||[r_3(1 + \beta_2) + \beta_2\beta_3 R_L]||\beta_1 r_{o1}||\beta_2 r_{o2} \tag{5.88}$$

$$R_{out} \approx \frac{1}{g_{m3} + g_{m4}} \tag{5.89}$$

$$V_{out} = V_{in} + \frac{V_T}{2} ln[\frac{I_1 I_3 I_{s2} I_{s4}}{I_2 I_4 I_{s1} I_{s3}}] \tag{5.90}$$

where $\beta_i$ is the current gain, $r_i$ is the base-emitter resistance, $r_{oi}$ is the output resistance, $g_{mi}$ is the transconductance, $I_i$ is the collector current, and $I_{si}$ is the base-emitter junction saturation current of the ith transistor. Fig. 5.19(a) shows the dc-transfer characteristic and $V_{in} - V_{out}$ versus $V_{in}$, where the drain currents of M1 and M2 are 100 $\mu$A, the collector currents of Q3 and Q4 are approximately 800 mA, the supply voltages are $\pm 4V$ and the circuit load resistance is 1 K$\Omega$. The curve in Fig. 5.19(b) is the ac-response using the device parameters in Table 5.2, $R_L$= 1 K$\Omega$, and $C_L$=100 pF. These curves show that the circuit is capable of achieving both low distortion and high speed.

The output stage circuit in Fig. 5.18(b) provides some dc-gain, practically infinite dc-input impedance, and very low output impedance. Transistors M7 and M8 provide the dc-gain and the infinite input impedance, and transistors M5 and M6 act as a current sink and current source, respectively. Transistors Q1-Q4 act as a unity gain buffer, where the desired output impedance and current drive capability are obtained by appropriately ratioing the emitter areas of Q3 and Q4 with respect to the emitter areas of Q1 and Q2. High dc-gain and low output impedance are easily achievable with this output stage. The expressions for these are:

$$R_{out} \approx \frac{R_L(R_x + R_y)}{R_L + R_x + R_y + R_L R_y(g_{m3} + g_{m4})} \tag{5.91}$$

$Adc \approx$
$$(g_{m6} + g_{m7}) \times [r_{o7}||r_{o8}||[[[r_3 + R_L(1 + \beta_3)]||r_{o6}]||[[r_4 + R_L(1 + \beta_4)]||r_{o5}]]]$$

$$= g_{m78}[r_{o7}||r_{o8}||(r_{34}(2 + \frac{1}{g_{m34}R_L} - \frac{1}{g_{m12}r_{34}}))] \tag{5.92}$$

where $g_{m1} = g_{m2} \gg \frac{1}{r_{12}}$, $R_x = r_{o5}||r_{o6}||r_{o7}||r_{o8}$, $R_y = r_3||r_4$, $g_{mij}$ and $r_{ij}$ are the transconductance and the base-emitter resistance of transistors i and j. Note that PMOS devices can be substituted for the PNP devices in this output stage. The output stage in Fig. 5.18(c) exhibits a similar performance to the one in Fig. 5.18(b). The expressions for its output impedance and dc-gain is the subject of Prob. 5.7, at the end of the chapter.

### 5.3.4  Voltage-to-Current Converters (VICs)

Wide input voltage range and high-speed VICs (also called transconductors) play important roles in the design of high-performance continuous-time filters, linear automatic control circuits, high-speed, four-quadrant multipliers, wide input voltage range differential difference amplifiers [20], and current-mode signal processor interface circuits. There are many different circuit techniques [21] available for design of low-to-medium speed ($< 10$ MHz), wide input range (linearized) VICs; however, high-speed linear VICs are not so readily available. Fig. 5.20 presents two possible BiCMOS VIC solutions. Currents $i1$ and $i2$ denote the small signal quantities only.

The VIC in Fig. 5.20(a) [19] is a common-drain (M1, M2), common-base (Q1, Q2) configuration. Transistors M1, M2, Q1, and Q2 operate in their active regions (MOSFETs in saturation and BJTs in active) and form the core of

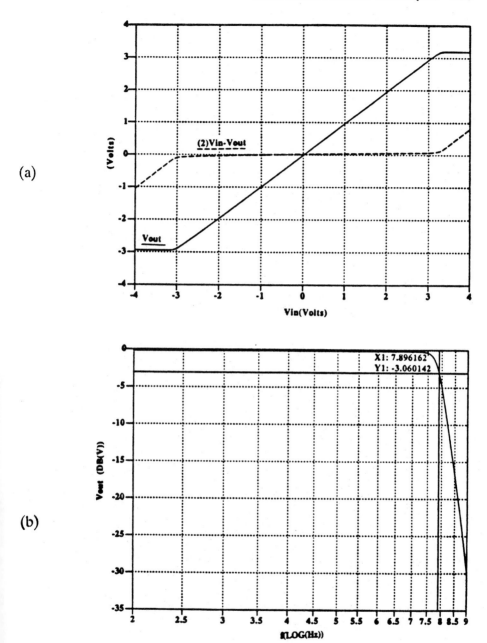

**FIGURE 5.19**
Performance of the circuit in Fig. 5.18(a) with $R_L$=1K and $C_L$=100PF, (a) dc-transfer characteristic and $V_{in} - V_{out}$ vs. $V_{in}$, (b) ac-response.

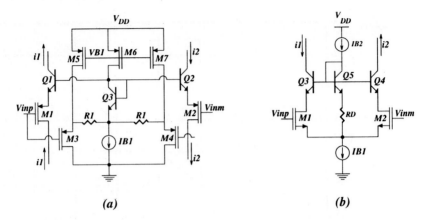

**FIGURE 5.20**
High speed BiCMOS VICs with MOSFETs (a) in saturation, (b) in triode region.

the VIC. The other devices provide proper common-mode dc-biasing and gain programmability. This circuit converts a differential input voltage (Vinp-Vinm) to a differential output current. The dc-transfer characteristic becomes mathematically linear (assuming square-law characteristic for the MOSFETs) when the bipolar transistors Q1-Q3 are replaced by NMOS transistors [19]. This BiC-MOS transconductor offers wider input voltage range (assuming $V_{BE} < V_{GS}$ for the same bias current) and higher bandwidth than its all-MOS counterpart. We now derive the dc-transfer function of this VIC using Eqs. (5.1)-(5.3) with $V_{in} = V_{inp} = -V_{inm}$ and $K_p = \frac{\mu_{eff}C_{ox}W}{2L}$. M3 and M4 is biased with a fixed current from the upper mirrors, and the source of M4 and M3 will thus simply follow their respective inputs, $V_{inm}$ and $V_{inp}$. Thus the emitter, as well as the base voltage of Q3, will simply follow the common mode input voltage. Straightforward analysis yields

$$- V_{in} - V_{GS1} - V_{BE1} + V_{CQ3} = 0 \tag{5.93}$$

$$+ V_{in} - V_{GS2} - V_{BE2} + V_{CQ3} = 0 \tag{5.94}$$

$$V_{GSi} = V_{TPi} - \sqrt{\frac{I_i}{K_{Pi}}} \tag{5.95}$$

Adding Eqs. (5.93) and (5.94) we get

$$V_{CQ3} = \frac{V_{GS1} + V_{GS2}}{2} + \frac{V_{BE1} + V_{BE2}}{2} \tag{5.96}$$

where

$$V_{GS1} = V_{GS3} = V_{TP} - \sqrt{(I_{B1} + I_{R1})/K_P}$$

$$V_{GS2} = V_{GS4} = V_{TP} - \sqrt{(I_{B1} - I_{R1})/K_P}$$

$$V_{BE3} = (V_{BE1} + V_{BE2})/2 \tag{5.97}$$

Substituting back into Eq. (5.96) we get

$$V_{CQ3} = V_{BE3} + \sqrt{\frac{I_{B1} + I_{R1}}{4K_{P_i}}} + \sqrt{\frac{I_{B1} - I_{R1}}{4K_{P_i}}} - V_{TP_i} \qquad (5.98)$$

where $V_{TP_i}$ is the threshold voltage of the ith PMOS transistor, $V_{CQ3}$ is the voltage at the collector terminal of Q3, and $I_{R1}$ is the current flowing in resistors R1. Assuming $K_{P_i} = K_P$, $V_{TP_i} = V_{TP}$, $V_{BE3} \approx \frac{V_{BE1}+V_{BE2}}{2} = V_{BE}$, $\Delta V_{BE} = V_{BE1} - V_{BE2} = V_t ln(\frac{I_1}{I_2})$ and $I_{B1} \gg I_{R1}$, the expressions for $I_1 - I_2$ and $I_1 + I_2$ are

$$i_1 - i_2 = -(2V_{in} + \Delta V_{BE}) \left[ \sqrt{(I_{B1} + I_{R1})K_P} + \sqrt{(I_{B1} - I_{R1})K_P} \right]$$

$$\approx -4\sqrt{I_{B1}K_P}V_{in} - 2\sqrt{I_{B1}K_P}\Delta V_{BE} \qquad (5.99)$$

$$i_1 + i_2 \approx K_P \left[ 2V_{in}^2 + 2\left(V_{BE} + \sqrt{\frac{I_{B1}}{K_P}}\right)^2 - 2V_{BE3}V_{BE4} \right.$$

$$\left. +2\Delta V_{BE}V_{in} - 4V_{BE}\sqrt{\frac{I_{B1}}{K_P}} \right] = \qquad (5.100)$$

$$K_P \left[ 2V_{in}^2 + 2V_{BE}^2 + \frac{I_{B1}}{K_P} - 2V_{BE3}V_{BE4} + 2\Delta V_{BE}V_{in} \right]$$

We note from Eq. (5.99) that the gain can be varied by the current source $I_{B1}$. Also, the voltage VB1 controls the current in M5-M7 and has the same effect on the gain of the transconductor as $I_{B1}$ does. The second term in the right side of Eq. (5.99) includes nonlinear components. This nonlinear term becomes insignificant relative to the linear term when the transconductance ratio of M1 (M2) to Q1 (Q2) is small. The smaller this ratio is the smaller $\Delta V_{BE}$ (with respect to the differential input voltage) becomes. Of course, a good design of this VIC should consider the tradeoffs between linearity, speed, and signal-to-noise ratio.

The VIC described above has been processed in a standard 2 $\mu$m BiCMOS process. Its design used the device sizes shown in Fig. 5.20(a) for the PMOS devices, a 2X emitter area for the bipolar transistors, and 5 K$\Omega$ poly resistors. Typical measured performance of this VIC is shown in Fig. 5.21. The curves in Fig. 5.21(a) are the large-signal dc-transfer characteristics, where the differential input voltage was varied from -1 V to 1 V and the differential output current was passed through two equal 1 K$\Omega$ resistors which were inserted between each collector terminal of transistors, Q1 and Q2, and the positive supply. The measurements were for $V_{DD} = 4$ V, $V_{SS} = -1$ V, $I_{B1} = 215$ $\mu$A, $V_{B1}$ was varied such that the currents in M5-M7 changed from 130 $\mu$ A to 190 $\mu$ A. The curves in Fig. 5.21(b) show $I_1 + I_2$, from which the square-law behavior is clearly evident.

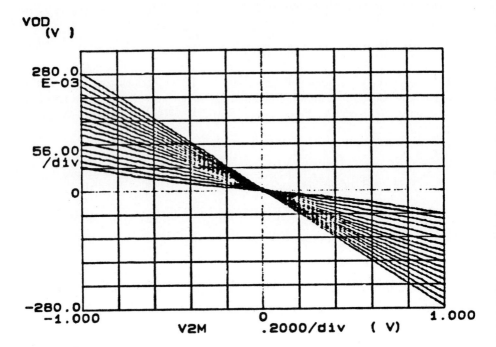

**FIGURE 5.21**
(a) Typical measured dc-transfer curves $(I_1 - I_2)$ of the BiCMOS VIC in Fig. 5.20.

Fig. 5.21(c) provides the nonlinearity measurements, where the differential in-put voltage was varied from -1.4 V to 1.4 V and the differential output and its difference from the ideal curve were plotted. One observes that the nonlinearity is about 1% or less over a 2.8 V input voltage range with a 5 V supply voltage. The left vertical axis in Fig. 5.21(c) pertains to the dc-transfer curve (straight line) and the vertical axis to the right pertains to another curve, which is the difference between this dc-transfer curve and its ideal curve. From Fig. 5.21(c), the nonlinearity was calculated to be the ratio of the peak error reading (from the right vertical axis) to the peak output voltage reading (from the left vertical axis).

The small-signal analysis of the VIC in Fig. 5.20(a) reveals that the gain-bandwidth is approximately equal to the ratio of the transconductance of the transistor M1(M2) to the load capacitance at the collector of Q1(Q2). The de-tailed result of the small-signal analysis is provided as follows.

$$A_{dc} = \frac{V_{od}}{V_{ind}} = \frac{(g_{m1} + g_{mb1})g_{m2}r_{o2}R_L}{(g_{m1} + g_{mb1} + g_{m2})r_{o2} + R_L(g_{m1} + g_{mb1} + \frac{1}{R_1})} \tag{5.101}$$

$$Z = \frac{g_{m1}}{C_{gs1}} \tag{5.102}$$

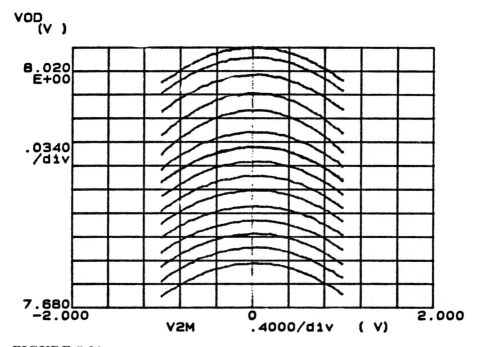

**FIGURE 5.21**
(b) $(I_1 + I_2)$ of the BiCMOS VIC in Fig. 5.20.

$$P_1 \approx -\frac{g_{m1} + g_{mb1} + g_{m2}}{C_1 + C_{gs1}} \qquad (5.103)$$

$$P_2 \approx -[\frac{1}{R_L C_L} + \frac{g_{m1} + g_{mb1} + \frac{1}{R_1}}{(g_{m1} + g_{mb1} + g_{m2})r_{o2}C_L}] \qquad (5.104)$$

where $V_{od} = (I_1 - I_2)R_L$, $V_{ind} = V_{in}^+ - V_{in}^-$, $R_L$ is the load resistance, $R_1 = r_{o1}\|r_1$, $C_1 = C_{gs1} + C_{cd1} + C_2$, $C_L = C_{load} + C_2 + C_{cs2}$, $g_{m1} + g_{mb1}$ and $g_{m2}$ are, respectively, the transconductance of M1(M2) and Q1(Q2) and $r_{o2}$ is the output resistance of Q1 (Q2). Assigning conservative parameters for the devices, $g_{m1} + g_{mb1} = 1.5$ m℧, $g_{m2} = 15$ m℧, $r_{o2} = 220$ KΩ, $R_L = 1$ MΩ and $R_1 = 6.5$ KΩ, the gain bandwidth product is computed as,

$$GBW = A_{dc}P_2 = \frac{1.364 \times 10^{-3}}{C_L} \approx \frac{g_{m1}}{C_L} \qquad (5.105)$$

The non-dominant pole, $P_1$, can be much larger than the unity gain bandwidth, GBW. Equation (5.105) shows that the VIC provides as high a unity gain bandwidth as the input PMOS transistor M1(M2). Therefore, the speed can be controlled by the geometry and the drain current of the input transistors.

The top and bottom waveforms in Fig. 5.22(a) are the measurements of each output terminal current across two equal 50 W resistors. The differential

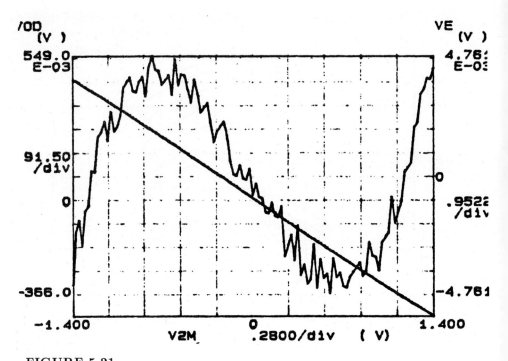

**FIGURE 5.21**
(c) Typical low frequency THD measurement of the BiCMOS VIC in Fig. 5.20.

input signal is 1Vpeak with a frequency of 90 MHz. The middle waveform is the differential output. Fig. 5.22(b) shows the spectrum of the differential output signal due to a 1 MHz, 1 Vpeak, differential input signal. From this spectrum the THD measures 0.63%.

The BiCMOS transconductor stage in Fig. 5.20(b) [22] is a common-source, common-base structure, which is suitable for high-speed analog signal processing. The input devices M1 and M2 are biased in the linear region of operation and determine the transconductance of the stage. The voltage drop across $R_D$ determines the drain-to-source voltage of M1 and M2 in quiescent conditions. The relationship between the input differential voltage and the output differential current is obtained as follows:

$$I_{di} = \frac{\mu_{neff}C_{ox}W}{L}[(V_{GS} - V_T)V_{DS} - \frac{V_{DS}^2}{2}], \tag{5.106}$$

$$2V_{in} = V_{GS1} - V_{GS2} \tag{5.107}$$

$$\frac{I_{d1}}{V_{DS1}} - \frac{I_{d2}}{V_{DS2}} = \frac{\mu_{neff}C_{ox}W}{L}(2V_{in} + \frac{V_{DS2}}{2} - \frac{V_{DS1}}{2}) \tag{5.108}$$

In Eq. (5.108), we have assumed equal device sizes for M1 and M2. As a differential input signal, $V_{in}$, is applied to the circuit, the drain-to-source voltages of M1 and M2 become unequal and nonlinear components are generated in the differential output current. The unity gain bandwidth of this transconductor is

$$GBW \approx \frac{g_m}{C_L} = \frac{\mu_{neff} C_{ox} W V_{DS}}{L \times C_L} \qquad (5.109)$$

where $g_m$ is the transconductance of M1 and M2, and $C_L$ is the load capacitance at the collector terminals of Q3 and Q4. Good linearity may be achieved when the input transistors have a small $V_{DS}$, which reduces GBW. From Eq. (5.108), the transconductance may be varied by changing $V_{DS}$ of the input devices. This is achieved by varying $I_{B2}$. The linearity of this transconductor improves with an increase in $I_{B1}$, which also determines the power consumption.

Another BiCMOS/CMOS transconductor stage suitable for medium speed applications is shown in Fig. 5.24 [19, 31]. This circuit is composed of two identical subcircuits. Transistors M1, M2, Q1, Q2, M5, and M7 make up one subcircuit and transistors M3, M4, Q3, Q4, M6, and M9 make up the other one. There are two feedback loops (Q1, M1, M5, Q2 and Q4, M4, M6, Q3) whose loop gain and bandwidth, respectively, determine the accuracy and the speed of this transconductor.

In order to determine the operation of this circuit we will take one step back and look at a pair of matched NMOS transistors biased as shown in Fig. 5.23. The differential output current will be given as

$$I_1 - I_2 = \frac{K}{2}(V_1 - V_{S1} - V_T)^2 - \frac{K}{2}(V_2 - V_{S2} - V_T)^2 \qquad (5.110)$$

$$= \frac{K}{2}(V_1 - V_2 - V_{S1} + V_{S2})(V_1 + V_2 - V_{S1} - V_{S2} - 2V_T)$$

In order to obtain the desired linear relationship between the differential input voltage and the differential output current, one might bias the sources such that they satisfy the following restriction

$$V_{S1} = V_2 - V_C - V_T$$
$$V_{S2} = V_1 - V_C - V_T \qquad (5.111)$$

where $V_C$ is a constant voltage. This can be accomplished using the circuit in Fig. 5.24. $V_C$ is simply $V_{GS} - V_T$ of transistors M1 and M4, which are biased with a constant current source. Thus the source voltages will simply follow the input voltages, and the difference is kept constant equal to $\sqrt{2I_C/K}$. Equations (5.112) and (5.113) give the difference and the sum of the two output currents.

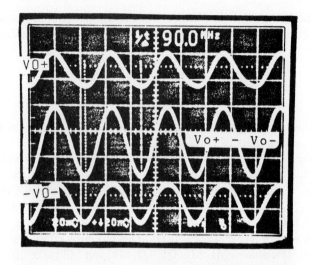

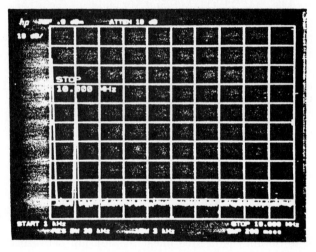

**FIGURE 5.22**
Typical performance of the BiCMOS VIC in Fig. 5.20(a), (a) input and output waveforms at f=90MHz and $V_{in} = 1.0V_{peak}$ (b) high frequency THD measurements.

$$I_1 - I_2 = 2KV_C(V_1 - V2)$$

$$= \sqrt{\frac{8\mu_{neff}C_{ox}WI_C}{L}}(V_1 - V_2) \qquad (5.112)$$

$$I_1 + I_2 = \frac{\mu_{neff}C_{ox}W}{2L}(V_1 - V_2)^2 + 2I_C \qquad (5.113)$$

We note that the transconductance is a function of the square root of the current $I_C$, which might be used for tuning. We note from Eq. (5.112) that the

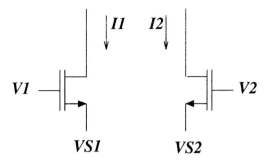

**FIGURE 5.23**
Two matched transistors operating in the saturation region.

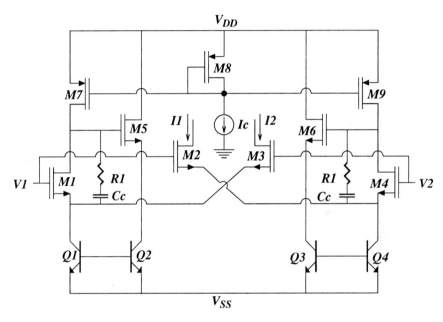

**FIGURE 5.24**
A medium speed transconductor gain stage.

difference between the two input signals, $V_1$ and $V_2$, is converted to an output current. Therefore, the linear operation of this transconductor does not necessarily depend on balanced differential input signals. The resistors and the capacitors in this VIC are used to increase the loops' phase margin. Speed measurements of this VIC from fabricated parts indicated that operation up to 50 MHz is achievable with a reasonable value of THD.

A typical large-signal dc-transfer characteristic of the transconductor stage is shown in Fig. 5.25. These curves are based on a $\pm 4$ V supply voltage, $\frac{W}{L} = \frac{10}{2}$ for M1-M6, $\frac{W}{L} = \frac{50}{2}$ for M7-M9, and the current IC was varied from $30\mu A$ to $300\mu A$.

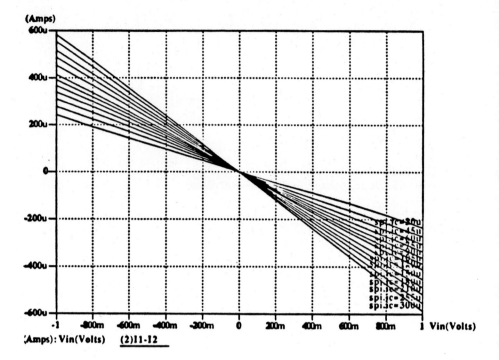

**(Amps)**

(Amps): Vin(Volts)    (2)I1-I2

## FIGURE 5.25
DC transfer characteristics of the transconductor in Fig.5.24.

### 5.3.5  Analog Signal Processing Circuits

Analog signal processing circuits have a wide range of applications in instrumentation, function generators, control systems, and neural networks. Computational circuits are designed to perform multiplication, squaring, square-rooting, inverting, inverse of squares, sum/difference of squares, square root of sum/difference of squares, log and anti-log of a variable, and special functions such as sine or cosine of a variable. Here, we present a few BiCMOS circuits that perform some of these functions. The circuit in Fig. 5.26(a) performs current squaring and inverting and then squaring of an input voltage [19]. The inputs of this circuit are the current, $i_x$, and the voltage, $V_1$. Its operation follows the same feedback principle described for the VIC in Fig. 5.24.

The output current expression is found by writing two KVL equations, one for the loop containing M1 and M3, and the other for the loop containing M2 and M3, which give

$$I_{out} = \frac{[i_x + V_c\sqrt{\frac{K(i_r+i_x)}{2}} - V_c\sqrt{\frac{K(i_r-i_x)}{2}}]^2}{2KV_1^2} = \frac{i_r^2}{2KV_1^2}\bigg|_{V_c=0} \qquad (5.114)$$

where $K = \frac{n_{eff}C_{ox}W}{2L}$ and $i_r$, is a static current. The differential input currents

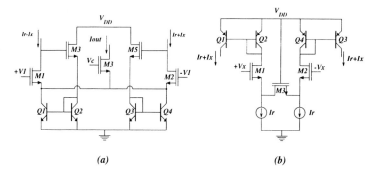

**FIGURE 5.26**
(a) Current squaring/square of inverse voltage. (b) Differential current generator circuit.

$(i_r - i_x)$ and $(i_r + i_x)$ may be produced by the differential amplifier in Fig. 5.26(b). In this circuit, the differential input voltage, $V_x$, is transferred across M3, which operates in the linear mode. Note that the ac-current, $i_x$, has a nonzero value only when a differential input signal is applied to the gate terminals of M1 and M2 of Fig. 5.26(b). In Fig. 5.26(b), the ratio of the differential input ac-signal to the on-resistance of the M3 constitutes the value of $i_x$.

Squaring of a variable may be accomplished using the circuit of Fig. 5.24 [19]. In Eq. (5.113) we found the square of the difference of two signals. Any values within allowable ranges (we must assure that all devices operate in the saturation region) can be chosen for the two variables $V_1$ and $V_2$.

Taking the square root of a nonzero variable is also possible with the circuit of Fig. 5.24. The variable is the current $I_C$ in Eq. (5.112), and the gain of the circuit may now be controlled by the difference between the two input voltages.

The circuit in Fig. 5.27 [19, 31] is a four-quadrant multiplier. It comprises three feedback networks connected in parallel and uses a circuit similar to that discussed in Prob. 5.16 as a basic building block. Input signals $V_1$, $V_2$, $V_3$, and $V_4$ are applied to the gate terminals of M1-M9 to perform analog multiplication on the differential voltages $V_1 - V_2$ and $V_3 - V_4$. The large-signal dc-transfer function may be obtained by writing the current equations for transistors M2-M4 and M6-M8. This can easily be done by looking at a simple loop where one of the transistors are biased with the fixed current Ic. In order to determine $I_2$ we take the KVL loop containing M2 and M9. By inspection we get $V4 - V_{GS(M2)} = V3 - V_{GS(M9)}$, and substituting in for $V_{GS(M2)}$ and $V_{GS(M9)}$ and solving for $I_2$ we get Eq. (5.115). Similarily all the required drain currents can be determined.

$$I_2 = \frac{\mu_{neff}C_{ox}W(V_4 - V_3)^2}{2L} + I_c + \sqrt{\frac{2I_c\mu_{neff}C_{ox}W}{L}}(V_4 - V_3) \qquad (5.115)$$

$$I_3 = \frac{\mu_{neff}C_{ox}W(V_3 - V_2)^2}{2L} + I_c + \sqrt{\frac{2I_c\mu_{neff}C_{ox}W}{L}}(V_3 - V_2) \qquad (5.116)$$

$$I_4 = \frac{\mu_{neff}C_{ox}W(V_1 - V_4)^2}{2L} + I_c + \sqrt{\frac{2I_c\mu_{neff}C_{ox}W}{L}}(V_1 - V_4) \qquad (5.117)$$

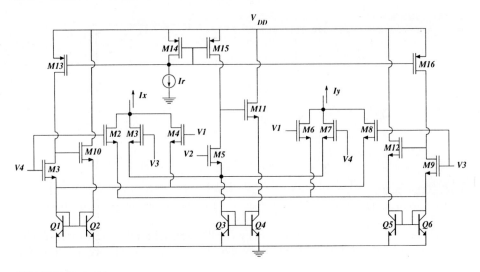

**FIGURE 5.27**
Four-quadrant multiplier based on the VIC cell in Fig.5.24.

$$I_6 = \frac{\mu_{neff}C_{ox}W(V_1 - V_3)^2}{2L} + I_c + \sqrt{\frac{2I_c\mu_{neff}C_{ox}W}{L}}(V_1 - V_3) \qquad (5.118)$$

$$I_7 = \frac{\mu_{neff}C_{ox}W(V_3 - V_4)^2}{2L} + I_c + \sqrt{\frac{2I_c\mu_{neff}C_{ox}W}{L}}(V_3 - V_4) \qquad (5.119)$$

$$I_8 = \frac{\mu_{neff}C_{ox}W(V_4 - V_2)^2}{2L} + I_c + \sqrt{\frac{2I_c\mu_{neff}C_{ox}W}{L}}(V_4 - V_2) \qquad (5.120)$$

The difference of the output currents $I_r$ and $I_y$ results in the desired four-quadrant multiplication function; that is

$$I_r - I_y = I_2 + I_3 + I_4 - I_6 - I_7 - I_8$$
$$= \frac{\mu_{neff}C_{ox}W(V_1 - V_2)(V_3 - V_4)}{L} \qquad (5.121)$$

Note from Eq. (5.121) that the nonlinear terms in $I_2$ through $I_8$ have been eliminated without requiring differential balanced inputs. Due to the parallel connection of input devices, the multiplier has a similar speed characteristic to that of the basic transconductor discussed in Prob. 5.16. It can be shown that the linear operation of the multiplier is directly proportional to the $\frac{W}{L}$ ratio of M10-M16 and is inversely proportional to the $\frac{W}{L}$ ratio of M1-M9. It is evident from Eqs. (5.115)-(5.120) that any mismatch among input transistors results in a second-order harmonic distortion. Typical dc-transfer curves of this multiplier are shown in Fig. 5.27. In these curves, the difference voltages $V_1 - V_2$ (62.5 mV increment) and $V_3 - V_4$ (horizontal axis) have been varied from -1 V to 1 V,

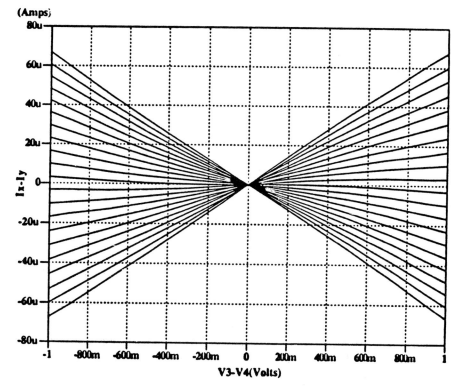

**(Amps)**

**FIGURE 5.28**
DC-transfer curves of the four-quadrant multiplier in Fig. 5.27.

while $\frac{W}{L}$ is $\frac{10}{5}$ for M1-M9, $\frac{10}{10}$ for M10-M12, $\frac{50}{2}$ for M13-M16, $I_C = 30$ $\mu$A, and the supply voltage is $\pm 4$ V.

The BiCMOS voltage-to-current converter in Fig. 5.20(a) can also be used to realize a combined four-quadrant multiplier/sum of squares functions. This is accomplished by connecting three of these VIC cells as shown in Fig. 5.29 [19], where transistors M1-M7 and Q1-Q5 represent the middle VIC cell. The input differential voltage to the middle cell is denoted by $V_1$ and the control signals are $V_{B1}$ and $I_{B1}$. The other two cells reside to the left and right of the middle VIC cell which provides their control signals. By writing similar loop equations as Eqs. (5.93)-(5.94), neglecting the effect of $\Delta V_{BE}$ terms and assuming equal device sizes for the input PMOS transistors ($K = \frac{\mu_{eff}C_{ox}W}{2L}$), we can obtain the following current equations for $I_1$, $I_2$, $I_a$, $I_b$, $I_c$, and $I_d$.

$$I_1 \approx KV_1^2 + I_{B1} + 2V_1\sqrt{I_{B1}K} \qquad (5.122)$$

$$I_2 \approx KV_1^2 + I_{B1} - 2V_1\sqrt{I_{B1}K} \qquad (5.123)$$

$$I_a \approx KV_2^2 + I_1 + 2V_2\sqrt{I_1K} \qquad (5.124)$$

$$I_b \approx KV_2^2 + I_1 - 2V_2\sqrt{I_1K} \qquad (5.125)$$

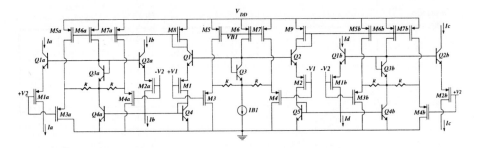

## FIGURE 5.29
Four-quadrant multiplier and sum of squares circuit.

$$I_c \approx KV_2^2 + I_2 + 2V_2\sqrt{I_2 K} \qquad (5.126)$$

$$I_d \approx KV_2^2 + I_2 - 2V_2\sqrt{I_2 K} \qquad (5.127)$$

Appropriate combinations of these current components will result in a four-quadrant multiplication and a sum of squares function as follows:

$$I_x - I_y = I_a + I_d - I_b - I_c = 8KV_1 V_2 \qquad (5.128)$$

$$I_x + I_y - 4I_{B1} = 4K(V_1^2 + V_2^2) \qquad (5.129)$$

Note that this circuit provides two copies of each current $I_a$, $I_b$, $I_c$, and $I_d$. If we assume that the bipolar transistors have high $\beta$, then Eqs. (5.128) and (5.129) can be realized with this circuit.

Finally we will introduce a BiCMOS implementation of a wide range four-quadrant multiplier which can operate successfully up to 100 MHz. The multiplier circuit is shown in Fig. 5.30 [28, 29]. All internal nodes within this multiplier have low impedance resulting in high operating speed. Unlike bipolar multipliers based on the Gilbert cell [32], the circuit does not employ feedback to extend the input voltage range at the expense of speed. Rather, it uses the four-transistor MOS circuit, M1 - M4, which was discussed in Chapter 3 to cancel MOS non-linearities. The four-transistor circuit operates linearly when all transistors are in saturation or when the cross-coupled pair operates in saturation while the other pair operates in triode, or visa versa. This extends input signal handling and allows operation from a low voltage supply. Furthermore, the cross-coupling of $M_1 - M_4$ reduces the effect of their Miller capacitances. Virtually, the maximum speed, $f_{MAX}$ is determined by the drain load capacitance, $C_L$ of $M_1 - M_4$ and their transconductance, $g_m$, that is $f_{MAX} = \frac{g_m}{2C_L}$. The circuit combines the linearity and high input impedance of MOS with the speed of Bipolar.

The circuit has two differential input voltages $(V_1 - V_2)$ and $(V_3 - V_4)$. The level shifters consisting of $M_5$ - $M_6$ and $Q_1$ - $Q_4$ provide a very low impedance at the sources of $M_1$ - $M_4$. With all MOS/bipolar devices operating in the saturation/active region the currents in $M_1$ - $M_4$ are given by:

$$I_1 = \mu_{neff}C_{ox}(\frac{W}{2L})_1(V_1 - V_{C1} - V_{T1})^2(1 + \lambda V_{ds1})$$

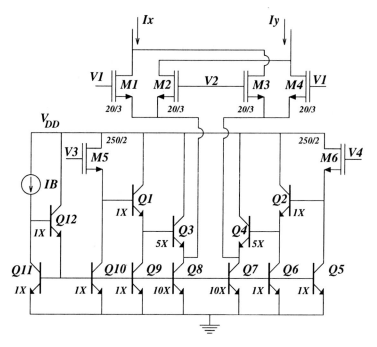

**FIGURE 5.30**
A wide range multiplier/mixer circuit.

$$I_2 = \mu_{neff}C_{ox}(\frac{W}{2L})_2(V_2 - V_{C1} - V_{T2})^2(1 + \lambda V_{ds2})$$

$$I_3 = \mu_{neff}C_{ox}(\frac{W}{2L})_3(V_2 - V_{C2} - V_{T3})^2(1 + \lambda V_{ds3})$$

$$I_4 = \mu_{neff}C_{ox}(\frac{W}{2L})_4(V_1 - V_{C2} - V_{T4})^2(1 + \lambda V_{ds4}) \qquad (5.130)$$

Also

$$V_{C1} = V_3 - V_{GS5} - V_{BE1} - V_{BE3}$$

$$V_{C2} = V_4 - V_{GS6} - V_{BE2} - V_{BE4} \qquad (5.131)$$

We assume that the emitter areas of $Q_7$ and $Q_8$ are scaled such that the static currents flowing in $Q_3$ and $Q_4$ are much larger than the currents in $M_1$-$M_4$. Thus $I_{EQ_3} = I_{EQ_4}$ and the base-emitter voltages of $Q_1$-$Q_4$ remain almost constant with $V_{BE_1} \approx V_{BE_2}$ and $V_{BE_3} \approx V_{BE_4}$. If we neglect the base currents in $Q_1$ and $Q_2$ (high $\beta$s) then $I_{d_5} = I_{d_6}$ and since $M_5$-$M_6$ are matched then $V_{GS_5} = V_{GS_6}$. As a result, from Eq. (5.131), we have $V_{C1} - V_{C2} = V_3 - V_4$ and from Eq. (5.130) assuming equal $V_T$ and $V_{ds}$ and matched $M_1$-$M_4$, we get

$$I_x - I_y = -\mu_{neff}C_{ox}\frac{W}{L}(V_1 - V_2)(V_3 - V_4)(1 + \lambda V_{ds}) \qquad (5.132)$$

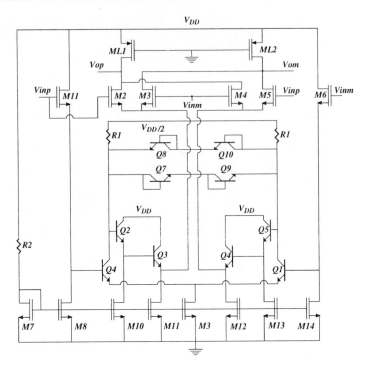

**FIGURE 5.31**
Precision BiCMOS absolute value circuit.

which indicates a threshold-independent linear multiplication of the two differential inputs.

The main sources of nonlinear errors are from device mismatches, unequal $V_{ds}$ for M1-M4 (or nonzero channel length modulation), finite emitter impedance of Q3-Q4 relative to the source impedance of M1-M4, and the MOS transistor mobility reduction. These errors can be minimized by choosing large gate length, small "on" voltage ($V_{GS} - V_T$), and good layout.

This circuit was implemented in the 2 $\mu$m BiCMOS process described in Table 5.2. A nonlinearity of 0.63% was measured for $V_1 = -V_2 = 2\,V_{p-p}$ and $V_3 = -V_4 = 2\,V_{p-p}$. The maximum operating frequency was about 100 MHz, which makes it suitable for medium frequency communication circuits.

Another widely used analog building block in most communication and signal processing systems is the absolute value circuit, or the full-wave rectifier. Most precision absolute circuits are op amp-based and have very narrow bandwidth. Here, we present a precision transistor-based BiCMOS absolute value circuit which features wide bandwidth. This circuit is shown in Fig. 5.31 [19]. Its differential input and output terminals are denoted by $V_{inp} - V_{inm}$ and $V_{op} - V_{om}$, respectively.

This circuit can be divided into four subcircuits. The first subcircuit is composed of M2-M5, whose source terminals are driven by the low impedance voltage sources provided by the emitter terminals of Q2 and Q4. Transistors M1-M6, Q1-Q10, and the two equal resistors, $R_1$, make up the second subcircuit, which performs high speed limiting on the differential input signal and provides a low output impedance. Transistors M7-M14 and the resistor $R_2$, which make up the third subcircuit, generate the necessary dc-currents for the circuit. The last subcircuit is a current-to-voltage converter which is composed of ML1 and ML2 operating in the linear region of operation. An expression for the output voltage is

$$V_{op} - V_{om} = \frac{\mu_n W_N(\frac{V_{dd}}{2} - 4V_{BE} + V_{TN})sgn(V_{in})}{\mu_p W_P(V_{dd} + V_{TP})} V_{in} \qquad (5.133)$$

where $W_N$ is the width of M2-M5, $W_P$ is the width of ML1 and ML2, $V_{BE}$ is the base-emitter voltage of the bipolar transistors, $V_{TN}$ is the threshold voltage of M2-M5, $V_{TP}$ is the threshold voltage of ML1-ML2, and $sgn(V_{in})$ is the sign function which takes on a value +1 when $V_{in}$ is above the mid-supply voltage and -1 when $V_{in}$ is below the mid-supply voltage. Typical performance of this circuit for a supply voltage of $\pm4$ V and resistor temperature coefficient of -1500 ppm is shown in Fig. 5.32. The plot in Fig. 5.32(a) is the sine-wave response of this circuit, and has unity dc-gain. The curves in Fig. 5.32(b) represent dc transfer curves at temperatures of $-10^\circ$C, $27^\circ$C, $65^\circ$C, and $100^\circ$C. The primary applications of this circuit include amplitude/frequency demodulation (for frequency demodulation, it must be preceded with an integrator or differentiator), high speed AGC, and signal detection.

### 5.3.6  Translinear Circuits

BiCMOS processes offer NPN and possibly PNP transistors. Translinear circuits [23, 24] are the type of circuits which use the exponential current-voltage characteristic of bipolar transistors to realize controlled nonlinear functions. A translinear circuit has all its inputs and outputs in the form of currents. Its unique operation arises from the exploitation of the logarithmic behavior of forward-biased PN junctions. It turns out that in any closed-loop of forward-biased PN junctions, in which for each clockwise facing junction there is another counterclockwise facing junction, it is possible to synthesize numerous translinear circuits which realize a wide variety of linear and nonlinear functions.

The circuits in Fig. 5.33 represent a translinear circuit where Q1 and Q3 face one direction while Q2 and Q4 face the opposite direction. We can write the following loop equation for this circuit:

$$V_{BE1} - V_{BE2} + V_{BE3} - V_{BE4} = 0 \qquad (5.134)$$

If we assume that base currents are negligible, transistors Q1-Q4 have equal geometry and are at the same temperature, then by applying Eqs. (5.3) in (5.134), we find

$$x \times I_2 = y \times I_1 \qquad (5.135)$$

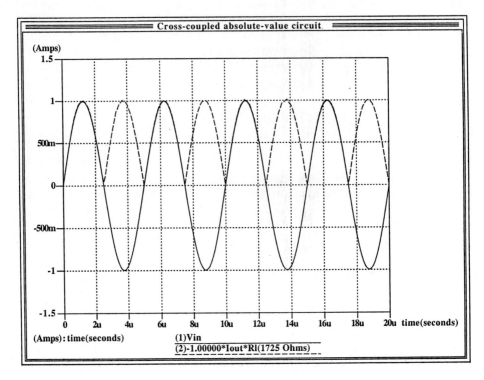

**FIGURE 5.32**
(a) The sine-wave response of the absolute value circuit in Fig. 5.31.

This is known as the translinear principle [23, 24], which states that the product of the collector currents of transistors having clockwise facing junctions is equal to the product of the collector currents of transistors having counterclockwise facing junctions. Using Eqs. (5.135) and $z = I_1 + I_2$ the expression for the difference of Q2 and Q3 collector currents is found to be

$$I_1 - I_2 = \frac{z(x - y)}{x + y} \qquad (5.136)$$

If one assumes that x, y, and z are temperature-independent, then this circuit is (to the first order) insensitive to temperature. Remember that we assumed Q1-Q4 are at the same temperature. The temperature insensitivity is inherent in all translinear bipolar circuits. This assumes that the transistors comprising the circuit are all at the same temperature. Of course, this neglects the fact that the input currents x, y, and z in Fig. 5.33 may have nonzero and/or different temperature coefficients.

This circuit may be used to realize useful functions. For instance, let $x = I + I_{in}$, $y = I - I_{in}$ and $Z = G \times I$, then $I_1 - I_2 = G \times I_{in}$, and the circuit functions

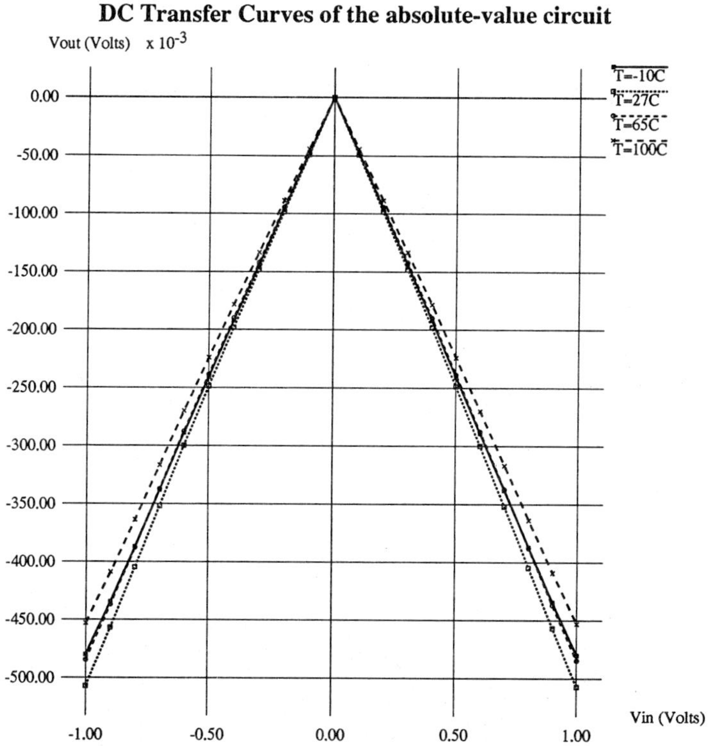

**DC Transfer Curves of the absolute-value circuit**

Vout (Volts) x 10$^{-3}$

FIGURE 5.32
(b) The dc-transfer curve at various temperatures of the absolute value circuit in Fig. 5.31.

as a variable gain cell (x modulates the amplitude of the input current $I_{in}$). For proper operation, $|I_{in}| < I$ and $x > 0$ must be assured. When $x = I_{in2}$ ($I_{in2} \geq 0$), then the circuit operates as a two-quadrant multiplier.

An efficient translinear four-quadrant multiplier is shown in Fig. 5.34. Assuming equal device sizes and neglecting the base currents, the expression for the difference of the output currents, $I_1 - I_2$, is

$$I_1 - I_2 = (x - y)(1 - \frac{z}{x + y}) \tag{5.137}$$

If $x = I + I_r$, $y = I - I_r$, and $z = 2I + I_z$, then $I_1 - I_2 = (I_r I_z)/I$. The limits for the inputs are: $|I_r| < I$ and $|I_z| < 2I$, which are the conditions for the circuit to properly function as a four-quadrant multiplier.

The input currents x, y, and z can be produced by using the voltage-to-current converter in Fig. 5.26(b). Although this current generator circuit should perform adequately for most applications, better linearity is obtained when M3 is replaced by a linear resistor.

Another simple translinear building block which performs multiplication/

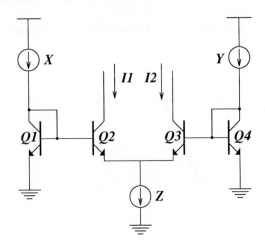

**FIGURE 5.33**
A translinear cell.

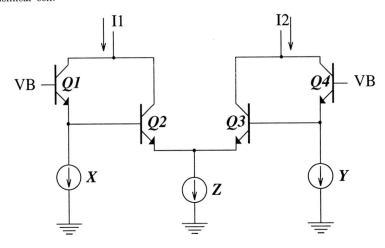

**FIGURE 5.34**
A translinear four-quadrant multiplier.

division in the current domain is shown in Fig. 5.35(a). The currents $I_x$, $I_y$, and $I_z$ are the inputs, and the current $I_o$ is the output. The input currents $I_y$ and $I_z$ can be generated by the voltage-to-current converter in Fig. 5.35(b), where the output current $I_{out} = \frac{V_{in}}{R_1}$. The current $I_x$ is generated by a complemented version of this VIC circuit.

By writing the following loop equation (5.137), an expression for $I_o$ is found,

$$V_{BE1} + V_{BE2} - V_{BE3} - V_{BE4} = 0 \tag{5.138}$$

$$I_o = \frac{I_x I_y}{I_z} \tag{5.139}$$

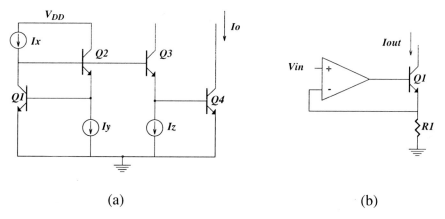

(a)                                           (b)

**FIGURE 5.35**
(a) Squaring/inverting/geometric mean and (b) VIC circuits.

If $I_x = I_y$ and $I_z = I$ (a constant current source), then the circuit performs a squaring function. Choosing constant currents for $I_x$ and $I_y$, the circuit provides the inverse of $I_z$. The collector current of Q3 is the geometric mean of $I_x$ and $I_y$, $I_{cQ3} = \sqrt{I_x I_y}$, when $I_z$ is set to zero and the collector terminal of Q4 is connected to its base.

The circuits in Fig. 5.36 function as current selectors. Consider the circuit in Fig. 5.36(a). Using the translinear principle, one finds that $I_{out} = I_x$. The current in Q5 is $I_{out} - I_y$ and is always greater than or equal to zero. When $I_y > I_x$, then $I_{out} = I_y$ (according to the translinear principle). But when $I_x > I_y$, no current flows in Q5, and $I_{out}$ is simply $I_x$. Therefore, the circuit selects the maximum of $I_x$ and $I_y$, or $I_{out} = max(I_x, I_y)$.

Now consider the circuit in Fig. 5.36(b). The current $I_{out}$ is the minimum of $I_x$ and $I_y$. Note that the collector currents in Q2 and Q4 are equal. When $I_x > I_y$, then Q2 is off and $I_{out} = I_y$. On the other hand, when $I_y > I_x$, the base voltage of Q3 is forced (by the currents in Q2 and Q4) to adjust such that a current equal to $I_x$ flows in Q3. Therefore, the circuit selects the minimum of $I_x$ and $I_y$, or $I_{out} = min(I_x, I_y)$.

To further illustrate the translinear technique, consider the circuit in Fig. 5.37(a). Applying the translinear principle to this circuit and assuming matched transistors, the expressions for $I_1$, $I_2$, their sum and difference are,

$$I_1 = \frac{I_x^2}{\sqrt{I_x^2 + I_y^2}} \qquad\qquad I_2 = \frac{I_y^2}{\sqrt{I_x^2 + I_y^2}}$$

$$I_1 + I_2 = \sqrt{I_x^2 + I_y^2} \qquad\qquad (5.140)$$

$$I_1 - I_2 = \frac{I_x^2 - I_y^2}{\sqrt{I_x^2 + I_y^2}} \qquad\qquad (5.141)$$

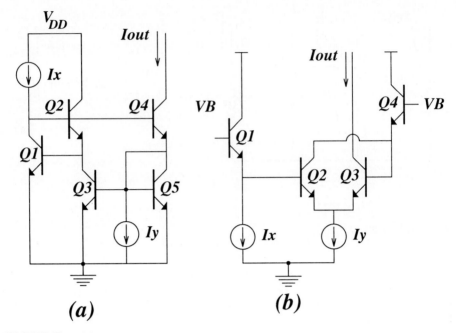

**FIGURE 5.36**
(a) Maximum selector $I_{out} = max(I_x, I_y)$ and (b) minimum selector $I_{out} = min(I_x, I_y)$
circuits.

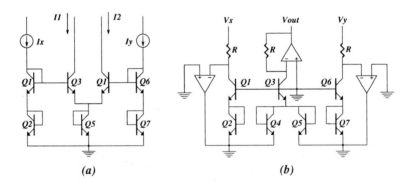

**FIGURE 5.37**
(a) Vector manipulation and (b) practical vector sum circuits.

Instrumentation and measurement systems use the vector-sum of Eq. (5.140)
One use of the vector-difference of Eq. (5.141) is to generate a double fre-
quency of an absolute-value sinusoidal signal. For example, if $I_x = |A \cos wt|$
and $I_y = |A \sin wt|$, then $I_1 - I_2 = A \cos 2wt$. The circuit in Fig. 5.37(b) rep-
resents a practical implementation of the vector-sum circuit. The expression for
the output voltage, $V_{out}$, in terms of the input voltages, $V_x$ and $V_y$, is similar to

Eq. (5.140) and is given below.

$$V_{out} = \sqrt{V_x^2 + V_y^2} \qquad (5.142)$$

The main shortcoming of this circuit is that the inputs have to be unipolar.

The above circuit examples are just some of the many functions realizable with the translinear technique. Other possible functions are: sign, modulus, root-mean square, harmonic mean, bridge linearizer and sine, cosine, and tangent/inverse of tangent approximations [19, 20].

## 5.4 CLOSING REMARKS

The importance of having both bipolar and MOS devices in a process can only be appreciated when designing high volume production ICs for high performance applications with relatively high levels of complexity. There are numerous practical issues associated with this design process including: design complexity, performance, testing, reliability, and cost. The addition of bipolar transistors to a CMOS process usually implies that the high CMOS performance must remain intact. The complexity of CMOS-based BiCMOS processes varies with the desired bipolar device characteristic. A bipolar-based BiCMOS process usually includes optimized bipolar devices. The desired packing density in both of these BiCMOS processes also dictates their complexity and cost. BiCMOS devices provide the best of both worlds. Judicious use of these devices in the design of high performance circuits at high speed can improve the cost by simplifying the circuit complexity, reducing power consumption (which lowers the packaging cost and increases the ICs' long-term reliability), improving yield by choosing appropriate devices for reducing design tolerances, and reducing the test time. Several BiCMOS building blocks were studied to illustrate the design flexibility with BiCMOS devices. More details, with additional new BiCMOS building blocks, including their experimental results, is provided in [19].

## PROBLEMS

**5.1.** Show that the transconductance of a bipolar transistor is at least ten times larger then that of an MOS transistor if they are biased with the same current and $V_{GS} - V_T$ is at least 0.6V. This is independent of the device sizes chosen for the two devices, as long as they are assumed to be operating in their proper regions; bipolar active, MOS saturation. What does this imply in terms of the unity gain frequency of a common-emitter and common-source amplifier, respectively?

**5.2.** Derive the current mismatch expressions of Eqs. (5.43) and (5.44). Assume $R_1 + R_2 = 2$ K$\Omega$, $R_1 - R_2 = \pm 10$ $\Omega$, $\frac{\Delta K_n}{K_n} = \pm.02$, $(V_{GS} - V_T) = .5$ V, $\Delta V_T = \pm 3$ mV, $I_D = I_C = 100$ $\mu$A, $\frac{\Delta A_E}{A_E} = \pm .02$, $\frac{\Delta Q_B}{Q_B} = \pm .02$, and $\frac{\Delta \alpha_F}{\alpha_F} = \pm .001$. Find the current mismatches when (a) R= 0 $\Omega$ and (b) R=2 K$\Omega$. Which current source (bipolar or MOS) would you use with and without resistor degeneration? Why?

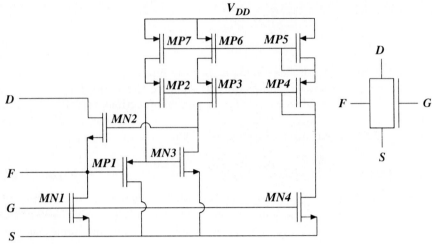

**FIGURE P5.5** An n-type super MOS circuit.

**5.3.** Compute the input-referred noise of the amplifiers in Fig. 5.11 when (a) $R_g = R_b = 1$ K$\Omega$ and (b) $R_g = R_b = 1000$ K$\Omega$, using the parameters in Table 5.2. Assume $K_f = 4 \times 10^{-23} V_2$ F, $I_c = I_d = 1 \times 10^{-4}$ A, $R_s = R_e = 1$ K$\Omega$, $R_d = R_c = 10$ K$\Omega$, $\Delta f = 1$ MHz and $\frac{W}{L} = \frac{200\mu}{3\mu}$. Compare the results for both circuits. Repeat the calculation for the case (a) with $\frac{W}{L} = \frac{1000\mu}{3\mu}$. Explain the results.

**5.4.** Derive an expression for the output resistance of the current mirror in Fig. 5.14(a). Assume that there exists a resistor, $R_E$, between the emitter terminal of Q3 and ground. Discuss whether any advantage is gained by the inclusion of this emitter degeneration resistor. Using the parameters in Table 5.2 determine the output impedances if (a) $R_E = 0$ $\Omega$ and (b) $R_E = 10$ K$\Omega$.

**5.5.** The transresistance stages presented in this chapter employ gain-boosting techniques to achieve high output resistance. It is possible to convert these stages to super MOS transistors similar to those discussed in Chapter 2 by adding a biasing stage. Fig. P5.5 shows the circuit for one such super MOS transistor formed from the transresistance stage of Fig. 5.14(a).

(a) Derive an approximate expression for the output resistance of the super MOS transistor.

(b) Now using the transistor parameters given in Table 5.2 calculate the $\frac{W}{L}$ ratio for the various transistors so that they are all in saturation while delivering a drain current of 150 $\mu$A for a gate voltage of -3.8 V. Let $V_{ss} = -5$ V and $V_{dd} = 5$ V.

(c) Simulate the above design and then adjust the aspect ratios of the transistors so that $V_{DS}$ of $M_{N1}$ is just 50 to 100 mV above its saturation voltage and the output resistance is greater than 150 M$\Omega$.

(d) Draw the corresponding p type super MOS circuit. Repeat part (c) for the p-type design. Note that you may have to place the n transistor corresponding to $M_{P1}$ in a separate well ($V_{SB} = 0$) in order to bias all the transistors in the proper regions of operation.

**5.6.** Determine the collector current for Q3 such that it causes an exact pole-zero cancellation in the basic amplifier of Fig. 5.15(c). What is the gain-bandwidth product at this current level? Use the base transit time curve in Fig. 5.2 ($V_{CE} = 3$ V), the BiCMOS device parameters in Table 5.2, $g_{m1} = 5$ m℧, $g_{mb1} = 2$ m℧, $I_{CQ2} = 200$ μA (note that the current in M1 can be chosen larger than the current flowing in Q2, see Fig. 5.17), $R_L = 10^7 \Omega$, $C_{gs1} = 0.16$ pF, $C_{gs1} + C_{sb1} + C_{\pi 2} + C_p$ (bias circuit and other parasitics) $= 0.5$ pF, $C_{\mu 2} + C_{cs2} + C_{\mu 4} + C_{cs4} = .44$ pF. Assume room temperature, and use $r_{oQ} = \frac{V_A}{I_C}$, $g_{mQ} = \frac{qI_C}{KT}$ for the bipolar transistors.

**5.7.** Derive expressions for the dc-gain and the output resistance of the BiCMOS output stage in Fig. 5.18(c). Compare the results with those from that in Fig. 5.18(b). What performance changes do you expect when the PNP transistors are replaced with PMOS devices (use the parameters in Table 5.2)?

**5.8.** In the advanced BiCMOS processes, bipolar transistors offer high speed. On the other hand, PMOS devices are usually the slowest among the BiCMOS devices. The circuit in Fig. P5.8 is a dual of the circuit in Fig. 5.20(a). Derive an expression for the large-signal dc-transfer function $\frac{I_1 - I_2}{V_{in}}$. Use the small-signal approach to find the poles/zeros for this voltage-to-current converter. Insert a feedforward capacitor between the source and drain terminals of each of M1 and M2 and repeat the derivations. Compare the results in both cases (with and without the feedforward capacitors).

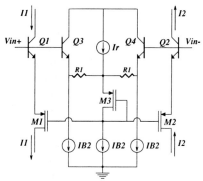

**FIGURE P5.8** High speed BiCMOS VIC.

**5.9.** Derive the large-signal dc-transfer equation as in Eq. (5.112). Obtain an expression for the maximum input range. Compare the maximum input range and the speed of the BiCMOS transconductor in Fig. 5.24 with its CMOS counterpart. (Hint, assure that all devices operate in their normal region. Use the results in Eqs. (5.17)-(5.18)).

**5.10.** Refer to the circuit in Fig. P5.10. The emitter-area ratio of transistors Q1-Q2 to Q3-Q4 is A. Compute the third and fifth harmonic components of the input Vin which exist in $(I_1 - I_2)$ for the cases 1) $A = 2 + \sqrt{3}$, and 2) A=4. What practical value(s) for the emitter-area ratios of Q1-Q4 minimize the magnitude of the third harmonic component? (Hint: Use a Taylor expression to derive the large-signal dc-transfer equation $\frac{i_1 - i_2}{V_{in}}$.)

**5.11.** Consider the circuit in Fig. P5.11. Determine the expression for $I_c$ such that the emitter of Q1 is at a virtual ground. Find an expression for $I_1 - I_2$ in terms of $V_x$ and $V_y$. Assume matched transistors and $|V_x| < I_oR$.

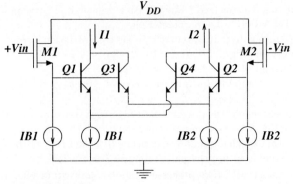

**FIGURE P5.10** BiCMOS VIC.

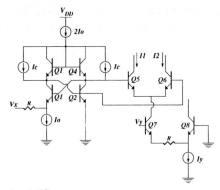

**FIGURE P5.11** Bipolar VIC.

**5.12.** Assuming finite $\beta$ and finite Early voltage for the matched bipolar transistors in the circuit of Fig. P5.12, derive an expression for $V_{out}$ in terms of $V_1$, $V_2$, $V_3$, and $V_4$. What is the limitation on the input signals for proper operation? Assume that PNP transistors are available, and they have comparable $\beta$, Early voltage, and speed to those of the NPNs. How would you use the PNPs and NPNs transistors to correct for the input limitation (what other problems may now exist?) Assume the op amps are ideal.

**5.13.** Most BiCMOS processing lines are optimized for high speed, high performance vertical NPN transistors and MOS devices. The analog designer will typically only have slower lateral PNP transistors available. For some applications a "pseudo-PNP" device can be applied as a substitute [25]. Obtain the small signal model for this structure shown in Fig. P5.13.

**5.14.** Obtain the open circuited voltage gain of a common-emitter amplifier implemented with the pseudo-PNP transistor given in Fig. P5.13. Simplify this expression, and compare it to that known for a regular common-emitter amplifier. Is it a valid approximation to look at the pseudo-PNP as a regular MOS transistor where the output impedance is reduced by a factor of $\beta + 1$, and the transconductance is increased by a factor of $\beta + 1$? Comment upon the stability of a circuit utilizing this "pseudo-PNP" structure.

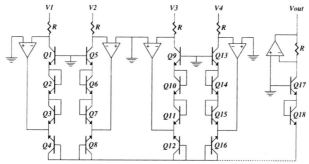

**FIGURE P5.12** Bipolar vector manipulation cell.

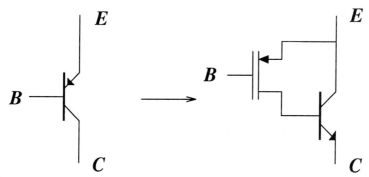

**FIGURE P5.13** Pseudo-PNP transistor structure.

**5.15.** Derive and compare the dc-transfer equations of circuits Fig. P5.15. Explain the ac-response behavior of the two circuits. Simulate the circuit responses in SPICE to determine their linear range. What characteristic change is expected when M1 in Fig. P5.15 is replaced by an NPN device?

**5.16.** In Fig. P5.16 assume M1 - M3 are operating in the saturation region. Using the simple drain current equation derive expressions for the difference and the sum of i1 and i2 as a function of the differential input, $\pm Vin = A\cos(\omega t)$. Derive expressions for the harmonic components and dc-offset voltage caused by mismatch in M1 and M2, and by mobility degradation. Calculate errors caused by mismatch given $K_{n1} - K_{n2} = 0.02K_n$, $K_{n3} = (1 \pm 0.01)K_n$, $\frac{W}{L} = \frac{50\mu}{3\mu}$, $K_n = 50$ $\mu A/V^2$, $I_C = 300\mu A$, and A = 1 V. Calculate errors caused by mobility degradation for $\theta$ equal to 0.01, 0.05, and 0.25.

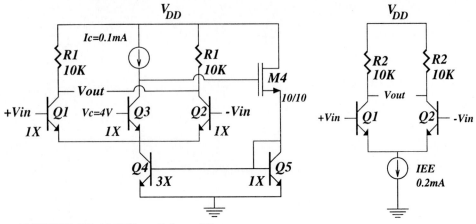

**FIGURE P5.15** Differential amplifier.

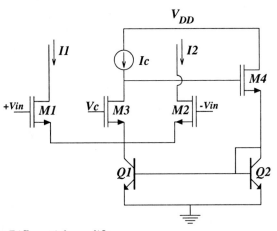

**FIGURE P5.16** Differential amplifier.

**5.17.** Consider the low-input capacitance unity gain buffer given in Fig P5.17.

(a) Determine the dc-transfer equation of this circuit. Use SPICE to validate your result (assume $C_{in} = \infty$ and the biasing current sources have very high output impedance).

(b) Assume that the buffer has a gain of 0.99, and the device capacitances associated with M1 are $C_{GS} = 0.5$ pF, $C_{GD} = 0.15$ pF, and $C_{GD} = 0.003$ pF. What should be the lower limit for the value of $C_{in}$ such that $\frac{V_{out}}{V_{in}} = \frac{1}{2}$?

**5.18.** The basic transconductance blocks given in Fig. P5.18 can be used to implement high frequency OTA-C filters [33], obtain the transfer characteristics for circuit (a) using the simple square-law current equation. What is the required constraint for Vb if a linear relationship is to be obtained between the differential input voltage and the differential output current? (Hint: express Vin+ and Vin- in terms of a common-mode and differential-mode input, and make Vb equal to

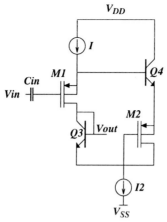

**FIGURE P5.17** Capacitance unity-gain buffer.

the common-mode signal.)

Perform the same analyses for circuit (b) using SPICE. How does the performance of the two transconductors compare?

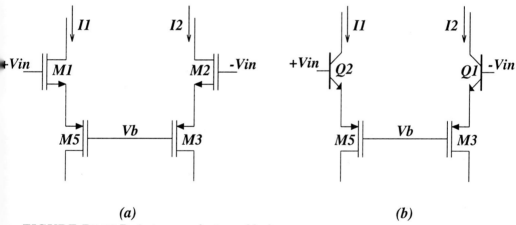

*(a)*                                        *(b)*

**FIGURE P5.18** Basic transconductance blocks.

**5.19.** Consider the circuit in Fig. P5.18. It can be used as a variable transconductor and a four-quadrant multiplier. This circuit is constructed by two BiCMOS unity gain buffer/level-shifter (dashed boxes) and an MOS cross-coupled network (M1-M4). Transistors M5-M6 provide a high input resistance and Q1-Q4 provide the low impedance nodes reached at the common-source nodes of M1-M4.

(*a*) Derive an expression for the dc-transfer equation of this circuit using the simple MOS square-law MOS current equation.

(*b*) SPICE simulate the dc-transfer curves using $V_{DD} = 8V$, $V_{SS} = 0V$, $I_A = 50\mu A$, $I_B = 200\mu A$, and $\frac{A_{E1}}{A_{D1}} = \frac{A_{E2}}{A_{D2}} = \frac{A_{E3}}{A_{D3}} = \frac{A_{E4}}{A_{D4}} = 3$. $\Delta E_i$ is the emitter area of the ith bipolar device. $\Delta D_i$ is the diode area of the ith diode device.

$\Delta V_{3-4} = \Delta V_{1-2} = 2V$ (centered at mid supply). Determine the input voltage range of this circuit.

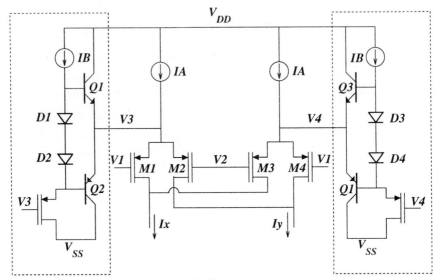

**FIGURE P5.19** Four-quadrant multiplier structure.

**5.20.** For the transconductance stage given in Fig. P5.20, determine the differential output current as a function of the differential input voltage. You can assume the voltage across RB to be small enough so that M1 and M2 are both in the linear region. The base-emitter voltage can be assumed to be constant and equal to 0.7V for all transistors.

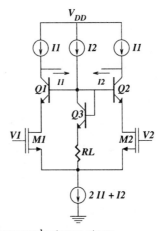

**FIGURE P5.20** Linear transconductance stage.

**5.21.** Consider the multiplier shown in Fig. P5.21.

 (a) Assume that $V_{BE}$ for the bipolar transistors is constant and neglect the base currents. Further, assume that $V_b$ is sufficiently low to force $M1$ and $M2$ into the triode region. Develop an expression for the output current $I_{out} = I_1 - I_2$ as a function of the input voltages, $V_{in1}$ and $V_{in2}$, and the DC control voltage $V_b$. What is the value of the transconductance, $g_m = I_{out}/(V_{in1} - V_{in2})$? What is the upper bound on the value of $V_b$ for which $M1$ and $M2$ will remain in the linear region? If an AC signal is applied to $V_b$, can the circuit be used as a multiplier?

 (b) Choose the $W/L$ of transistors $M1$ and $M2$ to give an approximate transconductance of $g_m = 65/muA/V$ with $V_b = 1.1V$. Simulate the DC transfer characteristic of $I_{out}$ vs. $V_{in1} - V_{in2}$ to estimate the transconductance. Explain any discrepancies between the calculated and simulated transconductance values. Use the circuit in the multiplication scheme suggested above to multiply a 1kHz and a 5kHz signal.

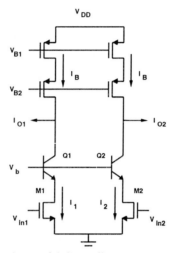

**FIGURE P5.21** Linear-region multiplier cell.

**5.22.** The circuit in Fig. P5.22 is a temperature-compensated voltage reference and current reference source. Use $I_C = CT^{2.4}exp\left[-\frac{E_{gapo}}{kT}\right]exp\left[\frac{qV_{BE}}{kT}\right] = I_{CS}exp\left[\frac{qV_{BE}}{kT}\right]$ for the bipolar devices, where C is a constant, T is the temperature in Kelvin, and $E_{gapo} = 1.205$ eV. At 300 degrees Kelvin we have $\frac{kT}{q} = 26$ mV, $I_{CS1} = 0.3$ pA, $V_{BE2} = 600$ mV, and the ratio $\frac{A_{E1}}{A_{E2}} = n$ is equal to four. Compute:

 (a) The nominal value of $V_{out}$.

 (b) The value of R1, R2, and R3 such that $I_{out}$ is 100 $\mu$A. Assume R3 has negligible temperature coefficient (TC).

 (c) The deviation of $V_{out}$ from the nominal value if the temperature is increased by 30 degrees.

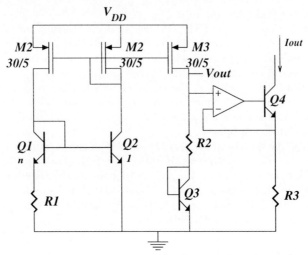

**FIGURE P5.22** Temperature-compensated voltage and current reference.

## REFERENCES

[1]  R. Gregorian and G. C. Temes, *Analog MOS Integrated Circuits for Signal Processing*, Wiley, 1986.

[2]  A. R. Alvarez, *BiCMOS Technology and Applications*, Kluwer Academic Publishers, 1989.

[3]  M. Kubo, I. Masuda, K. Miyata, and K. Ogiue, "Perspective on BiCMOS VLSI's," *IEEE J. of Solid-State Circuits*, vol. SC-23, no. 1, pp. 5-11, February 1988.

[4]  W. Heimsch et al., "A 3.8-ns 16K BiCMOS SRAM," *IEEE J. of Solid-State Circuits*, vol. SC-25, no. 1, pp. 48-54, February 1990.

[5]  A. Ohba et al., "A 7-ns 1-Mb BiCMOS ECL SRAM with shift redundancy," *IEEE J. of Solid-State Circuits*, vol. SC-26, no. 4, pp. 507-512, April 1991.

[6]  H. Kato et al., "A 6-ns 4Mb ECL I/O BiCMOS SRAM with LV-TTL Mask Option," *ISSCC Digest of Technical Papers*, pp. 212-213, February 1992.

[7]  T. Kawahara et al., "A circuit technology for sub-10-ns ECL 4-MB BiCMOS DRAM's," *IEEE J. of Solid-State Circuits*, vol. SC-26, no. 11, pp. 1530-1537, November 1991.

[8]  P. J. Lim and B. A. Wooley, "An 8-b 200-MHz BiCMOS comparator," *IEEE J. of Solid-State Circuits*, vol. 25, no. 1, pp. 192-199, February 1990.

[9]  A. N. Karanicolas et al., "A high-frequency fully-differential BiCMOS operational amplifier," *IEEE J. of Solid-State Circuits*, vol. 26, no. 3, pp. 203-208, March 1991.

[10]  K. Tsugaru et al., "A single-power supply 10-b video BiCMOS sampled-and-hold IC," *IEEE J. of Solid-State Circuits*, vol. 25, no. 6, pp. 653-659, June 1991.

[11]  M. Nayebi and B. Wooley, "A 10-b video BiCMOS track-and-hold amplifier," *IEEE J. of Solid-State Circuits*, vol. 24, no. 12, pp. 1507-1516, December 1990.

[12]  Z-Y Chang et al., "Design considerations of high-dynamic-range wide-band amplifiers in BiCMOS technology," *IEEE J. of Solid-State Circuits*, vol. 26, no. 11, pp. 1681-1688, November 1991.

[13]  C. Michael and M. Ismail., "Statistical modeling of device mismatch for analog MOS integrated circuits," *IEEE J. of Solid-State Circuits*, vol. 27, no. 2, pp. 154-166, February 1992.

[14]  P. R. Gray and R. G. Meyer, *Analysis and Design of Analog Integrated Circuits*, Second Edition. New York: Wiley, 1984.

[15] B. J. Hosticka, "Improvement of the gain of MOS amplifiers," *IEEE J. of Solid-State Circuits,* vol. SC-14, No. 6, pp. 1111-1114, December 1979.

[16] H. C. Yang and David Allstot, "An active-feedback cascode current source," *IEEE Trans. Circuits Syst.,* vol. 37, no. 5, pp. 644-646, May 1990.

[17] E. Sackinger and W. Guggenbuhl, "A high-swing, high-impedance MOS cascode circuit," *IEEE J. of Solid-State Circuits,* vol. SC-25, no. 1, pp. 289-298, February 1990.

[18] S. R. Zarabadi and M. Ismail, "Very high output impedance cascode current sources / current mirrors / transresistance stages and their applications," *International Journal of Circuit Theory and Applications,* vol. 20, pp. 639-648, 1992.

[19] S. R. Zarabadi, "Design of analog VLSI circuits in BiCMOS/CMOS technology," *Ph.D. dissertation,* The Ohio State University, June 1992.

[20] S. R. Zarabadi, S. C. Huang, and M. Ismail, "A wide dynamic range differential difference amplifier and its applications," SRC TECHCON'90, October. 1990

[21] S. T. Dupuie and M. Ismail, "High Frequency CMOS Transconductors,', in *Analog IC Design: The Current Mode Approach,* Ch. 5, edited by C. Toumazou, F. J. Lidgey, and D. G. Haigh, London, Peter Peregrinus Ltd., 1990.

[22] R. Alini, A. Baschirotto, and R. Castello, "8-32MHz tunable BiCMOS continuous-time filters," *Proc. European Solid-State Circuits Conference,* Milano, Italy, pp. 9-12, September 1991.

[23] B. Gilbert, "Translinear circuits: a proposed classification," *Electron. Lett.,* vol. 11, pp. 14-16. January 1975.

[24] E. Seevinch, "Analysis and synthesis of translinear integrated circuits," *D.Sc. dissertation,* Univ. Pretoria, S. Africa, May 1981.

[25] S. Sen and B. Leung, "A Low-Power Class-AB BiCMOS Op amp Using 'Pseudo-PNP' Transistors", *Proc. International Symposium on Circuits and Systems,* vol. 2, pp. 1136-1139, May 1993.

[26] P. Antognetti and G. Massobrio, "Semiconductor Device Modeling with Spice," Mc Graw-Hill, 1988.

[27] I.E. Getreu, "Modeling the bipolar transistor," Elsevier publisher, Tektronix Inc., 1978.

[28] S.R. Zarabadi, M. Ismail, and N.I. Khachab, "High frequency BiCMOS Linear V-I Converter, Voltage Multiplier, Mixer," U.S. Patent No. 5,151,625, filed Nov. 8, 1990, issued Sept. 29, 1992.

[29] S. Zarabadi and M. Ismail, "A 100Mhz Wide-Range BiCMOS Four-Quadrant Multiplier/Mixer," *Proc. European Solid-State Circuits Conference,* pp. 130-133, September 1993.

[30] K. Tsugaru et al., "A Single-Power-Supply 10-b Video BiCMOS Sample-and-Hold IC", *IEEE J. of Solid-State Circuits,* vol. 25, pp. 653-659, June 1990.

[31] S.R. Zarabadi and M. Ismail, "Linear Voltage to Current Converter Including Feedback Network", U. S. Patent pending, filed December 31, 1992.

[32] B. Gilbert, "A new High-Performance Monolithic Multiplier using Active Feedback", *IEEE J. of Solid-State Circuits,* pp. 364-373, December 1974.

[33] S. Zarabadi and M. Ismail, "A BiCMOS Transconductance-C Intermediate Frequency (IF) Filter," *Proc. European Solid-State Circuits Conference,* pp. 186-189, September 1993.

# CHAPTER
# 6

# CURRENT-MODE
# SIGNAL
# PROCESSING

## 6.1  INTRODUCTION

The developments of the current conveyor building block [1, 2] and the translinear principle [3] have been the catalysts for much of the recent interest in current-mode circuit activity [5]. These early circuit inventions offered an alternative to conventional voltage-mode design techniques. While fundamentally any design technique is limited by the device characteristics, there may be specific applications where current-mode circuits provide advantages. For example, because of the simplicity of current-mode circuits, they are easily adapted to systems requiring low power-supply voltages. Additionally, this simplicity may yield significant area savings for systems requiring moderate accuracy [4].

In this chapter, the design of current-mode signal processing circuits is presented. We first introduce the basic circuit building blocks, performance characteristics, and circuit topologies of continuous-time, current-mode circuits. Using these circuit blocks, the design of active ladder filters is described. Next, we describe the design and implementation of sampled-data, current-mode circuits. Finally, a current-mode implementation of a touch-tone phone keypad (DTMF) decoder is given.

## 6.2 CONTINUOUS-TIME SIGNAL PROCESSING

The most basic current-mode circuit building block is the MOS current mirror (Fig. 6.1). This circuit consists of two bias current sources, $I_1$ and $I_2$, and two transistors, $M_1$ and $M_2$, with the drain and gate of $M_1$ shorted or *diode-connected*. Assuming $M_1$ and $M_2$ are biased in the saturation region, neglecting channel-length modulation, and setting $i_{in} = 0$, the current flowing into the drain of $M_1$ is:

$$I_1 = I_{DS1} = \frac{k'}{2}\frac{W}{L}(V_{GS1} - V_T)^2 \tag{6.1}$$

This current is converted into a gate-source potential obtained by rearranging Eq. (6.1):

$$V_{GS1} = \sqrt{\frac{2I_1}{k'\frac{W}{L}}} + V_T \tag{6.2}$$

Since the gates and sources of $M_1$ and $M_2$ are connected, $V_{GS1} = V_{GS2}$, causing a current proportional to $I_{DS1}$ to flow into the drain of $M_2$. When the aspect ratios $(W/L)$ of transistors $M_1$ and $M_2$ are identical, $I_{DS2} = I_{DS1}$. Similarly, when $(W_2/L_2) = A(W_1/L_1)$, the output current is scaled by the factor $A$,

$$I_{DS2} = AI_{DS1} \tag{6.3}$$

If we now apply a small-signal current $i_{in}$ at the input, $I_{ds1}$ is the sum of the bias and signal currents:

$$I_{ds1} = I_1 + i_{in} \tag{6.4}$$

and the current in $M_2$ is,

$$I_{ds2} = A(I_1 + i_{in}) \tag{6.5}$$

Applying Kirchhoff's current law at the drain of $M_2$

$$i_{out} = I_2 - I_{ds2}\Big|_{I_2=AI_1} = A_1I_1 - AI_1 - Ai_{in} = -Ai_{in} \tag{6.6}$$

and therefore, the current mirror produces an output current that is an inverted and scaled version of the input current. To sum many input signals, the signals are all directly connected to the input. Thus, the current mirror performs the operations of signal inversion, scaling, and summation. These signal-processing operations may be combined to obtain more complex signal-processing operations including filtering and data conversion.

In this basic current-mode building block, summation of input signals is accomplished without requiring any additional circuitry. Conversely, generation of more than one output current requires adding branches identical to the $I_2$-$M_2$ branch. These two characteristics are the dual of the voltage-mode system where additional circuitry is required to sum signals but a single output voltage may be connected to many input terminals.

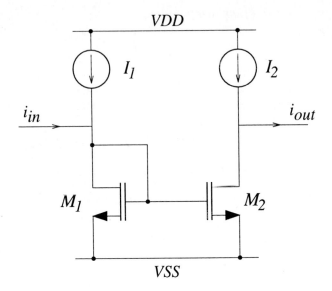

**FIGURE 6.1**
Simple current mirror with input current $i_{in}$, output current $i_{out}$ and $(W_2/L_2) = A(W_1/L_1)$.

## 6.2.1 Current Mirror Performance Characteristics

Accurate analog signal processing requires ideal circuit operation. In current-mode circuits, the current mirror – a current-controlled current source (CCCS) [6, 7]– determines the overall performance of the system. While the simple current mirror (Fig. 6.1) is suitable for some applications, it is far from ideal. In this section, we present the characteristics of the ideal CCCS and contrast these characteristics to those of practical circuit implementations.

1. Input and Output Resistance
    An ideal CCCS has zero input resistance and infinite output resistance. Consequently, the input voltage does not vary with the input current and the output current is independent of the output voltage. However, actual CCCS implementations have non-zero input resistance and non-infinite output resistance which cause errors in the output current. These errors can be illustrated by examining two cascaded current mirrors where the output of one mirror is connected to the input of the next mirror (Fig. 6.2). In this configuration, the error due to the non-ideal input and output resistances can be estimated. Let $r_{in}$ and $r_{out}$ be the input and output resistances of the cascaded mirrors, respectively [8]. The input to the second mirror, $i_{out}$ in terms of the input to the first mirror, $i_{in}$, is:

$$i_{out} \approx \frac{i_{in}}{1 + \frac{r_{in}}{r_{out}}}\bigg|_{r_{in} \ll r_{out}} \approx i_{in}\left(1 - \frac{r_{in}}{r_{out}}\right) \qquad (6.7)$$

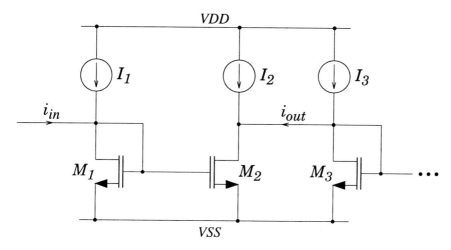

**FIGURE 6.2**
Cascaded current mirrors typical of a signal processing system with input current $i_{in}$ and output current $i_{out}$.

The error in the output current may be expressed as:

$$\Delta i = i_{in} - i_{out} \approx i_{in} \left( \frac{r_{in}}{r_{out}} \right) \tag{6.8}$$

Therefore, the error current is reduced by minimizing the ratio of the input resistance to the output resistance.

2. Input Linear Range

For the accurate reproduction of the signal current at the current mirror output (Fig. 6.1), the total input current, $I_1 + i_{in}$, must be in the range where both devices operate in saturation at all times. If $I_{ds1}$ is too small or negative, $M_1$ will cut off. Alternately, if $I_{ds1}$ is too large, $V_{gs1}$ may increase to the point that the bias current source, $I_1$, (circuit implementation not shown) will no longer be active. A peak signal current that is 50% of the bias current level generally provides a reasonable compromise between power dissipation and dynamic range [9].

3. Output Voltage Swing

An ideal CCCS produces an accurate output current regardless of the voltage at the output. In reality, a current source has a minimum voltage requirement to ensure that the devices operate in saturation. For the simple current mirror (Fig. 6.1), the minimum output voltage requirement is $V_{ds(sat)2}$ or no less than 0.25 V for operation in strong inversion. While the output current of the simple current mirror may be in error due to the nonideal input and output resistances, the current mirror has the capability of operating with low power supply voltages. Increasing the accuracy of the current mirror typically increases the supply voltage requirement. This places a challenge on the designer to develop high-performance current-mode circuits that operate with reduced power supply voltages. Several current mirror structures and their voltage requirements are covered in Sec. 6.2.2.

4. **DC-Balance**

The drain-source voltages of the mirror transistors $M_1$ and $M_2$ (Fig. 6.1) also affect the accuracy of the output current. If the devices are biased in saturation, $V_{ds}$ is related to the current by:

$$I_{ds} = \frac{k'}{2}\frac{W}{L}(V_{gs} - V_T)^2(1 + \lambda V_{ds}) \tag{6.9}$$

The ratio of the drain-source currents in the two devices is then

$$\frac{I_{ds2}}{I_{ds1}} = \frac{K_2/2(V_{gs2} - V_T)^2(1 + \lambda_2 V_{ds2})}{K_1/2(V_{gs1} - V_T)^2(1 + \lambda_1 V_{ds1})} \tag{6.10}$$

where $K = k'W/L$. Assuming all the process-dependent parameters are identical for $M_1$ and $M_2$ ($K = K_1 = K_2$, $V_{T1} = V_{T2} = V_T$, $\lambda = \lambda_1 = \lambda_2$), the ratio of the currents is:

$$\frac{I_{ds2}}{I_{ds1}} = \frac{(1 + \lambda V_{ds2})}{(1 + \lambda V_{ds1})} \tag{6.11}$$

Thus for $\lambda V_{ds1} \ll 1$, the error in the output current due to $V_{ds}$ mismatch is:

$$I_{error} \approx \lambda(V_{ds2} - V_{ds1}) \tag{6.12}$$

Clearly, if the drain-source voltages of $M_1$ and $M_2$ are not balanced (i.e., unequal), there will be an offset in the output current.

5. **Finite Bandwidth**

The amplification of current in current-mode circuits is typically less than a factor of 100 and, due to the constant gain-bandwidth product, high frequency circuit operation is possible. Consider the circuit in Fig. 6.1. The -3 dB bandwidth is approximately the total input resistance divided by the capacitance at that same node or $BW_{(-3dB)} \approx \frac{g_{m1}}{C_{gs1} + C_{gs2}}$. Current-mode circuits provide wide bandwidth operation at low gain values. Interestingly enough, voltage-mode circuits, operated with similar closed-loop gains, produce comparable bandwidths.

6. **Dynamic Range**

The dynamic range is determined by the ratio of the maximum signal level (at a specified level of distortion) to the minimum detectable signal level. The maximum signal current is determined by the input linear range as discussed above. In current-mode circuits, the input-referred noise current determines the minimum detectable signal. Calculating $\overline{i_{in}}^2$ in Fig. 6.1 involves translating the transistor noise voltage into a noise current by multiplying by the square of the transconductance, i.e.,

$$\overline{i_{in}}^2 = g_{m1}^2(\overline{v}_{n1}^2 + \overline{v}_{n2}^2) \tag{6.13}$$

where $\overline{v}_{n1}^2$ and $\overline{v}_{n2}^2$ are the mean square values of the noise voltages of $M_1$ and $M_2$, respectively [6].

**7.** Device Matching

Accurate mirroring of the signal current requires perfect matching of the current mirror transistors, $M_1$ and $M_2$ (Fig. 6.1). However imperfections in the IC manufacturing process lead to random and systematic errors in the devices (see Chapter 14 for a complete description of analog statistical modelling) [10]. Using large devices with common-centroid and unit-cell layout techniques (see Chapter 16 for a description of layout techniques), current matching of up to 0.1% may be obtained without special process trimming steps [11, 65, 59].

## 6.2.2   Current Mirror Topologies

**Simple Current Mirrors**   The performance characteristics described above must be taken into consideration when designing a current mirror building block. In this section, several CMOS current mirror topologies are contrasted with the ideal CCCS.

Consider first the simple current mirror of Fig. 6.1. Using the MOSFET low-frequency, small-signal model in Fig. 6.3 assumes all devices are biased in saturation and the input resistance is approximately $1/(g_{m1} + /g_{ds1})$. Typically $g_{ds1} \gg g_{m1}$, and

$$r_{in} \approx \frac{1}{g_{m1}} \tag{6.14}$$

Similarly, the output resistance is determined by evaluating the resistance seen looking into the drain of $M_2$, or $r_{out} = 1/g_{ds2}$. For a typical $2\mu$ CMOS process, the input resistance is on the order of a few k$\Omega s$ whereas the output resistance is in the M$\Omega$ range.

Let's examine the distortion in the output current due to dc-imbalance and non-zero input and finite output resistances. We will use the simple current mirror loaded at the output by an identical current mirror (Fig. 6.2). All bias currents are $100\mu A$ and all devices are $\frac{100}{10}$. Using SPICE to simulate the circuit and applying a peak sinusoidal input current of $50\mu A$, the gain error, offset current and total harmonic distortion (THD) are 0.74%, $0.23\mu A$, and 0.44%, respectively for a typical 2 $\mu$ process. In this example, the current mirror transistors are dc-balanced ($V_{ds1} = V_{ds2}$) and therefore, the distortion in the output is caused by the non-zero input resistance and finite output resistances. As shown in Eq. (6.8), the input-to-output resistance ratio determines the current error and, for the simple current mirror, this ratio is typically a few hundred $\Omega/\Omega$. Reduction of this error is obtained by employing cascoded devices in the current mirrors or by lowering the input resistance.

We next perform the same simulation, but replace $M_3$ and $I_3$ with a diode-connected p-channel transistor connected to the $V_{DD}$ and a bias current source connected from its drain to $V_{SS}$. The gain error, offset current and THD are now 0.88%, $2.8\mu A$, and 0.51%, respectively. Notice that the offset current increases by an order of magnitude due to the voltage difference in $V_{ds1}$ and $V_{ds2}$.

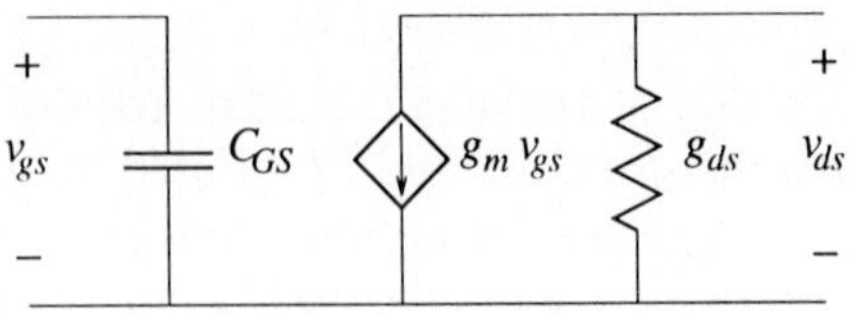

**FIGURE 6.3**
Low-frequency small-signal representation of the MOS transistor.

The simple current mirror has a very low power supply voltage requirement, i.e., $V_{GS1(max)} + V_{DS(sat)bias1}$. Thus, on the order of 1.2 V to 1.8 V is required to maintain the devices operating in saturation in a standard CMOS process. For this reason, the simple mirror is ideal for low supply voltage applications. However, other current mirror structures provide higher output current accuracy as discussed below.

**Cascoded Current Mirrors**    A current mirror which provides much higher output resistance than the simple current mirror is the cascode current source (Fig. 6.4(a)). The input resistance is approximately the same as the simple current mirror. The output resistance is increased by approximately the gain of the cascode device, $g_{m4}/g_{ds4}$. Cascoding additional devices will increase the output resistance by one transistor gain for each additional device. Unfortunately, the increase in the output resistance is at the expense of the output voltage swing. The minimum output voltage of the cascode current mirror is $V_{T2} + V_{DS(sat)2} + V_{DS(sat)4}$. Reduction in the minimum output voltage is obtained by using the high-swing cascode current mirror (Fig. 6.4(b)). The bias voltage, $V_{bias}$, of $M_3$ and $M_4$ is chosen so that $M_1$ and $M_2$ are biased on the edge of saturation when the total input current is at its maximum value. The output voltage can then swing as low as $V_{DS(sat)2} + V_{DS(sat)4}$ from $V_{SS}$. In addition to the increase in the output voltage swing, there is also an increase of $V_{GS}$ in the input linear range over the cascode current mirror.

While the high-swing cascode mirror has a reasonable input linear range, output voltage swing, and dc-balance, it has one significant drawback for current-mode circuits. If the input signal current is greater than about 10% of the bias current, the fixed bias of the cascode devices will tend to drive the lower mirror devices into the triode region. This introduces nonlinearities into the output current. The solution is to use an adaptive $V_{bias}$ that enables all the current mirror transistors to operate in the saturation region and the nonlinearities are eliminated.

**Regulated-Gate Cascode**    The regulated-gate cascode current mirror is a self-biasing current mirror (Fig. 6.5) [12, 14, 15, 16]. Active feedback is used to increase the output resistance. The feedback amplifier formed by $M_4$, $M_6$, and its biasing current source hold the drain of the mirror transistor, $M_1$, at a constant voltage. This eliminates channel-length modulation and the corresponding

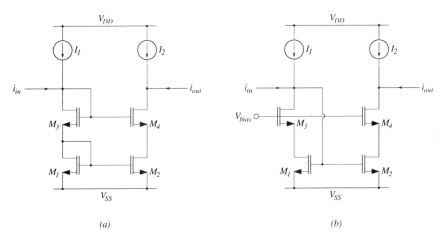

**FIGURE 6.4**
Cascode current mirrors (a) cascode and (b) high-swing cascode current mirror.

variation in drain current. The output resistance of this structure is typically hundreds of mega-ohms and calculation of the output impedance of this structure is left as an exercise. Transistor $M_6$ simply provides the active feedback and does not contribute to the current mirroring. Therefore, it can be operated in the sub-threshold region without severely affecting the circuit operation. DC balance is achieved in this current source by designing it such that $V_{DS1} = V_{DS2}$ or equivalently $V_{GS5} = V_{GS6}$.

Active feedback may also be applied to reduce input resistance. Replacing the diode-connection with an amplifier from drain to gate will reduce the input resistance by the gain of the amplifier [17].

Note that all of the above structures improve the output resistance at the expense of output voltage swing. Thus, as CMOS circuit power supply voltages are continually reduced to accommodate smaller device geometries, it becomes more difficult to obtain high accuracy.

### 6.2.3 Continuous-Time Current-Mode Filters

Employing the current-mode building blocks above, higher-order signal processing operations may be implemented. In this section, the design of a continuous-time, current-mode integrator is presented [19]. The differential integrator is the basic building block for many filter types. The design example given in this section uses this integrator to construct a low-pass active ladder filter.

A continuous-time, current-mode integrator (Fig. 6.6) is constructed by cascading two current mirrors and connecting the output to the input node (labeled $X$ in the figure) [19]. The two input signals, $i_1$ and $i_2$, are connected to the drains of $M_1$ and $M_3$, respectively. A transistor, configured with its drain and source shorted and the gate connected to $V_{SS}$, provides a large capacitance that creates the dominant pole of the integrator. The simplified expression for

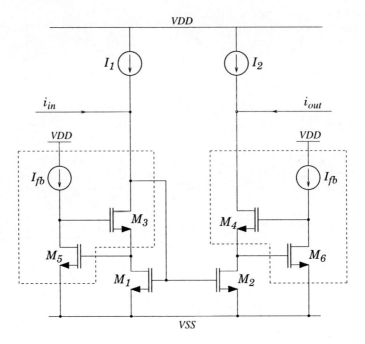

**FIGURE 6.5**

Regulated-gate cascode current mirror. The dotted line indicates the cascode devices, $M_3$ and $M_4$, and the active feedback devices, $M_5$ and $M_6$.

the output current is derived and analysis of the second-order effects is left as an exercise. Neglecting the transistor output conductances and applying KCL to nodes $X, Y$, and $Z$ of the small-signal equivalent of Fig. 6.6 yields:

$$i_1 + i_f = g_{m1}v_1 \tag{6.15}$$

$$g_{m2}v_1 + (g_{m3} + sC_{int})v_2 = i_2 \tag{6.16}$$

$$g_{m4}v_2 = -i_f \tag{6.17}$$

Typically $g_{m1} = g_{m2}$ and $g_{m3} = g_{m4}$. Substituting Eqs. (6.15) and (6.17) into Eq. (6.16) yields

$$i_1 + i_f + (g_{m3} + sC_{int})(-i_f/g_{m3}) = i_2 \tag{6.18}$$

Rearranging Eq. (6.18) and solving for $i_f$

$$i_f = (g_{m3}/sC_{int})(i_1 - i_2) \tag{6.19}$$

The output current is a scaled version of $i_f$ since $(W_5/L_5) = K(W_3/L_3)$

$$i_{out} = Ki_f \tag{6.20}$$

and

$$i_{out} = K(g_{m3}/sC_{int})(i_1 - i_2) \tag{6.21}$$

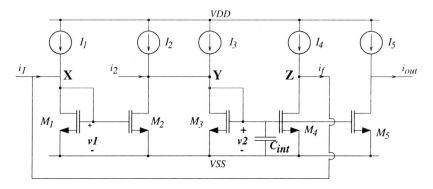

**FIGURE 6.6**
Continuous-time, current-mode integrator used to perform filtering with $(W/L)_5 = K(W/L)_3$.

From this expression, we see that $i_1$ and $i_2$ are applied to the non-inverting and inverting inputs, respectively, and the integrator's dominate pole frequency is determined by $g_{m3}/C_{int}$. The first current mirror creates a second pole at approximately $g_{m1}/(C_{gs1} + C_{gs2}) = g_{m1}/C_1$. The integrator quality factor, $Q$, is the ratio of these two poles:

$$Q \approx (g_{m1}/C_1)/(g_{m3}/C_{int}) \qquad (6.22)$$

Choosing $C_{int} \gg C_1$, a $Q$ greater than 20 may be obtained. As demonstrated in the example below, this integrator becomes the basic building block for active ladder (or leapfrog) filters. The term $K(g_{m3}/sC_{int})$ is the integrator coefficient or time constant. It determines the filter pole and zero locations, and thus, it is ideally very accurate. While $K$ depends on a *transistor ratio* and so it may be very accurate, $g_{m3}/C_{int}$ may vary widely with random and systematic process variations.

>    **Example 6.1.  Continuous-time Fifth-Order Chebyshev Filter.** Passive LC ladder networks were used extensively before the advent of integrated circuit technology due to their superior insensitivity to component values. Since then, much effort has been expended to find methods of simulating the characteristics of these networks with combinations of active and passive elements. Since the current-mode implementation contains only MOS transistors, the filter can be fabricated using a standard digital process.
>
>    Design a fifth-order current-mode Chebyshev low-pass filter with 0.5 dB ripple and a ripple bandwidth of $950kHz$.
>
>    *Solution:* The design procedure is nearly identical to that used in the synthesis of other active ladder implementations (for a complete description refer to Chapter 9) and begins with a passive LC prototype (Fig. 6.7). The choice of element values in this prototype are made with any one of a number of *normalized* design

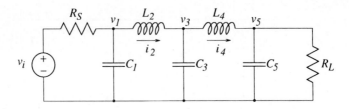

**FIGURE 6.7**
Passive prototype for the fifth-order, low-pass filter. Component values are determined by frequency- and impedance-scaling normalized values to meet the desired filter requirements.

tables [18]. The values are then frequency- and impedance-scaled to meet the desired filter requirements. Capacitances are scaled by $1/(Z \cdot 2\pi f_c)$, and inductances are scaled by $Z/(2\pi f_c)$, where $Z$ is the termination impedance, here chosen to be $2\Omega$, and $f_c$ is the desired ripple bandwidth. Once the passive prototype filter is scaled to meet the design specifications, our objective is to map this design to an all-transistor implementation.

We begin by writing the node and loop equations for each of the state variables in the passive prototype.

$$v_1 = \frac{1}{sC_1}\left[\frac{v_i - v_1}{R_S} - i_2\right] \tag{6.23}$$

$$i_2 = \frac{1}{sL_2}\left(v_1 - v_3\right) \tag{6.24}$$

$$v_3 = \frac{1}{sC_3}\left(i_2 - i_4\right) \tag{6.25}$$

$$i_4 = \frac{1}{sL_4}\left(v_3 - v_5\right) \tag{6.26}$$

$$v_5 = \frac{1}{sC_5}\left[i_4 - \frac{v_5}{R_L}\right] \tag{6.27}$$

The node and loop equations have been expressed as integrations in each case. To reproduce the frequency response of the passive network, the current-mode integrators must perform each of the integrations described by these equations. Therefore, each inductor and each capacitor of the passive prototype is "simulated" by a current-mode integrator. Equations (6.24), (6.25), and (6.26) are in the same form as Eq. (6.21). Equating these active and passive integrator expressions we obtain

$$K\frac{g_{m(int)}}{sC_{(int)}}\left(i_{(+)} - i_{(-)}\right) = \frac{1}{sC_3}\left(i_2 - i_4\right) \tag{6.28}$$

To produce the same integrator time constant as the passive structure,

$$K\frac{g_{m(int)}}{C_{(int)}} = \frac{1}{C_3} \tag{6.29}$$

**TABLE 6.1**

Low-pass filter component values: Fifth-order Chebyshev, 0.5 dB passband ripple, 950 kHz ripple bandwidth using 44 $\mu S$ transconductance current mirrors.

| Component | Normalized | Scaled | Integrator Cap. |
|:---------:|:----------:|:------:|:---------------:|
| $C_{1,5}$ | $1.8068F$ | $151.3nF$ | $6.690pF$ |
| $L_{2,4}$ | $1.3025H$ | $436.4nH$ | $19.29pF$ |
| $C_3$ | $2.6914F$ | $225.4nF$ | $9.965pF$ |

where $C_3$ is the frequency- and impedance-scaled capacitance in the passive prototype, $K$ is the integrator transistor ratio (Fig. 6.6), $g_{m(int)}$ is the transconductance of $M_3$, and $C_{(int)}$ is the capacitance setting the integrator dominate pole. Choosing values to satisfy this equation produces an integration identical to the original integration. This is repeated for Eqs. (6.24) and (6.26).

The expressions for the resistively terminated integrators in Eqs. (6.23) and (6.27) are rearranged to obtain a lossy integration, i.e.:

$$v_1 = \left( \frac{v_i}{R_S} - i_2 \right) \left( \frac{1}{sC_1 + \frac{1}{R_S}} \right) \qquad (6.30)$$

To obtain an expression for the current-mode integrator in this form, the second current mirror, $M_3$-$M_4$, in Fig. 6.6 is scaled by a factor $J$. Replacing $g_{m4}$ in (6.17) with $Jg_{m3} = Jg_{m(int)}$ yields the lossy integrator transfer function

$$i_{out} = (i_1 - i_2) \left( \frac{K}{s\frac{C_{(int)}}{g_{m3(int)}} + (1 - J)} \right) \qquad (6.31)$$

Choosing $C_{int}$ as in (6.29), the feedback coefficient is determined by

$$J = 1 - \frac{1}{R_S} \qquad (6.32)$$

Since the derivation for the load termination is identical for the integration in Eq. (6.27), $R_S$ is replaced with $R_L$.

From Eq. (6.29), it is apparent that in order to minimize the area consumed by the integration capacitors, the transconductance of the current mirror should also be minimized. Decreasing the transconductance, however, increases $(V_{GS} - V_T)$ of the mirror transistors, and hence, increases the power supply voltage requirements. This technique also aggravates the passband edge peaking due to the lowering of the second pole's frequency.

For the simulations in this example, a regulated cascode mirror (Fig. 6.5) was designed using a typical 2 $\mu$m CMOS process. The n-channel mirror transistors, $M_1$ and $M_2$, are $15\mu/22\mu$. Biasing the current mirror at $50\mu A$ yields a transconductance of $44\mu S$. Table 6.1 shows the integration capacitance values used.

As with other active ladder implementations, the integrator outputs must be scaled to ensure maximum dynamic range. This involves finding the peak

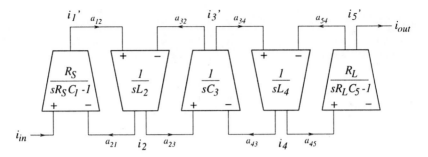

**FIGURE 6.8**

Fifth-order continuous-time, current-mode filter topology showing the interconnection of the integrators (Fig. 6.6). The dynamic range scaling factors ($a_{12}$, $a_{21}$, ...) are the ratios of the peak signal currents between adjacent integrators (i.e., $a_{12} = \hat{v}_1/\hat{i}_2$).

signal current for each integrator and then scaling the outputs as illustrated in Fig. 6.8. The most convenient method of finding these peak values, especially for high-order filters, is by simulation of the passive prototype filter.

With the element values determined, the filter is constructed by connecting the five current-mode integrators in the active leapfrog configuration of Fig. 6.8. Note that all but the first integrator have two outputs. As mentioned before, multiple outputs in current-mode circuits must be generated by additional output stages. For dynamic range scaling, this turns out to be a benefit, since a given integrator can easily have different feedback and feedforward coefficients.

The simulation results are shown in Fig. 6.9. Superimposed on the figure are the responses of the current-mode filter (solid) and the passive prototype (dashed). Note that the current-mode implementation exhibits significant pass-band edge peaking. The additional high-frequency pole of the first stage of each integrator causes excess phase shift at the integrators' unity-gain frequencies. Pre-distortion techniques can be used to reduce this peaking [19].

## 6.3  SAMPLED-DATA SIGNAL PROCESSING

With the development of switched-capacitor circuit techniques, analog sampled-data signal processing techniques gained widespread acceptance and usage in the late 1970's. Compared to continuous-time implementations, SC circuits offer superior accuracy and are more area efficient, especially for low frequency applications. Recently, the current-mode dual of SC circuits, known as switched-current (SI) circuits, has been developed. They also exhibit advantages over their continuous-time counterpart. In this section, we present the design and application of SI circuits.

We begin our discussion of SI circuits by describing the basic circuit building blocks. Sampled-data signal processing requires four basic operations: inversion, scaling, summation, and delay. In the previous section, we described how the first three of these are easily implemented with a current mirror. In this section, two delay blocks, the current-mode track-and-hold (T/H) [20] and the dynamic current mirror [21]-[24] are presented. Using these building blocks, the

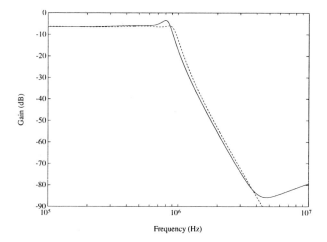

**FIGURE 6.9**
Simulated responses of the fifth-order Chebyshev continuous-time, current-mode (solid), and passive LC prototype (dashed) low-pass filters. Note peaking in the current-mode filter's response due to the integrators' non-dominant poles.

design of a high-Q bandpass ladder filter is described.

### 6.3.1   Current Delay Topologies

The current track-and-hold is constructed by placing a transistor between the gates of the mirror transistors $M_1$ and $M_2$ (Fig. 6.10). The added transistor is switched on and off by a precise digital clock signal. When on, the switch shorts the gates of the two mirror transistors. In this state, the circuit functions identically to the current mirror and, with an input applied to the drain of $M_1$, the output, $i_{out}$, tracks the input signal. When the switch is turned off, the gates of $M_1$ and $M_2$ are disconnected. The gate voltage of $M_1$, corresponding to the value of the input current at the instant the switch is opened, is sampled onto $C_{gs2}$. While the switch is open, $V_{gs2}$ remains constant; consequently, the output current is held at a constant value corresponding to the input current at the instant when the switch was opened.

The T/H circuit introduced above is a very versatile building block. However, it has an important disadvantage. The exact reproduction of the input current at the output depends on the precise matching of transistors $M_1$ and $M_2$ and the two bias current sources $I_1$ and $I_2$. The dynamic current mirror, or "current copier" [21]-[24], eliminates this problem by employing a single transistor and bias source for both input and output of current (Fig. 6.11). Two-phase non-overlapping clocks control the switches that configure the circuit first as a tracking amplifier and then as a holding amplifier. When $\phi_1$ is active ($S_1$ and $S_3$ closed) and $\phi_2$ is inactive ($S_2$ open), $M_1$ is diode-connected and $V_{gs1}$ tracks the

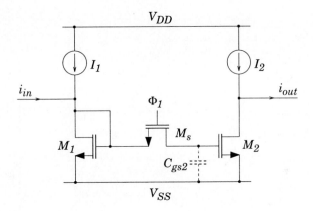

**FIGURE 6.10**
CMOS simple current track-and-hold.

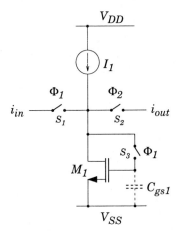

**FIGURE 6.11**
CMOS dynamic mirror or current copier. Circuit mirrors current in time rather than space. $S_1$ through $S_3$ are MOS switches.

total input current $I_1 + i_{in}$. To configure the dynamic mirror as a hold amplifier, $\phi_2$ is active ($S_2$ closed) and $\phi_1$ is inactive ($S_1$ and $S_3$ open). The voltage corresponding to the input current just before $S_3$ is opened is held on $C_{gs1}$ and, with $S_2$ closed, the held signal current is sensed at the output.

While the dynamic current mirror eliminates the errors due to transistor and bias source mismatches present in the current T/H, it produces only one copy of the output current. In a current-mode system, each output connects to only one input. An additional output branch must be generated for each input connection. To obtain multiple output currents in the dynamic mirror, additional devices can be connected to the gate of $M_1$ (Fig. 6.12) and, as with the simple track-and-hold, the dynamic mirror output current is scaled by scaling the device

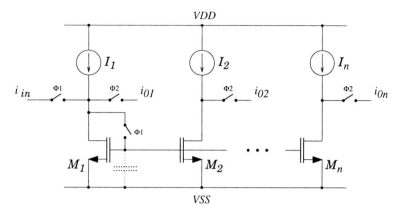

**FIGURE 6.12**
Dynamic current mirror with multiple output currents, $i_{o1}$, $i_{o2}$...$i_{on}$. Transistors $M_2$, ..., $M_n$ are added to produce additional, possibly scaled outputs.

aspect ratios and output bias currents. Unfortunately, the additional devices re-introduce errors due to device mismatches. Another way to scale the output current and produce multiple outputs is by time multiplexing [5].

Before describing how these building blocks are combined to obtain more advanced functions such as integrators and filters, we will contrast the current T/H and dynamic mirror characteristics. Note that the simple description of the basic circuit operation given above does not include many practical design considerations such as clock-feedthrough effects, transistor mismatch, noise, etc. These practical design issues will be discussed in detail in Sec. 6.5.

- The current T/H performs mirroring in space (silicon area) while the dynamic mirror performs the mirroring operation in time (clocked).
- The dynamic mirror consumes approximately half the power and half the chip area of the current T/H.
- In the dynamic mirror, the bias current is summed with the input signal current and then subtracted from it when the output is sensed. Thus, the exact value of the bias current is not critical. Mismatches in the current T/H bias sources, on the other hand, produce a dc-offset in the output current.
- The current T/H is a more flexible circuit building block since the dynamic mirror only produces one unscaled copy of the current at the output.

## 6.3.2   Switched-Current Filter Design

Using the dynamic mirror and current T/H, more complex operations may be performed. To illustrate this point, we will present the design of a switched-current integrator and, as done with the continuous-time integrator, show how it may be used in the design of active ladder filters.

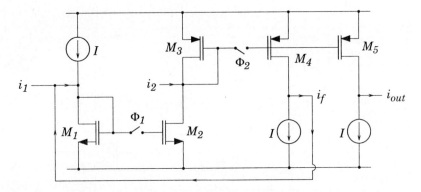

**FIGURE 6.13**
Switched-current integrator.

The switched-current integrator is composed of two cascaded current T/Hs (Fig. 6.13). The output of the second current T/H is connected to the input of the first T/H. Two-phase non-overlapping clocks control the switches in the two T/Hs. Breaking the feedback loop at the output and writing the expression for the output current at the drain of $M_4$:

$$i_f = (i_f + i_1)z^{-1} - i_2 z^{-12} \tag{6.33}$$

Both the feedback current, $i_f$, and the input current, $i_1$, are delayed by a full clock period to the output. Recall that the current T/H is an inverting amplifier and both $i_f$ and $i_1$ are applied to the noninverting input while $i_2$ is connected to the inverting input and delayed by one-half clock period. Rearranging Eq. (6.33),

$$i_f = \frac{i_1 z^{-1} - i_2 z^{-12}}{1 - z^{-1}} \tag{6.34}$$

To scale the integrator output, an output branch ($M_5$, $I_5$) is added resulting in

$$i_{out} = K \frac{i_1 z^{-1} - i_2 z^{-12}}{1 - z^{-1}} \tag{6.35}$$

where $(W/L)_5 = K(W/L)_3$. This expression corresponds directly to the sampled-data differential integration of SC circuits. Thus, all the design techniques established for SC filter design apply directly to the design of SI filters.

As previously demonstrated for the continuous-time differential integrator, the differential integrator may be used to design active ladder filters. The well-established methods of transforming an analog prototype ladder filter into a sampled-data one can be directly applied to the design of switched-current filters. As an example, we will construct a sixth-order bandpass filter using the lossless discrete integrator (LDI) transformation (refer to Chapter 9 for a detailed description of the LDI transformation).

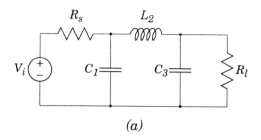

(a)

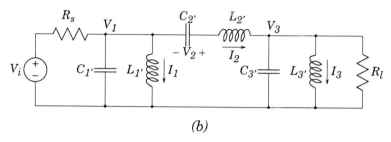

(b)

**FIGURE 6.14**
Low-pass and bandpass LC ladders: (a) third-order low-pass structure; (b) sixth-order bandpass structure produced by low-pass to band-pass transformation.

**Example 6.2.    High-Q Bandpass Filter.** Design a sixth-order Chebyshev bandpass SI filter. Determine the integrator gains and interconnections.

*Solution:* Consider the third-order low-pass LC ladder structure of Fig. 6.14(a). The values of $C_1$, $L_2$, and $C_3$ can be found using filter tables or numerical algorithms. The tables express component values normalized to (1 rad/sec), so the component values must be *denormalized* to the desired bandwidth of the bandpass filter by the relations $C = C_0/(2\pi bw \cdot R)$ and $L = (L_0 \cdot R)/(2\pi bw)$, where $C_0$ and $L_0$ are the normalized table values and $R$ is the termination resistance (assuming $R_s = R_l = R$ for convenience). The filter is then transformed to a bandpass structure by placing an inductor in parallel with each capacitor and a capacitor in series with each inductor. Each branch of the ladder is tuned to the center frequency using the relations $L' = 1/[(2\pi f_0)^2 \cdot C]$ and $C' = 1/[(2\pi f_0)^2 \cdot L]$. Note that $f_0$ is the *geometric* center frequency, $f_0 = \sqrt{f_l f_h}$, where $f_l$ and $f_h$ are the low and high breakpoints, respectively.

Once the component values are determined, the passive ladder prototype is described as a system of first-order differential equations in the complex frequency variable $s$:

$$V_1 = \frac{1}{sC_{1'}}\left(\frac{V_i - V_1}{R_s} - I_1 - I_2\right) \tag{6.36}$$

$$I_1 = \frac{1}{sL_{1'}}V_1 \tag{6.37}$$

$$V_2 = \frac{1}{sC_{2'}}(-I_2) \tag{6.38}$$

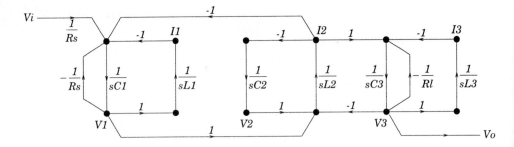

**FIGURE 6.15**
Signal flow-graph for the continuous-time system of Fig. 6.14(b).

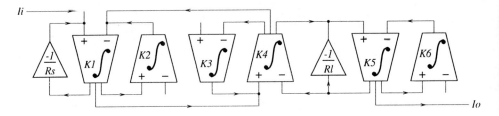

**FIGURE 6.16**
Block diagram of the bandpass filter resulting from LDI transformation of Fig. 6.15.

$$I_2 = \frac{1}{sL_{2'}}(V_1 + V_2 - V_3) \tag{6.39}$$

$$V_3 = \frac{1}{sC_{3'}}(I_2 - I_3 - \frac{V_3}{R_l}) \tag{6.40}$$

$$I_3 = \frac{1}{sL_{3'}}V_3 \tag{6.41}$$

The equations for $V_1$, $V_2$, and $V_3$ are obtained by applying Kirchhoff's current law at the appropriate nodes, and $I_1$, $I_2$, and $I_3$ are the result of KVL across $L_{1'}$, $L_{2'}$, and $L_{3'}$, respectively. Note that $V_2$ is defined as the voltage *across* $C_2$, and not as the voltage at the node between $C_2$ and $L_2$. This is done out of convenience rather than necessity, as any algebraically equivalent method of describing the system will result in the same transfer function.

Using the set of equations derived above, a signal flow-graph (Fig. 6.15) is then constructed. The vertical branches containing the $1/s$ terms represent integration with a time constant determined by the various component values. The two termination resistors are shown by the $1/R_s$ and $1/R_l$ branches.

LDI transformation involves replacing each integration branch in the continuous-time flow-graph with an appropriately scaled discrete-time integrator, in this case, a switched-current integrator. The scaling factors for each discrete-time integrator are related directly to the continuous-time component values.

The integrator scaling constants are found from the component values of the original circuit and the SI scaled branch $K$ is:

$$K = \frac{1}{X} \cdot \frac{1}{f_s} \tag{6.42}$$

where X is the component value ($C$ or $L$), and $f_s$ is the sampling frequency in Hertz. Combining this with Eqs. (6.36) - (6.41) yields

$$K_1 = \frac{1}{C_1}\frac{1}{f_s} = \frac{R \cdot 2\pi bw}{C_{10}} \cdot \frac{1}{f_s} \qquad (6.43)$$

$$K_2 = \frac{1}{L_1}\frac{1}{f_s} = \frac{(2\pi f_0)^2 \cdot C_{10}}{R \cdot 2\pi bw} \cdot \frac{1}{f_s} \qquad (6.44)$$

$$K_3 = \frac{1}{C_2}\frac{1}{f_s} = \frac{R \cdot L_{20}(2\pi f_0)^2}{2\pi bw} \cdot \frac{1}{f_s} \qquad (6.45)$$

$$K_4 = \frac{1}{L_2}\frac{1}{f_s} = \frac{2\pi bw}{L_{20} \cdot R} \cdot \frac{1}{f_s} \qquad (6.46)$$

$$K_5 = \frac{1}{C_3}\frac{1}{f_s} = \frac{R \cdot 2\pi bw}{C_{30}} \cdot \frac{1}{f_s} \qquad (6.47)$$

$$K_6 = \frac{1}{L_3}\frac{1}{f_s} = \frac{(2\pi f_0)^2 \cdot C_{30}}{R \cdot 2\pi bw} \cdot \frac{1}{f_s} \qquad (6.48)$$

where each $K$ is the integrator transistor aspect ratio $(W/L)_5 = K(W/L)_3$, $bw$ is the filter bandwidth, $C_{10}$ is the normalized capacitance, and $f_s$ is the clock frequency. Once each of these values is determined, the integrators are interconnected to satisfy the signal flow-graph (Fig. 6.16).

## 6.4 SWITCHED-CURRENT DATA CONVERTERS

Both A/D and D/A converters have been designed using the switched-current technique. Using this approach, high resolution data conversion is obtained without requiring precision linear capacitors. In this section, we will briefly discuss the implementation of SI data converters.

The sigma-delta A/D converter has received considerable attention because of the high-resolution, high-linearity conversion possible [57]. This high resolution is obtained by highly oversampling the input and trading off resolution in time for resolution in amplitude. The block diagram of a second-order sigma-delta modulator architecture is shown in Fig. 6.17. As in the design of the filters already discussed, the integrator is the basic building block of the converter. However, because this is a mixed analog-digital system (analog modulator and digital filtering), a fully-differential integrator is needed for common-mode rejection of the digital switching noise [58]. A fully-differential SI integrator is described in a later section. Note also that the conversion is performed on a current signal. To test the performance of the converter, a very linear current signal must be supplied. Thus, for testing purposes either a voltage-to-current converter must be included on chip or it can be built off-chip.

Other A/D architectures are also amenable for current-mode implementation. These include such architectures as the algorithmic and pipelined A/D converters [59]-[64]. In each case the dynamic mirror is optimized to perform the primary operation of signal delay (see Prob. 6.2).

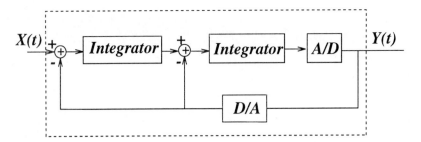

**FIGURE 6.17**
Second-order sigma-delta analog modulator.

There is increasing demand for very high resolution digital-to-analog converters for measurement equipment and digital audio applications. Often the resolution is constrained to 10 bits or less because of the limited component precision within a standard CMOS process. A 16-bit current-mode D/A converter overcomes this limitation by employing improved dynamic mirrors which produce current sources that match to within 0.02 % [65].

A common technique for implementing a D/A converter is to use many reference currents or current segments that are controlled by the digital input, Fig. 6.18. The total output current is determined by the number of reference current sources that are active. This output current is subsequently converted to a voltage. The accuracy of the currents determine the overall accuracy of the D/A. Because the dynamic mirror does not depend on matching, it is ideally suited for this application. A reference current is copied into each of the devices in the current segment. Each device will obtain the voltage corresponding to the reference current. These currents then can be output by controlling the switch to the output node.

While device mismatches are not a concern with the dynamic mirror, errors in the current will be produced due to finite output impedance and clock-feedthrough effects. Using a high quality dynamic mirror, such as the regulated-gate cascode, will virtually eliminate any errors due to finite output impedance. If the clock-feedthrough is exactly the same for each device it is copied into, it does not present a problem.

This dynamic mirror is then set up to be continuously calibrated such that no separate calibration cycle is required. This is done by having an extra dynamic mirror such that calibration is always performed on one mirror while the other mirrors are operating.

## 6.5 PRACTICAL CONSIDERATIONS IN SI CIRCUITS

Up to this point, we have not considered the practical circuit design issues in the application of SI circuits. In this section, we evaluate the accuracy, bandwidth capability, dynamic range, and low voltage characteristics. Additionally,

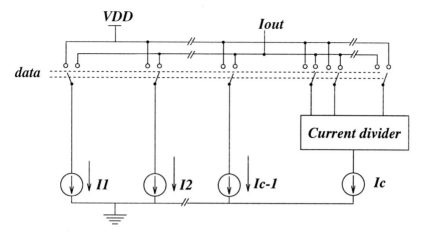

**FIGURE 6.18**
Basic diagram of D/A converter constructed of identical current segments.

simulation issues are discussed for current-mode circuits.

### 6.5.1 Accuracy

In the example above, we have not yet considered the obtainable frequency response accuracy. The inaccuracies introduced by all sources may be characterized by the dc-offset error, the ac-gain error, and harmonic distortion. DC offsets add or subtract current to the signal current, ac-gain errors change the amplification factor, and harmonic distortion reduces the dynamic range of the circuit. The source of these errors is the dc-current mirror biasing, charge injection, mismatch, and finite output impedance effects. Each of these errors are described in detail below.

**Current Mirror Device Biasing**   Highly linear current mirror operation is required to obtain accurate SI circuit operation. Whether the SI circuits are implemented using current T/Hs or dynamic mirrors, the dc-biasing of these building blocks determines the linearity, and therefore, the accuracy of the SI system. An example of a current mirror which is difficult to operate linearly over a large signal range (greater than about 10% of the bias current) is the high-swing cascode current mirror (Fig. 6.4(b)). With a fixed cascode bias voltage, $V_{bias}$, and a peak signal level that is 50% of the bias current, the output signal is distorted relative to the input current. The distortion is particularly evident at the peak value of the signal current. The large current produces a correspondingly large saturation voltage for transistors $M_1$ and $M_2$. This voltage exceeds the fixed drain voltage on $M_1$ and $M_2$ set by the constant bias voltage. Consequently, the mirror transistors, $M_1$ and $M_2$, operate in the triode region for large current inputs and the output current is no longer able to track the input current accurately.

Adaptively biasing the gates of the cascode devices $M_3$ and $M_4$ (Fig. 6.4(b)) eliminates the nonlinearity due to the improper biasing but requires two additional transistors [26]. The regulated-cascode current mirror is a self-biased current source and uses fewer transistors. It can be designed to operate linearly with large signal levels because, as the signal current changes, the cascode devices' voltage bias also changes.

**Device Matching**   Current-mode sampled-data systems rely heavily on accurate matching for high precision operation. In the current T/H and the ratioed dynamic mirror, mismatched devices will distort the output current. In this section, we will first examine the sources of transistor mismatches and the effect of these mismatches on the output current. Next, design techniques are presented that reduce the mismatch-induced errors.

Transistor parameter mismatch is a result of variations in the fabrication process [11, 10, 28]. Understanding the origin and magnitude of these variations gives insight into developing designs insensitive to these mismatches. The transistor parameters that vary and therefore cause the most significant errors in the current mirror output current are the threshold voltage, $V_T$, the transconductance parameter, $k'$, and the aspect ratio, $(W/L)$ [10]. For typical gate-source bias voltages ($\approx 1.2$ to $1.8$ V), the threshold voltage is the dominant source of error. These mismatches are due to process charge variations in ion implantations, fixed oxide charge, and substrate doping. Mobility variations cause mismatches in the transconductance parameter which is in part due to the threshold voltage implant variations. Because n-channel transistors have less variations in mobility than p-channel transistors, they exhibit better matching than identically designed and biased p-channel transistors. Matching of two transistors is also affected by the orientation of the two devices in the layout (see Chapter 16). As much as a 2% higher current mismatch results when one device is rotated 90 degrees with respect to the other [10]. Finally, with careful layout and device size selection, 8-bits of matching accuracy can be obtained with transistors and up to 10-bits is possible [11].

Pelgrom et al., [10] experimentally verified a model that accounts for the mismatches in $V_T$, $k'$, $W$, and $L$ between two transistors (see Chapter 13 for more detail). The graph of Fig. 6.19 illustrates the current error within one standard deviation of the mean as a function of the transistor area for a $2.5\mu$ n-well CMOS process technology. As seen in the graph, increasing $(V_{GS} - V_T)$ and/or the transistor area reduces the error in the output current. Designing a current T/H with high accuracy requires large devices biased at high gate-source voltage levels. The current error decreases for scaled processes as a result of the threshold voltage mismatch decreasing with thinner oxides. The transconductance parameter mismatch, on the other hand, remains approximately constant.

In current-mode circuits, we are interested in the effect of mismatches on the signal current accuracy. Using the simple current mirror of Fig. 6.1, the errors in the output signal can be expressed in terms of the mismatch parameters described above. Consider the simple current mirror of Fig. 6.1 with a signal

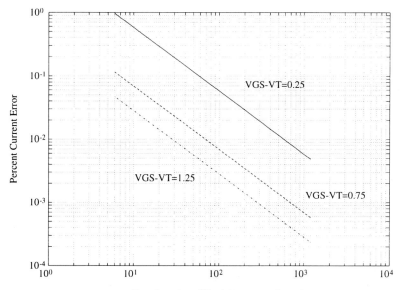

**FIGURE 6.19**
Percent current error versus the transistor area for several gate bias levels.

current $i_{in}$. If the mirror transistors $M_1$ and $M_2$ are perfectly matched and neglecting the finite output impedance effects, the output current is an exact replica of the input current. Mismatches in any one of the device parameters of $M_1$ and $M_2$ including $V_T$, $k'$, $\lambda$, $\frac{W}{L}$, will introduce errors in the output current [9]. The errors in the output current due to threshold voltage mismatches are evaluated below and the errors due to the other parameters will be left as an exercise.

Threshold voltage mismatch, $\Delta V_T = V_{T1} - V_{T2}$, is the major source of mismatch error in MOS transistors for typical gate-source bias voltages [10]. In order to analyze the effect of these mismatches, we will assume $M_1$ and $M_2$ of Fig. 6.1 are identical with the exception of the threshold voltage mismatches. Also assume that the dc-bias currents supplied to $M_1$ and $M_2$ is identical, i.e., $I_1 = I_2 = I$. The expression for the drain current in $M_2$ is

$$I_{ds2} = (K_2/2)(V_{gs2} - V_{T2})^2 \qquad (6.49)$$

where $K = k'(W/L)$. Since the gates of $M_1$ and $M_2$ are tied together, $V_{gs1} = V_{gs2}$, and we define

$$V_{gs1} = V_{T1} + \sqrt{\frac{2(I + i)}{K_1}} \qquad (6.50)$$

where $i = i_{in}$. Substituting Eq. (6.49) into Eq. (6.50) and applying the binomial expansion yields

**TABLE 6.2**
Distortion in the output current due to W, L, and k' mismatches.

| Parameter | DC Offset | AC Gain Error |
|---|---|---|
| $\Delta W = W_1 - W_2$ | $(\Delta W/L)I$ | $\Delta W/W$ |
| $\Delta L = L_1 - L_2$ | $-(\Delta L/L)I$ | $-\Delta L/L$ |
| $\Delta k' = k'_1 - k'_2$ | $(\Delta k'/k')I$ | $\Delta k'/k'$ |

$$I_{ds2} = I\left[1 + 2\frac{\Delta V_T}{(V_{gs} - V_{T1})} + \left(\frac{\Delta V_T}{V_{gs} - V_{t1}}\right)^2\right]$$

$$+ I\left\{(i/I) + \frac{2\Delta V_T}{V_{gs} - V_{T1}}\left[\frac{(\hat{i}/I)}{2} - \frac{(\hat{i}/I)^2}{8} + \frac{(\hat{i}/I)^3}{16} - ...\right]\right\} \quad (6.51)$$

where ideally $I_{ds2} = I + i_{in}$. The actual output current in Eq. (6.51) has a dc-offset indicated by the second and third terms in the first line and it depends on the magnitude of the bias current $I$. Ideally the second line of Eq. (6.51) should equal the signal input $i$, but due to the threshold voltage mismatches, ac-gain errors and harmonic distortion are present. The output current ac-gain errors are approximately equal to

$$\Delta G \approx \frac{2\Delta V_T}{V_{GS} - V_T}\hat{i} \quad (6.52)$$

The total harmonic distortion is approximately

$$THD \approx \frac{\Delta V_T}{I(V_{GS} - V_T)}\hat{i} \quad (6.53)$$

From Eqs. (6.52) and (6.53) it is evident that the harmonic distortion and ac-gain errors are reduced for signal currents that are small relative to the bias current. It is also desirable to make the excess gate voltage as large as possible to reduce the gain errors and harmonic distortion. The errors due to the width, length, and transconductance parameter mismatches are shown in Table 6.2. Unlike the threshold voltage mismatch errors, these parameter mismatches do not produce harmonic distortion because they are linearly related to the drain-source current.

**Charge Injection**   Clock-feedthrough induced charge injection onto the data-holding node is another major source of error in sampled-data circuits. The non-ideal characteristics of the MOSFET switch enable charge to couple onto the data-holding capacitance as the switch is being turned off. This injected charge produces errors in the held signal and limits the overall circuit accuracy. In this section, we describe the clock-feedthrough mechanisms and the techniques for modeling this phenomenon. We will also present design techniques for reducing the clock-feedthrough effects in current-mode sampled-data circuits.

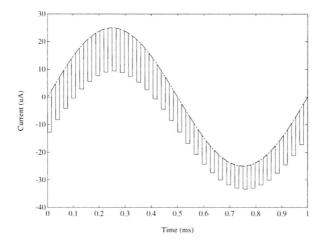

**FIGURE 6.20**
SPICE simulation of current T/H clocked at 20 times the signal frequency.

To illustrate the effects of clock-feedthrough on the output current, consider the current T/H of Fig. 6.10. A sinusoidal current is input to the T/H and sampled at 20 times the frequency of the input signal. The SPICE simulated output of the current T/H is shown in Fig. 6.20. The input current is represented by the dotted line and the solid line illustrates the current output. The ideal output waveform would be held at the value corresponding to when the switch opened. In the actual signal output, the held signal is offset due to clock-feedthrough voltage. This offset, due to charge injection, produces signal distortion in the form of ac-gain errors, dc-offset errors, and harmonic distortion. The errors due to charge injection effects are, to the first order, similar to threshold voltage mismatches. The difference is that the charge injection voltage varies as the signal level varies. In Fig. 6.20 note that the offset is higher for a peak signal level and lower for a minimum signal level.

To gain a thorough understanding of the charge injection mechanism, we will begin by describing the equivalent circuit model as the switch is being turned off for the current T/H of Fig. 6.10. Initially assume that the switch transistor, $M_s$, is conducting. That is, $V_{gs(switch)}$ is greater than the threshold voltage. The switch is modeled by the channel resistance, $R_{sw}(t)$, and the capacitances from gate-to-drain and gate-to-source, Fig. 6.21(a). The source resistance, $1/g_{m1}$, and capacitance, $C_{gs1}$, represent the equivalent circuit of the diode-connected transistor $M_1$. The model is completed by including the signal-holding capacitance which is the gate-source capacitance, $C_L$. Therefore, as the clock voltage decreases but the switch is still conducting, there are two sources of charge injection onto the data-holding node: channel charge and charge coupled through the gate-source and gate-drain overlap capacitances. Partial differential equations

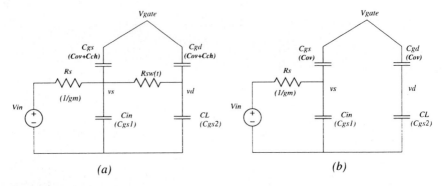

**FIGURE 6.21**
Clock-feedthrough model for switched current mirror when (a) the switch is conducting and (b) the switch is not conducting.

can be written to describe the voltage change on each of the circuit nodes [30].

$$\frac{\partial v_s}{\partial t} = \frac{(v_d - v_s)}{R_{sw}(t)C_{in}} - \frac{v_s}{R_s(t)C_{in}} - \frac{i_{inj}}{C_{in}} \tag{6.54}$$

$$\frac{\partial v_d}{\partial t} = \frac{(v_s - v_d)}{R_{sw}(t)C_L} - \frac{i_{inj}}{C_L} \tag{6.55}$$

$$i_{inj} = \frac{\partial(v_s - v_{gate})}{\partial t}C_{g(sw)} \tag{6.56}$$

$$C_{g(sw)} = \alpha(C_{ov} + 1/2C_{ox}W_{eff}L_{eff})_{sw} \tag{6.57}$$

where $\alpha$ is the clock slope in V/sec, and $V_{gate}$ is the voltage on the gate, $i_{inj}$ is the capacitive coupling through the gate capacitance including both the channel capacitance $(1/2C_{ox}W_{eff}L_{eff})$ and the overlap capacitance, $C_{ov}$. These expressions are valid until the switch is turned off, i.e., $V_{gate} = (V_T + V_s)_{sw}$. The dc-voltage applied to the input is assumed constant while the switch is being turned off – this assumption is valid for slowly varying signals relative to the clock turn-off rate.

The solutions to these expressions are found numerically and are used to determine the clock-feedthrough voltage induced onto the data-holding node, $v_d$. Once the switch is no longer conducting, the channel is eliminated and the channel resistance becomes infinite. The circuit model for the current T/H with the switch off is Fig. 6.21(b). With the channel eliminated, the gate-drain and gate-source coupling capacitance is just the gate overlap capacitance. The expression describing the voltage change on the drain at this time is:

$$\frac{\partial v_d}{\delta t} = -\frac{C_{ov}(sw)\alpha}{C_L} \tag{6.58}$$

$$\frac{\partial v_s}{\partial t} = -\frac{v_s}{R_s(t)C_{in}} + \frac{\partial v_d}{\partial t}\frac{C_L}{C_{in}} \tag{6.59}$$

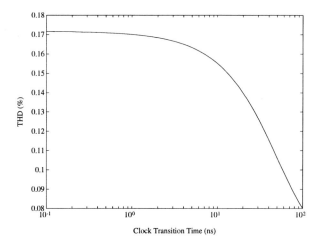

**FIGURE 6.22**
THD in output current of SI T/H due to clock-feedthrough versus clock turn-off rate.

The solution to these two equations yields the total charge injected onto the gate of $M_2$. This voltage is then transformed into current through the current-voltage relationship and for transistors operating in saturation this is:

$$I_{ds} = K/2(V_{gs} + v_d - V_T)^2 \qquad (6.60)$$

Solving the equations just described, we determine the gain error, offset error, and harmonic distortion. A plot of the total harmonic distortion, THD, versus clock rate is shown in Fig. 6.22. If the clock slope is slow relative to the $C_{in}R_{sw}$ time constant, some of the charge injected onto the load will leak off and reduce the error $v_d$ during this portion. Thus, in Fig. 6.22, for a clock slope of $100ns$ there is 0.08% THD compared to 0.16% with a $10ns$ turnoff time (see Prob. 6.15).

Based on the previous analysis, there are several design guidelines that should be observed in order to reduce the charge injection in SI circuits.

1. $C_{ov}(sw) \ll C_L$: Choosing the load capacitance to be very large with respect to the switch area will reduce the coupling of the clock signal into the data-holding node significantly. The load capacitance is made large by using a large mirror transistor, $M_2$, or by including an additional capacitance on the gate of $M_2$. A distinct advantage of current-mode circuits is that the capacitance need not be linear and, in some CMOS process technologies, the gate capacitance is more area-efficient than the highly-linear, double-poly capacitance. The trade-off of using a larger capacitance for improved accuracy is the reduction in the system bandwidth.

2. $V_{GS}$ *large:* Designing $V_{GS}$ as large as possible will result in larger absolute clock-feedthrough voltage. However, the larger value of voltage is a smaller

percentage of the total gate-source voltage, and therefore, contributes less overall error. The gate-source voltage is made large by designing with higher currents or choosing devices tending to be more square. A secondary benefit of increasing $V_{GS}$ is that there are less effects from errors due to threshold voltage mismatches.

3. *Slow clock slope relative to $C_{in}R_{sw}$ time constant:* As illustrated in Fig. 6.22, if the clock turns the switch off slowly, the charge on the data-holding node may leak off to the source side and this reduces the total error. The switch resistance may be increased instead of increasing the clock slope to produce charge leakage from the load capacitance. However, the switch resistance is increased by increasing the switch size which ultimately results in much higher capacitive coupling of the clock voltage onto the data-holding node.

4. *Small clock voltage swing:* Once the clock voltage is to the point that the switch in the current T/H is no longer conducting, reducing it further only contributes more clock-feedthrough voltage. Reducing the clock voltage swing to the point where the switch just turns on and off reduces the excess clock voltage coupling.

5. *Reduce the signal to bias current ratio:* For large signal excursions the variation in the clock-feedthrough voltage is larger resulting in more harmonic distortion. For maximum power efficiency, generally a 50% signal swing is desirable.

**Clock-feedthrough Reduction Schemes**   There are several circuit techniques used to reduce the charge injection errors in switched-current circuits. Two cancellation techniques used widely in SC circuits are the dummy switch and the transmission gate (Fig. 6.23). A dummy switch is one-half the size of the main switch and the drain and source terminals are shorted. It is controlled by an inverted clock so that, as the main switch is turned off, the dummy switch is turned on and collects the injected charge. Precise cancellation of the charge requires clocks which are identical (but inverted). The transmission gate shown in Fig. 6.23 consists of two complementary devices controlled by a clock and an inverted clock. Again, clock phasing is critical to obtain complete cancellation of the charge injection. In current-mode circuits, a combination of both cancellation schemes ensures the best performance. In general, current-carrying switches are best compensated by employing complementary devices. This is because the parallel combination of the two device types results in a lower total resistance and therefore less distortion in the signal path. Switches that short the gate and drain carry negligible currents and are compensated for with the area-efficient dummy switch.

Other circuit techniques have been developed for cancellation of the clock-feedthrough effects in switched currents such as an adaptive clock [23], offset cancellation [9], and signal-dependent cancellation [32]. In general, the additional circuitry required by each of these techniques does not provide improvement significant enough to warrant the increased area.

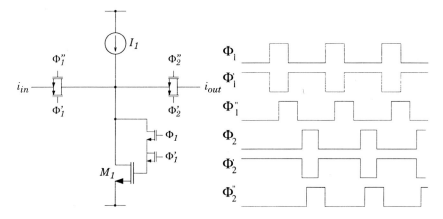

**FIGURE 6.23**
Dynamic mirror with a dummy switch and transmission gates employed for clock-feedthrough cancellation. Primed clock signals are inverted, and doubly-primed signals are inverted twice to add delay.

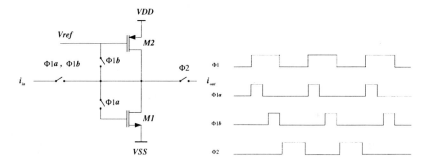

**FIGURE 6.24**
Two-step switched-current circuit and clock waveforms.

A circuit recently introduced SI delay cell that compensates for the signal dependent clock-feedthrough is the two-step approach [37]. The circuit and the three controlling clocks are shown in Fig. 6.24. On $\phi$1a, $M_1$ is diode-connected and $M_2$ is connected to a bias voltage. During this time, a bias current, $I$, is produced in $M_2$ and the total current in $M_2$ is the bias plus the signal current, $I + i_{in}$. Next, $\phi$1b is active and $M_2$ is diode-connected while the gate of $M_1$ floats. The signal current is still flowing into the input so that the current in $M_1$ is $I + i_{in} + i_{cft}$ and the current in $M_2$ is $I + i_{cft}$ where $i_{cft}$, is the signal-dependent clock-feedthrough. Finally when $\phi$2 is active, the current in $M_2$ is $I + i_{cft} + \Delta i$, the current in $M_1$ is $I + i_{in} + i_{cft}$, and the output is $-i_{in} + \Delta i$ (where $\Delta i$ is the clock-feedthrough current). Thus, a small, nearly constant offset remains at the output. This SI circuit may be particularly well suited for low voltage applications.

In switched-capacitor circuits, fully-differential integrators are employed to cancel the clock-feedthrough effects. In fully-differential SC integrators, the clock-feedthrough voltage appears, to a first order, as a common-mode signal and is therefore rejected. In SI circuits, fully-differential delay cells provide similar cancellation [34, 35, 38, 36].

The fully-differential dynamic-mirror structure is composed of two dynamic mirrors placed on top of a tail current source. Figure 6.25 shows the differential dynamic-mirror with the most important parasitic capacitances included.

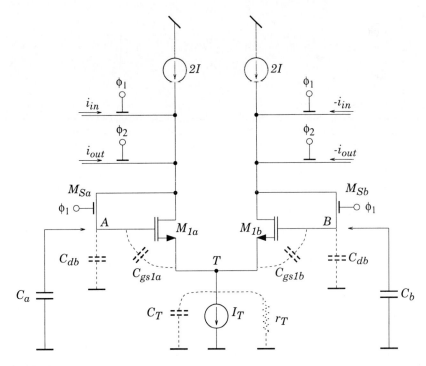

**FIGURE 6.25**
The fully-differential dynamic mirror.

The circuit is stimulated with purely differential signal currents $+i_{in}$ and $-i_{in}$ which, during $\phi_1$, pass through the diode-connected transistors $M_{1a}$ and $M_{1b}$. When the clock $\phi_1$ goes low, the switches $M_{Sa}$ and $M_{Sb}$ open and the information is stored on the gate-source capacitances of $M_{1a}$ and $M_{1b}$. Turning off the switches injects charge packets that make a voltage error $\delta V$ appear at the gate of $M_{1a}$ and $M_{1b}$. Since $M_{Sa}$ and $M_{Sb}$ switch at the same time, are the same size, and see the same load at the drain and source contacts, they will generate the same $\delta V$ that will be treated as a common-mode voltage input by the source-coupled differential pair. Any common-mode voltage input will produce variations in the voltage at node $T$ (the top of the tail current source). The high output impedance of the tail current source, $I_T$, prevents $\delta V$ from generating a current error through the data-holding devices. As the output information is differential, any even-ordered residual harmonics will be canceled.

It is important to make the distinction now between a fully-differential implementation and a fully-balanced configuration. The fully-balanced circuit consists of two single-ended circuits with opposite sign input signals where the difference of the signals is taken at the output. Both circuits are incapable of completely eliminating the clock-feedthrough (CFT) errors because this phenomenon is signal dependent. The fully-differential circuit, however, virtually cancels all errors due to the common part of the CFT (the signal independent part) before the signal is transformed into a distorted current. The balanced circuit processes the distorted current and cancels only the even harmonics at the output. Therefore, the accuracy of the fully-differential circuit is much higher. Furthermore, the degree of cancellation obtained by the fully-differential configuration depends on the matching of the data- holding FETs ($M_{1a}$ and $M_{1b}$) and of the switches ($M_{Sa}$ and $M_{Sb}$), while the balanced circuit would also depend on the accuracy of the current-mirrors that perform the final differencing.

Figure 6.25 reveals an unexpected problem. The clock-feedthrough injected charge is a common-mode charge that sees only the effective common-mode input capacitances at the gates of $M_{1a}$ and $M_{1b}$. This common-mode capacitance is the equivalent capacitance seen at nodes $A$ or $B$ to ground and is given by:

$$C_{com} = \left( \frac{C_{gs_a} \cdot \frac{C_t}{2}}{C_{gs_a} + \frac{C_t}{2}} \right) \parallel C_{db} \qquad (6.61)$$

Since $C_t$ is mainly a drain-gate overlap capacitance, it is two orders of magnitude less than $C_{gs}$. $C_{db}$ is the drain-to-bulk junction capacitance of the switch transistors, $M_{sab}$, and is also very small. The CFT charge, $q$, will be injected onto just a small $C_{com}$ capacitance and thus, the voltage jump $\delta V$ will be very large and the whole floating differential pair will move toward one of the supply rails. For example, in the case of N-channel switches, the injected charge is negative and the differential pair will move toward ground, pushing the tail current source out of saturation. This is essentially the problem of limited common-mode voltage range of a differential amplifier. The common-mode voltage jump at the gates of $M_{1a}$ and $M_{1b}$ has the value $\delta V = q/C_{com}$, immediately after the switches are turned-off. Although most of the voltage variation, $\delta V$, is absorbed by the tail current source, there will be a small variation, $\delta V_{gs}$, in the gate-to-source voltages of $M_{1a}$ and $M_{1b}$.

$$\delta V_{gs} = \delta V \frac{C_t/2}{C_{gs} + C_t/2} \qquad (6.62)$$

This small variation, $\delta V_{gs}$, causes a variation of the drain currents of the data-holding transistors and thus, $I_{d1a} + I_{d1b} \neq I_T$ and KCL is not satisfied. To satisfy it, the circuit shifts its common-mode voltage until the voltage variation on the tail current source compensates for the extra current. All the common-mode voltage variations are reflected at node $T$. But a shift of the common-mode voltage implies a voltage variation on the $C_{db}$ capacitors which can be accomplished only if the charge on them changes. This means that there will be charge redistribution between the gate of $M_{1ab}$ and the top plate of $C_{db}$. This charge redistribution will modify the original gate-to-source voltage variation given by Eq. (6.62). The equations that govern the steady-state voltage are the following:

$$\left\{ \begin{array}{ll} \Delta V\, C_{db} + \Delta V_{gs}\, C_{gs} = q & \text{– the charge conservation} \\ \Delta V_{gs} \cdot g_m = \frac{\Delta V - \Delta V_{gs}}{r_T} & \text{– the KCL at node } T \end{array} \right\} \qquad (6.63)$$

$\Delta V$ is the final common-mode voltage variation at the gates of $M_{1ab}$ and represents the amount of common-mode voltage that the fully-differential circuit has to be able to accommodate. $\Delta V_{gs}$ is the final gate-to-source voltage variation and represents the amount of error voltage that will affect the information stored during the hold period. After solving this system of equations, the solutions are:

$$\Delta V_{gs} = \frac{q}{g_m r_T C_{db} + C_{gs}} \qquad (6.64)$$

$$\Delta V = \frac{q}{C_{db} + \frac{C_{gs}}{g_m r_T}} \qquad (6.65)$$

Note that, in Eq. (6.64), the total resistance seen at node $T$, $r_T$, has a very important role. Ideally, $r_T = \infty$. For a finite $r_T$, the final gate-to-source voltage error can be much smaller than the single-ended corresponding error $(q/C_{gs})$ if $C_{db}$ can be made of the same order of magnitude as $C_{gs}$ due to the large value of $g_m r_T$. A straightforward way to increase $C_{db}$ is to add additional capacitors in parallel.

The final common-mode voltage is given by Eq. (6.65) and it is dependent on the value of $C_{db}$. If $C_{db}$ is small, the common-mode voltage can be very large.

There are several ways to correct this situation. One way is to design the structure to accept very large common-mode voltages. This implies designing a tail current source that withstands large voltage variations. This may not always be possible, especially with 3V power supply voltages.

Another way to correct for the small effective common-mode capacitance is to add physical capacitors in parallel with the inputs. These capacitors, $C_a$ and $C_b$, need not be very large, just on the order of $C_{gs}$. This would reduce the CFT-induced voltage jump to a manageable magnitude. $C_{gs_a}$ and $C_{gs_b}$ play a negligible role in the common-mode charge cancellation, so the area of these two transistors may be minimized. Matching of the input devices plays a small role in the integrator's accuracy, too. $C_a$ and $C_b$ are generated by using the gate-to-channel capacitance of supplementary transistors that have the source, drain and bulk contacts shorted to ground. As the circuit can handle a large amount of CFT (until we drive $I_T$ out of saturation), the transistors generating $C_a$ and $C_b$ can be much smaller than the usual T/H transistors and so a clear savings in real estate is created.

## 6.5.2 Low-Voltage Design

For low power applications, the transistors in the switched-current system may be operated in the subthreshold region (weak inversion) [27]. Although the current is related to the exponential of the voltage rather than the square of the voltage as in saturation, the subthreshold current T/H and dynamic mirror still perform all of the signal operations already described. For the same current level, the SI building block has a higher bandwidth when operated in the subthreshold region. There are two major disadvantages in operating the transistors in weak inversion rather than strong inversion: the flicker noise is higher and the mismatches are more significant. In subthreshold, $g_m/I$ is maximum and thus, the same threshold voltage mismatch between the mirror transistors produces as much as an order of magnitude higher error in the output current than for devices biased in saturation. For this reason, subthreshold transistor operation is reserved for applications where accuracy and dynamic range can be sacrificed to obtain lower power. A complete description of the subthreshold operation is given in Chapter 8.

## 6.5.3 Simulation

Simulation is a critical part of SI circuit design. Two types of simulators are currently available: circuit-level and system-level. The circuit-level simulators model the device non-ideal characteristics. Using this type of simulator, the circuit design may be optimized. However, when many SI cells are combined to form a system, i.e., a filter, the simulation-time becomes excessive. This is due to the sampled-data nature of the circuit. In general, many time-consuming transient simulations are required to obtain the frequency domain information. In this case, an efficient system-level simulator is usually used to evaluate the frequency domain system performance. While SI system-level design is very similar to SC system design, there are many unique characteristics of SI simulation. These unique characteristics are briefly described below.

SPICE or a SPICE-type simulator may be used to simulate the SI basic building blocks [42, 43, 44]. For example, to simulate the errors due to clock-feedthrough in the dynamic current mirror, a current signal is applied to the input and the current is sensed at the output. The input and output currents are compared to determine the error in the output current. If an ideal current source is applied to the input of the dynamic current mirror (Fig. 6.11), errors in the simulation will result. When the gate of the switch transistor $S_1$ is low (NMOS), the transistor is cut off. When this occurs, the voltage between the current source and the switch transistor will go to an artificially high (or low) value (such as 10,000 V) which will cause the switch to turn on. To overcome this situation, an MOS current mirror is placed between the ideal current source and the input to the dynamic current mirror. By doing this, the current source is isolated from the switch transistor. When the input switch is opened, the voltage at the input (drain of the switch) will not exceed the supply voltage since active devices are employed. To ensure that a very accurate reproduction of the current is obtained from the current mirror, the bias current is made large compared to the signal level and a high performance current mirror, such as the RGC, is used. To sense the current output, either a bias voltage is placed at the output or the actual load is applied (see Prob. 6.11).

Some simulators may not yield accurate transient simulation results for SI circuits. The tolerances in the simulator may be adjusted to improve the accuracy. However, adjusting the tolerances can significantly increase the simulation time. Some SPICE-type simulators do not perform accurate transient simulations because they do not conserve charge. This means that either charge is either created or destroyed rather than being conserved. It has been shown that a flaw in the MOS models causes the incorrect characterization of the charge.

The first widely used model for MOS devices was proposed by Meyer [45]. Meyer's model expresses the four-terminal MOS transistor I-V (current-voltage) characteristics in terms of the device physical parameters. The model equations describe the three-dimensional distribution of electrons and holes in the semiconductor under the gate oxide. Additionally, stray capacitances, such as wiring capacitance, coupling characteristics between terminals in the devices and the channel capacitance, are modeled so that the device behavior over frequency can be determined. The Meyer model provides a good description of the MOS transistor characteristics for most applications. As a result, it was incorporated into early versions of Berkeley SPICE and subsequent releases including SPICE3E2.

The Meyer model may not accurately predict the parasitic capacitances in the MOS transistor [46, 47, 48]. For sampled-data circuit applications the inaccuracies can lead to significant errors. These inaccuracies are particularly evident when simulating the clock-feedthrough effects in both switched-capacitor and switched-current circuits. The Meyer model inaccurately models the values of the capacitances between the gate and drain, leading to errors in the prediction of clock-feedthrough effects. This is illustrated by a simulation of the current track and hold, Fig. 6.26. A sinusoidal current is input to the circuit. The irregular jumps in the held portion are caused by the inaccuracies in the Meyer model capacitance prediction.

In the Meyer model the capacitances from the gate-to-source ($C_{gs}$) and the gate-to-drain ($C_{gd}$) are treated as lumped two-terminal nonlinear capacitors. It is assumed that the charge stored on $C_{gd}$ ($C_{gs}$) varies only as a function of $V_{gd}$ ($V_{gs}$). Since the charge on each capacitor is controlled by only one voltage, the capacitors are reciprocal, i.e.,

$$C_{gs} = C_{sg} \tag{6.66}$$

and

$$C_{gd} = C_{dg} \tag{6.67}$$

In reality, the charge stored on each capacitor is controlled by both $V_{gs}$ and $V_{gd}$. Therefore, the Meyer model erroneously models the charge stored on the capacitors and the capacitance values. In the literature, the problem is formally known as the Meyer Reciprocity Problem and the result of assuming reciprocity is that charge is not conserved in the MOSFET.

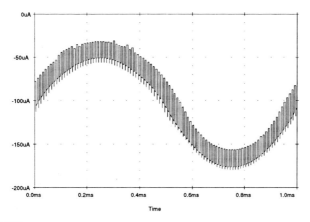

**FIGURE 6.26**
Plot showing problems with charge conservation.

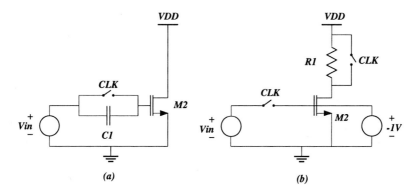

**FIGURE 6.27**
Charge conservation test circuits. (a) This circuit should be used to determine the analog simulators ability to conserve charge while the MOS transistor remains in the saturation region of operation. (b) This circuit should be used to determine the analog simulators ability to conserve charge while the MOS transistor jumps between linear and saturation regions of operation.

Two circuits are given to evaluate a simulators ability to conserve charge. The first circuit, shown in Fig. 6.27(a) tests for charge conservation of an MOS device in the saturation region. The combination of capacitor $C_1$ and the gate source capacitance of $M_2$ are in effect a capacitive voltage divider. The switch which is closed so there is no voltage across $C_1$, is opened at time $t = 0^+$ while $V_{in}$ is held at 3V. Then the voltage source is pulsed down to 0V and back to 3V ten times. If charge is conserved at the gate of $M_2$, the initial drain current should be equal to the current after the tenth cycle.

The second circuit, Fig. 6.27(b), begins with the switch across $R_1$ initially closed. The switch connected to the gate of $M_2$ opens at time $t = 0^+$ locking a charge, $Q$, on the gate of $M_2$. The switch across the resistor is subsequently cycled ten times, moving transistor $M_2$ from saturation to the linear region of operation and back again. If charge on the gate, $Q$, is conserved, the drain current after the tenth cycle should be the same as the drain current at time $t = 0^-$. If there are errors in the output currents, the simulator tolerances may be adjusted to compensate for these thus increasing the simulation time (see Prob. 6.14).

SC simulators may be used to simulate SI systems [49, 50, 51]. These simulators do not contain current-controlled current sources necessary to model the SI system. Therefore, in using these tools, only the signal flowgraph is verified. More recently, other simulators that include the nonideal circuit effects in the SI system-level simulation have been developed [54, 56, 53, 52]. Unfortunately, these tools are not widely available as yet.

## 6.6   DESIGN EXAMPLE: SI DTMF DECODER

The application of sampled-data current-mode circuits to system design has brought about further improvements in the basic circuit technique. In this section, the design of a current-mode, dual-tone, multiple-frequency decoder (DTMF) used for "touch-tone" telephone keypad detection is described.

A DTMF decoder interprets the tone signals associated with pressing the keys on a touch-tone telephone. The decoder processes an analog signal containing the tone (key) information and releases a digital code corresponding to the appropriate tone pair (key) is produced. Since the tones are transmitted via the same means as voice and other voice-band information (dial-tone, busy signal, etc.), the decoder must be able to 1) recognize tones in the presence of this other information, and 2) avoid interpreting the extraneous information as valid tone pairs. In addition, telephone transmission quality may vary depending on factors beyond the decoder's control. It must be able to recognize tone pairs which arrive at a range of levels.

### 6.6.1   Decoder Architecture

Each key on the touch-tone keypad corresponds to a row tone and a column tone. When a key is pressed, the DTMF transmitter generates the appropriate two tones simultaneously. The receiver senses which tone from each row and column is present and converts (decodes) it back to a number representing the key that was pressed. The basic structure of the decoder is shown in Fig. 6.28 [66, 67]. The idea behind decoding tone pairs is straightforward: simply apply the incoming signal to a set of bandpass filters and then determine which two filters have outputs that exceed some threshold. These two outputs correspond to the row and column of the key pressed on the keypad.

The DTMF decoder consists of several functional blocks including the following.

*Pre-filter.* The incoming signal is first pre-filtered to remove unwanted components. These include 60Hz line interference and dial tone, as well as compo-

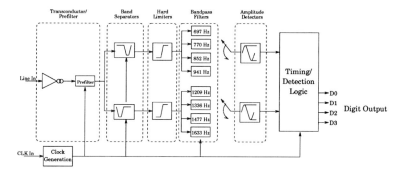

**FIGURE 6.28**
DTMF decoder block diagram.

nents above the Nyquist rate of the successive filters. Included in the pre-filtering is a voltage-to-current conversion.

*Band-Separating Filters.* The resulting band-limited signal is then split into the low (row) and high (column) tone groups. The low tone group spans 697Hz to 941Hz, and the high tone group spans 1209Hz to 1633Hz. Since these two bands are only 268 Hz apart and the system dynamic range is specified as 30dB, sixth-order Chebyshev filters are required. Extraneous signals will cause the zero crossing detectors (described below) to produce an erratic output, thereby increasing the immunity of the decoder to speech.

*Zero Crossing Detectors.* The band-separated signals are then converted into a hard-limited square wave by passing them through hard limiters. If there is only one valid tone present in the signal, then the output of the zero crossing detector will be a 50% duty cycle square wave at the tone frequency. If there are any other tones or extraneous information present (such as speech), then the output will be an erratic square wave due to the additional zero crossings. Since the signal may have dc-offset due to the preceding stages, the detector must be able to account for this and still output a 50% duty cycle signal. In addition, hysteresis is needed to reject extraneous signals below the 30dB.

*Bandpass Filter Bank.* Each of the hard-limited, band-separated signals is applied to a bank of four bandpass filters (i.e., one for each possible tone). These filters represent by far the highest level of technical challenge in the entire decoder. Their center frequencies must be accurate to $\pm(1.5\% + 2\text{Hz})$, and the quality factor, Q, required to obtain the desired selectivity is on the order of 20. The design must be immune enough to device mismatch and clock-feedthrough errors to ensure that the poles remain close enough to their designed locations.

*Amplitude Detectors.* The outputs of the bandpass filter bank are sensed by the amplitude detectors to determine which tone from each group is present. This is then compared to a threshold value. The output is a digital signal indicating the presence of a valid tone which corresponds to the key that was pressed.

*Digital Logic.* The digital functions include clock generation, timing and control logic, and output decoders and registers.

### 6.6.2   The SI DTMF Decoder Design

The SI DTMF decoder has been designed and integrated in a 2-micron p-well CMOS process. The design of each of the circuit blocks is briefly described.

*Sixth-Order Band-Separation Filters.* The two sixth-order Chebyshev band-reject filters are implemented using low-sensitivity coupled-biquad structures [68]. Beginning with an arbitrary low-pass filter prototype and transforming it to a band-reject filter, we are left with an even-ordered filter that can be separated into distinct biquadratic sections. Each section is built using a general biquad block, and the entire filter is obtained by coupling these in a leap-frog configuration.

As with the filters discussed in the previous examples, these filters are constructed by interconnecting switched-current integrators. As such, they are the most critical circuit building block in the system. Each integrator is composed of regulated-gate cascode (RGC) dynamic current mirrors (Fig. 6.29). In addition to dramatically increasing the output resistance, the RGC feedback structure provides a means of eliminating the permanent charge injection errors introduced on the data-holding node as the voltage spikes (toward the negative supply) in response to a current imbalance. During the time when $\Phi_1$ and $\Phi_2$ are both inactive, the drain of the dynamic mirror is in a high impedance state. Any current mismatch between the mirror and its bias source will therefore be reflected as a large voltage swing. A negative-going spike, if unsuppressed, will cause the drain voltage of $M_m$ to fluctuate, injecting charge onto the data-holding node and corrupting the sample. Adding the minimum-sized device $M_{ss}$ (dotted) will suppress this effect. Normally, $M_{ss}$ is off and does not affect the circuit operation. As $M_c$ begins to enter the triode region due to the decreasing output voltage, the feedback loop will cause its gate voltage to increase sharply. This will turn $M_{ss}$ on, feeding in enough extra current from the supply to maintain a constant voltage on the drain voltage of $M_m$ and minimize the impact on the held data.

With the current mirror chosen for the design, we next examine the integrator structure. Consider the integrator of Fig. 6.30(a). Due to the use of dynamic mirrors as the delay elements, the integration is free from device mismatch effects. The scaled output, however, has as much sensitivity to mismatch as a standard current mirror. In addition, the structure is lossless. Generating a feedback coefficient other than unity requires the addition of extra output stages. Output scaling ratios are also limited in resolution, since they must be chosen by scaling $W_3$ relative to the fixed width of the second dynamic mirror ($W_2$).

Figure 6.30(b) is a solution to this problem. The structure is similar to that of Fig. 6.30(a) in that dynamic mirrors are used to reduce area and eliminate mismatch effects. The operation, however, is slightly different. During $\Phi_2$, the current memorized by $M_1$ is divided between $M_{2a}$ and $M_{2b}$. The ratio of this

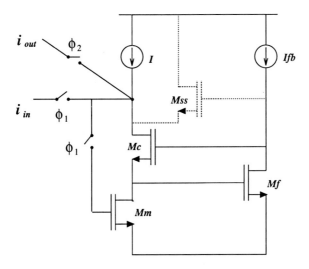

**FIGURE 6.29**

Regulated cascode structure. This high output impedance circuit is used to increase current mirror performance. Adding the minimum-sized device $M_{ss}$ eliminates the effects of voltage spiking in the dynamic current mirror.

division depends on the weighting of the aspect ratios of $M_{2a}$ and $M_{2b}$. During $\Phi_1$, only the current produced by $M_{2a}$ is fed back to $M_1$. Assuming $M_{2a}$ and $M_{2b}$ have equal gate lengths, the feedback coefficient, $A$, is determined by

$$A = \frac{W_{2a}}{W_{2a} + W_{2b}} \qquad (6.68)$$

The feedback coefficient, in addition to being less than unity, is now determined by two variables instead of one. By choosing $W_{2a}$ and $W_{2b}$ appropriately, almost any (rational) coefficient can be realized.

Once the coefficient is realized by choosing the width ratios, the actual sizes of the devices must be chosen. The results of Pelgrom et al. were extended to include two devices of different sizes as used to obtain the filter coefficients. Computer programs were used to obtain the filter coefficients. Taking into account the effects of transistor mismatch and round-off effects, a linear search algorithm then determined the transistor widths and width ratios to realize these coefficients to within 0.5%. The simulation of the ideal band-reject filters and the filters with the quantized coefficients is shown in Fig. 6.31.

*Bandpass Filters.* The second-order bandpass filter bank represents the highest degree of challenge in the DTMF implementation. As mentioned above, the selectivity and accuracy requirements dictate careful design. The implementation chosen is the "coupled" form for producing a pair of complex conjugate poles (Fig. 6.32) [69]. This structure was chosen based upon its simplicity and its similarity to ladder-type structures, which inherently have superior sensitivity properties. From Fig. 6.32, one can see that the structure requires only two coef-

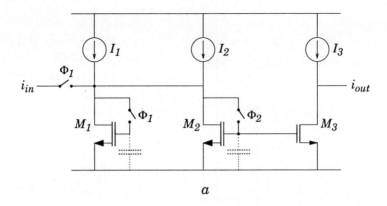

*a*

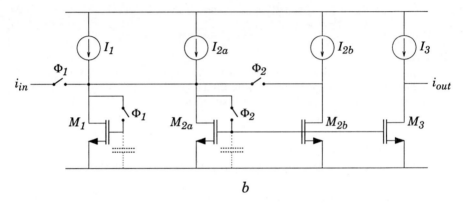

*b*

**FIGURE 6.30**
Dynamic integrator topologies. Structure of (a) lacks a variable feedback coefficient and has poor resolution of the scaled output. Improved structure (b) allows variable feedback coefficients less than unity and increases resolution.

ficients, thus simplifying the layout. In addition, these coefficients are simply the real and imaginary components of the z-plane pole locations. Given a sampling frequency, $f_s$, a center frequency, $f_0$, and the radius, $r$, of the poles, the filter coefficients are determined by

$$x = r\,cos(\theta) \tag{6.69}$$

$$y = r\,sin(\theta) \tag{6.70}$$

where

$$\theta = 2\pi\frac{f_0}{f_s} \tag{6.71}$$

To obtain the required quality factor of at least 15 in the bandpass filters, the pole radius needs to be approximately 0.97. This, in combination with the stringent tolerance on center frequency, gives rise to the need for an integrator topology possessing improved matching and resolution capabilities as described

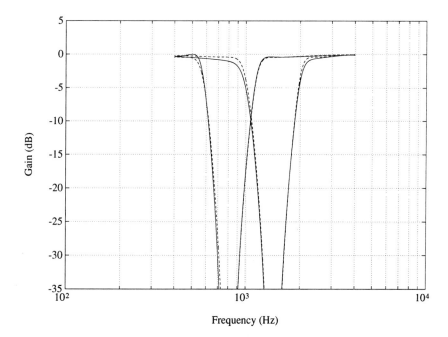

**FIGURE 6.31**
Coefficient quantization. Computer programs were written to initially quantize the coefficients to within 0.5% accuracy, and then iterate to minimize the error in the filter response using Pelgrom's model. Shown are the actual (solid) and ideal (dashed) transfer functions of both the high- and low-group reject filters.

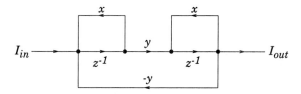

**FIGURE 6.32**
"Coupled-form" for the realization of complex conjugate poles. Structure contains only two coefficient values, representing the real and imaginary components of the pole locations.

above. Again, the transistor sizes used to realize the coefficients have been chosen based on the results of Pelgrom et al. [10]. To illustrate the effect of variations in the transistor ratios, a simulation of the filter with 1 % errors in the coefficients is shown (Fig. 6.33). The filter center frequency remains accurate while the filter bandwidth is reduced due to mismatches.

Another source of error that will contribute to the reduction in the Q is the charge injection and clock-feedthrough from the sampling switches. To minimize these effects, transmission gate compensation was used for the current-carrying

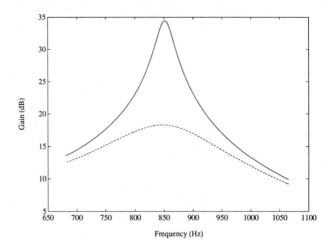

**FIGURE 6.33**

Effect of $1\%$ gain error introduced to illustrate the error in the frequency response of a high-Q bandpass filter where the solid line is the ideal response and the dotted line is the response with the gain error.

branches and dummy switch compensation (half sized devices) was used for the diode-connections. This compensation strategy provides an area-efficient and accurate cancellation scheme.

The filter was designed for a Q of 25 to ensure at least 15 was achieved over process variations. The measurements show that a Q of 25 was obtained and that the center frequency is accurate to within 0.1 %. Table 6.3 summarizes the measured results of the bandpass filters. The measured signal-to-noise-plus-distortion was 34.5 dB which meets the system requirements of 30 dB.

In addition to building a single-ended version of these filters, we also constructed a fully-differential bandpass filter designed to meet the same specifications. However, it was intended to allow evaluation of the low-voltage and high-frequency capability. The circuit was designed to operate with a 3 V power supply voltage and to be clocked 5 MHz. The advantage of the fully-differential structure is that the clock-feedthrough offset becomes common-mode and is effectively cancelled by the circuit. The input impedance of the integrator is designed to be less than 10 Ω. Since the input impedance is so low, the voltage variations at the input due to the signal are small. Thus, as the switches are turned off, the clock feedthrough and charge injection are reduced. And due to the differential structure, these constant errors are absorbed by the differential pair. In this design approximately 10 dB improvement is obtained due to the reduction in the charge injection effects.

*Amplitude Detection.* The output of each of the bandpass filters is monitored to determine if a particular tone is active. This operation is performed by a current-mode amplitude detector. To describe this circuit, we begin with the

**TABLE 6.3**
Measured performance and area statistics for the DTMF bandpass filter test chip implementation.

| Parameter | Bandpass Filter |
|---|---|
| Area per pole | $280\mu$m $\times$ $280\mu$m |
| Power per pole | 3mW |
| Center Freq. Accuracy | 0.1%* |
| Quality Factor | 25 |
| Clock Frequency | 9.321kHz |
| Input-Referred Noise | 475pA/$\sqrt{\text{Hz}}$ |
| PSRR | 2.17$\mu$A/V |
| Distortion | -35dB |
| Signal/(Noise+THD) | 34.5dB |

peak detection circuit of Fig. 6.34(a) [4]. The input section ($M_1$ and switch) is similar to the dynamic mirror except that the gate of the switch is connected to the drain of $M_1$. The circuit operates as follows. Initially, assume that the switch transistor is off, and that $M_1$ has memorized some current. The input node, $V_x$ (the drain of $M_1$), is in a high impedance state. Any difference between $i_{in}$ and the memorized current, therefore, will be reflected by a large voltage change at $V_x$. As $i_{in}$ exceeds the memorized value, $V_x$ will swing toward the positive supply, turning the switch on and diode-connecting the input. Any further increase in $i_{in}$ will cause a corresponding change in $V_{GS1}$, and $M_1$ will track the input. When $i_{in}$ begins to fall, the input voltage will decrease, turning the switch off and forcing the circuit into the "hold" configuration. The output, $i_{peak}$, will therefore reflect the peak value of the input signal. Figure 6.34(b) shows the SPICE-simulated operation of the peak detection circuit.

Ideally, the proposed circuit will respond instantaneously to an arbitrarily small difference between the sampled current and the input current. In reality, however, the performance is limited by the total parasitic capacitance and resistance associated with node $V_x$. The effects of these non-ideal parameters are quantified below as dc-sensitivity, dynamic range, and bandwidth [7].

For the detector to trigger, the input current must exceed the sampled value by some minimum amount to drive $V_x$ high enough to turn $M_s$ on. This minimum incremental current is expressed as

$$I_{min} = \frac{V_{Tsw}}{R_{out}} \tag{6.72}$$

where $V_{Tsw}$ is the threshold voltage of $M_s$. This may have a small signal-dependence if the bulk of the $M_s$ is connected to a potential other than its source. $R_{out}$ is the equivalent small-signal resistance at node $V_x$.

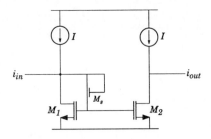 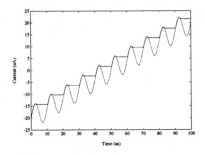

## FIGURE 6.34

Current-mode peak detector (a) circuit and (b) SPICE-simulated peak detector operation. Input (dashed) is a 100kHz sine wave with increasing dc-offset. The circuit detects each peak and holds that value (solid). Note a small dc-offset is present in the output due to the turn-off time of the switch.

The minimum detectable incremental current and the maximum input signal determine the dynamic range of the circuit. This is simply

$$DR = (\frac{\hat{I}}{I_{min}}) \qquad (6.73)$$

where $\hat{I}$ is the maximum input amplitude specified for a given level of distortion.

As the input current exceeds the sampled value, the total capacitance at $V_x$ will begin to be charged, causing $V_x$ to rise until $M_s$ is turned on. Any additional increase in $I_{in}$ will then produce a corresponding increase in the drain current of $M_1$. When $I_{in}$ begins to decrease, the capacitance will be discharged and $V_x$ will fall at a rate described by

$$\frac{dV_x}{dt} = \frac{1}{C_x} \cdot I_{diff}(t) \qquad (6.74)$$

where $I_{diff}$ is the difference between $I_{in}$ and the sampled value, and $C_x$ is the total capacitance associated with node $V_x$. To determine the bandwidth, consider an input described by $\hat{I}cos(2\pi ft)$. Defining $\tau_{DR}$ as the time when $|I_{diff}|$ reaches $I_{min}$ and integrating Eq. (6.74),

$$V_x(t) = \frac{1}{C_x} \int_0^{DR} \hat{I}[1 - cos(2\pi ft)]dt =$$

$$\frac{\hat{I}}{2\pi fC_x}[\theta_{DR} - sin(\theta_{DR})] \qquad (6.75)$$

where

$$\theta_{DR} = 2\pi f\tau_{DR} = cos^{-1}(1 - \frac{I_{min}}{\hat{I}}) = cos^{-1}(1 - \frac{1}{DR}) \qquad (6.76)$$

Assuming that the switch turns on abruptly when $V_x = V_{Tsw}$, Eqs. (6.75) and (6.76) become

$$f_{max} = \frac{1}{2\pi C_x V_{Tsw}}[\theta_{DR} - sin(\theta_{DR})] \qquad (6.77)$$

Conversely, when $M_s$ is off, the voltage stored on the gate of $M_1$ will decrease due to leakage of the reverse-biased substrate diode forming the drain of $M_s$. Again assuming that the input is a periodic sine function, the minimum operating frequency for the desired dynamic range is given by

$$f_{min} = \frac{g_{m1} J_s A_{D2}}{C_{GS1} I_{min}}$$   (6.78)

where $g_{m1}$ is the transconductance of $M_1$, $J_s$ is the reverse saturation current density of the drain junction diode, and $A_{D2}$ is the area of the drain junction of $M_2$.

The performance of the peak detector can therefore be optimized by increasing the output resistance by using a RGC. The complete circuit is shown in Fig. 6.35. An additional regulated cascode stage (Fig. 6.35, $M_4$ - $M_6$) is necessary to output the result of the peak detection operation. Assuming $M_1$ and $M_4$ have identical $\frac{W}{L}$ ratios,

$$I_{out} = -\hat{I}_{in}$$   (6.79)

Note the inverting nature of the current mirror structure. Scaled outputs are obtained by weighting the aspect ratios of $M_1$ and $M_4$. The detector is reset by turning on transistor $M_9$, added in parallel with $M_8$. This allows for the arbitrary diode connection of the mirror by an external control signal.

To construct an amplitude detector, two peak detector circuits are combined, one that is n-type and one that is p-type. Together the total amplitude is detected and the output current held at these values. The measured results of the peak detector are shown in Fig. 6.36.

*Limiter.* The hard-limiter is a critical part of the analog DTMF decoder, and is performed by a current-mode comparator with hysteresis (Fig. 6.37). The output of current mirror $X_m$ charges the gates of the four switch transistors in either the negative or the positive direction, depending upon the sign of the input current. If the input is negative, the output of $X_m$ will be positive, and the lower switches will turn on. Conversely, if the input is positive, the upper switches will turn on. This causes the output to be switched between $+I$ and $-I$ through $M_{p1}$ and $M_{n1}$, respectively. At the same time, $I_{hys}$ is added to or subtracted from the input to realize hysteresis. The amount of hysteresis determines the system's dynamic range by setting the minimum input signal amplitude that the receiver will respond to.

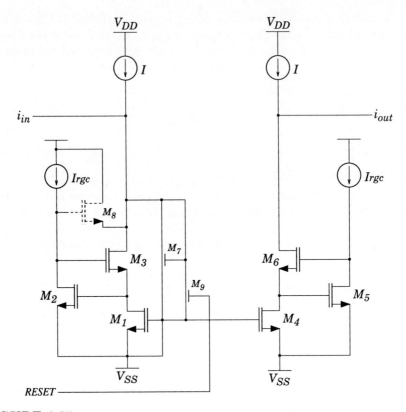

**FIGURE 6.35**
Complete peak detector with output generation ($M_4$ - $M_6$), reset ($M_9$), and spike suppression ($M_8$).

**Summary**   The layout of the analog IC in shown in Fig. 6.38. The power supplies, bias voltages, and clock signals run vertically just to the right of center. To the right of these are the eight bandpass filters and the amplitude detector and hard-limiter. To the left of center are the two sixth-order Chebyshev band-reject filters. The total chip area is 4 mm$^2$. To facilitate comparison to switched-capacitors, the filters were designed with the same minimum data-holding capacitance as a commercially available receiver. This makes both noise ($kT/C$) and clock feedthrough similar for both systems. The power consumption of the SI receiver is approximately $7.5mW$ per pole for the band-reject filters and $5.0mW$ per pole for the bandpass filters. Compared to the layout of a commercially available SC DTMF decoder, the SI implementation requires approximately 30% less area with the same minimum capacitance. The area comparison was based on actual circuit area scaled by process size and by unit-sized capacitance.

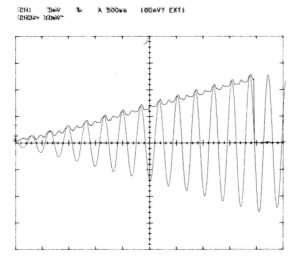

**FIGURE 6.36**
Measured results from current-mode amplitude detection circuit.

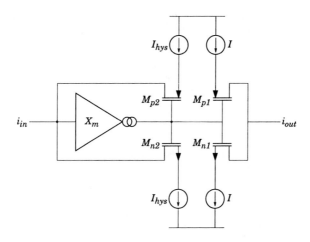

**FIGURE 6.37**
Current-mode comparator with hysteresis. The output current will be $\pm I$ and the hysteresis loop width is set by $I_{hys}$. $X_m$ is a current mirror.

## 6.7 CONCLUSION

Switched-current signal processing represents a complementary technique to the switched-capacitor technique. We have demonstrated that an approximate 30% area savings over a SC implementation with similar performance can be attained.

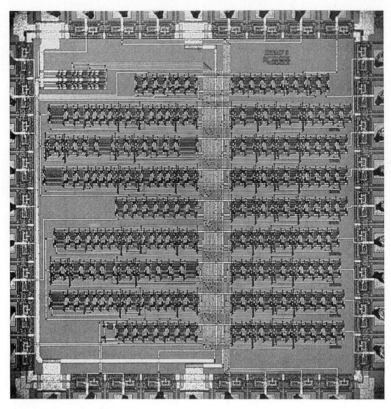

**FIGURE 6.38**
Die photograph of DTMF decoder.

SI circuit techniques complement the highly accurate SC circuit technique and are ideal for systems requiring moderate accuracy. Additionally, SI techniques do not require precision linear capacitors and may be fabricated in a standard digital CMOS process.

## PROBLEMS

Where required, use the $2\mu$ MOS transistor process parameters in Appendix A.

**6.1.** Design a 1 : 1 simple NMOS current mirror as shown in Fig. 6.1. Assume the following parameters: $V_{DD} = 5$V, $V_{SS} = 0$V, $I = 50\mu A$, $\hat{i}_{in} = 0.5 \cdot I = 25\mu A$. Connect the n-channel transistor bulks to $V_{SS}$.

   (a) Determine the aspect ratio $(W/L)$ required such that the maximum input voltage (when $I + \hat{i}_{in} = 75\mu A$) does not exceed half the supply voltage.

   (b) Using Pelgrom's model (Fig. 6.19), determine the gate area required for 0.5% matching between $i_{in}$ and $i_{out}$ for three standard deviations with a nominal current of $50\mu A$.

   (c) Determine the device widths and lengths in $\mu$m. Find $r_{in}$ and $r_{out}$ of the mirror.

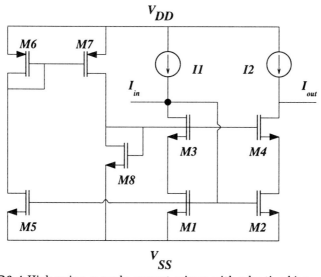

**FIGURE P6.4** High-swing cascode current mirror with adaptive bias.

(d) Using this ratio, what is the *minimum* input voltage?

(e) Simulate this circuit in SPICE to determine the output current total harmonic distortion due to the finite output impedance. Use a stage identical to this mirror as a load and apply a 1 kHz sinusoidal current of $25\mu$A.

**6.2.** Design a 1:1 regulated cascode current mirror (Fig. 6.5) using a procedure similar to that of Prob. 6.1(a) and (b). Determine the widths and lengths of M1 and M2 given $(W/L)_5 = (W/L)_6 = 9\mu/4\mu$, $(W/L)_3 = (W/L)_4 = 40\mu/4\mu$ and $I_{fb} = 5\mu$A. Determine $r_{in}$, $r_{out}$, and $V_o(min)$ for this current mirror?

**6.3.** The input resistance of the simple mirror of Prob. 6.1 can be reduced by placing an amplifier between the drain and gate of $M_1$ (Fig. 6.1). Determine the input resistance if an amplifier with a gain of 10 is inserted between the drain and gate of $M_1$.

**6.4.** Apply a $50\mu$A 1 kHz sine wave current signal to the cascode current mirror of Fig. 6.4(b) with $W_1/L_1 = W_2/L_2 = W_3/L_3 = W_4/L_4 = 50\mu/4\mu$, $I_1 = I_2 = 100\mu$A, $V_{SS} = 0$V and $V_{DD} = 5$V. Bias the output with a voltage source connected to 2.5 V and apply a fixed 2.6V bias to the gates of $M_3$ and $M_4$. Connect all bulks to $V_{SS}$. Using SPICE, plot one period of the output current versus time and determine the THD. Observe the distortion in the output current due to the fixed bias voltage. Next, use an adaptive voltage bias for the gates of $M_3$ and $M_4$ as shown in Fig. P6.4. Again plot the output current versus time and determine the THD. Compare the results.

**6.5.** Use the small signal model of Fig. 6.3 to find the expression for the transfer function $(i_{out}/i_{in})$ of the simple current mirror (Fig. 6.1) in terms of the device and bias parameters.

**6.6.** Deriving the complete transfer function of the regulated cascode current mirror is a difficult task. Fortunately, the dominant pole frequency is similar to that of

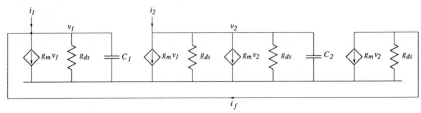

**FIGURE P6.8** Complete small-signal model for the continuous-time current-mode integrator.

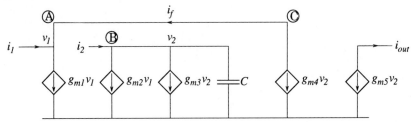

**FIGURE P6.9** Simplified small-signal model for the continuous-time current-mode integrator.

the simple current mirror. Verify this claim by simulating the circuit in Prob. 6.2. Compare the -3dB frequency with that obtained by a first-order hand analysis.

**6.7.** To scale the output current of a current mirror, the widths of the transistors are usually scaled. The lengths of the transistors may also be scaled to scale the output current but this introduces errors. Determine the percentage error in the output current of a simple current mirror (Fig.6.1) for $L_1 = 10L_2$, $L_2 = 2\mu$, $V_{DS1} = V_{DS2} = 2V$ and $\lambda = 0.027$ for $L = 2\mu$. Assume all other process and device parameters are identical.

**6.8.** In Sec. 6.2.3, a first-order derivation of the continuous-time current-mode integrator transfer function was performed. Find the more complete transfer function using the small signal model of Fig. P6.8.

**6.9.** By changing the feedback coefficient in the continuous-time integrator of Sec. 6.2.3 to less than unity, the dominant pole can be moved off of the origin. Using the simplified small signal model of Fig. P6.9 with $g_{m4} = kg_{m3}$, find the transfer function of the lossy integrator.

**6.10.** Design a third-order continuous-time current-mode elliptic low-pass filter:
  (a) Use a passive prototype as in Fig. P6.10 and element values of $C_1 = C_3 = 1.0512$, $C_2 = 0.2019$ and $L_2 = 0.9612$ to determine the scaled values for a $1MHz$ cut-off frequency.
  (b) With the state variables defined as in the figure, write the differential equations describing the system.
  (c) With the state variables defined as in the figure, write the differential equations describing the system.

**6.11.** Simulate a current T/H using the design in Prob. 6.2 and adding a NMOS switch between the gates of $M_1$ and $M_2$ with W/L=$3\mu/2\mu$. Apply a clock frequency

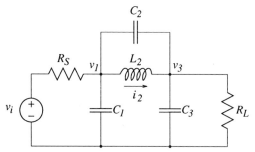

**FIGURE P6.10** Passive prototype filter.

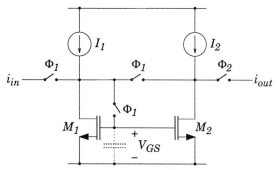

**FIGURE P6.13** Generating a coefficient by current division. On $\Phi_1$, $i_{in}$ is divided between $M_1$ and $M_2$. On $\Phi_2$, only the portion stored in $M_2$ is connected to the output.

of 100 kHz (0-5 Volts) and a signal frequency (sine wave) of 1 kHz. Simulate in SPICE to determine the THD.

**6.12.** Derive the equation for the output current in terms of the width, length, and transconductance parameter mismatch.

**6.13.** Derive the expression for the coefficient of variation for the current divider (Fig. P6.13). Compare the area required using a current divider and a current mirror for a scale factor of 0.1 to 0.9.

**6.14.** Simulate the circuit in Fig. 6.28(a) as described in the chapter. First simulate the circuit using the default settings in SPICE. Next, set RELTOL=1x10$^{-6}$ in the SPICE options card (in SABER set the density to 60) and resimulate the circuit. How do the results differ?

**6.15.** Clock-feedthrough induced errors in the current T/H are a major concern in the design of switched current circuits. For this problem use the circuit in Fig. 6.10 with $(W/L)_1 = (W/L)_3 = 100\mu/10\mu$, $(W/L)_2 = 3\mu/2\mu$ and the bias current is $100\mu$A.

(a) Verify clock feedthrough equations (6.54)-(6.59) by applying KCL to Fig. 6.21.

(b) Assuming that the clock voltage on the gate of $M_3$ in Fig. 6.10 varies from 5V to 0V, write a program to calculate the clock feedthrough voltage using the lumped model with $i_{in} = 0$. What is the amount of clock feedthrough voltage when the fall time of the clock is 1 nsec, 10 nsec, 100 nsec, and 1 $\mu$sec? Why does the clock feedthrough voltage reduce as the fall time

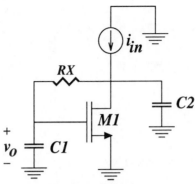

**FIGURE P6.16** Current copier in sampling mode.

increases? (Hint: Use the Euler method for numerical integration.)

(c) Simulate the circuit of Fig. 6.10 using SPICE. Use a diode-connected transistor as the load that is the same size as the input transistor. Use the option RELTOL=1E-6 in the SPICE option card to reduce the errors in numerical integration. What is the amount of clock feedthrough voltage for the fall time of 1 nsec, 10 nsec, 100 nsec, and 1 $\mu$sec? Do the results of the SPICE simulation agree with part (b) above?

(d) Use the program in part (b) to calculate the clock feedthrough voltage for different fall times and plot the *log* of the clock feedthrough as a function of the *log* of the fall time.

(e) For fast clock edges, the channel charge splits evenly between the source and the drain (i.e. half the charge is injected to the drain side and half is injected to the source side). Write a program to calculate the harmonic distortion in the output current due to clock feedthrough with a $50\mu$A sinusoidal input current.

(f) One method of reducing the clock feedthrough harmonic distortion is to use a fully differential structure. Use the program in part (e) to calculate the harmonic distortion for a differential approach. (Hint: Use two circuits, one with $i_{sig} = 25\mu$A $sin(\omega\tau)$ and one with $i_{sig} = -25\mu$A $sin(\omega\tau)$. Find the difference of the two outputs to determine the harmonic distortion.)

**6.16.** The settling time of the dynamic current mirror is one of the factors that determines the maximum clock frequency.

(a) Assume all the transistor capacitances are lumped into $C_1$ and $C_2$ in Fig. P6.16. Use the circuit shown in Fig. P6.16 to determine the expression for the small-signal transresistance, $\frac{v_o}{i_{in}}$.

(b) Write the expression for the settling time, $\tau_{sett} = 2\tau_1 + \tau_2$ where $\frac{1}{\tau_1}$= dominant pole and $\frac{1}{\tau_2}$ = second pole. Assume the poles are real.
    (Hint: $\left(\frac{C_1+C_2}{R_xC_1C_2}\right)^2 > \frac{4g_m}{R_xC_1C_2}$ )

**6.17.** (a) For the current copier circuit shown in Fig. P6.17(a), determine the error voltage $\Delta v_{gs1}$ introduced when transistor $M_s$ turns off. Use the small-signal circuit in Fig. P6.17(b) to calculate this error voltage. Note that Miller's theorem has been used to simplify the analysis. (Hint: $\Delta vgs = \frac{q_{inj}}{C_g}$ where $q_{inj} = 0.7 \times C_{ox}W_sL_{eff_s}(V_{clk} - V_{TH} - V_{Vgs1}) + C_{gd_s}V_{clk}$ and 0.7 is used since 70% of the charge is injected onto the gate of $M_1$.)

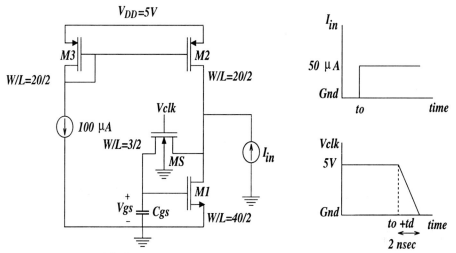

FIGURE P6.17(a) Complete current copier cell.

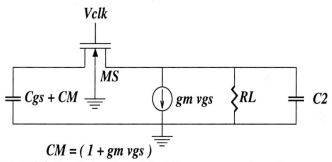

$$CM = ( 1 + gm \ vgs )$$

FIGURE P6.17(b) Small-signal model of the current copier cell.

(b) Calculate the relative current error $\frac{\Delta I_o}{I_o}$ after the switch is turned off where $I_o = I_{ref} + i_{in}$.

(c) Compare these results with SPICE. Do these results differ? why?

**6.18.** To reduce the $\frac{\Delta I_o}{I_o}$ error down to 0.1% in the previous problem, a capacitor may be inserted across $C_{gs1}$.

(a) What value of capacitance is required to meet this specification?

(b) What is the resulting settling time for the current copier?

(c) Compare these calculations with SPICE simulations.

**6.19.** In order to reduce the charge injection from the switch, a dummy switch is used as shown in Fig. P6.19.

(a) Determine the size of the dummy switch so that it cancels the charge injection.

(b) Compare this settling time with that obtained in the previous problem.

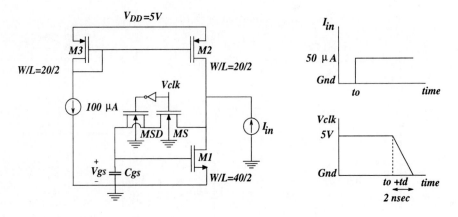

MS: W/L=3/2

MSD: W/L=6/2

**FIGURE P6.19** Current copier cell with dummy switch.

**6.20.** In the dynamic current mirror, charge is stored on the parasitic gate-source capacitance of the transistor ($M1$ in Fig. P6.20). When the switch ($S1$) is open, there exists a reverse biased diode from the drain of $S1$ to the substrate. This provides a leakage path for charge from the gate of $M1$.

   (a) Given that the reverse biased diode leakage current is 0.15 $\left(\frac{pA}{\mu m^2}\right)$ at room temperature, what is the minimum clock frequency required to obtain no more than $0.1\mu A$ reduction in the drain current? Use $g_m = 564\ \mu S$.

   (b) What is the minimum clock frequency at $100^\circ C$.

      (Hint: The leakage doubles for every $10^\circ C$ increase in temperature.

   (c) Perform this same calculation including the leakage effects due to the dummy switch.

**6.21.** The circuit in Fig. P6.21 is proposed to cancel the charge injection without increasing errors due to charge leakage.

   (a) Determine the channel and overlap charge for each transistor.

   (b) What condition is required for precise cancellation of the channel charge.

**6.22.** Calculate the input-referred noise voltage in the regulated-gate cascode current mirror (Fig. 6.5).

**6.23.** Simulate the two-step switched-current circuit in Fig. 6.24. Determine the error in the output current. Next, replace each transistor with a RGC structure. Again simulate the circuit. Why is there a reduction in the error in the output current?

**6.24.** Design a transconductor that converts the voltage at the input to a current. The transconductance should be 50 $\mu S$ and provide 1 % linearity from +/-1 V with a +/-2.5 V power supply.

**6.25.** Using the component values given in Prob. 6.10, design a switched-current elliptic filter with a cutoff frequency of 5 kHz and a clock frequency of 100 kHz.

   (a) Show the interconnection of the integrators and the integrator coefficients.

   (b) Determine values for $K_1$-$K_3$.

   (c) If a continuous-time current-mode version of this same filter were designed with $C_{int}$ = 2pF and $g_{m(int)} = 400\mu S$, what transistor ratios, $K_1$-$K_3$, would

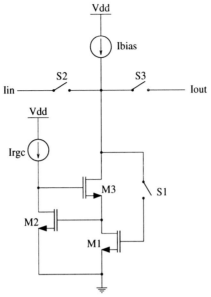

**FIGURE P6.20** Regulated-gate cascode current mirror.

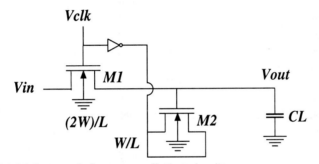

**FIGURE P6.21** Improved charge cancellation circuit.

be required? Compare these transistor ratios with that of the sampled-data version.

**6.26.** The quarter-square principle is a well known technique to realize a multiplier. For signals in the form of currents, the principle can be written as: $I_o = (I_x + I_y)^2 - (I_x - I_y)^2 = 4I_xI_y$. To multiply the input currents $I_x$ and $I_y$, one can square both the sum and the difference of these currents and then subtract the result.

(a) The circuit in Fig. P6.26(a) is proposed as a current squarer. Determine the expression for $I_{out}$ in terms of $I_{in}$. (Hint: Apply Kirchhoff's voltage law on the loop formed by the gates and sources of transistors $M1$, $M2$, $M3$ and $M4$.)

(b) Suppose a second input current is injected into node $B$ in Fig. P6.26(a). Show that the effective input current is the difference of the two currents.

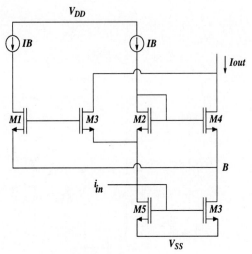

**FIGURE P6.26(a)** Current squaring circuit.

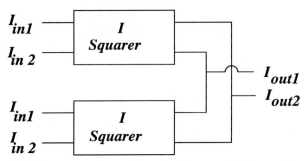

**FIGURE P6.26(b)** Two current squaring circuits connected to perform current multiplication.

    (c) By using two current squaring circuits and two input currents, construct a four-quadrant current multiplier of Fig. P6.26(b). Simulate the circuit with $+/-50\mu A$ input currents and plot the output current versus the input current. Assume all transistors are $\frac{100\mu}{10\mu}$ with $100\mu A$ bias currents.

**6.27.** Figure P6.27 represents a current-mode pipelined analog to digital converter which has a symmetrical input range of $-500\mu A$ to $+500\mu A$. The input current is fed through the sample/hold circuit and compared with zero to determine if it is greater or less than zero. The most significant bit $B_1$ is set high if the input current is greater than zero and it is set low otherwise. Depending on the result of comparison, $I_{ref}$ is subtracted (if $I_{in} > 0$) or added (if $I_{in} < 0$) to $I_{in}$. All the following stages multiply the output current of the previous stage by two (performed by the two current copiers), compare it to zero which generates the corresponding bit $B_i$ and subtract(add) $I_{ref}$ from the result depending on wether $B_i$ is active(inactive).

    (a) Derive the expression for the input current in the $n$-th stage of the converter in terms of input current and $B_i$, $i < n$. Recall both the sample/hold and the copier output currents are inverted relative to the input signal. (Hint: The input current to the second stage can be expressed as $I_2 = -I_{in} - [2(-1)^{B_1} + 1]I_{ref}$.)

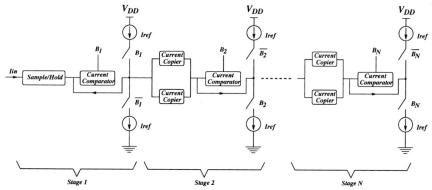

**FIGURE P6.27**

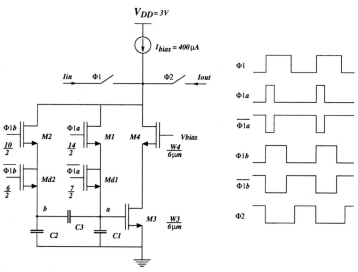

**FIGURE P6.28**

**6.28.** In Figure P6.28 is represented an enhanced dynamic current mirror with clock feed-through cancellation. The circuit shows a two-step cancellation scheme which employs a negative feedback loop. Assume that $V_{DD} = 3V$ and the switches are controlled by a $5V$ clock signals as represented in the figure. The input current range is $-100\mu A$ to $+100\mu A$ and the ideal current source is $I_{bias} = 400\mu A$.

(a) Determine the sizes of $M_4$ and $M_4$ to obtain an output resistance larger than $10M\Omega$ and to keep both $M_3$ and $M_4$ in the saturation region. (Hint:

Look at the voltage change of $V_{gs,3}$ and $V_{gs,4}$ with the maximum and the minimum input current.)

(b) The circuit provides negative feedback through $M_3$, $M_4$, $M_2$ when $M_1$ is turned off and a clock feed-through voltage variation appears on $C_1$. This error voltage is then cancelled when $M_2$ is turned off. The voltage variation on $C_1$ is no more than 0.5mV when $M_2$ is turned off (before the dummy switch $M_{d2}$ is turned on) and the voltage variation on $C_1$ should preserve the error current to within $100nA$. Write the equations for $\Delta V_a$ and $\Delta V_b$ at the end of $\Phi_{1a}$ and $\Phi_{1b}$ in terms of $C_1$ and $C_3$. Determine the value of $C_3$ to meet the voltage variation requirement at nodes $a$. How much charge is injected at node $b$ at the end of $\Phi_{1a}$ if the voltage variation at node $a$ is $\Delta V_a = 0.8mV$. Consider $C_1 = C_2 = 1pF$.

(c) Simulate the circuit using SPICE. Use an input current pulse with the amplitude in the range $\pm 150 \mu A$ and a diode-connected transistor load. (Hint: Use a cascoded current mirror for $I_{bias}$.)

**6.29.** Figure P6.29 represents a regenerative comparator with offset cancellation also used as a current comparator in Prob. 6.27. Switch $S_d$ is closed all the time and is used to balance the current mirror $M_5 - M_8$. When $S_1$ and $S_2$ are closed and zero input current is applied, the current copier samples and memorizes the offset current. Then $S_2$ is opened, the input current is applied and the regenerative process makes the voltage at the drain of $M_8$ converge to a high or a low value depending on the comparison result. Simulate the circuit using SPICE to determine the speed when $(W/L)_2 = (W/L)_1 = \frac{60\mu}{2\mu}$ and $(W/L)_2 = 10(W/L)_1 = \frac{60\mu}{2\mu}$. Notice the change in the offset voltage. Explain which aspect ratio is more suitable.

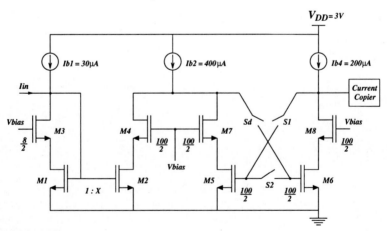

**FIGURE P6.29**

# REFERENCES

[1] K.C. Smith and A.S. Sedra, "The current conveyor–a new circuit building block," *Proc. IEEE*, vol. 56, pp. 1368-1369, August 1968.

[2] A.S. Sedra and K.C. Smith, "A second-generation current conveyor and its applications," *IEEE Trans. Circuit Theory*, vol. CT-17, pp. 132-134, February 1970.

[3] B. Gilbert, "Translinear circuits: a proposed classification," *Electronics Letters*, vol. 11, pp. 14-16, 1975.

[4] R. Croman, "High-accuracy switched-current filter design," M.S. thesis, Washington State University, May 1993.

[5] C. Toumazou, F.J. Lidgey and D.G. Haigh editors, "*Analogue IC design: the current-mode approach,*" Peter Peregrinus Ltd., London, 1990.

[6] P.R. Gray and R.G. Meyer, *Analysis and design of analog integrated circuits*, 2 ed., John Wiley & Sons, Inc., 1984.

[7] A.S. Sedra and K.C. Smith, "*Microelectronics circuits*" 2 ed., Holt, Rinehart & Winston, Inc., 1987.

[8] T.S. Fiez and D.J. Allstot, "CMOS switched-current ladder filters," *IEEE J. Solid-State Circuits*, vol. 25, no. 6, pp. 1360-1367, December 1990.

[9] H.C. Yang, T.S. Fiez, and D.J. Allstot, "Current-feedthrough effects and cancellation techniques in switched-current circuits," *Proc. of IEEE Intl. Symp. Circuits and Systems*, pp. 3186-3188, May 1990.

[10] M.J.M. Pelgrom, A.C.J. Duinmaijer, and A.P.G. Welbers, "Matching Properties of MOS Transistors," *IEEE J. Solid-State Circuits*, vol. 24, No. 5, pp. 1433-1439, October 1989.

[11] J.B. Shyu, G.C. Temes, and F. Krummenacher, "Random error effects in matched MOS capacitors and current sources," *IEEE J. Solid-State Circuits*, vol. SC-19, no. 6, pp. 948-955, December 1984.

[12] B. J. Hosticka, "Improvement of the gain of MOS amplifiers," *IEEE J. Solid-State Circuits*, vol. SC-14, no. 6, pp.1111-1114, December 1979.

[13] C. Toumazou, J.B. Hughes, and C.M. Pattullo, "Regulated cascode switched-current memory cell," *IEE Electron. Lett.*, vol. 26, no. 5, pp. 303-305, March 1, 1990.

[14] H.C. Yang and D.J. Allstot, "An active-feedback cascode current source," *IEEE Trans. Circuits Syst.*, vol. 37, no. 5, pp. 644-646, May 1990.

[15] E. Sächinger and W. Guggenbühl, "A high-swing, high-impedance MOS cascode circuit," *IEEE J. Solid-State Circuits*, vol. 25, no. 1, pp. 289-298, February 1990.

[16] K. Bult and G. Geelen, "A fast-settling CMOS op amp for SC circuits with 90-dB DC gain," *IEEE J. Solid-State Circuits*, vol. 25, no. 6, pp. 1379-1384, December 1990.

[17] D.G. Nairn and C.A.T. Salama, "High-resolution current mode A/D converters using active current mirrors," *Electron. Lett.*, vol. 24, pp. 1331-1332, 1988.

[18] A.B. Williams and F.J. Taylor, *Electronic filter design handbook,* 2 ed., McGraw-Hill, Inc., 1981.

[19] R. Zele, S.S. Lee, G. Liang, and D.J. Allstot, "A continuous-time current-mode integrator," *IEEE Trans. Circuits Syst.*, vol. 38, no. 10, pp.1236-1238, October 1991.

[20] J.B. Hughes, N.C. Bird, and I.C. Macbeth, "Switched currents – A new technique for analog sampled-data signal processing," *IEEE Intl. Symp. Circuits Syst.*, May 1989, pp. 1584-1587.

[21] E.A. Vittoz, "Dynamic analog techniques," in *VLSI Circuits for Telecommunications*, edited by Y.P. Tsividis and P. Antognetti, Englewood Cliffs, NJ: Prentice Hall, 1985.

[22] S.J. Daubert, D. Vallancourt, and Y.P. Tsividis, "Current copier cells," *Electon. Lett.*, vol. 24, no. 25, pp. 1560-1562, December 8, 1988.

[23] G. Wegmann and E.A. Vittoz, "Analysis and improvements of accurate dynamic current mirrors," *IEEE J. Solid-State Circuits*, vol. 25, no. 3, pp. 699-706, June 1990.

[24] G. Wegmann, "Design and analysis techniques for dynamic current mirrors," *Ingenieur Electricien Diplome EPFL*, De Nationalite Suisse Et Francaise, These NO. 890, 1990.

[25] E.A.M. Klumperink and E. Seevinck, "MOS current gain cells with electronically variable gain and constant bandwidth," *IEEE J. Solid-State Circuits*, vol. 24, no. 5, pp. 1465-1467, Oct. 1989.

[26] M.G. Degrauwe, J. Rijmenants, E.A. Vittoz, and H.J. DeMan, "Adaptive biasing CMOS Amplifiers," *IEEE J. Solid-State Circuits*, vol. 19, no. 3, pp. 522-528, June 1982.

[27] E.A. Vittoz "Micropower Techniques," *VLSI Circuits for Telecommunications*, pp. 104-144, ed. Y. P. Tsividis and P. Antiognetti, Englewood Cliffs, NJ: Prentice Hall, 1985.

[28] K.R. Lakshmikumar, R.A. Hadaway and M.A. Copeland, "Characterization and modeling of mismatch in MOS transistors for precision analog design," *IEEE J. Solid-State Circuits*, vol. SC-21, no. 6, pp. 1057-1066, December 1986.

[29] J.H. Shieh, M. Patil, and B.J. Sheu, "Measurement and analysis of charge injection in MOS analog switches," *IEEE J. Solid-State Circuits*, vol. SC-22, no. 2. pp. 277-281, April 1987.

[30] C. Eichenberger and W. Guggenbühl, "On charge injection in analog MOS switched and dummy switch compensation techniques," *IEEE Trans. Circuits and Syst.*, vol. 37, no. 2, pp. 256-264, February 1990.

[31] T.S. Fiez, "Design of CMOS switched-current filters," Ph.D. dissertation, Oregon State University, June 1990.

[32] T.S. Fiez, D.J. Allstot, G. Liang, and P. Lao, "Signal-dependent clock-feedthrough cancellation in switched-current circuits," *1991 China Intl. Conf. Circuits and Syst.*, June 1991, pp. 785-788.

[33] T.S. Fiez, G. Liang, and D.J. Allstot, "Switched-current design issues," *IEEE J. Solid-State Circuits*, vol. 26, no. 3, pp. 192-202, March 1991.

[34] R.H. Zele, D.J. Allstot, and T.S. Fiez, "Fully-differential CMOS current-mode circuits," *IEEE Custom Integ. Circuits Conf. (CICC)*, May 1991, pp. 24.1.1-24.1.4.

[35] J.B. Hughes and K.W. Moulding, "Switched-current signal processing for video and beyond," *IEEE J. Solid-State Circuits*, March 1993.

[36] R.H. Zele, D.J. Allstot and T.S. Fiez, "Fully balanced CMOS current-mode circuits," *IEEE J. Solid-State Circuits,* vol. 28, no. 5, pp. 569-575, May 1993.

[37] J.B. Hughes and K.W. Moulding, "$S^2I$: A two-step approach to switched-currents," *Proceeding of the IEEE Intl. Symp. Circuits Syst.*, May 1993, pp. 1235-1238.

[38] M. Goldenberg, "Low-voltage fully-differential switched-current integrators," M.S. thesis, Washington State University, December 1992.

[39] S.J. Daubert and D. Vallancourt "Operation and analysis of current copier circuits," *IEE Proceedings*, vol. 137, Pt. G, no. 2, pp. 109-115, April 1990.

[40] J.H. Fischer, "Noise sources and calculation techniques for switched capacitor filters," *IEEE J. Solid-State Circuits*, vol. SC-17, no. 4, pp. 742-752, Aug. 1982.

[41] G.M Jacobs, D.J. Allstot, R.W. Brodersen, and P.R. Gray, "Design techniques for MOS switched capacitor ladder filters," *IEEE Trans. Circuits Syst.*, vol. CAS-25, no. 12, pp. 1014-1021, December 1978.

[42] D.O. Pederson, "A historical review of circuit simulation," *IEEE Trans. Circuits Syst.*, vol. CAS-31, no. 1, pp. 103-111, January 1984.

[43] B.J. Sheu, D.L. Scharfetter, P. Ko, and M. Jeng, "BSIM: Berkeley short-channel IGFET model for MOS transistors," *IEEE J. Solid-State Circuits*, vol. SC-22, no. 4, pp. 558-565, August 1987.

[44] P. Antognetti and G. Massobrio, *Semiconductor Device MOdeling with SPICE*, McGraw-Hill, Inc., New York, 1988.

[45] J.E. Meyer, "MOS models and circuit simulation," *RCA Review*, vol. 32, pp. 42-63, March 1971.

[46] K.A. Sakalla, Y.T. Yen and S.S. Greenberg, "A first-order charge conserving MOS capacitance model," *IEEE Trans. on Computer-Aided Design*, vol. 9, no. 1, pp. 99-109, January 1990.

[47] D.E. Ward and R.W. Dutton, "A charge-oriented model for MOS transistor capacitances," *IEEE J. Solid-State Circuits*, vol. SC-13, no. 5, pp. 703-708, October 1978.

[48] P. Yang, B.D. Epler and P.K. Chatterjee, "An investigation of the charge conservation problem for MOSFET circuit simulation," *IEEE J. Solid-State Circuits*, vol. SC-18, no. 1, pp. 128-138, February 1983.

[49] M.L. Liou, Y.L. Kuo, and C.F. Lee, "A tutorial on computer-aided analysis of switched-capacitor circuits," *Proceedings of the IEEE*, vol. 71, no. 8, pp. 987-1005, August 1983.

[50] K. Suyama, S.C. Fang, and Y.P. Tsividis, "Simulation of mixed switched-capacitor/digital networks with signal-driven switches," *IEEE J. Solid-State Circuits*, vol. 25, no. 6, pp. 1404-1413, December 1990.

[51] S.E. Belter, S.C. Bass, and S.N. Stevens, "DINAP: digital filter analysis program short form user's manual and guide," West Lafayette, Indiana, Circuits and Systems Group, School of EE, Purdue University, 1985.

[52] A.C. De Queiroz, P.R.M. Pinheiro, and L.P. Caloba, "Systematic nodal analysis of switched-current filters," *Proceedings of the IEEE Intl. Symp. Circuits Syst.*, June 1991, pp. 1805-1808.

[53] J.A. Barby, "Switched-current filter models for frequency analysis in the continuous-time domain," *Proceedings of the IEEE Intl. Symp. Circuits Syst.*, vol. 2, May 1993, pp. 1427-1430.

[54] E.M. Schneider, "Simulation of Current-Mode Sampled-Data Signal Processing Systems," M.S. thesis, Washington State University, April 1992.

[55] E.M. Schneider and T.S. Fiez, "Simulation of Switched-Current Systems," *Proceedings of the IEEE Intl. Symp. Circuits Syst.*, vol. 2, May 1993, pp. 1420-1423.

[56] Analogy Inc., *Saber Users Guide* and *Sampled-Data Systems Library*, Beaverton, OR., Analogy Inc., 1991.

[57] J.C. Candy and G.C. Temes, "Oversampling methods for A/D and D/A conversion," in *Oversampling Delta-Sigma Data Converters*, J.C. Candy and G.C. Temes eds., IEEE Press, New Jersey, 1991.

[58] S.J. Daubert and D. Vallencourt, "A transistor-only current-mode sigma-delta modulator," *IEEE Custom Integ. Circuits Conference*, May 1991, pp. 24.3.1-24.3.4.

[59] D.G. Nairn and C.A.T. Salama, "Current-mode algorithmic analog-to-digital converters," *IEEE J. Solid-State Circuits*, vol. 25, no. 4, pp. 997-1004, August 1990.

[60] D.G. Nairn and C.A.T. Salama, "A ratio-independent algorithmic analog-to-digital converter combining current-mode and dynamic techniques," *IEEE Trans. Circuits Syst.*, pp. 319-325, March 1990.

[61] P. Deval, J. Robert and J.J. Declercq, "A 14-bit CMOS A/D converter based on dynamic current mirrors," *Proceedings of the IEEE Intl. Symp. Circuits Syst.*, May 1991, pp. 997-1004.

[62] A.M. Cujec, C.A.T. Salama and D.G. Nairn, "An optimized bit cell design for a pipelined current-mode algorithmic A/D converter," *Analog Integrated Circuits and Signal Processing*, pp. 137-140, 1993.

[63] Ligang Zhang, "A 2 V 12-bit pipelined A/D converter using current-mode techniques," M.S. thesis, Washington State University, May 1993.

[64] W.R. Krenik, R.K. Hester and R.D. DeGroat, "Current-mode flash A/D conversion based on current-splitting techniques," *IEEE Intl. Symp. Circuits Syst.*, pp. 585-588, May 1992.

[65] D.W.J. Groeneveld, H.J. Schouwenaars, H.A.H. Termeer and C.A.A. Bastiaansen, "A self-calibration technique for monolithic high-resolution D/A converters," *IEEE J. Solid-State Circuits*, vol. 24, no. 6, pp. 1517-1522, December 1989.

[66] R. N. Battista, C. G. Morrison, and D. H. Nash, "Signaling System and Receiver for Touch-Tone Calling," *IEEE Transactions on Communications Electronics*, pp. 9-17, March 1963.

[67] B. J. White, G. M. Jacobs, and G. F. Landsburg, "A Monolithic Dual Tone Multifrequency Receiver," *IEEE Journal of Solid-State Circuits*, vol. SC-14, no. 6, pp. 991-997, December 1979.

[68] K. Martin and A.S. Sedra, "Exact Design of Switched-Capacitor Bandpass Filters Using Coupled-Biquad Structures," *IEEE Trans. on Circuits and Systems*, vol. CAS-27, no. 6, pp. 469-475, June 1980.

[69] A. V. Oppenheim and R. W. Schafer, *Discrete-Time Signal Processing*, Englewood Cliffs, NJ: Prentice Hall, Inc., 1989.

# CHAPTER
# 7

# NEURAL
# INFORMATION
# PROCESSING I

## 7.1  INTRODUCTION

Rapid advances in computer and communication networking technology make it possible to integrate the widely distributed information from various types of data fields. Different kinds of media such as image, voice, and text are linked and shared among the information processing systems supporting multi-media capabilities. High performance computation based on data processing algorithms in real time is indispensable for large-scale information processing systems. The demand for intensive computational power for information processing applications is increasing. In the future, these systems are anticipated to support tera operations-per-second in 1996 and beyond [90]. Remarkable advances of very large-scale integration (VLSI) technologies have made it possible for the compact electronic systems to possess high functional capabilities.

Artificial neural networks represent one of the approaches to enhance the computational capabilities in real-time information processing. Unlike conventional methods based on a single powerful central processing unit, an artificial neural network consists of a large collection of simple processing elements which are efficiently interconnected as shown in Fig. 7.1. Inspired by the physiology of the human brain, these simple processing elements execute mathematical algorithms to carry out the information processing through their responses to stimuli. Artificial neural networks have demonstrated great ability to deliver simple, powerful solutions in areas that have challenged conventional computing approaches

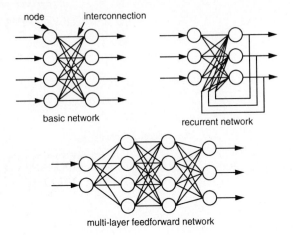

**FIGURE 7.1**
Various artificial neural network models.

during the past decade [28]. Key operations include classification, pattern matching, pattern completion, noise removal, optimization, and control [85]. According to the network topology and the types of the learning and retrieving schemes, many different classes of artificial neural networks exist [37, 77]. Some of the major networks are summarized in Table 7.1.

The fastest way to execute artificial neural network algorithms is to develop a neurocomputing processor (or neurocomputer, neuroprocessor). It supports the fully massive parallelism. Such neurocomputing processors have been used in various scientific and engineering applications such as machine vision, speech and pattern recognitions, robotics, and expert systems. Some design factors for implementing neural systems in hardware are [11] :

- Flexibility of working for several neural processing algorithms,
- Number of neurons which can be monolithically integrated,
- Storage of programmable weights and support of learning schemes,
- Small silicon area and low power consumption, and
- Stability, reproducibility, expandability, and affordability.

Digital neurocomputers [23, 20, 45, 43, 44, 10, 62, 66, 65] are based upon the multiprocessor architecture. Each processing element consists of a digital multiplier, adder, and cache memory. Many processing elements are connected through a common bus. The network is fully programmable and reconfigurable to facilitate on-chip learning. The operational accuracy is related to usage of the hardware resources. Each digital processing element consumes a large silicon area and significant electrical power. Analog or mixed-signal neural network processors have been widely developed and preferred for dedicated applications. Various circuit techniques are available for implementation of synapses and neurons in analog circuits. Since silicon area and power dissipation of analog elements are small, a significantly large number of components can be monolithically inte-

Table 7.1: Major artificial neural network models.

| • neural models | • strengths | • limitations | • primary applications |
|---|---|---|---|
| perceptron | simple structure | cannot recognize complex patterns | typed-character recognition |
| Hopfield | simple structure | weights are pre-determined. | retrieval of data/image from fragments |
| multi-layer percetron with delta rule | more genral than the perceptron | cannot recognize complex patterns | pattern recognition |
| back-propagation | most popular, work well, and is simple to adapt | required for large volume of examples in advance | wide range: speech synthesis to loan application scoring |
| Boltzmann machine | simple network using noise function to reach global minimum | long training time | pattern recognition for radar/sonar |
| self-organizing map | better performance than many algorithmic techniques | extensive learning | mapping one geometrical region onto another |

grated. Thus, the fully parallel configuration provides very high computational power. The accuracy, however, is limited due to the imperfect analog device characteristics.

In the following sections, various analog circuit design techniques to implement artificial neural network models are described. In Sec. 7.2, hardware implementations of biologically-inspired neural network models are presented: silicon retina, silicon cochlea, and vision chips. Pulse-stream approaches to process a neural signal is also described. In Sec. 7.3, artificial neural network components using transistors biased in the strong-inversion region are described in detail. Various synapse cells and neurons are discussed. In Sec. 7.4, one of the important technologies to circumvent the analog synapse storage problem, the floating-gate MOSFET design, is explained with an example using the commercial ETANN chip. In addition, the new $\nu$MOSFET circuit technique is introduced. It can be used to efficiently build neural network chips and use voltage-mode operation for low-power consumption.

## 7.2  BIOLOGICALLY-INSPIRED NEURAL NETWORKS

In this section, the biologically-inspired neural networks are described. Many neural network models have been studied in order to implement them in software-hardware co-design approaches [29, 7, 69].

### 7.2.1  Subthreshold Conduction of MOS Transistors

Since implementation of a biologically-inspired neural network might require a significantly large number of transistors, power consumption and chip area are

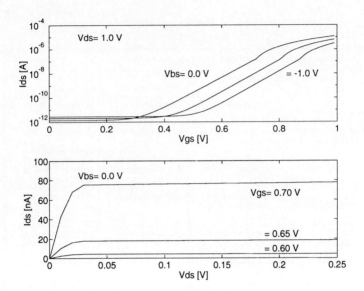

**FIGURE 7.2**
SPICE simulation results of the subthreshold MOSFET.

critical design issues. Design methodology using the transistors biased in the subthreshold region is an ideal approach to greatly reduce the power dissipation. Several advantages come from the subthreshold conduction-approach [69, 63, 87].

### 7.2.1.1   MOS TRANSISTOR IN SUBTHRESHOLD REGION.

The drain current of an $n$-channel metal-oxide-semiconductor (MOS) transistor biased in the subthreshold region can be expressed as,

$$I_{DS} = I_0 e^{-\frac{qV_G}{kT}} \left( e^{\frac{qV_S}{kT}} - e^{\frac{qV_D}{kT}} \right) \tag{7.1}$$

where $V_G$, $V_D$, and $V_S$ are the gate voltage, the drain voltage, and the source voltage with respect to the local substrate potential. The quantity $kT/q$ is the thermal voltage, and the slope factor $\alpha$ results from the capacitive voltage division which relates the variation of the gate voltage to that of the surface potential. Its value is approximately equal to 1.3 for a typical VLSI technology. The current-voltage characteristics are shown in Fig. 7.2, where the SPICE simulation results are shown based upon the BSIM_plus model [21]. The drain current has an exponential dependence on the gate voltage over many orders of the operating range. The small-signal transconductance value is easily determined as,

$$G_m \equiv \frac{\partial I_{DS}}{\partial V_G} = \frac{I_{DS}}{\alpha KT/q} \tag{7.2}$$

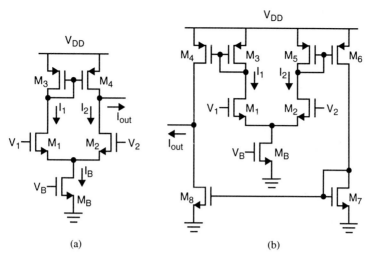

**FIGURE 7.3**
Transconductance (TC) amplifiers. (a) Basic TC amplifier. (b) Wide-range TC amplifier.

The output conductance is approximately proportional to the drain current and can be characterized by an extrapolated voltage parameter $V_E$ as follows,

$$G_o \equiv \frac{I_{DS}}{V_E} \qquad (7.3)$$

When all terminal voltages are normalized to the quantity of $kT/q$, then Eq. (7.1) is changed into

$$I_{DS} = I_0 e^{\frac{1}{q}V_G}(e^{-V_S} - e^{-V_D}) \qquad (7.4)$$

For an $p$-channel transistor, signs of all voltage notations are reversed.

#### 7.2.1.2  TRANSCONDUCTANCE AMPLIFIER.

The basic building component for a large neural network is the transconductance amplifier. The basic circuit schematic is shown in Fig. 7.3(a) [69]. Here, all voltages are assumed to be normalized with respect to the thermal voltage $kT/q$. The output current $I_{out}$ is expressed as,

$$I_{out} = I_1 - I_2 = I_B \tanh(\frac{V_1 - V_2}{2\alpha}) \qquad (7.5)$$

The transconductance, $g_m$, is defined as the slope of the output current curve around the zero input voltage and is expressed as

$$g_m \equiv \frac{\partial I_{out}}{\partial (V_1 - V_2)}|_{V_1 - V_2 = 0} = \frac{I_B}{2\alpha}. \qquad (7.6)$$

Since the transconductance, $g_m$, is proportional to the tail current, $I_B$, the gate terminal of the transistor $M_B$ is used for controlling the characteristics of the

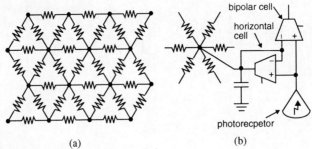

(a)                                          (b)

**FIGURE 7.4**
Conceptual diagram of the silicon retina. (a) Hexagonal resistive network. (b) Block diagram of one pixel node.

transconductance amplifier. When the input voltage difference is small, then the tanh($\cdot$) function can be approximated by its argument so that the output current in Eq. (7.5) is simplified to

$$I_{out} = g_m(V_1 - V_2), \tag{7.7}$$

which is obvious from the definition of Eq. (7.6). The circuit schematic of an improved transconductance amplifier for wide-range operations is shown in Fig. 7.3(b).

### 7.2.2   Electronic Retina, Cochlea, and Vision Chips

#### 7.2.2.1   SILICON RETINA.

Silicon implementation of the biological retina consists of three main building blocks: the photoreceptor, the horizontal cell, and the bipolar cell [69]. A conceptual diagram of the silicon retina is shown in Fig. 7.4(a), where the hexagonal resistive network is constructed for spatial averaging. Figure 7.4(b) shows the block diagram of one pixel node. At each node of Fig. 7.4(a), the following functions are performed:

- to receive the light intensity and convert it into electrical signal (**photoreceptor**),
- to average spatially and temporally (**horizontal cell**), and
- to provide the difference signal between the received input signal and the local signal (**bipolar cell**).

Figure 7.5 shows the detailed circuit schematic of one pixel of the silicon retina. The photoreceptor consists of the p-channel MOS transistors, $M_1$, $M_2$, and a pnp-bipolar transistor $Q_0$ with the base terminal floating. The parasitic pnp-bipolar transistor, which can be easily formed in a BiCMOS process or can be formed vertically in a conventional CMOS process, receives the light and produces the photocurrent $I_{ph}$. The output voltage of the photoreceptor can be expressed as

$$V_{ph} = \alpha^2 V_{DD} - (\alpha^2 + \alpha)\ln\frac{I_{ph}}{I_0} \tag{7.8}$$

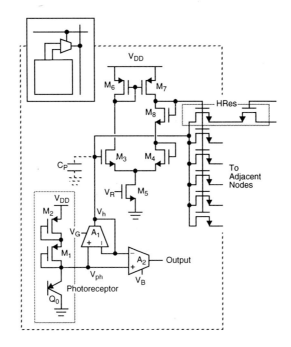

## FIGURE 7.5

One pixel node consists of a photoreceptor, a horizontal cell, and a bipolar cell.

Since the response of the photoreceptor exhibits a logarithmic behavior, the input signal range of many orders of magnitude can be compressed into a moderate output range. In addition, the contrast ratio between two points can be represented in terms of a voltage difference.

The transconductance amplifier, $A_1$, and its output parasitic capacitor, $C_p$, form the follower-integrator circuit. The following small-signal differential equation can be used to describe the behavior of the output node voltage, $V_h$, of the horizontal cell [69],

$$C_p \frac{dV_h}{dt} = g_{m1}(V_{ph} - V_h) \tag{7.9}$$

where $g_{m1}$ is the transconductance of the amplifier $A_1$. By linear superposition, we can obtain the temporal-smoothing (or averaging) characteristics [69],

$$V_h(t) = \int_0^\infty V_{ph}(t - m)e^{-m} dm \tag{7.10}$$

where $\tau$ is the characteristic time constant defined as $C_p/g_{m1}$.

At each pixel, six resistors are used to construct the hexagonal resistive network for a high degree of network symmetry and connectivity. Each horizontal resistor (*Hres*) consists of two series transistors which connect two adjacent

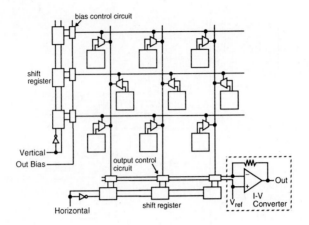

**FIGURE 7.6**
Floorplan of the silicon retina chip.

pixels. The effective resistance is

$$R_{eff} = \frac{kT/q}{I_0 e^{V_G}/2} \qquad (7.11)$$

Transistors $M_3$ to $M_8$ form the bias circuitry shared by six horizontal resistors. Solutions to the hexagonal network are given by the forward-recursive formula as [69]

$$V_{n+1} = \frac{n(RG+4)V_n - (2n-1)V_{n-1}}{2n+1} \qquad (7.12)$$

where $R$ and $G$ are the connecting resistance value and the shunting conductance value, respectively. Here $V_0$ is the voltage at the node where the light input is applied, and $V_1$ is the voltage in the periphery of the the first hexagon around $V_0$-node, and so on.

A bipolar cell consists of the transconductance amplifier $A_2$. It provides the output current which is proportional to the difference between the incoming signal $V_{ph}$ and the averaged horizontal cell signal $V_h$.

Figure 7.6 shows the floorplan of a silicon IC chip. By alternating the two rows which contains the pixels with the opposite input/output directions, the layout of the hexagonal matrix can be efficiently achieved in a small silicon area. The shift registers and the control circuitry at the vertical edge and the horizontal edge are used for transmitting the desired pixel signal to the output node. The output current signal is converted into the voltage format by a current-to-voltage converter.

Continued improvements of silicon retina chips have been attempted. Mead presented an initial design of an adaptive silicon retina using the floating gate technology to compensate component variations [84]. Boahen and Andreou developed a contrast-sensitive silicon retina with reciprocal synapses [81]. Delbrück

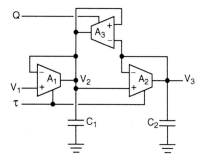

**FIGURE 7.7**
The second-order section [18].

implemented a silicon retina chip with correlation-based, velocity-tuned pixels [18].

### 7.2.2.2   SILICON COCHLEA.

A cochlea is a traveling-wave structure that creates the first-level representation in the auditory system. It converts time-domain information into spatially encoded information by spreading out signals in space according to their time scale. The basic structure, basilar membrane, can be implemented in silicon as a transmission line with a velocity of propagation which is electrically tunable [38]. Output taps are located at intervals along the line to observe the signal. This silicon model is intended to approximate the dynamics of a second-order system of fluid mass and membrane stiffness.

The building element of the transmission line is a second-order section [69]. Its block diagram is shown in Fig. 7.7. Three transconductance amplifiers and two capacitors are used. The amplifier-capacitor circuits ($A_1$, $C_1$ and $A_2$, $C_2$) are in the follower-integrator configuration. The amplifier $A_3$ is the destabilizing element due to the positive feedback scheme. By setting $C_1 = C_2 = C$ and $g_{m1} = g_{m2} = g_m$, the $s$-domain transfer function can be expressed as [69]

$$H(s) \equiv \frac{V_3(s)}{V_1(s)} = \frac{1}{\tau^2 s^2 + 2(1 - \eta)\tau s + 1} \tag{7.13}$$

where $\tau = C/g_m$ and $\eta = g_{m3}/(g_{m1}+g_{m2}) = g_{m3}/2g_m$. To ensure the stability of the circuit, $\eta$ has to be less than 1. By rearranging Eq. (7.13) in terms of the $Q$-factor, $H(s)$ becomes [69]

$$H(s) = \frac{1}{\tau^2 s^2 + \frac{1}{Q}\tau s + 1} \tag{7.14}$$

where the $Q$-factor is defined as

$$Q = \frac{1}{2(1 - \eta)} \tag{7.15}$$

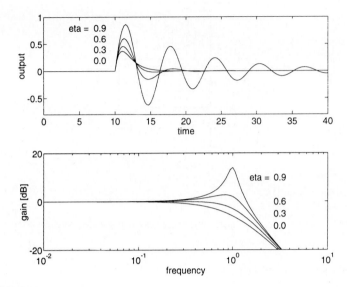

**FIGURE 7.8**
Calculated results of a second-order section.

Note that $Q$ starts from 0.5 for $\eta = 0$ and grows toward infinity as $\eta$ approaches 1.

Figure 7.8 shows the transient behavior and the frequency characteristics of the second-order section. Here, all values are normalized to the value of $\tau$. The impulse input is applied at the $x$-axis value of 10. The sinusoidal behavior exponentially decays and is inversely proportional to the value of $\eta$. If $\eta = 0$, which corresponds to no feedback, there is no oscillation. The frequency characteristics shows that if $Q$ is less than 0.707, $i.e.$ $\eta < 0.2928$, the gain of the transfer function is less than unity. The peak occurs at the frequency $\omega_{max}$,

$$\omega_{max}^2 = 1 - \frac{1}{2Q^2}, \tag{7.16}$$

and the peak gain is

$$|H(\omega)|_{max} = \frac{Q}{\sqrt{1 - \frac{1}{4Q^2}}} \tag{7.17}$$

If $Q^2 = 0.5$, the response is maximally flat with the peak gain of 1. If $Q = 1$, the peak gain is 1.16 at the normalized peak frequency of 0.707.

The block diagram of the silicon cochlea with 80 second-order sections is shown in Fig. 7.9. When two different voltages are applied to the two ends of the time constant $\tau$-control line, a linear gradient in voltage can be achieved because the control line is made of the long polysilicon line of resistance. Each linear gradient voltage is applied to the gate terminal of a bias transistor in the transconductance amplifier, producing the exponential tail current which is

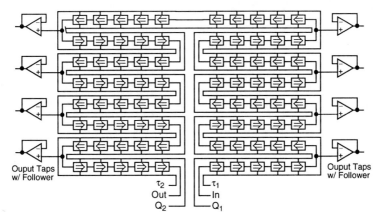

**FIGURE 7.9**
Block diagram of a silicon cochlea chip.

proportional to the transconductance value. Thus the propagation velocity and cutoff frequency, which are determined by the transconductance value, can have an exponential dependence on the distance along the control line. The $Q$-factor control line is used to adjust all sections so that they have the same $Q$ value.

The calculated transfer function of 80 second-order sections is shown in Fig. 7.10. Responses of the 10-th, 20-th, 40-th, and 80-th output taps are observed. The value of the $Q$-factor is 0.8, the open-loop dc-gain of the amplifier is 1000, and the time constant value increases by the factor of 1.01 as the sections increase. The resultant transfer function of cascading multiple second-order sections can be expressed as [38],

$$H_M(\omega) = \prod_{i=0}^{M-1} H_i(\omega) = \exp\left[\sum_{i=0}^{M-1} \ln(H_i(\omega))\right] = \exp\left[j\sum_{i=0}^{M-1} k(\omega, i\Delta x)\Delta x\right]$$
(7.18)

where $M$ is the number of stages and $k(\omega, i\Delta x)$ is the complex wave number. Several other IC chips for auditory signal processing have been reported in [41, 36, 70], and circuit techniques based on switched-currents have recently been described in [71].

### 7.2.2.3   VISION CHIPS.

Computational vision is the field of visual information processing [47, 83]. The main objectives of computer vision are to develop image understanding systems and to understand the human vision. Early vision is the set of visual modules which can extract such physical properties of the object under view as distance, orientation, reflectance, color, texture, and so on. Several obstacles have been circumvented and efficient algorithms have been developed in implementing the hardware of the early vision system. Image acquisition and the spatio-temporal smoothing for noise removal can be performed in the previously-mentioned silicon retina chip. The feature-extraction processing is the next step

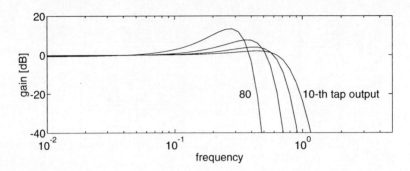

**FIGURE 7.10**
Transfer functions of the silicon cochlea model consisting of 80 second-order sections. Four
curves correspond to the $80^{th}$, $40^{th}$, $20^{th}$, $10^{th}$ tap outputs from the left to the right.

which emphasizes some important properties of the image such as the edge-
detection and motion-detection processing.

*Edge Detection*

Edge detection can be achieved with the Laplacian of the Gaussian $\nabla^2 G$
followed by the zero-crossing technique. The Difference of the Gaussian (DOG) is
successful in approximating the Laplacian of the Gaussian. The chip implements
the difference of exponentials (DOE) and was shown to be adequate for edge-
detection purposes [82].

Figure 7.11 shows the block diagram of two pixels in the edge-detection
chip. Light is applied to the network through the photoreceptor and is converted
into the input voltage. There are two rails of the resistive network, $R_1$'s and
$R_2$'s. Two shunting conductances are implemented by the two transconductance
amplifiers $A_1$ and $A_2$. Each resistive network provides the decaying exponential functions with different characteristics. The outputs from the wide-range
transconductance amplifiers are used to produce the difference between two node
voltages of the pixel.

The zero-crossing detection and threshold circuit are included in the final
row of the figure. Two neighboring currents charge or discharge the inputs of
the exclusive-OR gate and the difference of currents is applied to the input of
the NAND gate. If the slope of the zero-crossing is greater than the threshold
current, $I_{th}$, then the input of the NAND gate is charged to the high-state.
Typical operational waveforms of the chip are shown in Fig. 7.12.

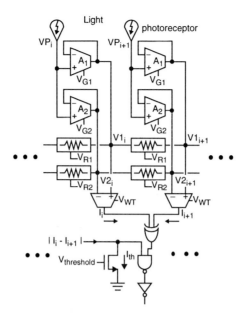

**FIGURE 7.11**
Edge-detection circuitry using zero-crossing.

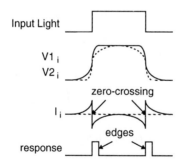

**FIGURE 7.12**
Typical waveforms of the edge-detection chip.

*Optical Motion Sensor*

An algorithm for two-dimensional velocity detection should solve the problem of inherent ambiguity between motions along the two axes resulting from the fact that the field of view is limited, known as the aperture problem [86, 24]. The relationship between the time-derivative of the intensity and the velocities can be expressed as [86]

$$\frac{\partial I}{\partial t} = -\frac{\partial I}{\partial x} v_x - \frac{\partial I}{\partial y} v_y \qquad (7.19)$$

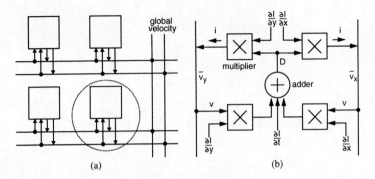

(a)                                         (b)

## FIGURE 7.13

Implementation of motion detection algorithm. (a) Computing cells containing local variables and one global line. (b) Block diagram of one cell.

where $I$ is the optical intensity, $v_x$ and $v_y$ are the $x$- and $y$-components in the velocity plane, respectively. Rearranging (7.19) into a line equation, we can get

$$\frac{\partial I}{\partial x}v_x + \frac{\partial I}{\partial y}v_y + \frac{\partial I}{\partial t} = 0 \tag{7.20}$$

Thus, each local set of three derivatives defines a constraint line in the velocity plane along which the actual velocity must lie.

In order to address the ambiguity problem, a global solution is required which is constrained by each pixel point. Each pixel contains only local variables such as $\partial I/\partial x$, $\partial I/\partial y$, and $\partial I/\partial t$. The block diagram of the motion-computation chip is shown in Fig. 7.13(a). The global velocity is represented by the voltage on the global line.

Each cell checks whether this global velocity satisfies its constraint. If there is an error, the circuit provides a "force" to move the global velocity toward the local constraint by charging or discharging the global line through the current. The force term "$F$" can be expressed as [86]

$$F_x = \left(\frac{\partial I}{\partial x}v_x + \frac{\partial I}{\partial y}v_y + \frac{\partial I}{\partial t}\right)\frac{\partial I}{\partial x} \tag{7.21}$$

$$F_y = \left(\frac{\partial I}{\partial x}v_x + \frac{\partial I}{\partial y}v_y + \frac{\partial I}{\partial t}\right)\frac{\partial I}{\partial y} \tag{7.22}$$

Figure 7.13(b) shows the detailed block diagram of one pixel. The simulated response curve as a function of $\partial I/\partial x$ is shown in Fig. 7.14 for different input frequencies.

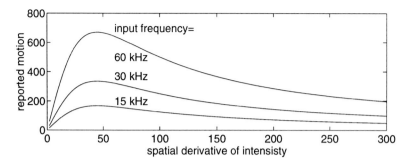

**FIGURE 7.14**
Calculated response with respect to the value of $\partial I/\partial x$ for different frequencies.

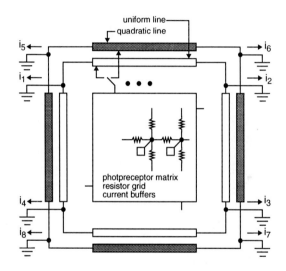

**FIGURE 7.15**
Block diagram of the object-orientation chip.

*Object Orientation and Segregation*

Resistive networks for computer vision can also be used in object-related applications. In addition to the basic photoreceptor-resistor network, additional simple circuitry is required for implementing some specific functions. In Fig. 7.15, a block diagram is shown for obtaining the orientation of an object [51]. The main portion consists of the photoreceptor matrix and the resistor grids. The current buffers are located around the perimeter of this network holding each parameter node at a common ground. The output currents from the current buffers flow into the linear and quadratic resistor lines in order to produce the second moments. These second moments are used to uniquely determine the angle of the object with respect to the axis.

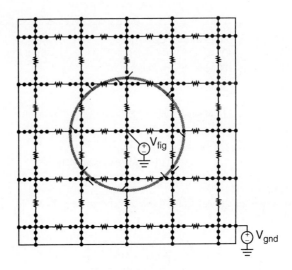

**FIGURE 7.16**
Block diagram of the figure-ground segregation chip.

Figure 7.16 shows the schematic diagram of the figure-ground segregation. The resistive grid is combined with four switches at each node [35]. All points inside the contour of the given objects are labeled with a certain voltage which is different from that for the background. The input to the chip consists of a binary edge map, which specifies the presence or absence of edges in the image. At every grid point where the edges are found, four switches are opened to isolate that node from its four neighbors. If the center node is connected to $V_{fig}$ and other nodes are connected to $V_{gnd}$, then all interior points inside the contour rise to $V_{fig}$ and the points outside the contour settle to $V_{gnd}$.

### 7.2.3 Pulse-Stream Method

In this subsection, another approach inspired by the biological neuron model is described. A pulse-stream is used for representation of the neuronal signal [60, 19, 17, 39, 22, 16]. In the nervous system, a neuron which is in the on-state fires the pulses, and a neuron in the off-state ceases firing the pulses. The axon of each neuron connects to all other neurons through the synapses. Each synapse functions by gating the signals from the transmitting neuron to the receiving one. These signals are either excitatory or inhibitory in order to turn on or turn off the state of the receiving neuron. The neural network implementation based on the pulse-stream scheme has several advantages: the network can be easily supported by digital VLSI technology, the input/output are easily interfaced with the host controller, and it is based on the understanding of the biological neurons.

The schematic diagrams of the neuron and the synapse cells are shown in Fig. 7.17(a) and (b), respectively. The neural activity (input potential) is

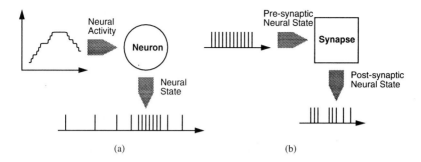

**FIGURE 7.17**
Neural network implementation in the pulse-stream approach. (a) Neuron. (b) Synapse.

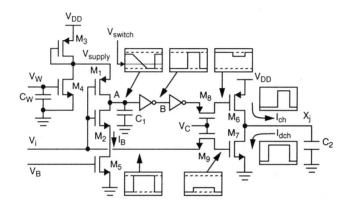

**FIGURE 7.18**
Analog synapse cell in the pulse-stream approach.

determined by the combination of the excitatory and the inhibitory inputs. The output pulses representing the neuron states are generated in proportion to this value. The synapse cell modifies the pre-synaptic neuron state into the post-synaptic neuron state by the amount of the stored weight value. In the digital storage [60, 19, 17], the weight value is stored in the internal $D$-flip flops and the post-synaptic signal is produced by the reference clocks and several logic gates. In an analog signal processing scheme [39, 22, 16], the weight value stored on the dynamic capacitor controls the pulse width through the $RC$ discharge circuitry. The digital weight approach occupies a large silicon area and might require separate excitatory and inhibitory lines. The analog weight scheme occupies a small silicon area and still can maintain full programmability.

Figure 7.18 shows the circuit schematic diagram of one analog synapse cell [39]. The weight value stored on the capacitor $C_W$, controls the supply voltage of the inverter formed by transistors $M_1$ and $M_2$. Thus, when the pre-synaptic signal, $V_i$, is in the low-state, the "high" output voltage of the inverter is determined from the weight value. When the input goes to the high-state, transistor

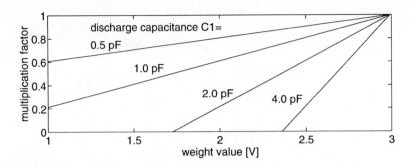

**FIGURE 7.19**
Multiplication factor with respect to the synapse value.

$M_1$ turns off and the node $A$ starts to discharge through transistors $M_2$ and $M_5$. As the voltage of node $A$ decreases, it crosses the switching threshold voltage $V_{switch}$ of the inverter and produces a pulse at node $B$. Since the $RC$ discharging rate is determined by the bias voltage, $V_B$, and the capacitor, $C_1$, which are constant during the operation, the width of the pulse generated at the node $B$ is determined by the supply voltage, $V_{supply}$, or effectively the weight value, $V_W$. The inputs to transistors $M_6$ and $M_7$ are excitatory and inhibitory, respectively. From these two inputs, the output potential $X_j$ is determined by the charge/discharge currents into/out of the output capacitor, $C_2$, from all synapse cells relevant to the specific output neuron. Transistors $M_8$ and $M_9$ attenuate the inputs to transistors $M_6$ and $M_7$ so that they can operate in the subthreshold region.

In Fig. 7.19, the calculated pulse-width multiplication factor is shown as a function of the weight voltage. It is defined as

$$F_{PW} = \frac{T_W - C_1(V_{supply} - V_{switch})/I_B}{T_W} \tag{7.23}$$

where $T_W$ is the pulse width of the input pulse (pre-synaptic state).

The neuron function is implemented by a voltage-controlled oscillator (VCO) and can operate in the asynchronous mode. Its circuit schematic diagram is shown in Fig. 7.20 [73]. The comparator and resistors $R_1$, $R_2$ form a Schmitt trigger. Its output controls the charge/discharge action at the comparator input through pass transistors $M_{10}$ and $M_{11}$. The oscillation frequency can be expressed as

$$f_o = \frac{1 + R_2/R_1}{4C} \cdot \frac{I}{V_{DD}} = \frac{1 + R_2/R_1}{4C} \cdot \frac{v_{in} - V_{DC}}{R_{in}V_{DD}} \tag{7.24}$$

## 7.3  ANALOG STRONG-INVERSION NETWORKS

One of the most important objectives in implementing the neural operation is to compute weighted summation in a fully-paralleled scheme to achieve very high

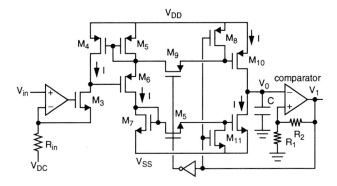

**FIGURE 7.20**
Neuron in the pulse-stream approach.

computational capability. Feedforward operation of synapses and neurons can be expressed as,

$$o_j^l = f(\sum_{i=1}^{M} w_{ji}^l o_j^{l-1}) \quad for \ j = 1 \ldots N \tag{7.25}$$

where $l$ is the layer number and $f(\cdot)$ is a nonlinear function which is usually a sigmoid function. The $o_j$ is an output value of the neuron activity. There have been significant research activities on the efficient implementation of this operation.

Khachab and Ismail presented nonlinear signal processing building blocks and discussed their use in the VLSI implementation of the Hopfield network [31]. Robinson, Yoneda, and Sánchez-Sinencio presented a modular design of a Hamming network [48] and Linares-Barranco, Sánchez-Sinencio et al. implemented a modular T-mode bidirectional associative memory (BAM) [33]. Sheu and Choi et al. implemented a neuroprocessor for self-organization mapping [49]. Satyanarayana and Tsividis et al. developed a reconfigurable analog VLSI neural chip for general-purpose applications [55]. Van der Spiegel et al. presented a prototype analog neural computer for speech processing [53]. Boser et al. presented an analog neural processor for high-speed character recognition [3]. Tsay and Newcomb designed an adaptive resonance theory (ART1) memories [61]. Salam et al. constructed a real-time 50-neuron chip for pattern recognition [52]. Arima et al. [1] presented a self-learning neural network chip for the Boltzmann machine. Shima et al. [56] presented a general-purpose neural chip which supports the on-chip back-propagation/Hebbian learning schemes. Many cellular neural network (CNN) implementations have been achieved in [64, 4]. The power for cellular neural networks and their potential applications are tremendous. In addition, various design and fabrication technologies such as BiCMOS [46], field-programmable analog arrays (FPAA) [34], and charge-coupled devices (CCD) [9] have been used for efficient construction of neural computers. Switched-capacitor circuit designs were also employed for building a specific neural network in hardware [25, 40].

Analog current-mode signal processing [88] has great potential to construct efficient signal processing systems. In this section, analog current-mode signal processing schemes are used in implementing neural networks with transistors not biased in the subthreshold region. The main advantages of using strong-inversion transistors are to achieve higher operational speeds and a better noise immunity at a cost of moderate power consumption.

### 7.3.1   Synapse Cells

The main function of the synapse cell is to achieve a linear multiplication and to provide reliable storage of the weight value. Since the number of the synapse cells is a dominant factor in a VLSI neural network chip, careful design of the synapse cell is crucial in achieving compact silicon area, minimized power consumption, and large dynamic range.

#### 7.3.1.1   MULTIPLYING DIGITAL-TO-ANALOG CONVERTER (MDAC) WITH DIGITAL WEIGHTS.

The weight value of a digital synapse cell can be stored in the internal $D$-flip flops. The digital weight is converted into an analog current using the binary-weighted current sources. Current sources are connected by pass transistors. The gate voltage of each pass transistor is controlled by the corresponding bit of the digital weight. Since the weight value is stored in the regenerative $D$-flip flops, the synapse content can be maintained as long as the power of the chip is on without the need of any external refreshing. In addition, the digital bus for synapse contents serves as a simple interface for the neural chip to communicate with the host machine or the main data processor.

Figure 7.21(a) and (b) show two examples of synapse cell implementations presented in [27] and [53], respectively. In Fig. 7.21(a), the input voltage, $V_{in}$, is applied to the differential pair. The $N$-bit weight stored on the $N$-stage shift register is used to represent the digital tail current, $I_{SS}$, through the pass transistors and the binary-weighted current sources of $M_0$ to $M_{N-1}$. The bias voltage, $V_{BB}$, is set to be around $V_T + 200mV$ in order to achieve small power consumption. Within the small input range, the output current of the differential pair can be expressed as

$$I_{out} \equiv I_{D1} - I_{D2} \simeq \frac{K}{2} V_{in} \sqrt{4 I_{SS}/\beta} \tag{7.26}$$

By replacing $I_{SS}$ with the digital representation of the weight value, the output current is given by

$$I_{out} = \frac{\beta}{2} V_{in} \sqrt{4 \sum_{i=1}^{n} I_0 2^{i-1} b_i / \beta} \tag{7.27}$$

where $b_i$ is either "0" or "1" according to the binary state of $B_i$. The operational range of $V_{in}$ is dependent on the synapse value. Multiplication is done in the first and third quadrants.

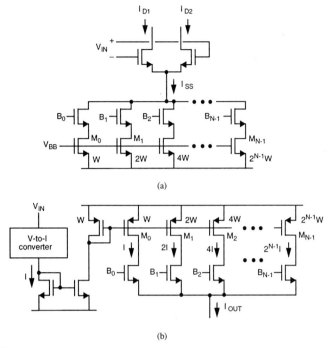

**FIGURE 7.21**
Synapse cells with digital weight representations.

In the synapse cell of Fig. 7.21(b), the applied input is first converted into a current which is used as the unit current for the binary-weighted current mirror. The output current is determined by,

$$I_{out} = \sum_{i=1}^{n} I_0 2^{i-1} b_i = G_m V_{in} \cdot \sum_{i=1}^{n} 2^{i-1} b_i \qquad (7.28)$$

where $G_m$ is the proportionality constant of the voltage-to-current converter.

### 7.3.1.2  MULTIPLIER WITH ANALOG WEIGHTS.

An analog weight value can be stored as the voltage across a capacitor. Since the capacitor is a basic component in a MOS VLSI technology, the synapse weight storage is very straightforward. Thus, the savings in silicon area is very apparent. For example, the area of one digital synapse cell in [27] is 27,000 $\lambda^2$ while the area of one analog synapse cell in [30] is 35 $\lambda^2$. The charge retention, however, is a major design issue in analog weight storage. It needs to be carefully addressed because the weight value might be reduced due to the leakage current in a reverse-biased $pn$-junction. A periodic refreshing is required. All weight values are maintained in the main memory of the digital processor and a digital-to-analog converter is used to produce the corresponding analog value. It is written into the synapse capacitors of the network in every refresh cycle.

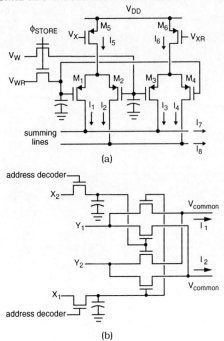

**FIGURE 7.22**
Synapse cells with analog weight representations.

*Fully Functional Multiplier*

Circuit schematic diagrams of two analog synapse designs are shown in Fig. 7.22(a) and (b), which were described in [30] and [31], respectively. In Fig. 7.22(a), the modified Gilbert multiplier cell is used for linear multiplication of the input voltage and the synapse value. The differential output current can be expressed as,

$$I_7 - I_8 \simeq \sqrt{\frac{K_l K_u}{2}}(V_W - V_{WR})(V_X - V_{XR}) = \sqrt{\frac{K_l K_u}{2}}\Delta V_W \Delta V_X \qquad (7.29)$$

where $K_l = K_1 = K_2 = K_3 = K_4$ and $K_u = K_5 = K_6$. The circuit is capable of four-quadrant multiplication. Since the weight value is stored in the differential format of $\Delta V_W = V_W - V_{WR}$, it can have a much larger charge retention time than the single-ended format. In addition, the common-mode noise rejection is also better.

In Fig. 7.22(b), the circuit schematic diagram of a multiplier with transistors biased in the linear region is shown. Four transistors are used to cancel the nonlinear terms in the drain current expression of the linear-region transistor. The output current is

$$I_1 - I_2 = -K(X_1 - X_2)(Y_1 - Y_2) \qquad (7.30)$$

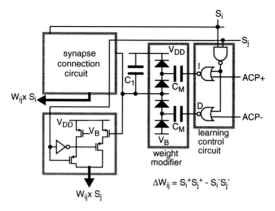

**FIGURE 7.23**
On-chip charge-modifiable synapse.

where $X_i$'s represent the differential synapse value, $Y_i$'s represent the differential input value in the synapse cell, and $K$ is the transconductance coefficient of the transistor with the same $W/L$ ratio.

*Charge Modifiable Synapse*

In some synapse cell designs, the weight value is stored on the capacitor and can be modified by injecting (or extracting) some charge onto (or from) the capacitor. One such synapse implementation is shown in Fig. 7.23 [2]. The electrical charge on the capacitor $C_1$ represents the weight value of $W_{ji}$. When the input signal $S_j$ is "low", the output current is produced by the bias voltage (corresponding to the reference zero). When the input is "high", the output current is dependent on the synapse value, $W_{ji}$. Thus, the current output is the product of the synapse value, $W_{ji}$, and the binary input, $S_j$. Summation of these output signals is compared with the reference in the next neuron stage.

If the weight modifier consists of two charge-pump circuits, the weight value can be changed within the synapse cell. Figure 7.23 shows an example design. Nodes $I$ and $D$ are the control terminals which increase and decrease the stored weight value, respectively. Pulses on these control lines are capacitively coupled to the weight node through the modifier capacitor $C_M$'s and diodes. The learning control circuit performs the simplified Boltzmann machine learning rule as,

$$\Delta W_{ji} = \eta(S_i^+ S_j^+ - S_i^- S_j^-) \qquad (7.31)$$

where $\eta$ is the learning rate constant. When both $S_i$ and $S_j$ are high, learning can take place. The $ACP^+$ and $ACP^-$ are the corresponding pulses of increasing and reducing the magnitude of weight values, respectively.

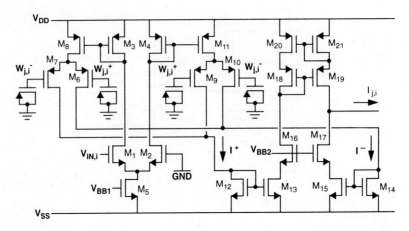

**FIGURE 7.24**
Wide-range Gilbert multiplier synapse cell.

**7.3.1.3   DETAILED DESIGN OF THE SYNAPSE CELL.**

*Linear Multiplication*

The synapse cell based on the wide-range Gilbert multiplier structure is shown in Fig. 7.24 [12]. The $p$-channel MOS transistor mirrors, $M_3$-$M_8$ and $M_4$-$M_{11}$, implement the upper differential pairs of the basic Gilbert multiplier cell with the $p$-channel MOS transistors $M_6$-$M_7$ and $M_9$-$M_{10}$. By stacking a fewer number of transistors between the power supply lines, a larger operational range can be achieved. Transistors $M_{12}$ through $M_{21}$ are used to convert the differential output current into a single-ended value with a cascode current mirror. When the single-ended input voltage and the differential weight voltage are applied, the synapse current $I_{ji}$ from the $(j,i)^{th}$-synapse cell can be expressed as

$$I_{ji} = G_M(W_{ji}^+ - W_{ji}^-)V_{INi} = G_M W_{ji} V_{INi} \qquad (7.32)$$

where $G_M$ is the transconductance parameter of the multiplier. Transistor sizes are listed in Tab. 7.2. The SPICE model parameters of a 2-$\mu m$ CMOS technology used in the simulation are listed in Fig. 7.25. Figure 7.26 shows the simulated dc-characteristics of the multiplication operation by this synapse cell.

*Weight Storage*

The weight value is stored on a capacitor which is made of the intrinsic gate-capacitance of an MOS transistor. An additional capacitor can be added to increase the equivalent capacitance value for a larger charge retention capability. A single common signal line for weight modification/refresh is connected to the desired synapse cell through pass transistors which are controlled by the row/column address decoders. Since the weight update process is performed by a sample-and-hold operation, there could be an error voltage due to the charge feedthrough effect [54]. This problem can be significantly reduced by using dif-

Table 7.2: Transistor sizes of a synapse cell in Fig. 7.24.

| Transistor | W/L [μm / μm] | Transistor | W/L [μm / μm] |
|---|---|---|---|
| $M_1$ , $M_2$ | 4 / 40 | $M_{12}$ , $M_{14}$ | 8 / 4 |
| $M_3$ , $M_4$ | 20 / 2 | $M_{13}$ , $M_{15}$ | 24 / 4 |
| $M_5$ | 30 / 2 | $M_{16}$ , $M_{17}$ | 60 / 2 |
| $M_6$ , $M_7$, $M_9$ , $M_{10}$ | 4 / 30 | $M_{18}$ , $M_{19}$ | 16 / 4 |
| $M_8$ , $M_{11}$ | 20 / 2 | $M_{20}$ , $M_{21}$ | 14 / 6 |

```
* SPICE LEVEL-2 MODEL PARAMETERS
*
.MODEL CMOSN NMOS LEVEL=2 LD=0.240081U TOX=395.000000E-10
+ NSUB = 2.246064E+16 VTO=0.949471 KP=5.193000E-05 GAMMA=0.9877
+ PHI=0.6 UO=594 UEXP=0.212155 UCRIT=127137
+ DELTA=1.72793 VMAX=72575.8 XJ=0.250000U LAMBDA=2.712315E-02
+ NFS=2.697586E+12 NEFF=1 NSS=1.000000E+10 TPG=1.000000
+ RSH=26.260000 CGDO=3.148231E-10 CGSO=3.148231E-10
+ CGBO=4.230883E-10
+ CJ=3.945100E-04 MJ=0.457442 CJSW=4.704500E-10 MJSW=0.335833
+ PB=0.80
* Weff = Wdrawn - Delta_W
* The suggested Delta_W is 0.31 um
*
.MODEL CMOSP PMOS LEVEL=2 LD=0.247330U TOX=395.000000E-10
+ NSUB=5.638600E+15 VTO=-0.79371 KP=2.482000E-05 GAMMA=0.4949
+ PHI=0.6 UO=283.941 UEXP=0.272137 UCRIT=21057.1
+ DELTA=1.27611 VMAX=34646.6 XJ=0.250000U LAMBDA=5.908812E-02
+ NFS=8.367052E+11  NEFF=1.001 NSS=1.000000E+10 TPG=-1.000000
+ RSH=94.190000 CGDO=3.243289E-10 CGSO=3.243289E-10
+ CGBO=4.355993E-10
+ CJ=1.808400E-04 MJ=0.390981 CJSW=2.422600E-10 MJSW=0.293202
+ PB=0.70
* Weff = Wdrawn - Delta_W
* The suggested Delta_W is 0.36 um
```

## FIGURE 7.25
SPICE model parameters of a 2-$\mu m$ CMOS technology.

ferential weight signals and minimum-sized pass transistors.

The weight value is stored as the voltage across the capacitor and its magnitude is continuously decayed due to the small leakage current through the reverse-biased $pn$-junction of the pass transistor. In order to maintain accurate synapse values, periodic refresh is required [75]. The refresh cycle is determined by several factors such as the storage capacitance value, computational accuracy, the leakage current, and the junction temperature.

The requirement on the refreshing cycle for 8-bit accuracy is determined by,

$$T_{cycle} = N \cdot T_{write} < 10^{-3} \qquad (7.33)$$

where $N$ is the number of the synapse cells connected to one common signal line and $T_{write}$ is the time needed for writing the analog value into one synapse cell. Since learning is performed in the digital processor, the final weight value is stored in the main memory. This digital weight value is converted into an analog

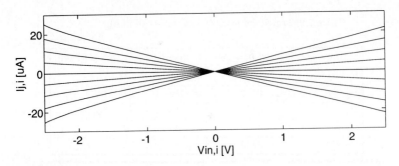

**FIGURE 7.26**
DC-characteristics of the multiplication in the synapse cell. The differential weight values increase from -2 $V$ by the step size of 0.5 $V$.

value by a digital-to-analog converter and written onto the synapse capacitor. Thus, $T_{write}$ consists of the following components,

$$T_{write} = T_{data} + T_{dac} + 6 \cdot R_{on} C_W, \qquad (7.34)$$

where $T_{data}$ is the time needed for fetching the digital data from the system memory, $T_{dac}$ is the digital-to-analog conversion time, $R_{on}$ is the on-resistance of the access-switch transistor, and $C_W$ is the effective capacitance for holding the analog value of the synapse weight.

### 7.3.2 Input and Output Neurons

#### 7.3.2.1 INPUT NEURON.

An input neuron buffers the input signal voltage and provides the driving capability for a large capacitive load. It consists of an operational amplifier configured as a unity-gain buffer. Since there could exist a large number of synapse cells to be driven by one input neuron, the load capacitance of the input neuron might be quite huge. Thus, a fast settling response of the input neuron is highly desired. In addition, the input neuron should occupy a compact silicon area and consumes low power because many input neurons are required in a large network. Figure 7.27 shows the circuit schematic diagram of an input neuron. In order to reduce the power consumption in the output branch while maintaining a high speed operation, a class-AB output stage is used.

#### 7.3.2.2 OUTPUT NEURON.
*Linear Current-to-Voltage Conversion*

In Fig. 7.28(a), the circuit diagram of an output neuron is shown. Current-to-voltage conversion occurs through the transimpedance amplifier consisting of an operational amplifier and a feedback resistor. Since the output impedance of the synapse cell is finite, the output current is strongly dependent on the voltage, which appears as the input node voltage of the output neuron. In one specific

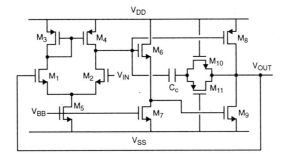

**FIGURE 7.27**
Input neuron.

design, this node is connected at virtual ground to avoid the effect of the output
current variation due to the finite output resistance of the synapse cells.

The large array of synapse cells may contribute to the summed current.
The transimpedance amplifier ought to have a sufficient capability to handle
a large magnitude of current for proper linear conversion. In a specific design,
the operational amplifier includes a source follower as an output stage and a
distortion-compensated active resistor is used. The feedback resistor is implemented by six active transistors for high accuracy and occupies a much smaller
silicon area than a passive resistor. The circuit schematic is shown in Fig. 7.28(b)
[5]. The summed current is converted into a voltage as

$$V_{LINj} = R_{eq} \cdot \sum_{i=1}^{M} I_{ji} = \frac{\sum_{i=1}^{M} I_{ji}}{2K_{r1}(V_C - V_{Tp})} \tag{7.35}$$

where $V_C$ is the gate voltage controlling the equivalent resistance value $R_{eq}$, which
is designed to be from 2 $k\Omega$ to 8 $k\Omega$. Here $K_{r1}$ and $V_{Tp}$ are the transconductance
parameter and the threshold voltages of the $p$-channel transistor pair $M_{R1}$-$M_{R2}$,
respectively. This $I$-to-$V$ converter has a fairly good linearity for a wide range
of operation.

*Sigmoid Function Generation*

Sigmoid function generation is required for performing the conventional
back-propagation learning algorithm. The sigmoid function is realized by cascading simple inverting amplifiers with the input and feedback resistors as shown
in Fig. 7.28. The voltage gain is determined by the amplifier gain and resistance
ratio as,

$$\frac{V_{SIGj}}{V_{LINj}} = \left[ \frac{R_{eq2}/R_{eq1}}{1 + \frac{1}{A}(1 + R_{eq2}/R_{eq1})} \right]^2 \tag{7.36}$$

where $A$ is the voltage gain of the simple inverting amplifier. The resistors are
realized by transistors biased in the triode region. Their equivalent resistance

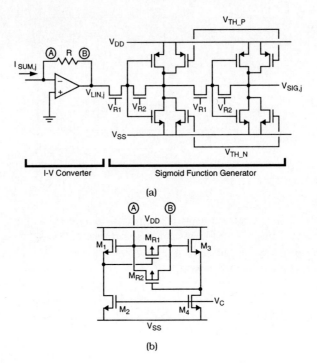

**FIGURE 7.28**
Output neuron. (a) Circuit diagram of the output neuron. (b) Floating linear resistor.

values are determined as

$$R_{eq1(2)} = \frac{1}{K(V_{R1(2)} - V_T)} \tag{7.37}$$

where $K$, $V_T$ are the transconductance parameter and the effective threshold voltage of the transistor, respectively. The control voltages, $V_{TH\_P}$ and $V_{TH\_N}$ are used to reduce the offset voltage of the sigmoid function generator induced by fabrication process. Their values are set during the chip-initialization phase. The entire gain of the output neuron is tunable by the gate voltages of the $p$-channel transistors of the feedback resistor in the transimpedance amplifier, i.e. $V_C$ in (7.35) and those of the input and the feedback resistors in the sigmoid function generator, i.e. $V_{R1}$ and $V_{R2}$ in (7.37). Figure 7.29 shows the SPICE simulation results of the dc characteristic of the output neuron with various neuron gains.

*On-Chip Learning Neuron*
    In one on-chip learning implementation [56], a single complex neuron includes multiple functional blocks. The block diagram of the single neuron for back-propagation and/or Hebbian learning is shown in Fig. 7.30. There are two input terminals and two output terminals. One input terminal is for the forward signal and the other one is for the backward signal. A transimpedance amplifier

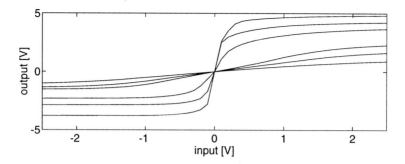

**FIGURE 7.29**
Simulated dc-characteristics of the sigmoid function generation with variable gains.

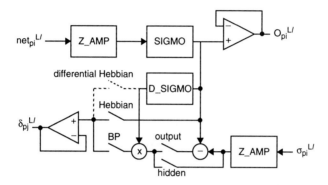

**FIGURE 7.30**
On-chip neuron.

converts the input current into voltage. Two output terminals are used for the neuron output and the learning signal output, respectively. Both output terminals are driven by buffers. Three switches are used to activate a specific learning algorithm. Two other switches are used for identifying the hidden neuron or the output neuron.

## 7.4 FLOATING-GATE, LOW-POWER NEURAL NETWORKS

Neural network processor design with floating-gate transistors has been one of the most important approaches in implementing reliable analog memory with permanent synapse weight storage [15, 50, 8]. In this section, two design techniques are described. The first is a commercial analog chip from Intel Corp. and the second is based on the $\nu$MOS design technique.

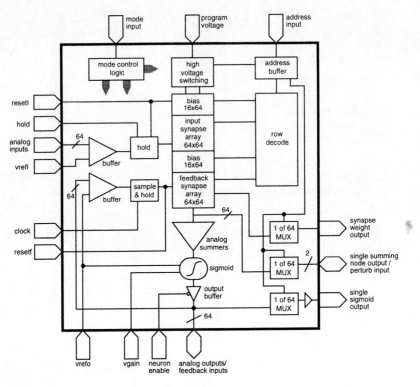

**FIGURE 7.31**
Block diagram of the ETANN chip.

## 7.4.1 The ETANN Chip

The first commercial chip implementation for general-purpose neural network applications was the Electrically Trainable Analog Neural Network (ETANN) chip (80170NW) from Intel Corp. [26, 89]. The block diagram of the ETANN chip is shown in Fig. 7.31. It consists of 64 neurons and 10,240 synapses. A total of 160 synapses are connected to each output neuron. Sixty-four synapses are from the external input group, and 64 synapses are for the feedback group. There are 16 synapses for the fixed bias from each group. Thus, 128 configurable inputs are available. The synapse value can be stored and updated at the floating gate of an MOS transistor in an EEPROM technology. The neuron performs the sigmoid function for the dot product of the input signal and the weight value from the synapse array.

Fully-paralleled processing achieves high performance in excess of 2 billion multiply and accumulate operations, or connections per second (XPS). In addition to this high feedforward processing rate, the chip can support a 100 KXPS learning rate for the individually-addressable weight update. Learning is implemented by an off-chip approach for maximizing flexibility in order to support

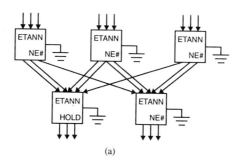

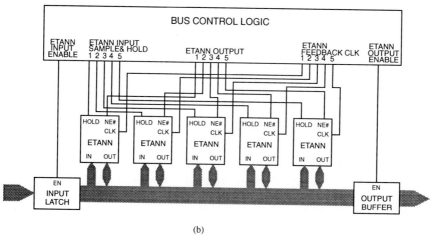

**FIGURE 7.32**

Interconnection schemes of the ETANN chip. (a) Direct pin-to-pin interconnection. (b) Bus connection.

various learning algorithms such as the back-propagation learning and competitive learning. The network also has the hold and feedback capability as shown in Fig. 7.31, for the configuration of various operating modes.

The chip was fabricated with the fast, low-power CHMOSIII EEPROM technology. It uses a single power supply voltage of 5 $V$ and dissipates an electrical power of 1.5 $W$ in the active mode. The chip is packaged in a dense 208-pin PGA and supports the TTL-compatible input/outputs levels. Since the input and the output voltage levels of the chip are fully-compatible, multi-chip processing is possible through the use of direct pin-to-pin interconnection and the bus connection as shown in Fig. 7.32(a) and (b), respectively.

### 7.4.1.1   BUILDING BLOCKS - SYNAPSE AND NEURONS.

The feedforward operation for each neuron can be summarized as

$$v_i = f_s[\sum_j W_{ij} \cdot v_j + \sum_k I_{ik} \qquad (7.38)$$

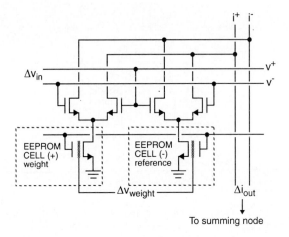

**FIGURE 7.33**
Synapse cell of the ETANN chip.

Where $v_i$ is the output voltage of the $i^{th}$ neuron, $v_j$ is the voltage from the $j^{th}$ neuron, and $I_{ik}$'s are the currents from the 16 bias synapses. Here $f_s$ is a sigmoid function. The dot product calculation consists of the linear multiplication of the input signal and the synapse value.

Figure 7.33 shows the circuit diagram of one synapse cell. It is based on a modified Gilbert multiplier. The differential input is applied to the upper differential pairs and the weight value is stored in the lower differential pair. Each synapse contribution appears as a differential output current which is summed with other synapse output currents at the neuron. The weight value is stored as a difference in the floating-gate threshold voltages of a pair of EEPROM transistors. Weight update involves changing the charge stored on the floating gates. Synapse weight value can be modified continuously for either polarity. Typical resolution of the analog input signal and the output signal in EEPROM technology is around 6 bits. The main advantage of using the floating-gate storage technique is its capability of very long data-retention time, which can be longer than 10 years, at room temperature.

The block diagram of the neuron is shown in Fig. 7.34. The summed current from all contributing synapses is converted into the voltage format with load resistors of about 2 $k\Omega$. This voltage is transformed by the sigmoid function to produce the output voltage. For each neuron, the differential summing nodes and the final neuron output voltage can be available in the pads of SMO+, SMO- and NO. Addresses $A_5$-$A_0$ select the desired pair of summing nodes among 64 neurons.

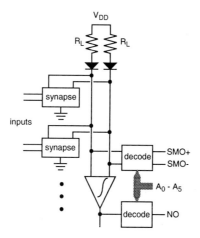

**FIGURE 7.34**
Neuron of the ETANN chip.

## 7.4.2 The MOSFET Technology

The basic structure of the functional MOS transistor (a neuron MOSFET or abbreviated as $\nu$MOSFET) is shown in Fig. 7.35 [57]. It consists of $n$-channel MOS transistors with electrically floating gate electrodes. The $N$-input gates are capacitively coupled to the floating-gate. In Fig. 7.35(c), the net charge in the floating-gate $Q_F$ can be expressed as,

$$Q_F = Q_0 - \sum_{i=1}^{N} Q_i = \sum_{i=0}^{N} C_i(\Phi_F - V_i), \tag{7.39}$$

where $\Phi_F$ is the floating-gate potential and $V_i$'s are the input signal voltages. Here $Q_F$ is the initial charge in the floating gate if no charge injection takes place during the operation. For simplicity, $Q_F$ and $V_0$ can be set to zero. The floating-gate potential $\Phi_F$ can be expressed as,

$$\Phi_F = \frac{C_1 V_1 + C_2 V_2 + \cdots + C_N V_N}{C_{TOT}}, \tag{7.40}$$

where

$$C_{TOT} = \sum_{i=0}^{N} C_i \tag{7.41}$$

According to (7.40), the floating-gate potential $\Phi_F$ is determined by a linear sum of all input signals weighted by the capacitive coupling coefficients $C_i/C_{TOT}$. This voltage-mode summation is an extremely attractive feature as compared to the wired summation of currents. Significant savings in power consumption can be achieved.

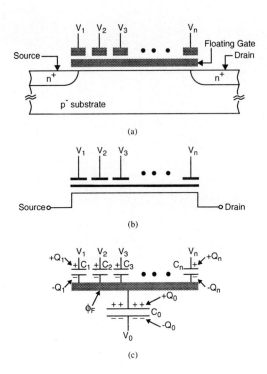

**FIGURE 7.35**

$\nu$MOSFET. (a) Cross-section view. (b) Circuit symbol. (c) Equivalent circuit model.

The effective threshold voltage of the transistor with respect to the floating gate is $V_T^*$. The transistor can turn on if the floating-gate potential $\Phi_F$ is larger than this effective threshold voltage,

$$\Phi_F = \frac{\sum_{i=1}^{N} C_i V_i}{C_{TOT}} > V_T^* \qquad (7.42)$$

In other words, when the linear weighted sum of all input signals exceeds a certain threshold voltage, the transistor turns on. This behavior is highly similar to the neural operation, the functional MOSFET is also called the neuron-MOSFET or a neuMOS ($\nu$MOSFET).

### 7.4.2.1 VARIABLE THRESHOLD TRANSISTOR.

The effective threshold voltage of the transistor can be changed by arranging the input voltages $V_i$'s so that $V_1$ is the input of the transistor and $V_2$ to $V_N$ are used to control the effective threshold voltage value. Equation (7.42) can be rearranged

$$V_1 > \frac{C_{TOT}}{C_1} V_T^* - \frac{\sum_{i=2}^{N} C_i V_i}{C_1} \qquad (7.43)$$

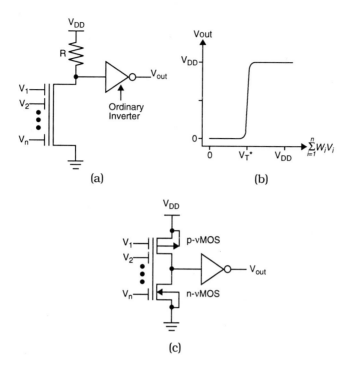

**FIGURE 7.36**
$\nu$MOSFET inverters. (a) $n$-channel $\nu$MOS inverter. (b) DC-transfer curve. (c) Complementary $\nu$MOS inverter.

and the effective threshold voltage with respect to gate 1, $V_T^{(1)}$, can be expressed as

$$V_T^{(1)} = \frac{C_{TOT}}{C_1} V_T^* - \frac{\sum_{i=2}^{N} C_i V_i}{C_1} \tag{7.44}$$

Thus, by proper combination of $C_i V_i$ products, various threshold voltages can be achieved.

### 7.4.2.2 MOS INVERTER.

The circuit schematic of an inverter based on the $n$-channel $\nu$MOS transistor and its dc-transfer curve are shown in Fig. 7.36(a) and (b), respectively. Figure 7.36(c) shows the complementary $\nu$MOS inverter circuit to implement a ratioless circuit for very low power consumption. If the switching threshold of the inverter with respect to the floating gate is defined as the voltage at which the inverter output is $V_{DD}/2$, it can be determined as

$$V_{inv}^* = \frac{V_{DD} + \sqrt{K_R} V_{Tn}^* + V_{Tp}^*}{\sqrt{K_R} + 1} \tag{7.45}$$

where $V_{Tn}^*$ and $V_{Tp}^*$ are the $n$-channel and $p$-channel $\nu$MOSFET threshold voltages with respect to the floating gate, respectively. $K_R$ is the ratio of $K_n$ to

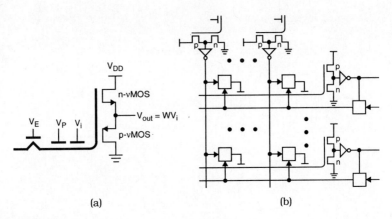

(a)                                                              (b)

**FIGURE 7.37**
On-chip learning $\nu$MOSFET neural network. (a) One synapse cell. (b) Block diagram of the neural network.

$K_p$. From the variable threshold expression of (7.44), the complementary $\nu$MOS inverter threshold as seen from gate 1 is

$$V_{inv}^{(1)} = \frac{C_{TOT}}{C_1}\left(V_{inv}^* - \frac{C_{0p}}{C_{TOT}}V_{DD}\right) - \frac{\sum_{i=2}^{N} C_i V_i}{C_1} \tag{7.46}$$

where $C_{0p}$ is the gate oxide capacitance of a $p$-channel $\nu$MOS transistor. Here, the second term in the parenthesis is due to the fact that in the $p$-channel MOS transistor, the substrate is connected to $V_{DD}$ instead of ground. By changing the $C_i V_i$ products properly, the inverter switching threshold can be adjusted.

### 7.4.3 MOS Circuits for Neural Networks

The complementary $\nu$MOS inverter of Fig. 7.36(c) performs the linear weighted summation and thresholding operation which are the generic process of a neuron model.

#### 7.4.3.1 SELF-LEARNING SYNAPSE CELL.

The synapse cell should store the information about the weight value and update its value during the learning process. The circuit diagram of the synapse cell using the complementary $\nu$MOS source follower is shown in Fig. 7.37(a) [58]. Here, $V_i$ is the voltage from the previous neuron layer, which has the binary states. $V_P$ is the synapse programming control voltage and $V_E$ is the terminal for the programming electrode where the thin oxide allows the charge transfer between the floating gate and this node through tunneling mechanism.

Because the output voltage is obtained by subtracting the threshold, $V_T$, from the input voltage, the resulting weight factor is $W = 1 - V_T/V_{DD}$ for the input value of $V_{DD}$. By programming $V_T$, the weight value can be changed. The

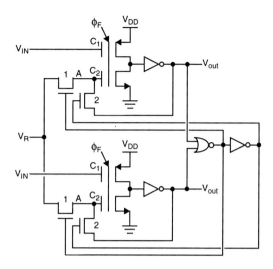

**FIGURE 7.38**
Winner-take-all circuit with $\nu$MOS transistors.

portion of the gate oxide under the $V_E$ electrode is especially thin such that
tunneling can easily occur. If $V_E$ electrode is connected to ground and $V_i$ and
$V_P$ electrodes are connected to positive values, then electrons are injected from
the $V_E$ electrode to the floating gate. If the net charge in the floating gate $Q_F$ is
not zero, the threshold voltage is changed from $V_T$ to $V_T$ - $Q_F/C_{TOT}$, performing
the weight modification process. The schematic of a multi-layer neural network
implementation using the $\nu$MOS inverters and the synapse cells is shown in
Fig. 7.37(b).

### 7.4.3.2    WINNER-TAKE-ALL CIRCUIT.

A winner-take-all (WTA) circuit receives multiple analog input signals and
selects the largest one as the winning input. The output corresponding to the
winning input will appear as "1" to signify a winner and other outputs will be
"0". The circuit diagram of a two-input WTA circuit is shown in Fig. 7.38 [67].
Each $\nu$MOS inverter receives two inputs: one inverter input voltage $V_{IN}$ and
one common ramp input $V_R$. The potential of the floating gate $\Phi_F$ is $(C_1V_{IN} +
C_2V_R)/C_{TOT}$ which becomes $(V_{IN}+V_R)/2$ when $C_1 = V_2 > V_0$. At the beginning
when $V_R$ is quite small, $\Phi_F$ is smaller than $V_{inv}^*$ in (7.45) and all output voltages
are zero. Hence, all $M_1$ transistors turn-on and $M_2$ transistors turn-off. When
the ramp voltage increase, $\Phi_F$ exceeds the inverter threshold voltage for the first
time at the winning cell where the largest input is applied. At this condition, the
corresponding output voltage goes to "1" which makes all $M_1$ transistors turn-
off and $M_2$ transistors turn-on. When the feedbacks are established through $M_2$
transistors, the high output voltage increases the node $A$ toward $V_{DD}$ while all
other low outputs result in a low voltages at their node $A$'s.

For further readings of neural network applications and VLSI implementations, consult the following books [68, 79, 74, 78, 76, 80, 72] and Chapters 3 and 8 of this book.

## PROBLEMS

**7.1.** The voltage amplification factor of an MOS transistor in the saturation region is defined as the ratio of the transconductance to the output conductance, $g_m/g_o$. Compare the voltage amplification factors in both the subthreshold and the strong-inversion regions. Also, verify your results by SPICE circuit simulation.

**7.2.** Derive the dc-voltage gain of the two voltage amplifiers in Fig. P7.2. Here $V_{DD} < V_{Tn} + |V_{Tp}|$.

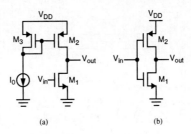

(a)                    (b)

**FIGURE P7.2** Two voltage amplifiers.

**7.3.** Describe the operation of the rectifier circuit in Fig. P7.3. by identifying the signal path for $V_1 \geq V_2$ and $V_1 < V_2$, respectively.

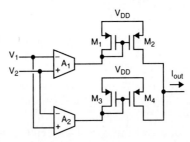

**FIGURE P7.3** Rectifier circuit.

**7.4.** Consider the synapse and the neuron circuits in Fig. P7.4 [14]. All transistors operate in the subthreshold region.
  (a) Derive an expression of the output current of the synapse cell in terms of the voltages and determine the weight value. The differential inputs are applied at $V_1$ and $V_2$.
  (b) Derive an expression of the output voltage of the neuron in terms of the currents. The differential inputs are applied at $I_5$ and $I_6$.

**7.5.** The photoreceptor circuit shown in Fig. P7.5 receives the light intensity and converts the photocurrent into voltage.
  (a) Derive the voltage expression of Eq. (7.8).

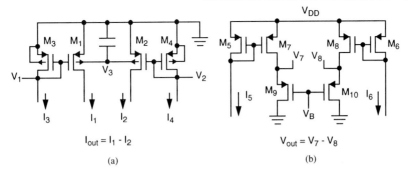

FIGURE P7.4 Synapse and neuron circuits.

(b) Derive the voltage expression when the number of stacked MOS transistors is 3.

**7.6.** Consider the analog synapse cell of the pulse-stream neural network implementation shown in Fig. 7.18
  (a) Obtain an expression for the curve in Fig. 7.19. (*Hint:* In the steady state, there is no current flowing in the path of transistors $M_1$ and $M_2$.)
  (b) What are the upper/lower limit synapse values for which the linear relation can be maintained?
  (c) Perform SPICE simulations to verify the obtained curve.

**7.7.** The circuit in Fig. P7.7 is a self-resetting neuron used in a pulse-stream neural network [16]. Derive the quantitative timing diagrams of $V_{in}$ and $V_{out}$. (*Hint:* Consider the charge conservation law in the capacitance path of $C_1$ and $C_2$ for the case when $V_{out}$ changes to either $0\ V$ or $V_{DD}$.)

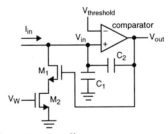

FIGURE P7.7 Self-resetting neuron cell.

**7.8.** The analog counter shown in Fig. P7.8 is useful for implementing the learning algorithms in the analog signal format [42]. Consider the increment operation when $I/\bar{D}$ line is high.
  (a) Derive an expression of the output voltage $V_W(n)$ after the $n$-th $\phi$ period as a function of $n$.
  (b) Derive the expression of the amount of the one counter step and plot the error curve between the one counter step results and results from the desired linear step of $V_{DD}(C_I/C_W)$.
  (c) Calculate the maximum error voltage due to non-zero rising time $t_r$. $\phi$ increases linearly from zero to $V_{DD}$ during $t_r$. (*Hint:* The maximum error occurs when both transistors $M_1$ and $M_2$ have the saturation current.)
  (d) Verify your results by SPICE circuit simulation.

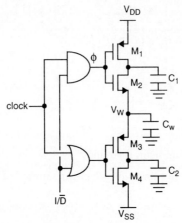

**FIGURE P7.8** Analog counter circuit.

**7.9.** Derive the output current expressions of Eqs. (7.29) and (7.32).

**7.10.** Consider the implementation of a neural network as shown in Fig. P7.10(a) [33]. The combination of the resistor $R$ and the sigmoid function generator can be implemented using the circuits in (b). Plot the transfer characteristics of $J_i$ versus $y_i$.

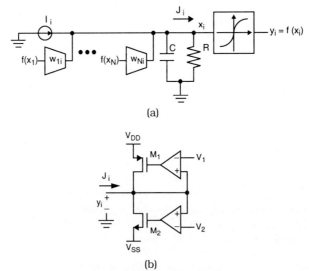

**FIGURE P7.10** T-mode neural network implementation.

**7.11.** Verify that the synapse cell shown in Fig. P7.11 [3] performs the multiplication of the digital inputs $X_0$, $X_1$ and the reference current $I_{ref}$ represented by the voltages stored on the capacitor $C_W$. (*Hint:* Separate the multiplication operation in terms of the magnitude and the sign.)

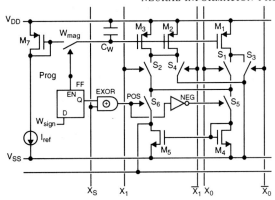

**FIGURE P7.11** Synapse cell with digital inputs and analog weights.

**7.12.** The circuit in Fig. P7.12 approximately generates the derivative of the sigmoid function in the neuron block of Fig. 7.30 [56].
   (*a*) Derive an expression of the output voltage.
   (*b*) Verify your results by SPICE circuit simulation.

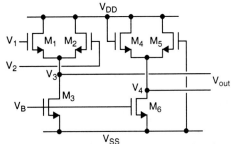

**FIGURE P7.12** Sigmoid-derivative function generator circuit.

**7.13.** The circuit diagram of Fig. P7.13 shows the implementation of the input neuron; synapse cell; output neuron combination in a compact hardware [32].
   (*a*) Derive an expression of the synapse current $I_{ij}^s$ for the binary input voltage.
   (*b*) Derive an expression of the synapse current $I_{ij}^s$ for the analog input voltage.
   (*c*) Derive the expression of the input voltage of the output neuron when there are $N$ synapse cells connected to the output neuron.
   (*d*) Derive the expression of the output voltage of the output neuron.
   (*e*) Use SPICE circuit simulator to verify your derivations.
**7.14.** The circuit shown in Fig. P7.14 approximately implements a Gaussian function [13]. Discuss the programmability of the magnitude, mean value, and the variance value. Verify your results by SPICE circuit simulation.
**7.15.** The circuits in Fig. P7.15(a) and (b) are synapse cells with floating-gate MOS transistors [6]. Derive an expression of the differential output current in each cell.
**7.16.** Consider a 3-input variable-threshold complementary $\nu$MOS inverter as shown in Fig. P7.16. Here $V_I$ is the input terminal, $V_A$ and $V_B$ are the control inputs. Determine the conditions for the variable effective threshold voltage with respect

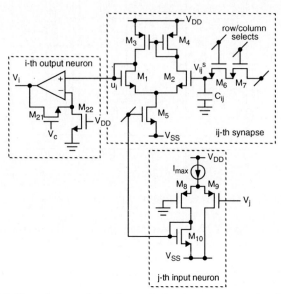

**FIGURE P7.13** Neural network implementation in a compact hardware.

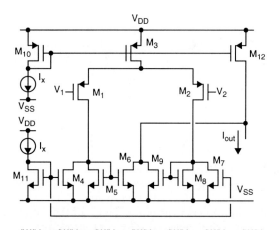

$(W/L)_4 = (W/L)_5 = (W/L)_6 = (W/L)_7 = (W/L)_8 = (W/L)_9 = (W/L)_{11}$
$(W/L)_{10} = (W/L)_{12} = 0.5(W/L)_3$

**FIGURE P7.14** Gaussian function generator circuit.

to gate 1, $V_T^{(1)}$, to become $V_{DD}$, $(2/3)V_{DD}$, $(1/3)V_{DD}$, and 0, respectively.

## REFERENCES

[1] Y. Arima, K. Mashiko, K. Okada, T. Yamada, A. Maeda, H. Kondoh, and S. Kayano, "A self-learning neural network chip with 125 neurons and 10 K self-organization synapses," *IEEE J. Solid-State Circuits*, vol. 26, no. 4, pp. 607–611, April 1991.
[2] Y. Arima, K. Mashiko, K. Okada, T. Yamada, A. Maeda, H. Notani, H. Kondoh, and S. Kayano, "A 336-neuron, 28 K-synapse, self-learning neural network chip with branch-

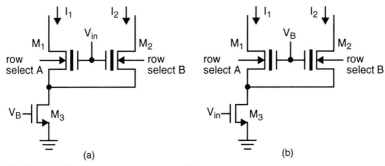

**FIGURE P7.15** Floating-gate synapse cells.

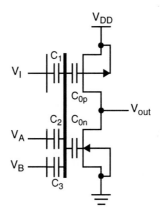

**FIGURE P7.16** Three-input $\nu$MOSFET inverter.

neuron-unit architecture," *IEEE J. Solid-State Circuits,* vol. 27, no. 11, pp. 1637–1644, November 1991.

[3] B. Boser, E. Säckinger, J. Bromley, Y. LeCun, and L. D. Jackel, "An analog neural network processor with programmable topology," *IEEE J. Solid-State Circuits,* vol. 26, no. 12, pp. 2017–2025, December 1991.

[4] I. A. Baktir and M. A. Tan, "Analog CMOS implementation of cellular neural network," *IEEE Trans. Circuits and Systems, II,* vol. 40, no. 3, pp. 200–206, March 1993.

[5] M. Banu and Y. Tsividis, "Floating voltage-controlled resistors in CMOS technology," *Electronic Letters,* vol. 18, pp. 678–679, 1982.

[6] T. H. Borgstrom, M. Ismail, and S. B. Bibyk, "Programmable current-mode neural network for implementation in analogue MOS VLSI," *IEE Proceeding, Part G,* vol. 37, no. 2, pp. 175–184, April 1990.

[7] T. W. Berger, T. P. Harty, X. Xie, G. Barrionueve, and R. J. Sclabassi, "Modeling of neuronal networks through experimental decomposition," *Proc. IEEE 34th Midwest Symp. Circuits and Systems,* pp. 91–97, 1992.

[8] T. Blyth, S. Khan, and R. Simko, "A non-volatile analog storage device using EEPROM technology," *Tech. Digest IEEE Intl. Solid-State Circuits Conf.,* San Francisco, CA, February 1991.

[9] A. M. Chiang and M. L. Chuang, "A CCD programmable image processor and its neural network applications," *IEEE J. Solid-State Circuits,* vol. 26, no. 12, pp. 1894–1901, December 1991.

[10] C.-F. Chang and B. J. Sheu, "Digital VLSI multiprocessor design for neurocomputers," *Proc. IEEE/INNS Inter. Joint Conf. Neural Networks,* vol. II, no. 1-6, Baltimore, MD, June 1992.

[11] D. R. Collins and P. A. Penz, "Considerations for neural network hardware implementations," *Proc. IEEE Inter. Symp. Circuits and Systems,* pp. 834–837, Portland, OR, May 1989.

[12] J. Choi, S. H. Bang, and B. J. Sheu, "A programmable analog VLSI neural network processor for communication receivers," *IEEE Trans. Neural Networks,* vol. 4, no. 3, pp. 484–495, May 1993.

[13] J. Choi, B. J. Sheu, and C.-F. Chang, "A Gaussian synapse circuit for analog VLSI neural networks," *IEEE Trans. Very Large Scale Integration (VLSI) Systems,* to be published, 1994.

[14] M. H. Cohen and A. G. Andreou, "Current-mode subthreshold MOS implementation of the Herault-Jutten autoadaptive network," *IEEE J. Solid-State Circuits,* vol. 27, no. 5, pp. 714–727, May 1992.

[15] D. A. Durfee and F. S. Shoucair, "Comparison of floating gate neural network memory cells in standard VLSI CMOS technology," *IEEE Trans. Neural Networks,* vol. 3, no. 3, pp. 347–353, May 1992.

[16] J. Donald and L. Akers, "An adaptive neural processing node," *IEEE Trans. Neural Networks,* vol. 4, no. 3, pp. 413–426, May 1993.

[17] J. A. Dickson, R. D. Mcleod, and H. C. Card, "Stochastic arithmetic implementations of neural networks with in situ learning," *Proc. IEEE/INNS Inter. Joint Conf. Neural Networks,* vol. II, pp. 711–716, San Francisco, CA, March 1993.

[18] T. Delbrück, "Silicon retina with correlation-based, velocity-tuned pixels," *IEEE Trans. Neural Networks,* vol. 4, no. 3, pp. 529–541, May 1993.

[19] G. Erten and R. M. Goodman, "A digital neural network architecture using random pulse trains," *Proc. IEEE/INNS Inter. Joint Conf. Neural Networks,* vol. I, pp. 190–105, Baltimore, MD, June 1992.

[20] M. Griffin, G. Tahara, K. Knorpp, R. Pinkham, B. Riley, D. Hammerstrom, and E. Means, "An 11 million transistor digital neural network execution engine," *Tech. Digest IEEE Inter. Solid-State Circuits Conf.,* pp. 180–181, San Francisco, CA, February 1991.

[21] S. M. Gowda, B. J. Sheu, and C.-H. Chang, "Advanced VLSI circuit simulation using the BSIM_plus model," *Proc. IEEE Custom Integrated Circuits Conf.,* pp. 14.3.1–14.3.5, San Diego, CA, May 1993.

[22] A. Hamilton, A. F. Murray, D. J. Baxter, S. Churcher, H. M. Reekie, and L. Tarassenko, "Integrated pulse stream neural networks: results, issues, and pointer," *IEEE Trans. Neural Networks,* vol. 3, no. 3, pp. 385–393, May 1992.

[23] D. Hammerstrom, "A VLSI architecture for high-performance, low-cost, on-chip learning," *Proc. IEEE/INNS Inter. Joint Conf. Neural Networks,* vol. 2, pp. 537–543, 1990.

[24] J. Hutchinson, C. Koch, J. Luo, and C. Mead, "Computing motion using analog and binary resistive networks," *IEEE Computer,* vol. 21, pp. 52–64, March 1988.

[25] J. E. Hansen, J. K. Skelton, and D. J. Allstot, "A time-multiplexed switched-capacitor circuit for neural network applications, *Proc. IEEE Inter. Symp. Circuits and Systems,* vol. 3, pp. 2177–2180, 1989.

[26] M. Holler, S. Tam, H. Castro, and R. Benson, "An electrically trainable artificial neural network (ETANN) with 10240 floating gate synapses," *Proc. IEEE/INNS Inter. Joint Conf. Neural Networks,* vol. 2, pp. 191–196, 1989.

[27] P. W. Hollis and J. J. Paulos, "Artificial neural networks using MOS analog multiplier," *IEEE J. Solid-State Circuits,* vol. 25, no. 3, pp. 849–855, June 1990.

[28] R. Hecht-Nielsen, "Neural-computing: picking the human brain," *IEEE Spectrum,* vol. 25, no. 3, pp. 36–41, March 1988.

[29] B. N. Krieger, T. W. Berger, and R. J. Sclabassi, "Instantaneous characterization of time-varying nonlinear systems," *IEEE Trans. Biomedical Engineering,* vol. 39, pp. 420–424, 1992.

[30] F. J. Kub, K. K. Moon, I. A. Mack, and F. M. Long, "Programmable analog vector-matrix multipliers," *IEEE J. Solid-State Circuits,* vol. 25, no. 1, pp. 207–214, February 1990.

[31] N. I. Khachab and M. Ismail, "A nonlinear CMOS analog cell for VLSI signal and information processing," *IEEE J. Solid-State Circuits,* vol. 26, no. 11, pp. 1689–1699, November 1991.

[32] B. W. Lee and B. J. Sheu, "General-purpose neural chips with electrically programmable synapses and gain-adjustable neurons," *IEEE J. Solid-State Circuits,* vol. 27, no. 9, pp. 1299–1302, September 1992.

[33] B. Linares-Barranco, E. Sánchez-Sinencio, A. Rodríguez-Vázquez, and J. L. Huertas, "A modular T-mode design approach for analog neural network hardware implementation," *IEEE J. Solid-State Circuits,* vol. 27, no. 5, pp. 701–713, May 1992.

[34] E. K. F. Lee and P. G. Gulak, "A CMOS field-programmable analog array," *IEEE J. Solid-State Circuits,* vol. 26, no. 3, pp. 1860–1867, December 1991.

[35] J. Luo, C. Koch, and B. Mathur, "Figure-ground segregation using an analog VLSI chip," *IEEE Micro,* vol. 12, pp. 46–57, December 1992.

[36] J. Lazzaro, J. Wawrzynek, M. Mahowald, M. Sivilotti, and D. Gillespie, "Silicon auditory processors as computer peripherals," *IEEE Trans. Neural Networks,* vol. 4, no. 3, pp. 523–528, May 1993.

[37] R. Lippmann, "An introduction to computing with neural nets," *IEEE Acoustic, Speech, and Signal Processing Magazine,* pp. 4–22, April 1987.

[38] R. F. Lyon and C. A. Mead, "An analog electronic cochlea," *IEEE Trans. Acoustics, Speech, and Signal Processing,* vol. 36, no. 7, pp. 1119–1134, July 1988.

[39] A. F. Murray, "Pulse arithmetic in VLSI neural networks," *IEEE Micro Magazine,* pp. 64–74, December 1989.

[40] B. J. Maundy and E. I. El-masry, "Feedforward associative memory switched-capacitor artificial neural networks," *Jour. Analog Integrated Circuits and Signal Processing,* vol. 1, pp. 321–338, 1991.

[41] C. A. Mead, X. Arreguit, and J. P. Lazzaro, "Analog VLSI models of binaural hearing," *IEEE Trans. Neural Networks,* vol. 2, pp. 230–236, March 1991.

[42] K. Madani, P. Garda, E. Belhaire, and F. Devos, "Two analog counters for neural network implementation," *IEEE J. Solid-State Circuits,* vol. 26, no. 7, pp. 966–974, July 1991.

[43] M. S. Melton, T. Phan, D. S. Reeve, and D. E. Van den Bout, "The TInMANN VLSI chip," *IEEE Trans. Neural Networks,* vol. 3, no. 3, pp. 375–384, May 1992.

[44] N. Mauduit, M. Duranton, J. Gobert, and J.-A. Sirat, "Lneuro 1.0: A piece of hardware LEGO for building neural network systems," *IEEE Trans. Neural Networks,* vol. 3, no. 3, pp. 414–422, May 1992.

[45] R. Mason, W. Robertson, and D. Pincock, "An hierarchical VLSI neural network architecture," *IEEE J. Solid-State Circuits,* vol. 27, no. 1, pp. 106–108, January 1992.

[46] T. Morishita, Y. Tamura, and T. Otsuki, "A BiCMOS analog neural network with dynamically updated weights," *Tech. Digest IEEE Inter. Solid-State Circuits Conf.,* pp. 142–143, San Francisco, CA, February 1990.

[47] T. Poggio, V. Torre, and C. Koch, "Computational vision and regularization theory," *Nature,* vol. 317, no. 6035, pp. 314–319, September 1985.

[48] M. E. Robinson, H. Yoneda, and E. Sáchez-Sinencio, "A modular CMOS design of a Hamming network," *IEEE Trans. Neural Networks,* vol. 3, no. 3, pp. 444–456, May 1992.

[49] B. J. Sheu, J. Choi, and C.-F. Chang, "An analog neural network processor for seld-organizing mapping," *Tech. Digest IEEE Inter. Solid-State Circuits Conf.,* pp. 136–137, San Francisco, CA, February 1992.

[50] C.-K. Sin, A. Kramer, V. Hu, R. R. Chu, and P. K. Ko, "EEPROM as an analog storage device, with particular applications in neural networks," *IEEE Trans. Electro Devices,* vol. 39, no. 6, pp. 1410–1419, June 1992.

[51] D. L. Standley, "An object position and orientation IC with embedded images," *IEEE J. Solid-State Circuits,* vol. 26. no. 12, pp. 1853–1859, December 1991.

[52] F. M. A. Salam and Y. Wang, "A real-time experiment using a 50-neuron CMOS analog silicon chip with on-chip learning," *IEEE Trans. Neural Networks,* vol. 2, no. 4, pp. 461–464, July 1991.

[53] J. V. der Spiegel, P. Mueller, D. Blackman, P. Chance, C. Donham, R. Etienne-Cummings, and P. Kinget, "An analog neural computer with modular architecture for real-time dynamic computations," *IEEE J. Solid-State Circuits,* vol. 27, no. 1, pp. 82–91, January 1992.

[54] J.-H. Shieh, M. Patil, and B. J. Sheu, "Measurement and analysis of charge injection in MOSanalog switches," *IEEE J. Solid-State Circuits,* vol. 22, no. 2, pp. 277–281, April 1987.

[55] S. Satyanarayana, Y. Tsividis, and H. P. Graf, "A reconfigurable VLSI neural network," *IEEE J. Solid-State Circuits,* vol. 27, no. 1, pp. 67–81, January 1992.

[56] T. Shima, T. Kimura, Y. Kamatani, T. Itakura, Y. Fujita, and T. Iida, "Neuro chips with on-chip back-propagation and/or Hebbian learning," *IEEE J. Solid-State Circuits,* vol. 27, no. 12, pp. 1868–1876, December 1992.

[57] T. Shibata and T. Ohmi, "A functional MOS transistor featuring gate-level weighted sum and threshold operations," *IEEE Trans. Electron Devices,* vol. 39, no. 6, pp. 1444-1455, June 1992.

[58] T. Shibata and T. Ohmi, "A self-learning neural-network LSI using neuron MOSFETs," *Tech. Digest IEEE 1992 Symp. VLSI Technology,* pp. 84–85, Seattle, WA, June 1992.

[59] C. Tomovich, "MOSIS - A gate way to silicon," *IEEE Circuits and Devices Magazine,* vol. 4, no. 2, pp. 22–23, March 1988.

[60] J. E. Tomberg and K. K. K. Kaski, "Pulse-density modulation technique in VLSI implementations of neural network algorithms," *IEEE J. Solid-State Circuits,* vol. 26, no. 5, pp. 1277–1286, October 1990.

[61] S. W. Tsay and R. W. Newcomb, "VLSI implementation of ART1 memories," *IEEE Trans. Neural Networks,* vol. 2, no. 2, pp. 214–221, March 1991.

[62] K. Uchimura, O. Saito, and Y. Amemiya, "A high-speed digital neural network chip with low-power chain-reduction architecture," *IEEE J. Solid-State Circuits,* vol. 27, no. 12, pp. 1862–1867, December 1992.

[63] E. A. Vittoz and J. Fellrath, "CMOS analog integrated circuits based on weak inversion operation," *IEEE J. Solid-State Circuits,* vol. 12, no. 3, pp. 224–231, June 1977.

[64] J. E. Varrientos, E. Sáchez-Sinencio, and J. Ramírez-Angulo, "A current-mode cellular neural network implementation," *IEEE Trans. Circuits and Systems, II,* vol. 40, no. 3, pp. 147–155, March 1993.

[65] T. Watanabe, K. Kimura, M. Aoki, T. Sakata, and K. Itho, "A single 1.5-V digital chip for a $10^6$-synapse neural network," *IEEE Trans. Neural Networks,* vol. 4, no. 3, pp. 387–393, May 1993.

[66] M. Yasunaga, N. Masuda, M. Yagyu, M. Asai, K. Shibata, M. Ooyama, M. Yamada, T. Sakaguchi, and M. Hashimoto, "A self-learning digital neural network using wafer-scale LSI," *IEEE J. Solid-State Circuits,* vol. 28, no. 2, pp. 106–114, February 1993.

[67] T. Yamashita, T. Shibata, and T. Ohmi, "Neuron MOS winner-take-all circuit and its application to associative memory," *Tech. Digest IEEE Inter. Solid-State Circuits Conf.,* pp. 236–237, San Francisco, CA, February 1993.

[68] C. Lau, (Ed.), "Neural Networks: Theoretical Foundations and Analysis," IEEE Press, 1992.

[69] C. A. Mead, "Analog VLSI and Neural Systems," Addison-Weslley Publishing Company, 1989.

[70] A.G. Andreeou and W. Liu, "BiCMOS circuits for silicon chochleas," *Proc. 11th European Conference on Circuit Theory and Design, Part I,* pp. 503-508, Elsevier, September 1993.

[71] J.C. Park, C. Abel and M. Ismail, "Design of a silicon cochlea using MOS switched-current techniques," *Proc. 11th European Conference on Circuit Theory and Design, Part I,* pp. 269-274, Elsevier, September 1993.

[72] T. McKenna, J. L. Davis, and S. F. Zornetzer, (Editors), "Single Neuron Computation," Academic Press: New York, NY, 1992.

[73] F. Gregorian and G. C. Temes, "Analog MOS Integrated Circuits for Signal Processing," A Wiley-Interscience Publication, 1986.

[74] R. Hecht-Nielsen, "Neurocomputing," Addison-Weslley Publishing Company, 1990.

[75] B. W. Lee and B. J. Sheu, "Hardware Annealing in Analog VLSI Neurocomputing," Kluwer Academic Publishers, 1991.

[76] C. Mead and M. Ismail (Eds.), "Analog VLSI Implementation of Neural Systems," Kluwer Academic Publishers, 1989.

[77] D. E. Rumelhart, J. L. McClelland, and the PDP Research Group, "Parallel Distributed Processing," vol. 1, The MIT Press, Cambridge, MA, 1989.

[78] U. Ramacher and U. Ruckert (Eds.), "VLSI Design of Neural Networks," Kluwer Academic Publishers, 1991.

[79] E. Sáchez-Sinencio and C. Lau (Eds.), "Artificial Neural Networks: Paradigms, Applications, and Hardware Implementations," IEEE Press, 1992.

[80] S. F. Zornetzer, J. L. Davis, and C. Lau (Eds.), "An Introduction to Neural and Electronic Networks," Academic Press: New York, NY, 1990.

[81] K. Boahen and A. Andreou, "A contrast sensitive silicon retina with reciprocal synapses," *Advances in Neural Information Processing Systems,* vol. 4, J. E. Moody, S. J. Hanson, and R. P. Lippmann (Eds.), pp. 764–774, Morgan Kaufmann, San Mateo, CA, 1992.

[82] W. Bair and C. Koch, "An analog VLSI chip for finding edges from zero-crossing," *Neural Information Processing Systems,* vol. 2, R. P. Lippmann, J. E. Moody, and D. S. Touretzky (Eds.), pp. 399–405, Morgan Kaufmann, Palo Alto, CA, 1991.

[83] B. K. P. Horn, "Parallel analog networks for machine vision," *Artificial Intelligence at MIT: Expanding Frontiers,* vol. 2, P. H. Winston and S. A. Shellard (Eds.), pp. 437–471, The MIT PRESS, Cambridge, MA, 1990.

[84] C. Mead, "Adaptive retina," *Analog VLSI Implementations of Neural Systems,* C. Mead and M. Ismail (Eds.), pp. 239–246, Kluwer Academic Publishers, Norwell, MA, 1989.

[85] P. K. Simpson, "Foundations of Neural Networks," *Artificial Neural Networks: Paradigms, Applications, and Hardware Implementations,* E. Sánchez-Sinencio and C. Lau (Eds.), IEEE Press, Piscataway, NJ, 1992.

[86] J. Tanner and C. A. Mead, "An integrated analog optical motion sensor," *VLSI Signal Processing II* S.-Y. Kung, R. E. Owen, and J. G. Nash (Eds.), pp. 59–76, IEEE Press, New York, NY, 1987.

[87] E. A. Vittoz, "Micropower Techniques," *Design of MOS VLSI Circuits for Telecommunications,* Y. Tsividis and P. Antognetti (Eds.), pp. 104–144, Prentice-Hall, Englewood Cliffs, NJ, 1985.

[88] *IEE Proceedings, Part G,* Special issue on current-mode analogue signal processing circuits, April 1990.

[89] "User Manual of 80170 NW ETANN Chip," Intel Corporation, 1990.

[90] "Grand challenges: High performance computing and communications, The FY 1992 U.S. Research and Development Program," National Science Foundation, Washington D.C., 1991.

# CHAPTER
# 8

# NEURAL INFORMATION PROCESSING II

## 8.1  INTRODUCTION

A "neural network" is a computational paradigm founded on the functional and organizational principles of biological information processing systems. For a historical perspective and a comprehensive literature survey on neural networks, the reader should refer to the review article by Cowan and Sharp [1]. The recent excitement about neural networks stems from the belief that hardware embodiments [2] of this computational paradigm would provide effective solutions to problems in the fields of robotics, human/computer interfaces, and adaptive information processing [3].

Biological systems excel at sensory perception, motor control and sensorimotor coordination by sustaining high computational throughput with minimal energy consumption and heat production. Understanding information processing in biological systems will provide insight into the most effective signal representation and architectures for these tasks. We believe that these insights will be equally applicable to human-engineered systems.

In this chapter, the emphasis is on analog VLSI systems [4] that are designed for *low-energy, high-functionality computation* with applications in *real-time, adaptive, sensory-level processing*. High functionality is achieved by exploiting the physics of the computational substrate—rather than abstract mathematical algorithms tailored for digital hardware.

358

The difference between the conventional computational paradigm and one which abstracts principles from biology, may be illustrated with an example from machine/computer vision [5]. Most computer vision algorithms use the light intensity recorded by a CCD camera as the input signal. This intensity field is subsequently processed outside the camera to discard any absolute luminance information and form a representation where only relative illumination, i.e. contrast, is retained. In contraposition, motivated by the known organization of early vision processing in biological systems [6], one could make the *architectural*, as opposed to *algorithmic*, decision to integrate some low–precision analog processing with the transducer elements to extract contrast information at the focal plane [7].

The neuromorphic solution is attractive from a computational perspective because *contrast*, an invariant representation of the visual world, has been obtained with a front-end that is robust, small, and extremely low power (a few mW). There is also an engineering benefit because subsequent processing stages are not burdened with handling and processing signals of wide dynamic range.

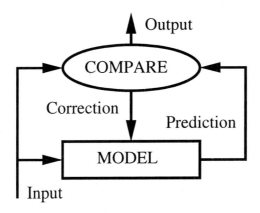

**FIGURE 8.1**

An adaptive system. External inputs to the system are compared with its internal state (model), to produce an output. Information from the input and from the comparison process is used to adjust the parameters of the internal model. (Adapted from [2]).

Adaptation is a fundamental property of neural systems. Each level of processing has an internal model that is refined by experience as shown by the functional–block-level description in Fig. 8.1. The neuromorphic focal-plane processor (or silicon retina) we described in the preceding paragraph is adaptive, but only in a limited sense. It has an *a-priori* model of the world; one that assumes that the intensity is either uniform or, in the case of non-uniform illumination, is a linear function of space. The output of the system is the difference between the input intensity field and the model. As such, the output is a measure of the second spatial derivatives (or the Laplacian) of the intensity field. However, this information is not used to automatically adjust the parameters of the model.

In the field of computer vision, linear methods based on regularization theory are used to impose a second-order smoothness constraint [8, 9]. In the adaptive system framework, their model also assumes second and higher order derivatives are zero. However, embodiments of the algorithm are in the form of software simulations run on general purpose digital hardware.

In biological retinae, where computation has to be performed using the inherent physics of the underlying substrate, with limited energy resources, and in a limited physical space, the embodiment of such information processing strategies take a different form. To illustrate this, consider the diffusion of ions through intra-cellular fluid, a natural physical process. If there is no net flux into a particular region in space, the distribution of ions there will be linearly graded. The second and higher order spatial derivatives at a particular location are proportional to the net flux of particles introduced there. Biological retinae exploit this property to compute deviations from the spatial intensity model. Contrast sensitivity is achieved by having the local diffusivities of the medium be functions of the injected current. This *local gain control* mechanism makes the system non-linear. Silicon retinae can also exploit this property since diffusion is the dominant charge transport mechanism in the subthreshold MOS transistor [7].

Overall, there are three hierarchical levels of system integration at which analog circuits will play an important role. At the lowest level, there are adaptive sensors for transducing environmental signals and extracting invariant representations of the external world; silicon retinas [4, 7] and silicon cochleas [10, 11, 12] are examples of such systems. At the intermediate level, there is adaptive processing through self-organization [13, 14] and unsupervised learning. Pattern processing at this level involves coordinate transformations to facilitate information processing by subsequent stages [15]. Decorrelators, independent component analyzers [16, 17] and principal component analyzers belong to this class of processing. At the highest level, nonlinear mappings are computed for classification and decision making. Systems for computing Hamming [18, 20, 19, 21], Euclidean [22] or other similarity measures are necessary.

Space limitations do not permit extensive coverage of this rapidly developing field; however, there are some excellent references. A discussion of system prototypes and different circuit design styles can be found in [4, 23, 24, 25], and in Chapter 17 of [26]. Furthermore, journals such as IEEE Transactions on Circuits and Systems, IEEE Transactions on Neural Networks, IEEE Journal of Solid–State Circuits, Analog Integrated Circuits and Signal Processing, Neural Computation and Neural Networks carry regular articles and have special issues on hardware implementations of neural networks.

In this chapter, the focus is on *circuits* and *design methodologies*. A *minimalist* design style is adhered to; the MOS device itself is our basic module [27]. This style efficiently exploits the physical properties of devices, electronic or otherwise, to effectively compute desirable functions. The design constraints, namely, high integration density (VLSI) and micropower operation, are satisfied by using current-mode (CM) techniques based on the translinear properties of subthreshold MOS transistors in the saturation [18, 28, 29] and ohmic regions [7].

The requirements of high-volume, low-cost, application-specific systems are satisfied by using devices that are readily available in standard CMOS processes.

The objective here is to introduce the reader to a consistent design framework for analog VLSI circuits, beginning with the basic devices and following through to the system level. In many ways, the material here is complementary to that in the book by Carver Mead [4] and it is assumed that the reader is familiar with that monograph.

The chapter is organized into five sections. The next section reviews CMOS technology. A charge-based model for the MOS transistor is introduced and the large signal device equations are derived. This derivation emphasizes the symmetry between source and drain. The analysis and modeling of the floating gate MOSFET is presented next together with a brief discussion of large signal models for bipolar transistors which are available in standard CMOS processes. The third section covers the current-mode design methodology. Basic circuit techniques, the current-conveyor concept, and the translinear principle are introduced. The fourth section shows how the design methodology is applied to the design of small networks. A bidirectional communication technique for programmable fully-connected current-mode networks is described. Networks that perform global normalization, local normalization, and Winner-Takes-All conversion from amplitude-to-place coding are described. In addition, transistor-based implementations of diffusive media are presented. The fifth section focuses on the implementation of a 46,000 pixel contrast-sensitive silicon retina; a current-mode analog VLSI system that employs over 500,000 transistors operating in the subthreshold region.

## 8.2  CMOS TECHNOLOGY AND MODELS

CMOS technology and, in particular, subthreshold MOS operation has long been recognized as the technology of choice for implementing digital VLSI and analog LSI circuits that are constrained by power dissipation requirements [30, 31]. The advantages of using standard digital CMOS processes for cost-effective engineering solutions to analog signal processing problems are surveyed in [32] and are also discussed in [33]. CMOS has the highest integration density attainable today, making it especially attractive for *analog VLSI* models of neural computation [4]. Moreover, the physical properties of silicon and its native oxides, together with recent advances in micromachining of electromechanical elements [34], make silicon-based technologies the prime candidate for highly integrated, truly complex, analog computational systems. A final advantage is that CMOS silicon technologies are readily available for experimentation and rapid prototyping through foundry services at relatively low cost. This advantage of CMOS technologies accelerates the research and evolution of complex systems.

## 8.2.1   Charge-Based MOS Model Formulation

We begin by deriving the basic equations for an NMOS transistor; the derivation for an PMOS device can be done in an analogous manner. We seek to obtain an equation for the drain current that is charge-based, valid in all regions of operation, and is useful for scaled-down devices with thin oxides, short channels, and velocity-saturated charge transport. Our approach is similar to Maher and Mead's (see Appendix B of [4]) except that we do not ignore the role of the surface potential in determining the effective channel depth. This effect is important at the onset of weak inversion and must be taken into consideration to describe the condition at threshold precisely.

  The general outline of the derivation is as follows. First we consider the electrostatics and obtain expressions for the semiconductor charge in terms of the potential at the surface. We proceed with an examination of the solutions for the classical drift-diffusion equation which relates the current to the charge. This approach yields a simple, physically-intuitive, closed-form description valid over the entire operating range of the transistor. The equations that relate charge and current are derived in a set of natural units as in Appendix B of [4]. Finally, we make some approximations that are valid over a limited range of operation in order to obtain simple expressions for the current in terms of the terminal voltages. All quantities with the little square ($^\circ$) as superscript (charges and capacitances) denote amount per unit area.

**Physical System**   The physics of the MOS transistor can be best understood if we consider it as a two-component system. The first component is the gate-oxide-semiconductor structure that creates a depletion region under the gate and controls the potential at the surface. The second component consists of two p-n junctions that exist between the heavily-doped source and drain areas and the depleted region under the gate.

  The potential energy of carriers traveling along the channel is shown in Fig. 8.2. All the potentials are referenced to the potential deep in the bulk where the bands are flat—this is the same as the bulk terminal voltage if the contact potential is zero. The potential difference $\psi_s$ between the intrinsic Fermi level in the bulk ($E_i$)—the middle of the band gap—and its level at the surface of the channel is referred to as the *surface potential*, for short. The potential difference between the intrinsic Fermi level in the bulk and the Fermi level ($E_F$) is denoted by $\phi_F$. Since the drain and the source of the device are heavily doped (often degeneratively doped), their Fermi level roughly coincides with the edge of the conduction band. At the same time, their existence in the neighborhood of the channel affects the local concentrations of the mobile charge density. To include this effect in our formulation, the potential $\phi_c(x)$ is defined as the quasi-Fermi level for the mobile charge carriers (minority carriers) at any particular point in the channel.

  Externally applied voltages can be related to internal potentials using the *quasi-Fermi levels* or *imref*s. The reference level for both potentials and voltages

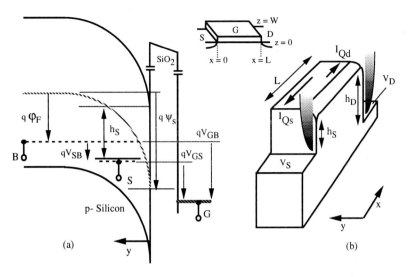

(a)

(b)

## FIGURE 8.2

Energy band diagrams for an NMOS device. The gate voltage $V_{GS}$ is set so that the surface potential $\psi_s$ at the source end of the channel equals $2\phi_F + V_{SB}$ and the drain and source voltages are set such that $V_{DB} > V_{GB} > V_{SB} > 0$. (a) Along a 1-D path from the gate deep down into the bulk at the source-end of the channel. (b) Over a 2-D surface extending across the length of the channel and down into the bulk. The mobile charge at the source and drain boundaries is coded with shades of gray—darker shades correspond to higher charge densities. $h_S$ and $h_D$ are the effective barrier energy at the source and drain, respectively, and $L$ is the channel length.

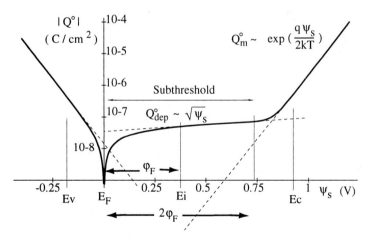

## FIGURE 8.3

Variation of $Q^\circ$, the total charge density under the gate of an MOS capacitor, as a function of the surface potential $\psi_s$. The position of the conduction band $E_c$, valence band $E_v$, Fermi level $E_F$, and the intrinsic Fermi level $E_i$ are also shown (in eV). This plot is computed by solving the one-dimensional Poisson equation at a temperature of $300^\circ K$ for an acceptor dopant concentration of $2.1 \times 10^{16}/cm^3$—a typical value for a 40nm gate-oxide process.

is the Fermi level in the substrate (for an NMOS transistor this can be thought as the quasi–Fermi level for holes.) Hence, near the ends of the channel, $\phi_c$ is equal to the source and drain voltages, i.e., $\phi_c(0) = V_{SB}$ and $\psi_c(L) = V_{DB}$, and deep in the bulk it is zero.

Therefore, the source and drain voltages determine the energy barriers, $h_S$ and $h_D$, at the channel boundaries, which, in turn, determine the charge concentrations there.

**Electrostatics**   We begin by obtaining an expression for the total semiconductor charge in terms of $\psi_s$, the potential at the surface of the channel. First, we consider the two terminal MOS capacitor structure, integrate the depletion layer charge from deep down in the bulk to the surface to obtain the electric field—invoking Gauss' Law—and then integrate the electric field to get the surface potential. The result is plotted in Fig. 8.3. In addition to the depletion layer charge term which is proportional to the square-root of $\psi_s$, there is a decaying exponential term due to accumulation of holes as the surface potential becomes negative and a growing exponential term due accumulation of electrons as the surface potential becomes positive. These electrons constitute the mobile charges and invert the surface, making it n-type.

There are two regions here that are of primary interest. When $\psi_s = \phi_F$, the exponential hole/electron accumulation terms are equal; this is the onset of *weak inversion*. When $\psi_s = 2\phi_F$, the conduction band is as close to the Fermi-level at the surface as the valence band is to the Fermi-level in the bulk. Therefore, the electron concentration at the surface equals the hole concentration deep in the p-type bulk; this is the onset of *strong inversion* and we will call it the threshold. $\phi_F$ depends on the acceptor dopant concentration, $N_A$, and can be calculated from:

$$\phi_F = (kT/q) \ln \left( \frac{N_A}{n_i} \right) \tag{8.1}$$

where $n_i$ is the intrinsic carrier concentration, $T$ is the temperature in Kelvin, $k$ is Boltzmann's constant, and $q$ is the electronic charge. Higher doping concentrations shift the Fermi-level closer to the valence band level and increases $\phi_F$.

When the surface potential exceeds a few times the thermal voltage, the hole accumulation term is negligible and the total charge per unit area is given by [36].

$$Q^\diamond(\psi_s, \phi_c) = Q^\diamond_{dep}(\psi_s) \sqrt{1 + \frac{kT/q}{\psi_s} \exp \left( \frac{\psi_s - (2\phi_F + \phi_c)}{kT/q} \right)} \tag{8.2}$$

Note that it is the difference between the surface potential $\psi_s$ and the quasi-Fermi potential, $\phi_c(x)$, that determines the mobile charge (exponential term)—not the Fermi level in the bulk. Taking this into account, the condition at threshold is $\psi_s = 2\phi_F + \phi_c(x)$. The calculation of the charge in Fig. 8.3 was performed for a two-terminal MOS capacitor where $\phi_c = 0$.

The depletion charge, $Q_{dep}^{\diamond}$, is related to the surface potential, $\psi_s$, and the acceptor concentration, $N_A$, in the substrate through the following expression:

$$Q_{dep}^{\diamond} = -\sqrt{\psi_s} \sqrt{2q\,\epsilon_{Si}\,N_A} = -\sqrt{\psi_s}\,\gamma\,C_{ox}^{\diamond} \qquad (8.3)$$

where $C_{ox}^{\diamond}$ is the oxide capacitance per unit area, $\epsilon_{Si}$ is the permittivity of silicon, $q$ is the electronic charge and $\gamma$ is defined as:

$$\gamma \equiv \frac{\sqrt{2q\,\epsilon_{Si}\,N_A}}{C_{ox}^{\diamond}} \qquad (8.4)$$

The mobile charge per unit area may be obtained simply by subtracting the depletion charge from the total charge. The above expression for the total charge is valid for weak, moderate, and strong inversion; these conditions characterize the so-called subthreshold, transition, and above-threshold regions.

The dependence of the surface potential on the gate voltage is fundamentally different in these three regions. In the subthreshold region, i.e., $5kT/q < \psi_s < 2\phi_F - 5kT/q$, the depletion charge is the dominant term. Field lines from charges added to the gate find acceptor atoms deep in the bulk, increasing the depletion layer charge and the surface potential. Therefore, the surface potential $\psi_s$ is constrained to follow changes in the gate potential. At threshold, the mobile charge equals $C_{dep}^{\diamond}kT/q$—changing the surface potential by exactly one thermal voltage if we ignore the gate-oxide capacitance. This result is obtained simply by setting $\psi_s = 2\phi_F + \phi_c$ in Eq. (8.2), subtracting $Q_{dep}^{\diamond}$, and using the following expression for the depletion layer capacitance

$$C_{dep}^{\diamond} \equiv \frac{\partial(-Q_{dep}^{\diamond})}{\partial \psi_s} = \frac{\gamma}{2\sqrt{\psi_s}}\,C_{ox}^{\diamond} \qquad (8.5)$$

obtained by differentiating Eq. (8.3).

Above threshold, i.e., $\psi_s > 2\phi_F + 5kT/q$, the mobile charge can no longer be ignored because it is comparable to the depletion charge. However, field lines terminate on the thin sheet of conduction electrons at the surface and thus, the surface potential is more or less constant. In the transition region, i.e., $2\phi_F - 5kT/q < \psi_s < 2\phi_F + 5kT/q$, neither term is dominant and both need to be taken into account to determine the surface potential.

An expression relating the gate voltage, $V_{GB}$, to the total semiconductor charge $Q^{\diamond}$ can be derived as follows. Summing all the voltage drops encountered as we go from the bulk to the gate, we find that

$$V_{GB} = \psi_s + \psi_{ox} + \phi_{MS}$$

where $\psi_{ox}$ is the voltage across the gate-oxide capacitor, $C_{ox}^{\diamond}$, an $\phi_{MS}$ is the contact potential between the gate material and the bulk material. $\psi_{ox} = -(Q^{\diamond} + Q_0^{\diamond})/C_{ox}^{\diamond}$ where $Q_0^{\diamond}$ is the charge due to electrons trapped at the oxide interface and ions implanted at the channel surface.

The *flat-band voltage*, $V_{FB} \equiv \phi_{MS} - Q_0^\diamond/C_{or}^\diamond$, is introduced to account for the constant voltage offset due to the fixed charges and the contact potential. Hence

$$V_{GB} = V_{FB} + \psi_s - Q^\diamond/C_{or}^\diamond \qquad (8.6)$$

When $V_{GB} = V_{FB}$, the semiconductor charge and the surface potential are both zero, i.e., the surface potential is the same as the potential at the bulk terminal. Under these conditions, the bands remain flat from the surface deep into the bulk. By substituting the expression for $Q^\diamond$ given in Eq. (8.2) into Eq. (8.6), a relation between the gate voltage and the surface potential can be obtained.

**8.2.1.1  CHARGE TRANSPORT.** For the derivation of the transport equations, it is convenient to introduce a self-consistent set of natural units that will be used to normalize voltage, charge, current, and length. This will yield dimensionless equations—allowing terms representing different physical measures to be readily compared [4].

The natural units are $V_T = kT/q$, $Q_T = (C_{dep}^\diamond + C_{or}^\diamond)V_T$, $I_0 = Q_T v_0$, and $l_0 = \mu V_T/v_0$, for voltage, charge, current, and distance, respectively. $v_0$ is the carrier saturation velocity and $\mu$ is the mobility in the channel. $l_0$ is the distance over which a thermal voltage will yield a large enough electric field to accelerate carriers to the saturation velocity. Quantities that are expressed in these units will be denoted by lower-case letters. In these natural units, the classical drift-diffusion charge-transport equation is restated [4]:

$$i = \frac{q_m'}{1 + q_m'}q_m + q_m' \qquad (8.7)$$

for a device of unit width. $q_m$ is the mobile charge and $q_m'$ is its spatial derivative with respect to distance along the channel. The term proportional to $q_m$ is the drift term. The factor $q_m'/(q_m' + 1)$ models the velocity saturation. It asymptotes approaches unity at high electric fields since $q_m'$ is proportional to the derivative of the surface potential—assuming a constant channel capacitance—which is, in turn, proportional to the electric field strength.

As shown in [4], the diffusion term, $q_m'$, may be replaced with $q_m'/(q_m' + 1)$ since velocity-saturated transport is not achieved by the diffusion process, i.e., $q_m' \ll 1$. For larger values of $q_m'$, the drift term dominates and hence the error incurred is negligible. This makes it possible to integrate the differential equation with the boundary conditions $q_m = q_s$ and $q_m = q_d$ at $x = 0$ and $x = l$, respectively:

$$i = \frac{q_s - q_d}{q_s - q_d + l}\left(\frac{q_s + q_d}{2} + 1\right) \qquad (8.8)$$

For long-channel devices (no velocity saturation effects), the denominator $(q_s - q_d + l) \approx l$ and Eq. (8.8) becomes:

$$i \approx \frac{(q_s + q_d)}{2}\frac{(q_s - q_d)}{l} + \frac{(q_s - q_d)}{l}$$

The first term corresponds to drift and the second to diffusion. Both terms are proportional to $(q_s - q_d)/l$—this represents the concentration gradient as well as the electric field (assuming constant capacitance). However, the drift component is also proportional to the average number of carriers, i.e., $(q_s + q_d)/2$. Hence, drift will dominate when the mobile charge is large compared with $Q_T$ and diffusion will dominate when the mobile charge is small compared with $Q_T$. The point where $q_m = 1$, i.e., $Q_m^\circ = Q_T$, that divides these two regimes is used by Maher and Mead [4] as the threshold condition. We have adopted a definition, i.e., $\psi_s = 2\phi_F + V_{SB}$, which is qualitatively different but quantitatively very similar. For this value of $\psi_s$, the mobile charge in natural units is $q_m = C_{dep}^\circ/(C_{ox}^\circ + C_{dep}^\circ)$ which is fairly close to one.

Collecting like terms, we find that the current has two independent opposing components:

$$i \equiv i_{q_s} - i_{q_d} = \frac{\frac{1}{2}q_s^2 + q_s}{l} - \frac{\frac{1}{2}q_d^2 + q_d}{l}$$

The forward component is a function of the charge at the source and the reverse component is a function of the drain charge. When the charge at the drain, $q_d$, is much less than the charge at the source, $q_s$, the latter is negligible and the current saturates—no longer increasing as the drain charge decreases. Note that this decomposition is only possible in the absence of velocity saturation and channel-length modulation effects.

A conversion of Eq. (8.8) into regular engineering units and introduction of $W$, the width of the transistor, yields:

$$I = \mu \, \frac{W}{L + \frac{1}{v_0} \frac{Q_s^\circ - Q_d^\circ}{C_{ox}^\circ + C_{dep}^\circ}} (Q_s^\circ - Q_d^\circ) \left( \frac{1}{2} \frac{Q_s^\circ + Q_d^\circ}{C_{ox}^\circ + C_{dep}^\circ} + \frac{kT}{q} \right) \tag{8.9}$$

Notice that velocity saturation has the same effect as changing the effective channel length. It may be characterized by a parameter similar to the Early voltage which is used to characterize the effect of the source and drain depletion layers on the channel length.

**Approximations**  Simple expressions can be obtained for the charge and the gate voltage if we limit the region of operation to subthreshold or above-threshold.

In subthreshold, the exponential term in Eq. (8.2) (due to mobile charge) is negligible. Dropping that term and substituting the charge $Q^\circ = Q_{dep}^\circ$ in Eq. (8.6) yields:

$$V_{GB} = V_{FB} + \psi_s + \gamma\sqrt{\psi_s} \tag{8.10}$$

and, solving for $\psi_s$,

$$\psi_s = \left[ -\frac{\gamma}{2} + \left( \frac{\gamma^2}{4} + V_{GB} - V_{FB} \right)^{\frac{1}{2}} \right]^2$$

We can obtain a simpler result by linearizing Eq. (8.10). This is a good approximation in the region corresponding to values of $\psi_s$ between $(\phi_F + \phi_c)$ and

$(2\phi_F + \phi_c)$ and therefore, two parameters, a slope, $\kappa$, (pronounced 'kappa') and an intercept, $V_{GB}^*$, may be introduced to relate the gate voltage to the surface potential:

$$\psi_s(V_{GB}) = (1.5\phi_F + \phi_c) + \kappa(V_{GB} - V_{GB}^*) \tag{8.11}$$

where $V_{GB}^*$ is the gate voltage at which the surface potential is at the mid-point of the range, i.e., $\psi_s(V_{GB}^*) = (1.5\phi_F + \phi_c)$. From Eq. (8.10)

$$\kappa \equiv \left.\frac{\partial \psi_s}{\partial V_{GB}}\right|_{=15_F+_c(x)} = \frac{1}{1 + \gamma/(2\sqrt{1.5\phi_F + \phi_c(x)})} \tag{8.12}$$

The physical significance of $\kappa$ is apparent if we express it in terms of the oxide capacitance and the depletion capacitance[1]. Rewriting Eq. (8.12) and using Eq. (8.5)

$$\kappa = \frac{C_{ox}^\diamond}{C_{ox}^\diamond + C_{dep}^\diamond}$$

it becomes clear that the oxide and depletion capacitances form a capacitive divider between the gate and bulk terminals that determines the surface potential. Lighter doping reduces $\gamma$, reduces $C_{dep}^\diamond$, and pushes the divider ratio closer to unity. A larger surface potential also reduces $C_{dep}^\diamond$.

All that is needed now is an expression for the mobile charge in subthreshold.

$$Q_m^\diamond(\psi_s, \phi_c) = Q^\diamond - Q_{dep}^\diamond$$

$$= Q_{dep}^\diamond(\psi_s)\left(\sqrt{1 + \frac{kT/q}{\psi_s}\exp\left[\frac{\psi_s - (2\phi_F + \phi_c)}{kT/q}\right]} - 1\right)$$

$$\approx -\frac{kT}{q}\frac{\gamma}{2\sqrt{\psi_s}}C_{ox}^\diamond\exp\left[\frac{\psi_s - (2\phi_F + \phi_c)}{kT/q}\right] \tag{8.13}$$

The inverse–square-root pre-exponential factor comes from integrating all the mobile charge down the depth of the channel, i.e., from the surface down into the bulk. When $\psi_s$ is large, the bands are fully bent, so the mobile charge concentration dies rapidly away from the surface and hence, the integral is small and does not change much with $\psi_s$. On the otherhand, when $\psi_s$ is close to zero, this integral is large and changes a lot. Therefore, it must be taken into account to predict very low-level currents accurately.

We can now express the subthreshold current in terms of the terminal voltages by using Eqs. (8.9), (8.11), and (8.13), which are repeated here for convenience.

---

[1]Note that $\kappa = 1/\alpha$, where $\alpha$ is defined in Chap. 2

$$Q_s^\diamond(\psi_s, V_{SB}) \approx -\frac{kT}{q}C_{dep}^\diamond \exp\left[\frac{\psi_s - (2\phi_F + V_{SB})}{kT/q}\right]$$

$$Q_d^\diamond(\psi_s, V_{DB}) \approx -\frac{kT}{q}C_{dep}^\diamond \exp\left[\frac{\psi_s - (2\phi_F + V_{DB})}{kT/q}\right]$$

$$\psi_s(\phi_c, V_{GB}) \approx 1.5\phi_F + \phi_c + \kappa_n(V_{GB} - V_{GB}^*(\phi_c))$$

$$I(Q_s^\diamond, Q_d^\diamond) \approx \mu\,\frac{kT}{q}\frac{W}{L}(Q_s^\diamond - Q_d^\diamond) \tag{8.14}$$

We have set the quasi-Fermi potential $\phi_c(x)$ to $V_{SB}$ and $V_{DB}$ at the source and drain boundaries of the channel, respectively, to obtain the mobile charge concentrations there. The same thing is done to obtain the surface potential at the ends of the channel. In subthreshold, the surface potential does not change along the channel because the mobile charge does not affect the electrostatics. Hence, we need only calculate it at the source-end.

We have dropped the $Q^2$ terms in the current equation since $Q^\diamond \ll (kT/q)(C_{ox}^\diamond + C_{dep}^\diamond)$ in subthreshold. Velocity saturation effects have also been dropped. Note that $\psi_s$ does not change along the channel because the depletion charge and the gate charge are uniformly distributed and thus, the equations at the boundaries are now written in terms of the depletion capacitance, $C_{dep}^\diamond$.

Thus, we obtain the familiar result for an NMOS device;

$$I_{SD} = -I_{DS} = -\,S\,I_{no}e^{\frac{\kappa V_{GB}}{kTq}}\left(e^{-\frac{V_{SB}}{kTq}} - e^{-\frac{V_{DB}}{kTq}}\right) \tag{8.15}$$

$$I_{n0} = \mu\left(\frac{kT}{q}\right)^2 C_{dep}^\diamond \exp\left[-\frac{\phi_F/2 + \kappa_n V_{GB}^*}{kT/q}\right]$$

the PMOS equation is similar except for a sign-flip on the terminal voltages due to a change in sign of the charge carriers. For long-channel devices, $S$ is simply the geometrical factor of the device, i.e., $W/L$. This form of the device equation is most useful for analyzing circuits that exploit the drain/source symmetry of MOS transistors.

The adopted conventions for positive voltages and currents in PMOS and NMOS transistors are shown in Fig. 8.4. For example, in an $n$-well process, the local substrate voltage for the NMOS transistors is the p-substrate voltage $V_{NBB}$ (usually $V_{SS}$), and for the PMOS transistors, the voltage $V_{PBB}$ of the n-well (usually $V_{DD}$). Our notation is consistent with the notation in [4] and also with the subthreshold MOS literature [31].

The process-dependent parameters, $I_{no}$ and $\kappa_n$, can be measured experimentally. Note that $I_{no}$ is not simply the point where the extrapolated $\log(I)$

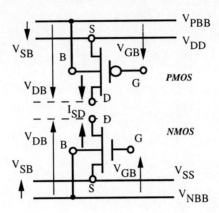

**FIGURE 8.4**
MOS transistor sign conventions for voltages and currents.

vs. $V_{GS}$ line intercepts the $V = 0$ axis. Fitting the complete expression above, to take into account the $1/\psi_s$ pre-exponential factor, is the only way to determine $I_{no}$ accurately. Care should also be exercised in obtaining the subthreshold slope coefficient, $\kappa$, because it depends on the surface potential and thereby on $V_{SB}$. For a given value of $V_{SB}$, $\kappa$ may be obtained from the slope of the $\log(I)$ vs. $V_{GS}$ curve.

Often, it is more convenient to have the terminal voltages referenced to the source instead of the bulk:

$$I_{DS} = SI_{no} \ \exp\left[\frac{(1 - \kappa_n)V_{BS} + \kappa_n V_{GS}}{kT/q}\right]\left(1 - e^{-\frac{V_{DS}}{kTq}}\right) \qquad (8.16)$$

This shows explicitly the dependence on $V_{BS}$ and the role of the bulk as a *back-gate* that underlies this.

For devices that have the source and substrate at the same potential and biased with $V_{DS} \geq 4\,(kT/q)$, (saturation) the drain current is given by:

$$I_{DS} = S\,I_{no} \ \exp\left[\frac{\kappa_n V_{GS}}{kT/q}\right] \qquad (8.17)$$

This equation, having only the dependence on $V_{GS}$, is used for circuit designs where devices operate in saturation. However, channel-length modulation—which we have ignored completely—becomes significant in saturation. So the device equations must be augmented with terms that model this effect to accurately predict the output conductance.

### 8.2.2  Symmetric MOS Model and CM Circuit Design

In the previous subsections, we derived expressions that relate the voltages at the terminals of a MOS transistor to the current. A charge-based formulation

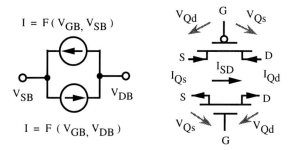

**FIGURE 8.5**
Large signal model for an NMOS transistor (left). Adopted conventions for current decomposition in PMOS and NMOS devices (right).

that preserves the symmetry between its source/drain terminals was used. Which terminal of the device actually serves as the source or the drain is determined by the circuit, the bias conditions—and even the input signals. Therefore, it is more sensible to use a symmetric model for the device which preserves the inherent symmetry of its source/drain terminals and does not assign them different roles. Furthermore, the obvious symmetry in our device model can be exploited in our circuit design.

The MOS device has a very simple current-charge relationship because diffusion and drift are both proportional to the concentration gradient. As shown in previous subsections, this yields a quadratic expression for the current that consists of two independent opposing components—in the absence of velocity saturation and channel-length modulation effects (Fig. 8.5). A key property of the MOS device that makes this possible is lossless channel conduction. Unlike a bipolar transistor, the charge on the gate is confined there by the almost infinite gate-oxide resistance. Therefore, there is no recombination between the current-carrying charge in the channel and the current-modulating charge on the gate.

The familiar ohmic/saturation dichotomy introduced in voltage-mode design can be reformulated in terms of these opposing components. In the saturation region, $|I_{Q_d}| \ll |I_{Q_s}|$ and $I \approx I_{Q_s}$ and therefore, the current is independent of the drain voltage. In the ohmic region, $I_{Q_d} \simeq I_{Q_s}$ and $I = I_{Q_s} - I_{Q_d}$ and therefore, the current depends on the drain voltage as well as the source and gate voltages. The functional dependence of the current components on the terminal voltage is fixed and remains the same throughout the ohmic and saturation regions.

The charge-voltage relationship is much more complicated than the current-charge one because in general, both the mobile charge and the depletion charge are involved in the electrostatics. Eqs. (8.2), (8.3), and (8.9) yield the following general functional form for the current-voltage relationship that was first introduced by [41] for above threshold operation:

$$I_{SD} \propto \mathcal{F}(V_{GB}, V_{SB}) - \mathcal{F}(V_{GB}, V_{DB}) \qquad (8.18)$$

The device current is decomposed into two opposing components that are func-

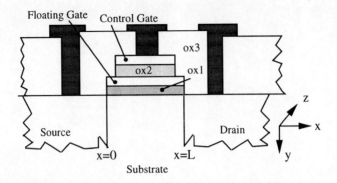

**FIGURE 8.6**

Simplified structure for an FGMOS device. The properties of the different dielectric layers (ox1, ox2, and ox3) that surround the two polysilicon gates depend on the particulars of the manufacturing process. Injecting charge on the floating gate involves current flow through these dielectric layers and thus programming the device is highly process dependent.

tions, $\mathcal{F}$, of the terminal voltages [41, 36]. For an n-type device, $\mathcal{F}$ is a nonpositive, monotonically decreasing function of $V_{GB}$ (see Eqs. (8.2) and (8.6)) and a monotonically increasing function of $V_{SB}$.

In subthreshold operation, the special form of Eq. (8.15) suggests the following factorization of $\mathcal{F}$ :

$$I \propto \mathcal{G}(V_{GB}) \left[ \mathcal{H}(V_{SB}) - \mathcal{H}(V_{DB}) \right] \tag{8.19}$$

where $\mathcal{G}$ and $\mathcal{H}$ are exponential functions. This shows that the source-driven and drain-driven components are controlled independently by $V_{SB}$ and $V_{DB}$. However, $V_{GB}$, acting through the surface potential, also controls both components in a symmetric and multiplicative fashion.

### 8.2.3 The Floating-Gate MOS Transistor

The MOS device model is extended to describe floating-gate MOS transistors in this subsection. The Floating-Gate MOS transistor (FGMOS) is an important element in neuromorphic electronic systems. It can be used for long term non-volatile storage of neural network model parameters [22, 42, 43, 44] and for compensating parametric variations in device characteristics [45, 46, 47]. The floating-gate can also be employed as a summing node to perform mathematical operations in the charge domain with high-linearity [48].

In an FGMOS transistor fabricated in a standard double-poly CMOS process, the floating gate is first polysilicon and the control gate is second polysilicon (see Fig. 8.6). The floating gate controls the current in the channel beneath it and is capacitively-coupled to the control gate above it. The voltage on the floating-gate is determined by the amount of charge deposited on it, as well as the voltages on the control gate, drain, source and substrate.

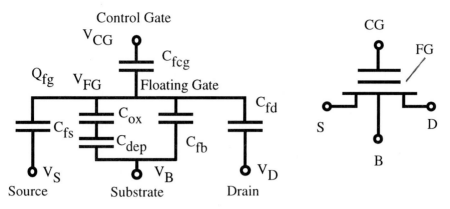

**FIGURE 8.7**
Equivalent circuit model, and symbol for an n-type FGMOS transistor

Several methods have been reported in the literature for metering charge onto or off from a floating gate. Lance Glasser was the first to demonstrate the feasibility of using unfocused UV-light to write-enable binary information on floating-gate structures fabricated in a standard NMOS process [49]. Experimental results with UV-writable non-volatile analog storage in standard double-polysilicon CMOS processes have been reported [50, 51]. Other researchers have reported programming devices that have tunneling injectors using a standard CMOS process [47, 52, 53, 54]. The review article by Horio and Nakamura [55] offers a comprehensive account on this subject. There is also a discussion of other approaches for analog memory and its applications to neural networks.

The best way to incorporate a floating-gate transistor as part of a system is in a *closed-loop* configuration. The relaxation phenomena in the different dielectrics [51] introduce hysteresis in the device characteristics. The time constants depend on the details of the manufacturing process. Thus open-loop control of the charge on the floating gate may be difficult.

### 8.2.4   Large Signal FGMOS Model

The electrostatics of the MOS transistor, with the addition of a floating gate, will now be revisited. First, the voltage on the floating gate is obtained as a function of the voltages on the nodes that are capacitively-coupled to it. Then, the large-signal FGMOS equation is obtained by substituting this voltage for $V_{GB}$ in the equation for the regular MOS transistor. The resulting model is not much different except for two extra terms arising from capacitive-coupling to the source and drain. This effect is significant and cannot be ignored in the design of analog circuits using FGMOS transistors.

The voltage on the floating gate of the FGMOS can be calculated with the aid of the simple circuit model in Fig. 8.7. If transient oxide interface states are ignored, the voltage on the floating-gate is given by:

$$V_{FGB} = \frac{C_{fcg}V_{CGB} + C_{fs}V_{SB} + C_{fd}V_{DB} + C_{ox}\psi_s + Q_{fg}}{C_{sum}} \qquad (8.20)$$

where

$$C_{sum} = C_{fcg} + C_{fs} + C_{fd} + C_{fb} + C_{ox},$$

$Q_{fg}$ is the charge on the floating gate, $C_{fcg}$ is the capacitance between the control gate and the floating gate, $C_{fs}$ and $C_{fd}$ are the capacitances between the floating gate and source and drain, respectively, $C_{ox}$ is the oxide capacitance between the floating gate and the channel, and $C_{fb}$ is the capacitance between the floating gate and the substrate along the edge of the channel. $V_{CGB}$ is the control gate voltage, $V_{FGB}$ is the floating gate voltage, $V_{SB}$ is the source voltage, $V_{DB}$ is the drain voltage, and $\psi_s$ is the channel potential—all referenced to the bulk.

If Eq. (8.11) is substituted into Eq. (8.20), $\psi_s$ is eliminated and an expression for $V_{FGB}$ is obtained in terms of the capacitances and the terminal voltages

$$V_{FGB} = \frac{C_{fcg}V_{CGB} + C_{fs}V_{SB} + C_{fd}V_{DB} + C_{ox}(1.5\phi_F + V_{SB} - \kappa V_{GB}^*) + Q_{fg}}{C_{sum}^*}$$

$$(8.21)$$

where $C_{sum}^* = C_{sum} - \kappa C_{ox}$.

The above equation for the floating gate voltage is now inserted in Eq. (8.15) to obtain:

$$I = S\, I_{fno}\, e^{\zeta \frac{V_{CGB}}{kTq}}\, e^{\frac{\zeta_S V_{SB} + \zeta_D V_{DB}}{kTq}} \left( e^{\frac{-V_{SB}}{kTq}} - e^{\frac{-V_{DB}}{kTq}} \right) \qquad (8.22)$$

where $I_{fno}$ is given by

$$I_{fno} = I_{no} e^{\frac{Q_{fg} + C_{ox}(15_F + V_{SB} - V_{GB}^*)}{C_{sum}^* kTq}} \qquad (8.23)$$

and $\zeta_C \equiv \kappa C_{fcg}/C_{sum}^*$, $\zeta_S \equiv \kappa C_{fs}/C_{sum}^*$, and $\zeta_D \equiv \kappa C_{fd}/C_{sum}^*$. The capacitive divider between the control gate and the floating gate reduces its log-lin slope coefficient $\zeta_C$ (pronounced 'zeta'). Note that the current is also exponentially dependent on the drain and source voltages with log-lin slope coefficients of $\zeta_S$ and $\zeta_D$ due to capacitive coupling to the fringes of the floating-gate.

Ideally, we would like to have $\zeta_C = 1, \zeta_S = \zeta_D = 0$. This may be achieved by adding some extra capacitance between the control and floating-gates. The floating-gate charge, $Q_{fg}$, is the only variable that can be changed after the chip is fabricated; it changes $I_{fno}$ and shifts the I-V curve.

### 8.2.5   Small Signal MOS and FGMOS Models

A simplified small-signal model for the MOS transistor is shown in Fig. 8.8. For completeness, this diagram should be augmented with the capacitances in the equivalent circuit model (Fig. 8.7). In the saturation region, the MOS transistor is a voltage-controlled current source with transconductance from the gate:

$$g_m \equiv \frac{\partial I_{ds}}{\partial V_{gs}} = \frac{\kappa I_{ds}}{kT/q} \qquad (8.24)$$

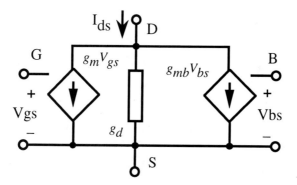

**FIGURE 8.8**
Small-signal MOS transistor model.

and from the local substrate terminal:

$$g_{mb} \equiv \frac{\partial I_{ds}}{\partial V_{bs}} = \frac{(1 - \kappa)I_{ds}}{kT/q} \tag{8.25}$$

The conductance $g_s$ at the source is given by:

$$g_s \equiv \frac{\partial I_{ds}}{\partial V_s} = \frac{I_{ds}}{kT/q} \tag{8.26}$$

Linear relationships between conductance and current are the basis of the *Translinear Principle* [56] which certainly applies to this device. Circuits that exploit this principle will be introduced in the next section. For an FGMOS transistor the parameter $\kappa$ is replaced with $\zeta_C$.

When the device is operating as a controlled current source (saturation), its characteristics are not ideal. The variation of the current with drain voltage results in an output conductance. The depletion layers created at the source and drain junctions by the voltage across them reduces the effective channel length. Charge is extracted from the channel as soon as it reaches the edge of the depletion layer by the large electric field. Therefore, the dependence of the current on the channel length and the dependence of the depletion layer width on the voltage must be taken into account to determine the output conductance. The depletion layer width may be obtained by dividing the charge (Eq. (8.3)) by the acceptor concentration $N_A$.

For a regular MOS transistor

$$g_d \equiv \frac{\partial I_{ds}}{\partial V_d} = \frac{\partial I_{ds}}{\partial L} \frac{\partial L}{\partial V_d} = \left(-\frac{I_{ds}}{L}\right)\left(-\sqrt{\frac{q\epsilon_{Si}}{2N_A}}\frac{1}{\sqrt{V_{DB}}}\right) \approx \frac{I_{ds}}{V_0} \tag{8.27}$$

The inverse square-root dependence of the depletion layer width is replaced by a nominal value $V_0$ which has units of volts and is proportional to $L$. This voltage is analogous to the Early voltage in bipolars and is determined by experiment.

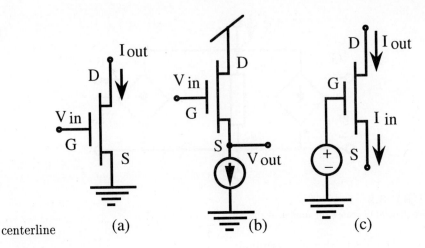

centerline      (a)         (b)         (c)

**FIGURE 8.9**
Different bias configurations for an n-MOS transistor: (a) transconductance amplifier (common source), (b) voltage follower (common drain), and (c) current buffer (common gate).

The output conductance of an FGMOS transistor has an additional term due to capacitive-coupling between the drain and the floating gate. This term is exponentially related to the drain voltage, with slope coefficient $\zeta_D$. The output conductance is therefore given by:

$$g_d^* = g_d + \zeta_D \frac{I_{ds}}{kT/q} \tag{8.28}$$

The extra term increases the output conductance significantly and therefore cascoding the FGMOS is strongly advised.

From the small-signal point of view, the MOS transistor itself can perform several useful circuit functions:

1. In the *common-source* mode (Fig. 8.9(a)), with the drain connected to an ideal voltage source, the device is a **transconductance amplifier** with transconductance gain $G \equiv \Delta I_{out}/\Delta V_{in} = g_m$. Alternatively, with the drain connected to an ideal current source, it is an **inverting amplifier** with high voltage gain: $|A_v| \equiv \Delta V_{out}/\Delta V_{in} = g_m/g_d$. The intrinsic gain of our devices are $|A_{vo}| = \kappa V_o/(kT/q) \approx 400$.

2. In the *common-drain* mode (Fig. 8.9(b)), the MOS transistor is a **voltage follower**, i.e., $\Delta V_{out} \approx \Delta V_{in}$, with a low output resistance: $1/g_m$.

3. In the *common-gate* mode (Fig. 8.9(c)), the device is a **current buffer**, i.e., $\Delta I_{out} = \Delta I_{in}$, with a low output conductance: $g_d$.

Typical design parameters for a 2-micron $n$-well fabrication process are listed in Tab. 8.1. The parameter $V_o$ is listed as a function of the actual (not drawn) channel length $L$.

| | $I_o$ | $\kappa$ | $V_o$ | $C_{fs}^I, C_{fd}^I$ | $C_{ox}^\diamond, C_{fcg}^\diamond$ | $C_{fb}^\diamond$ |
|---|---|---|---|---|---|---|
| | $(fA)$ | | $(V)$ | $(fF/\mu m)$ | $(fF/\mu m^2)$ | $(fF/\mu m^2)$ |
| $P$ | 0.3 | 0.6 | $4.36 \times L^{1.33}$ | 0.42 | 0.5 | 0.05 |
| $N$ | 2 | 0.6 | $9.55 \times L^{1.1}$ | 0.24 | 0.5 | 0.05 |

**TABLE 8.1**
Typical MOS transistor parameters in a 2-micron n-well CMOS process. The channel length $L$ is given in $\mu m$. Note that the parametric equations for the Early Voltage $V_o$ are accurate for devices with $L > 10\mu m$.

## 8.2.6   Bipolar Transistor Model

Bipolar transistors are available in standard CMOS processes in the form of parasitic lateral devices [59, 60] or vertical transistors with a common collector connection (the substrate). The most common application of these devices is in Proportional To Absolute Temperature (PTAT) voltage references [61]. More recently, trends towards digital BiCMOS technologies have endowed standard CMOS processes with a base implant in the well. This enhancement provides for a vertical bipolar transistor, albeit with a high collector resistance. We will use bipolar transistors available in a standard CMOS process to implement transducers such as phototransistors and photodiodes, and in circuits that demand a translinear element operating at higher currents than are attainable by small area subthreshold MOS devices.

A large-signal model useful for the analysis of circuits that includes bipolar transistors is presented in this section. The model will also be used to study the similarities and the subtle—but important—differences between subthreshold MOS and bipolar device operation.

The Ebers-Moll model [57, 58] for an npn bipolar transistor is:

$$
\begin{aligned}
I_E &= -I_F + \alpha_R I_R \\
I_C &= \alpha_F I_F - I_R
\end{aligned}
\tag{8.29}
$$

$$
\begin{aligned}
I_F &= I_{ES}(e^{\frac{qV_{BE}}{kT}} - 1) \\
I_R &= I_{CS}(e^{\frac{qV_{BC}}{kT}} - 1)
\end{aligned}
\tag{8.30}
$$

$$
\alpha_F I_{ES} = \alpha_R I_{CS}
\tag{8.31}
$$

where $I_C$ and $I_E$ are the collector and emitter currents respectively and

$V_{BE}$  is the base to emitter voltage,

$V_{BC}$  is the base to collector voltage,

$I_{ES}$  is the saturation current of emitter junction with zero collector current,

$I_{CS}$  is the saturation current of collector junction with zero emitter current,

$\alpha_F$  common-base current gain.

$\alpha_R$  common-base current gain in inverted mode, i.e. with the collector functioning as an emitter and the emitter functioning as a collector.

By convention, the currents for bipolars are positive when flowing into its terminals.

Combining Eqs. (8.29), (8.30), and (8.31), the collector current can be expressed as:

$$I_C = \alpha_F I_{ES} \left( (e^{\frac{qV_{BE}}{kT}} - 1) - \frac{1}{\alpha_R}(e^{\frac{qV_{BC}}{kT}} - 1) \right) \tag{8.32}$$

For an ideal device which has both common-base current gain, $\alpha_F$, and common-base current gain in inverted mode, $\alpha_R$, very close to unity the above equation becomes:

$$I_C = I_{ES}(e^{\frac{qV_{BE}}{kT}} - e^{\frac{qV_{BC}}{kT}}) \tag{8.33}$$

Regular bipolar transistors do not have both $\alpha_F$ and $\alpha_R$ near unity.

When the collector to base voltage equals zero or the collector is reverse biased with respect to the base, the above equation simplifies to the familiar:

$$I_C = A_E J_{ES}(e^{\frac{qV_{BE}}{kT}} - 1) \approx A_E J_{ES}e^{\frac{qV_{BE}}{kT}} \tag{8.34}$$

where $A_E$ is a design parameter, the area of the emitter junction, and $J_{ES}$ the saturation current density for the emitter, a process dependent parameter. In this case, $I_R \ll I_F$ and the equations above give $I_C = -\alpha_F I_E$. Using the relation $I_E + I_C + I_B = 0$ (KCL) we get the familiar result

$$I_C = \frac{\alpha_F}{1 - \alpha_F}I_B \equiv \beta_F I_B \tag{8.35}$$

where $\beta_F$ is the common-emitter current gain.

In a standard n-well CMOS process, a vertical pnp transistor is available for circuit design, but only in the common collector configuration since it has the p-substrate as the collector—the n-well forms the base and a p-diffusion the emitter. This device is useful as a light sensor; the smallest possible phototransistor permitted by the 2-micron design rules has base dimensions of $16\mu m \times 16\mu m$ when the emitter has an area $A_E = 6\mu m \times 6\mu m$. The dark current in these minimum size phototransistors is approximately 100fA. These sensors show a linear response over at least eight orders of magnitude in light intensity. Experimentally determined responsively of a device with $A_E = 100\mu m \times 100\mu m$ was $73.8A/W$

at $\lambda = 632.8nm$ and $118.5A/W$ at $\lambda = 834nm$. The $\beta$ is approximately 200 with Early voltage $V_o = 48V$. The frequency response is limited to a few hundred KHz by the large base-collector capacitance.

In some processes, an npn transistor is offered through an extra implant in the n-well to form the base. Typical forward $\beta$ for these devices are 60 for emitter area $A_E = 8\mu m \times 8\mu m$. The performance of such bipolar devices is limited by the collector resistance $r_c$ which is in the $k\Omega$ range if there is no buried collector implant. The Early voltage of these devices is approximately $45V$. At high collector currents (high injection conditions) the characteristics of bipolars deviate from exponential and their current gain $\beta$ is also reduced. For the npn vertical bipolars, with a minimum emitter area, this high current effect becomes important at current levels above a few hundred microamps. At low collector currents, the $\beta$ is also limited by recombination in the base. Typical betas range from 20 at current levels of a few nanoamps, to their maximum at current levels of a few hundred microamps.

### 8.2.7   MOS Device Matching in Subthreshold

Traditional MSI and LSI MOS analog integrated circuits employ large-size devices operating above threshold to minimize the effects of transistor mismatch. By operating above threshold and at high values of $V_{GS}$ the transconductance of the device per unit current is dramatically reduced and thus the effects of threshold voltage mismatch are minimized. By using large areas, the random spatial fluctuations in doping and fixed interface charges are averaged out. Better matching is thus achieved at the expense of power dissipation and die size.

In VLSI systems, small geometry transistors must be used, typically $4\mu m \times 4\mu m$ or $6\mu m \times 6\mu m$, to achieve high densities. Furthermore, it is preferable to operate the devices in the region where the transconductance per unit current is maximum, i.e., the subthreshold and transition regions. Small device geometries and high transconductance per unit current makes the drain current strongly dependent on spatial variations of process-dependent parameters, particularly $I_o$. Characterization of the fabrication process and the matching properties of the basic devices is thus of paramount importance because it provides the necessary information for designing working systems. Some of the rules developed over a periods of years by veteran analog designers [31] are applicable to analog VLSI. More importantly, *adaptation* [46] and robust architectures/algorithms [7] must be employed to provide reliable computation in the presence of parametric device variations and defects in the fabrication process.

In a recent experimental study, dense transistor arrays have been fabricated and characterized in the subthreshold region of operation [62, 63, 64]. More than 150,000 transistors have been tested using an automated data acquisition system. The results show that there are three factors affecting matching: edge effects, striation effects, and random variations (Fig. 8.10). The experimental data presented here are typical of 40nm gate-oxide CMOS processes.

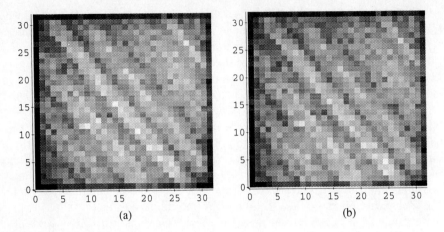

**FIGURE 8.10**

Density plots of currents in a $32 \times 32$ array of $4\mu m \times 4\mu m$ N-MOS transistors. Each transistor is represented by a square pixel. Current level is coded by the shade of gray, where the minimum and maximum values are represented by black and white, respectively. The current at two different nominal current levels of (a) 100pA and (b) 100nA is obtained by setting the same $V_{GS}$ for the entire array. The transistors are biased in saturation.

The edge effect is due to the dependence of a transistor's current on its position with respect to surrounding structures. N-type transistors surrounded by other n-type transistors have a larger drain current than identically designed and biased transistors on the edges of the arrays. "Edge" in this context is defined as the region that is within approximately $35\mu m$ from the array periphery and is not limited to a single device. The opposite is true for p-type devices. Variations in transistor currents due to the edge effect typically range from 5% to 15% for n-type transistors and from 20% to 50% for p-type transistors.

The striation effect is thought to be due to spatial variations in the surface properties of the startup material that result in a sinusoidal spatial variation in transistor current. The amplitude is typically about 30% of the average current and the spatial period typically varies gradually between $100\mu m$ to $200\mu m$. Edge effects and striation effects present significant problems in large scale analog computational systems since they cannot be reduced by increasing transistor area—unless the transistor area is about the size of one period. If the striations are startup material related, their presence or absence may be known apriori. However, their orientation is unknown and thus, this effect could be potentially damaging.

After discounting these deterministic effects, we are left with the random variations. Random mismatch in the subthreshold region can be characterized in terms of the simple model parameters $I_o$ and $\kappa$. The parameter $\kappa$ is very stable with a typical normalized standard deviation of $\sigma(\kappa)/\langle\kappa\rangle \approx 0.3\%$, where $\langle.\rangle$ denotes mean value. The small variations in $\kappa$ suggest that doping and gate-oxide thickness are extremely uniform. The fuzziness in the data points in Fig. 8.11 is

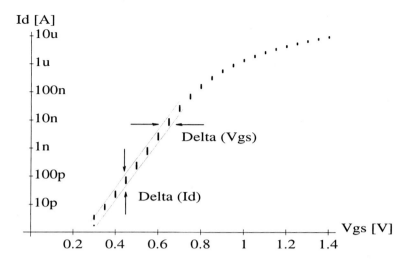

**FIGURE 8.11**

Superimposed I-V characteristics for 32 transistors ($4 \times 4\mu m$). The fuzziness in the current is constant in subthreshold (on a $\log(I)$ scale) and decreases as the device enters the transition and above threshold regime.

therefore not due to changes in slope. This implies that characteristics are displaced from one transistor to the next, implicating the flat-band voltage which depends on the contact potential, $\phi_{MS}$, and the fixed interface charge (implanted ions and trapped electrons), $Q_0$. Most probably, it is the fixed charges that are fluctuating. The variations in the current can of course be related to the fixed-charge distribution through the transconductance and the gate-oxide capacitance.

$$\frac{\sigma(I)}{\langle I \rangle} = \kappa \frac{\sigma(Q_0)/C_{ox}}{kT/q} \tag{8.36}$$

Figure 8.12 shows the dependence of the normalized standard deviation $\sigma(I)/\langle I \rangle$ of the drain current on transistor size. Each data point represents measurements from approximately 1000 transistors. The normalized standard deviation of the current is inversely proportional to the square root of the device area, $A$, (square geometry) and is given by:

$$\frac{\sigma(I_D)}{\langle I_D \rangle} = \sigma_0 \frac{1}{\sqrt{W \times L}} = \sigma_0 \frac{1}{A} \tag{8.37}$$

where $\sigma_0$ is the proportionality constant for a given device type and process. This is exactly as expected for independent random variables: when you sum $N$ variables, the sum and standard deviation increase by factors of $N$ and $N$, respectively, and therefore, the normalized standard deviation decreases by $N$. Since $A \propto N$, this explains the area dependence and allows us to confidently extrapolate our results to rectangular devices.

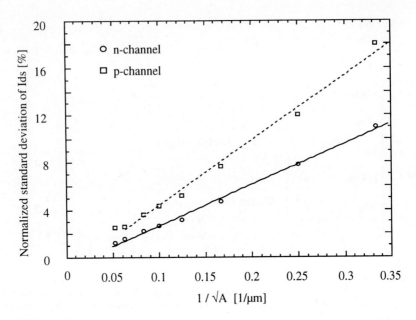

**FIGURE 8.12**

Dependence of normalized standard deviation of subthreshold drain current on transistor size; the lines are best fits to the data. All devices have square geometries with area $A = L^2$. Notice that the normalized standard deviation of $I_{ds}$ saturates at large transistor geometries.

## 8.3  DESIGN METHODOLOGY

Neuromorphic information processing is based on functions that arise directly from the intrinsic *physics* of the computational substrate. In our technology, these are continuous functions of *time, space, voltage, current* and *charge*. To help manage the complexity in VLSI systems, these functions will be considered at three hierarchical levels: *device level, circuit level* and *architectural level*.

At the *device* level, we have seen in an earlier section how subthreshold MOS device physics yields the following functional form for the drain current

$$I = S \, \mathcal{G}(c \, V_{GB}) \, [\mathcal{H}(V_{SB}) - \mathcal{H}(V_{DB})] \tag{8.38}$$

where $\mathcal{G}$ and $\mathcal{H}$ are growing and decaying exponential functions, respectively. This results in the excellent circuit properties of the MOS transistor as a voltage-input, current-output device with good fan-out capabilities (high transconductance $\mathcal{G}$) and good fan-in capability (almost zero conductance at the input). The exponential form is also a very versatile computational primitive, as we will show in the subsection on translinear circuits. Its inverse function, the logarithm, is readily available using the exponential function in a negative feedback loop. For above threshold operation, the device follows a quadratic law (Eq. (8.18)) and therefore, its computational properties are different.

At the *circuit* level, conservation laws, that is conservation of charge (Kirchoff's Current Law) and conservation of energy (Kirchoff's Voltage Law) may be used to realize simple constraint equations.

At the *architecture level,* differential equations from mathematical physics will be employed to implement useful signal processing functions. For example, the *biharmonic* equation

$$\lambda \nabla^2 \nabla^2 \Phi + \Phi = \Phi_{in} \qquad (8.39)$$

where $\nabla^2 \equiv \partial^2/\partial x^2 + \partial^2/\partial y^2$ is the Laplacian operator, constrains the sum of the fourth derivatives of $\Phi$ and $\Phi$ itself to be equal to a fixed input $\Phi_{in}$. From a signal processing view-point, solutions to this equation represent an optimal estimation of the underlying smooth continuous function, given a set of noisy discrete samples $\Phi_{in}$. The solution is optimal in the sense that it simultaneously minimizes the squared error and the energy in the second derivative—the parameter $\lambda$ is the relative cost associated with the derivative term. A large value for $\lambda$ will therefore favor smooth solutions while a small value favors a closer fit. As we will show in Sec. 4, a network that solves the biharmonic equation may be implemented by using a diffusive, or resistive, grid to compute a discrete-approximation of the Laplacian.

### 8.3.1   The Current-Mode Approach

VLSI system considerations and requirements for a high degree of integration make the subthreshold MOS transistor the device of choice for complex neuromorphic systems. Subthreshold operation offers the highest processing rates per unit power. Current-mode (CM) operation yields large dynamic range, simple and elegant implementations of both linear and nonlinear computations, and low power dissipation without sacrificing speed.

Our choice of subthreshold operation is based on the principle: *"Active devices should be used in the region where their transconductance per unit current is maximized.".* This is the way to minimize the energy per operation and maximize the speed per unit power consumed:

$$\frac{\text{speed}}{\text{power}} = \frac{1/\tau}{I\Delta V} = \frac{g_m/C}{I^2/g_m} = \frac{1}{C}\left(\frac{g_m}{I}\right)^2$$

A squared factor is obtained because both voltage swings ($\Delta V$) and propagation delays ($\tau$) are inversely proportional to the transconductance for a given current level. However, only a linear factor is realized if the power supply voltage is not reduced to match the voltage swings $\sim I/g_m$. The transconductance per unit current increases as the current decreases—throughout the above-threshold and transition regions—and reaches a maximum in the subthreshold region.

By taking advantage of the high subthreshold transconductance per unit current, voltage swings are kept to a few thermal voltages, and reasonable processing bandwidths are achieved. Dynamic power dissipation and supply noise are also reduced as a result of the smaller voltage swings. Smaller voltage swings eliminate the current that is wasted in charging and discharging parasitic capacitances, thereby allowing us to use smaller current signals and cut quiescent power dissipation as well. Thus, this approach yields relatively fast analog circuits with power dissipation levels compatible with future trends in system integration. Fast digital circuits can also be designed using source-coupled logic gates and current steering (ECL-like circuits).

The essence of CM VLSI signal processing is signal representation and normalization in terms of a unit current. Since signals are represented by currents, the CM approach enables the design of systems that function over a wide dynamic range. The low-end is limited by leakage currents in the junctions and by noise in the system; the high-end is limited by degrading transconductance per unit current above threshold. Large dynamic range is essential in neuromorphic systems that receive inputs from real world environments (for example, silicon retinas). The value of the unit current ultimately determines the overall power dissipation and controls the temporal response of the system. Thus, power consumption can be managed and *adaptively* controlled to satisfy the temporal response requirements of the system.

In the remainder of this section, we will introduce the translinear principle and the current conveyor concept. The **translinear principle** provides a powerful design and analysis framework for linear and non-linear computational circuits. The **current conveyor** concept facilitates the analysis and synthesis of CM circuits. In addition, we will expose the reader to three important CM functions that are often encountered in neuromorphic computational systems. These are: Normalization, Scaling, Multiplication, and Bidirectional to Unidirectional Current Conversion (Absolute value).

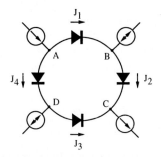

**FIGURE 8.13**
A Translinear loop using ideal p-n junctions.

## 8.3.2 The Translinear Principle

The translinear (TL) principle [56] can be stated as follows:

In a closed loop containing an equal number of oppositely-connected translinear elements, the product of the current densities in the elements connected in the clockwise (CW) direction is equal to the corresponding product for elements connected in the counter clockwise (CCW) direction.

A translinear element is a physical device with a linear relationship between conductance and current. This implies an exponential dependence of the current on the voltage. Summing the voltages around a loop consisting of translinear elements is therefore equivalent to taking the product of the currents. Thus, the translinear principle is simply a restatement of Kirchoff's voltage law or the principle of conservation of energy. The p-n junction, with its exponential I-V characteristics, is a translinear element.

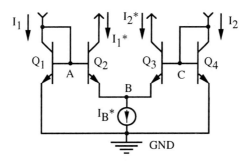

**FIGURE 8.14**
The Gilbert gain stage. This circuit rescales the differential input current.

As an example of the application of the TL principle, consider the four-diode loop shown in Fig. 8.13. Applying the translinear principle around the loop A-B-C-D-A yields:

$$\prod_{CCW} J = \prod_{CW} J \quad \text{or} \quad J_1 J_2 = J_3 J_4$$

This relationship holds between the diode currents inspite of the amount of current injected into or drawn from the loop.

Three-terminal devices are also translinear if we consider the relationship between the current and the controlling voltage. All that is required is that the relationship is exponential *and* the two terminals across which the controlling voltage is applied exhibit true diode-like behavior, i.e., increasing the voltage on one terminal is exactly equivalent to decreasing the voltage on the other terminal by the same amount. In this case, the loop consists of voltage drops across pairs of control terminals and we exploit the linear transconductance-current relationship—hence, the term *translinear.*

Translinear circuits have traditionally been designed using bipolar transistors, exploiting the exponential nonlinearity between the collector current and base-emitter voltage [56] (see Eq. (8.34)) and the equivalent effectiveness of the base and the emitter. Several circuit examples have been discussed in Chapter 5. The simple translinear bipolar circuit in Fig. 8.14 is excellent for rescaling differential current signals. The analysis of this circuit is typical for translinear circuits that involve differential current signals. Application of the translinear principle around the loop A-B-C-GND-A yields:

$$I_1 \, I_2^* = I_1^* \, I_2 \quad \Rightarrow \quad \frac{I_1^*}{I_1} = \frac{I_2^*}{I_2}$$

Using elementary algebra,

$$\frac{I_1^* - I_2^*}{I_1 - I_2} = \frac{I_1^* + I_2^*}{I_1 + I_2}$$

and

$$\Delta I^* = \frac{I_B^*}{I_B} \, \Delta I$$

where $\Delta I \equiv I_1 - I_2$, $I_B \equiv I_1 + I_2$, and similarly for $\Delta I^*$, $I_B^*$. The differential output current $\Delta I^*$ is a scaled version of the differential input current $\Delta I$. The voltage between node $B$ and GND should be such that the current source $I_B^*$ stays in saturation.

This circuit was employed in a BiCMOS transconductor for silicon cochleas [66]. Translinear circuits form a large subclass of CM circuits (Chapter 2 in [26]). Designs based on the translinear principle can perform complex computations on current inputs. Some examples are: products, quotients, and power terms with fixed exponents [56, 65].

### 8.3.3 Translinear Circuits in Subthreshold MOS

The translinear principle is equally applicable to MOS transistors in subthreshold–where the I-V characteristic is exponential (see Eq. (8.24)). However, one must be aware of certain nonidealities of the MOS transistor.

The gate and the source are not equally effective in controlling the current. If their voltages are measured by the same amount, the same currents are not obtained because the transconductances are different (compare Eqs. (8.24) and (8.26)). Whereas 100% of the source voltage goes to change the barrier height, only a certain fraction ($\kappa$) of the gate voltage is effective. However, this fraction can be pushed very close to unity by biasing the device with a large surface potential $\psi_s$ (and thus large $V_{SB}$ and $V_{GB}$) as predicted by Eq. (8.12). Then the subthreshold MOS device becomes an ideal translinear element.

Another way to circumvent this problem is to move the local substrate voltage together with the source voltage—which is exactly equivalent to increasing $V_{GB}$ by the same amount and does not change $V_{SB}$. In practice, this may be achieved simply by shorting the source to the local substrate. Clearly, this is only possible with devices in separate wells. For this reason, it is preferable to bias the device at large values of $\psi_s$, as described in the previous paragraph, since that works for both types of transistors and is more area efficient (there is no need for separate wells).

Other areas of concern are the poor matching characteristics of MOS transistors in subthreshold (much worse than bipolars) and the high impedance of the gate terminal which requires special frequency compensation techniques. Also, circuit techniques that exploit the current-input/current-output operational mode of the bipolar transistor are not amenable to MOS devices. For a detailed discussion of subthreshold MOS Translinear circuits please refer to [40].

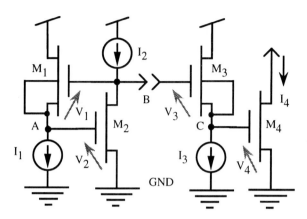

## FIGURE 8.15
A translinear circuit that performs a one-quadrant multiply and divide. The multiplicands are $I_1$ and $I_2$ and the divisor is $I_3$; $I_4 = I_1 I_2 / I_3$ is the result.

Our first example of a TL MOS circuit is the one-quadrant multiply-divide circuit shown in Fig. 8.15. Applying the translinear principle to the loop GND–A–B–C–GND, we find a total of four equivalent diode junctions and obtain

$$I_1 I_2 = I_3 I_4 \quad \text{or} \quad I_4 = \frac{I_1 I_2}{I_3} \tag{8.40}$$

The above relationship can also be derived by summing the voltages around the loop (conservation of energy)

$$V_1 + V_2 - V_3 - V_4 = 0$$

Replacing the gate-source voltages for $M_1, M_2, M_3, M_4$ with their respective drain-source currents using Eq. (8.16) (assuming all devices are in saturation, have negligible drain conductance, have $\kappa = 1$ and the same $I_0$), we obtain:

$$\frac{kT}{q} \left( \ln \left( \frac{I_1}{I_0} \right) + \ln \left( \frac{I_2}{I_0} \right) - \ln \left( \frac{I_3}{I_0} \right) - \ln \left( \frac{I_4}{I_0} \right) \right) = 0$$

or

$$\ln \left( \frac{I_1 I_2}{I_0} \right) = \ln \left( \frac{I_4 I_3}{I_0} \right)$$

from which Eq. (8.40) readily follows.

Yet another way of capturing the function of this circuit is to view it as a *log-antilog* block. Transistors $M_1$, $M_2$ and $M_3$ do the *log-ing* and $M_4$ does the

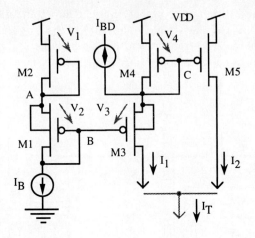

**FIGURE 8.16**

A CM circuit that converts a bidirectional current on a single line into two unidirectional currents on separate wires. This is essentially an absolute value circuit. The sign of the bidirectional input current is assumed to be positive when it adds positive charge to node $C$.

*antilog-ing.* A large number of these CMOS multipliers have been employed in the implementation of a correlation-based motion-sensitive silicon retina [90].

Our second example addresses the problem of converting a bidirectional current to unidirectional currents. This is a common operation in CM designs [17]. A translinear circuit that computes this highly nonlinear function is shown in Fig. 8.16. The bidirectional current $I_{BD}$ is steered through transistor $M_3$ when $I_{BD} > 0$ and through transistors $M_{45}$ when $I_{BD} < 0$. Concentrating on transistors $M_{1234}$, we identify a loop (VDD-A-B-C-VDD) and apply the translinear principle to it

$$I_B I_B = I_1 \left( I_1 - I_{BD} \right)$$

A second equation is obtained from the trivial loop VDD-C-VDD:

$$I_2 = I_1 - I_{BD}$$

which can also be obtained by observing that transistors $M_4$ and $M_5$ form a simple current mirror.

These equations may be solved for $I_1$ and $I_2$ in terms of $I_B$ and $I_{BD}$:

$$I_1 = \tfrac{1}{2} \left( I_{BD} + \sqrt{4 I_B{}^2 + I_{BD}{}^2} \right)$$

$$I_2 = \tfrac{1}{2} \left( -I_{BD} + \sqrt{4 I_B{}^2 + I_{BD}{}^2} \right)$$

$$(8.41)$$

Which shows that $I_1 \simeq |I_{BD}|$ and $I_2 \simeq 0$ when $I_{BD} > 0$ (or vise versa) if $I_{BD} \gg I_B$. The absolute value is obtained by connecting the two output wires together in which case:

$$I_T \equiv I_1 + I_2 \simeq |I_{BD}| \quad \text{if } I_{BD} \gg I_B$$

An operating point corresponding to the bias current $I_B$ can also be set by a voltage source driving the gate of transistor $M3$. This is not advisable, since the operating point in the system will be ultimately controlled by a temperature-compensated current reference.

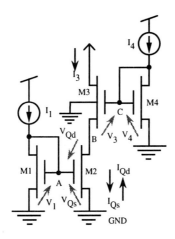

**FIGURE 8.17**
A translinear circuit that that exploits subthreshold MOS ohmic characteristics and computes correlation between two current signals.

### 8.3.4   TL Circuits in Subthreshold Ohmic Region

We extend the translinear principle to MOS transistors operating in the ohmic region in this subsection. In the previous section, we saw how the source-drain current of a MOS transistor can be decomposed into a source component $I_{Q_s}$ and a drain component $I_{Q_d}$ and that these components superimpose linearly to yield the actual current $I_{SD} \equiv I_{Q_s} - I_{Q_d}$ (see Fig. 8.5). In the ohmic region, these components are comparable and decomposition and linear superposition may be used to exploit the intrinsic translinearity of the gate-source and gate-drain "junctions." This is the basis for extending the TL principle to the ohmic region. On the otherhand, in the saturation region, we can exploit the translinearity of the gate-source "junction" directly because the drain component is essentially zero and decomposition is of no consequence.

To demonstrate the application of the translinear principle to circuits that include MOS transistors in the ohmic regime, consider the one-quadrant current-correlator circuit in Fig. 8.17. Transistor $M_2$ operates in the ohmic region. Proper circuit operation requires that the output voltage is high enough to keep $M_3$ in saturation. This circuit was first introduced by Delbrück [67] and later incorporated in a larger circuit that implements the non-linear Hebbian learning rule in an auto-adaptive network [68].

An expression relating the output current, $I_3$, to the input currents, $I_1$ and $I_4$, can be derived by treating the source-gate and the drain-gate "junctions" of the ohmic device as separate translinear elements and applying the translinear principle. For the two loops formed by nodes GND-A-GND and GND-A-B-C-GND in Fig. 8.17, we obtain

$$I_1 = I_{Q_s}$$

$$I_3 I_{Q_s} = I_4 I_{Q_d}$$

(8.42)

In writing Eq. (8.42), we have tacitly assumed that the source-drain current of the MOS transistor can be decomposed into a source component $I_{Q_s}$ and a drain component $I_{Q_d}$—controlled by their respective "junction" voltages $V_{Q_s}$ and $V_{Q_d}$. These opposing components superimpose linearly to give the actual current passed by $M_2$, i.e.,

$$I_{DS2} = I_{Q_s} - I_{Q_d}$$

Furthermore, both $M_2$ and $M_3$ pass the same current and therefore:

$$I_3 = I_{Q_s} - I_{Q_d}$$

(8.43)

Using Eqs. (8.42) and (8.43), the output current is given by:

$$I_3 = \frac{I_1 I_4}{I_1 + I_4}$$

(8.44)

One could argue that decomposition is also possible with bipolar devices (see Eq. (8.34)). However, while the difference of two exponentials is the *exact* form for MOS devices, it is only an *approximation* for bipolars (see Eq. (8.33)). This is due to the fact that the forward and inverse alphas of the device are never unity. This distinction is a fundamental and important difference between MOS and bipolar transistors arising from lossless transport in a MOS channel which is not the case in the bipolar device due to recombination in the base.

### 8.3.5   The Current Conveyor

Sedra and Smith originated the notion of a current-conveyor [70]—a hybrid voltage/current three-port device (Fig. 8.18). It is a versatile building block for analog signal processing applications. The current conveyor is characterized by small-signal relationships between the voltages and currents at different nodes.

Analogous to the operational amplifier, there is a virtual short for voltage from node $Y$ to node $X$ and a virtual short for current from node $X$ to node $Z$. Therefore, node $X$ can transmit a current signal to node $Y$ while simultaneously receiving a voltage signal from node $Y$. This original current-conveyor implementation (CC-I) has been exploited effectively in linear analog LSI circuit design [26]. The high functionality embodied in this very simple—yet elegant— circuit makes it a good candidate for analog VLSI systems where it satisfies the

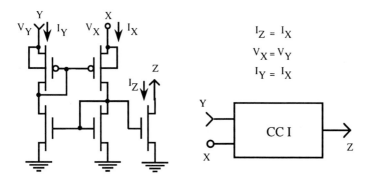

**FIGURE 8.18**
The original Sedra and Smith five-transistor (5T) current-conveyor circuit (CC-I) implemented using MOS transistors. In this conveyor, $Y$ is a hybrid voltage-input/current-output node, $X$ is a hybrid voltage-output/current-input node and $Z$ is a current-output node.

need for large fan-in and large fan-out. In the context of the CM systems described here, node $X$ is used as a *communication* node, node $Y$ as a *control* node, and node $Z$ as an *output* node.

Although the original implementation of the current conveyor concept used five transistors, it can infact be realized with just one. A MOS transistor, in saturation, can transfer a current from its high-conductance source terminal to its low-conductance drain terminal or a voltage from its high-impedance gate terminal to its low-impedance source terminal (see Fig. 8.9(b) and (c)). These two actions can be exploited to simultaneously achieve a voltage short from gate-to-source and a current short from source-to-drain (see Fig. 8.19(a)). This dual role—obtained with a single device—captures the essence of the current-conveyor. As such, the gate is the control voltage ($Y$) node, the drain is the output current ($Z$) node, and the source is the hybrid input-current/output-voltage communication ($X$) node. In this minimalist implementation, we forsake the redundant current-output at node $Y$ and introduce a nominally constant voltage offset between nodes $X$ and $Y$.

The large-signal behavior for the 1T current-conveyor is described by the following set of equations:

$$I_Z = I_X$$

$$V_X = \kappa V_Y - \frac{kT}{q} \ln\left(\frac{I_X}{SI_o}\right)$$

(8.45)

assuming the drain conductance is zero. Note that, unlike Sedra's CC-I circuit, the 1T conveyor is highly nonlinear. However, the compressive log function does not seriously degrade the voltage following property. The small-signal behavior of the 1T conveyor is due to the combined action of a single-transistor voltage follower and a single-transistor current buffer—realized by the same device. That is, if the current source at node $X$ has an output conductance $g_i$, then the output conductance at node $Z$ is approximately $g_i/A_{v_0}$ and the input conductance

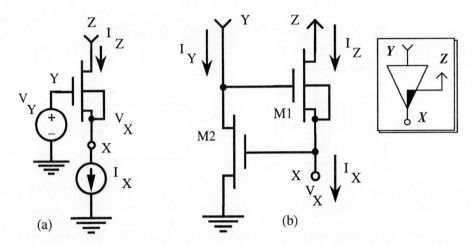

**FIGURE 8.19**
(a) A one-transistor (1T) current conveyor on (b) a two-transistor (2T) current-controlled current conveyor circuit and conveyor symbol. The white triangle represents the voltage buffering action from $Y$ to $Z$ while the in-laid black triangle represents the current buffering action from $X$ to $Z$.

at node $X$ is $g_m$, where $A_{v_0}$ is the intrinsic gain of the device and $g_m$ is its transconductance.

A two-transistor (2T) current-controlled current conveyor is shown in Fig. 8.19(b) This circuit has a hybrid current-input/voltage-output communication node $X$, a current-input control node $Y$, and a current-output node $Z$. The voltage $V_X$ at the communication node $X$ is determined by the current supplied to node $Y$— instead of using a control voltage. The current-controlled conveyor's large-signal behavior is described by the following equations:

$$I_Z = I_X$$

$$V_X = \frac{kTq}{2} \ln\left(\frac{I_Y}{S_2 I_o}\right) \tag{8.46}$$

$$V_Y = \frac{kTq}{21} \ln\left(\frac{I_Y}{S_2 I_o}\right) + \frac{kTq}{1} \ln\left(\frac{I_X}{S_1 I_o}\right)$$

The voltage buffering between nodes $Y$ and $X$ has been replaced with a generalized transimpedance ($M_2$) that generates a voltage $V_X$ proportional to $\log(I_Y)$. As will be seen in a later section, in a current-mode circuit design style where all signals are implicitly represented by "log I" voltages, this nonlinearity is transparent. The log-ing action is achieved by diode-connecting $M_2$—using the voltage-following action of $M_1$ from node $Y$ to node $X$. $M_1$ sets $V_X$ so as to make $M_2$'s current equal $I_Y$. $M_2$ inverts and amplifies small changes in $V_X$ to generate the requisite voltage at node $Y$.

Compared to the 1T conveyor, the small-signal conductance at the current-output node ($Z$) is lower by a factor of $A_{v_0}$ and the small-signal conductance at the communication node $(X)$ is increased by the same factor. More specifically, the conductance seen at node $Z$ is approximately $g_i/(A_{v_1} A_{v_2})$, i.e., the buffering action of $M_1$ is improved by the gain $A_{v_2}$ of $M_2$. The conductance seen at node $X$ is approximately $(A_{v_2} g_{m_1})$, i.e., the source conductance of $M_1$ times $M_2$'s gain.

The 2T current-controlled conveyor was used in a two-way communication scheme for a CM, clamped bit-line, associative memory design [27, 71]. In that application, the dual role of the 2T current-controlled current conveyor was fully exploited to simultaneously fan-in currents and fan-out voltage over the same wire. Recently, Säckinger independently proposed this arrangement to increase the output impedance of a MOS transistor current source, as we have shown here. He used this circuit in a high-performance op-amp design and dubbed it the *regulated cascode* [72]. Indeed, the negative feedback arrangement used in the 2T conveyor is not new and is well-known to veteran analog circuit designers. One of its most elegant applications is the three-transistor Wilson current mirror [73]. Having completed the discussion of the translinear principle and the current conveyors we are now ready to employ these concepts in neuromorphic networks.

## 8.4   NETWORKS

The computational capabilities of neural networks emerge from the dynamics of large pools of richly-interconnected neurons [1] (see Fig. 8.20). Analog VLSI neuroprocessors reflect this. They can be divided into two broad classes: globally-connected or locally-connected. The contrast-sensitive silicon retina [7] described in the next section is locally-connected—each cell communicates only with its immediate neighbors. The associative memory [18, 20, 19] and the Herault-Jutten network [17] are globally-connected—each cell communicates with all cells in the other layer.

With local connectivity, $N$ neurons require $O(N)$ synapses. Therefore, the area taken up by synapses grows at the same rate as that used for neurons. This allows a 2-D array of neurons covering the entire chip area to be built. With global connectivity, $O(N^2)$ synapses are required. Therefore, only a 1-D array of neurons is possible, since the $N$ synapses required by each neuron take up the second dimension. In this case, neurons are placed on the periphery or along the diagonal.

Neuromorphic architectures are much more similar to memory systems than they are to logic systems since they have large numbers of identical units. There-fore, the core cell—either a synapse or a neuron with local synapses—**must** employ circuits of minimal complexity and be carefully layed-out. Support cir-cuitry (a. k. a. sense amplifiers and address decoders in a memory design) must be designed to match the pitch of the core cell. This is totally unlike traditional analog MSI/LSI circuit design.

In this section we show by example how the device principles and circuit techniques we have presented are applied to neuromorphic system design. The first example is an associative memory design using current-controlled current

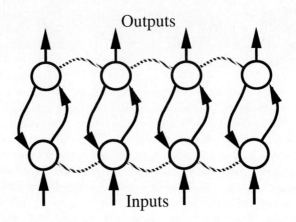

**FIGURE 8.20**

Network architecture. A typical neural network has a layered topology with multiple inputs and outputs. The nodes and arcs are abstractions of neurons and synapses. A modifiable coupling strength (weight) is associated with each arc and is set according to a prescribed algorithm (learning rule). For the sake of clarity, a simple one-dimensional array with only nearest-neighbor connections is shown here.

conveyors interconnected through a simple bidirectional junction. The conveyor's hybrid voltage-out/current-in communication node results in very efficient use of wires. The second example is a one-dimensional network for current signal normalization. This is an application of the translinear principle to a vector of current signals. The third example is a network for local aggregation; it is based on carrier diffusion in subthreshold MOS transistors. The last example is a non-linear network that performs Winner-Takes-All competition; it consists of interacting current-conveyors.

### 8.4.1   Globally-Connected Networks

A scheme for implementing global connectivity is shown in Fig. 8.21. The neurons use current conveyors to communicate bidirectionally with other conveyor-based neurons [27]. A simple two-transistor circuit—called a bidirectional junction—connects one neuron to another (see Fig. 8.22(a)). Bidirectional interaction is possible through this junction; it models reciprocal chemical synapses. Each device receives a voltage-input from one node and transforms it into a current-output at the other node. This is activated by grounding node $S$, deactivated by bringing node $S$ to $V_{dd}$, or modulated by changing the local substrate voltage $V_c$. The transconductance gain of both junctions are changed equally in this fashion if they share the same well. Experimental data, including the modulation of the well voltage, was reported in [27, 29].

The bidirectional junction's devices must be in saturation, i.e., $V_{n_1}, V_{n_2} > V_{dsat}$, to prevent cross-talk between in-coming voltage signals and out-going current signals on the same wire. A small change in $V_{n_1}$, caused by a corresponding

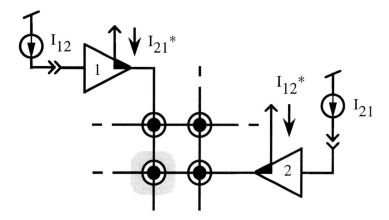

**FIGURE 8.21**
A simple network of neurons that use 2T current-conveyors and bidirectional junctions to communicate. The two neurons shown, labeled 1 and 2, interact through the junction in the shaded area. They send current signals $I_{12}$ and $I_{21}$ and receive currents $I_{21}^{*}$ and $I_{12}^{*}$, respectively.

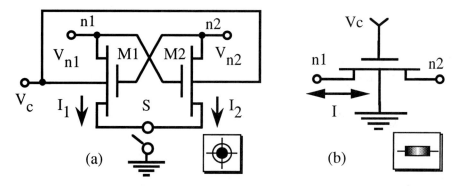

**FIGURE 8.22**
Circuits for bidirectional (two-way) communication between neurons. (a) Bidirectional junction. The switch activates or deactivates the connection. The cross-hair in the symbol represents the two lines, $n_1$ and $n_2$, between which the junction is formed. (b) Diffusor. Charge diffuses between nodes $n_1$ and $n_2$; $V_C$ controls the effective diffusivity.

change in the neuron $n_1$'s output current $I_{12}$, results in a desired change in the post-junction current $I_{12}^{*}$ and an undesired change in the in-coming current $I_{21}^{*}$. These changes are related by:

$$\frac{\Delta I_{21}^{*}}{\Delta I_{12}^{*}} = \frac{g_d(I_{21}^{*})}{g_m(I_{12}^{*})} = \frac{1}{A_{vo}} \frac{I_{21}^{*}}{I_{12}^{*}}$$

where $A_{vo}$ is the intrinsic gain of the device–assuming it is in saturation. The high intrinsic gain of subthreshold MOS devices makes bidirectional communication possible with less than $-50$dB cross-talk comparing the fractional changes

$(\Delta I/I)$ in the currents.

Cross-talk can also occur at the current-conveyor end of the connection due to its finite communication node conductance. In this case, a small change in the out-going voltage $V_{n_1}$, caused by a corresponding change in the in-coming current $I_{21}{}^*$, results in an undesirable change in the post-junction current $I_{12}{}^*$. These variations are related by:

$$\frac{\Delta I_{12}{}^*}{\Delta I_{21}{}^*} = \frac{g_m(I_{12}{}^*)}{A_{vo}g_m(I_{21}{}^*)} = \frac{1}{A_{vo}}\frac{I_{12}{}^*}{I_{21}{}^*}$$

This demonstrates that large current–fan-in and large voltage–fan-out is possible with the simple 2T current conveyor.

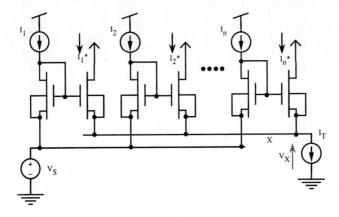

**FIGURE 8.23**

A CM array normalizer. This is a network of current–mirrors with the sources of their output devices connected to a common global line.

### 8.4.2  Global Normalization Network

Vector normalization is an important operation in biological systems. Elaborate structures and mechanisms in the sensory periphery are dedicated to the transduction and normalization of real-world stimuli that have a wide dynamic range. These operations match the dynamic range of input signals to levels compatible with the needs of subsequent processing stages. In this subsection, we present a translinear CM network due to Barrie Gilbert that performs global normalization [69]. Unlike the translinear circuits presented earlier, this circuit operates on a vector of input current signals. See Fig. 8.23.

The analysis of this circuit follows the same general method used in analyzing the Gilbert gain stage, thus:

$$\frac{I_1^*}{I_1} = \frac{I_2^*}{I_2} = \frac{I_3^*}{I_3} = \frac{I_n^*}{I_n} \tag{8.47}$$

And hence the input and output currents are related by:

$$I_n^* = I_T \frac{I_n}{\sum_m I_m} \tag{8.48}$$

Thus any output current, $I_n^*$, in the array is proportional to the corresponding input current $I_n$—normalized to the sum of all input currents. The sum of all the output currents must equal the tail current, $I_T$; this current serves as a scaling parameter that can be controlled externally.

This circuit performs global normalization very elegantly—just one wire is used for communication. The voltage on this wire follows the highest input voltage and the corresponding device gets most of the tail current, $I_T$. The array normalizer can also be viewed as a network of 1T current conveyors with their communication nodes tied together. Each conveyor attempts to set the voltage there but the one with the highest $Y$-node voltage wins because it can supply the most current. *Local* normalization is complementary to global normalization; the diffusive networks discussed in the following subsection are used to perform this computation.

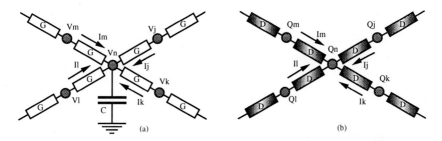

**FIGURE 8.24**
Simulating diffusion with (a) conductances and voltage/current variables or (b) diffusors and charge/current variables.

### 8.4.3 Diffusive Networks

*Local aggregation*, i.e., the linear addition of signals over a confined region of space, is a computation that occurs throughout the nervous system. A voltage-mode technique for peforming this extremely useful computation is presented in Chapter 6 of [4]; it is the basis for many neuromorphic systems described therein. In this subsection, we take a close look at *diffusion*, the physical process that underlies local aggregation in the nervous system, and come up with a more efficient current-mode technique based on diffusion in subthreshold MOS devices.

The diffusion process is described by the following equation:

$$\frac{dN}{dt} = D\nabla^2 N(x, y) \tag{8.49}$$

$N$ is the concentration of the diffusing species and $D$ is their diffusivity. This applies to the 2-D case where the concentration is assumed uniform in the third dimension and $N$ is the number of particles per unit area. Two alternative analog simulations of this process on a discrete grid are shown in Fig. 8.24.

The first network uses voltages and currents (Fig. 8.24(a)). Its node equation is

$$\frac{dV_n}{dt} = \frac{4G}{C} \left( \frac{1}{4}(V_j + V_k + V_l + V_m) - V_n \right) \tag{8.50}$$

which is homologous with Eq. (8.49) since the term in large parenthesis is a first-order approximation to the Laplacian. However, this solution is not amenable to VLSI integration because linearized transconductances ($G$) with large linear range are very area consuming and power hungry.

The second network uses charges (positive) and currents (Fig. 8.24(b)). Its node equation is

$$\frac{dQ_n}{dt} = 4D \left( \frac{1}{4}(Q_j + Q_k + Q_l + Q_m) - Q_n \right) \tag{8.51}$$

Note that $dQ_n/dt$ is the same as the current supplied to node $n$ by the network. This solution is easily realized by exploiting diffusion in subthreshold MOS transistors. As shown in the device physics section, the current is linearly proportional to the charge difference across the channel—in the absence of velocity saturation and channel-length modulation. The diffusivity obtained with a MOS device is

$$\frac{W}{L} \frac{kT}{q} \mu$$

where $\mu$ is the surface mobility of the semiconductor (see Eq. (8.14)). Therefore, the diffusion process may be modeled using devices with identical $W/L$ ratios and identical gate voltages. The former guarantees they have the same diffusivity and the latter guarantees that the charge concentrations at all the source/drains connected to node $n$ are the same and equal $Q_n$.

In both of these networks, the boundary conditions may be set up by injecting current into the appropriate nodes. In the voltage-mode network, the solution is the node voltages. They are easily read without disturbing the network. On the otherhand, the other network represents the solution by charge concentrations $Q_s$ and $Q_d$ at source/drains—not the charge on the node capacitance. The source/drain charge cannot be measured directly without disturbing the network. It may be inferred from the node voltage.

Diffusive media in biological cell syncytia are hardly ever isotropic like our simple initial example. Nerve cells make gap junctions of varying area. Neuromodulators like dopamine can vary the pore permeability. Thus, nerve cells can control the diffusivity between them and neighboring cells or the extracellular fluid. We can control the diffusivity of the MOS network by exploiting the factorization property (see Eqs. (8.15), (8.19)). This allows us to extract the gate-voltage term in the charge expression and lump it with the geometrical factor ($W/L$). Now we use the gate to set the *effective width* ($W_{\text{eff}}$) of the channel.

$Q_n$ is redefined to be the source/drain charge for $\psi = 2\phi_F$ or $V_{GS} = V_{TO}$. Thus, the current through a NMOS diffusor connecting nodes $n$ and $m$ is given by

$$I_{nm} = \frac{W_{\text{eff}}}{L} \frac{kT}{q} \mu \ (Q_n - Q_m) \qquad (8.52)$$

$$Q_{nm} = -\frac{1}{2} \frac{kTC_{ox}}{q} \frac{\gamma}{\sqrt{\psi}} e^{-\frac{V_{nm}}{kTq}} \qquad (8.53)$$

$$W_{\text{eff}} = W e^{\frac{-2_F}{kTq}} \qquad (8.54)$$

where the effective width $W_{\text{eff}}$ accounts for the physical width of the channel as well as the surface potential (or gate-voltage) which determines the barrier height.

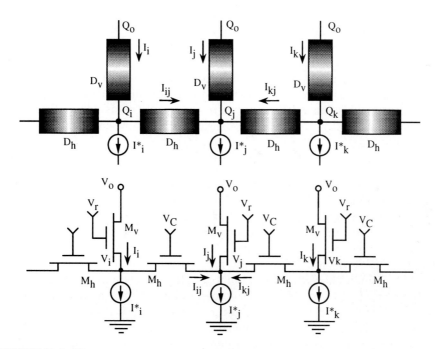

**FIGURE 8.25**
A one-dimensional diffusive network for local aggregation. The lateral diffusors model gap junctions between cells and the vertical ones model the membrane leakage. (a) Schematic representation. (b) MOS transistor implementation.

The network shown in Fig. 8.25 may be analyzed with the help of a diffusor model. The node equation is

$$\frac{dQ_j}{dt} = I_j^* = I_j + I_{ij} + I_{kj} = D_v(Q_0 - Q_j) + D_h(Q_i + Q_k - 2Q_j) \qquad (8.55)$$

where the effective diffusivities ($D_h$ and $D_v$) are given by $\frac{kT}{q}\mu(\frac{W_{eff}}{L})$ (see Eq. (8.54)). This is a discrete approximation to the 1-D diffusion equation

$$I^*(x) = \frac{dQ(x)}{dt} = D_v(Q_0 - Q(x)) + D_h\frac{d^2Q(x)}{dx^2}$$

The extra term arises from the vertical diffusors which shunt charge to ground. This equation yields the solution to the following optimization problem: Find the smooth function $Q(x)$ that best fits the data $I^*(x)$ with the minimum energy in its first derivative. The ratio $D_h/D_v$ is the cost associated with the derivative energy—relative to the squared-error of the fit. Note that in this example $Q_0 > Q(x)$ and thus the extra term is positive.

The vertical elements afford us the opportunity to read the solution, i.e., the source/drain charge $Q_j$. This is directly proportional to the vertical current $I_j$ if $Q_0$ is zero. Eq. (8.53) predicts that $Q_0$ is negligibly small if $V_0$ is a few thermal voltages above $V_j$ (saturation). Therefore, we can measure $Q_j$ by setting $V_0$ at a high voltage and measuring the current there. The actual value used does not matter because $Q_0$ only serves to set the quiescent (zero input) drain/source charge.

For circuit analysis, it is convenient to use the translinear principle together with channel current decomposition to analyze these diffusor circuits. However, we need to include the voltage sources in the loop. This is done following the approach suggested by [74] and [26]. Using their result, we have:

$$I_{Q_j} = I_j e^{\frac{(V_c - V_r)}{kTq}}$$

assuming $I_j$'s drain-driven component is zero, i. e. $M_v$ is in saturation. Similar results are obtained for $I_{Q_i}$ and $I_{Q_k}$; simply replace $I_j$ with $I_i$ and $I_k$, respectively.

This is immediately obvious if we observe that $M_h$ and $M_v$ are a differential pair operating subthreshold. These devices act as a current-divider for current driven by the charge at their common node. The divider ratio is set by their effective widths which depend on the geometrical width as well as the surface potential. Here, we have used the $\kappa$ approximation to relate the surface potential to the gate-bulk voltage. The surface potential is constant as long as the gate and bulk voltages are fixed—assuming the mobile charge is negligible. Therefore, the divider ratio is constant and linear division occurs. However, as we enter the transition and above-threshold regions this assumption fails and the surface potential starts to follow the source voltage. Consequently, the divider-ratio is no longer independent of the current level. This limits the dynamic range of the diffusor.

The node equation may be rewritten as

$$I_j^* = I_{ij} + I_{kj} + I_j = I_{Q_i} - I_{Q_j} + I_{Q_k} - I_{Q_j} + I_j$$

and substituting the expressions for the current components, we get

$$I_j^* = I_j + e^{\frac{(V_c - V_r)}{kTq}}(I_i + I_k - 2I_j) \tag{8.56}$$

This is identical to Eq. (8.55) if we replace $I_j$ with $D_v Q_j$, $I_k$ with $D_v Q_k$, etc., and $\exp(\kappa q(V_c - V_r)/kT)$ by $D_h/D_v$. The area efficiency and controlled coupling strength available using the diffusor this circuit particularly attractive for implementing the local aggregation.

The diffusive network in Fig. 8.25 was recently described in terms of "pseudo-conductances" [75]. We prefer the charge-based formulation using diffusors first proposed in [7] and elaborated here because it is a *physical* model. This provides an intuitive understanding of the device and yielded the insight that enabled us to extend the translinear principle to subthreshold MOS transistors in the ohmic region and diffusors. We have provided a comprehensive *current-mode* approach for analyzing subthreshold MOS circuits. The essence of this approach is the representation of variables and parameters by charge, current, and diffusivity. Voltages and conductances are not used explicitly.

Bult and Geelen proposed an identical network for linear current division above-threshold and used it in a digitally-controlled attenuator [76]; they also analyzed its subthreshold behavior. However, they stipulate that all gate voltages must be identical and control the division by manipulating the geometrical factor $W/L$ of the devices. We have shown here, and previously in [7], that this constraint may be relaxed in subthreshold without disrupting linear operation. This is a real bonus because it allows us to modify the divider ratio or space constant of the network after the chip is fabricated by varying $V_c - V_r$. Tartagni et al. have demonstrated a current-mode centroid network [77] using subthreshold MOS devices whose operation is described by the diffusors discussed here.

### 8.4.4   Winner-Takes-All Networks

Winner-Takes-All (WTA) competition involves the allocation of a finite resource to one member of a set of competing cells. The cell with the largest input takes everything. A voltage-input WTA can be implemented using 1T current convey-

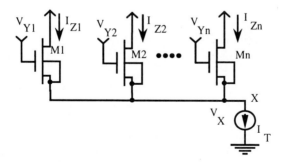

**FIGURE 8.26**
Winner-take-all with 1T current–conveyors. $N$ cells, driven by voltage inputs, compete through their communication nodes $(X)$ for the tail current $I_T$. The two-input version of this circuit is the well-known differential pair.

ors interacting with one another as shown in Fig. 8.26. Here, node $X$ follows the largest input voltage $V_{Yi}$, turning off all other current-conveyors. The tail current $I_T$ is then entirely conveyed to the output node $Z_i$ identifying the $i^{th}$ input voltage as the largest and the $i^{th}$ cell as the winner.

Current-controlled current-conveyors can also interact with each other through the communication node (Fig. 8.27) to implement MAX or WTA functions. Current inputs are supplied to their control nodes $Y$. Output currents are obtained from the output nodes $Z$, and the communication nodes $X$ are connected together. Thus, $n$ current conveyors compete for the current supplied to the common communication line. This current, $I_T$, is steered to the output of the conveyor with the largest input current; all other outputs are zero. Lazzaro et al. [78] first proposed this circuit for the WTA function and used the $Y$ node voltages as the outputs.

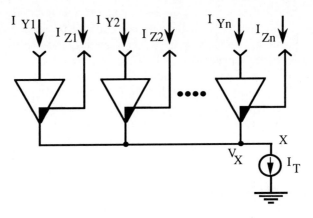

**FIGURE 8.27**
Winner-takes-all with 2T current-conveyors. $N$ current signals are applied to $N$ conveyors. The conveyor that has the largest input conveys the tail current $I_T$ supplied to the common line; the other conveyors have zero output.

The circuit works as follows. Each conveyor sees a voltage source at its communication node—not a current source. Let $\mathcal{I}(V_X)$ be the current in a device with gate voltage $V_X$. This is also the control current required to set the $X$-node voltage to $V_X$. If $I_{Y_i} < \mathcal{I}(V_X)$, then transistor $M_2$ discharges the $i^{th}$ control node and enters the linear region, turning $M_1$ off (refer to Fig. 8.19). Otherwise, node $i$ is charged up and turns $M_1$ on harder. Therefore, $M_1$ pulls $V_X$ up and sets $\mathcal{I}(V_X) = I_{Y_i}$. The end result is the node with the largest input current will set $V_X$, turn off all the other conveyors, and take all the tail current.

When the inputs are very similar, the conversion from input to output is *exponential*, with the sum of the outputs normalized to $I_T$. In this case $M_1$ remains on and $M_2$ stays in saturation. The inputs develop voltage signals across $M_2$'s drain conductance, $g_{d2}$; these voltages are converted exponentially to current by $M_1$. For example, a one percent input difference produces a voltage difference of 0.15V, so the corresponding outputs differ by a factor of 75. A detailed discussion of the dynamics for current conveyors and current-conveyor–based network circuits can be found in [19].

## 8.5    A CONTRAST SENSITIVE SILICON RETINA

In this final section, we will discuss an example of an analog VLSI neuromorphic system that was designed using the methodology described here [7]. It is a specialized "camera" for early vision applications. Several versions of this system have been fabricated. The original implementation [7] has $90 \times 92$ pixels on a $6.8 \times 6.9$ mm die in a $2\mu m$ n-well double-metal, double-poly, digital-oriented CMOS process. The same design was scaled-down and fabricated in a $1.2\mu m$ process on a $1 \sim$cm square die.

This chip employs 590,000 transistors in a 48,000 pixel array and is fully functional. Alternative approaches and design methodologies for analog VLSI vision chips can be found in the work of Harris [79], Standley [80] and De-Weerth [81]; these are summarized in [9].

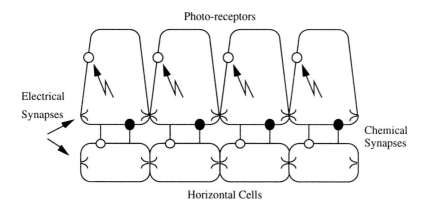

**FIGURE 8.28**
One-dimensional model of neurons and synapses in the outer-plexiform layer. Based on the red-cone system in the turtle retina.

### 8.5.1    System organization

The neuromorphic chip is modeled after neurocircuitry in the distal part of the vertebrate retina—called the outer-plexiform layer. Fig. 8.28 illustrates interactions between cells in this layer [6]. The well-known center/surround receptive field emerges from this simple structure, consisting of just two types of neurons. These neurons are non-spiking, unlike the ganglion cells in the inner retina and neurons in the rest of the nervous system. Thus, this system is well-suited to analog VLSI.

The photoreceptors are activated by light; they produce activity in the horizontal cells through excitatory chemical synapses. The horizontal cells, in turn, suppress the activity of the receptors through inhibitory chemical synapses. The receptors and horizontal cells are electrically coupled to their neighbors by electrical synapses. These allow ionic currents to flow from one cell to another, and are characterized by a certain conductance per unit area.

*Shunting inhibition* is employed to compute a normalized output that is proportional to a local measure of contrast. The horizontal cells compute the local average intensity and modulate a conductance in the cone membrane proportionately. Since the current supplied by the cone outer-segment is divided by this conductance to produce the membrane voltage, the cone's response will be proportional to the ratio between its photoinput and the local average, i.e., to contrast. This is a very simplified abstraction of the complex ion-channel dynamics involved. The advantage of performing this complex operation at the focal plane is that the dynamic range is extended (local-automatic gain control).

At any particular intensity level, however, the outer-plexiform behaves like a linear system that realizes a powerful second-order regularization algorithm for edge detection. Regularization algorithms have traditionally been employed in computer vision to deal with noisy, discretely-sampled images [8]. In the wetware implementation, regularization mitigates the inherent random variations in the neuronal characteristics as well, leading to very robust performance. The same benefits accrue to analog hardware implementations using the poor devices available in garden-variety digital VLSI CMOS technology. Thus, we see how architectures and organization for physics-based computational systems must account for the properties of the computational substrate, something that is irrelevant when implementing algorithms on general purpose, digital, computers.

## 8.5.2 Details of the Implementation

The neurocircuitry may be mapped onto silicon as shown in Fig. 8.29. Chemical synapses are modeled as a current source driving the postsynapse, controlled by a voltage in the presynapse. A MOS transistor in saturation serves this purpose. The cone outer-segment is also modeled as a controlled current source using a light-sensitive bipolar transistor in saturation. The phototransistor's emitter current is proportional to the incident light intensity. The controlled current source model results in a simple implementation but does not capture the shunting effects of the synapses. These effects are negligible in the horizontal cells because they have a relatively low input resistance due to tight coupling. However, they play an important role in the cones where they control the gain.

Electrical gap junctions are modeled by MOS diffusors. The channel is directly analogous to a porous membrane where carrier concentration is the analog of ionic species concentration. In addition to offering a compact gap junction with electronically adjustable "area," the diffusor has a dynamic range of at least three decades. The effective diffusivity is set globally by the bias voltages $V_{cc}$ and $V_{hh}$ for the cone and horizontal cell syncytia, respectively (see Fig. 8.29).

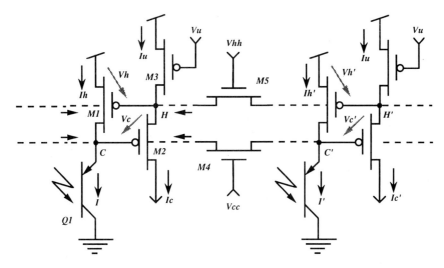

## FIGURE 8.29

One-dimensional current-mode implementation of outer-plexiform retinal processing. There are two diffusive networks coupled together by controlled current-sources. Nodes in the upper layer correspond to horizontal cells while those in the lower layer correspond to cones. Devices $M_1$ and $M_2$ model chemical synapses while $M_4$ and $M_5$ model gap junctions. $Q_1$ models the outer segment of the cone and $M_3$ models a leak in the horizontal cell membrane.

The leak, $I_u$, from the horizontal cell node counterbalances the synaptic input from the cone and is controlled by $V_u$. This current sets the operating point of the system and the average cone current ($I_c$) (output normalization).

The circuit operation is as follows. The currents $I_c$ and $I_h$ represent the responses of the cone and the horizontal cell, respectively. These signals are actually in the postsynaptic circuit—the actual presynatic signals $V_c$ and $V_h$ are not used because they encode the logarithm of the response. Increasing the photocurrent $I$ causes $V_c$ to increase, turning on $M_2$ and increasing its current $I_c$. This is excitation, i.e., an increase in the input signal results in an increase of the output signal. $I_c$ increases $V_h$, turning on $M_1$ and increasing its current $I_h$; another excitatory effect. However, $I_h$ opposes the photocurrent $I$, decreasing $V_c$, turning off $M_2$, and reducing $I_c$; this is inhibition.

We can analyze this circuit using Eq. (8.51)

$$I_h(x,y) + D_c \nabla^2 Q_c(x,y) = I(x,y)$$

$$I_u + D_h \nabla^2 Q_h(x,y) = I_c(x,y)$$

where we have used the continuous form. Assuming $Q_c \propto I_c$ and $Q_h \propto I_h$, we can eliminate $I_c$ and show that

$$\lambda \nabla^2 \nabla^2 I_h(x,y) + I_h(x,y) = I(x,y)$$

where $\lambda = (\alpha_c D_c)(\alpha_h D_h)$ and $\alpha_c, \alpha_h$ are the proportionality constants. Thus, $I_h$ is the solution to the biharmonic equation and $I_c$ is the Laplacian of the solution.

We may derive the actual relationships between the charges and the currents by finding the surface potential corresponding to the currents $I_c$ and $I_h$ in $M_2$ and $M_1$, substituting the surface potential into Eq. (8.10) to obtain their gate voltages, and then substituting these voltages into Eq. (8.53) to get the charge. This procedure is simplified by using the $\kappa$ approximation, i.e.,

$$I_c = I_{0_p} e^{\frac{(2(V_c+V_h)-V_h)}{kTq}} \; ; \quad I_h = I_{0_p} e^{\frac{1}{kTq}V_h}$$

Replacing $V_{nm}$ by $V_{dd} - V_h - V_c$ and $V_{dd} - V_h$ in Eq. (8.53) and substituting the above expressions for the exponentials, we find

$$Q_c = Q_0 \left(\frac{I_c}{I_{0_p}}\right)^{1_2} \left(\frac{I_h}{I_{0_p}}\right)^{1_{12}} \; ; \quad Q_h = Q_0 \left(\frac{I_h}{I_{0_p}}\right)^{1_1}$$

respectively. Therefore, if $\kappa_1, \kappa_2$ are close to unity and $I_h(x,y)$ varies slowly, $Q_c$ and $Q_h$ will be approximately proportional to $I_c$ and $I_h$, respectively. In that case, the network is linear and described by the biharmonic equation.

To operate in the linear regime, the surface-bulk voltage must be large enough to push $\kappa_1$ and $\kappa_2$ towards unity. This may be accomplished by tying $M_3$'s source a volt or two below $V_{dd}$. Also the diffusitivity of the horizontal cell network must be high to guarantee that $I_h(x,y)$ varies slowly. However, $I_h$ will track the overall input intensity level since $I_h(x,y) \approx \langle I(x,y)\rangle$. Hence, the proportionality constant $\alpha_c$ increases as the intensity increases, increasing $\lambda$ and resulting in bigger receptive fields.

This effect is really a manifestation of the gain-control property of the circuit which makes it only sensitive to contrast. By increasing the cone network's effective diffusivity, it shunts the input current to neighboring nodes and thereby reduces the gain. If the gain is inversely proportional to the local average intensity, the output will be proportional to contrast. Biological retinae avoid the undesirable enlargement of the receptive field by shunting the input current to ground instead of to neighboring nodes as this circuit does.

The core of the silicon retina is an array of pixels with either four-neighbor or six-neighbor connectivity (see Fig. 8.30). The wiring is included in the layout of the cell so that they may be tiled to form the retina. The two-layer architecture can be accommodated in a cell area of $80\mu m \times 94\mu m$ using the 2-micron single-poly two-metal technology. In the implementation reported in [7], a 2-micron double-poly, double-metal technology was used and the cell area was only $66\mu m \times 73\mu m$. First metal and polysilicon wires are used for interconnects; second metal is used to cover the entire array, shielding the substrate from undesirable photogenerated carriers. Support circuitry in the periphery extracts signals from the imager and interfaces with the display. Mead and Delbrück discuss the details of a synchronous communication interface that can drive a multisync video monitor [82].

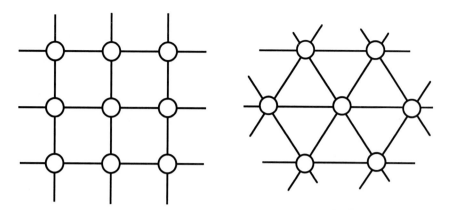

**FIGURE 8.30**
Cellular connecitvity of retina processors. (a) Square grid with four-neighbor connectivity. (b) Hexagonal grid with six-neighbor connectivity.

## 8.6   CONCLUDING REMARKS

This chapter has focused on subthreshold MOS operation. The device physics, circuit concepts, and basic mathematical models were introduced. We also discussed briefly the changes in behavior as we enter the transition and above threshold regions. Above threshold behavior is discussed extensively in earlier chapters of this book. One should be aware of the inferior transconductance characteristics and the deviation from the exact exponential characteristics in both ohmic and saturation regions outside the subthreshold region.

Neuromorphic computation with time-varying functions, was not discussed in this chapter. In the examples presented here, we ignored the dynamics introduced by the node capacitances. As related to analog VLSI vision chips, time-domain processing using minimal complexity focal plane circuits was reported in [87, 88, 89, 90]. Other researchers have used time in pulse signaling schemes for communication and for implementing spiking neurons [83, 11, 84, 85, 86].

Over the last ten years, advances in process technology and CAD tools have enabled the realization of highly complex digital VLSI logic systems for fast, general-purpose, high-precision calculations on symbolic information [91]. In this chapter, we attempt to leverage these advances to synthesize truly complex low-power, special-purpose, massively-parallel systems for *perceiving* information from the natural environment in real-time.

Nature has long discovered ways of solving efficiently and effectively hard problems in sensory perception. Perhaps, principles of organization and function can be abstracted from our understanding of biological information processing systems [4]. Other researchers with long and distinguished careers in microelectronics research related to medicine and biology share the same belief:

...... sensors hold the key to better understanding of biological/cellular systems, which may contribute to long term fundamental improvements in the electronic art ......

## PROBLEMS

**8.1.** Outline a procedure to determine the parameter $\kappa$ and its dependence on the surface potential (parameter $\gamma$). (*Hint:* When the device is operating under the conditions where $Q_s \gg Q_d$, and the current in the device is fixed by an external constant current source, the device current is determined only by the charge at the source.)

**8.2.** A CM *squareroot function* is a useful compressive non-linearity in CM designs. Modify the circuit in Fig. 8.15 to obtain a normalized square root function. Using the translinear principle verify that indeed this is the correct function. (*Hint:* The minimum design involves only four transistors.)

**8.3.** A CM *square function* is also a useful expansive non-linearity in CM designs. Modify the circuit in Fig. 8.15 to obtain a normalized square function. Using the translinear principle verify that indeed this is the correct function. (*Hint:* The minimum design involves only four transistors.)

**8.4.** The absolute value circuit in Fig. 8.16 has an asymmetry in the way the output signals are obtained. Suggest a method to remedy this. (*Hint:* Add one transistor.)

**8.5.** Design a circuit which produces an output current that represents the Euclidean distance of two input currents. This is defined as $I_{out} = \sqrt{I_1^2 + I_2^2}$ and is a useful metric in neuromorphic information processing systems. The minimum design involves only five transistors. (*Hint:* Take a careful look at the absolute-value circuit in Fig. 8.16.)

**8.6.** When the bidirectional junction is switched off, and $V_{n1} > V_{n2}$, transistor $M_1$ is shut-off but $M_2$ sources some amount of current $I_{off}$. Give an equation for $I_{off}$ and determine its value if we know that the currents communicated on the line range between $100pA$ and $100nA$.

**8.7.** Redesign the silicon retina in Chapter 15 of [4], using current-mode circuits. (*Hint: Use diffusors in a diffusive network for local aggregation.*)

**8.8.** Use the partial differential equations that describe the state of the two networks (one dimensional) to solve analytically for the current $I_h(x)$ when $I(x) = \delta(x)$ (a delta function spatial excitation). Plot the solution and verify that it has the characteristic on-center/off-surround shape.

**8.9.** (Project) Do a layout for a contrast sensitive silicon retina pixel in a single-poly, double-metal, 2 micron process. Compare your cell area with design that employs two polysilicon layers (cell area $66 \times 73$ micron. All transistors in the cell should be $6 \times 6$ microns and all switching devices can be minimum size. Use a hexagonal tesselation with six neighbours per cell. The phototransistor should have the minimum size permitted by the design rules. The spatial response of the contrast sensitive silicon retina discussed in this chapter is dependent on the absolute illumination level. This is not a desirable feature. Redesign the circuit so that the receptive fields do not depend on the absolute illumination.

**8.10.** (Project) The temporal response of the centroid computation system in Reference [81] depends on the absolute value of the illumination. That is, when the light is dim, the system is slow, and when the light is bright, the system is fast! Redesign the system so that the centroid computation is invariant to absolute level changes in the illumination. (*Hint: Do some kind of global normalization*

*before the centroid computation.*)
**8.11.** (Project) Design and layout a memory cell for a Hamming distance classifier. Much like a standard RAM cell, it will be necessary to store vector (word) bits in flip-flops. The word address can be selected using word lines. To save space, you will need to design the cell so that two bit lines can be used to both program the cell and encode matches/mismatches as currents. Do the floor planning for a chip. The system will do analog computation in subthreshold/transition region but the I/O will be digital. Outline procedures and safeguards so that this memory intensive system can survive catastrophic faults in a row or column and in the analog Winner Takes All circuit (MAX select circuitry).

# REFERENCES

[1] J. D. Cowan and D. H. Sharp, "Neural Nets," *Quarterly Reviews of Biophysics,* vol. 21, no. 3, pp. 365–427, 1988.
[2] C. A. Mead, "Neuromorphic electronic systems," *Proceedings IEEE,* vol. 78, no. 10, pp. 1629–1636, October 1990.
[3] S. Haykin, "Adaptive Filter Theory," Prentice-Hall, Inc. Englewood Cliffs, NJ, 1991.
[4] C. A. Mead, *Analog VLSI and Neural Systems,* Reading, MA, Addison-Wesley, 1989.
[5] A. Rosenfeld, "Computer vision: basic principles," *Proceedings IEEE,* vol. 76, no. 8, pp. 857–1056, August 1988.
[6] J. E. Dowling, "The retina: an approachable part of the brain," The Belknap Press of Harvard University, Cambridge, MA, 1987.
[7] K. A. Boahen and A. G. Andreou, "A contrast sensitive silicon retina with reciprocal synapses," *Advances in Neural Information Processing Systems 4,* Moody, J. E., Hanson, S. J. and Lippmann, R. P. (eds.), Morgan Kaufmann Publishers, San Mateo, CA, 1992.
[8] T. Poggio, V. Torre, and C. Koch, "Computational vision and regularization theory," *Nature,* vol. 317, pp. 314–319, 1985.
[9] C. Koch, "Seeing Chips: analog VLSI circuits for computer vision," *Neural Computation,* vol. 1, no. 2, pp. 184–200, 1989.
[10] R. F. Lyon, "Analog VLSI hearing systems," in *VLSI Signal Processing III,* Brodersen and Moscovitz (eds.), IEEE Press, 1988.
[11] J. Lazzaro, "A silicon model of an auditory neural representation of spectral shape," *IEEE J. Solid-State Circuits,* vol. SC-26, pp. 772–777 May 1991.
[12] W. Liu, A. G. Andreou, and M. H. Goldstein, "Voiced-speech representation by an analog silicon model of the auditory periphery," *IEEE Transactions on Neural Networks,* vol. 3, no. 3, May 1992.
[13] T. Kohonen, *Associative Memory: A System Theoretic Approach,* Springer Verlag, Berlin, Heidelberg, New York, 1977.
[14] T. Kohonen, *Self-Organization and Associative Memory,* Springer Verlag, (2nd edition), Berlin, Heidelberg New York, 1988.
[15] B. E. Boser, E. Säckinger, J. Bromley, Y. Le Cun, and L. Jackel, "An analog neural network processor with programmable topology," *IEEE J. Solid-State Circuits,* vol. SC-26, pp. 2017–2025, November 1991.
[16] E. A. Vittoz and X. Arreguit, "CMOS Integration of Herault-Jutten Cells for Separation of Sources," Ch. 3 in Analog VLSI Implementation of Neural Systems, C. Mead and M. Ismail (Eds.), Kluwer Academic Publisher, 1989.
[17] M. Cohen and A. G. Andreou," Current-mode subthreshold MOS implementation of the Herault–Jutten autoadaptive network," *IEEE J. Solid-State Circuits,* vol. 27, no. 5, pp. 714–727, May 1992.

[18] K. A. Boahen, P. O. Pouliquen, A. G. Andreou, and R. E. Jenkins, "A heteroassociative memory using current–mode MOS analog VLSI circuits," *IEEE Trans. Circuits and Systems,* vol. CAS-36, no. 5, pp. 747–755, May 1989.

[19] K. A. Boahen and A. G. Andreou, "Design of a bidirectional associative memory chip," Associative Neural Memories: Theory and Implementation, edited by M. Hassoun, Chapter 17, Oxford University Press, New York, 1993.

[20] P. O. Pouliquen, K. Strohbehn, A. G. Andreou, and R. E. Jenkins, "A Haming Distance Classifier based on a Winner-Takes-All Circuit," Proceedings of the 36th Midwest Symposium on Circuits and Systems, Detroit, MI, August 1993.

[21] Y. He, U. Cilingiroglu, and E. Sanchez-Sinencio, "A high density and low-power charge-based Hamming network," *IEEE Transactions on Very Large Scale Integration (VLSI) Systems,* vol. 1, no. 1, pp. 56–62, March 1993.

[22] C.-K. Sin, A. Kramer, V. Hu, R. R. Chu, and P. K. Ko, "EEPROM as an analog storage device, with particular application in neural networks," *IEEE Trans. Electron Devices,* vol. ED-39, no 6, pp. 1410–1419, June 1992.

[23] C. A. Mead and M. Ismail (eds.), *Analog VLSI Implementation of Neural Systems,* Norwell, MA, Kluwer Academic Publishers, 1989.

[24] U. Ramacher and U. Rückert (eds.), *VLSI Design of Neural Networks,* Kluwer Academic Publishers, Boston, MA, 1991.

[25] E. Sánchez Sinencio and C. Lau (eds.), *Artificial Neural Networks: Paradigms, Applications, and Hardware Implementations,* New York, NY, IEEE Press, 1992.

[26] C. Toumazou, F. J. Lidgey, and D. G. Haigh (eds.), "Analogue IC Design: The Current–Mode Approach," *IEE Circuits and Systems Series,* vol. 2, Peter Peregrinus Ltd, London, 1990.

[27] A. G. Andreou and K. A. Boahen, "Synthetic neural circuits using current domain–signal representations," *Neural Computation,* vol. 1, pp. 489–501, 1989.

[28] A. G. Andreou, "Synthetic neural systems using current–mode circuits," Proceedings of the IEEE 1990 International Symposium on Circuits and Systems, New Orleans, pp. 2428–2432, May 1990.

[29] A. G. Andreou, K. A. Boahen, P. O. Pouliquen, A. Pavasović, R. E. Jenkins, and K. Strohbehn, "Current–mode subthreshold MOS circuits for analog VLSI neural systems," *IEEE Trans. on Neural Networks,* vol. 2, no. 2, pp. 205–213, March 1991.

[30] E. A. Vittoz and J. Fellrath, "CMOS analog integrated circuits based on weak inversion operation," *IEEE J. of Solid-State Circuits,* vol. SC-12, no. 3, pp. 224–231, June 1977.

[31] E. A. Vittoz, "Micropower techniques," in *VLSI Circuits for Telecommunications,* edited by Y. P. Tsividis and P. Antognetti, Prentice Hall, 1985.

[32] E. A. Vittoz, "The design of high-performance analog circuits on digital CMOS chips," *IEEE J. of Solid-State Circuits,* vol. SC-20, no. 3, pp. 657–665, June 1985.

[33] Y. P. Tsividis, "Analog MOS integrated circuits–certain new ideas, trends and obstacles," *IEEE J. Solid-State Circuits,* vol. SC-22, pp. 317–321, June 1987.

[34] "Microsensors," edited by R. S. Muller, R. T. Howe, S. D. Senturia, R. L. Smith, and R. M. White, IEEE Press, 1991.

[35] T-Spice Program, Tanner Research, Inc., 180 North Vinedo Avenue, Pasadena, CA 91107, USA.

[36] Y. Tsividis, "Operation and Modeling of the MOS Transistor," McGraw-Hill, New York, NY, 1987.

[37] E. S. Yang, "Microelectronic Devices," McGraw-Hill, New York, NY, 1988.

[38] S. M. Sze, "Physics of Semiconductor Devices," John Wiley and Sons, New York, NY, 1969 (First Edition).

[39] W. Fichtner and H. W. Pötzl, "MOS modelling by analytical approximations," *Intl. J. Electronics,* vol. 46, no. 1, 1979.

[40] K. A. Boahen, A. G. Andreou, and C. A. Mead, "A generalization of the translinear principle to ohmic region of MOS transistor operation," to appear in *Electronics Letters,* 1993.

[41] J. E. Meyer, "MOS Models and Circuit Simulation," RCA Review, vol. 32, pp. 42–63, March 1971.

[42] M. Holler, S. Tam, H. Castro, and R. Benson, "An electrically trainable artificial neural network ETANN with 10420 'floating gate' synapses," Proc. Intl. Joint Conference on Neural Networks, Washington D.C., vol. II, pp. 191–196, June 1989.

[43] T. H. Borgstrom, M. Ismail, and S. B. Bibyk, "Programmable current–mode neural network for implementation in analog MOS VLSI," IEE Proceedings, vol. 137, pt. G, no. 2, pp. 175–184, April 1990.

[44] E. Vittoz, H. Oguey, M. A. Maher, O. Nys, E. Dijkstra, and M. Chevroulet, "Analog storage of adjustable synaptic weights," VLSI Design of Neural Networks, U. Ramacher and U. Rückert (eds.), Kluwer Academic Publishers, Boston, MA, 1991.

[45] E. Sackinger and W. Guggenbuhl, "An Analog Trimming Circuit based on a Floating Gate Device," *IEEE J. of Solid-State Circuits,* vol. SC-23, no. 6, pp. 1437–1440, December 1988.

[46] C. A. Mead, "Adaptive retina," in *Analog VLSI Implementation of Neural Systems,* C. A. Mead and M. Ismail (eds.), Kluwer Academic Publishers, Boston, MA, pp. 239–246, 1989.

[47] L. R. Carley, "Trimming Analog Circuits Using Floating-Gate Analog Memory," *IEEE J. of Solid-State Circuits,* vol. SC-24, pp. 1569–1575, December 1989.

[48] K. Yang and A. G. Andreou, "Multiple Input Floating-Gate MOS Differential Amplifier and Applications for Analog Computation," Proceedings of the 36th Midwest Symposium on Circuits and Systems, Detroit, MI, August 1993.

[49] L. Glasser, "A UV write-enabled PROM," Proceedings of the 1985 Chapel Hill Conference on Very Large Scale Integration, pp. 61–65, Computer Science Press, 1985.

[50] D. A. Kerns, J. E. Tanner, M. A. Sivilotti, and J. Luo, "CMOS UV-writable non-volatile analog storage," Proceedings of Advanced Research in VLSI: International Conference 1991, Santa Cruz CA, MIT Press.

[51] G. Cauwenberghs, C. F. Neugebauer, and A. Yariv, "Analysis and verification of an analog VLSI incremental outer-product learning systems," *IEEE Transactions on Neural Networks,* vol.3, no. 3, May 1992.

[52] A. Thomsen and M. A. Brooke, "A Floating-Gate MOSFET with Tunneling Injector Fabricated Using a Standard Double-Polysilicon CMOS Process," *IEEE Electron Device Letters,* vol. EDL-12, no. 12, pp.111–113, March 1991.

[53] H. Yang, B. J. Sheu, and J.-C. Lee, "A nonvolatile analog neural memory using floating-gate MOS transistors," *Analog Integrated Circuits and Signal Processing,* vol. 2, no. 1, February 1992.

[54] K. Yang and A. G. Andreou, "Subthreshold Analysis of Floating Gate MOSFET's," Proceedings of the 10th Biennial University Government Industry Microelectronics Symposium, Research Triangle, NC, pp. 141–144, May 1993.

[55] Y. Horio and S. Nakamura, "Analog memories for VLSI neurocomputing," in *Artificial Neural Networks: Paradigms, Applications, and Hardware Implementations,* New York, NY, IEEE Press, pp. 344–366, 1992.

[56] B. Gilbert, "Translinear circuits: A proposed classification," *Electronics Letters,* vol. 11, No. 1, pp. 14–16, 1975.

[57] J. J. Ebers and J. L. Moll, "Large-signal behavior of junction transistors," *Proceedings IRE,* vol. 42, pp. 1761–1772, December 1954.

[58] R. S. Muller and T. I. Kamins, "Device electronics for integrated circuits," New York, John Wiley and Sons, 1977.

[59] E. A. Vittoz, "MOS transistors operated in the lateral bipolar mode and their application in CMOS technology," *IEEE J. of Solid-State Circuits,* vol. SC-18, no. 3, pp. 273–279, June 1983.

[60] X. Arreguit, "Compatible lateral bipolar transistors in CMOS technology: model and applications," D.Sc. thesis, no. 817, EPFL, Lausanne, 1989.

[61] Y. Tsividis and R. Ulmer, "A CMOS Voltage Reference," *IEEE J. of Solid-State Circuits,* vol. SC-13, pp. 774–778, December 1978.

[62] A. Pavasović, "Subthreshold Region MOSFET Mismatch Analysis and Modeling for Analog VLSI Systems," Ph.D. dissertation, The Johns Hopkins University, Baltimore, 1990.

[63] A. Pavasović, A. G. Andreou, and C. R. Westgate, "Characterization of CMOS process variations by measuring subthreshold current," *Non-Destructive Characterization of Materials,* vol. IV, C. O. Ruud and R. E. Green (eds.), Plenum Press, 1991.

[64] A. Pavasović, A. G. Andreou, and C. R. Westgate, "Characterization of subthreshold MOS mismatch in transistors for VLSI systems," *Journal of VLSI Signal Processing,* also *Analog Integrated Circuits and Signal Processing,* 1994.

[65] E. Seevinck, "Analysis and synthesis of translinear integrated circuits," Studies in Electrical and Electronic Engineering 31, Elsevier, Amsterdam, 1988.

[66] A. G. Andreou and W. Liu, "BiCMOS circuits for silicon cochleas," Proceedings of the 1993 European Conference on Circuit Theory and Design, Davos, Switzerland, September 1993.

[67] T. Delbrück, " "Bump" circuits for computing similarity and dissimilarity of analog voltages," Proc. Intl. Joint Conf. on Neural Networks, Seattle WA, pp. 475–479, 1991.

[68] M. Cohen and A. G. Andreou," MOS circuit for nonlinear Hebbian learning," *Electronics Letters,* vol. 28, no. 9, pp. 809–810, April 1992.

[69] B. Gilbert, "A Monolithic 16-Channel Analog Array Normalizer," *IEEE J. of Solid-State Circuits,* vol. SC-19, no. 6, 1984.

[70] K. C. Smith and A. S. Sedra, "The Current Conveyor—a new circuit building block," *IEEE Proc.,* vol. 56, pp. 1368–1369, August 1968.

[71] K. A. Boahen, A. G. Andreou, P. O. Pouliquen, and A. Pavasovic̀, "Architectures for associative memories using current-mode analog MOS circuits," Proceedings of the Decennial Caltech Conference on VLSI, C. Seitz (ed.), MIT Press, 1989.

[72] E. Säckinger and W. Guggenbühl, "A High–Swing, High Impedance MOS Cascode Circuit," *IEEE J. Solid-State Circuits,* vol. 25, no. 1, pp. 289–298, February 1990.

[73] G. R. Wilson, "A Monolothic Junction FET–npn Operational Amplifier," *IEEE J. Solid-State Circuits,* vol. SC-3, no. 4, pp. 341–348, December 1968.

[74] B. L. Hart, "Translinear circuit principle: a reformulation," *Electronics Letters,* vol. 15, no. 24, pp. 801–803, November 22, 1979.

[75] E. Vittoz and X. Arreguit, "Linear networks based on transistors," *Electronics Letters,* vol. 29, pp. 297–299, February 4, 1993.

[76] K. Bult and G. J. G. M Geelen, "An inherently linear and compact MOST-only current division technique," *IEEE J. Solid-State Circuit,* vol. SC-27, no.12, pp. 1730–1735, December 1992.

[77] M. Tartagni and P. Perona, "Computing centroids in current-mode technique," *Electronics Letters,* vol. 21, October 1993.

[78] J. Lazzaro et al., "Winner-Take-All networks of O(n) complexity," in *Advances in Neural Information Processing Systems,* vol. 1, pp. 703–711, D. S. Touretzky (ed.), Morgan Kaufmann, San Mateo, CA, 1988.

[79] J. Harris, "Analog Models of Early Vision," Ph.D. dissertation, Caltech, Pasadena, CA , 1991.

[80] D. L. Standley, "Analog VLSI implementation of smart vision sensors: stability theory and an experimental design," Ph.D. dissertation, MIT, Cambridge, MA, 1991.

[81] S. P. DeWeerth, "Analog VLSI circuits for stimulus localization and centroid computation," *International Journal of Computer Vision,* vol. 8, no. 2, pp. 191–202, 1992.

[82] C. A. Mead and T. Delbrück, "Scanners for visualizing activity in analog VLSI circuitry," *Analog Integrated Circuits and Signal Processing,* vol. 1, no. 2, October 1991.

[83] N. El-Leithy, M. E. Zaghloul, and R. W. Newcomb, "A basic MOS neural-type junction; A perspective on neural-type microsystems," IEEE First International Conference on Neural Networks, San-Diego, CA , pp. III-469–III-477, June 1987.

[84] B. Linares-Barranco, E. Sanchez-Sinencio, A. Rodriguez-Vazquez, and J. Huertas, "A CMOS implementation of FitzHugh-Nagumo neuron model," *IEEE J. Solid-State Circuits,* vol. SC-26, pp. 956–965, July 1991.

[85] A. F. Murray and A. V. W. Smith, "Asynchronous arithmetic for VLSI neural systems," *IEEE Electronics Letters,* vol. 23, no. 12, June 1987.

[86] J. L. Meador, A. Wu, C. Cole, N. Nintunze, and P. Chintrakulchai, "Programmable impulse neural circuits," *IEEE Trans. on Neural Networks,* vol. 2, no. 1, pp. 101–109, January 1991.

[87] T. Delbrück and C. A. Mead, "Silicon adaptive photoreceptor array that computes temporal intensity derivatives," in Proc. SPIE 1541, vol. 1541-12, pp. 92–99, San Diego, CA, July 1991.

[88] R. G. Benson and T. Delbrück, "Direction sensitive silicon retina that uses null inhibition," *Advances in Neural Information Processing Systems 4,* J. E. Moody, S. J. Hanson, and R. P. Lippmann (eds.), Morgan Kaufmann Publishers, San Mateo, CA, 1992.

[89] C. P. Chong, C. A. T. Salama, and K. Smith, "A sub-nanoampere current differentiator," Proceedings of the IEEE 1991 International Symposium on Circuits and Systems, Singapore, pp. 2593–2596, June 1991.

[90] A. G. Andreou, K. Strohbehn, and R. Jenkins, "A silicon retina for motion computation," Proceedings of the IEEE 1991 International Symposium on Circuits and Systems, Singapore, pp. 1373–1376, June 1991.

[91] C. A. Mead and L. Conway, *Introduction to VLSI Systems,* Reading, MA, Addison-Wesley, 1980.

# CHAPTER
# 9

## SAMPLED-DATA ANALOG FILTERS

### 9.1  INTRODUCTION

Sampled-data analog filters are discrete-time systems that operate on periodic samples of the input signal and provide an output signal as a sequence of samples. They are analog systems because the magnitudes of the signals are continuous; unlike digital systems where the magnitude also is discretized or digitized.

Central to the operation of a discrete-time filter is a clock, and the analog filters studied in this chapter are no exception. They all require a two-phase non-overlapping clock such as that whose waveform is shown in Fig. 9.1.

At the the present time there are two classes of discrete-time analog filters that are suitable for implementation in integrated-circuit form. These are the switched-capacitor (SC) filter and the switched-current (SI) filter. Switched-capacitor filters have been in large-volume commercial use since the late 1970's [1, 2] and currently represent a mature filter design technology. Switched-current filters, on the other hand, are only a few years old and are still in the research stage [3, 4, 5, 6].

The switched-capacitor filter technique is based on the idea that a capacitor that is periodically switched can be arranged to cause packets of charge to be transferred between two circuit nodes. If the switching frequency is much greater than the frequency of the signals at the circuit nodes, the charge transfer process appears to be almost continuous. The switched capacitor then gives the effect of

414

a resistor that connects the two circuit nodes. It follows that an active-RC filter can be converted to a switched-capacitor filter by simply replacing each resistor with a periodically switched capacitor. Thus SC filters are composed of active elements (op amps), capacitors, and analog switches. Since all three types of components are realizable in MOS IC technology, SC filters can be implemented in any of the MOS technologies, with CMOS being the technology of choice.

While SC filters process samples of the signal *voltage*, switched-current (SI) filters operate on the signal in *current* form as discussed in Chapter 6, for SI filters the active element is not an op amp but rather a current mirror. Although charge storage is an essential operation for both filter types, as it provides the means of realizing signal delay, it is performed differently in the two technologies. In SC filters, charge is transferred from one capacitor to another under the influence of the clock with the result that the time-scale of the circuit is determined by the ratio of two capacitances as well as the frequency of the clock. By contrast, in SI filters the input current causes charge to be stored on the input capacitance of a current mirror. This charge in turn produces an output current of the mirror. It can be seen that the signal transfer process is governed by the current transfer ratio of the mirror and, of course, the clock frequency and is independent of the exact value of the charge-storage capacitor.

This chapter provides an introduction to the subject of sampled-data analog filter design. Specifically, we shall study methods for the synthesis of SC and SI filter circuits. This will be done by considering the SC case in detail and then presenting a method for converting SC circuits to SI circuits.

## 9.2 FIRST-ORDER SC CIRCUITS

In this section we shall study the basic building blocks of SC filters. These are circuits that perform a form of time-integration and have first-order transfer functions.

### 9.2.1 A Pair of Stray-Insensitive Integrators

Figure 9.2 shows a pair of circuits that have become the standard building blocks of SC filters [7, 8]. The operation of these circuits can be analyzed in the time domain to obtain a first-order difference equation that represents the operation of each circuit over one clock period. For the circuit in Fig. 9.2(a) we obtain:

$$v_O(n) = v_O(n-1) - \frac{C_1}{C_2}v_I(n) \qquad (9.1)$$

where $v_I(n)$ and $v_O(n)$ denote the values of the input and output voltages at the end of clock phase, $\phi_2$ (refer to the clock waveform in Fig. 9.1), and $v_O(n-1)$ denotes the value of $v_O$, a clock period earlier. To evaluate the frequency response of this circuit we apply the $z$ transformation to Eq. (9.1), thus obtaining

$$V_o(z) = z^{-1}V_o(z) - \frac{C_1}{C_2}V_i(z) \qquad (9.2)$$

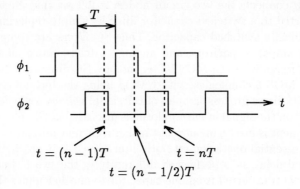

**FIGURE 9.1**
The two-phase nonoverlapping clock waveform used in SC filters.

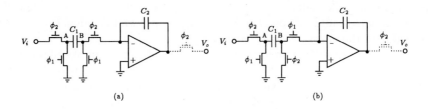

(a)              (b)

**FIGURE 9.2**
Pair of stray-insensitive SC integrator circuits: (a) inverting integrator (b) non-inverting integrator.

which can be rearranged to obtain the z-domain transfer function

$$\frac{V_o(z)}{V_i(z)} = -\frac{C_1}{C_2}\frac{1}{1-z^{-1}} \tag{9.3}$$

which can be written in the alternate form

$$\frac{V_o(z)}{V_i(z)} = -\frac{C_1}{C_2}\frac{z^{12}}{z^{12}-z^{-12}} \tag{9.4}$$

To obtain the response for physical frequencies $\omega$ we substitute $z = e^{jT}$, where $T$ is the clock period. This corresponds to evaluating the transfer function in Eq. (9.4) along the unit circle in the complex $z$ plane (see Fig. 9.3). The result is:

$$\frac{V_o}{V_i}(\omega) = -\frac{C_1}{C_2}\frac{e^{jT2}}{j2\sin(\omega T/2)} \tag{9.5}$$

Now if the clock frequency $\omega_c = 2\pi/T$ is much greater than the signal frequency $\omega$ then $\omega T << 1$ and $\sin(\omega T/2) \approx \omega T/2$ with the result that the transfer function in Eq. (9.5) can be approximated as

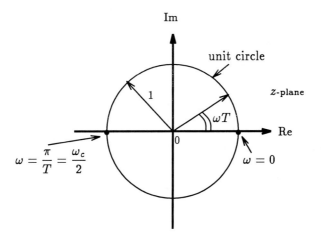

**FIGURE 9.3**
The frequency response of discrete-time circuits is found by evaluating the transfer function along the unit circle in the $z$ plane.

$$\frac{V_o}{V_i}(\omega) \approx -\frac{C_1}{C_2} \frac{e^{jT2}}{j\omega T} \tag{9.6}$$

Apart from the excess phase lead represented by the numerator term $e^{jT2}$, Eq. (9.6) is that of an inverting integrator having a time constant $T(C_2/C_1)$. Thus the circuit of Fig. 9.2(a) performs an approximate integration function and is known as an *inverting integrator*. The approximation involved is twofold: there is an excess phase lead, and the transfer function is inversely proportional to $\sin(\omega T/2)$ rather than to $\omega$. It will be shown that if we arrange things properly the excess phase lead can be rendered immaterial. The second approximation, however, can result in significant errors unless it is either taken into account in the synthesis process (in what is known as *exact* design) or the filter is clocked at a frequency much higher than the signal frequency (so that $\omega T \ll 1$).

A similar analysis can be performed on the circuit of Fig. 9.2(b) and results in the transfer function

$$\frac{V_o(z)}{V_i(z)} = +\frac{C_1}{C_2} \frac{z^{-12}}{z^{12} - z^{-12}} \tag{9.7}$$

which for physical frequencies, $z = e^{jT}$, yields

$$\frac{V_o}{V_i}(\omega) = +\frac{C_1}{C_2} \frac{e^{-jT2}}{j2 \sin(\omega T/2)} \tag{9.8}$$

This transfer function is similar to that in Eqs. (9.4) and (9.5) except that here the excess phase is lag and the transfer function has a positive sign. Thus the circuit of Fig. 9.2(b) performs an approximate *non-inverting integration*.

It will be seen in subsequent sections that SC filters of order 2 and greater are formed of feedback loops each containing one inverting and one non-inverting integrator. It follows that in each such *two-integrator loop* the excess phase lead ($z^{12}$ or $e^{jT2}$) of the inverting integrator cancels the excess phase lag ($z^{-12}$ or $e^{-jT2}$) of the non-inverting integrator, with the result that the excess phase around the loop is zero. This phase cancellation occurs provided that the clock phasing is that shown in Fig. 9.2. Observe that this clock phasing is consistent in the sense that the input to each integrator is provided during $\phi_2$ and the output of each integrator also is sampled during $\phi_2$. The cancellation of the integrators' excess-phase in the two-integrator loop is the reason we shall not concern ourselves here any further with the numerator terms ($z^{\pm12}$) in Eqs. (9.4) and (9.7).

It can be shown that the operation of the integrator circuits of Fig. 9.2 is not affected by the values of the parasitic capacitances present between the drain and source diffusions of the MOS switches and the substrate (signal ground); that is, between the nodes labeled A and B and ground. This stray insensitivity property is a crucial one for the practical viability of these circuits; it means that one can design the circuits with very small capacitances $C_1$ and $C_2$ while still obtaining precise frequency response. Here it should be noted that the precision of the filter response depends on the precision to which the time constants of the integrators are realized. The integrator time constant, in turn, depends on two things: the clock frequency and the capacitor ratio. Thus response precision is determined by the precision to which capacitor ratios can be realized rather than on the accuracy of the values of individual capacitors; another reason why SC filters are ideally suited for IC implementation.

### 9.2.2   The Lossless Digital Integrator (LDI) Frequency Variable

In the above we have shown that the circuits of Fig. 9.2 perform an *approximate* integration function in terms of the physical frequency $\omega$. They function as ideal integrators, however, in terms of a new complex-frequency variable $\gamma$ which is defined as

$$\gamma = \frac{1}{2}(z^{12} - z^{-12}) \tag{9.9}$$

To see how this comes about, we replace ($z^{12} - z^{-12}$) in Eqs. (9.4) and (9.7) by $2\gamma$ and thus express the transfer function of the circuit in Fig. 9.2(a) as

$$\frac{V_o}{V_i} = -\frac{C_1}{2C_2}\frac{z^{12}}{\gamma} \tag{9.10}$$

and the transfer function of the circuit in Fig. 9.2(b) as

$$\frac{V_o}{V_i} = +\frac{C_1}{2C_2}\frac{z^{-12}}{\gamma} \qquad (9.11)$$

Now if we ignore the excess phase factors $(z^{\pm 12})$ we see that the transfer functions in Eqs. (9.10) and (9.11) are those of ideal integrators in the $\gamma$ plane. Since $\gamma$ is known as the lossless digital integrator (LDI) variable [9], the two circuits of Fig. 9.2 are known as LDI integrators. Finally, it is worthwhile to note that physical frequencies $\omega$ map to locations on the imaginary axis of the $\gamma$ plane according to

$$\text{Im}(\gamma) = \sin(\omega T/2) \qquad (9.12)$$

Observe that $\omega = 0$ maps to $\gamma = 0$ and $\omega = \pi/T$ (half the sampling frequency) maps to $\text{Im}(\gamma) = 1$; thus the useful frequency band of interest maps onto the range $0 \leq \text{Im}(\gamma) \leq 1$.

### 9.2.3 Settling Time of the Integrator Circuits

Throughout this chapter the synthesis of SC filter circuits is performed assuming ideal op amps. Nevertheless, the designer must consider the nonideal effects of op amps, namely the effects of the op amp finite gain and bandwidth on the operation of SC filters. For space limitations, we shall not deal with this topic [10] here except for making frequent observations on the relationship between clock phasing and the effect of the op amp dynamics (i.e., finite bandwidth or settling time). Taking these observations into account in SC filter design can result in circuits whose operation depends minimally on the op amp settling characteristics [11, 12].

Consider the inverting LDI integrator of Fig. 9.2(a). We see that during clock phase $\phi_2$, charge is being transferred from the input capacitor $C_1$ to the op amp feedback capacitor $C_2$, thus the output voltage of the op amp will be changing during $\phi_2$. Unfortunately, the output voltage is also read (or sampled) during $\phi_2$. The situation for the non-inverting LDI integrator of Fig. 9.2(b) is quite different. Here, charge is transferred from $C_1$ to $C_2$ during $\phi_1$ and thus the output voltage is not read until the next clock phase $\phi_2$. In other words, the op amp in the non-inverting integrator is allowed an additional half-clock cycle for its output to settle. We, therefore, expect the non-inverting integrator circuit of Fig. 9.2(b) to be less dependent on the op amp finite bandwidth and settling characteristics than the inverting integrator circuit of Fig. 9.2(a). This has been verified by detailed analysis [10].

### 9.2.4 Differential Integrator Circuits

In the integrator circuits of Fig. 9.2 signals are single-ended, that is referenced to ground. Although these single-ended circuits are economical in their use of capacitors they suffer from some practical nonideal effects. Chief among these

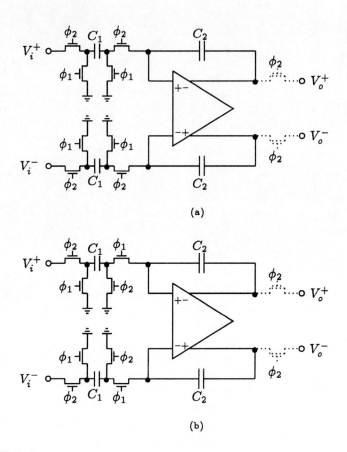

**FIGURE 9.4**

Differential form of the standard SC integrator circuits: (a) inverting integrator (b) non-inverting integrator.

problems are amplifier offset voltage and clock feedthrough. Briefly, when the MOS switches turn off, the inversion charge under their gate is injected into the circuit. This charge injection can cause output offset voltages, often on the order of 100 mV. Because clocks switch to the power supply, the effect can also increase the dependence of the output voltage on the power supply; that is, reduce the power-supply rejection. Further, when the injected charge depends on the signal, transfer function errors and non-linear distortion result [13, 14].

Although various schemes have been proposed to reduce the effect of charge injection and clock feedthrough on the operation of SC filters, the most effective approach involves the use of differential circuits [15]. Indeed, this is now a standard strategy and it reduces the influence of a wide range of other nonidealities including noise. In differential circuits, signals are represented by the difference between voltages in two symmetrical circuits and thus any common-mode disturbance has no effect. Figure 9.4 shows the differential form of the inverting and non-inverting integrators of Fig. 9.2. The differential circuits utilize a differential-in/differential-out op amp with $V_o^+ = -V_o^-$ and $V_o^+ - V_o^- = A(V^+ - V^-)$ where $A$ is the open-loop gain. The transfer function of each differential integrator is defined as

$$T(s) \equiv \frac{V_o^+ - V_o^-}{V_i^+ - V_i^-} \tag{9.13}$$

and it can be easily shown that their transfer functions are identical to those of their single-ended counterparts, given by Eqs. (9.4) and (9.7). Note, however, that here we have an additional degree of freedom: By simply interchanging the input or output terminals the polarity of the transfer function is reversed. This makes possible new and better filter circuits as will be demonstrated in later sections.

## 9.3  BILINEAR TRANSFORMATION

The area of analog continuous-time filters is a mature one with a wealth of design methods and design aids. It would therefore be extremely useful if one were able to apply these methods to the discrete-time case. The link that makes this possible is the bilinear transformation [16, 17]

$$\lambda = \frac{z - 1}{z + 1} \tag{9.14}$$

where $\lambda$ is a new complex-frequency variable called the bilinear transform variable.

To see how the bilinear transformation enables one to apply directly continuous-time design methods and data to the discrete-time case we find the mapping of physical frequencies $\omega$ from the $z$ plane to the $\lambda$ plane. Toward that end we write Eq. (9.14) as

$$\lambda = \frac{z^{12} - z^{-12}}{z^{12} + z^{-12}} \tag{9.15}$$

and then substitute $z = e^{jT}$ to obtain

$$\lambda = j \tan(\omega T/2) \tag{9.16}$$

Thus the unit circle of the $z$ plane, which is the contour of physical-frequency points, maps onto the entire imaginary axis of the $\lambda$ plane. Equation (9.16) indicates that $\omega = 0$ maps to $\mathrm{Im}(\lambda) = 0$ and $\omega = \pi/T$ (half the clocking frequency) maps to $\mathrm{Im}(\lambda) = \infty$. This is a one-to-one mapping with half the unit circle ($\omega T = 0$ to $\omega T = \pi$) being mapped onto the positive half of the imaginary axis in the $\lambda$ plane. Now, since continuous-time analog filters are designed to meet specifications posed along the imaginary axis of a complex plane (the $s$ plane) these design methods can be directly applied to the discrete-time case,

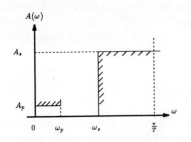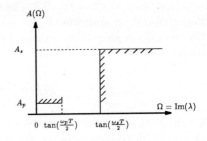

**FIGURE 9.5**
Prewarping frequency specifications of discrete-time filter according to the bilinear transformation.

first by renaming $s$ to $\lambda$ and then by applying the bilinear transform of Eq. (9.14).

To illustrate, let it be required to find a transfer function $T(z)$ whose magnitude response $|T(e^{jT})|$ meets the low-pass specifications given in Fig. 9.5(a). First, we use the relationship in Eq. (9.16) to find the specs versus $\text{Im}(\lambda)$ which we denote $\Omega$,

$$\Omega = \text{Im}(\lambda) = \tan(\omega T/2) \tag{9.17}$$

This results in the tolerance plot shown in Fig. 9.5(b). Since the relationship between $\Omega$ and $\omega$ is a nonlinear one, the frequency axis is said to be *warped* and the process we have just performed to recast the specs versus $\Omega$ is known as *prewarping*. Next we use continuous-time filter approximation methods to obtain a transfer function $T_c(\lambda)$ whose magnitude function $|T_c(j\Omega)|$ meets the specifications of Fig. 9.5(b). This can be done using standard filter design tables [18] or filter design programs [19]. For illustration purposes we assume that a fifth-order elliptic function provides the best approximation to the prewarped specs; that is, $T_c(\lambda)$ is a fifth-order elliptic function. Finally the discrete-time transfer function required to meet the original specs in Fig. 9.5(a) can be found by applying the bilinear transform as follows:

$$T(z) = T_c(\lambda)\Big|_{=\frac{z-1}{z+1}} \tag{9.18}$$

Note that the attenuation pole at $\Omega = \infty$ will map to half the sampling frequency.

The example above illustrates the use of the bilinear transformation in obtaining transfer functions for discrete-time filters (both analog and digital). Let's now consider its use in filter synthesis. To do that assume we are able to find a block diagram realization of the continuous-time transfer function $T_c(\lambda)$ with the blocks being integrators, i.e., having transfer functions of the form $(1/\lambda)$. Such a block diagram can be implemented in switched-capacitor form provided we have SC blocks having transfer functions of the form $\frac{1}{\cdot} = \frac{z+1}{z-1}$.

Simply replacing each block in the block-diagram realization with one of these *bilinear SC integrators* results in an SC circuit realization of the filter.

The problem then reduces to that of finding a circuit for the SC integrator. Unfortunately it turned out that it is *not possible* to find a single-amplifier, single-ended, stray-insensitive SC bilinear integrator [20]. Other indirect means had to be found to utilize the bilinear transform in the synthesis of SC filters, as will be demonstrated in subsequent sections. There is, however, a differential bilinear SC integrator circuit that can be used in the direct synthesis of SC filters. We now discuss this circuit.

The transfer function of the bilinear SC integrator is

$$T(z) = k\frac{z+1}{z-1} \tag{9.19}$$

where $k$ is a constant. This transfer function can be written as

$$T(z) = k\frac{z}{z-1} + k\frac{1}{z-1}$$
$$= k\frac{z^{12}}{z^{12} - z^{-12}} + k\frac{z^{-12}}{z^{12} - z^{-12}} \tag{9.20}$$

which is the sum of the transfer function of an LDI integrator with excess phase lead and the transfer function of an LDI integrator with excess phase lag. Based on this observation we obtain the circuit realization of the bilinear integrator as in Fig. 9.6(a). Observe that the input branches with capacitors $C_1$ realize the term with $z^{+12}$ in the numerator and those with input capacitors $C_1'$ realize the term with $z^{-12}$ in the numerator. Of course we have to select $C_1' = C_1$ and the resulting transfer function of the circuit of Fig. 9.6(a) is

$$\frac{V_o^+ - V_o^-}{V_i^+ - V_i^-} = \frac{C_1}{C_2}\frac{z+1}{z-1} \tag{9.21}$$

Figure 9.6(b) shows a refinement on the circuit of Fig. 9.6(a). Here only two capacitors as opposed to four are needed, reducing the matching requirements of capacitors $C_1$ and $C_1'$ in the circuit of Fig. 9.6(a). It can be shown that as long as the input signal does not change during $\phi_1$, the circuit of Fig. 9.6(b) implements a bilinear integrator with transfer function of Eq. (9.21) [21]. Note that the output of the circuit also does not change during $\phi_1$, thus the circuit can be used to feed another identical one (for instance, to form a two-integrator loop).

The bilinear integrator circuit of Fig. 9.6 has a major disadvantage from a settling-time point of view. Specifically note that charge is transferred to the feedback capacitors during $\phi_2$ and the output is sampled also during $\phi_2$. Worse still, if the output is fed to an identical circuit then the op amp of this latter circuit will be settling during the same clock phase in which the first op amp is settling. The result is an increased dependence on the op amp dynamic response which

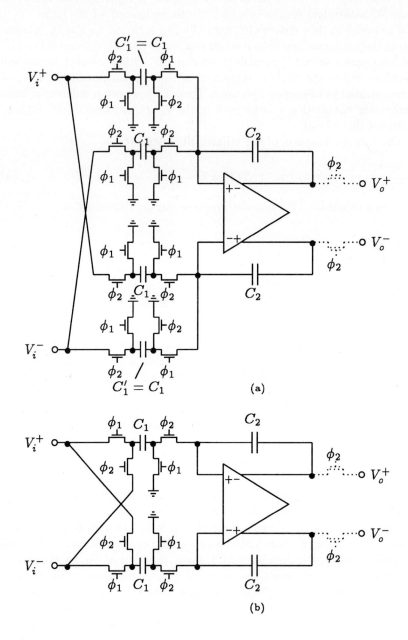

**FIGURE 9.6**
(a) A bilinear SC integrator (b) Refined bilinear integrator.

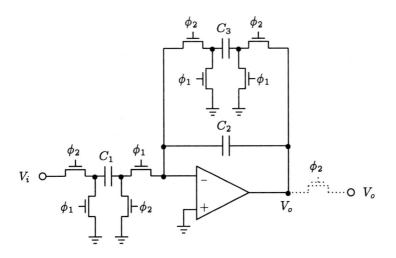

**FIGURE 9.7**
Damped non-inverting integrator.

manifests itself in transfer function error especially when the clocking frequency is sufficiently high in comparison with the op amp's unity-gain bandwidth.

### 9.3.1 Damping the SC Integrator

Damping an SC integrator can be accomplished by connecting a switched capacitor across the integrator capacitor as shown in Fig. 9.7. The phasing of the switched capacitor in the feedback should be noted; this phasing results in negative feedback and thus positive damping. The transfer function of the damped integrator is most easily derived by considering it as an undamped integrator with two inputs: $V_i$ through the switched capacitor $C_1$ and $V_o$ through the switched capacitor $C_3$. Thus we can express $V_o$ as

$$V_o(z) = \frac{C_1}{C_2}\frac{z^{-12}}{z^{12}-z^{-12}}V_i(z) - \frac{C_3}{C_2}\frac{z^{12}}{z^{12}-z^{-12}}V_o(z) \qquad (9.22)$$

from which the transfer function is found as

$$\frac{V_o(z)}{V_i(z)} = \frac{C_1 z^{-12}}{C_2(z^{12}-z^{-12})+C_3 z^{12}} \qquad (9.23)$$

For physical frequencies,

$$\frac{V_o}{V_i}(\omega) = \frac{C_1 e^{-jT2}}{j2C_2 \sin(\omega T/2) + C_3\left[\cos(\omega T/2) + j\sin(\omega T/2)\right]} \qquad (9.24)$$

We note that the damping term in the denominator is frequency dependent. The denominator terms can be regrouped to yield

$$\frac{V_o}{V_i}(\omega) = \frac{C_1 e^{-jT2}}{j2\left(C_2 + \frac{C_3}{2}\right)\sin(\omega T/2) + C_3 \cos(\omega T/2)} \tag{9.25}$$

Thus damping the SC integrator changes its time constant or integrator frequency. Furthermore, the damping term is frequency dependent. For $\omega T \ll 1$,

$$\frac{V_o}{V_i}(\omega) \approx \frac{C_1 e^{-jT2}}{j\left(C_2 + \frac{C_3}{2}\right)\omega T + C_3} \tag{9.26}$$

indicating that the time constant is $T\frac{C_2 + \frac{C_3}{2}}{C_1}$ and the 3-dB frequency is $\frac{1}{T}\frac{C_1}{C_2 + \frac{C_3}{2}}$.

In terms of the LDI variable $\gamma$, the transfer function of the damped integrator is

$$\frac{V_o(z)}{V_i(z)} = \frac{C_1 z^{-12}}{2\gamma C_2 + C_3 z^{12}} \tag{9.27}$$

Introducing yet another complex-frequency variable $\mu$,

$$\mu \equiv \frac{1}{2}(z^{12} + z^{-12}) \tag{9.28}$$

we can express $z^{12}$ as

$$z^{12} = \gamma + \mu \tag{9.29}$$

and thus the transfer function in Eq. (9.27) becomes

$$\frac{V_o}{V_i} = \frac{C_1 z^{-12}}{\gamma(2C_2 + C_3) + \mu C_3} \tag{9.30}$$

This represents the transfer function of a damped integrator in the $\gamma$ plane. Once again we note that the the damping capacitance $C_3$ modifies the coefficient of $\gamma$ in the denominator, and that the damping term is dependent on frequency through $\mu$ which for physical frequencies is

$$\mu = \cos(\omega T/2) \tag{9.31}$$

These effects do not have counterparts in continuous-time active-RC integrators.

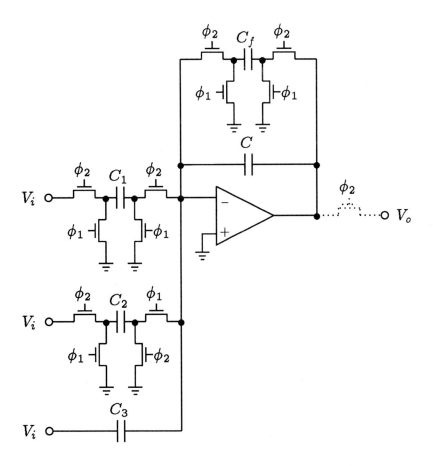

**FIGURE 9.8**
General first-order building block.

### 9.3.2 General First-Order Building Block

We conclude this section by presenting in Fig. 9.8 a general first-order SC build-ing block in the single-ended form. The generality stems from the fact that all three possible feed-in branches are included; namely, an inverting branch, a non-inverting branch, and an unswitched branch. The transfer function can be derived by expressing $V_o$ as the sum of four components due to the three feed-in branches and the feedback branch as follows:

$$V_o(z) = -\frac{C_1}{C} \frac{z^{12}}{z^{12} - z^{-12}} V_i(z) + \frac{C_2}{C} \frac{z^{-12}}{z^{12} - z^{-12}} V_i(z)$$

$$-\frac{C_3}{C} V_i(z) - \frac{C_f}{C} \frac{z^{12}}{z^{12} - z^{-12}} V_o(z) \tag{9.32}$$

The transfer function can then be obtained as

$$\frac{V_o(z)}{V_i(z)} = \frac{-C_1 z^{12} + C_2 z^{-12} - C_3(z^{12} - z^{-12})}{C(z^{12} - z^{-12}) + C_f z^{12}} \tag{9.33}$$

which can be expressed in terms of $\gamma$ and $\mu$ as

$$\frac{V_o(z)}{V_i(z)} = \frac{-C_1 z^{12} + C_2 z^{-12} - 2\gamma C_3}{\gamma(2C + C_f) + \mu C_f} \tag{9.34}$$

## 9.4   SECOND-ORDER SC CIRCUITS AND CASCADE DESIGN

In this section we utilize the basic SC integrators of the previous section to form second-order circuits of the two-integrator-loop type. These biquadratic circuits, or *biquads* for short, can be connected in cascade so as to realize a high-order filter function. Also, it will be seen in the next section that SC filters that are designed on the basis of simulating the operation of LC ladder networks, consist of two-integrator loops. Thus, a thorough understanding of the two-integrator loop is essential to the synthesis of SC filters.

### 9.4.1   Structure of the Two-Integrator Loop

The two-integrator loop consists of an inverting and a noninverting integrator connected in a feedback loop [17]. Figure 9.9 shows such a loop formed using the pair of single-ended LDI integrators of Fig. 9.2. Observe that during clock phase $\phi_2$ charge is transferred from $C_3$ to $C_1$, and the output of the op amp $A_1$ is sampled by charging capacitor $C_4$. Then, during clock phase $\phi_1$ the charge of $C_4$ is transferred to $C_2$ and the output of op amp $A_2$ is allowed to settle until the next clock phase $\phi_2$. Thus, in this loop integrator $A_1$ will be more dependent on op amp dynamics than $A_2$. Nevertheless, from the point of view of the dependence on op amp dynamics this loop is the best that can be designed using single-ended integrators [10]. Finally, note that there is one clock period of delay around the loop.

Figure 9.10 shows the differential equivalent of the two-integrator loop of Fig. 9.9. Recalling, however, that differential circuits provide an extra degree of freedom, namely that signal inversion can be obtained by simply interchanging the op amp output terminals, we can obtain a loop that is more tolerant of amplifier settling time [11] than in Fig. 9.10. Such a loop is shown in Fig. 9.11.

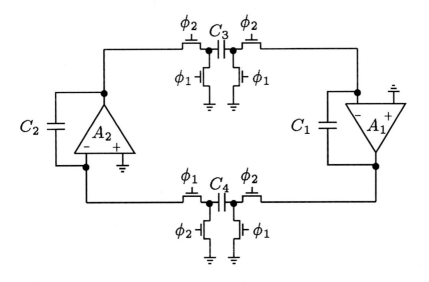

**FIGURE 9.9**
A two-integrator loop of single-ended non-inverting and inverting SC integrators.

Observe that during $\phi_1$, $C_4$ is charged and its charge is transferred to $C_2$. Op amp $A_2$ is then allowed to settle until the next phase, $\phi_2$. During $\phi_2$, $C_3$ is charged and its charge is transferred to $C_1$. Op amp $A_1$ is then allowed to settle until the next clock phase, $\phi_1$. Thus this two-integrator loop has the best performance in the presence of finite-bandwidth op amps, assuming that the op amps have low output impedances [11, 12].

Many practical SC filters utilize transconductance op amps that have high output impedance. In this case, the two-integrator loop shown in Fig. 9.12 has been found to yield better high-frequency response [11] than that of Fig. 9.11. The reason is that the input capacitors in the loop of Fig. 9.12 are half the size of the corresponding ones in the circuit of Fig. 9.11. This halving of capacitance values is possible because the capacitors are switched between $+V_i$ and $-V_i$, rather than between $V_i$ and ground. The circuit of Fig. 9.12 is said to utilize differencing integrators.

Figure 9.13 shows yet another two-integrator loop. This loop is formed from two non-inverting integrators with the signal inversion obtained by interchanging the output terminals of $A_2$. This loop has worse settling-time dependence than the loop of Fig. 9.11 since the output of each amplifier is sampled during the same clock phase in which charge is transferred to its feedback capacitor. Nevertheless, this loop has a very attractive feature: In each phase only one amplifier is active. Thus we can dispense with one of the amplifiers and multiplex the other amplifier to obtain a single-amplifier biquad [22, 12], as shown in Fig. 9.14. This circuit is useful when clocking is at moderate speeds.

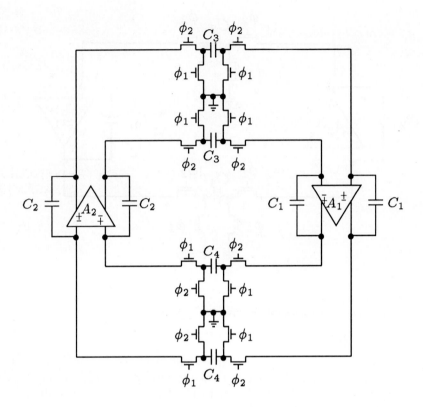

**FIGURE 9.10**
A two-integrator loop of fully-differential non-inverting and inverting SC integrators.

The final two-integrator loop we show is that utilizing the bilinear integrator of Fig. 9.6(b). This loop, shown in Fig. 9.15, has the advantages that it can be directly employed in designs using the bilinear transformation process outlined in the previous section. Unfortunately, however, it has poor settling-time performance because during $\phi_2$ the entire loop is active!

### 9.4.2 A Simple Biquad Circuit

Having studied a variety of two-integrator-loop structures we now consider their application in biquad design. The simplest situation is shown in Fig. 9.16 where the input signal is fed through a switched capacitor to the two-integrator loop of Fig. 9.9, and one of the integrators is damped by placing a switched-capacitor across its integrating capacitor. By analogy to active-RC filters one can see that this biquad circuit realizes a bandpass function at the output of $A_1$ and a low-pass function at the output of $A_2$. The realized functions, however, are approximate; for instance, it can be shown that the bandpass function has one transmission zero at $z = 1$ and the other at $z = 0$. While the first zero corresponds to

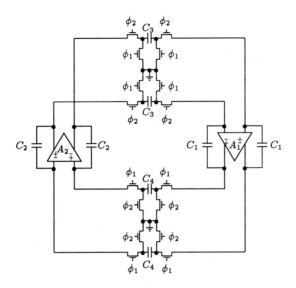

**FIGURE 9.11**
Fully-differential two-integrator loop for high frequency applications.

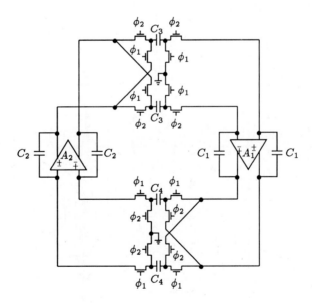

**FIGURE 9.12**
A two-integrator loop of two fully-differential SC differencing integrators. This structure has
the best high-frequency behavior when the op amps have high output impedances.

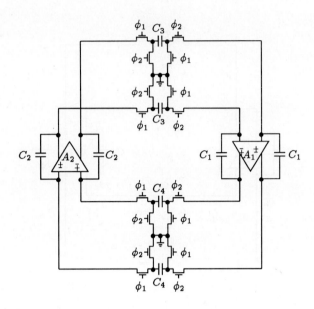

**FIGURE 9.13**
Fully-differential two-integrator loop suitable for a single op amp time-sharing application. Note that only one op amp is active during any one clock phase.

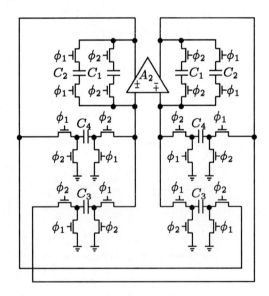

**FIGURE 9.14**
Fully-differential two-integrator loop consisting of a single op amp based on the circuit of Fig. 9.13.

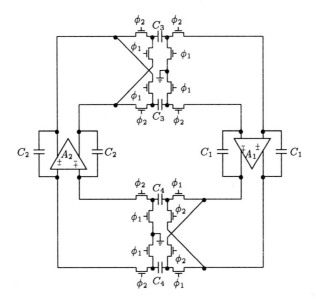

**FIGURE 9.15**
A two-integrator loop consisting of two bilinear integrators.

$\omega = 0$, the second has no corresponding physical frequency. The second zero therefore contributes little to the selectivity of the bandpass filter response. In other words, while the active-RC circuit has a zero at $\omega = 0$ and the other at $\omega = \infty$, both contributing to selectivity, the discrete-time SC filter has only one physical-frequency zero (at $\omega = 0$). A more selective bandpass response can be obtained by placing a transmission zero at half the clock frequency ($\omega = \pi/T$). Such is the case in the biquad circuit that is based on the bilinear transformation, discussed next.

Before leaving the simple biquad circuit of Fig. 9.16, we note that there is an alternative method for damping it to place the poles at their proper locations: Rather than placing a switched capacitor across one of the integrator capacitors, an unswitched capacitor can be placed across one of the switched-capacitor branches in the two-integrator loops. This "capacitive damping" leads to better capacitor values in certain applications [23].

### 9.4.3 A Biquad Circuit Based on the Bilinear Transformation

Recall from Section 9.2 that the bilinear transformation provides a method for obtaining a transfer function $T_c(\lambda)$ whose corresponding $z$-domain transfer function meets exactly the given specifications. Let us now go a step further and

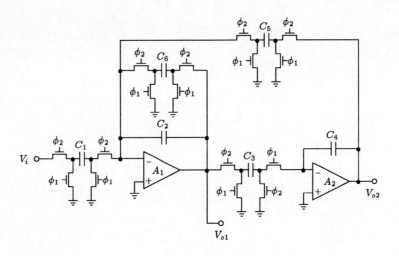

**FIGURE 9.16**
A simple two-integrator loop SC biquad circuit.

consider the realization of $T_c(\lambda)$. Towards that end, assume that $T_c(\lambda)$ is of even order. Thus, it can be factored into the product of biquadratic transfer function of the general form

$$t(\lambda) = \pm \frac{a_2\lambda^2 + a_1\lambda + a_0}{\lambda^2 + b_1\lambda + b_0} \tag{9.35}$$

Now if we have an SC circuit that can realize $t(\lambda)$ then the SC filter can be realized exactly by simply cascading a number of these biquad circuits, each realizing one of the biquadratic functions in the factorization of $T_c(\lambda)$. SC biquad circuits capable of realizing the general biquadratic function $t(\lambda)$ in Eq. (9.35) have been reported in [8, 24, 23]. As an example, we show one such circuit in Fig. 9.17. Note that although the circuit is shown in single-ended form, it can be converted to differential form easily. Analysis of this circuit results in the $z$-domain transfer function

$$\frac{V_o(z)}{V_i(z)} = -\frac{K_3z^2 + (-2K_3 + K_1K_5 + K_2K_5)z + (K_3 - K_2K_5)}{z^2 + (-2 + K_4K_5 + K_5K_6)z + (1 - K_5K_6)} \tag{9.36}$$

The $\lambda$-plane transfer function is obtained by substituting $z = \frac{(1+)}{(1-)}$ which is the inverse of the transformation in Eq. (9.14). Design equations for the SC biquad can be then obtained by equating coefficients in the resulting transfer function to the corresponding coefficients in Eq. (9.35). Equating corresponding denominator coefficients results in two equations in the three capacitor ratios that determine the biquad poles,

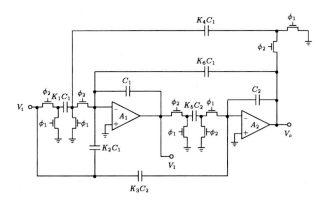

**FIGURE 9.17**
A multi-purpose SC biquad circuit.

$$K_4 K_5 = 4mb_0$$

$$K_5 K_6 = 2mb_1$$

$$\text{where} \qquad m = \frac{1}{(1 + b_1 + b_0)}$$

Because we have only two equations in the three unknowns $K_4$, $K_5$, and $K_6$, one of the capacitor ratios can be chosen arbitrarily. This choice is usually done with a view to maximizing the biquad dynamic range. For high-$Q$ biquads, this is approximately obtained when the two-loop time-constants are made equal, that is $K_4 = K_5$.

Design equations for the feed-in capacitor ratios, which determine the biquad transmission zeros, are obtained by equating coefficients of corresponding numerator terms. The result for some special cases are:

$(a)$ Low $-$ Pass $(a_1 = a_2 = 0)$
$\qquad K_2 = 0 \quad K_1 K_5 = 4ma_0 \quad K_3 = ma_0$

$(b)$ Bandpass $(a_0 = a_2 = 0)$
$\qquad K_1 = 0 \quad K_2 K_5 = 2ma_1 \quad K_3 = ma_1$

$(c)$ High $-$ Pass $(a_0 = a_1 = 0)$
$\qquad K_1 = K_2 = 0 \quad K_3 = ma_2$

$(d)$ Notch $(a_1 = 0)$
$\qquad K_2 = 0 \quad K_1 K_5 = 4ma_0 \quad K_3 = m(a_0 + a_2)$

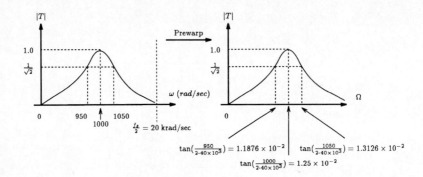

**FIGURE 9.18**
Prewarping SC bandpass specifications into the $\lambda$-plane for Example 9-1.

The circuit of Fig. 9.17 realizes a negative bandpass function. If a positive band-pass is needed, it can be obtained by taking the output at $V_1$. While the pole design equations remain unchanged, the zeros' equation becomes

$$K_1 = 0 \quad K_3 K_4 = 2ma_1 \quad K_2 = ma_1\left(1 + \frac{b_1}{b_0}\right)$$

The reader is urged to note that the low-pass and bandpass functions realized in the circuit of Fig. 9.17 have more selective frequency responses than the corresponding functions realized in the circuit of Fig. 9.16. This is due to the transmission zero at half the sampling frequency ($\omega = \pi/T$ which is the mapping of $\lambda = \infty$).

**Example 9.1.** Using the general SC biquad circuit shown in Fig. 9.17, design a negative-gain bandpass filter for which the center frequency is located at 1 krad/s and has a lower and upper 3-dB frequency of 950 rad/s and 1050 rad/s, respectively. The center frequency gain should be set to unity. The sampling rate is assumed equal to 10 krad/s.
*Solution:*
Prewarping the SC specifications as illustrated by Fig. 9.18, we arrive at the following $\lambda$-plane specifications: $\omega_0 = 0.3249$ rad/s and $\omega_0/Q = BW_{3-dB} = 0.0348$ rad/s. Using standard notation for a bandpass transfer function, i.e.,

$$t(\lambda) = \frac{(\omega_0/Q)\lambda}{\lambda^2 + (\omega_0/Q)\lambda + \omega_0^2}$$

and comparing it with the general form of the transfer function for the biquad SC circuit of Fig. 9.17 in Eq. (9.35) we can immediately write

$$a_2 = 0 \qquad a_1 = \omega_0/Q = 0.0348 \qquad a_0 = 0$$

$$b_1 = \omega_0/Q = 0.0348 \quad b_0 = \omega_0^2 = (0.3249)^2$$

Using the above design rules described for a BP filter function, allows us to compute

$$m = 1/(1 + b_1 + b_0) = 1/(1 + 0.0348 + (0.3249)^2) = 0.87692$$

$$K_4 K_5 = 4mb_0 = 4(0.87692)(0.3249)^2 = 0.37027$$

$$K_5 K_6 = 2mb_1 = 2(0.87692)(0.0348) = 0.061033$$

$$K_1 = 0$$

$$K_2 K_5 = 2ma_1 = 2(0.87692)(0.0348) = 0.061033$$

$$K_3 = ma_1 = (0.87692)(0.0348) = 0.030517$$

To solve for individual coefficients, consider setting $K_4 = K_5$. Therefore,

$$K_4 = \sqrt{4mb_0} = \sqrt{0.37027} = 0.60850$$

$$K_5 = \sqrt{4mb_0} = \sqrt{0.37027} = 0.60850$$

$$K_6 = 2mb_1/K_5 = 0.061033/0.60850 = 0.10030$$

$$K_2 = 2ma_1/K_5 = 0.061033/0.60850 = 0.10030$$

The actual value of each capacitor will therefore depend on the absolute value of capacitance chosen for $C_1$ and $C_2$. We leave this situation for the reader to taylor to his/her particular application.

The bilinear-based biquad circuit of Fig. 9.17 can be used also in the design of filters utilizing any of the coupled-biquad structures [24, 25]. The design method is very simple: $T_c(\lambda)$ is realized using the chosen coupled-biquad structure in exactly the same way employed for active-RC filters and then each of the biquad blocks is realized with the circuit of Fig. 9.17 using the design equations presented above.

## 9.5 SWITCHED-CAPACITOR LADDER FILTERS

Doubly terminated LC ladder networks that are designed to effect maximum power transfer from source to load in the filter passband feature very low sensitivities to variations in the component values [26]. This fact has over the years spurred considerable interest in finding active-RC, digital, and switched-capacitor filter structures that simulate the internal workings of LC ladder prototype networks. Essentially these structures consist of a connection of first-order blocks that implement the integral $I - V$ relationships of the $L$ and $C$ elements in the prototype; that is, the basic building blocks are integrators.

In this section we present a method for SC filter design on the basis of simulating the operation of an LC ladder prototype [27]. We note at the outset that since the LDI integrators are not perfect integrators (in the sense that their transfer functions are not proportional to $1/\omega$) they cannot be used *directly* in the design of SC ladder filters. Rather, in our method, which is an *exact* synthesis, one will employ the LDI integrators in a rather indirect way.

To illustrate the method, let it be required to synthesize an SC ladder circuit whose attenuation function meets standard low-pass specifications such as those depicted in Fig. 9.5(a). First we prewarp the specifications according to $\Omega = \tan(\omega T/2)$, thus obtaining the modified specifications shown in Fig. 9.5(b) where $\Omega = \text{Im}(\lambda)$ and $\lambda$ is the bilinear complex-frequency variable. Next we use filter design tables or computer programs to synthesize an LC ladder network whose attenuation function meets the prewarped specifications of Fig. 9.5(b). This ladder is synthesized in the $\lambda$ plane and thus its inductors will have impedances $\lambda l$ and its capacitors will have impedances $1/\lambda c$. The synthesis process, however, is the same as that used in the continuous-time case (i.e., in the $s$ plane).

For illustration purposes, assume that a fifth-order elliptic filter meets the given specs. The corresponding LC ladder network is shown in Fig. 9.19(a). Direct simulation of this ladder network using active-RC techniques requires integrators with transfer functions of the form $1/\lambda$. As mentioned in Sec. 9.2, these bilinear integrators are impossible to implement in single-ended form using stray-insensitive, single-amplifier SC circuits. Although a differential bilinear-integrator circuit exists (Fig. 9.6) we have found in Sec. 9.3 that its clock phasing results in two-integrator loops that are highly sensitive to the limited bandwidth of the op amps. We shall therefore employ an indirect method for the simulation of the LC network of Fig. 9.19(a).

The method is based on scaling the impedances of the LC ladder of Fig. 9.19(a) by the factor $\mu$ (defined in Eq. (9.28)). Specifically, if we divide all impedances by $\mu$, the voltage transfer function remains unchanged. An inductor in the original circuit with impedance $(\lambda l)$ transforms into an element with impedance $(l)$, a capacitor with impedance $(1/\lambda c)$ transforms into an element with impedance $(1/\lambda \mu c)$, and the ladder termination resistors, $r_s$ and $r_l$, transform into impedances $(r_s/\mu)$ and $(r_l/\mu)$. To see what the transformed elements are, we use the defining equations for $\gamma$, $\mu$, and $\lambda$,

$$\gamma = \frac{1}{2}(z^{12} - z^{-12}) = \frac{1}{2}(e^{sT2} - e^{-sT2}) = \sinh(sT/2) \tag{9.37}$$

$$\mu = \frac{1}{2}(z^{12} + z^{-12}) = \frac{1}{2}(e^{sT2} + e^{-sT2}) = \cosh(sT/2) \tag{9.38}$$

$$\lambda = \frac{z-1}{z+1} = \frac{z^{12} - z^{-12}}{z^{12} + z^{-12}} = \tanh(sT/2) \tag{9.39}$$

to obtain the following two relationships:

$$\lambda = \frac{\gamma}{\mu} \tag{9.40}$$

$$\mu^2 = 1 + \gamma^2 \tag{9.41}$$

Now, the transformed capacitor with impedance $(1/\lambda\mu c)$ can be seen using Eq. (9.40) to have an impedance $(1/\gamma c)$, which is a $\gamma$-plane capacitor. The transformed inductor with impedance $(\,l\,)$ has an admittance $(\frac{1}{l})$ which can be written as

$$\frac{\mu}{\lambda l} = \frac{\mu^2}{\lambda\mu l} = \frac{1+\gamma^2}{\gamma l} = \frac{1}{\gamma l} + \frac{\gamma}{l} \qquad (9.42)$$

which is the parallel connection of a $\gamma$-plane capacitor of value $1/l$ and a $\gamma$-plane inductor of value $l$.

With these results in hand, we can sketch the transformed LC ladder as in Fig. 9.19(b). Next we combine the parallel capacitors and convert the input signal source to its Norton's form to obtain the circuit in Fig. 9.19(c). Utilizing a technique from active-RC filters [28], the bridging capacitors can be replaced with voltage-controlled current sources, as shown in Fig. 9.19(d). The resulting ladder network consists of $\gamma$-plane shunt capacitors, and $\gamma$-plane series inductors, and controlled sources proportional to $\gamma$. It follows that this ladder can be simulated using LDI blocks. The only apparent problem is that the terminations are not pure resistors but the complex impedances $(r_s/\mu)$ and $(r_l/\mu)$. This turns out to be not a problem but rather a welcome result: Recall that a damped LDI integrator has a damping term that includes $\mu$ [see Eq. (9.30)]. Thus, as will be seen shortly, these $\mu$-dependent terminations will be simulated naturally by the damped integrators that simulate the operation of the two end capacitors.

The operation of the ladder in Fig. 9.19(d) can be fully described by the following five equations, each describing the operation of one of the reactive components:

$$V_1 = \frac{\mu(V_i/r_s) - I_2 + \gamma c_2 V_3}{\gamma c_1' + (\mu/r_s)} \qquad (9.43)$$

$$I_2 = \frac{V_1 - V_3}{\gamma l_2} \qquad (9.44)$$

$$V_3 = \frac{I_2 - I_4 + \gamma c_2 V_1 + \gamma c_4 V_5}{\gamma c_3'} \qquad (9.45)$$

$$I_4 = \frac{V_3 - V_5}{\gamma l_4} \qquad (9.46)$$

$$V_5 = \frac{I_4 + \gamma c_4 V_3}{\gamma c_5' + (\mu/r_l)} \qquad (9.47)$$

Except for the $\mu$ term that multiples $V_i$ in Eq. (9.43), these equations are in the form of the transfer function of the general first-order block of Fig. 9.8 (Eq. (9.34)). From this analogy we can directly sketch the SC circuit realization shown in Fig. 9.20. This circuit includes an input branch that multiples $V_i$ by the variable $\mu$. The transfer functions of the five blocks in Fig. 9.20 can be written by inspection as

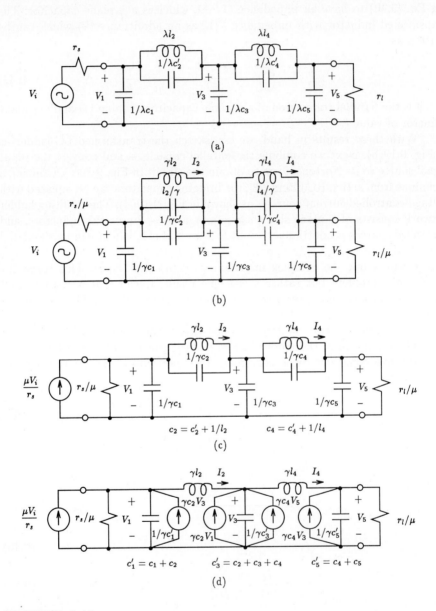

**FIGURE 9.19**

(a) $\lambda$-plane LC ladder network realization of a fifth-order elliptic filter. (b) Impedance scaling the $\lambda$-plane LC ladder network shown in (a) by $\mu$. (c) Simplifying the $\gamma$-plane LC ladder and converting the input voltage source to its Norton's form. (d) Norton's theorem is used to replace $c_2$ and $c_4$ with voltage-controlled current sources.

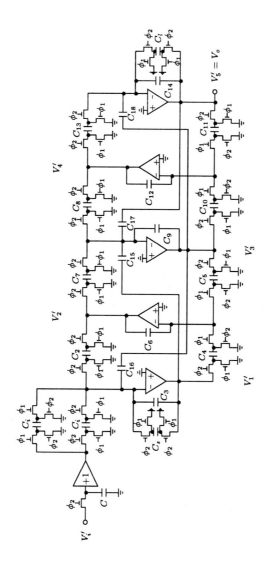

**FIGURE 9.20**
Switched-capacitor circuit that simulates the operation of the fifth-order LC ladder network of
Fig. 9.18(d).

$$V_1' = \frac{-2C_i\mu V_i' - C_2 z^{12} V_2' - \gamma 2C_{16} V_3'}{\gamma(2C_3 + C_s) + \mu C_s} \tag{9.48}$$

$$V_2' = \frac{C_4 z^{-12} V_1' + C_5 z^{-12} V_3'}{\gamma 2C_6} \tag{9.49}$$

$$V_3' = \frac{-C_7 z^{12} V_2' - C_8 z^{12} V_4' - \gamma 2C_{15} V_1' - \gamma 2C_{17} V_5'}{\gamma 2C_9} \tag{9.50}$$

$$V_4' = \frac{C_{10} z^{-12} V_3' + C_{11} z^{-12} V_5'}{\gamma 2C_{12}} \tag{9.51}$$

$$V_5' = \frac{-C_{13} z^{12} V_4' - \gamma 2C_{18} V_3'}{\gamma(2C_{14} + C_l) + \mu C_l} \tag{9.52}$$

Comparing corresponding equations between the equation set (9.43) – (9.47) and the set (9.48) – (9.52) we obtain the following correspondences between the ladder currents and voltages (Fig. 9.19(d)) and the op amp voltages (Fig. 9.20):

$$(-z^{-12} V_i') \Leftrightarrow V_i \quad (z^{-12} V_1') \Leftrightarrow V_1 \quad V_2' \Leftrightarrow I_2$$

$$(-z^{-12} V_3') \Leftrightarrow V_3 \quad (-V_4') \Leftrightarrow I_4 \quad (z^{-12} V_5') \Leftrightarrow V_5$$

With these correspondences we see that the SC circuit in Fig. 9.20 exactly simulates the operation of the LC ladder network of Fig. 9.19(d). Initial values for the capacitors of the SC circuit are obtained by equating the coefficients of corresponding terms between the set (9.43) – (9.47) and the equation set (9.48) – (9.52). The result is

$$C_i = \frac{1}{2r_s} \qquad C_2 = 1 \qquad C_s = \frac{1}{r_s} \qquad C_3 = (c_1' - \frac{1}{r_s})/2$$

$$C_4 = C_5 = 1 \qquad C_6 = l_2/2 \qquad C_7 = C_8 = 1 \qquad C_9 = c_3'/2$$

$$C_{10} = C_{11} = 1 \qquad C_{12} = l_4/2 \qquad C_{13} = 1 \qquad C_l = \frac{1}{r_l}$$

$$C_{14} = (c_5' - \frac{1}{r_l})/2 \quad C_{15} = C_{16} = c_2/2 \quad C_{17} = C_{18} = c_4/2$$

Using these capacitor values gives the SC filter a dc-gain equal to that of the LC ladder. If the gain of the SC filter is to be a factor $K$ greater than that of the LC ladder then the value of $C_i$ must be changed to $KC_i$.

The initial capacitor values above must then be scaled so that the SC filter has the widest possible dynamic range. Dynamic range scaling is usually based on equalizing the peaks obtained at the outputs of the op amps as the frequency of an input sinusoid is swept over a desired band. To perform scaling one needs the values of the signal peaks, denoted $\hat{V}_1'$, $\hat{V}_2'$, $\hat{V}_3'$, $\hat{V}_4'$, and $\hat{V}_5'$. These can be found from a simulation of the SC circuit [29] with the capacitor values given above. Alternatively, the peak values can be obtained from an analysis of the LC ladder network.

Once the spectral peaks have been determined the capacitor values can be scaled as follows: capacitor $C_{ij}$ which is connected between the output of op amp $i$ and input of op amp $j$ is changed to $C_{ij}(\hat{V}_i'/\hat{V}_j')$. As an example, capacitor $C_2$ changes from the initial value of unity to $(\hat{V}_2'/\hat{V}_1')$. Note that this scaling process preserves the magnitude of loop transmission of every two-integrator loop, and hence the filter transfer function (except for a gain constant) remains unchanged. To keep the overall gain unchanged, $C_i$ must be scaled by the factor $(\hat{V}_5'/\hat{V}_1')$.

The next and final step in the design involves scaling the capacitor values to minimize the total capacitance. This is achieved by considering the five blocks one at a time. For block #$i$, assume that the smallest capacitance connected to the virtual ground input of the op amp is of value $\tilde{C}_i$. We wish to scale so that this capacitor is equal to the minimum capacitor value possible with the given technology; call it $C_{min}$ (this may be 0.1 pF or so). It follows that for block #$i$ the scaling factor $k_i = C_{min}/\tilde{C}_i$. Every capacitor of block #$i$ is multiplied by $k_i$. The transfer function of block #$i$ obviously remains unchanged. The process is then repeated for the other blocks.

This completes the design procedure of SC ladder filters. As noted earlier this design is exact and has been used for clock frequencies as low as five times the passband edge. Also, since filter approximation and ladder synthesis are performed in terms of the bilinear ($\lambda$) variable, the transmission zero at $\pi/T$ is preserved. Finally we note that this exact synthesis method has been extended to high-pass and bandpass filters [30, 31].

**Example 9.2.** Let it be required that a fifth-order low-pass SC filter circuit be designed to have an elliptic transfer function having a passband edge located at 1.5 kHz and a stopband edge located at 3 kHz. The maximum passband ripple should not exceed 0.5 dB, and in addition, the filter should have a dc-gain of unity. The minimum stopband attenuation should exceed 60 dB. Due to system requirements, the clock frequency has been assigned to be 8 kHz and that a single-ended design as opposed to a fully-differential design is to be chosen. *Solution:*
Prewarping the SC specifications according to the bilinear transform we arrive at the $\lambda$-plane specifications shown in Fig. 9.21(a). Subsequently, using a filter design computer program [19] we found the LC ladder network shown in Fig. 9.21(b) with component values

$$r_s = 1.000 \quad c_1 = 2.5238 \quad c_2' = 0.033619 \quad l_2 = 1.8124$$
$$c_3 = 3.7041 \quad c_4' = 0.088493 \quad l_4 = 1.7612 \quad c_5 = 2.4707$$
$$r_l = 1.000$$

had an attenuation characteristic that satisfied the $\lambda$-plane specifications. Following the synthesis method outlined above, we then convert the $\lambda$-plane ladder network into the $\gamma$-plane by performing the following set of capacitor calculations:

$$c_2 = c_2' + \frac{1}{l_2} = 0.033619 + \frac{1}{1.81241} = 0.5854$$

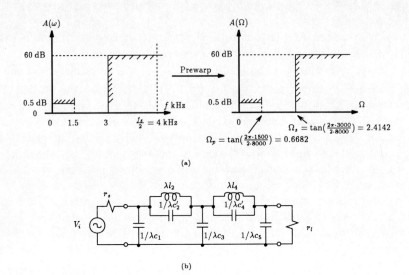

**FIGURE 9.21**
(a) Prewarped filter specifications. (b) fifth-order LC ladder network whose attenuation characteristic satisfy the prewarped specifications of part (a).

$$c_4 = c_4' + \frac{1}{l_4} = 0.088493 + \frac{1}{1.7612} = 0.6563$$

Subsequently, the capacitor values of the $\gamma$-plane LC ladder are renormalized for its Norton form, resulting in:

$$c_1' = c_1 + c_2 = 2.5238 + 0.5854 = 3.1092$$

$$c_3' = c_2 + c_3 + c_4 = 0.5854 + 3.7041 + 0.6563 = 4.9458$$

$$c_5' = c_4 + c_5 = 0.6563 + 2.4707 = 3.1270$$

Then, according to the formulae provided, the capacitor values of the fifth-order SC circuit (Fig. 9.19) are computed as follows:

$C_i = \frac{1}{2r_s} = 0.5$ $\qquad$ $C_2 = 1.0$ $\qquad$ $C_s = \frac{1}{r_s} = 1.0$

$C_3 = (c_1' - \frac{1}{r_s})/2 = 1.0546$ $\qquad$ $C_4 = C_5 = 1.0$ $\qquad$ $C_6 = l_2/2 = 0.9062$

$C_7 = C_8 = 1.0$ $\qquad$ $C_9 = c_3'/2 = 2.4729$ $\qquad$ $C_{10} = C_{11} = 1.0$

$C_{12} = l_4/2 = 0.8806$ $\qquad$ $C_{13} = 1.0$ $\qquad$ $C_l = \frac{1}{r_l} = 1.0$

$C_{14} = (c_5' - \frac{1}{r_l})/2 = 1.0635$ $\quad$ $C_{15} = C_{16} = c_2/2 = 0.2927$ $\quad$ $C_{17} = C_{18} = c_4/2 = 0.3281$

Analyzing the resulting SC circuit using the computer program SWITCAP [29], we find that the peak value of the magnitude of each op amp output is as follows (node voltages corresponds to those in Fig. 9.20):

$$\hat{V}_1' = 0.793 \quad \hat{V}_2' = 1.428 \quad \hat{V}_3' = 0.788 \quad \hat{V}_4' = 1.198 \quad \hat{V}_5' = 0.5$$

Using these peak values we can then scale the various components of the SC circuit for maximum dynamic range as follows:

$$C_i \rightarrow C_i \times \hat{V}_5'/\hat{V}_1' = 0.3152 \quad C_4 \rightarrow C_4 \times \hat{V}_1'/\hat{V}_2' = 0.5553$$

$$C_2 \rightarrow C_2 \times \hat{V}_2'/\hat{V}_1' = 1.8008 \quad C_5 \rightarrow C_5 \times \hat{V}_3'/\hat{V}_2' = 0.5518$$

$$C_s \rightarrow C_s \times 1 = 1.0 \qquad C_6 \rightarrow C_6 \times 1 = 0.9062$$

$$C_3 \rightarrow C_s \times 1 = 1.0546 \quad C_7 \rightarrow C_7 \times \hat{V}_2'/\hat{V}_3' = 1.8122$$

$$C_8 \rightarrow C_8 \times \hat{V}_4'/\hat{V}_3' = 1.5203 \qquad C_l \rightarrow C_l \times 1 = 1.0$$

$$C_9 \rightarrow C_9 \times 1 = 2.4729 \qquad C_{14} \rightarrow C_{14} \times 1 = 1.0635$$

$$C_{10} \rightarrow C_{10} \times \hat{V}_3'/\hat{V}_4' = 0.6578 \quad C_{16} \rightarrow C_{16} \times \hat{V}_3'/\hat{V}_1' = 0.2909$$

$$C_{11} \rightarrow C_{11} \times \hat{V}_5'/\hat{V}_4' = 0.4174 \quad C_{15} \rightarrow C_{15} \times \hat{V}_1'/\hat{V}_3' = 0.2946$$

$$C_{12} \rightarrow C_{12} \times 1 = 0.8806 \quad C_{18} \rightarrow C_{18} \times \hat{V}_3'/\hat{V}_5' = 0.5171$$

$$C_{13} \rightarrow C_{13} \times \hat{V}_4'/\hat{V}_5' = 2.3960 \quad C_{17} \rightarrow C_{17} \times \hat{V}_5'/\hat{V}_3' = 0.2082$$

At the present time, the SC circuit of Fig. 9.20, together with the above capacitor values, simulates the input–output behavior of the doubly-terminated LC ladder in Fig. 9.21(b) and as a result has a 6 dB loss in the passband region (consider, at dc, a voltage-divider network is formed between the equal-valued resistors $r_s$ and $r_l$). The passband gain can be increased to unity by simply doubling the value of the input capacitor $C_i$ from 0.3152 capacitor units to 0.6304 capacitor units.

Finally, we scale the above capacitor values to minimize the total capacitance. Let us assume that the minimum capacitor value possible with this technology $C_{min}$ is 0.1 pF. On doing so, the following capacitor values result:

| Block#1 : | $C_i$ | $C_s$ | $C_2$ | $C_3$ | $C_{16}$ |
|---|---|---|---|---|---|
| Before Scaling | 0.6304 | 1.0 | 1.8008 | 1.0546 | 0.2909 |
| After Scaling | 0.217 pF | 0.344 pF | 0.619 pF | 0.363 pF | 0.1 pF |

| Block#2 : | $C_4$ | $C_5$ | $C_6$ |
|---|---|---|---|
| Before Scaling | 0.5553 | 0.5518 | 0.9062 |
| After Scaling | 0.101 pF | 0.1 pF | 0.164 pF |

| *Block#3* : | $C_7$ | $C_8$ | $C_9$ | $C_{15}$ | $C_{17}$ |
|---|---|---|---|---|---|

Before Scaling  1.8122  1.5203  2.4729  0.2946  0.2082
After Scaling  0.870 pF  0.730 pF  1.188 pF  0.1415 pF  0.1 pF

| *Block#4* : | $C_{10}$ | $C_{11}$ | $C_{12}$ |
|---|---|---|---|

Before Scaling  0.6578  0.4174  0.8806
After Scaling  0.1576 pF  0.1 pF  0.211 pF

| *Block#5* : | $C_{13}$ | $C_{14}$ | $C_l$ | $C_{18}$ |
|---|---|---|---|---|

Before Scaling  2.396  1.0635  1.0  0.5171
After Scaling  0.463 pF  0.206 pF  0.193 pF  0.1 pF

## 9.6  SYNTHESIS OF SWITCHED-CURRENT FILTERS

The switched-current (SI) technique is an alternate method for realizing mono-lithic sampled-data analog filters [3, 5]. Its main advantage over the SC technique is that it does not require linear capacitors, and as such, SI filters can be con-structed using MOS transistors only. As a result, the SI approach can implement accurate signal processing functions in a standard digital CMOS process. This, therefore, enables digital IC manufacturers to realize mixed-signal ICs, i.e., im-plement both digital and analog circuits on the same IC with existing low-cost CMOS processes.

By exploiting the *inter-reciprocal* property of linear networks, the SC syn-thesis methods of the previous sections can be used directly to synthesize SI filter circuits [6]. The basic idea is to simply realize branches of a transposed version of the signal-flow-graph (SFG) that represents the SC filter circuit with SI integrator circuits.

Before we discuss this SI filter synthesis method, let us briefly describe the basic building blocks used in SI filters. A more detailed discussion of SI circuit techniques, together with experimental results, was presented in Chapter 6.

### 9.6.1   The Switched-Current Technique

As described in the previous sections, precision SC filters are realized using integrator circuits, usually a pair of integrator circuits, one inverting and one non-inverting. In SI technology, integrator circuits can be constructed with transfer functions similar to those of SC integrators except that the signals involved are currents not voltages. Each integrator circuit consists of a cascade of two current track-and-hold circuits [32] of the type shown in Fig. 9.22. With switch $\phi$ closed, the current that appears at the output ($I_{out}$) simply tracks the input current $I_{in}$. However, when the switch is opened, a constant current is delivered to the output at a value equal to the input current at the moment the switch was opened. This

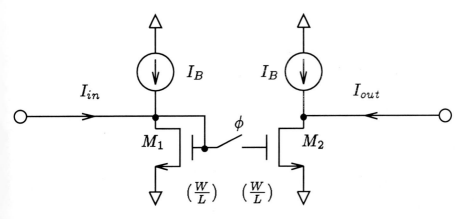

**FIGURE 9.22**
A current track-and-hold circuit.

constant output current is maintained by the voltage $V_{GS}$ of $M_2$ which in turn is maintained by the charge stored on the gate-source capacitance of $M_2$.

Cascading two track-and-hold stages in a master-slave arrangement, together with an external feedback path, enables present output currents to be combined with past or present inputs to implement either the inverting or non-inverting integrator transfer functions. For instance, in Fig. 9.23(a) we show an SI circuit that implements the non-inverting integration operation. It is straightforward to write a set of difference equations in terms of the circuit currents, transform them to the z-domain via the z-transform, and express the transfer function of the input to the first output $I_{o_1}$ according to the following

$$\frac{I_{o_1}}{I_{in}}(z) = K_1 \frac{1}{z-1} \qquad (9.53)$$

or using our alternative notation, we can write

$$\frac{I_{o_1}}{I_{in}} = K_1 \frac{z^{-12}}{2\gamma} \qquad (9.54)$$

Other current outputs are generated by connecting other current mirrors to the gate of transistor $M_4$ as is also shown in Fig. 9.23(a). Thus, this circuit is able to generate various positively-scaled integrated versions of the input current signal. Graphically, we can represent the flow of signals from the input to the multiple outputs of this circuit with the SFG shown in Fig. 9.23(b). To simplify our diagrams, we shall denote this integrator circuit with the symbol or short-

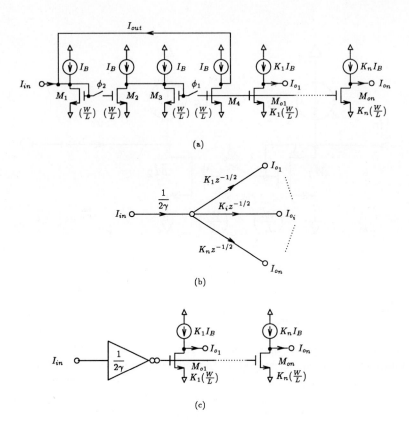

**FIGURE 9.23**
(a) A multiple-output SI non-inverting integrator circuit. (b) An SFG representation of the circuit shown in (a). (c) Short-hand notation.

hand notation shown in Fig. 9.23(c). The triangle with two circles represents the sample-and-hold portion of the integrator (i.e. transistors $M_1$, $M_2$, $M_3$, and $M_4$). We attach two small circles at the output of this triangle to denote that it is a non-inverting integrator.

An inverting SI integrator circuit is realized with the exact same circuit structure as the non-inverting integrator; however, the output is derived from the other track-and-hold circuit as is depicted in Fig. 9.24(a). The transfer function from the input to one output, for instance $I_{o_1}$, is

$$\frac{I_{o_1}}{I_{in}}(z) = -K_1 \frac{z}{z-1} = -K_1 \frac{z^{12}}{2\gamma} \tag{9.55}$$

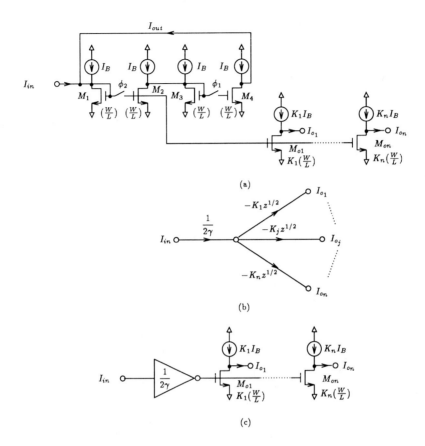

**FIGURE 9.24**
(a) A multiple-output SI inverting integrator circuit. (b) A SFG representation of the inverting SI integrator. (c) Short-hand notation.

The SFG representation for the inverting integrator is shown in Fig. 9.24(b) and its short-hand notation is provided in Fig. 9.24(c). Note that a single circle is attached to the triangular symbol to denote that it is an inverting integrator.

## 9.6.2 Comparing the SFGs of SI and SC Integrator Circuits

An important relationship exists between a multiple-output SI integrator and a multiple-input SC integrator of the same polarity. To see this, we show in Fig. 9.25(a) a multiple-input SC non-inverting integrator circuit, and in Fig. 9.25(b) its SFG. Now, if we reverse the direction of all the branches in this SFG while maintaining the same branch transmittances and convert the summing nodes into branch nodes, and vise-versa, then the SFG shown in Fig. 9.25(c) results.

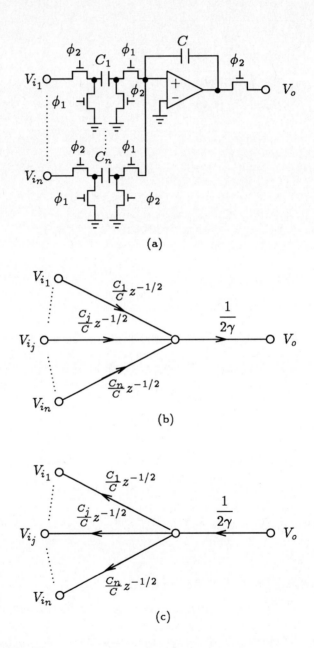

**FIGURE 9.25**

(a) A multiple-input SC non-inverting integrator circuit. (b) An SFG representation of the circuit in (a). (c) A transposed version of the SFG shown in (b).

Observe that in this transposed SFG the transfer function $\frac{V_{i_1}}{V_o}$ is the same as the corresponding transfer function $\frac{V_o}{V_{i_1}}$ in the original SFG; $\frac{V_{i_2}}{V_o}$ is the same as $\frac{V_o}{V_{i_2}}$ and so on. Also observe that the SFG of Fig. 9.25(c) is identical to that in Fig. 9.23(b) provided that

$$K_1 = C_1/C, \quad K_2 = C_2/C, \quad etc.,$$

and that the following signal correspondences are made:

$$V_o <=> I_{in}, \quad V_{i_1} <=> I_{o_1}, \quad etc.$$

Thus, from the theory of SFGs [33], we can state that the SFG for a multiple-input SC non-inverting integrator is equivalent to the transpose of the SFG representation of a multiple-output SI non-inverting integrator. Clearly, the same holds true for the inverting integrator case.

The procedure just described is known as *SFG transposition*. Although, this procedure can be applied to any SFG, we are particularly interested in its application to the SFG of SC filters constructed of multiple-input single-output integrators of the type shown in Fig. 9.25(a) (and of course the inverting type). We have shown in the above that the transpose of the SFG of each SC integrator is an SFG of a single-input, multiple-output SI integrator. Now, since the transposed SFG will have the same input-to- output transfer function as that of the original SFG, it constitutes a realization of the original filter. It follows that by realizing the individual integrator SFGs with SI integrator circuits we obtain an SI realization of the filter. Thus we have a method for converting any integrator-based SC filter realization into an SI filter realization.

In the network theory literature, two networks that are topologically identical and that realize the same transfer function when the excitation and response of one network are interchanged with the response and excitation of the other are said to be *inter-reciprocal* [34]. We can therefore conclude that SI circuits generated from SC circuits using SFG transposition form inter-reciprocal pairs. As a consequence of their inter-reciprocity, they will also possess identical component sensitivities [16]. This is important because it suggests that the transformation of low sensitivity SC filter circuits will lead to low sensitivity SI filter circuits.

To illustrate our synthesis procedure, we show in Fig. 9.27(a) a third-order LDI lowpass SC filter circuit. This SC circuit structure is slightly different than the SC circuit of Fig. 9.20 that resulted from the exact design method of Sec. 9.4. It consists of only LDI integrators. As a result, this type of SC circuit can only approximate the magnitude behavior of an LC ladder owing to the frequency dependence of the two damping sections of this circuit as outlined in Sec. 9.2. Nonetheless, if the clock frequency is much larger than the input signal frequencies then a filter response quite close to the intended response can result.

Let us now transform the SC circuit of Fig. 9.27(a) into an SI circuit using the method just outlined. We shall assume that the value of each capacitor of the SC circuit is known. Based on our knowledge of SC circuits we can draw the

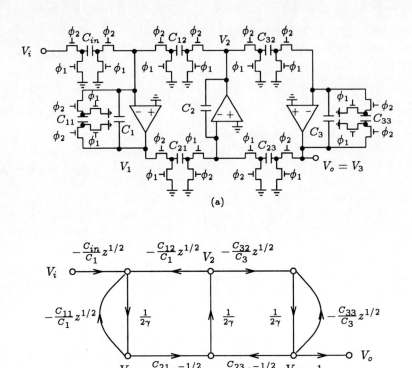

**FIGURE 9.26**
(a) A third-order LDI low-pass SC filter circuit. (b) An SFG representation of SI circuit implementation of the SFG shown in (c).

SFG for this SC circuit in terms of the basic integrator blocks. This is shown in Fig. 9.27(b). The transpose of this SFG is that shown in Fig. 9.27(c). Notice from this SFG that the integrators pointing downward on the page together with the branches that span outward from these integrators with transmittances proportional to $z^{-12}$ can be replaced with the single-input, multiple-output SI non-inverting integrator of Fig. 9.23(a). Conversely, the integrators pointing upward, together with fan-out branches with transmittances proportional to $z^{12}$ can be replaced with the single-input, multiple-output SI inverting integrator of Fig. 9.24(a). On doing so, the circuit shown in Fig. 9.27(d) results. We have simplified the circuit schematic by representing the sample-and-hold portion of each SI integrator by its short-hand notation.

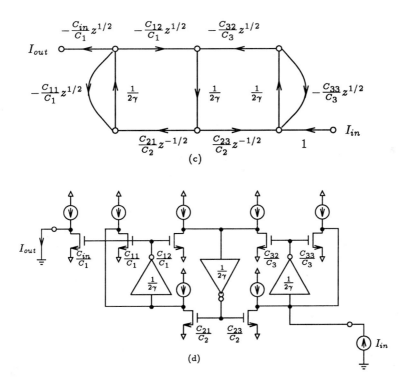

**FIGURE 9.26**
(c) A transposed version of the SFG shown in (b). (d) And the circuit in (a).

### 9.6.3  Switched-Current Ladder Filters

We conclude our discussion of SI filter synthesis by demonstrating how SC filter circuits synthesized using the exact design method outlined in 9ection 9.4 can be transformed into corresponding SI circuits. To illustrate this process, consider the SFG of the fifth-order SC filter circuit of Fig. 9.20, shown in Fig. 9.27(a). The basic structure of this SFG is similar to that seen previously in Fig. 9.6.2(b) with several exceptions. The most notable ones are the branches that interconnect the various integrators having transmittances proportional to $2\gamma$. These branches correspond to the unswitched capacitors in the SC circuit. In addition, one other branch that did not appear in the SFG seen previously in Fig. 9.6.2 is the branch $-\frac{C_i}{C_3}z^{-12}$ that interconnects the sampled-and-held input to the first integrator. Up to this point we have not yet seen any SI circuits that realize these transmittances; however, as we shall see below, these branches pose no difficulty for SI technology.

Let us proceed and transpose the SFG corresponding to the SC circuit. The result is shown in Fig. 9.27(b). On inspection, we see that each downward-

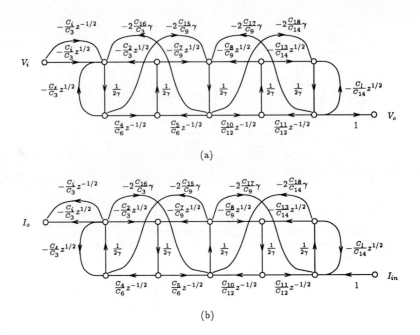

(a)

(b)

**FIGURE 9.27**
(a) An SFG representation of the fifth-order SC circuit shown in Fig. 9.19. (b) A transposed version of the SFG shown in (a).

pointing integrator branch, together with the $z^{-12}$ branches that fan out from its output, can simply be realized with the multiple-output, non-inverting SI integrator of Fig. 9.23(a). In contrast, the integrators that point upward have various types of transmittances that fan outward. Only the branches that have transmittances proportional to $z^{12}$ can be replaced with the multiple-output inverting SI integrator of Fig. 9.24(a). In order to realize the remaining branch transmittances, we extend the structure of the basic SI integrator such that it includes these additional transmittances. This is shown in Fig. 9.28(a). It consists of a single current input and three different types of outputs: an inverting integrator, a non-inverting integrator, and an inverting amplifier. The transfer function from the input to each output is found by writing a set of difference equations, and transforming the result into the $z$-domain. The results are:

$$\frac{I_{o1}}{I_{in}}(z) = K_1 \frac{1}{z-1} = K_1 \frac{z^{-12}}{2\gamma} \tag{9.56}$$

$$\frac{I_{o2}}{I_{in}}(z) = -K_2 \tag{9.57}$$

$$\frac{I_{o3}}{I_{in}}(z) = -K_3 \frac{z}{z-1} = -K_3 \frac{z^{12}}{2\gamma} \tag{9.58}$$

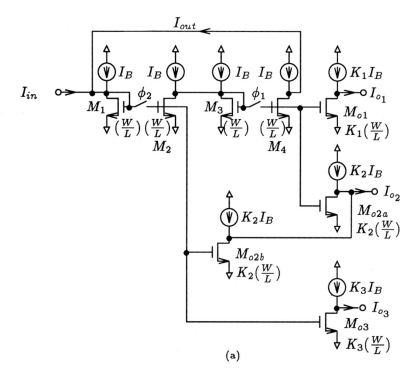

(a)

**FIGURE 9.28**
(a) General first-order SI building block.

Extension to multiple outputs of each type is achieved by simply attaching additional current mirror transistors. Inverted versions of any one of these outputs can be obtained by simply placing an additional current mirror in cascade with the output. The SFG representation of this general first-order SI building block is shown in Fig. 9.28(b). We denote the sample-and-hold portion of the building block with another triangular symbol having three separate outputs as shown in Fig. 9.28(c). Now that we can realize all branches of the SFG in Fig. 9.27(b), we can replace each one of them with the appropriate SI circuit and generate the fifth-order SI filter circuit shown in Fig. 9.29. Issues that pertain to the scaling of this SI circuit as to maximize its dynamic range and its efficient use of silicon area are unclear at this stage of development and represents an active area of research.

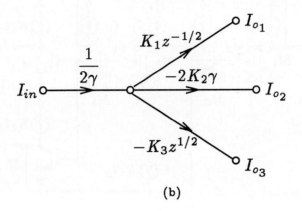

(b)

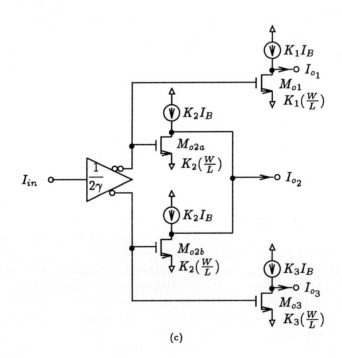

(c)

**FIGURE 9.28**
(b) SFG representation. (c) Short-hand notation.

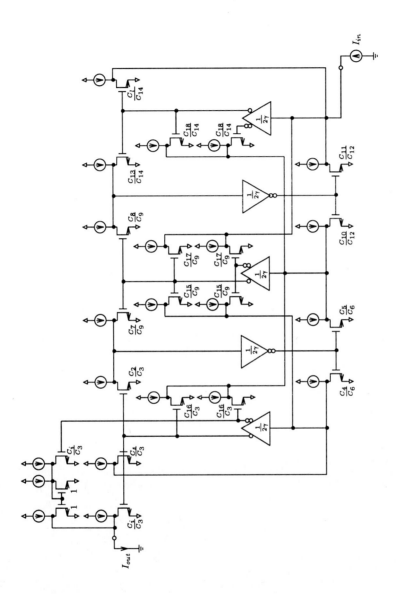

**FIGURE 9.29**

Switched-current circuit that simulates the operation of the fifth-order LC ladder network of Fig. 9.19. This circuit was obtained by transposing the SFG of the SC filter circuit shown in Fig. 9.20 and replacing the appropriate branches of the transposed SFG with SI circuits.

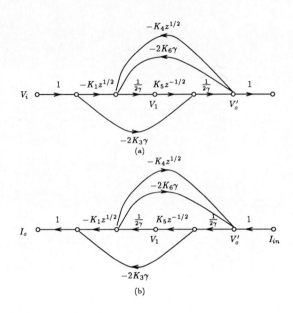

**FIGURE 9.30**
(a) SFG for SC circuit of Fig. 9.17 with $K_2 = 0$. (b) Transposed SFG of that shown in part (a).

**Example 9.3.**

For the second-order SC filter circuit shown in Fig. 9.17 the following capacitor values have been derived to provide a second-order, low-pass filter function:

$$C_1 = 1.0 \qquad C_2 = 1.0 \qquad K_1 = 1.0824 \qquad K_2 = 0$$
$$K_3 = 0.2929 \quad K_4 = 1.0824 \quad K_5 = 1.0824 \quad K_6 = 0.7654$$

Create an SFG representation of this circuit, transpose the resulting SFG and create a corresponding SI filter circuit. The sampling rate is assumed equal to 100 kHz.
*Solution:*
The SFG for the second-order SC filter circuit of Fig. 9.17 with $K_2 = 0$ is shown in Fig. 9.30(a). The corresponding transposed SFG is then shown in part (b) of Fig. 9.30. The input and output nodes of the transposed SFG have been re-labelled as current variables to indicate its reference to a current-mode circuit. Finally, the SI circuit implementation shown in Fig. 9.31 is obtained directly from the SFG of Fig. 9.30(b).

## 9.7 CONCLUSION

This chapter has provided methods for the synthesis of two different types of sampled-data analog filters for monolithic implementation: those using the well-

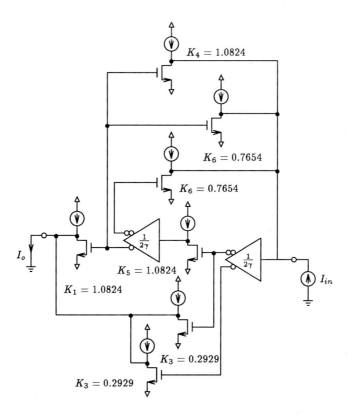

**FIGURE 9.31**
SI circuit corresponding to the SFG of Fig. 9.30(b).

established switched-capacitor technique, and filters utilizing the more recently proposed switched-current technique.

Beginning with a set of first-order SC building blocks, various two-integrator-loops structures were proposed, with the advantages of each described. Subsequently, a general biquadratic filter circuit was presented with which any second-order filter function can be realized. Such second-order circuits can be connected in cascade to realize an arbitrary high-order filter function.

An alternative approach to realizing high-order filter functions is by simulating the behavior of an LC ladder network whose attenuation characteristics meet the given specifications. This approach generally leads to filter circuits with low sensitivity to manufacturing component variations. An exact design method was given which makes use of the two-integrator loop sections proposed earlier.

Finally, we concluded the chapter by demonstrating how SC filter circuits synthesized using the methods given here can be transformed into corresponding SI filter circuits with exactly the same sensitivities to component variations.

## PROBLEMS

**9.1.** The simple logic circuit of Fig. P9.1 consisting of two NOR gates is used to generate a two-phase nonoverlapping clock signal for SC filter circuits. Assuming that each NOR gate has a constant propagation delay of 5 ns and the input is driven by a 10 kHz square-wave input signal, sketch the voltage signal that appears at the output of each gate. Assume each gate has an output voltage swing of 5-volts.

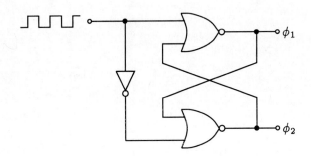

**FIGURE P9.1**

**9.2.** Using first principles, derive the first-order difference equation that describes the input–output behavior of a noninverting SC integrator circuit. Repeat for the inverting integrator case.

**9.3.** Compare the frequency response behavior (both magnitude and phase) of a SC non-inverting and inverting integrator over a normalized frequency interval $(\omega T)$ between 0 to $\pi$.

**9.4.** If a non-inverting integrator is connected in cascade with an inverting integrator with all capacitors of equal size, plot the phase behavior of the two-integrator cascade as a function of normalized frequency $(\omega T)$ over the interval between 0 to $\pi$. How does this phase behavior compare with a corresponding arrangement of active-RC integrators?

**9.5.** The time-constant associated with an SC integrator circuit is formed as the ratio of two capacitors. Let us denote this time-constant as $C_1/C_2$. Consider the effect of capacitor variations on the integrator time-constant assuming: (a) capacitor $C_1$ varies by 20 percent and $C_2$ is ideal; (a) capacitor $C_2$ varies by 20 percent and $C_1$ is ideal; (c) both $C_1$ and $C_2$ vary by 20 percent. What general conclusions can you state about the effect of correlated capacitor errors on the time-constant of SC integrator circuits?

**9.6.** Physical frequencies in the $z$-domain represent points on the unit circle. Sketch corresponding physical frequencies in the $\gamma$-plane. Indicate critical frequencies such as dc and half the sampling frequency.

**9.7.** If a new frequency variable is created called *delta* ($\Delta$) and is related to the $z$-variable according to the following,

$$\Delta = z - 1$$

sketch the contour in the $\Delta$-plane that corresponds to physical frequencies.

**9.8.** If a new frequency variable is created called *beta* ($\beta$) and is related to the $z$-variable according to the following,

$$\beta = \frac{z-1}{z}$$

sketch the contour in the $\beta$-plane that corresponds to physical frequencies.

**9.9.** Using first principles, derive the input–output transfer function for the two fully-differential SC integrator circuits of Fig. 9.4.

**9.10.** Assuming an ideal op amp, what dc-voltage relative to analog ground appears at the input terminals of the op amp in a fully-differential SC integrator circuit? *Note: This is a trick question and requires careful thought!*

**9.11.** A low-pass SC filter circuit is to be designed to have passband and stopband edges located at 1 kHz and 5 kHz, respectively. The clock frequency is set at 20 kHz. Prewarp these specifications using the bilinear transform and determine the frequencies at which the passband and stopband edges are located.

**9.12.** If a bandpass filter is designed in the $\lambda$-plane to have a lower and upper passband edge of 5 rad/sec and 6 rad/sec, respectively, determine the location of the passband region in the $z$-plane assuming that the clock rate is 20 kHz. What is the resulting bandwidth?

**9.13.** A fifth-order, unity-gain discrete-time filter transfer function is specified in terms of its pole and zero locations according to the following:

| *Poles* | *Zeros* |
|---|---|
| $0.99998588 \pm j6.56958682 \times 10^{-5}$ | $0.999999972 \pm j2.34553678 \times 10^{-4}$ |
| $0.99999581 \pm j9.75357925 \times 10^{-5}$ | $0.999999987 \pm j1.56506231 \times 10^{-4}$ |
| $0.99997948$ | $-1.00000$ |

Determine the location of the pole and zero locations in the $\lambda$-plane.

**9.14.** A fifth-order, $\lambda$-plane transfer function has the following pole and zero locations:

| *Poles* | *Zeros* |
|---|---|
| $-0.1685 \pm j0.4730$ | $\pm j1.436$ |
| $-0.0553 \pm j0.7216$ | $\pm j2.222$ |
| $-0.2294$ | $\infty$ |

Determine the location of the pole and zero locations in the $z$-plane.

**9.15.** Using first principles, derive the input–output transfer function of the bilinear SC integrator shown in Fig. 9.6(b). Assume that the input signal is sampled and held on clock phase $\phi_2$ (i.e., the input signal does not change during clock phase $\phi_1$).

**9.16.** Repeat problem 9.15 above but this time assume that the input signal is sampled and held on clock phase $\phi_1$ (i.e., the input signal does not change during clock phase $\phi_2$).

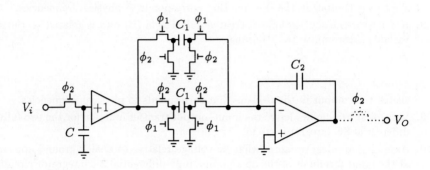

**FIGURE P9.17**

**9.17.** Figure P9.17 displays a two-op amp, single-ended version of a bilinear SC integrator circuit. Derive the input–output transfer function for this circuit.

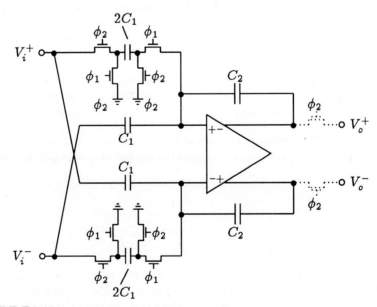

**FIGURE P9.18**

**9.18.** Figure P9.18 illustrates another circuit implementation for the bilinear SC integrator. Show that the input–output transfer function for this circuit is:

$$\frac{V_o^+ - V_o^-}{V_i^+ - V_i^-} = \frac{C_1}{C_2}\frac{\mu}{\gamma}$$

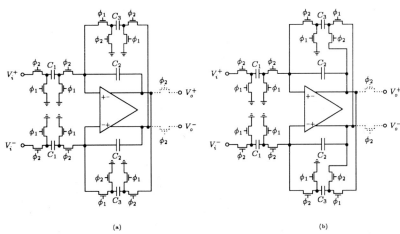

(a)           (b)

**FIGURE P9.19**

**9.19.** SC circuits have traditionally been damped with negative feedback and a "positive resistor" combination as shown in Fig. 9.7. Using fully-differential circuits other damping arrangements can be created, such as those seen in Figure P9.19. Derive the input–output transfer function for these two damped integrator circuits.

**9.20.** Repeat Ex. 9-1 but this time realize a positive version of the bandpass function.

**9.21.** Design a second-order, low-pass SC filter circuit with a 3-dB bandwidth of $10^3$ rad/sec and a dc-gain $= 1$. Assume a sampling rate of $5 \times 10^3$ rad/sec.

**9.22.** An application requires a notch filter to reject a 60 Hz signal appearing at the output of some pre-amp circuit. Using the second-order SC circuit of Fig. 9.17, select the capacitor values such that the center frequency of the notch occurs at 60 Hz, has a 3-dB bandwidth of 10 Hz, and the low- and high-frequency gain is unity. Let us assume that the sampling rate is 8 kHz.

**9.23.** Derive the design equations outlined in Section 9.3 for the second-order biquad circuit of Fig. 9.17.

**9.24.** Prove the following relationships:

$$\lambda = \frac{\gamma}{\mu}$$

and

$$\mu^2 = 1 + \gamma^2$$

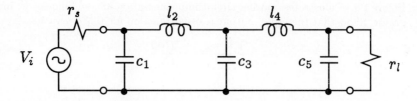

**FIGURE P9.25**

**9.25.** The doubly-terminated LC ladder of Fig. P9.25 is a realization of a fifth-order Chebyshev low-pass filter with a frequency normalized passband edge of 1 rad/sec. The passband has an attenuation ripple of 1 dB. The component values for the LC ladder are found from a computer program to be as follows:

$$r_s = 1.000 \quad c_1 = 2.1349 \quad l_2 = 1.0911 \quad c_3 = 3.0009$$
$$l_4 = 1.0911 \quad c_5 = 2.1349 \quad r_l = 1.000$$

Following the method outlined in Section 9.4 design an SC filter circuit using bilinear integrators that simulates the operation of the above LC ladder network assuming the clock frequency is 10 kHz. The passband edge is to be located at 1 kHz. Create a neat sketch of your circuit.

**9.26.** Repeat Prob. 9.25 above, but this time, design the SC filter circuit to minimize the effect of op amp finite gain and bandwidth.

**9.27.** Create a fifth-order SI filter realization of the SC filter circuit designed in Ex. 9-2.

**9.28.** Create an SI filter realization of the Chebyshev low-pass SC filter circuit designed in Prob. 9.25.

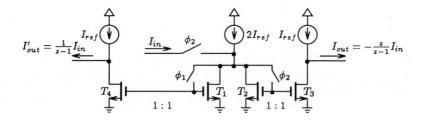

**FIGURE P9.29**

**9.29.** An improved SI integrator circuit is provided in Fig. P9.29. It is referred to in the literature as a second-generation SI integrator. Using first principles, derive the input-output transfer function for the two current outputs.

# REFERENCES

[1]  J. T. Caves, M. A. Copeland, C. F. Rahim, and S. D. Rosenbaum, "Sampled analog filtering using switched capacitors as resistor equivalents," *IEEE J. Solid-State Circuits,* vol. SC-12, pp. 592-600, December 1977.

[2]  B. J. Hosticka, R. W. Brodersen, and P. R. Gray, "MOS sampled-data recursive filters using switched-capacitor integrators," *IEEE J. Solid-State Circuits,* vol. SC-12, pp. 600-608, December 1977.

[3]  J. B. Hughes, N. C. Bird, and I. C. Macbeth, "Switched currents – a new technique for analog sampled-data signal processing," *Proc. of IEEE Intl. Symp. on Circuits and Systems,* pp. 1584-1587, May. 1989.

[4]  J. B. Hughes, I. C. Macbeth, and D. M. Pattullo, "Switched current filters," IEE Proceedings (Part G), Special issue on current-mode analogue signal processing circuits, Vol. 137, No. 2, pp. 156-162, April 1990.

[5]  T. Fiez and D. Allstot, "CMOS switched-current ladder filters," *IEEE J. Solid-State Circuits,* vol. SC-25, pp. 1360-1367, December 1990.

[6]  G. W. Roberts and A. S. Sedra, "Synthesizing switched-current filters by transposing the SFG of switched-capacitor filter circuits," *IEEE Trans. Circuits Syst.,* vol. CAS-38, pp. 337-340, March 1991.

[7]  K. Martin, "Improved circuits for the realization of switched-capacitor filters," *IEEE Trans. Circuits Syst.,* vol. CAS-27, pp. 237-244, April 1980.

[8]  K. Martin and A. S. Sedra, "Stray-insensitive switched-capacitor filters based on the bilinear z transform," *Electronics Letters,* vol. 19, pp. 365-366, June 1979.

[9]  L. T. Bruton, "Low-sensitivity digital ladder filters, *IEEE Trans. Circuits Syst.,* vol. CAS-22, pp. 168-176, March 1975.

[10]  K. Martin and A. S. Sedra, "Effects of the operational amplifier gain and bandwidth on the performance of switched-capacitor filters," *IEEE Trans. Circuits Syst.,* vol. CAS-28, pp. 822-829, August 1981.

[11]  D. B. Ribner and M. A. Copeland, "Biquad alternatives for high-frequency switched-capacitor filters," *IEEE J. Solid-State Circuits,* vol. SC-20, no. 6, pp. 1085-1095, December 1985.

[12]  G. W. Roberts, D. G. Nairn, and A. S. Sedra, "On the implementation of fully differential switched-capacitor filters," *IEEE Trans. Circuits Syst.,* vol. CAS-33, no. 4, pp. 452-455, April 1986.

[13]  W. Wilson, H. Massoud, E. Swanson, R. George, and R. Fair, "Measurement and modeling of charge feedthrough in n-channel MOS analog switches," *IEEE J. Solid-State Circuits,* vol. SC-20, pp. 1206-1213, December 1985.

[14]  C. Eichenberger and W. Guggenbuhl, "On charge injection in analog MOS switches and dummy compensation techniques," *IEEE Trans. Circuits Syst.,* vol. CAS-37, no. 2, pp. 256-264, February 1990.

[15]  K. C. Hsieh, P. R. Gray, D. Senderowicz, and D. G. Messerschmitt, "A low-noise chopper-stabilized differential switched-capacitor filtering technique," *IEEE J. Solid-State Circuits,* vol. SC-16, pp. 708-715, December 1981.

[16]  A. V. Oppenheim and R. W. Schafer, *Digital Signal Processing,* Prentice-Hall, Englewood Cliffs, N.J., 1975.

[17]  A. S. Sedra and P. O. Brackett, *Filter Theory and Design: Active and Passive,* Matrix Publishers, Portland, Oregon, 1978.

[18]  A. I. Zverev, *Handbook of Filter Synthesis,* John Wiley and Sons, Inc., New York, 1967.

[19]  W. M. Snelgrove, "FILTOR-2: A computer aided filter design program," Dept. of Elec. Engrg., Univ. of Toronto, Toronto, Canada, 1981.

[20] M. S. Lee, G. C. Temes, C. Chang, and M. G. Ghaderi, "Bilinear switched-capacitor ladder filters," *IEEE Trans. Circuits Syst.*, vol. CAS-28, no. 8, pp. 811-822, August 1981.

[21] D. Senderowicz, S. F. Dreyer, J. H. Huggins, C. F. Rahim, and C. A. Laber, "A family of differential NMOS analog circuits for a PCM CODEC filter chip," *IEEE J. Solid-State Circuits*, vol. SC-17, pp. 1014-1023, December 1982.

[22] F. Montecchi, "Time-shared switched-capacitor ladder filters insensitive to parasitic effects," *IEEE Trans. Circuits Syst.*, vol. CAS-31, pp. 349-353, April 1984.

[23] P. E. Fleischer and K. R. Laker, "A family of active switched-capacitor biquad building blocks," *Bell Syst. Tech. J.*, vol. 58, pp. 2235-2269, December 1979.

[24] K. Martin and A. S. Sedra, "Exact design of switched-capacitor bandpass filters using coupled-biquad structures," *IEEE Trans. Circuits Syst.*, vol. CAS-27, pp. 469-475, June 1980.

[25] G. W. Roberts and A. S. Sedra, "Switched-capacitor filter networks derived from general parameter bandpass ladder networks," *Proc. IEEE Intl. Symp. Circ. Syst.*, pp. 1005-1008, May 1988.

[26] H. J. Orchard, "Inductorless filter," *Electronics Letters*, vol. 2, pp. 224-225, June 1966.

[27] R. B. Datar and A. S. Sedra, "Exact design of stray-insensitive switched-capacitor ladder filters," *IEEE Trans. Circuits Syst.*, vol. CAS-30, pp. 888-898, December 1983.

[28] A. S. Sedra, "Active-RC Filters," in *Miniaturized and Integrated Filters*, S. K. Mitra and C. F. Kurth, (Eds), New York, Wiley, 1989.

[29] S. C. Fang, Y. Tsividis, and O. Wing, "SWITCAP: A switched-capacitor network analysis program," Parts I and II, *IEEE Circuits and Systems Magazine*, September and December 1983.

[30] R. B. Datar, "Exact design of stray-insensitive switched-capacitor ladder filters," Ph.D. thesis, Dept. of Elec. Engrg., Univ. of Toronto, October 1983.

[31] R. B. Datar and A. S. Sedra, "Exact design of stray-insensitive switched-capacitor high-pass ladder filters," *Electronics Letters*, vol. 19, no. 29, pp. 1010-1012, November 1983.

[32] E. A. Vittoz, "Dynamic analog techniques," in *MOS VLSI Circuits for Telecommunications*, Y. Tvisidis, and P. Antognetti, (Eds), Englewood Cliffs, N.J.: Prentice-Hall, 1985.

[33] S. J. Mason and H. J. Zimmermann, *Electronic Circuits, Signals, and Systems,* John Wiley & Sons, Inc., New York, 1960.

[34] J. L. Bordewijk, "Inter-reciprocity applied to electrical networks," Applied Science Research, vol. B6, pp. 1-74, 1956.

# CHAPTER
# 10

## OVERSAMPLED
## A/D CONVERTERS

## 10.1 INTRODUCTION

As the cost of VLSI circuits decreases there is a greater demand on the analog-digital interfaces between these VLSI circuits and the analog world. The importance of developing low cost analog-digital interfaces in applications ranging from instrumentation, voice, audio, and video coding cannot be overemphasized. Due to the large number of analog-digital interfaces in VLSI systems, the development of low-cost monolithic MOS analog-digital conversion techniques has been an important element in reducing the cost of the total system. As MOS technology feature sizes continue to fall, the high speed, programmability, and low production cost offered by VLSI technologies are making digital signal processors a more attractive alternative to conventional analog signal processing. To exploit the significant computational power offered by these processors, high resolution analog-to-digital interfaces must be developed.

This chapter reviews the oversampling technique as a method of performing high resolution data conversion. The technique is especially attractive for VLSI implementations because it can take advantage of the inherent speed advantage of the digital VLSI technology [1-11]. In Sec. 10.2, conventional ADCs (Analog-to-Digital Converters) operating at the Nyquist frequency will be reviewed together with a summary of their limitations. Two broad classes of oversampled ADCs, based on prediction and noise shaping, will be compared in Sec. 10.3. The material in Sec. 10.3.2 concentrates on a particular class of oversampled ADC, the

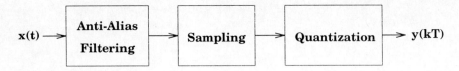

**FIGURE 10.1**
A/D converter block diagram.

sigma-delta modulator. Various techniques of realizing the accompanying decimation filters are presented in Section 10.8. Finally, a silicon implementation of
a multi-channel voiceband coder using sigma-delta modulators is presented in
Section 10.9.

## 10.2 NYQUIST RATE A/D CONVERTERS

Shown in Fig. 10.1 are the primary functions of an Analog-to-Digital (A/D)
converter which converts a continuous-time analog input signal, $x(t)$, into a sequence of digital codes, $y(kT)$, at a rate $f_N$. To prevent aliasing, the analog filter
in front of the sampling block in Fig. 10.1 limits the bandwidth of $x(t)$ to half
the sampling rate. The sampled signal is then converted to digital codes, $y(kT)$,
by the quantizer.

A/D converter architectures can be subdivided into two groups based on the
sampling rate of the converter. Nyquist rate converters are defined as converters
that sample the analog signal at its Nyquist frequency $f_N = 2W$, where $W$ is the
input signal bandwidth of the converter input. Oversampled converters denote
converters that sample at a much higher rate than the Nyquist, $f_s \gg f_N$. The
ratio of the sampling frequency to the Nyquist rate is defined as the oversampling
ratio, $D$. The key is that for many applications, the bandwidth of the signal is
small compared to the typical speed of operation available in VLSI circuits. In
these cases, oversampled A/D converters offer the potential to trade this excess
speed for improved resolution.

### 10.2.1 Classifications

The Nyquist rate architectures can be classified according to the number of clock
cycles required to perform the quantization, which may vary from 1 to $2^N$, where
N is the resolution of the converter in bits. In a parallel A/D converter, commonly
referred to as flash A/D converter, all of the possible quantization levels are
simultaneously compared to the sampled analog input. To achieve the parallel
operations $2^N - 1$ comparators are required for an N-bit converter. In practical
implementations, the limitations in the resolution result from the exponential
increase in the number of comparators as a function of the number of bits. For
an additional bit of resolution, the matching and accuracy requirements in the
reference circuits will double. To date, most flash A/D converters have been
limited to approximately 10 bits of resolution.

To avoid the resolution limitations of fully parallel converters, the quantization operation may be distributed over extra clock cycles. For example, in a two-stage pipelined A/D converter, two steps are included in a conversion cycle. In the first pass, the conversion consists of a coarse quantization of the most significant bits (MSBs). The MSBs from the flash A/D subconverter are converted back to analog form by a D/A subconverter and subtracted from the input signal. The difference, or residue, is then subject to a fine quantization during the second pass, thus producing the least significant bits (LSBs). Another method, called subranging, includes performing coarse quantization during the first clock cycle to zoom in on the range of the input signal. In the second clock cycle this information is used to provide a suitable set of reference levels for the fine quantization process.

As a trade-off, the number of comparators can be further reduced by performing the data conversion with a cascade of low-resolution stages. This reduction is accomplished at the expense of increased conversion time. In the extreme case, the stages can be pipelined to allow a one-bit conversion per stage, but with latency equal to the total number of bits.

In architectures in which one bit is resolved per clock cycle, only one comparator is required. Two distinct types of converters fall into this category. In successive approximation converters, starting from the MSB, a feedback voltage $V_f$ is used to approximate the input analog voltage, $V_{in}$, in a sequence of successive steps. During each step, $V_f$ is changed in accordance with the result of the previous comparison between $V_{in}$ and $V_f$. At the same time, the value of each bit is determined based on whether or not it causes the approximation to exceed the input. In algorithmic converters, the successive approximation algorithm is carried out serially in time. The difference between the converter input and a weighted value of the reference voltage, as determined by the previous bits, is multiplied by two. The difference is then compared with the reference voltage to determine the next bit.

At the lowest end of the data converter speed ladder are Nyquist-rate A/D converter architectures that perform the comparison between the input and each of $2^N$ reference levels sequentially. They are sometimes referred to as indirect converters, as a quantity like frequency or time is used as an intermediate variable to obtain the final digital output code. The number of clock cycles it takes to resolve N bits is then $2^N$.

## 10.2.2 Limitations of Nyquist Rate A/D Converters

**Anti-Aliasing**   Nyquist rate converters and oversampled converters differ in two fundamental aspects: anti-alias filtering and quantization. In the Nyquist rate converters, the filtering has to be done before the A/D converter, thus demanding the use of complex analog filters. In the case of oversampled converters, the filtering can be done in multiple stages [3, 12]. The analog anti-aliasing fil-

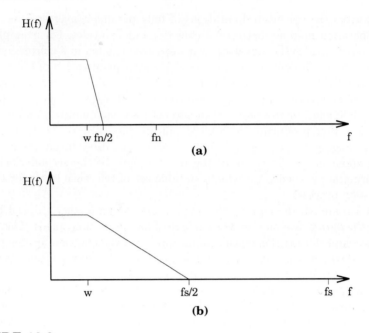

**FIGURE 10.2**
Requirements on the anti-alias filter for (a) Nyquist rate ADC, (b) Oversampled ADC where $f_n$ is the Nyquist frequency, $\omega$ is the input signal BW, and $f_s$ is the sampling frequency.

ter preceding the A/D conversion has a relaxed requirement due to the large "don't care" regions. The latter stages of filtering for subsequent decimation can be implemented digitally, thereby taking advantage of the enormous amount of digital signal processing capability provided by the VLSI technology. Shown in Fig. 10.2 is a comparison of the requirements on the anti-aliasing filter of Nyquist rate and oversampled A/D converters. The Nyquist rate converter requires the use of an anti-aliasing filter with a very sharp transition in order to provide adequate aliasing protection. The transition band of the anti-aliasing filter of an oversampled A/D converter, on the other hand, is much wider than its passband, since anti-aliasing suppression is required only for a small band of frequencies around half the sampling rate, $\frac{f_s}{2}$. This simplification results in a considerably simpler anti-aliasing filter for oversampled A/D converters.

**Quantization**   Another difference between the two converters takes place in quantization. If the analog samples can be represented exactly, there is no apparent advantage to sampling at a rate higher than the Nyquist rate. However, during the analog-to-digital conversion, quantization has to be performed, resulting in quantization errors. If more samples are taken, these errors can be reduced. For example, if the number of samples taken per unit time is doubled and each pair of samples produced is averaged to produce new samples at the original sample rate, the signal component of the samples will add linearly while the quantization noise component in the samples will add as random variables. If the noise is uncorrelated then the signal-to-noise ratio will improve by 3dB. Consequently, in the case of oversampled A/D converters, a series of rough amplitude quantizations is made for every analog sample. This redundancy is combined with the averaging provided by a digital decimation filter to obtain a more precise estimate for the analog input [13]. On the other hand, for a Nyquist rate A/D converter, quantization is performed for every sampling instant at the full precision of the converter. This full precision is determined by the accuracy of the analog circuit components. To appreciate more fully the advantage of the oversampled approach over the Nyquist approach the quantization process in actual implementations will now be discussed.

In VLSI implementations, the signal quantization is achieved by comparing the discrete-time representation of the analog signal to a set of reference voltages generated by passive elements. The result of the comparisons is to generate a digital code with a nearest lower level to the input level. The resolution of a converter is defined by the spacing of the reference levels that can be generated and resolved. For a converter with 16-bit or higher resolution, the spacing of these levels will have to be in the microvolt range. From the implementation viewpoint, the spacing is determined by the matching properties of either the resistors or capacitors employed to generate the reference voltages. It has been shown in the past that the matching tolerance of these passive components is substantially lower than the above requirement [14]. In the past, techniques such as laser trimming or self-calibration have been utilized to improve the resolution of A/D converters. Oversampling can be viewed as another technique that will push the resolution of the converter beyond the component tolerances imposed by the technology.

## 10.3   MODULATORS FOR OVERSAMPLED A/D CONVERSION

The quantizer in an oversampled A/D converter performs quantization by approximating the analog input with the nearest lower level from a set of discrete reference levels. The quantization error, $e$, is defined as the difference between the quantizer input and the value of the output. If the input to the quantizer is varying sufficiently, this error can be modeled as uncorrelated noise. The variance of the noise is a measure of the quantizer's resolution and in a Nyquist rate converter it is reduced by increasing the number of reference levels. In comparison, the quantization noise within the signal baseband of oversampled A/D converters with a quantizer with a given resolution is reduced through the use of feedback. Two approaches toward achieving this have been proposed: prediction and noise shaping [15, 16].

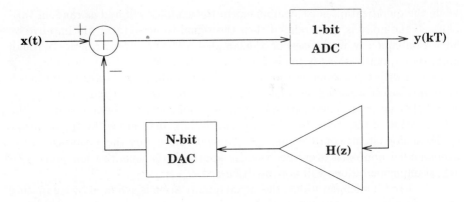

**FIGURE 10.3**
Block diagram of predictive coder.

## 10.3.1   Predictive Coders

The block diagram in Fig. 10.3 illustrates the basic architecture of a predictive or differential modulator [17]. For a first-order predictive coder, the transfer function H(z) will be an integrator. It works on the principle that if one knows the course of a signal up to a certain point in time, it is possible to make some inference about its continuation. This extrapolation process is called prediction. To achieve predictive encoding, the value of the present sample is used to calculate the next sample. The difference between the actual value and the estimation is quantized by the 1-bit ADC. The quantized value is then stored in the integrator. Since the error is encoded in digital form, to calculate the analog predicted value, a DAC (Digital-to-Analog Converter) is required in conjunction with either an analog or digital integrator. As an analog-to-digital interface, the advantage offered by prediction is that it reduces the dynamic range of the ADC by subtracting an estimate $\hat{X}(t)$ from the modulator input $x(t)$. The step size of the quantizer is adjusted to this reduced range, which results in a decrease of the quantization noise.

## 10.3.2   Noise Shaping Coders

Noise shaping modulators do not reduce the magnitude of the quantization noise, but instead shape the power density spectrum of this error, moving the energy toward high frequencies. Provided that the analog input to the quantizer is oversampled, the high-frequency quantization noise can be eliminated with a digital lowpass filter without affecting the signal band. Figure 10.4 shows the noise shaping function of a first-order coder.

A wide selection of implementations of both predictive and noise shaping modulators have been reported. The former, commonly known as delta modulators, perform an implicit differentiation of the input signal. If there is a mismatch between the integrators in the encoder and decoder, the signal can be distorted.

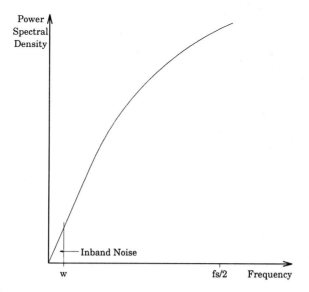

**FIGURE 10.4**
Power spectral density of a first-order noise shaping coder.

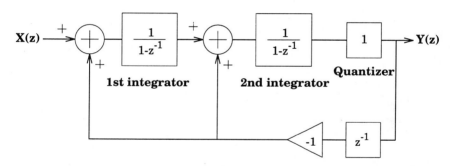

**FIGURE 10.5**
Second-order noise shaping coder (A).

In addition delta modulators exhibit slope overload for rapidly rising input signals and their performance is therefore dependent on the frequency of the input signal. Another potential shortcoming of the delta modulator is that, with the integrator in the feedback loop, any nonideality of the integrator will appear in the output [18]. Modulators that combine both the prediction and noise shaping techniques have been proposed. A multi-bit DAC is needed in the feedback loop that must be accurate because it determines the overall linearity of the converter. As in the case of predictive coding, the performance of these modulators is susceptible to the slope of the input signal.

Noise shaping modulators encode the signal itself and their performance is insensitive to the rate of change of the signal. When the internal ADC and DAC are implemented with one-bit resolution, the modulator is called delta-sigma

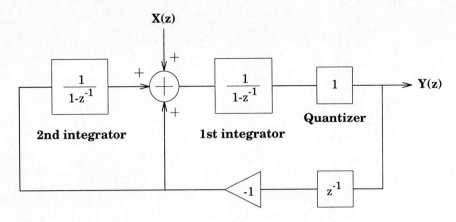

**FIGURE 10.6**
Oversampled coder with first-order noise shaping, first-order prediction (B).

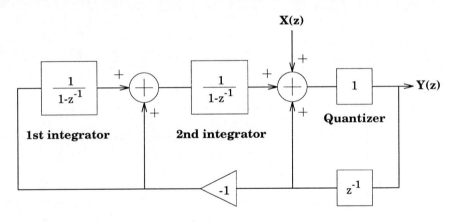

**FIGURE 10.7**
Second-order predictive coder (C).

modulator [19] or sigma-delta modulator [20]. The following discussion summarizes the implementation issues that differentiate the predictive coders from
the noise shaping coders [21]. A second-order oversampled modulator shown in
Fig. 10.5 has been selected as an example. Three configurations are shown: a
second-order noise shaping (Fig. 10.5), a first-order predictive with first-order
noise shaping (Fig. 10.6), and a second-order predictive converter (Fig. 10.7).
They will henceforth be denoted as structures A, B, and C respectively. The
signal transfer function (STF) and noise transfer function (NTF) for each structure is shown in Table 10.1. Theoretically each structure has no major difference
in terms of performance. However significant differences arise in actual VLSI
implementation.

First of all integrators in A and the first integrator in B are switched
capacitor integrators realized with analog circuits. The second integrator in B

**TABLE 10.1**
Transfer functions of the three structures.

| Structure | STF | NTF |
|-----------|-----|-----|
| A | 1 | $(1 - z^{-1})^2$ |
| B | $(1 - z^{-1})$ | $(1 - z^{-1})^2$ |
| C | $(1 - z^{-1})^2$ | $(1 - z^{-1})^2$ |

**TABLE 10.2**
Comparison among three structures.

| Structure | A | B | C |
|-----------|---|---|---|
| Number of analog stages | 2 | 1 | 0 |
| DAC resolution | 1 bit | medium | high |
| Complexity of decimation filter | large | medium | small |
| High speed multiplier | no | yes | yes |
| Decimation filter attenuation | high | medium | low |

and both integrators in C can be replaced by a digital integrator followed by a DAC. With this arrangement, the number of operational amplifiers required will be 2, 1, and 0 for A, B, and C, respectively. The resolution or number of bits required for the DAC is different for each type of topology. As an example if we assume the oversampling ratio is 128, then the resolution needed for the DAC will be 1-, 6-, and 12-bits for A, B, and C, respectively [21, 22]. In addition, in cases B and C, nonidealities introduced by the integrators and the multi-bit DACs will limit the linearity of the modulator.

For the decimation filter, multipliers operating at the oversampling frequency $f_s$ are required for the cases in B and C, but not in A, since the output contains only a single bit. With a second-order high-pass quantization noise spectrum, the highest out-of-band attenuation is required for the decimation filter in A. The differences between the three cases are now summarized in Table 10.2.

## 10.4  FIRST- AND SECOND-ORDER SIGMA-DELTA MODULATORS

The most widely used oversampled modulator architecture consists of a noise shaping modulator with a one-bit internal quantizer, called the sigma-delta modulator. A block diagram of the first-order sigma-delta modulator is shown in Fig. 10.8. It consists of a summing node, an integrator, and one-bit A/D and Digital-To-Analog, (D/A) converters in a feedback loop. Since the integrator has infinite gain at dc, the loop gain is infinite at dc, and therefore the dc-component or the average of the error signal is zero. Consequently, the dc-component or the

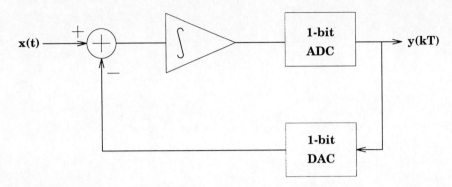

**FIGURE 10.8**
First-order sigma-delta modulator.

average of the output from the DAC will be identical to the dc-component of the input signal. This means that even though the quantization error at every sample is large because a quantizer with only two levels is being used, the average of the quantized signal, and therefore the modulator output, $y(kT)$, tracks the analog input signal, $x(t)$. This average is computed by the decimation filter that follows the converter.

In general the integrator output ramps up and down according to the value of the DAC. Correspondingly, the 1-bit ADC outputs a bit stream of ones and zeroes which is a pulse density modulated representation of the input dc-value. As an example, when the input is $\frac{1}{7}$ and the initial condition of the integrator is zero, the output sequence for the first 20 cycles is: "0," "0," "0," "0," "1," "0," "0," "0," "0," "0," "0," "1," "0," "0," "0," "0," "0," "0," "1," "0". The average value approaches $\frac{1}{7}$. The resolution of the converter increases when more samples are included in the averaging process, or as the oversampling ratio $D$ increases. However, the signal bandwidth is decreased at the same time. Consequently, the resolution of oversampled A/D converters is a function of the ratio of the sampling rate to the bandwidth of the converter. The principle of operation for oversampled A/D converters relies on this fundamental trade-off between resolution and time.

To analyze the operation of a first-order sigma-delta converter, the proper model will now be developed. The basic model of a typical first-order coder is shown in Fig. 10.9. The integrator usually has the delay in the forward path for ease of implementation using the switched capacitor technique. The input of the coder is defined as $x(t)$ and the output as $y(kT)$. The ADC in the forward loop can be modeled as a unity gain block in conjunction with an additive noise source having error, $e$, and the DAC in the feedback loop can be represented by a unity gain element. The difference equation describing the entire system is

$$y(kT) = x(kT - T) + e(kT) - e(kT - T) \qquad (10.1)$$

Note that when compared to the Nyquist rate converter whose input-output

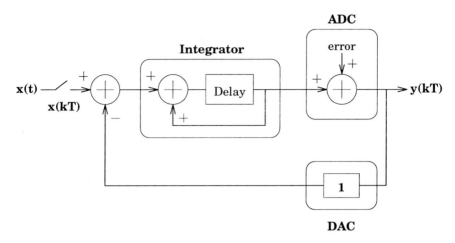

**FIGURE 10.9**
Model for first-order sigma-delta modulator.

relationship is described by

$$y(kT) = x(kT) + e(kT) \qquad (10.2)$$

the error becomes a differential error. This is a first-order difference relationship. In essence the coder is trying to cancel the error by subtracting quantization errors from two adjacent samples. This principle of reducing the error source by exploiting the statistics between them can be extended to higher-order coders, where more past error samples are involved in the cancellation process to reduce the overall error. Viewed from the frequency domain, this difference operation acts to attenuate the quantization noise at low frequencies, thus shaping the noise. Modulators with a second-order transfer function involve the cancellation of the two past samples and thus exhibit stronger attenuation at low frequencies. The noise shaping function of a first- and second-order modulator are compared in Fig. 10.10 [23]. Note that as the order of the system increases the quantization error is decreased.

The above analysis can be formalized to yield quantitative results for the resolution of sigma-delta modulators, provided that the spectral distribution of the quantization error, $e$, can be assumed to be uncorrelated [24]. Then the modulator can be regarded as a linear system for which the power spectral density of the noise can be calculated [25]. For a first-order system, the noise transfer function $NTF$ in discrete-time domain:

$$NTF = (1 - z^{-1}) \qquad (10.3)$$

The power spectral density of the noise $S_{nn}$ is given by

$$S_{nn} = 4\sigma_c^2 \sin^2 \left( \pi \frac{f}{f_s} \right) \qquad (10.4)$$

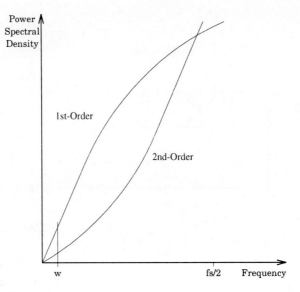

**FIGURE 10.10**
Power spectral density for the first and second-order sigma-delta modulators.

where $\sigma_c^2$ is the variance of the quantization noise due to the 1-bit quantizer and $f_s$ is the sampling frequency. If the passband is denoted by $f_c$ where $f_s \gg f_c$, the variance of the output noise is given by

$$\sigma_n^2 = \frac{8}{3}\pi^2 \sigma_c^2 \left(\frac{f_c}{f_s}\right)^3 \tag{10.5}$$

In general the dynamic range of an A/D converter is defined as the ratio of the output power for a full scale sinusoidal input to the output signal power for a small input whose SNR is unity (0dB). If the 1-bit quantizer in the sigma-delta modulator is modeled as an additive white noise source, then the dynamic range $DR$ is

$$DR = \frac{3}{2}\frac{2L+1}{\pi^{2L}} D^{2L+1} \tag{10.6}$$

where L is the order of the sigma-delta coder and D is the oversampling ratio. Figure 10.11 shows the dynamic range of first-, second-, and third-order sigma-delta modulators. The resolution is expressed as the equivalent number of bits of a Nyquist rate converter having the same dynamic range.

**Influence of Circuit Parameters**    Because the one-bit quantizer introduces gross nonlinearities, an analytical expression for predicting the influence of circuit nonidealities is difficult to obtain. Most past efforts have resorted to time-domain simulations [26]. As an example, let us take a look at a second-order sigma-delta modulator as shown in Fig.10.5, which is perhaps the most widely used configuration to date. An important nonideality arises from the circuit noise

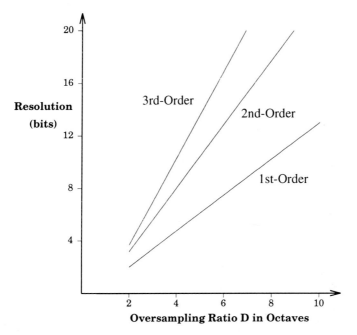

**FIGURE 10.11**
Dynamic range of sigma-delta modulators.

contribution. This places a lower limit on the achievable dynamic range. Noise can come from power-supply or substrate coupling, clock-signal feedthrough, 1/f and thermal noise generated by the transistors. All of these, except for the thermal noise, can be reduced by circuit and layout techniques, such as using a fully-differential configuration. The fundamental limitation imposed by thermal noise has been shown to be [26]

$$SNR = \frac{V_{dd}}{4} \sqrt{\frac{D \times C}{kT}} \tag{10.7}$$

where $V_{dd}$ is the power supply voltage, D is the oversampling ratio, and C is the size of the sampling capacitor.

Op amps used in the implementation of integrators can also limit the performance of the overall converter. The requirement on the op amp open-loop gain is relatively low, on the order of the oversampling ratio. In-band quantization noise starts to increase dramatically with lower open-loop gain values. The bandwidth of the op amp, on the other hand, determines the maximum speed at which the modulator can operate. Simulation results have confirmed that the loop is very tolerant of linear settling errors. For example, a 10% settling error introduces a negligible effect in the SNR at the 14-bit level.

Finally, the capacitor matching requirement is very relaxed and is set mainly to ensure a stable modulator. From simulations, the ratio of the gains

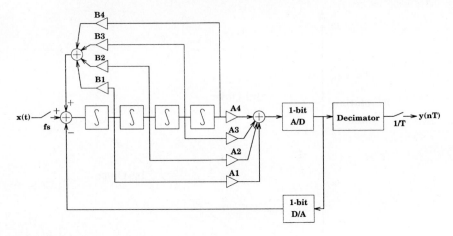

**FIGURE 10.12**
Third-order topology with feedback and feedforward coefficients.

of the two feedback paths needs to match to within 5% to yield satisfactory performance.

## 10.5 INTERPOLATIVE MODULATORS

As one moves to higher order modulators [4, 9, 27], the spectral density of the quantization noise can be modified to improve the stability and enhance quantization noise reduction. The spectral density may no longer have a $\sin^L$ form, where L is the order of the modulator. An interesting example of these coders is the interpolative modulator which incorporates both feedforward and feedback in the transfer function as shown in Fig. 10.12 [9]. The system function where the ADC is modeled as an additive noise source with a delay $z^{-1}$ is shown to be:

$$Y(z) = H_X(z)X(z) + H_E(z)E(z) \tag{10.8}$$

The quantization noise to output transfer function is shown to be

$$|H_E(f)| = \left| \left( \frac{\pi f_b}{f_s} \right)^N \left( \frac{2}{A_N} \right) T_N \left( \frac{f}{f_b} \right) \right| \quad \text{for } f \le f_b < f_s \tag{10.9}$$

Here $f_b$ is the baseband frequency, $f_s$ is the sampling frequency, $A_N$ is the feedforward coefficient as shown in Fig. 10.12 where N is 3. $T_N(x)$ is the Chebyshev polynomials defined by

$$T_0(x) = 1 \tag{10.10}$$

$$T_1(x) = x \tag{10.11}$$

$$T_N(x) = 2xT_{N-1}(x) - T_{N-2}(x) \tag{10.12}$$

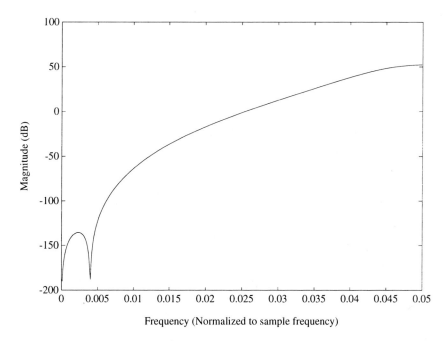

**FIGURE 10.13**
Filter seen by the quantization noise for the interpolative modulator.

In this design the poles are distributed throughout the signal band in order to lower the in-band noise. The zeros are chosen to flatten the filter response at high frequency in order to reduce the high-frequency noise and prevent it from reducing the dynamic range. The noise spectrum obtained from such a modulator is given in Fig. 10.13. Note the suppression provided by the Chebyshev zeros throughout the passband.

## 10.6   CASCADED ARCHITECTURE

While high-order sigma-delta modulators offer the potential to achieve higher resolution for a given oversampling ratio, modulators with more than two integrators exhibit the potential of being unstable. An architecture where several first-order or second-order modulators are cascaded in order to achieve a performance level that is comparable to that of higher-order modulators has been suggested as a means to overcome the stability problem. Another advantage of cascading is that it has been shown theoretically that if the number of stages is large, then the output is free from undesired limit cycles [28, 29]. A two-stage modulator configuration is shown in Fig. 10.14 [30, 31]. Each stage consists of a first-order sigma-delta modulator. The first stage quantizes the input signal in the same way as a conventional sigma-delta modulator. The first stage quantiza-

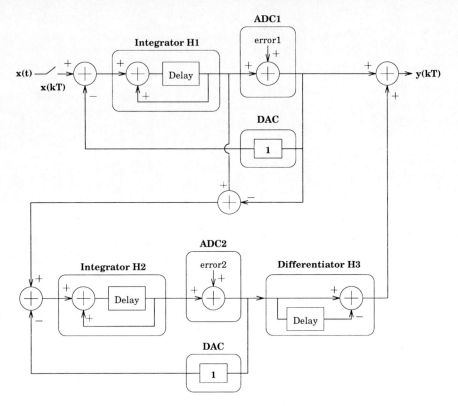

**FIGURE 10.14**

Model for two-stage noise shaping modulator.

tion noise is fed into the second stage to be quantized and shaped. In conjunction with the differentiator, the quantization noise coming out from the second stage would be identical to that of the first stage. Cancellation is achieved by subtracting this from the first stage output. If $H_1(z)$ is assumed equal to $H_2(z)$, the converter output $Y(z)$ is given by the equation:

$$Y(z) = \frac{H_1(z)z^{-1}}{1 + H_1(z)z^{-1}}X(z) + \frac{[H_3(z) - H_1(z)]\,z^{-1}}{[1 + H_1(z)z^{-1}]\,H_3(z)}\sigma_{n1} +$$

$$+ \frac{\sigma_{n2}}{[1 + H_1(z)z^{-1}]\,H_3(z)} \tag{10.13}$$

If we concentrate on baseband frequencies, then $|H_1(z)| \gg 1$

$$Y(z) = X(z) + \frac{[H_3(z) - H_1(z)]\,z^{-1}}{[1 + H_1(z)z^{-1}]\,H_3(z)}\sigma_{n1} + \frac{\sigma_{n2}}{[1 + H_1(z)z^{-1}]\,H_3(z)} \tag{10.14}$$

where $\sigma_{n1}$ and $\sigma_{n2}$ are standard deviations of the quantization noise of ADC1 and ADC2, respectively, and $\frac{1}{H_3(z)}$ is the differentiator. If $H_1(z)$ is equal to $H_3(z)$, the second term is cancelled and the third term is suppressed by $H_1(z)$ and $H_3(z)$. Consequently, the resulting spectral density function of the noise is equivalent to embedding a second-order transfer function in the feedback loop. In general this can be extended to an arbitrary number of stages. However, as shown above, if $H_1(z)$ and $H_3(z)$ do not match precisely, then there will be residual first-order noise. This requirement on precise gain matching between the individual first-order sections runs contrary to the original intention of designing A/D converters that are insensitive to component mismatches.

## 10.7   MULTI-BIT SIGMA-DELTA MODULATORS

Another approach to increase the resolution of a sigma-delta modulator of a given order is to incorporate a multi-bit quantizer within the feedback loop. This approach can be applied to modulators of higher-order ($> 2$), or cascaded, architectures. It has been shown in the past that multi-bit internal quantizers can improve the stability characteristics of higher-order modulators, thus significantly lowering the quantization noise for a given oversampling ratio [10, 11, 32, 33]. One principal drawback of multi-bit sigma-delta converters has been the stringent absolute accuracy requirements placed on the low resolution feedback DAC. Past attempts to solve this problem resulted in trimming techniques. The use of digital error correction on the DAC has also been reported [11]. Another approach is the use of the dynamic element approach. This technique is applied to the DAC in the feedback loop by having different permutations on the arrangement of passive elements in generating a given voltage level. The use of random permutation involves randomly selecting the passive elements that generate the desired reference level, thereby translating the harmonic distortion due to mismatch into random noise [10, 32]. A variation of this technique consists of converting what would have been the distortion components caused by element mismatch in the feedback DAC to high-frequency components lying outside the passband. A simple realization of this technique is called the "clocked averaging approach," whereby the averaging is done by permutating the DAC elements in a periodic fashion, has been reported [34]. This technique, however, can generate undesirable tones inside the passband for certain signal frequencies [18]. If the oversampling ratio is denoted by D and the order of the modulator is L, then in order to ensure the tones do not fall inside the passband, the following inequality has to be observed:

$$D > \frac{2^L + 1}{2} 2^L \tag{10.15}$$

As an example, L is 3 for a third-order coder. Substituting this into inequality (13) yields a minimum oversampling ratio D of 36. In addition, even if the above inequality is observed, the tones can still become a problem from an anti-aliasing standpoint. To suppress the amplitude of the tone, anti-aliasing filtering has to be provided throughout the "don't care" regions in general. An improved

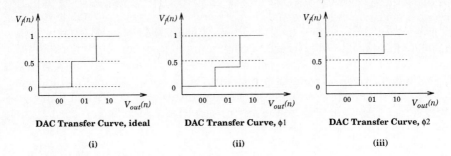

**FIGURE 10.15**
DAC transfer curves with ideal and nonideal matching.

technique, called the individual level averaging approach, achieves the averaging by permutating the DAC elements on a per level basis.

As an example, this approach is applied to a first-order modulator that incorporates a feedback DAC with 2 unit elements. If the unit elements are identical, then the DAC will have 3 levels as shown in Fig. 10.15(i). The input to the feedback DAC is the output of the sigma-delta converter, $v_{out}(n)$, and the DAC output is the feedback signal, $v_f(n)$. If there is a mismatch between the two elements such that their values are "1.1" and "1.2," then the DAC will have 3 levels corresponding to: "0," "$\frac{11}{23}$," and "1" as shown in Fig. 10.15(ii). If the elements are swapped periodically between two phases, $\phi_1$ and $\phi_2$, then the DAC transfer characteristics will alternate between Fig. 10.15(ii) and Fig. 10.15(iii) at the clock rate $f_{clk}$. The DAC levels corresponding to Fig.15a(iii) will be "0," "$\frac{12}{23}$," and "1." Fig. 10.16 shows the four different cases when a dc-input of "0.25" is applied to $v_{in}$. Here $v_{out}$ for the first four cycles are assumed to be "00," "01," "00," "01." Note that in the individual level averaging case, both the $\frac{11}{23}$ and $\frac{12}{23}$ levels that correspond to the code "01" are exercised. Consequently the error is averaged out. The output sequences, $v_f$, for the different cases are summarized in Table 10.3.

An experimental chip implementing a second-order multi-bit sigma-delta D/A converter using this technique has achieved an improvement of 6dB in SNR over the random averaging approach [35] .

## 10.8 DECIMATION FILTERS

The decimation filter in an oversampled A/D converter consists of a low-pass filter and a resampler. The purpose of the filter is to remove out-of-band quantization noise and to suppress spurious out-of-band signals. Upon filtering, the signal is resampled at the Nyquist frequency. The rate reduction, or decimation, is usually performed in two or more steps to increase the ratio of the width of the transition band of the filter to the sampling rate, thereby decreasing the order of the individual filters. This decimation filter design differs from traditional decimation filter design in that the desired objective is to suppress out-of-band quantization noise as opposed to meeting a certain frequency response specifica-

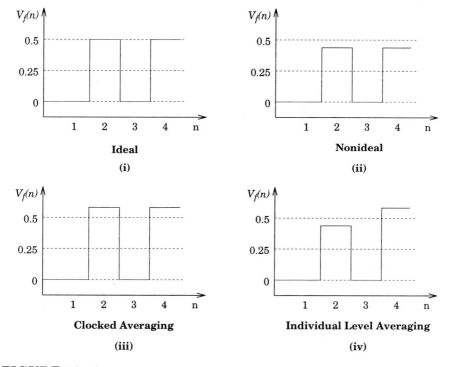

**FIGURE 10.16**
$V_f(n)$ for different dynamic element approaches, $V_{in}(n) = 0.25$.

**TABLE 10.3**
DAC output $v_f$ for different averaging algorithms $V_{in} = 0.25$, first-order $\Sigma - \Delta$ coder, 2 unit elements.

| Case | n=1 | n=2 | n=3 | n=4 | Avg Val. | Error |
|------|-----|-----|-----|-----|----------|-------|
| Ideal | 0 | 0.5 | 0 | 0.5 | .25 | 0 |
| No Averaging | 0 | $\frac{1.1}{2.3}$ | 0 | $\frac{1.1}{2.3}$ | .239 | -.011 |
| Clocked Averaging | 0 | $\frac{1.2}{2.3}$ | 0 | $\frac{1.2}{2.3}$ | .261 | .011 |
| Individual Level Avg. | 0 | $\frac{1.1}{2.3}$ | 0 | $\frac{1.2}{2.3}$ | .25 | 0 |

tion [36, 37, 38]. Furthermore, if the output code is 1-bit only, then no multiplier is needed.

### 10.8.1  Sinc Decimators

For the sigma-delta modulator whose power spectral density of the quantization noise has a sinusoidal response, it has been shown that the decimation filter can be implemented efficiently using a cascade of comb filters. The decimation

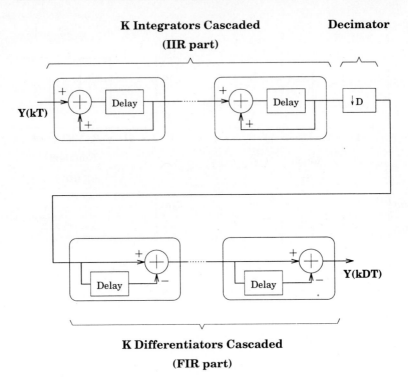

**FIGURE 10.17**
Decomposition of the comb filter with sinc-type frequency response.

filter will exhibit a sinc-type frequency response [12]. This type of filter can be implemented very efficiently because neither a ROM nor a multiplier is required. The filter transfer function is

$$\left(\sum_{i=0}^{i=D-1} z^{-i}\right)^k = \left(\frac{1}{D}\frac{1-z^{-D}}{1-z^{-1}}\right)^k \tag{10.16}$$

where $D$ is the decimation factor and $k$ is the order of the sinc filter. If the sigma-delta modulator is of order L, then setting $k$ to L+1 would provide sufficient suppression of the quantization noise [12]. Notice if the denominator is factored out, then the numerator can be implemented at the decimated frequency [39, 40, 41] and the computation speed is greatly reduced. A block diagram of this implementation is shown in Fig. 10.17. The attenuation of the filter in the passband can be found and compensated for in the second-stage decimation filter.

### 10.8.2 FIR Decimation Filters Based on Quadratic Programming

Recently another approach based on quadratic optimization of the out-of-band quantization noise has been reported [42]. The methodology tackles the FIR filter design problem by formulating a quadratic programming problem that minimizes the integral of the aliased noise subject to the passband and stopband constraints. This approach offers a design whose response is optimized to meet arbitrary quantization noise Power Spectral Density (PSD) and anti-alias requirements.

Conceptually the optimization process tries to minimize the noise energy that is the product of the FIR filter transfer function squared and the noise PSD, integrated over the desired bandwidth. The mathematical treatment will now be presented. There are four cases of FIR filters with exactly linear phase [43], of which two cases apply to low-pass filters: Case 1 for filters with odd length and Case 2 for filters with even length. Without loss of generality, Case 2 filters are assumed for the rest of the chapter. Given a Case 2 FIR filter with an impulse response, $\{h(n)\}$, the frequency response is

$$H(e^j) = \sum_{n=1}^{n=\frac{L}{2}} b(n) \cos(n - \frac{1}{2})\omega \tag{10.17}$$

where

$$b(n) = 2h(\frac{L}{2} - n), n = 1, ..., \frac{L}{2} \tag{10.18}$$

Defining the filter coefficient vector

$$X = \left[ b(1)b(2)...b\left(\frac{L}{2}\right) \right]^T \tag{10.19}$$

and the transform kernel vector

$$S(\omega) = \left[ \cos \frac{1}{2}\omega \cdots \cos \left(\frac{L}{2} - \frac{1}{2}\right)\omega \right]^T \tag{10.20}$$

the zero phase response can be written as an inner vector product

$$H(e^j) = X^T S(\omega) \tag{10.21}$$

Furthermore, if we define the spectral density of the quantization noise from the oversampled ADC to be $S_{nn}$, then the quantization noise energy over a frequency interval [a, b] is given by

$$E_s = \frac{1}{2\pi} \int_{=a}^{=b} |H(e^j)|^2 |S_{nn}(e^j)|^2 d\omega = \frac{1}{2\pi} X_T Q X \tag{10.22}$$

where the energy matrix

$$Q = \int_{=a}^{=b} S(\omega)S^T(\omega) |S_{nn}(e^j)|^2 d\omega \tag{10.23}$$

is symmetric.

The desired filter passband and stopband performance is attained by imposing magnitude constraints at a discrete set of points. The constraints are divided into equality and inequality constraints. The equality constraints would be $|H(e^j)|_{=_i} = k_i$ and the inequality constraints would be $|H(e^j)|_{=_j} < k_j$. Using vector notation, we have $ST|_{=_i}X = K$ and $ST|_{=_j}X < K_1$ where $K = [k_0, ..k_n]^T$ and $K_1 = [k_{10}, ..k_{1n}]^T$. The design of the *quadratic* filter can be formulated as a quadratic programming problem: Minimize $E_s$ subject to $ST|_{=_i}X = K$ and $ST|_{=_j}X < K_1$. The optimization adopts the method of active set constraints [44].

From a physical point of view the quantity $E_s$ represents energy and therefore is non-negative. Now $E_s$ is defined by

$$E_s = \frac{1}{2\pi} X^T Q X \qquad (10.24)$$

Therefore Q is positive definite. This implies that the projected Hessian matrix $Z_k^T Q Z_k$ is also positive definite at every iteration k [45], where $Z_k$ is the matrix whose columns form the basis for every feasible direction at iteration k. Therefore the search direction is well defined at every iteration and the quadratic function has a unique minimum in the subspace defined by $Z_k$. Furthermore, it has been shown that the function has a global minimum [46].

This approach offers design flexibility over the comb filter with a sinc-type response but it requires ROM to store the impulse response coefficients. Specifically it can be used to design filters with more complicated noise-shaping functions [4, 9, 47]. It also provides a better trade-off between filter length and the amount of noise suppression provided. Reduction in filter length up to a factor of two can be achieved when compared to filters with sinc-type frequency responses.

### 10.8.3   Polyphase IIR Decimation Filter Based on Optimization

FIR filters simplify the computation by exploiting the 1-bit output nature of the oversampled modulators. Furthermore, in the FIR decimation filter, the computation is done at the decimated frequency. However, in cascaded modulators or higher-order modulators involving multi-bit internal quantizers, the modulator output will be multi-bit and the 1-bit advantage can no longer be exploited. In addition, a method of transforming conventional IIR transfer functions to a form whose denominator is a polynomial in $Z^D$ only (i.e., polyphase structure), has been proposed [48]. Since $D$ is the decimation ratio, the computation can now be done at the decimated frequency, making the IIR filter equally efficient. The filter transfer function is of the form

$$H(z) = \frac{\displaystyle\sum_{i=0}^{i=L} b_i z^{-i}}{1 - \displaystyle\sum_{j=0}^{j=K} a_j z^{-D}} \tag{10.25}$$

Here $a_j$ and $b_i$ are the coefficients of the filter transfer function. The multiplication rate has been shown to be [49]

$$\frac{K + 1 + \frac{L}{2}}{D} \tag{10.26}$$

Based on this polyphase architecture, a design methodology has been proposed whose objective is to minimize the out-of-band quantization noise, as well as the in-band phase distortion [50]. In the optimization process, the degree of phase nonlinearity can be selected as one of the design specifications. The resulting polyphase IIR filters can achieve up to a factor of two reduction in multiplication rate over the quadratic programmed based FIR filters, with similar performance in quantization noise reduction.

## 10.9  DESIGN EXAMPLE: A MULTI-CHANNEL VOICEBAND CODER

The design of a 4-channel voiceband coder for telephony applications using sigma-delta modulators will now be presented as an example [3].

### 10.9.1  Multi-Channel Oversampled PCM Coders

For voiceband telephony applications, first- and second-order sigma-delta modulators described above are of the most practical interest. In voiceband telephony, per-line component cost is a critically important parameter. An important objective is to reduce the silicon area per channel in the coder function to the lowest possible value. While second-order sigma-delta coders are most frequently used in single-channel applications because a lower sample rate can be used, the realization of multi-channel PCM interfaces in a single chip brings about a different set of tradeoffs in silicon area minimization which tend to make a first-order implementation more attractive than would be the case for a single-channel coder. An area comparison is made to determine the relative area consumption of first-order and second-order coders in a multi-channel implementation.

A realistic comparison of the area required for the decimation filter required for first- and second-order coders is quite complex and must necessarily be made under a set of assumptions that might be violated in some innovative future design. In the present section, such a comparison will be made in order to compare the areas of such filters as they would be implemented in the prototype described later. These assumptions are:

1. In the telephony CODEC application, the first-order coder implementation utilizes a $sinc^2$ filter with an L/D of 2, a decimation ratio of 512, and a sampling rate of 4 MHz. The second-order coder utilizes a $sinc^3$ filter with an L/D of 3, a decimation ratio of 128, and a sampling rate of 1 MHz [12]. These filter configurations and sample rates meet the dynamic range and idle channel requirements of PCM telephony with comparable margins. These particular FIR filter configurations have been previously shown to reduce out-of-band aliased noise to a value that is small compared to the in-band noise for the two-coder cases. Other decimation filters that meet the telephony transmission requirements are certainly possible. For example, use of a second-order loop operating at a relatively high sampling rate with a $sinc^2$ filter gives a coder whose in-band noise is dominated by aliased out-of-band noise which is capable of meeting the noise requirements. Such configurations do not appear to have great promise for having minimum area and will not be investigated further.

2. It is assumed an FIR implementation of the decimation filter is used. Other configurations such as the polyphase IIR decimation filter are possible. Making this assumption will also enable the computation of the IIR filter to be performed at the decimated frequency also. However, in a sigma-delta loop the input signal to the decimation filter is a one-bit code so that only an AND gate is needed instead of a multiplier to form the product of the input and coefficients. This can be used in an FIR filter to eliminate the need for a multi-bit multiplier. However in the IIR filter, because there is feedback there are multiplications in the feedback loop that are between two multi-bit words. Multipliers will be needed that will lead to an increase in the filter complexity.

3. The overload point of the coder is defined as that point at which the large-signal S/(N+D) ratio is $40dB$. This impacts the dynamic range and required oversampling ratio of second-order implementations because such coders display significant degradations in SNR due to overload as the signal level approaches full scale.

4. The area of the ROM, PLA, up-down counter, or other circuitry used to develop the impulse response is negligible in the multi-channel case because it is shared over many channels. Thus the area of this function has little effect on the comparison of the area of first-order versus second-order filter in the multi-channel case.

5. In the $sinc^2$ filter, a datapath width of 14 bits is required because the dc-gain of a $sinc^2$ filter with an L of 256 is approximately $2^{14}$. Similarly in the $sinc^3$ filter, a width of 18 bits is required because the dc-gain of a $sinc^3$ filter with an L of 192 is approximately $2^{18}$.

6. The speed capability of the technology used is such that the first-stage of decimation required in the digital filter cannot easily be time shared over multiple channels.

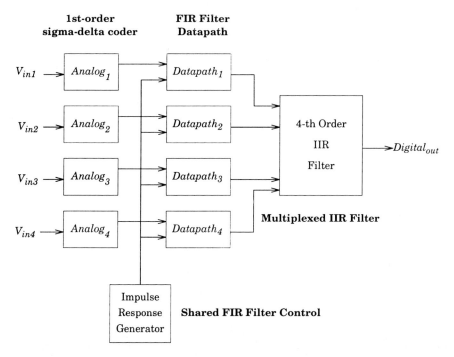

**FIGURE 10.18**

Multi-channel voiceband coder using sigma-delta modulators.

Under the above assumptions, the datapath dominates the area. The area of the datapath is primarily dependent of the number of additions per second that it must perform and the data width it must handle. This depends only on the L/D ratio and the datapath width. Both are smaller for the $sinc^2$ case and hence it is likely to take less area under the assumptions above. Consequently the first-order coder was chosen for use in the particular prototype described here.

## 10.9.2   Prototype Discussion

A prototype of a four-channel voiceband oversampled A/D converter with transmit filter realizing a dynamic range of $79dB$ and crosstalk of $-83dB$ has been fabricated in a 3-micron double metal, single-poly CMOS technology. The coder uses a first-order analog front end followed by a time multiplexed digital filter. The coder oversamples at 4 MHz and decimates to generate the output at 8 kHz. Shown in Fig. 10.18 is the overall block diagram of a four-channel oversampled A/D converter. There are separate analog front ends for each channel to minimize the channel-to-channel crosstalk. Each front end is working at 4MHz. The filter is divided into two stages, a high-speed FIR decimation filter followed by an IIR filter . The FIR filter has a shared control since the impulse response is the same for all four channels. The datapath is separate for each channel because of the speed limitation imposed by a $3\mu$ technology. The FIR filter is working at 8MHz. The IIR filter is time shared and has a clock frequency of 4MHz with an input sampling frequency of 16kHz.

An additional problem with a first-order loop is that the quantization noise is correlated with the input, giving rise to noise components in the passband. Specifically, at low input level, the quantization noise spectrum exhibits tones that may fall into the passband. To de-correlate the quantization noise at low input level, a dither signal is applied. By adding a square-wave dither signal at the input, the tones in the passband will be phase-modulated by the fundamental and harmonics of the square wave [51]. Consequently the tone energy inside the passband is reduced. The reduction of the tone energy depends in a complicated way on the modulation index, which in turn is a function of the dither amplitude. However the addition of this square-wave dither signal reduces the dynamic range of the coder and so there is a trade-off in the desirable amplitude of the square wave. The dither frequency is chosen to lie at the zeroes of the decimation filter so that the dither signal is removed by the filter. Under the above considerations and with the help of system simulations, a square-wave dither signal with a frequency of 250 kHz and an amplitude of -12dB from the full scale is used.

**Decimation Filter Implementation**    The implementation of the decimation filter will now be discussed. The filter needs to suppress the out-of-band quantization noise as discussed above. It also needs to have an anti-alias suppression of more than $33dB$ and a passband ripple of less than $0.25dB$. As discussed above an L/D of 2 is required to ensure that the out-of-band quantization noise will be sufficiently suppressed. To satisfy the other two requirements, the filter would need a higher L/D ratio. Since the FIR filter computes at the high input sampling frequency $f_s$, its complexity should be minimized. By breaking the filter into multiple stages, the requirements can be met while keeping the L/D ratio of the first-stage FIR filter at 2. The overall filter structure is shown in Fig. 10.19 where it takes a 4 MHz 1-bit input data through an FIR filter with a triangle impulse response. The 14-bit output is decimated to 32kHz and fed into a second-stage FIR filter. The output is further decimated to 16kHz and fed into the IIR filter where it is decimated to 8kHz. FIR1 is implemented in a custom structure. FIR2 and IIR are implemented in a time-shared microprogrammed structure.

Next the prototype custom FIR1 design will be discussed. Figure 10.20 is the block diagram of the filter FIR1. It is shown for a four-channel filter where one counter provides all the coefficients. Although FIR decimation filters that do not require impulse response generator exist [36, 37], these approaches do not have particular advantage in area because the generator is shared over different channels. Furthermore this approach does not lend itself easily to reduction in area by time multiplexing the datapath without introducing extra registers and multiplexers. As shown in the figure the signals X1 through X4 are the one-bit output code from the four analog front ends. The distinct feature of this approach is that the adder in each channel is time multiplexed. Careful design of the

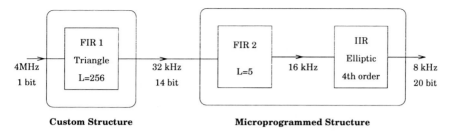

**FIGURE 10.19**
Filter structure.

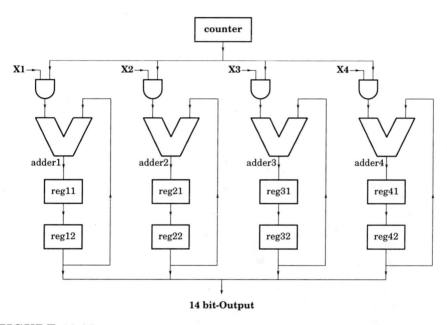

**FIGURE 10.20**
Block diagram of FIR1.

control structure allows this to be implemented in a simple fashion. The detailed implementation of the one-bit slice of the adder and the registers is shown in Fig. 10.21. Here $PC_0$ and $PC_8$ are the LSB and MSB from the program counter which is an up-counter. $PC_i$ is the $i$th-bit of the counter. The $PC_8$ signal is used to invert the counter output after it counts D samples. It converts the up-counter into an up-down counter. Since L/D equals 2, two sets of coefficients are needed. If two accumulators are used these two sets of coefficients have to be provided simultaneously. Since only one adder is used, they will have to be provided in consecutive clock cycles. Furthermore since the impulse response is triangular, the two sets of coefficients are just the inverse of one another, leading to further simplification. In the circuit the $PC_0$ signal is used to invert the counter output between every clock cycle to generate these two sets of coefficients. There are

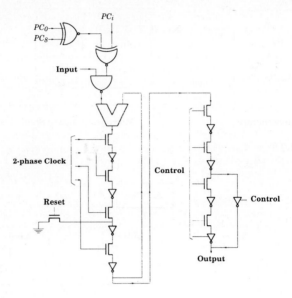

**FIGURE 10.21**
One-bit slice of FIR1 Filter.

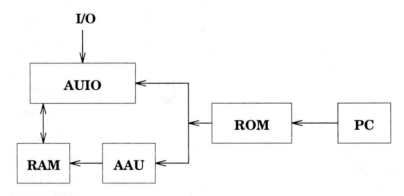

**FIGURE 10.22**
Filter FIR2 and IIR implementation.

altogether four registers. Two of the registers are used to hold state variables for the addition. The other two registers are used to hold the decimated outputs. These outputs are then latched and transferred to the next stage filter. The control signals are generated by ANDing a latch signal and the two-phase clock. The latch signal occurs once every D samples and it latches the new output as well as pushes the old output down like a FIFO. The reset signal is used to reset the FIR filter after the output has been latched.

The IIR filter and the FIR filter FIR2 are both implemented using a micro-programmed architecture. FIR2 is used to provide anti-aliasing around 16 kHz and the IIR filter is used to maintain the passband ripple as well as the stopband attenuation. The IIR filter has to compensate for the droop in the passband due to the filters FIR1 and FIR2 while maintaining a passband ripple of less than 0.25dB. In the stopband the suppression must be more than 33dB at 4.6kHz and beyond. The microprogram structure as shown in Fig. 10.22 is composed of the PC (program counter), the ROM for the program, the RAM for the state variables and the AUIO (arithmetic unit and I/O unit ) for computations, AAU (address arithmetic unit) for calculating addresses for state variables that belong to different channels. The basic hardware is generated by the LAGER system [52], a silicon compiler that accepts an assembler program for the filter as the input and generates the layout of the filter. The hardware is heavily pipelined to increase throughput. To save area no hardware multiplier is used, and multiplications are achieved by shifting and adding. In order to reduce the number of shift and add cycles, thus improving the speed of the IIR filter and allowing for more multiplexing, the IIR filter coefficients are optimized by a program Candi [53] that generates the CSD (canonical signed digit) code with the fewest number of bits in the filter coefficients. The filter uses a datapath of 20 bits which ensures that the idle channel noise is low enough even in the presence of limit cycles. The ROM is 50 words by 26 bits and the RAM is 40 words by 20 bits. The microprogrammed structure runs at 4 MHz. In addition special modifications are introduced to achieve area efficiency for the present application. In particular the AAU(address arithmetic unit) is replaced by simple logic to calculate the addresses of the state variables for the different channels.

**Analog Front End Implementation**   The switched capacitor first-order analog front end integrator is shown in Fig. 10.23. To minimize the crosstalk four separate analog front ends are used. With the digital filter being integrated on the chip, power supply rejection is also a primary consideration. A fully differential architecture is used to reduce crosstalk and improve power supply rejection. Since the process has only one layer of poly, a metal2-metal1-poly capacitor structure was used. The Dith+, Dith- are inputs for dithering, and Vf+, Vf- are outputs from the comparators. Furthermore the output digital pad buffers are switched only at the decimated frequency to minimize switching transients.

Next the op amp and comparator design will be described. The op amp-gain requirement in a sigma-delta loop is determined by its effect on the quantization noise suppression at low frequency. As discussed above the effect of the finite gain of the op amp is to modify the frequency response of the integrator from having a pole at dc to having a pole at some finite frequency. Moreover the dc-gain of an integrator is no longer infinite, but equals the op amp gain. The effect of changing the frequency response of the integrator is to modify the noise transfer function. This modification results in more quantization noise in the passband. For the prototype, the oversampling ratio is 512. Based on the above result, it is determined that a gain of about 1000 is sufficient.

A fully-differential folded-cascode configuration with dynamic common-mode feedback was chosen. The common-mode feedback [54] is connected directly to the output node to improve the speed at the expense of having a lower gain.

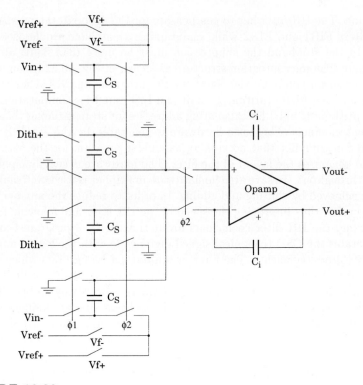

**FIGURE 10.23**
Prototype switched capacitor integrator.

From simulations the amplifier has a typical gain of 1000 and settles to 0.01% in 75 ns with a 1.2 V differential step into a 1.2 pF load. To reduce the harmonic distortion due to signal dependent charge injection the sampling switches are turned off sequentially, with the switches connecting to ground being turned off first.

Since the comparator offset does not affect the performance of the converter, a simple regenerative latch is used to achieve fast switching. In simulation the comparator latches correctly in 20 ns with an overdrive of 10 mV. The photomicrograph of the entire chip is shown in Fig. 10.24.

### 10.9.3   Experimental Results

The measured signal-to-noise ratio versus input amplitude curve is shown in Fig. 10.25 for a $V_{dd}$ of 4.75 V. The signal-to-noise ratio of the converter was evaluated by putting a sine wave of 1.024 kHz into the converter, uploading the digital output to a computer, and then running a 1024 point FFT to compute the ratio of the signal power to the noise power. The measured results meets the D3 specifications. The frequency response of the overall coder is measured. The response has more than 33dB suppression at 4.6kHz and beyond. The passband

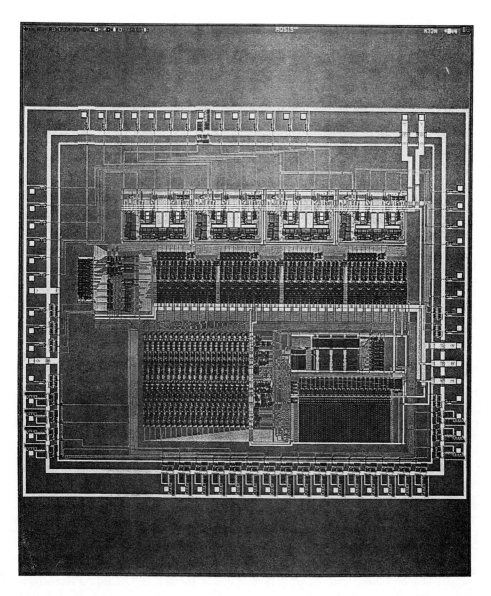

**FIGURE 10.24**
Chip photograph.

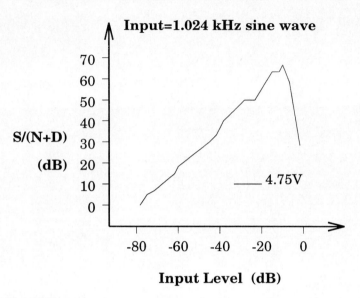

**FIGURE 10.25**
Measured signal-to-noise ratio versus input amplitude.

ripple is less than 0.25dB. The crosstalk was measured by putting a full scale sine wave at 1 kHz at the input of channel one, a zero input to channel two, and then measured the 1 kHz component at the output of channel two. The crosstalk was measured to be -83dB. Power supply rejection ratio at 1 kHz was measured by applying a 20 mV peak-to-peak sine wave at 1 kHz at the $V_{dd}$ of the chip, zero input to the chip and measured the 1 kHz component at the output. Measurements were repeated for different frequencies from 500 Hz to 20 kHz. The PSRR at 10 kHz is better than that at 1kHz because the power noise is suppressed by the digital filter. The idle channel noise was measured by putting zero input at the channel input, summing up the noise at the output, and taking the ratio between the measured noise with the input signal level at the overload point. The measured value was 12 dBrnC0 C-message weighted. In addition, the variation in idle channel noise as a function of dc-offset has been evaluated. A dc-offset in increments of 5 mV and ranges from -100 mV to +100 mV was applied to the coder and the idle channel noise was measured to vary less than 1 dB. This indicates the dithering signal is effective in decorrelating the quantization noise. The anti-alias requirement is also tested at higher frequency at 13 kHz and 29kHz to measure how effective the filter FIR1 and FIR2 are in removing frequency components around 16 kHz and 32 kHz, respectively. At 13 kHz the signal is attenuated by 42 dB and at 29 kHz the signal is attenuated by 39 dB, both attenuations larger than the 33 dB requirement. The performance is summarized in Table 10.4.

**TABLE 10.4**
Data summary for typical performance: 5 V $V_{dd}$, 25 deg C .

| Area | 8250 sq. mils |
|---|---|
| Dynamic Range | 79dB |
| Idle Channel Noise | 12dBrnC0 |
| Crosstalk | -83dB |
| PSRR(1 kHz)/ | 3dB |
| PSRR(10 kHz) | 60dB |
| Harmonic Distortion(2nd) | -76dB |
| Harmonic Distortion(3rd) | -82dB |
| Power | 50mW |

## PROBLEMS

**10.1.** Show that the signal transfer function (STF) and noise transfer function (NTF) of a second-order oversampled modulator, with first-order predictive and first-order noise shaping configuration shown in Fig. P10.1, are:

$$STF = 1 - z^{-1}$$

$$NTF = \left(1 - z^{-1}\right)^2$$

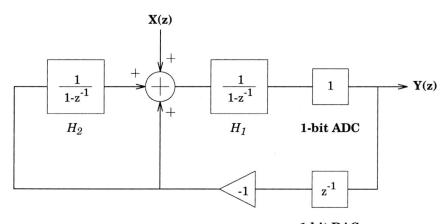

**FIGURE P10.1** Oversampled coder with first-order noise shaping and first-order prediction.

**10.2.** Show that the dynamic range (DR) of a sigma-delta modulator, assuming a white noise model, is given by

$$DR = \frac{3}{2} \frac{2L+1}{\pi^{2L}} D^{2L+1}$$

where $L$ is the order of the sigma-delta modulator and $D$ is the oversampling ratio.

**10.3.** Show that the fundamental limitation on the achievable SNR of an oversampled modulator, imposed by thermal noise is

$$SNR = \frac{V_{dd}}{4}\sqrt{\frac{D \times C}{kT}}$$

where $V_{dd}$ is the power supply voltage, $D$ is the oversampling ratio, and $C$ is the size of the sampling capacitor, $k$ is the Boltzmann constant, and $T$ is the absolute temperature.

**10.4.** The block diagram of a cascaded modulator, in which a second-order $\Sigma - \Delta$ A/D converter is cascaded with a first-order modulator, is shown in Fig. P10.4. Assume the quantization noise of the internal ADCs can be modeled as additive white noise and that the integrators have a transfer function of $\frac{z^{-1}}{1-z^{-1}}$.

(a) Derive an expression for the output $Y(z)$ in terms of the input $X(z)$ and the variance of the quantization noise. Simplify the final result for the case when $g_1 j_1 k_{1a} k_{1b} = 1$.

(b) What is the increase in the mean-square-error (MSE) of the output due to the presence of a mismatch characterized by the factor $\delta = g_1 j_1 k_{1a} k_{1b} = 1$ ?

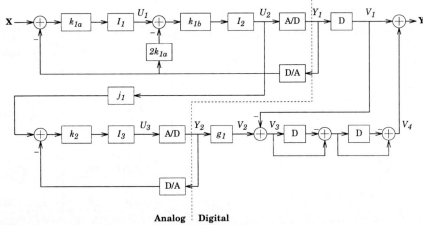

**Analog : Digital**

**FIGURE P10.4** Multi-stage $\Sigma - \Delta$ A/D converter.

**10.5.** For the architecture shown in Fig. P10.4, evaluate an expression for the increase in MSE due to the finite gain of the first integrator's op amp. The transfer function of the integrator with finite op amp gain is

$$I_1(z) = \frac{z^{-1}}{(1 - z^{-1})(1 + \mu) + \alpha\mu}$$

where $\mu = \frac{1}{A}$, $A$ being the gain of the op amp, and $\alpha$ is the ratio of the sampling capacitor the integrating capacitor.

**10.6.** A two-stage $\Sigma-\Delta$ A/D converter used for data acquisition application is shown in Fig. P10.6. The integrators are reset at the beginning of each conversion and then the converter runs for p+1 cycles. $V_R$ is the reference voltage.

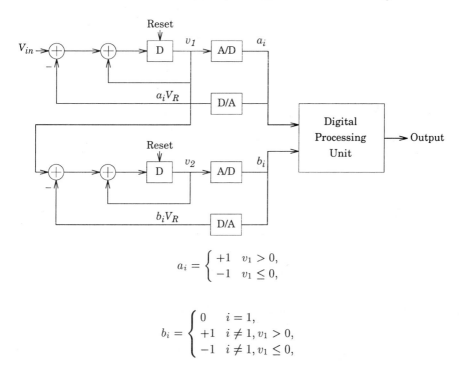

$$a_i = \begin{cases} +1 & v_1 > 0, \\ -1 & v_1 \le 0, \end{cases}$$

$$b_i = \begin{cases} 0 & i = 1, \\ +1 & i \ne 1, v_1 > 0, \\ -1 & i \ne 1, v_1 \le 0, \end{cases}$$

**FIGURE P10.6** Data acquisition $\Sigma - \Delta$ A/D converter.

Derive an expression for the output voltage of the second integrator $v_2$ at the end of (p+1)st cycle in terms of the constant input voltage $V_{in}$ (Note: $-V_R \le V_{in} \le V_R$) and the digital outputs $a_i$ and $b_i$. How should the final output of the ADC be calculated from $a_i$s and $b_i$s? Evaluate the equivalent resolution in bits. (Hint: $-V_R \le V_{in} \le V_R$ for i = 0, 1, ..., p+1)

**10.7.** For the third-order Lee-Sodini $\Sigma-\Delta$M A/D converter in Fig. P10.7, the transfer function of the integrators is $I(z) = (z-1)^{-1}$. Derive the input and quantization noise transfer functions.

**10.8.** Show that in a multi-bit modulator with order L using clocked averaging, in order that the tones do not fall inside the passband the following inequality has to be observed.

$$D > \frac{2^L + 1}{2} 2^L$$

Here $L$ is the order of the sigma-delta modulator and $D$ is the oversampling ratio.

**10.9.** In the design of the decimation filter with a $sinc^K$ frequency response, if $K$ is set to $L + 1$, it has been shown that for most design that is adequate. Here $L$ is the order of the modulator. Evaluate the rms spectral density of the output

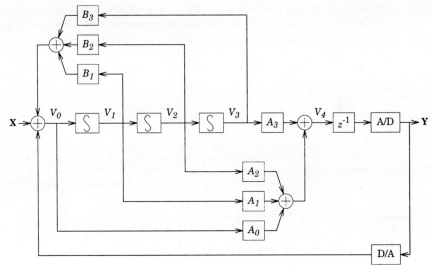

**FIGURE P10.7** Third-order Lee-Sodini $\Sigma - \Delta$ A/D converter.

noise $N(f)$, when $K$ is set to $L-1$, in terms of $D$, the decimation ratio, $f_s$ the sampling frequency, $L$ the order of the modulator and $\bar{e} = \frac{\sigma}{\sqrt{12}}$, where $\sigma$ is the quantization step.

**10.10.** For a polyphase IIR decimation filter, the transfer function is

$$H(z) = \frac{\displaystyle\sum_{i=0}^{i=L} b_i z^{-i}}{1 - \displaystyle\sum_{j=0}^{j=K} a_j z^{-D}}$$

where $D$ = decimation ratio. Show that the multiplication rate is

$$\frac{K + 1 + \frac{L}{2}}{D}$$

# REFERENCES

[1]  B. Boser and B. Wooley, "The Design Of Sigma-Delta Modulation Analog-to-Digital Converters," *IEEE J. of Solid-State Circuits,* vol. 23, no. 6, pp.1298-1308, December 1988.

[2]  M. Hauser, P. Hurst, and R. Brodersen, "MOS ADC-Filter Combination That Does Not Require Precision Analog Components," *International Solid-State Circuits Conference Digest,* vol. 28, pp. 80-81, February 1985.

[3]  B. Leung, N. Robert, P. Gray, and R. Brodersen, "Area-Efficient Multi-channel Oversampled PCM Voice-Band Coder," *IEEE J. of Solid-State Circuits,* vol. 23, no. 6, pp. 1351-1357, December 1988.

[4]  A. B. Signore, D. Kerth, N. Sooch, and E. Swanson, " A Monolithic 20b Delta-Sigma A/D Converter," *International Solid-State Circuits Conference Digest,* vol. 33, pp.170-171, February 1990.

[5] R. Koch, B. Heise, F. Eckbauer, E. Engelhardt, J. A. Fisher, and F. Parzefall, "A 12-bit Sigma-Delta Analog-To-Digital Converter With 15-Mhz Clock Rate," *IEEE J. of Solid-State Circuits,* vol. SC-21, no. 6, pp. 1003-1010, December 1986.

[6] P. Defraeye, D. Rabaey, W. Roggeman, J. Yde, and L. Kiss, "A 3-um CMOS Digital Codec with Programmable Echo Cancellation and Gain Setting," *IEEE J. of Solid-State Circuits,* vol. SC-20, no. 3, pp. 679-688, June 1985.

[7] V. Friedman, D. Brinthaupt, De-Ping Chen, T. Deppa, J. Elward, Jr., E. Fields, J. Scott, and T. Viswanathan, "A Dual-Channel Sigma-Delta Voiceband PCM Codec Using $\Sigma\Delta$ Modulation Technique," *IEEE J. of Solid-State Circuits,* vol. 24, no. 2, pp. 274-280, April 1989.

[8] S. Norsworthy, I. Post, and H. Fetterman, "A 14-bit 80kHz Sigma-Delta A/D Converter: Modeling, Design, and Performance Evaluation," *IEEE J. of Solid-State Circuits,* vol. 24, no. 2, pp. 256-266, April 1989.

[9] K. Chao, S. Nadeem, W. Lee, and C. Sodini, "A Higher-Order Topology for Interpolative Modulators for Oversampling A/D Converters," *IEEE Trans. Circuits and Syst.,* vol. 37, no. 3, pp. 309-318, March 1990.

[10] A. R. Carley, "A Noise-Shaping Coder Topology For 15+ Bit Converters," *IEEE J. of Solid-State Circuits,* vol. 24, no. 2, pp. 267-273, April 1989.

[11] A. R. Walden, G.C. Temes, et al., "Architectures For High-Order Multi-bit Sigma-delta Modulators," *Proceedings of the 1990 IEEE Intl. Symp. on Circuits and Syst.,* May 1990, vol. 2, pp. 895-898.

[12] J. C. Candy, "Decimation for Sigma Delta Modulation," *IEEE Trans. on Communications,* vol. COM-34, no. 1, pp. 72-76, January 1986.

[13] J. C. Candy, "A Use Of Limit Cycle Oscillations To Obtain Robust Analog-to-Digital Converters," *IEEE Trans. on Communications,* vol. COM-22, no. 3, pp. 298-305, March 1974.

[14] J. McCreary, "Matching Properties and Voltage and Temperature Dependence of MOS Capacitors," *IEEE J. of Solid-State Circuits,* vol. SC-16, no. 6, pp. 608-616, December 1981.

[15] S. Tewksbury and R. Hallock, "Oversampled, Linear Predictive And Noise-Shaping Coders Of Order N > 1," *IEEE Trans. Circuits and Syst.,* vol. CAS-25, no. 7, pp. 436-447, July 1978.

[16] R. Steele, "SNR Formula For Linear Delta Modulation With Band-Limited Flat And RC-Shaped Gaussian Signals," *IEEE Trans. on Communications,* vol. COM-28, no. 12, pp. 1977-1984, December 1980.

[17] K. Cattermole, "Principles Of Pulse Code Modulation," *Iliffe Books Ltd.,* 1969, p. 201.

[18] B. Leung and S. Sutarja, "Multi-bit Sigma-delta A/D Converter Incorporating, A Novel Class Of Dynamic Element Matching Techniques," *IEEE Trans. Circuits and Syst.,* vol. 39, no. , pp. 35-51, January 1992.

[19] H. Inose, Y. Yasuda, and J. Murakami, "A Telemetering System by Code Modulation – delta-sigma modulation," *IRE Trans. on Space Electronics and Telemetry,* vol. SET-8, pp. 204-209, September 1962.

[20] J. C. Candy, Y. C. Ching, and D. S. Alexander, "Using Triangularly Weighted Interpolation to Get 13-Bit PCM from a Sigma-Delta Modulator," *IEEE Trans. on Communications,* vol. COM-24, no. 11, pp. 1268-1275, November 1976.

[21] A. Yukawa, R. Maruta, and K. Nakayama, "An Oversampling A-to-D Converter Structure For VLSI Digital Codecs," *International Conference On Acoustics, Speech And Signal Processing Proceedings,* pp. 1400-1403, May 1985.

[22] A. Yukawa, "Constraints Analysis For Oversampling A-to-D Converter Structures On VLSI Implementation," *Proceedings of Intl. Symp. Circuits and Syst. Conference,* vol. 2, pp. 467-472, May 1987.

[23] J. C. Candy, "A Use Of Double Integration In Sigma Delta Modulation," *IEEE Trans. on Communications,* vol. COM-33, no. 3, pp. 249-258, March 1985.

[24] W. Bennett, "Spectra of Quantized Signals," *Bell System Technical Journal,* pp. 446-472, July 1948.

[25] B. Agrawal and K. Shenoi, "Design Methodology for Sigma-Delta Modulators," *IEEE Trans. on Communications,* vol. COM-31, no. 3, pp. 360-369, March 1983.

[26] M. Hauser and R. Brodersen, "Circuit and Technology Considerations for MOS Delta-Sigma A/D Converters," *Intl. Symp. Circuits and Syst. Conference Proceedings,* vol. 3, pp. 1310-1315, May 1986.

[27] R. Adams, "Design and Implementation of an Audio 18-bit A/D Converter Using Over-sampling Techniques," *J. Audio Engineering Society,* vol. 34, no.3, pp. 153-166, March 1986.

[28] J. C. Candy and O. Benjamin, "The Structure Of Quantization Noise From Sigma-Delta Modulation," *IEEE Trans. on Communication,* vol. COM-29, no. 9, pp. 1316-1323, September 1981.

[29] W. Chou, P. Wong, and R. Gray "Multi-Stage Sigma-Delta Modulation," *IEEE Trans. on Information Theory,* vol. 35, no.4, pp. 784-796, July 1989.

[30] K. Uchimura, et. al, "Oversampling A-to-D and D-to-A Converters with Multistage Noise Shaping Modulators," *IEEE Trans. on Acoustics, Speech and Signal Processing,* pp. 1899-1905, December 1988.

[31] Y. Matsuya, K. Uchimura, A. Iwata, T. Kobayashi, M. Ishikawa, and T. Yoshitome, "A 16-bit Oversampling A/D Conversion Technology using Triple Integration Noise Shaping," *IEEE J. of Solid-State Circuits,* vol. SC-22, no. 6, pp. 921-929, December 1987.

[32] R. Carley, "An Oversampling Analog-to-Digital Converter Topology for High Resolution Signal Acquisition Systems," *IEEE Trans. on Circuits and Syst.,* vol. CAS-34, no. 6, pp. 83-91, January 1987.

[33] T. Cataltepe, G. C. Temes, et al., "Digitally Corrected Multi-bit $\Sigma - \Delta$ Data Converters," *Proceedings of the 1989 IEEE Intl. Symp. Circuits and Syst.,* pp. 647-650, May 1990.

[34] Yashui Sakina and P. Gray, " Multi-bit $\Sigma - \Delta$ Analog-to-Digital Converters with Nonlinearity Correction Using Dynamic Barrel Shifting," *Master thesis,* pp. 24-32, June 1990.

[35] F. Chen and B. Leung, "A Multi-bit $\Sigma - \Delta$ DAC with Dynamic Element Matching Techniques," *Proceedings of IEEE Custom Integrated Circuit Conference,* May 1992, pp. 16.2.1-16.2.4.

[36] E. B. Hogenauer, "An Economical Class of Digital Filers for Decimation and Interpolation," *IEEE Trans. on Acoustics, Speech, and Signal Processing,* vol. ASSP-29, no. 2, pp. 155-162, April 1981.

[37] A. N. Netravali, "Optimum Digital Filters for Interpolative A/D Converters," *Bell System Technical Journal,* pp. 1629-1641, November 1977.

[38] A. Huber, et. al., "FIR Lowpass Filter For Signal Decimation with 15 MHz Clock Frequency," *Intl. Conf. on Acoustics, Speech, And Signal Processing,* pp. 533-1536, April 1986.

[39] T. Saramäki, T. Karema, T. Ritoniemi, and H. Tenhunen, "Multiplier-Free Decimator Algorithms For Superresolution Oversampled Converters," *Proceedings Intl. Symp. on Circuits And Syst.,* vol. 4, pp. 3275-3278, May 1990.

[40] T. Saramäki, "Design Of Optimal Multistage IIR and FIR Filters For Sampling Rate Alteration," *Proceedings Intl. Symp. on Circuits And Syst.,* vol. 1, pp. 227-230, May 1986.

[41] E. Dijkstra, et. al., "On The Use Of Modulo Arithmetic Comb Filters In Sigma Delta Modulators," *Proceedings Intl. Symp. on Circuits And Syst.,* pp. 2001-2004, May 1988.

[42]  B. Leung, "Design Methodology of Decimation Filters for Oversampled ADC Based on Quadratic Programming," *IEEE Trans. on Circuits and Syst.,* vol. 38, no. 10, pp. 1121-1132, October 1991.

[43]  L. Rabiner and B. Gold, *Theory and Application of Digital Signal Processing,* Prentice-Hall, 1975.

[44]  J. Bunch, "A Computational Method for the Indefinite Quadratic Programming Problem," *Linear Algebra and Its Applications,* pp. 341-371, 1980.

[45]  P. Gill, W. Murray, and M. Wright, *Practical Optimization,* Academic Press, 1981, pp. 177-178.

[46]  M. Avriel, *Nonlinear Programming, Analysis and Methods,* Prentice-Hall, 1976, p. 434.

[47]  R. Adams, "Companded Predictive Delta Modulation: A Low-Cost Conversion Technique For Digital Recording," *J. of Audio Engineering Society,* vol. 32, no. 9, pp. 659-672, September 1984.

[48]  M. Bellanger, G. Bonnerot, and M. Coudreuse, "Digital Filtering by Polyphase Network: Application to Sample-Rate Alteration and Filter Banks," *IEEE Trans. on Acoustic, Speech and Signal Processing,* vol. ASSP-24, no. 2, pp. 109-114, April 1976.

[49]  H. Martinez and T. W. Parks, "A Class of Infinite-Duration Impulse Response Digital Filter for Sampling Rate Reduction," *IEEE Trans. on Acoustic, Speech and Signal Processing,* vol. ASSP-27, no. 2, pp. 154-162, April 1979.

[50]  Z. Ma and B. Leung, "Polyphase IIR Decimator Filter Design for Oversampled A/D Converter with Approximately Linear Phase," *IEEE Trans. on Circuits and Syst.,* vol. 39, no. 8, pp. 497-505, August 1992.

[51]  J. E. Iwersen, "Calculated Quantizing Noise of Single-Integration Delta-Modulation Coders," *Bell System Technical Journal,* pp. 359-2389, September 1969.

[52]  J. Raebaey, et. al., "An Integrated Automated Layout Generation System for DSP Circuits," *IEEE Trans. CAD of Circuits and Syst.,* vol. 4, no. 3, pp. 285-296, July 1985.

[53]  R. Jain, F. Catthoor, J. Vanhoof, B. DeLoore, G. Goosens, N. Goncalvez, L. Claesen, J. Van Ginderdeuren, J. Vandewalle, and H. De Man, "Custom Design of a VLSI PCM-FDM Transmultiplexer from a System Specifications to Circuit Layout Using a Computer-Aided Design System," *IEEE J. of Solid-State Circuits,* vol. SC-21, no. 1, pp. 73-86, February 1986.

[54]  R. Castello and P. Gray, "A High-Performance Micropower Switched-Capacitor Filter," *IEEE J. of Solid-State Circuits,* vol. SC-20, no. 6, pp. 1122-1132, December 1985.

# CHAPTER
# 11

## ANALOG
## INTEGRATED
## SENSORS

### 11.1  INTRODUCTION

Many signal processing applications require inexpensive, reliable sensors compatible with analog circuitry. The challenge of combining sensors with the appropriate circuitry may be met by microsensors, i.e., silicon sensors with on-chip circuitry fabricated by using the same integrated circuit technology. Indeed, a variety of microsensors fabricated in standard industrial CMOS or bipolar IC technologies have been demonstrated for mechanical, magnetic field, temperature and radiation and chemical measurands [1]. This makes silicon a very strong candidate for the principal sensor material of the near future. Certain thermal, mechanical, and chemical sensors can also be obtained by combining established IC technologies with one or two additional processing steps, (i) film deposition or/and (ii) micromachining which are specific to the sensor function and compatible with the IC process.

Sensors, with respect to electronic circuits integrated on the same chip, can be divided into two broad groups. The first, active sensors, provide an output signal as voltage or current. For example, a resistive bridge, which is sensitive to pressure or to temperature produces a voltage output. A magnetic-field sensitive transistor provides an output current. Two circuit approaches may be used to interface to these sensors. In the first approach, the signal conditioning circuitry begins with a low-noise voltage or current amplifier or a low-noise I-V conversion stage. Alternately, in the second approach voltage or current controlled circuits, for example, voltage or current controlled oscillators may be used for sensor interfacing.

The second group of sensors are passive sensors. They respond to an external force by modification of their basic parameters. A typical example is a capacitive humidity sensor. Two kinds of circuits can be co-integrated with these sensors. The basic parameter, i.e., capacitance, can be measured by one of the circuits for direct capacitance measuring. Another possibility is co-integration of the sensor with a parameter controlled circuit: an oscillator (multivibrators are primarily used for this purpose) or a one-shot multivibrator.

Finally, in both active and passive sensors certain biasing or driving conditions are required for normal operation. These conditions can be provided by an electronic circuit. An example is a gas flow meter which requires constant power dissipation in a variable heating resistor.

This chapter reviews basic mechanical, temperature, and magnetic field sensors obtained by combinations of thin film and/or micromachining with industrial CMOS and bipolar IC process technologies. Electronic circuits are described which are used for signal processing and to support sensor operating conditions.

## 11.2   MECHANICAL SENSORS

Mechanical signal sensors are used to measure force-related measurands (strain, acceleration, weight, pressure, torque) or motion-related measurands (displacement, velocity, mass flow).

A silicon sensor reminiscent of a strain gauge bridge is a good representative example of a sensor obtained by adding one or two additional processing steps to a basic IC technology. Silicon is a good piezoresistive material. The gauge factor (the ratio of the relative change of resistance to the relative increase in the resistor length) is very high for p-type silicon (up to 175) [1]. Yet, the separate realization of silicon strain gauges is difficult. The accurate measurement of strain requires that the gauges be very thin. As silicon is brittle, such thin strain gauges are difficult to handle. They have to be supported by a substrate and sense the substrate strain. One possible solution is to create a very thin membrane under the strain gauge in the chip which includes the signal-processing circuitry. Combining a standard bipolar technology with post-processing anisotropic etching from the bottom side of the chip, one obtains such a strain gauge sensor (Fig. 11.1(a)). The piezoresistors are diffused or ion implanted on the upper surface of the diaphragm during the signal processing circuit fabrication.

Silicon piezoresistivity is temperature dependent. Therefore, in addition to the signal-conditioning circuit, it is desirable to include on the same chip a temperature-compensation circuit in order to reduce the temperature sensitivity. The temperature coefficient of sensitivity can then be reduced to less than 1% which facilitates sensor interfacing.

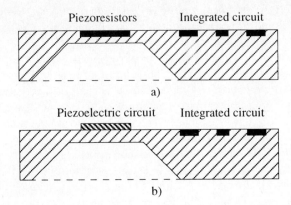

**FIGURE 11.1**

Strain gauge pressure sensors a) piezoresistive b) piezoelectric.

Silicon is non-piezoelectric and does not convert mechanical energy into electrical energy which is necessary for piezoelectric pressure sensors. Using thin-film deposition (combined with etching), one can deposit piezoelectric layers on top of the silicon substrate (Fig. 11.1(b)). The most frequently used material is $ZnO$ which can be deposited in a reliable way. The desired signal conversion then takes place in these layers. Silicon is used as the material in which electronic circuitry can be realized with standard IC processing, and by micromachining, diaphragms or cantilevers are obtained by a first postprocessing step.

Such a chip can be contained in a can (Fig. 11.2(a)). Exposed to a given pressure, the device becomes a pressure sensor. Pressure sensors with different operating pressure ranges can be obtained by varying the diameter and thickness of the diaphragm. One can find highly sensitive sensors with a range of 0-10 kPa and high-pressure sensors with a range of 0-200 MPa [1]. Bridge circuits require 5 or 10 V power supplies and the sensitivity varies between 10 mV/kPa for low pressure to 0.001 mV/kPa for high-pressure devices. The sensor encapsulation not only protects the circuit but plays an important role in the device operation.

An even more active role of encapsulation is seen in the capacitive pressure sensor. It can be easily converted into a weight or into a displacement sensor (Fig. 11.2(b)) where the second plate of the capacitor is deposited on a glass cover.

Typical acceleration sensors using cantilever beams (Fig. 11.3) are also obtained by anisotropic etching. A piezoresistor (or a piezoelectric layer) is bonded on the cantilever neck (from one of the surfaces) and the change of the resistance (which can be part of a bridge circuit) or the piezoelectric voltage is a measure of the acceleration. Such accelerometers can be manufactured with measurements ranges from 5 g to 10,000 g.

All these mechanical sensors can be obtained in bipolar as well as in CMOS standard technologies. In CMOS technologies, there is a strong tendency to obtain sensitive elements located on the free field oxide layer. For example, Fig.

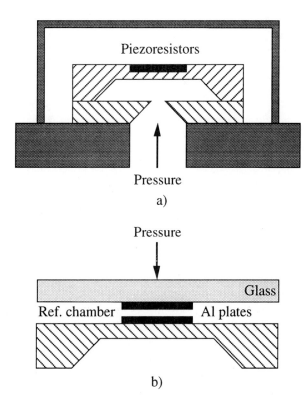

**FIGURE 11.2**
Encapsulated pressure sensors a) piezoresistive b) capacitive.

11.4 shows a mechanical device [2] where the etching solution was manipulated in such a way that a thin cantilever beam of $SiO_2$ results.

## 11.3  THERMAL SENSORS

Most of the commercially available temperature sensors are based on the well-known [3] relationship between the base-emitter voltage, $V_{BE}$, and the collector current, $I_C$, of a bipolar transistor,

$$V_{BE} = V_G(0) + (kT/q)ln[I_C - ln(kT^r/\eta)] \tag{11.1}$$

where $(kT)/q = V_t = 26$ mV at room temperature. The other constants in this equation depend on the semiconductor material and its impurity concentration. The bandgap voltage $V_G(0)$ varies between 1.12 and 1.19 V, $r$ varies between 3 and 5, and $\eta$ is equal to 1 above around 20° K for silicon.

When $I_C$ is kept constant, the voltage $V_{BE}$ drops at a constant rate of approximately -2 mV/°K as temperature increases. Based on Eq. (11.1), a transistor can be directly used as an electrical thermometer. In many cases the dif-

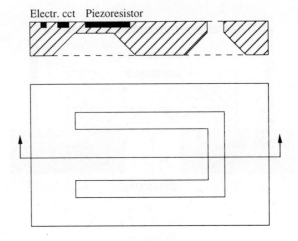

Electr. cct   Piezoresistor

**FIGURE 11.3**
Cantilever beam acceleration sensor.

ference between two base-emitter voltages of two identical transistors carrying different collector currents is used. Then the voltage difference is

$$\Delta V_{BE} = V_{BE1} - V_{BE2} = (kT/q)ln(I_{C1}/I_{C2}) \qquad (11.2)$$

The value $\Delta V_{BE}$ is proportional to absolute temperature. It does not depend on material or transistor parameters and, for normal sensor operations, only the current ratio $I_{C1}/I_{C2}$ has to be set and maintained. A similar result can be obtained if the currents through both transistors are equal but the ratio $N$ of the emitter areas is chosen to be different from unity. This is usually achieved by designing a large transistor as a parallel connection of $N$ small transistors.

Circuits using the temperature dependence of $V_{BE}$ and $\Delta V_{BE}$ can be realized in both bipolar and CMOS technologies. In the latter case, the substrate transistor is used as part of the sensor circuit.

Since the temperature dependence of the resistivity of metals and semiconductors is well known, temperature sensors can also be designed using metal films deposited on silicon chips. For metal film resistors, the temperature dependence of the resistance $R$ is expressed by

$$R(T) = R(0) + AT + BT^2 \qquad (11.3)$$

Here $R(T)$ and $R(0)$ are the resistances at temperature $T$ and at 0 degrees and $A$ and $B$ are constants. The constants $A$ and $B$ are rather small; yet the temperature dependence expressed in Eq. (11.3) is well established over a wide range of temperatures.

Finally, thermocouples are also compatible with standard microelectronic technologies. Figure 11.5 shows the design, in principle, of a silicon/aluminum

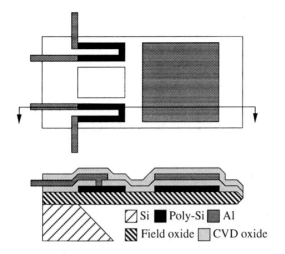

**FIGURE 11.4**
Multilayer oxide cantilever beam.

thermopile using p-type strips made by ion implantation or diffusion in an n-type epitaxial layer. The p-silicon strips are cross-connected by sputtered aluminum strips. A thermopile in CMOS technology can be obtained using cross-connected polysilicon and aluminum strips [4].

A temperature difference, $\Delta T$, between the two sets of contacts produces the thermoelectric voltage

$$\Delta V = \alpha_z \Delta T \tag{11.4}$$

between the thermopile ends, where $\alpha_z$ denotes the Seebeck coefficient. Such thermopiles can have high sensitivity. For p-Si/Al the Seebeck coefficient of 1.17 mV/°K is obtainable. If, for example, the thermopile includes 20 thermocouples, the sensitivity is typically 20 mV/°K. The most obvious application of silicon temperature sensors is the direct measurement of temperature and temperature differences. Silicon is suitable as long as the temperatures are measured in the range of -50 °C to +150 °C. In addition, silicon thermoelectric sensors can be used for indirect measurement of other physical measurands. A common method for measurement of gas or liquid fluid flow velocity is based on the measurements of the temperature difference, $\Delta T$, between the upstream and downstream ends of a (heated) temperature sensor (Fig. 11.6(a)). For small values of the flow velocity this temperature difference can be expressed as

$$\Delta T = C(T_S - T_F)v \tag{11.5}$$

Here $C$ is a constant, $T_S$ is the average temperature of the sensor, $T_F$ is the average temperature of the fluid or gas, and $v$ is the flow velocity. For larger values of the flow velocity, the temperature difference becomes approximately proportional to the square root of velocity.

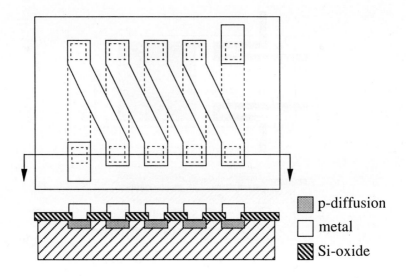

p-diffusion
metal
Si-oxide

**FIGURE 11.5**
Silicon/aluminum thermopile.

In CMOS technologies the resistive sensor and heater can be located on a bridge consisting of a sandwich of field oxide and CVD oxide and nitride (Fig. 11.6(b)). The thermocouple cold contacts are usually located on bulk silicon, while the heater is on the tip of a cantilever oxide-nitride beam (Fig. 11.6(c)).

For a sensor element one can use either a diffused resistor bridge (Fig. 11.7 (a)), or two transistors (one upstream, one downstream (Fig. 11.7(b)), or a thermopile (Fig. 11.7(c)) with the corresponding orientation of the thermopile contacts. In the last case, the temperature difference decreases when the flow velocity increases. Notice that two sensors include heating resistors. It was found [4] that to obtain a constant sensor sensitivity, it is desirable to keep a constant power dissipation in these heaters. The silicon resistors have variable resistance depending on temperature. The constant power can be provided by special electronic circuits, some of which are considered at the end of the chapter.

The device shown in Fig. 11.6(c) also can be used as a vacuum sensor. This sensor consists of a silicon chip in which, by electrochemical etching techniques, a cantilever beam is fabricated. Before the structure is etched, a heating resistor and a thermopile are placed at the tip of the beam on the cantilever. The heater maintains a temperature gradient, which depends on the gas pressure, between the tip and the rim where the beam is connected to the chip. When the gas pressure is low, the gradient is high and the thermopile output voltage is high as well. When the gas pressure increases, the temperature gradient decreases and the corresponding decrease of the thermopile output voltage is proportional to the gas pressure.

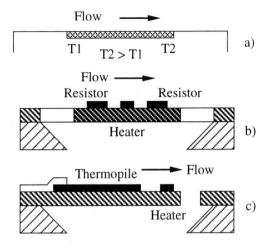

**FIGURE 11.6**
Flow sensor a) on chip, b) on oxide bridge, c) on cantilever beam.

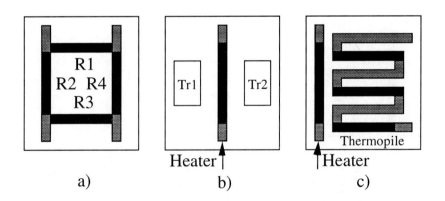

**FIGURE 11.7**
Temperature sensors a) resistive bridge, b) two-transistor, c) thermopile.

## 11.4   HUMIDITY SENSORS

Humidity is a chemical measurand which has attracted much attention. Many different types of humidity sensors have been studied recently. Here is how such a sensor can be fabricated in a standard double-poly CMOS technology [5]. In the sensor element (Fig. 11.8), a poly-1 layer is used as a ground plane and a pair of poly-2 electrodes arranged in an interdigitated fashion serve as the two electrodes of a lateral capacitor. Each poly-2 electrode consists of some fingers with a series of contact-cuts and pad-openings placed between the fingers to expose the region between the electrodes to the ambient air as the dielectric. Since the dielectric

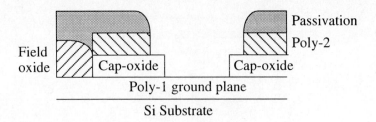

**FIGURE 11.8**
Lateral capacitance humidity sensor.

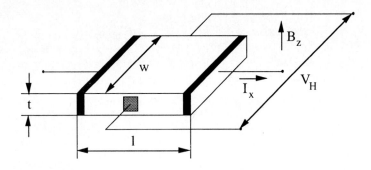

**FIGURE 11.9**
Hall plate structure.

permittivity of the ambient air is a function of the ambient relative humidity (r.h.) level, the capacitance measured between the poly-2 electrodes is modulated by the r.h. level of the ambient atmosphere.

Polyimides are known to absorb moisture. They offer the following crucial advantages: linear response to humidity changes, good absorption of water, resistance against chemical attack, mechanical strength, and compatibility with IC fabrication. As a result polyimide film deposition is frequently used as a post-processing step. The basic structure of the humidity sensitive capacitor is the same but instead of a passivation layer, one uses a polyimide film deposited on the interdigitated structure [6].

## 11.5   MAGNETIC SENSORS

We will discuss both Hall plate and magnetotransistor structures for sensing magnetic fields.

### 11.5.1   Hall Plates

If a thin semiconductor plate (Fig. 11.9) is subjected to a magnetic flux density, $B_z$, perpendicular to the plane of the plate, the so-called Hall voltage, $V_H$, occurs between the sides of the plates, if the plate is carrying a current, $I_r$. This voltage is proportional to the current and the magnetic induction and is expressed by

$$V_H = \frac{R_H I_r B_z}{t} \tag{11.6}$$

where $R_H$ is the Hall coefficient and $t$ is the thickness. This voltage does not depend directly on the length $l$ and width $w$. The Hall coefficient provided that $l \gg w \gg t$ is

$$R_H = \frac{\mu_p - \mu_n}{n_i q(\mu_p + \mu_n)} \tag{11.7}$$

Here $\mu_p$ and $\mu_n$ are the mobilities of holes and electrons, $n_i$ is the intrinsic carrier concentration and $q$ is electron charge. For an n-doped material

$$R_H = -\frac{r_n}{nq} \tag{11.8}$$

and for p-doped material

$$R_H = +\frac{r_p}{nq} \tag{11.9}$$

with the temperature dependent scattering factors $r_n$, $r_p$ of the order of 1.

As follows from Eqs. (11.8) and (11.9) weakly-doped materials have larger Hall coefficients, which explains the fact that in bipolar technology the epitaxial collector layer is often used for the production of Hall plates (Fig. 11.10(a)). The impurity concentration of this layer is about $4 \times 10^{15}$ cm$^{-3}$ and the Hall coefficient is 1600 cm$^3$/coulomb.

Hall plates are available in MOS technology as well. The MOS Hall plate (sometimes called MAGFET) shown in Fig. 11.10(b) looks like an ordinary MOS transistor. It consists of the source, drain, and gate but, in addition, shows two Hall contacts at the side edge of the channel region. These Hall contacts are simultaneously diffused with the source and drain regions. The device is used in the linear region of operation, when the channel is not pinched-off. The Hall voltage for this device is

$$V_H = \frac{I_D B}{C_{ox}(V_{GS} - V_T)} \tag{11.10}$$

in which $I_D$ is the transistor drain current, $V_{GS}$ is the gate-source voltage, $V_T$ is the threshold voltage, and $C_{ox}$ is the gate capacitance per unit area.

## 11.5.2 Magnetotransistors

The characteristics of many transistor structures, by virtue of the Lorentz force, can be influenced by a magnetic field. This fact is used in the construction of magnetotransistors which are especially designed to measure the magnetic field. While bipolar in function, such devices can be fabricated in both bipolar and MOS technologies. Most magnetotransistors are designed with a single emitter and two or more collectors to pick up the current.

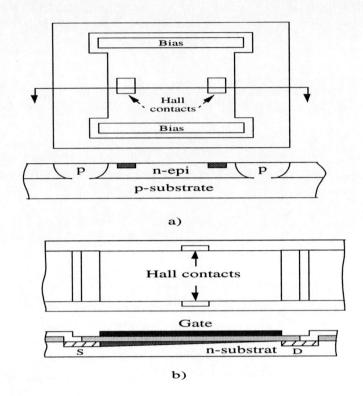

**FIGURE 11.10**
Hall plate realization a) bipolar, b) MOS.

One can have lateral magnetotransistors in which lateral currents (parallel to the surface of the chip) are influenced by the magnetic field, and vertical transistors in which the vertical component of the current is deflected by the magnetic field. In MOS technology, it is usually the lateral current that is affected, but the devices with a vertical component of the current can be obtained by special design of their emitters.

Two mechanisms are responsible for the magnetic sensitivity of magnetotransistors: carrier deflection, and modulation of emitter injection because the Hall voltage in the base region changes the emitter-base potential along the emitter length. As a result of either of these effects, the application of a magnetic field causes a difference in the collector currents of a multi-collector magnetotransistor.

A lateral PNP magnetotransistor (Fig. 11.11(a)) obtained in a standard bipolar technology consists of, e.g., a p-doped emitter, $E$, and two p-doped collectors, $C_1$ and $C_2$, diffused in an n-doped epilayer island. Two n-doped base contacts, $B_1$ and $B_2$, allow control of the emitter current and create a lateral drift field accelerating the injected holes in their passage to the collectors. When

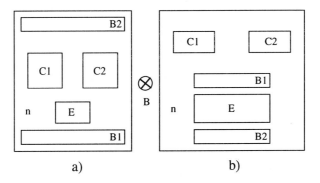

**FIGURE 11.11**
Lateral dual collector PNP magnetotransistors a) carrier deflection controlled, b) injection modulation controlled.

a magnetic field perpendicular to the device surface is applied, the hole emitter current injected into the base is deflected from its original path. Moreover, the base region acts as a Hall plate generating a Hall field causing an asymmetry in the emitter injection. Both effects result in a collector current difference, $\Delta I_C$, proportional to the magnetic induction $B$ and emitter current $I_E$, i.e.,

$$\Delta I_C = PBI_E \qquad (11.11)$$

Here $P$ is a constant which depends on the bias conditions, geometry, electrical properties of the device, and electron and hole mobilities.

When the base contacts are located close to the emitter area (Fig. 11.11(b)), injection modulation is considered the predominant mechanism which determines the variation of the collector currents. The difference of the collector currents is also given by the dependence in Eq. (11.11).

In vertical magnetotransistors (Fig. 11.12), the current deflected by the magnetic field is mainly vertical. The main current flow in a vertical dual-collector magnetotransistor is normal to the substrate. The device consists of two vertical NPN transistors with common emitter and base. A magnetic field parallel to the transistor surface causes deflection of the electrons toward one of the collectors at the expense of the other collector. The collector current difference is again proportional to the magnitude of the magnetic induction.

The sensitivity of magnetotransistors can be improved if the path of carriers from emitter to collector becomes longer. In a lateral device this is achieved by focusing the emitter injection in the center of the emitter junction. Such a device is called a suppressed sidewall injection magnetotransistor (SSIMT) [7]. SSIMT with focused emitter injection can be realized in a CMOS technology as well as in a bipolar technology. In a CMOS technology, the device is realized as a lateral transistor located in a p-well (Fig. 11.13), while in a bipolar technology, it is realized as an ordinary lateral transistor located in the epilayer. A characteristic feature of this device is the $p^+$-stripes placed between the emitter and each of

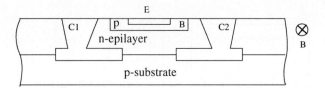

**FIGURE 11.12**
Dual-collector vertical magnetotransistor.

the two collectors. The stripes play a threefold role. First, they suppress carrier injection from the emitter in the lateral direction toward the collectors. Secondly, by applying a reverse potential to these stripes with respect to the emitter, carrier injection from the emitter into the base can be confined to the center of the bottom of the emitter-base junction. The current confinement increases with increasing reverse potential on the $p^+$-stripes. Current confinement can also be obtained by connecting $B_2$ directly to the emitter. Thirdly, applying a reverse potential to these stripes with respect to the emitter creates a lateral accelerating electric field in the neutral base region. Because of these three effects, both high sensitivity and a linear response to the magnetic field can be achieved simultaneously.

The device operation is as follows. In the forward active regime, the electrons are injected from the emitter into the base region. If $B_2$ is connected to the emitter or to a negative potential with respect to the emitter, the injection is restricted to the bottom of the emitter. Part of the injected electrons are collected by the substrate and part of them by the left and right collectors. In the absence of a magnetic field, the collector currents $I_{C1} = I_{C10}$ and $I_{C2} = I_{C20}$ are equal because of the structure's symmetry (assuming zero offset). When a magnetic field is applied parallel to the chip surface (perpendicular to the drawing plane in Fig. 11.13) an imbalance arises in the collector currents due to the following "double-deflection" effect. The Lorentz force acts on all three current components $I_S$, $I_{C1}$, and $I_{C2}$. The $I_S$ component is deflected in the $Y$ direction, contributing an increase in $I_{C2}$ and a decrease in $I_{C1}$. Moreover, the current components 1 and 2, $I_{C1}$ and $I_{C2}$, (Fig. 11.13) are deflected in the $-Z$ and $+Z$ directions, respectively, causing a further increase of $I_{C2}$ and a decrease of $I_{C1}$. When the field direction is reversed, $I_{C1}$ increases and $I_{C2}$ decreases.

The device is characterized in terms of the relative sensitivity

$$S_R = \frac{|I_{C1} - I_{C2}|}{I_{C10} + I_{C20}} \frac{1}{B} \tag{11.12}$$

Finally, this structure has the potential to annul the collector current offset (a current difference at zero magnetic field) using different potentials on the two p-stripes.

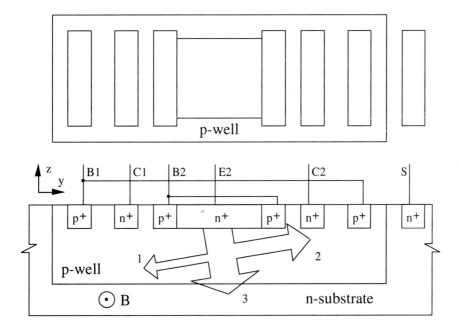

**FIGURE 11.13**
Suppressed sidewall injection magnetotransistor (SSIMT) in MOS technology.

## 11.6  SENSOR INTERFACES

A sensor made on a silicon substrate, using an IC technology, can be combined with interface electronic circuitry on the same chip. Most sensors produce a small signal in response to a measurand. This signal should be linearly amplified (see Chapter 3) and converted into a different format such as voltage, current, frequency, duty cycle, and pulse duration. The circuits considered below represent the approaches most frequently used for this purpose.

### 11.6.1  Bridge Circuits

Variable resistors appear in many sensors. To obtain a signal proportional to a small relative resistance variation, $\delta$, one can put such a resistor in a bridge (Fig. 11.14(a)), apply a reference voltage, $V_R$, to this circuit and measure the output voltage, $V_S$. For this particular configuration we obtain

$$V_S = \frac{V_R \delta}{2(2 + \delta)} \approx \frac{V_R \delta}{4} - \frac{V_R \delta^2}{8} \qquad (11.13)$$

The first term in Eq. (11.13) is the useful linear term, the second term is a nonlinearity error. The relative nonlinearity error is $\delta/2$. If one has two resistors

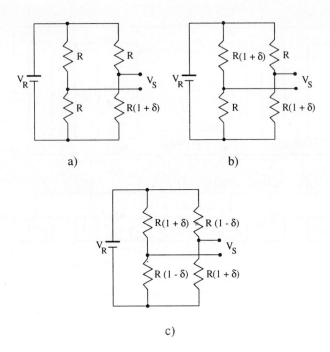

a)

b)

c)

**FIGURE 11.14**
Bridge circuits driven by a reference voltage.

which change simultaneously in the bridge configuration (Fig. 11.14(b)), then the output voltage is

$$V_S = \frac{V_R\delta}{(2+\delta)} \approx \frac{V_R\delta}{2} - \frac{V_R\delta^2}{4} \tag{11.14}$$

and still has a nonlinearity term. Both terms are twice as large as in Eq. (11.13) and the relative nonlinearity error is still $\delta/2$. Finally, if the bridge includes four resistors (Fig. 11.14(c)), two increasing and two decreasing, the bridge output voltage is

$$V_S = V_R\delta \tag{11.15}$$

In this case the bridge is linearized, i.e., its output voltage does not include a nonlinear term.

The same bridge configurations can be supplied by a reference current source $I_R$ (Fig. 11.15). For the first configuration, one obtains

$$V_S = \frac{I_R R\delta}{(4+\delta)} \approx \frac{I_R R\delta}{4} - \frac{I_R R\delta^2}{16} \tag{11.16}$$

One can see that the relative nonlinearity error is diminished and becomes $\delta/4$.

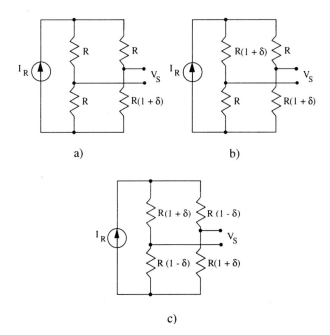

**FIGURE 11.15**
Bridge circuits driven by a reference current.

For the bridge of Fig. 11.15(b) one finds

$$V_S = \frac{I_R R \delta}{2} \tag{11.17}$$

and for the bridge of Fig. 11.15(c) we have

$$V_S = I_R R \delta \tag{11.18}$$

The results in Eqs. (11.17) and (11.18) do not include any nonlinear term; the bridge linearization has been achieved. The current sources frequently used in bridge linearization are shown in Fig. 11.16. In the case of Fig. 11.16(a), the reference current will be $V_R/R$, in the case of Fig. 11.16(b), it is $[V_{CC} - V_{BE(ON)}]/R$. More complicated circuits using temperature and supply independent voltage and current references are possible as well.

The bridge signal, $V_S$, should be amplified and an instrumentation operational amplifier (Fig. 11.17) is frequently used for this purpose. The output signal of this amplifier is

$$V_O = (1 + 2a)bV_S \tag{11.19}$$

with the resistance ratios $a$ and $b$ shown in Fig. 11.17.

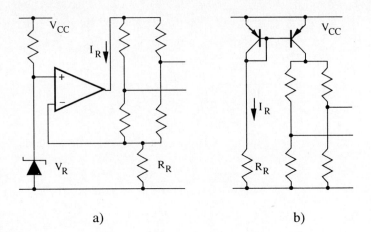

a)                                          b)

**FIGURE 11.16**
Current sources for bridges.

Simpler single op amp bridge configurations (Fig. 11.18) are also used. For the circuit of Fig. 11.18(a), a one obtains

$$V_O = \frac{V_R R_1}{R} \frac{\delta}{[(2+\delta)+(R/R_1)(1+\delta)]} \approx \frac{V_R R_1 \delta}{R(2+\delta)} \qquad (11.20)$$

where the approximation is valid when the gain $R_1/R$ is large. For the circuit of Fig. 11.18(b) the output voltage is

$$V_O = -\frac{V_R a \delta}{1+a+\delta} \approx -V_R \delta \qquad (11.21)$$

Finally, the bridge with one variable resistor can also be linearized if a circuit with two operational amplifiers [8] is used (Fig. 11.19). For this circuit

$$V_O = V_R \frac{R_1}{R} \delta \qquad (11.22)$$

### 11.6.2   Magneto-Operational Amplifier

In the bridge circuits discussed above, the operational amplifier is independent of the bridge. The magneto-operational amplifier (MOP) [9] considered in this section is an example of a circuit where the sensor is merged with an operational amplifier input stage. This results in an additional operational amplifier input corresponding to the measurand (in this case the magnetic field density). If the measurand is not applied to the circuit, one has an ordinary operational amplifier.

Figure 11.20(a) shows the conceptual diagram of this approach. The MOP has a conventional pair of input electrodes; the signals at these electrodes are

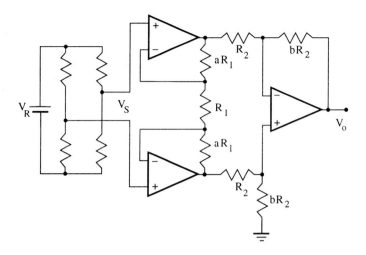

**FIGURE 11.17**
Instrumentation amplifier connected to bridge.

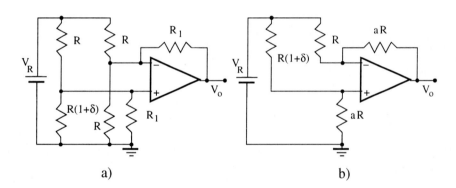

**FIGURE 11.18**
Single op amp bridge amplifiers.

applied to the summation point. The signal $V_m$ proportional to the applied measurand appears inside the operational amplifier and is applied to the summation point with a positive sign (thus, the interpretation of $V_m$ is similar to that used for calculations of the offset voltage effects [10]). Then, the MOP output voltage becomes

$$V_{out} = A(V_{in}^+ - V_{in}^- + V_m) \tag{11.23}$$

where, in this case,

$$V_m = S_b B \tag{11.24}$$

is referred to the input voltage proportional to the applied magnetic induction.

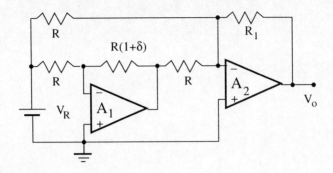

**FIGURE 11.19**
Transducer amplifier with linear response.

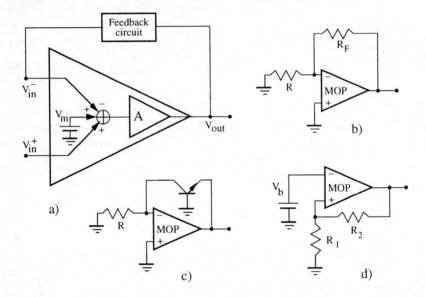

**FIGURE 11.20**
Magneto-operational amplifier.

The coefficient $S_b$ is the system sensitivity. When $S_b$ is known and $V_m$ is obtained, the MOP output voltage is calculated using general properties of operational amplifiers. For example, if a negative feedback circuit is added, the output voltage becomes

$$V_{out} = \frac{V_{in}^+ + V_m}{(1/A) + F} \qquad (11.25)$$

By using two external resistors $R_F$ and $R_S$ (Fig. 11.20 (b)), one realizes a

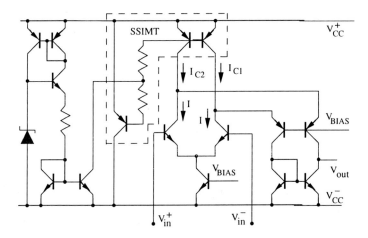

**FIGURE 11.21**
MOP bipolar realization.

linear magnetic sensor with

$$V_{out} = BS_b(1 + R_F/R_S) \tag{11.26}$$

If the feedback circuit includes a transistor (Fig. 11.20(c)), one obtains a logarithmic amplifier where

$$V_{out} = -V_t \ln(S_b B/I_s R) \tag{11.27}$$

Here $V_t = kT/q$ is the thermal voltage, and $I_s$ is the transistor saturation current. Finally, with positive feedback one can obtain a magnetically controlled Schmitt trigger where the switching points are determined by:

$$B_1 = S_b^{-1}[V_b - V_o^+ \frac{R_1}{(R_1 + R_2)}] \tag{11.28}$$

$$B_2 = S_b^{-1}[V_b - V_o^- \frac{R_1}{(R_1 + R_2)}] \tag{11.29}$$

The MOP realization is shown in Fig. 11.21. Similar to an ordinary system operational amplifier, it includes a differential input stage, and a folded-cascode stage with differential-to-single-ended signal conversion. The active load of the differential stage is realized using an SSIMT in the collector circuits. The difference of currents which enters the folded cascode stage can be forced either by the magnetic field or by application of the voltage difference between positive and negative terminals. Equating these effects one obtains

$$S_b = \frac{S_R(I_{C10} + I_{C20})}{g_m} \tag{11.30}$$

where $g_m = I/V_t$.

The relative sensitivity $S_R$ of the SSIMT decreases with increasing temperature and increases when the base current of transistors increase. Both dependencies are almost linear [9]. As a result one can compensate the thermodependence of $S_R$ including temperature dependent biasing shown in Fig. 11.21.

This particular circuit illustrates the concept of the integrated smart sensor: co-integration of the sensor with signal processing circuitry, and auxiliary circuits providing normal sensor operation on the same chip.

### 11.6.3 Controlled Multivibrators

Multivibrators represent a widely used group of electronic circuits. A lot of work has been done to investigate and design the circuits where the oscillation parameters (amplitude, frequency, phase, and duty cycle) provide information on the values of the passive (resistors, capacitors) and active (current and voltage sources) elements which are parts of the oscillating circuits. These values can be functions of external measurands (mechanical force, electric or magnetic field, temperature, humidity, etc.) Then, the modification of oscillation parameters gives information about such measurands.

The sensors included in multivibrators appear in the disguise of the above mentioned elements and multivibrators including sensors become important interface circuits [11]. Multivibrators with linearly controlled frequency are preferred. The frequency output, being noise immune, is required in most cases. A current trend in the realization of integrated sensors is the demand for communication with a microprocessor. The rectangular multivibrator output is more easily converted to the final digital form. Yet, the duty cycle modulation available in a sensor-driven multivibrator can be a very robust form of signal as well and is used in some interesting applications.

In this section the chosen circuits are mostly treated on the level of operational amplifiers, comparators and switches considered as building blocks. Besides describing the principle of operation of the circuits, it is our goal to present the control characteristics in a form allowing estimation of their nonlinearity. Special attention is given to slew-rate limitation of operational amplifiers and comparators. This nonideality, which is not "visible" in the ordinary self-oscillating circuits, can be a main source of errors in sensor applications.

Our scope is limited by continuous-time multivibrators. The application of switched-capacitor circuits to the conversion of RLC parameters into a frequency or digital signal can be found in [12].

The conversion of the frequency-modulated rectangular signal into useful analog or digital form can be made highly immune to noise and interference signals. Here phase-lock loop techniques are most important [13, 14]. They allow an easy transformation of the signal into analog form and also frequency conversion and multiplexing. The signal can be put in the frequency band suitable for telecommunication transmission.

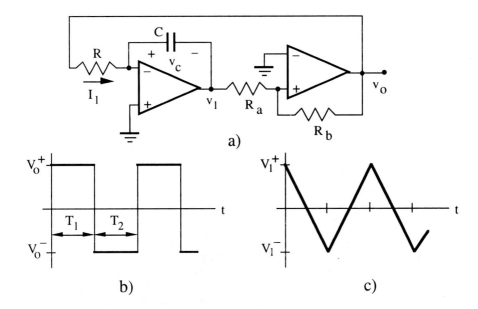

a)

b)                                        c)

**FIGURE 11.22**
Common two-amplifier multivibrator a) circuit b) output voltage, and c) integrator output voltage.

**Multivibrators with Operational Amplifiers**   One of the frequently used multivibrator circuits is shown in Fig. 11.22(a). It includes an integrator and a Schmitt trigger. The output voltage, $v_o$, (Fig. 11.22(b)) is a rectangular wave limited by the values $V_o^+$ and $V_o^-$. The integrator output voltage $v_1$ (Fig. 11.22(c)) is a triangular wave achieving the values $V_1^+$ and $V_1^-$. The transition of the multivibrator from one quasistable state to another is determined by the balance of voltage at the positive terminal of the Schmitt trigger. At the end of the interval $T_1$, one has

$$V_1^- \frac{R_b}{R_a + R_b} + V_o^+ \frac{R_a}{R_a + R_b} = 0 \qquad (11.31)$$

which gives $V_1^- = -V_o^+(R_a/R_b)$. By the same token, one will obtain at the end of $T_2$ that $V_1^+ = -V_o^-(R_a/R_b)$. Thus, the capacitor voltage, $v_c$, has the swing of $\Delta V_c = V_1^+ - V_1^- = (V_o^+ - V_o^-)(R_a/R_b)$. During the first interval the capacitor $C$ is recharged by the current $I_1 = V_o^+/R$, therefore, the interval duration is

$$T_1 = RC(\frac{V_o^+ - V_o^-}{V_o^+})\frac{R_a}{R_b} \qquad (11.32)$$

To obtain $T_2$, it is sufficient to interchange "+" and "−" in the superscripts of Eq. (11.32). Hence, the oscillation period is

$$T = T_1 + T_2 = RC \frac{R_a(V_o^+ - V_o^-)(V_o^- - V_o^+)}{R_b V_o^+ V_o^-} \tag{11.33}$$

If $V_o^+ = -V_o^- = V_o$, Eq. (11.33) is simplified and the oscillation frequency becomes

$$f_0 = \frac{1}{4RC} \left( \frac{R_b}{R_a} \right) \tag{11.34}$$

The oscillation frequency can be modulated by modifying $R$ or $C$.

The main source of dynamic frequency errors in this multivibrator is the finite slew rate, $S_{ri}$, of the integrator. The correct frequency value can be calculated in the following way. If $V_{IM}$ is the maximum useful input voltage of the op amp with respect to the slew rate [15] then the error voltage source

$$v_i = -\frac{V_{IM}}{S_{ri}} \frac{dv_1}{dt} \tag{11.35}$$

is introduced at the input terminal of the integrator (Fig. 11.23(a)). The integrator output voltage when the multivibrator output voltage is $V_o^+$, will be

$$v_1 = -\frac{1}{RC} \int V_o^+ dt + v_i + \frac{1}{RC} \int v_i dt \approx -\frac{1}{RC} \int V_o^+ dt + \frac{1}{RC} \int v_i dt \tag{11.36}$$

Differentiating Eq. (11.36) and using Eq. (11.35) one obtains that during this period of time

$$\frac{dv_1}{dt} = -\frac{V_o^+}{RC + \eta_i} \tag{11.37}$$

where $\eta_i = V_{IM}/S_{ri}$. From the other side, the capacitor voltage swing

$$\Delta V_c = \pm (V_o^+ - V_o^-) \frac{R_a}{R_b} \tag{11.38}$$

is not influenced by the presence of $v_i$. Thus, the correct oscillation period is

$$T = T_1 + T_2 = (RC + \eta_i) \frac{R_a(V_o^+ - V_o^-)(V_o^- - V_o^+)}{R_b V_o^+ V_o^-} \tag{11.39}$$

i.e., the source, $v_i$, introduces the relative frequency error

$$\frac{\delta f}{f_0} \approx -\frac{\eta_i}{RC} = -\frac{4 f_0 \eta_i R_a}{R_b} \tag{11.40}$$

which increases with frequency.

When the oscillation frequency becomes very high, the finite slew rate $S_{rs}$ of the Schmitt trigger operational amplifier must be considered as well. The Schmitt trigger has positive feedback and one can assume that its output voltage always

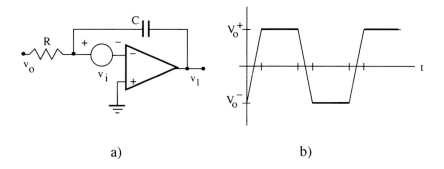

a)                                                b)

**FIGURE 11.23**
Slew-rate error voltage and multivibrator wave distortion a) integrator slew-rate error voltage
b) output waveform.

changes with a constant speed of $S_{rs}$ between quasistable states. The oscillation
period becomes

$$T = T_1 + T_2 = (RC + \eta_i)\frac{R_a(V_o^+ - V_o^-)(V_o^- - V_o^+)}{R_b V_o^+ V_o^-} + \frac{2(V_o^+ - V_o^-)}{S_{rs}} \qquad (11.41)$$

so that the frequency error is

$$\frac{\delta f}{f_0} \approx -\frac{\eta_i}{RC} = -\frac{4V_o f_0}{S_{rs}} \qquad (11.42)$$

The integrator output voltage will have visibly smoothed corners.

**Switches in the Basic Multivibrator Circuits**    Useful modifications occur when switches are incorporated in the basic multivibrator circuit. Switches
realized by active devices of minimal geometry are not a heavy burden, neither in terms of multivibrator complexity nor chip area. An example of such a
modification is shown in Fig. 11.24.

The output voltage in this multivibrator controls the switches, $P_1$ and $P_2$,
connecting the resistors, $R_1$ and $R_2$, to the integrator input. It is easy to find
that for the symmetric output wave ($V_o^+ = -V_o^- = V_o$ and $V_{CC}^+ = -V_{CC}^- = V_{CC}$)
the durations are

$$T_1 = \frac{2V_o C R_1 R_a}{V_{CC} R_b} \qquad (11.43)$$

and

$$T_2 = \frac{2V_o C R_2 R_a}{V_{CC} R_b} \qquad (11.44)$$

The circuit is used when the resistor change is transformed into a duty
cycle change. Then it is convenient to consider the quasistable state duration
ratio $D = T_1/T_2$. If $R_1 = R + \delta R$ and $R_2 = R - \delta R$, then the oscillation

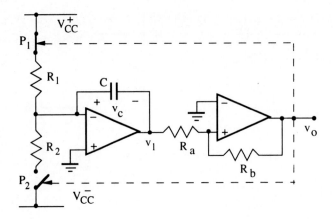

**FIGURE 11.24**
Multivibrator with switch-commutated resistors.

frequency for the circuit of Fig. 11.24 does not change and

$$D = \frac{T_1}{T_2} = \frac{R_1}{R_2} \approx 1 + \frac{\delta R}{R} \tag{11.45}$$

The errors caused by the slew-rate of the operational amplifiers can be calculated individually for $T_1$ and $T_2$ as outlined in the previous section. In addition, these durations should be augmented by the switching time of the switches.

**Bridge-to-Frequency Converter**   Using switches in a two-operational amplifier multivibrator one can obtain control of the oscillation frequency by detuning a resistive bridge (Fig. 11.25). For normal circuit operation, at least the resistors $R_1$ and $R_2$ should be different.

The conditions for the circuit transition from one quasistable state to another are $V_1^- = -V_o^+(R_a/R_b)$ and $V_1^+ = -V_o^-(R_a/R_b)$. Let the position of switches shown in Fig. 11.25 correspond to the first quasistable state and the output voltage be $v_o = V_o^+$ in this state. Then, at the end of this state, the capacitor voltage, $v_c$, is

$$v_c(T_1) = \frac{V_{CC}^+ R_4 + V_{CC}^- R_3}{(R_4 + R_3)} + V_o^+ \frac{R_a}{R_b} \tag{11.46}$$

The voltage $v_c(T_1+T_2)$ at the end of the second quasistable state is equal to $v_c(0)$ at the beginning of the first quasistable state. It is obtained by interchanging plus and minus signs in the superscripts of Eq. (11.46). This will give

$$v_c(0) = \frac{V_{CC}^- R_4 + V_{CC}^+ R_3}{(R_4 + R_3)} + V_o^- \frac{R_a}{R_b} \tag{11.47}$$

Thus, the capacitor voltage swing during the first quasistable state is

$$\Delta V_c = \frac{(V_{CC}^+ - V_{CC}^-)(R_4 - R_3)}{(R_4 + R_3)} + (V_o^+ - V_o^-)\frac{R_a}{R_b} \tag{11.48}$$

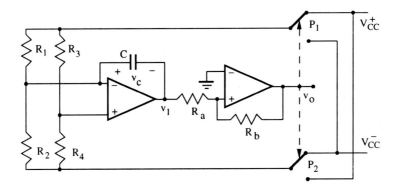

**FIGURE 11.25**
Switch-commutated bridge in the two-amplifier multivibrator.

During this quasistable state the capacitor is recharged by the current

$$I_1 = \frac{(V_{CC}^+ - v^+)}{R_1} + \frac{(V_{CC}^- - v^+)}{R_2} = \frac{V_{CC}^+ - V_{CC}^-}{(R_4 + R_3)} \left( \frac{R_3}{R_4} - \frac{R_4}{R_2} \right) \qquad (11.49)$$

Then one obtains the duration of the quasistable state

$$T_1 = T_2 = \frac{(R_4 - R_3)C}{[(R_3/R_1) - (R_4/R_2)]}$$

$$+ \frac{R_a(R_3 + R_4)C}{R_b[(R_3/R_1) - (R_4/R_2)]} \left( \frac{V_o^+ - V_o^-}{V_{CC}^+ - V_{CC}^-} \right) \qquad (11.50)$$

The oscillation frequency is

$$f_0 = \frac{1}{2C} \left( \frac{R_3}{R_1} - \frac{R_4}{R_2} \right)$$

$$\times \frac{(V_{CC}^+ - V_{CC}^-)R_b}{[R_b(R_4 - R_3)(V_{CC}^+ - V_{CC}^-) + R_a(R_4 + R_3)(V_o^+ - V_o^-)]} \qquad (11.51)$$

The switches also allow [15] elimination of the feedback resistances of the Schmitt trigger (Fig. 11.26). For this circuit the transition from the first quasistable state to the second is determined by the condition $v_1 = V_1^- = V_{CC}^-$ (the condition of the opposite transition is $v_1 = V_1^+ = V_{CC}^+$). The capacitor recharges from

$$v_c(0) = \frac{V_{CC}^- R_4 + V_{CC}^+ R_3}{(R_4 + R_3)} - V_{CC}^+ \qquad (11.52)$$

to

$$v_c(T_1) = \frac{V_{CC}^+ R_4 + V_{CC}^- R_3}{(R_4 + R_3)} - V_{CC}^- \qquad (11.53)$$

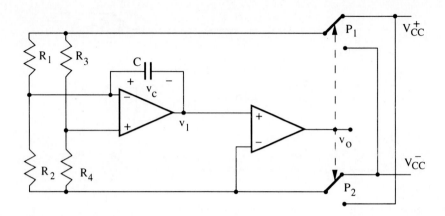

**FIGURE 11.26**
Switch-commutated bridge in the amplifier-comparator multivibrator.

so that

$$\Delta V_c = 2(V_{CC}^+ + V_{CC}^-)\frac{R_4}{(R_4 + R_3)} \qquad (11.54)$$

The recharge current is again given by Eq. (11.49) and one obtains

$$T_1 = T_2 = \frac{2CR_4}{[(R_3/R_1) - (R_4/R_2)]} \qquad (11.55)$$

and the oscillation frequency is

$$f_0 = \frac{1}{4CR_4}(\frac{R_3}{R_4} - \frac{R_4}{R_2}) \qquad (11.56)$$

The finite slew rate of the integrator can be a major source of the frequency error in the circuits of Figs. 11.25 and 11.26. For the error analysis, it is convenient, in this case, to use two separate error voltages $V_I^+$ and $V_I^-$ for each quasistable state and to represent them as

$$V_I^+ = -V_I^- = \frac{V_{IM}}{S_{ri}}(\frac{V_1^+ - V_1^-}{T/2}) = \frac{2f_0 V_{IM}(V_o^+ - V_o^+)R_a}{R_b S_{ri}} \qquad (11.57)$$

Figure 11.27 shows the integrator of the multivibrator with this source of error during the first quasistable state. Then one can write

$$v_c(T_1) = \frac{V_{CC}^+ R_4 + V_{CC}^- R_3}{(R_3 + R_4)} + V_I^+ + V_o^+ \frac{R_a}{R_b} \qquad (11.58)$$

Again, the initial value of the capacitor voltage can be obtained interchanging the signs in the superscripts. This gives

$$v_c(0) = \frac{V_{CC}^- R_4 + V_{CC}^+ R_3}{(R_3 + R_4)} + V_I^- + V_o^- \frac{R_a}{R_b} \qquad (11.59)$$

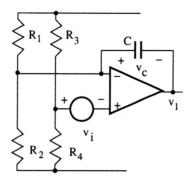

**FIGURE 11.27**
Slew-rate error voltage in the integrator with bridge.

so that

$$\Delta V_c = (V_{CC}^+ - V_{CC}^-)(\frac{R_4 - R_3}{R_4 + R_3}) + (V_I^+ - V_I^-) + (V_o^+ - V_o^-)\frac{R_a}{R_b} \qquad (11.60)$$

The recharge current in this case will be

$$I_1 = (\frac{V_{CC}^+ - V_{CC}^-}{R_4 + R_3})(\frac{R_3}{R_1} - \frac{R_4}{R_3}) - V_I^+(\frac{1}{R_1} + \frac{1}{R_1}) \qquad (11.61)$$

Using Eqs. (11.60) and (11.61) one can find the duration of $T_1$. Then, changing the signs of the subscripts in $T_1$, the duration of $T_2$ will be found. After this the oscillation frequency can be found. These complex expressions are omitted here. In the case of $V_{CC}^+ = -V_{CC}^- = V_{CC}$, $V_o^+ = -V_o^- = V_o$, and $V_I^+ = -V_I^- = v_i$, the result is

$$f = f_0 + \delta f = \frac{\frac{2V_{CC}}{(R_3 + R_4)}(\frac{R_3}{R_1} - \frac{R_4}{R_2}) - V_I(\frac{1}{R_1} + \frac{1}{R_2})}{4C[V_{CC}(\frac{R_4 - R_3}{R_4 + R_3}) + V_I + V_o\frac{R_a}{R_b}]} \qquad (11.62)$$

Neglecting the term $V_I$ in the denominator of Eq. (11.62) one obtains the relative frequency error

$$\frac{\delta f}{f_0} \approx -\frac{V_I}{2V_{CC}}F(R_B) = -\frac{f_0}{S_{ri}}\frac{V_{IM}V_o}{V_{CC}}\frac{R_a}{R_b}F(R_B) \qquad (11.63)$$

where $F(R_B) = [(R_1 + R_2)(R_3 + R_4)]/(R_2R_3 - R_1R_4)$.

It is easy to verify that for the circuit of Fig. 11.26, the error is [15]

$$\frac{\delta f}{f_0} \approx -\frac{f_0\eta_i}{S_{ri}}F(R_B) \qquad (11.64)$$

With increasing oscillation frequency the output comparator slew rate in the circuits of Figs. 11.25 and 11.26 augments the durations of $T_1$ and $T_2$, as for the circuit of Fig. 11.22. In addition, the switching time of $P_1$ and $P_2$ should be added to $T_1$ and $T_2$.

**Voltage-to-Frequency Converters**   In the multivibrators considered so far, it was assumed that the ratio $D = T_1/T_2$ is about 0.5. If $D$ is modified on purpose to strongly differ from 0.5, the circuit becomes a voltage-to-frequency converter (VFC). The output voltage of these multivibrators is a train of narrow pulses where the frequency can be controlled by the input voltage. The circuit should be designed so that the width of the pulses is as narrow as possible; in this case better linearity of the control characteristic is provided.

Modifying [16], for example, the circuit of Fig. 11.22, one obtain the VFC shown in Fig. 11.28. During the long part $T_1$ of the oscillation period, the switch, $P$, is open and the timing capacitor $C$ is charged from a positive input voltage $V_i$. During the short part $T_2$ the switch $P$ is closed and the timing capacitor is recharged from the voltage $V_d^-$. The integrator output voltage $v_1$ changes from $V_1^+ = V_o^+ \rho$ to $V_1^- = V_o^- \rho$ where $\rho = R_a/(R_a + R_b)$. The duration $T_1$ can be found from the charge balance equation,

$$T_1 \frac{V_i}{R} = C(V_1^+ - V_1^-) \tag{11.65}$$

and the duration $T_2$ is calculated from the equation

$$T_2 \left( \frac{V_i}{R} + \frac{V_d^-}{R_1} \right) = C(V_1^- - V_1^+) \tag{11.66}$$

The oscillation frequency, $f$, can be found as

$$f = \frac{1}{T_1 + T_2} = \frac{V_i}{RC(V_1^+ - V_1^-)} \left( 1 + \frac{V_i R_1}{V_d^- R} \right) \tag{11.67}$$

If the second term in the brackets of Eq. (11.67) is small and can be neglected, then the oscillation frequency

$$f_0 = \frac{V_i}{RC(V_1^+ - V_1^-)} \tag{11.68}$$

is proportional to the input voltage. When $V_i$ increases this second term increases as well and determines the relative frequency error

$$\frac{\delta f}{f} = \frac{V_i R_1}{V_d^- R} \tag{11.69}$$

The additional errors caused by the slew rate limitation and finite switching time are added (at the upper end of the frequency range) in the same fashion as in the circuit of Fig. 11.22.

The minimum value of the error term given by Eq. (11.69) is limited by the integrator maximum current. Aside from this, the auxiliary voltage source $V_d^-$ should provide a high current for the capacitor recharge. These deficiencies are eliminated [17] in the circuit shown in the Fig. 11.29(a). Here the active integrator is replaced by a passive $RC$-circuit. A buffer is introduced between this circuit and a dynamic Schmitt trigger.

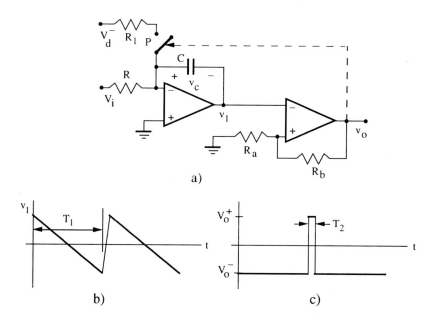

**FIGURE 11.28**
VFC with timing capacitor recharge a) circuit, b) integrator output, c) VFC output.

The circuit operation can be understood by examining the voltage $v^+$ (Fig. 11.29(b)). During the interval $T_1$, the switch $P$ is open and the large capacitor, $C$, is being charged by the input voltage, $V_i$. The voltage $v_1$ is equal to the voltage at $C$. The values of $R_a$, $R_b$, and $C_d$ (a small value capacitor) should be such that the voltage $v^+$ at the Schmitt trigger positive terminal follows $v_1$ as well. When $C_d$ is small the initial transient for $v^+$ is short and can even be completely eliminated (the required parameter relationship is given below). When $v^+$ attains the value of $V_{Ref}$ the output voltage $v_o$ jumps to $V_o^+$ and the switch $P$ will be closed. The voltage $v_1$ drops to zero, yet the voltage $v^+$ becomes much higher than $V_{Ref}$ because the voltage at $C_d$ and $V_o^+$ are operating in series. Then the capacitor $C_d$ starts to recharge and $v^+$ quickly diminishes. When $v^+$ drops to the value of $V_{Ref}$, the output voltage drops to $V_o^-$, the switch will be open again, and the cycle is repeated.

The durations $T_1$ and $T_2$ are calculated considering the voltages in the Schmitt trigger feedback circuit. When the switch $P$ is open (Fig. 11.29(d)) the buffer output voltage is

$$v_i = V_i[1 - exp(-t/RC)] \approx (V_i t)/(RC) \qquad (11.70)$$

If $C_d$ is small then

$$i_{cd} = C_d \frac{dv_1}{dt} \approx \frac{V_i C_d}{RC} \qquad (11.71)$$

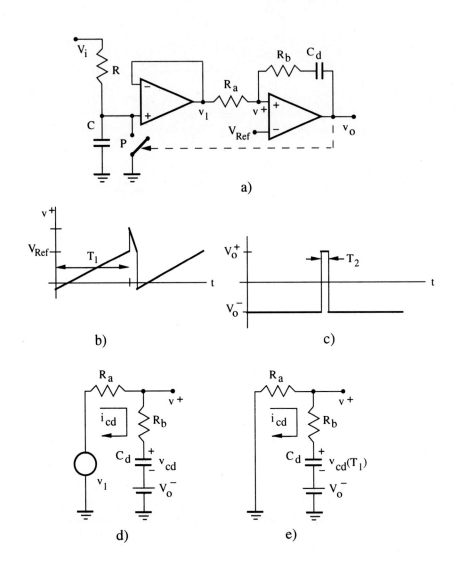

**FIGURE 11.29**
VFC with full discharge of the timing capacitor a) circuit, b) comparator input, c) VFC output,
d) charge of $C_d$, e) discharge of $C_d$.

At the end of the first quasistable state $v_1(T_1) = V_{Ref} + i_d R_a$, i.e.,

$$V_i[1 - exp(-t/RC)] = V_{Ref} + \frac{V_i C_d}{RC} \tag{11.72}$$

and from Eq. (11.72) one obtains that

$$T_1 = RC\ln\{[1 - (\frac{V_{Ref}}{V_i} + \frac{R_a C_d}{RC})]^{-1}\} \approx RC\frac{V_{Ref}}{V_i} + R_a C_d \tag{11.73}$$

At the end of this part of the oscillation period the voltage $v_{cd}$ at the feedback capacitor becomes $v_{cd}(T_1) = V_{Ref} - V_o^- - (V_i R_b C_d/RC)$.

When the circuit switches in the second quasistable state, the voltage $v^+$ can be calculated from the circuit of Fig. 11.29(e). First of all, it is seen that the condition

$$[V_o^+ + v_{cd}(T_1)]\rho > V_{Ref} \tag{11.74}$$

should be satisfied to obtain this quasistable state. Here $\rho = R_a/(R_a + R_b)$ again. Then one can write

$$v^+ = [V_o^+ + v_{cd}(T_1)]\rho \exp\{-(t - T_1)/[C_d(R_a + R_b)]\} \tag{11.75}$$

for $T_1 < t < (T_1 + T_2)$, and the condition $v^+(T_1 + T_2) = V_{Ref}$ gives

$$T_2 \approx C_d(R_a + R_b)\{1 - \frac{V_{Ref}/\rho}{[V_o^+ - V_o^- + V_{Ref} - (V_i R_b C_d)/(RC)]}\} \tag{11.76}$$

The oscillation frequency, which is

$$f = \frac{1}{T_1 + T_2} = \frac{V_i}{RCV_{Ref} + V_i(R_a C_d + T_2)} \tag{11.77}$$

includes for this circuit the error term increasing with $V_i$. Indeed, if one denotes

$$f_0 = \frac{V_i}{RCV_{Ref}} \tag{11.78}$$

then, as it follows from Eq. (11.77), the relative frequency error is

$$\frac{\delta f}{f} \approx - f_0(R_a C_d + T_2) \tag{11.79}$$

Finally, at the end of the second quasistable state $v_{cd}(T_1 + T_2) = v_{cd}(0) = V_{Ref}[1 + (R_b)/(R_a)] - V_o^+$ and, if the condition $V_{Ref} = V_o^+ \rho$ is satisfied, the voltage $v^+$ follows the voltage at the timing capacitor without a transient in the first quasistable state.

**Current-to-Frequency Converters**    All previously described VFCs could be transformed into current-to-frequency converters (CFCs) if the voltage source $V_i$ and resistor $R$ were replaced by a current source charging the timing capacitor $C$.

The CFCs considered in this section are based on a different approach. It is assumed here that the timing capacitor is periodically charged and discharged from two tightly matched current sources. The rest of the multivibrator circuit provides the redirection of these currents. A transistor current source can operate with a wide current range [18] and the redirection is realized by switches which can be very fast. The approach requires more complicated circuits but results in the circuits operating at high frequencies and having very linear control characteristics. The ratio $D = T_1/T_2$ of these multivibrators is normally about one which makes them very suitable for operation with phase-locked loops.

Most CFCs used in sensor applications have one plate of the timing capacitor grounded. This requirement resulted in the Schmitt trigger CFCs with three different circuits (Fig. 11.30) providing redirection of the charging current.

In the first circuit (Fig. 11.30(a)), the timing capacitor is charged by the top current source $I$. When the voltage $v_c$ attains the value of $V_{cH}$, the trigger turns off the top current source $I$, opens the switch $P$, closes the switch $P_1$, and turns on the bottom current source $I$. Now the timing capacitor is discharged and when $v_c$ attains the value of $V_{cL}$, the trigger turns off the current source $I$ and the switching operations are repeated in the inverse order.

In the second circuit (Fig. 11.30(b)), the current source $I$ permanently charges the timing capacitor from the top and the trigger redirects the current source $2I$. During the discharge of the timing capacitor, both current sources are connected to the timing capacitor, which is discharged by the difference of two currents.

Finally, in the third circuit (Fig. 11.30(c)), one has only one redirected current source. When the switch $P$ is ON, the current $I$ is directed into the current mirror and charges the timing capacitor from the top. When the switch $P_1$ is ON, the current source is connected to the capacitor and discharges it.

For all three circuits the oscillation frequency is

$$f_0 = \frac{I}{2C(V_{cH} - V_{cL})} \tag{11.80}$$

In the first circuit the switching and trigger subcircuits operate in series. The trigger can be realized using a comparator and a two-resistor positive feedback. When the top and bottom currents are changing simultaneously, one obtains a CFC. The relative frequency error for these converters can be estimated from

$$\frac{\delta f}{f} \approx -2f_0 T_s \tag{11.81}$$

where $T_s$ is the combined delay introduced by the comparator and the switch. If the currents are changing separately, but their sum is preserved, one obtains a duty-cycle modulation.

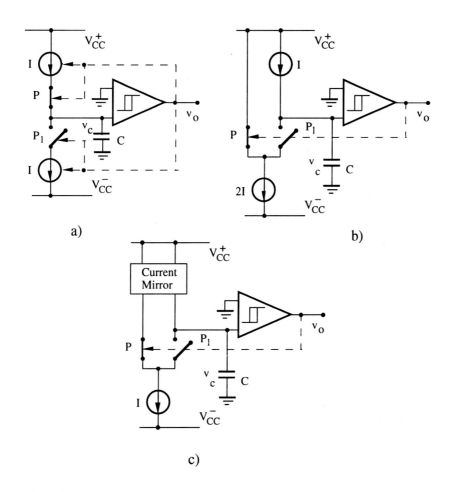

**FIGURE 11.30**
CFC's with Schmitt trigger.

In the second and third circuits, the switching and trigger subcircuits are driven in parallel. This results in faster circuits capable of operating at higher frequencies. In addition, these CFCs are structurally divided into two independent subcircuits, namely, the threshold circuit and the charge-discharge circuit. This makes the design more flexible and provides an abundance of circuits adapted for different design conditions [19, 20]. The charge-discharge circuit including the two complementary switches is typically realized using a differential pair, which is able to provide fast switching of its tail current. The Schmitt trigger designs are different. The fast converters rely on simpler structures of the Schmitt triggers also based on switching a differential pair so that the whole circuit becomes a fast nonsaturated multivibrator. These circuits show the widest range of control current and can be realized (in bipolar technology) using NPN transistors only [19]. For many of these circuits, the bipolar design can be topologically transferred into CMOS technology (with minor additional limitations).

It is feasible to design a multivibrator where the frequency of oscillation depends linearly on the control voltage. The three decades (four decades in the bipolar case) of frequency variation can be achieved with one capacitor value.

## 11.7 STABILIZATION OF POWER DISSIPATION

Stabilization of dissipation power is important for the correct operation of some integrated sensors [1, 4]. The design of a circuit for this purpose is very simple if an active device is used to provide constant power dissipation. The circuit is more complicated if a constant dissipation power is required in a variable resistor representing a sensor heater.

Two circuits are chosen here to solve this problem. In the first circuit, an active equivalent of the ballast resistor in parallel with the variable load is designed. The circuit uses linear voltage-controlled current sources and a current mirror subtracting circuit. The load power depends on one external voltage and two resistors on the same chip. The power is stabilized in the vicinity of the quiescent operating point and small variations of the load resistor are assumed. In the second circuit, a controlled current source is introduced in the translinear loop of four transistors. The load power is controlled by an external current source and one resistor on the chip.

**Power Stabilizer with Voltage Controlled Current Sources**   In the circuit shown in Fig. 11.31, the load current is

$$I_L = \left(\frac{V_S}{R_{Sn}} - \frac{V_L}{R_{0n}}\right) \tag{11.82}$$

where $R_{Sn} = R_S/n$ and $R_{0n} = R_0/n$. Using $V_L = I_L R_L$, one can find from Eq. (11.82) that the load current is

$$I_L = \frac{(V_S/R_{Sn})}{1 + (R_L/R_{0n})} \tag{11.83}$$

The power dissipated in the load is

$$P_L = I_L^2 R_L = \frac{4P_Q \sigma}{(1 + \sigma)^2} \tag{11.84}$$

where $P_Q = [(V_S/R_{Sn})^2 R_{0n}]/4$ and $\sigma = R_L/R_{0n}$.

The parameter $P_Q$ is the maximal power dissipated in the load for a chosen value of $V_S$ and circuit parameters. Then the function

$$W = 1 - \frac{P_L}{P_Q} = \left(\frac{1 - \sigma}{1 + \sigma}\right)^2 \tag{11.85}$$

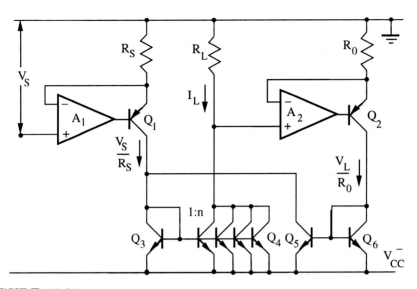

**FIGURE 11.31**
Power stabilizer for small resistor variations.

can characterize the power stabilization (Fig. 11.32). This function shows that even a 50% variation of the load resistor leads only to a 3% variation in the dissipated power.

The operational amplifiers, $A_1$ and $A_2$, being loaded only by the transistor base currents can be designed as system operational amplifiers. The design can be transferred to CMOS technology as well.

**Stabilizer with a Control Current Source in a Translinear Loop**  The design idea of this power stabilizer can be illustrated by the circuit shown in Fig. 11.33(a). For the translinear loop (see Chapter 5) including matched transistors $Q_1$ to $Q_4$ we find

$$I_R^2 = I_L I_C \tag{11.86}$$

If the current $I_C$ is made controllable in such a way that $I_C = V_L/R_0$, where $V_L$ is the voltage drop at the resistor $R_L$ and $R_0$ is a transresistance, then one obtains

$$I_R^2 R_0 = I_L V_L \tag{11.87}$$

Thus, the circuit will provide a constant power

$$P_Q = I_R^2 R_0 \tag{11.88}$$

independent of the variation in the resistor $R_L$. The circuit of Fig. 11.33(b) shows how the required control function can be realized. Transistor $Q_5$ is an emitter follower and transistor $Q_6$ compensates the $V_{BE}$ voltage drop in this follower. The voltage across $R_0$ becomes equal to $V_L$. The current $V_L/R_0$ is intercepted

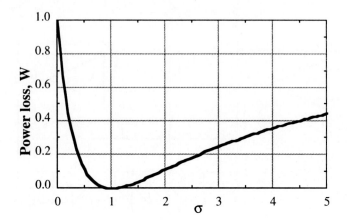

**FIGURE 11.32**
Characteristic of power stabilizer.

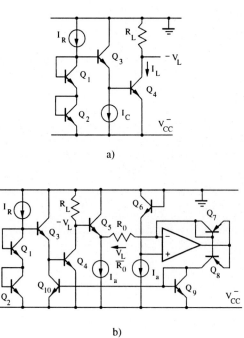

**FIGURE 11.33**
Power stabilizer with translinear loop.

by $Q_7$ and mirrored by $Q_8$. Then it is mirrored by $Q_9$ and $Q_{10}$ and introduced in the translinear loop providing the required control function.

One can use a general purpose operational amplifier in the circuit of Fig. 11.33(b). But in this case, it is necessary to have a positive power supply. This deficiency can be avoided if the gain provided by a differential pair with differential-to-single-ended conversion is sufficient.

## PROBLEMS

**11.1.** A piezoresistive bridge (Fig. 11.14(c)) operating in the range of 0 to 10 kPa has the sensitivity of 1mV/kPa. The bridge is driven by the reference voltage $V_R$=10 V and R(0)=100 k$\Omega$. Find the relative change of resistor length if the gauge factor is 150.

**11.2.** Design a Widlar current source (Fig. P11.2) to drive a Hall plate with $R_H$=1500 cm$^3$/coulomb and thickness $t$ =10 $\mu$m. The current should provide the sensitivity of 0.15 mV/gauss.

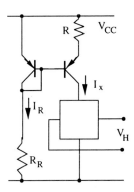

**FIGURE P11.2**

**11.3.** In the bridge shown in Fig. P11.3, the variable resistors change with temperature so that $R(T) = R_0[1 + \alpha T - \beta T^2]$ where $\alpha = 4.10^{-3}$ $\Omega/\Omega°$C and $\beta = 6.10^{-6}$ $\Omega/\Omega(°C)^2$ and $T$ is the temperature in degrees Celsius. The bridge nonvariable resistors are constant and have the value of $R_0$=100 k$\Omega$. Find the bridge sensitivity and nonlinearity error if $V_R$=10 V and the bridge is operating in the temperature range of 0 to 100 °C.

**11.4.** In the bridge shown in Fig. P11.4, the variable resistors change with temperature so that $R(T) = R_0[1 + \alpha T - \beta T^2]$ where $\alpha = 4.10^{-3}$ $\Omega/\Omega°$C and $\beta = 6.10^{-6}\Omega/\Omega(°C)^2$ and T is the temperature in °C. The bridge nonvariable resistors are constant and have the value of $R_0$ =100 k$\Omega$. Find the bridge sensitivity and nonlinearity error if $I_R$=100 $\mu$A and the bridge is operating in the temperature range of 0 to 100 °C.

**11.5.** Find the sensitivity of the temperature sensor using two matched transistors and two current sources (Fig. P11.5).

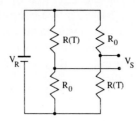

**FIGURE P11.3**

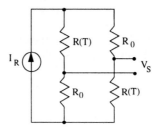

**FIGURE P11.4**

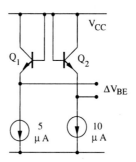

**FIGURE P11.5**

**11.6.** Find the sensitivity of the temperature sensors using two matched currents obtained by splitting the drive current (Fig. P11.6). The circuit transistors are matched.

**11.7.** A humidity dependent capacitor $C$ is used in the multivibrator circuit shown in Fig. 11.22(a). At the nominal operating point the capacitor has the value of 50 pF and changes with $\Delta C/C = 0.4\%/(\%\text{r.h.})$. Find the frequency variation if the humidity changes by 10% around the nominal point. Assume $R = 100$ k$\Omega$, $R_a/R_b = 1$, and the maximum useful voltage for the integrator is $V_{IM}=0.15$ mV. Find the required operational amplifier slew rate if the frequency error at the nominal point should not exceed more than 0.05%.

**11.8.** A capacitor $C = 50$ pF is used in the resistor bridge multivibrator circuit of Fig. 11.25. In this circuit $R_1 = R_2 =100$ k$\Omega$, $R_3 =130$ k$\Omega$, $R_4 =110$ k$\Omega$, and $R_a/R_b =1$. Find the relative frequency error if the maximum useful voltage

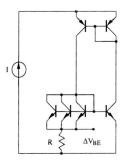

**FIGURE P11.6**

$V_{IM}$ =0.15 mV for the integrator operational amplifier and this amplifier has slew rate of 0.6 V/$\mu$sec.

**11.9.** The input voltage for the voltage-to-frequency converter (Fig. 11.29(a)) has the range 1 to 30 V, the reference voltage of the circuit is $V_{Ref}$ =0.5 V. The circuit has $R$ =10 k$\Omega$, $C$ = 0.5 $\mu$F, $C_d$=220 pF, $R_a$=1 k$\Omega$, $R_b$=10 k$\Omega$, $V_o^+ = V_o^-$ =5 V. Find the parameters of the waveform and evaluate the maximum frequency error. Give the voltage-to-frequency characteristic.

**11.10.** In the current-to-frequency converter shown in Fig. 11.30(c), the current changes from 1 to 100 $\mu$A. The capacitor has the value of 50 pF and the threshold circuit has $V_{cH} - V_{cL}$ =1 V. Calculate the current-to-frequency characteristic if the source of the frequency error is the delay introduced by the threshold comparator having a slew rate of 10 V/$\mu$sec and an output voltage swing from +2 V to -2 V.

**11.11.** Estimate the error introduced by the transistor base currents in the relationship of Eq. (11.86) for the circuit of Fig. 11.33(a). Assume that transistors are matched and their betas are equal. What will be the load power in this case if $I_R$ and $I_C$ are given?

# REFERENCES

[1] S. Middelhoek and S. A. Audet, *Silicon sensors*, San Diego, CA, Academic Press, 1989.

[2] H. Baltes, "Microtransducers by Industrial IC Technology and Micromachining," *Dig. Tech. Papers, Intl. Conf. Solid-State Sensors and Actuators (Transducer'91)*, Tokyo, Japan, 1991, pp.17-23.

[3] P. R. Gray and R. G. Meyer, *Analysis and Design of Analog Integrated Circuits*, second edition, J. Wiley, New York, 1984.

[4] D. Moser, R. Lenggenhager, and H. Baltes, "Silicon Gas Flow Sensors Using Industrial CMOS and Bipolar IC Technology," *Sensors and Actuators A*, vol. 25-27, 1991, pp. 577-581.

[5] H. Baltes, E. Charbon, M. Parameswaran, and A. M. Robinson, "Humidity-Sensitive Oscillator Fabricated in Double Poly CMOS Technology," *Sensors and Actuators B*, vol. 1, 1990, pp. 441-445.

[6] T. Boltshauser and H. Baltes, "Capacitive Humidity Sensors in SACMOS Technology with Moisture Absorbing Photosensitive Polyimide," *Sensors and Actuators A*, vol. 25-27, 1991, pp. 509-512.

[7] L. Ristic, H. P. Baltes, T. Smy, and I. M. Filanovsky, "Suppressed Sidewall Injection Magnetotransistor with Focused Emitter Injection and Carrier Double Deflection," *Electronic Device Letters*, vol. EDL-8, pp. 395-397, 1987.

[8] S. Franco, *Design with Operational Amplifiers and Analog Integrated Circuits,* McGraw-Hill, New York, 1988.

[9] K. Maenaka, H. Okada, and T. Nakamura, "Universal Magneto-Operational Amplifier (MOP)," *Sensors and Actuators*, vol. A21-A23, pp. 807-811, 1990.

[10] J. Dostal, *Operational amplifiers*, Elsevier, New York, 1981.

[11] S. Middelhoek, P. J. French, J. H. Huijsing, and W. J. Lian, "Sensors with Digital or Frequency Output," *Sensors and Actuators*, vol. 15, pp. 119-133, 1988.

[12] A. Cichocki and R. Unbehauen, "Application of Switched-Capacitor Self-Oscillating Circuits to the Conversion of RLC Parameters Into a Frequency or Digital Signal," *Sensors and Actuators A*, vol. 24, pp. 129-137, 1990.

[13] R. Best, *Phase-locked loops*, McGraw-Hill, New York, 1984.

[14] D. H. Sheingold, (Editor), *Transducer interfacing handbook*, Analog Devices, Norwood, MA, 1980.

[15] J. H. Huising, G. A. Van Rossum, and M. Van der Lee, "Two-wire Bridge-to-Frequency Converter," *IEEE J. Solid-State Circuits*, vol. SC-22, pp. 343-349, 1987.

[16] G. B. Clayton, *Operational amplifiers* (2nd edition), Newnes-Butterworths, London, 1979.

[17] I. M. Filanovsky, V. A. Piskarev, and K. A. Stromsmoe, "Nonsymmetric Multivibrators with an Auxiliary RC-Circuit," *IEE Proc.*, vol. 131, pt. G, no. 4, pp. 141-146, 1984.

[18] T. J. Van Kessel, and R. J. Van de Plassche, "Integrated Linear Basic Circuits," *Philips Techn. Review*, vol. 32, no. 1, pp. 1-12, 1971.

[19] I. M. Filanovsky, I. G. Finvers, L. Ristic, and H. P. Baltes, "Multivibrators with Frequency Control for Application in Integrated Sensors," *Proc. First Int. Forum on ASIC and Transducer Technology (ASICT'88)*, Honolulu, Hawaii, USA, February 7-10, pp.65-76, 1988.

[20] A. Nathan, I. A. McKay, I. M. Filanovsky, and H. P. Baltes, "Design of a CMOS Oscillator with Magnetic Field Frequency Modulation," *IEEE Journal of Solid-state Circuits*, vol. SC-21, pp. 230-232, 1987.

[21] G. C. M. Meijer, R. van Gelder, V. Nooder, J. van Drecht, and H. M. M. Kerkvliet, "A Three-Terminal Wide-Range Temperature Transducer with Microcomputer Interfacing," *Dig. Tech. Papers, European Solid-State Circuits Conf.*, Delft, The Netherlands, 1986, pp. 161-163, September 16-18.

# CHAPTER
# 12

## DESIGN FOR TESTABILITY

## 12.1  INTRODUCTION

With the increased complexity of analog VLSI circuits and the reduced access to internal circuit nodes, the task of properly testing these devices is becoming a major bottleneck during their prototyping, development, production, and maintenance. Analog circuits can easily turn out to be untestable, moreover quality is likely not to be guaranteed and the time-to-market period becomes dominated by test-program development and debugging. In addition, there is a strong tendency to integrate analog as well as digital parts onto one IC, where the analog (and mixed-signal ) testing of the part is likely to cause most of the test problems.

In this chapter, we will distinguish between pure analog VLSI (Fig. 12.1) and a mixed analog and digital (mixed-signal) VLSI (Fig. 12.2). In the latter case, we can again distinguish between devices having a majority of analog circuitry with some digital circuitry for control, or a large portion of (often microprocessor-based) digital circuitry with some peripheral analog circuitry. The latter is especially expected to dominate the market in the mid and late nineties. It will be assumed that the analog portion of the VLSI IC is synthesized from functional blocks or so-called macros, like analog-to-digital (A/D) and digital-to-analog (D/A) converters, operational amplifiers, filters, phase-locked loops, oscillators, etc. The overall design is assumed to have hierarchy. These blocks may originate from libraries or from parameterized module generators. This type of analog synthesis supports the conventional approach of functional

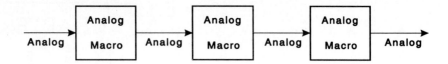

**FIGURE 12.1**
The assumed architecture of a pure analog VLSI IC partitioned into functional macros to support functional testing.

tests for analog devices. Although this approach can also be used for the digital portion of the VLSI chip, it is not required. In mixed-signal devices, the digital circuitry is assumed to be isolated from the analog circuitry in terms of testing because of their specific nature; however, digital circuitry may be used to control analog testing.

Although in the digital world many different Design For Testability (DFT) approaches already exist and are currently being used, like macro-test, boundary-scan techniques and self test, the analog design and test community lags behind. There are, however, good reasons for the differences in the level of analog test development.

The analog circuit must be designed so that all necessary tests may be carried out. The design should be realized with minimum effort in terms of cost and time, and with minimum loss in performance or increase in the circuit overhead. So, what is a properly tested analog circuit? For a digital IC, a well-known measure is used, the so-called fault coverage. It is usually defined as the percentage of "stuck-at model" faults that can be detected by a certain number of input test patterns, divided by the total number of "stuck-at" faults that can occur. Usually this fault coverage is in the high ninety-percent range. The generally accepted "diagnostic" granularity of testing is the logic-gate level and not the transistor level.

A quite different approach to test makes use of a definition for the testability of a circuit. Although many different definitions exist, most of these definitions are based on a combination of observability and controllability of a node. Controllability is the ease by which a node can be controlled and set to a certain value, while observability indicates how well a node value can be observed or measured. In the ideal case, a node is at one time a primary output (100% observability) and at another time a primary input (100% controllability). Unfortunately, 100% observability and controllability are not usually obtained.

Which faults can occur in analog circuits for which tests should be derived? Unfortunately no general fault models have been established; it is clear that this seriously hampers the development of tests for analog and mixed-signal ICs and also the associated DFT. The difficulty is that there are a much wider range of fault models in analog circuits than in digital circuits. Only the so-called catastrophic faults bear some resemblance to the conventional digital "stuck-at" fault modeling. The most common fault models that have been developed will be discussed in some more detail in Sec. 12.2.

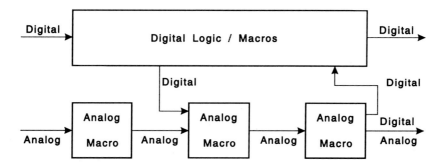

**FIGURE 12.2**
Suggested partitioning of a mixed-signal VLSI IC. The analog and digital parts are separated and both employ a macro design approach.

To quantify how many faults are covered by a set of test-signals or patterns, a fault simulation is required. In the case of catastrophic faults, the simulation can also be used to generate a fault dictionary for diagnostic purposes. In this case, the intention is to imitate the approach most commonly used in digital testable design. Although this approach is not yet widely used in the analog world, for obvious computational reasons, it is discussed in Sec. 12.2.

Based on the applied test patterns, a testability analysis can be carried out that will determine where DFT changes are required in the hardware. These test patterns can be derived manually or automatically. When the key parameters, usually of the transfer function (or nodes) of a circuit, are determined which can cause poor testability, measures can be taken to increase the testability. Usually this requires increasing the controllability and observability of nodes. Some suggestions for analog testability measures are provided in Sec. 12.3.

The improvement in testability of an analog circuit can be realized in an ad-hoc manner or in a structured way as explained in Sec. 12.4. Usually the latter is preferred, although from an efficiency point of view an ad-hoc approach may be desirable. Examples of both methods will be provided. Structural methods include scan methods (Sec. 12.5), boundary-scan techniques (Sec. 12.6) , built-in self tests (Sec. 12.7), and the use of analog test busses (Sec. 12.8).

Finally, some new approaches toward analog DFT will be presented which make use of e-beam measurement equipment linked to powerful computer-aided test and computer-aided design tools. In this particular case, design for testability translates into design for electron-beam testability as will be shown in Sec. 12.9.

In addition to the obvious advantages of analog design for testability, there are also several disadvantages. By adding additional hardware, the design time increases and the yield of the testable circuit is likely to decrease. The latter is caused by an increase in silicon area. Furthermore, there is usually an increase in the number of I/O pins required. Often the total cost of the IC may also increase due to the special CAD tools needed for the development of an analog testable design. Last, but certainly not least, it is likely that the overall performance of the circuit will decrease during normal operation, and this can mean losing the edge over the competitors. Hence, a very careful trade-off between the pros and cons of the different methods is necessary and some guidelines will be provided.

The impact of the rapid introduction of mixed-signal ICs on testing analog VLSI parts is considered to be of major importance in the near future. It is expected new approaches and benefits in analog DFT will continue to be developed. Furthermore, the new structured design methodologies and associated CAD and CAT tools for analog and mixed-signal VLSI will make the testable design and testing of these devices less a black art and significantly lower the time-to-market.

## 12.2   FAULT MODELING AND SIMULATION

A distinction is usually made in IC testing between fault detection (go/no go) and fault diagnosis. Very detailed knowledge about the physical causes of faults and their circuit-level modeling is required to obtain the physical causes of faults. More or less elaborate design-for-test methods have to be employed, depending on the required level of detail of the diagnostics. For instance, some faults can be found by using only dc-tests, thereby not requiring any wide bandwidth interconnection paths between a tester and the nodes of interest. Hence, the basics of analog fault modeling must be understood in order to apply them accurately to fault simulation.

### 12.2.1   Fault Modeling

During processing, a physical fault may occur, such as a pinhole in the oxide or an unacceptable change in substrate doping. This can result in a change or failure of the (transistor) component behavior as shown in Fig. 12.3. This, in turn, can result in a change or error in an entire circuit's analog behavior. Note that a fault does not always cause an error in the circuit behavior. It is, however, possible that the combination of several of these minor faults can cause an error.

Analog fault modeling lacks a general model, as will be discussed later in the chapter. Additionally, the statistical distributions of these faults are not well known yet.

Two basic fault categories are usually distinguished:

1. Catastrophic faults (hard faults)
2. Parametric/deviation faults (soft faults)

The first category of faults is usually randomly occurring and often caused by structural deformations. For instance, they can occur as a result of large dust particles (greater than minimum design dimensions). Examples include open or shorted interconnections or where very large deviations in component parameters occur. These faults are somewhat similar to the well-known stuck-at, stuck-open and stuck-on fault models sometimes used in digital fault modeling.

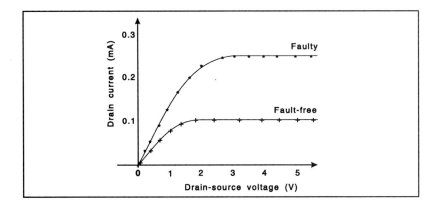

**FIGURE 12.3**
The result of an oxide pin hole on the drain current of an MOS transistor. The line with the
"+" symbol shows the fault-free behavior.

The second category of faults is much more difficult to test for. These faults
can be caused by small yet unacceptable variations in process parameters or tiny
dust particles on the wafer (less than minimum design dimensions). Although
a normal or Gaussian distribution is often assumed for the process parameters
involved, these faults, can exceed the allowed tolerances of the chip, wafer, or
lot. It is very difficult to identify the faults if several of these parameters are
correlated. There will be values in the set of parameters for which the circuit
may or may not satisfy all specifications. Moreover, distributed allowed deviations of different parameters might cause an unacceptable deviation from the
circuit specifications, although this should be observed during a worst case circuit
simulation.

One of the unsolved problems in analog testing is that there is no common agreement as to which fault models should be used. Some publications
state that approximately 80% of all analog faults are caused by catastrophic
shorts and opens of components, similar to digital circuits (CMOS) where single
catastrophic faults are assumed to dominate. On the other hand, there are also
examples (e.g., bipolar analog ICs) in which case multiple-fault, non-catastrophic
errors seem to dominate.

In BiCMOS processes, detailed failure studies have started. At this moment, a mix of multiple-fault catastrophic and deviation faults are anticipated
to exist depending on the implementation of the analog circuitry.

### 12.2.2 Fault Simulation

A fault simulation is required in order to verify whether or not faults are detected
in a circuit. Given that a set of assumed fault models of the set of input test
patterns is known, the fault coverage percentage can be calculated.

The fault coverage also indicates which faults are not detected at specific nodes, and hence provides important information about where potential testability problems may arise. This can lead to modification of the design to improve its testability (DFT). This approach and the corresponding terminology are similar to the digital counterpart, especially where only catastrophic faults are considered. In the analog case however, if it concerns very complex systems, fault simulation has been relatively uncommon until now.

One reason for this is that analog fault simulation is a very tedious and expensive task. The number of fault simulations that have to be carried out is estimated to be a power of two to six higher than a digital circuit of comparable complexity. There are also a large number of possible errors due to deviation errors. Furthermore, until recently use was made of conventional circuit simulators, like SPICE and other derivatives, which suffer from poor reliability, improper fault modeling insertion, inflexibility and inefficiency in terms of CPU time.

An important aspect in fault simulation is which faults are simulated and how they are modeled within the simulator. Usually, only catastrophic faults are considered. Furthermore, in almost all cases only single faults are assumed. The question still remains whether or not this assumption can be made in analog and mixed-signal BiCMOS circuits.

Catastrophic faults, like shorts and opens, can be modeled by low and high impedance paths, modeled by different resistor values or conducting or nonconducting transistors. The possible locations in the netlist of the faults should also be included. One option is to assume that all lines can have errors, but also special requests for fault locations would be possible. After the circuit simulation of the fault-free circuit, the error(s) are introduced, the circuit simulated and the relevant outputs observed. Hence, relationships exist between faults and are shown between occurring faults and the output responses. Parametric or deviation faults can also be modeled. This type of fault occurs if a parameter is outside the range of its accepted nominal value including its associated and accepted tolerances. These parameters can be low-level parameters such as the oxide thickness of transistors. However, in order to keep CPU times within practical limits, high-level parameter deviation errors are more likely to meet this requirement. The limited number of parameters involved in such a simulation should have a known probability distribution function.

Next, simulations can be carried out, and the performance of the system can be calculated under normal conditions including allowed tolerances. Unacceptable deviations of these parameters can be introduced, and the responses stored. Simple cases usually assume dominance of a particular parameter. It is obvious that fault simulation at a low hierarchical level, incorporating many different parameters and dependent errors will result in unacceptable CPU times and very large and probably unreliable fault dictionaries. Global deviation faults of important process parameters can usually be detected by measuring process-control modules on the wafer, and their joint effect on the performance of a circuit can be simulated and confirmed by measurements. Local deviation faults, especially if several occur simultaneously, are extremely difficult to handle and diagnose. With the recently developed hierarchical mixed-signal simulators, like SMASH and SABER, featuring macro modeling, and various types of fault analysis, including Monte Carlo analysis and data post-processing, new possibilities exist for fault dictionary preparation at the macro-level, test-pattern generation, and internal test-probe selection. It has been shown that in analog circuits there are a large number of different types of faults. They tend to be more fuzzy, and hence, much more difficult to detect than for their digital counterparts. This translates into the need for elaborate means of DFT. Analog fault simulators, especially the new hierarchical versions, could be used to identify areas that require DFT.

## 12.3   TESTABILITY-ANALYSIS TECHNIQUES

In contrast to fault simulation techniques, where test patterns are evaluated on their effectiveness of detecting faults in a circuit, or are used to construct fault dictionaries, testability analysis evaluates the relative degree of difficulty in testing circuit nodes, or key parameters for particular fault models and given test-patterns.

In the digital case, the testability of a node is considered to be some mathematical function of the controllability and observability of that node. Usually good controllability and observability results in good testability. However, the measure is soft and relative. If one has the netlist of the total circuit, one is able to calculate the above properties of the nodes based on the controllability and observability features of the logic gates (e.g., NAND) involved.

In the digital world, several proven testability analysis tools already exist, such as TICOP, SCOAP, TMEAS, and CAMELOT. They are, however, not yet widely accepted by designers. Critics claim that this is a result of the structured approach in digital design, the experience of most designers and the difficulty in interpreting the calculation results.

However, it is generally accepted that testability analysis tools are quite capable of guiding the test-pattern generation process and can also contribute to guided probing in the case of detailed fault diagnosis.

Although controllability and observability of analog waveforms at embedded nodes could be considered as testability measures in analog circuits, the problem is much more complicated. This is due to the fact that a simple (logic-like) control or propagation of signals (and the associated calculations/simulations) is usually not possible. Analog circuits exhibit 'soft' multiple deviation faults. In the past, several approaches toward analog testability analysis and testability measures have evolved. Some use heuristics while others describe the testability as a rather robust and quantitative property.

Most methods are based on a rank-test algorithm. Essentially, the degree of solvability of a set of (usually nonlinear) equations describing the relationships between measurements at the available nodes and the nonredundant parameters describing the functional behavior of a circuit, is used as a measure. Notice that

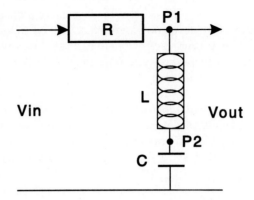

**FIGURE 12.4**
A simple RCL network with alternative test points P1 or P2 [19].

in this approach, the analog circuits are tested using the usual functional tests.

It can also be described as the information return of parameters in response to stimuli. Observe that in the analog case, one often talks about determining parameters, e.g., the gain of an amplifier. In a more mathematical sense, the testability is related to the maximum number of nearly independent columns of a matrix associated with these equations. This again has consequences for the rank of a matrix which is therefore often used as the basis for the testability. In this case, the testability, $T$, can be considered as a scalar measure of the average rate of information returned on the specific parameter, and hence, depends on the available test points and the input signal. This makes it suitable for use in automatic test-pattern generation and test-point allocation.

As an example, Fig. 12.4 shows a simple LRC network where one wants to determine the value of parameter R. It is an essential parameter in the transfer function of the circuit describing the relationship of the input, $V_{in}$, to the output, $V_{out}$ (test point P1).

Assuming a sinusoid voltage at the input with radian frequency, $\omega$, and an amplitude of $\sqrt{2}$, the testability $T(\omega)$ is:

$$T(\omega) = \frac{2(\omega RC)^2 \cdot (1 - \omega^2 LC)^2}{\{(1 - \omega^2 LC)^2 + \omega^2 \cdot (RC)^2\}^2} \tag{12.1}$$

A more detailed derivation of this equation can be found in [19]. The testability $T$ of parameter $R$ versus the radian frequency, with test points 1 (P1) and 2 (P2) is shown in Fig. 12.5. By measuring at test point 2, a much higher testability of parameter $R$ is obtained than by using test point 1.

One can also make use of sensitivity analysis of parameters which can sometimes be one or more components, e.g., a resistor. By looking at the deviations in the transfer function (notice the conventional functional testing aspect) of a system as result of a deviation in a key parameter, one is able to calculate the sensitivity of the transfer function with respect to that parameter. Difficulties

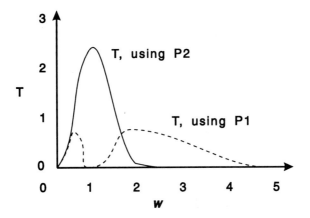

**FIGURE 12.5**
The testability T of the resistor R with the radian frequency and the test points P1 and P2 as variables [19].

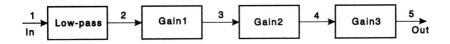

**FIGURE 12.6**
Basic set-up of a simple analog system with only one observable output node (5).

arise when parameters are dependent on each other, making them part of the same ambiguity group. Basically it means that the deviations in both parameters cause similar effects in the resulting measurements making it difficult to distinguish the origin.

Table 12.1 shows test results of an attenuating amplifier illustrating resistors and capacitors with their associated testability values. The measurement accuracies, the number of different input frequencies and the number of observable nodes are varied. Not surprisingly, it shows that increasing the number of input frequencies test patterns increased the number of observable nodes and provided better testability results.

The results of another testability-analysis approach (TASTE) based on relative determination accuracies and including possible measurement errors is shown in the Figs. 12.6 and 12.7, with the associated Tabs. 12.2 and 12.3. The system operates only on key parameters of high-order primitives or so-called macros like op amps and filters. These key parameters can be e.g., filter coefficients or ac-gain values.

In this case, the accuracies of the measurements are incorporated in the calculations. The system equations describing the behavior of the system in terms of the key parameters and input and measured output signals are used to determine the key parameters with a certain accuracy. However, sometimes

**TABLE 12.1**
The testability values of components as function of the test conditions.

| Test number | 1 | 2 | 3 | 4 | 5 | 6 |
|---|---|---|---|---|---|---|
| Test nodes | 1 | 1 | 1 | 1 | 3 | 3 |
| Test frequency | 6 | 41 | 4 | 41 | 9 | 41 |
| Measured error | 1E-5 | 1E-5 | 3E-4 | eE-4 | 1E-5 | 3E-4 |

| Component | Testability Values (*1E-3) | | | | | |
|---|---|---|---|---|---|---|
| R10 | 0.061 | 0.15 | 0.48 | 1.2 | 100 | 640 |
| R1 | 0.38 | 0.79 | 0.48 | 1.1 | 0.097 | 57 |
| R7 | 6.6 | 17 | 10 | 30 | | |
| C2 | 0.055 | 0.13 | 0.45 | 1.1 | 0.11 | 1.1 |
| R2 | 0.057 | 0.14 | | | 0.097 | |
| R3 | 0.059 | 0.12 | | | | |
| C5 | | | | | 0.15 | 17 |
| C6 | | | | | 0.10 | |
| **Change factor** | **1** | **2.3** | **4.7** | **6.1** | **410** | **3550** |

**TABLE 12.2**
The parameters, relative determination accuracies, and required non-faulty parameters as calculated from Fig. 11.6. The frequencies used at node 5 were 0.1, 1.0, and 1.7 Hz.

| Parameter | Relative determination accuracy | Required non-faulty parameters |
|---|---|---|
| Gain1 | 2.245 | Gain2, Gain3 |
| Gain2 | 2.245 | Gain1, Gain3 |
| Gain3 | 0.714 | Gain1, Gain2 |
| a | 0.181 | None |
| b | 0.566 | None |

key parameters are dependent of each other and are inseparable parameters. Matrix manipulations and the wel-known Gaussian elimination procedure with full pivoting are used to determine the dependencies of parameters and to provide the solution of the equations.

In the example shown, a low-pass filter is considered, followed by three amplifiers characterized respective to the filter coefficients $a$ and $b$, and small-signal amplifications of $gain1$, $gain2$, and $gain3$. The system requires a simple netlist description consisting of the names of the macros and their interconnections and

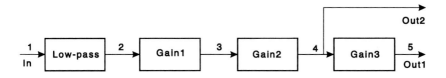

**FIGURE 12.7**
The extended set-up of an analog system with two observable output nodes (4) and (5).

**TABLE 12.3**
The parameters, relative determination accuracies, and required non-faulty parameters as calculated from Fig. 11.7. The frequency components used at node 5 were 0.1, 1.0, and 1.7 Hz and at node 4 the frequency used was 0.1 Hz.

| Parameter | Relative determination accuracy | Required non-faulty parameters |
|-----------|--------------------------------|-------------------------------|
| Gain1 | 1.077 | Gain2 |
| Gain2 | 1.077 | Gain1 |
| Gain3 | 0.224 | None |
| a | 0.181 | None |
| b | 0.566 | None |

the nominal, standard deviation and desired accuracies of the key parameters of the macros. Furthermore, an input signal with a number of frequency components and the accuracy of the measurements has to be provided.

The resulting calculations in Tab. 12.2 show that *gain1* and *gain2* cannot be measured very accurately and some non-faulty parameters are required for determining the key parameters. By the designer adding an additional measurement point (*out2*), recalculations show that *gain1* and *gain2* can now be determined much more accurately and less non-faulty parameters are required (Tab. 12.3). It illustrates that although a specific definition for testability is not used, a designer can still immediately see the consequences of alterations.

## 12.4   AD HOC METHODS AND GENERAL GUIDELINES

There are two fundamental approaches toward testable design. The first approach looks only for answers to the test problems of a specific circuit. It is usually carried out after the initial design and the solutions used are not applicable in general. It is referred to as the ad hoc approach.

The second approach starts immediately at the beginning of the design. General guidelines are followed for testable design or even very strict and actually sometimes design limiting rules. In this section, several ad hoc methods and

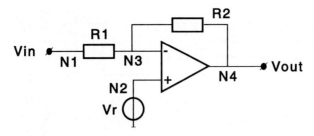

**Access: N1, N2, N4**

**Formula: Vout=(R2/R1+1).Vr-(R2/R1)Vin**

* **Closed-loop voltage gain**
* **Offset voltage**          **R L conditions**
* **Slew rate**

**FIGURE 12.8**

Requirements for testing an amplifier macro including key parameters to be tested and the required accessible nodes.

general guidelines are discussed.

When it is concluded after the initial design of a circuit that certain nodes are difficult to access for control and observation purposes, one is able to enhance the testability by modifying the design. The methods that can be employed in analog design are not fundamentally different from their digital counterparts. Basically, one tries to attain the same goal: an increase in internal controllability and observability. Both are increased by carrying out a partitioning and enabling direct or indirect access to a macro cell. Sometimes isolation of the macro cell is required because of specific loading conditions.

One way of modifying the design is by (re)partitioning of the chip. This can, however, be done in many different ways. One brute force method is by cutting the design in half until the remaining parts are expected to be easy to test. A more sophisticated approach attempts to distinguish between high-speed parts and low-speed parts usually requiring different test requirements, and hence, approaches. In the case of mixed-signal ICs, it can be the partitioning between the analog and the digital part.

The most frequently used approach of partitioning a complex chip is based on using the different functional modules that can be distinguished in a design. This method is usually referred to as 'macro testing.' The current hierarchical approaches toward design support this method. Furthermore, analog testing often involves functional testing and hence methods for testing specific macros are quite well known. Figure 12.8 shows an amplifier macro including the nodes to be accessed, as well as the key parameters to be determined and the associated test conditions. It is possible to have such data available for every macro in a library.

The problem still remains of obtaining observability, controllability, and isolation in order to carry out these tests. An increase in observability is obtained in the following ways. The most direct mechanical approach is quite straightforward. One can add a test point in the form of a small metal contact connected to the node in the layout; the contact area dimensions are smaller than conventional bonding contacts as bonding is usually not required. In order to reduce their number, one sometimes combines test points by means of multiplexing. Furthermore, buffered high-impedance micro probes are usually required for measurements. These probes can be automatically guided to the contact points by means of a software navigation system connected to the CAD database. Note that in this approach, the original load of the node is still connected and could be the cause of a failure.

In an extreme example of an ad hoc approach, it is sometimes required to physically add a probing point even if the chip has already been manufactured. This is accomplished by depositing new metal on the chip by using a focused ion-beam system. The disadvantage of this approach is clear: it takes a lot of chip area, increases the node capacitance, and usually will require some form of buffering. This buffering is accomplished by source-followers or operational amplifiers. As a result, one is able to reduce loading disturbances at the node. In the case of high-speed devices, impedance mismatch problems must also be taken into account. In addition, mechanical damage is likely to occur by using probe pins, and depending on the application, it can be a sensitive point for crosstalk. In the latter case, a shielding ring to ground around the contact is recommended. A more sophisticated method is to create additional very small contacts or vias to the top-layer metal and carry out measurements with a low-capacitance electron-beam machine. This approach is discussed in more detail in Sec. 12.9.

A less direct but more sophisticated approach makes use of built-in switches, implemented with transmission gates and analog multiplexers. In this case, several nodes can be accessed by the multiplexer or transmission gates controlled by digital control signals. Again these elements are usually buffered for the same reasons as discussed above. The disadvantages are the increase in hardware, the required control signals, the additional design time, the loss of signal due to the multiplexers and transmission gates, and the long wires to the multiplexers or primary outputs. A simple example is shown in Fig. 12.9. However, it requires additional digital control and analog pins.

Increased controllability is somewhat more difficult to realize. It is usually recommended that the input be isolated from the output of the previous stage. The most direct way to do this is by again adding a metal contact probing point for micro-probing, and destructively or non-destructively breaking the connection with the previous stage. A destructive way can be realized by using a laser-beam (like is often carried out in RAMs to remove faulty cells) or a focused ion-beam. A more sophisticated method uses transmission gates and analog multiplexers. It has the advantage that it automatically disconnects the inputs from the outputs of the previous stages. It is advised that buffers be used at the analog inputs. The same disadvantages of this approach hold as discussed

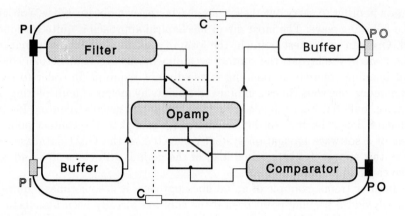

**FIGURE 12.9**
Example of using digital switches and multiplexers in order to increase observability and controllability of macros. Buffers are inserted to minimize the influence of additional loads.

earlier in the section on enhancing the observability. This is also illustrated in Fig. 12.9.

To achieve complete isolation of inputs and outputs of macros either a destructive or non-destructive method has to be applied. Examples of both approaches have already been discussed. Destructive methods use mechanical or chemical removal of metal layers, e.g., focused-ion beam. The use of analog multiplexers is clearly preferred. However, it must be kept in mind that the additional hardware should in no way influence the measurement results due to the additional loading or noise. Finally, analog circuits often make use of feedback, in which case measurements can be quite difficult. In order to break these loops, either analog switches controlled by digital signals or laser and focused ion-beam techniques are used for breaking and subsequently restoring feedback loops.

Although it is not a very elaborate design for testability technique, the ad hoc approach can be fast and cheap as compared to a structured approach. Sometimes it is the last resort for making the device testable. It is, however, preferred to use the suggested methods as guidelines at the beginning of a design.

## 12.5 SCAN TECHNIQUES

Most currently used design-for-testability methods in the digital world are based on scanning techniques. Although a number of different approaches exist, one basically enhances the controllability and observability of nodes by connecting them each to a shift register. If a serial-to-parallel-to-serial shift register approach is used, the number of additional input and output pins can be limited to only two, excluding a control pin and possibly one or two clocking lines. It is clear that if the shift register is large, because many internal nodes are to be controlled and observed, the scanning and hence testing time will be long. The latter is one of the reasons why several techniques have been developed to reduce this time.

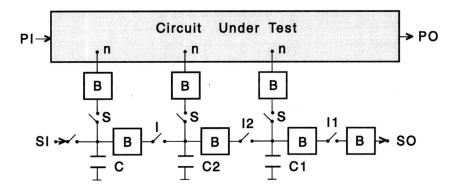

**FIGURE 12.10**
Basic set-up of the analog shift register during normal operation [31].

It is not surprising then that clocked scanning techniques have been explored for testing analog devices. Therefore, the fundamental question arises whether one can provide the proper analog clocked shift registers; i.e., good behavior in terms of signal, impedance, and bandwidth levels required for the measurements. In the analog case, the registers function as a sampling oscilloscope. Essentially two well-known approaches of analog shift registers can be used: bucket-brigade-like devices and charge-coupled devices.

### 12.5.1 Bucket-Brigade-Like Devices

The essential features required in observing analog nodes are the following. First, one requires a switch which is capable of isolating the analog shift register test circuit from the node $n$ to be tested during normal operation. Furthermore, during testing, the influence of the test hardware on the node to be measured should be minimized. This can be implemented by a conventional voltage follower (buffer $B$) which has a very high input impedance and a very low output impedance in series with a switching device $S$. The global architecture is shown in Fig. 12.10. For storing the analog information, a sample-and-hold circuit is used. It can be built by using a switch, $I$, for sampling, a capacitor, $C$, for storing the information, and a voltage follower, $B$, for impedance buffering between the capacitors.

The required clocking for the different switches can be provided by a simple digital shift register. The operation of the overall structure is similar to the digital scan counterpart and proceeds as follows.

There are essentially two test modes. In the normal mode, all switches are open and no connection exists between the test nodes and the analog shift register (Fig. 12.10). During the test mode, the switches $S$ are closed and the test data at the nodes $n$ are sampled and connected to the storage capacitors $C$. All nodes are loaded in parallel to the capacitors. In the mean time, all pass switches $I$ remain turned off, thus isolating all stages.

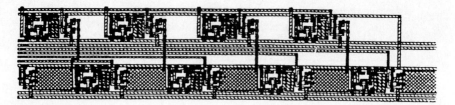

**FIGURE 12.11**
Layout of the analog scan register in CMOS technology including buffers.

After sampling the values, the pass switches $I$ are subsequently turned on, thereby enabling the signals on the capacitors to transfer their values to the scanning output $SO$. More specifically, initially the pass switch $I1$ is closed, and the signal stored on the capacitor $C1$ is passed to the output through the voltage followers $B$ connected to $SO$. Next, pass switch $I2$ is closed ($I1$ remains closed), and the signal on the next capacitor $C2$ is serially transferred to the output via the three voltage followers $B$. As soon as the serial passing process is completed, the whole cycle is repeated from the beginning by loading the node signals on the capacitors. Figure 12.11 shows a possible layout of four of the stages depicted in the previous figure.

## 12.5.2 Charge-Coupled Devices

Another well-known type of analog shift register is the CCD or Charge-Coupled Device. Very compact transistors and capacitors are used to shift discrete analog information in the form of charges. In this case, a sampling of the input data again takes place. A possible approach is shown in Fig. 12.12. Input voltages and currents can be handled. They are first linearly converted into distinct amounts of charge in the charge converter labeled $CC$. Next, these charges are dumped into a conventional CCD register and transferred to the output. At the output, the discrete charge packets providing information about the value of a voltage or current are converted back into the voltage domain labeled in the block $C{\rightarrow}V$.

If a current integrating circuit is used at the CCD input, it provides a small capacitive load on the test node, high linearity and accommodates high sampling frequencies. If a large input voltage range is required, a fill-and-spill input circuit is preferred. The floating diffusion output structure provides a good transformation of the charge packet into the voltage domain.

Depending on whether high measuring accuracies or high-speed devices are to be measured, either a single-node per single-well construction or a single-node and multiple-well construction can be used. As with the previous approach, a clocking mechanism has to be used to operate the charge-coupled device. It can be accomplished by digital shift registers. If buried-channel CCDs and overlapping poly-silicon gates are used, the manufacturing process will have to be equipped for this. Aside from this, the major disadvantages of this approach are the amount of required clocking circuits and the small pitch of CCD registers

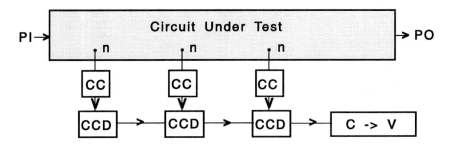

**FIGURE 12.12**
A charge-coupled device used as an analog signal observation register. The information is stored as discrete analog charge packets.

leading to relatively large measurement interconnection lines.

An embedded analog element can also be tested if it is preceded by a stage which provides sufficient gain and bandwidth for undisturbed passages of a continuous input signal to the element. A similar argument can also be given for a succeeding stage giving undisturbed passages of the output signal of the embedded element. Such circumstances can occur in several classes of analog circuits, for instance in active analog filters.

If one looks at the active analog filter shown in Fig. 12.13, one can distinguish three stages connected in cascade. This clearly resembles a conventional scan path. Now one is able to test stage 2 by widening the bandwidths in the test mode of stages 1 and 3. The latter processes will be referred to in a figurative way as *scanning*. Note that in this case no clocking or discrete storage is employed and that there is a continuous flow of signals.

The increased controllability and observability in the filter of Fig. 12.13 is now realized by dynamically broadening the bandwidth of each stage. This is accomplished by adding MOS transistors and associated control lines and signals to the passive elements like resistors, capacitors, and series and parallel combinations of these elements. This has already been carried out in Fig 12.13. The three MOS transistors have been added to the original filter structure for this purpose. Care has been taken that the original pole-zero locations were not significantly changed. The rules for adding the MOS transistors are:

1. Single resistor: MOS transistor in parallel
2. Single capacitor: an MOS transistor in series, and one in parallel with the series structure
3. Series RC branch: MOS transistor in series with the capacitor or in parallel with the branch
4. Parallel RC branch: MOS transistor in series with the capacitor or in parallel with the branch

Experiments have shown that the original filter characteristics above have been changed insignificantly by the additional hardware. Furthermore, the catas-

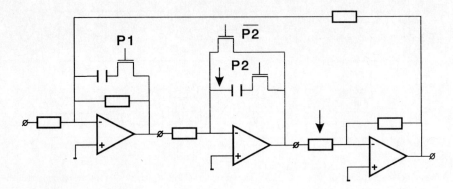

**FIGURE 12.13**

Example of an already modified analog active filter capable of allowing catastrophic as well as deviation failure diagnosis.

trophic and out-of-range faults were determined successfully in the modified filter. A stuck-open catastrophic error and a deviation error of a resistor have been indicated by arrows in Fig. 12.13. For fault diagnosis purposes, several SPICE simulations ($N + 2$, assuming $N$ stages) were carried out to relate faults with particular simulation results. The previous method has already been extended to include switch-capacitor filters. It is expected that it will be applied to more classes of analog circuits in the near future, such as sigma-delta ADCs implemented with switched-capacitor circuits.

## 12.6 BOUNDARY-SCAN

Although the boundary-scan concept was originally developed for debugging digital printed-circuit boards and chips mounted on these boards, analog and mixed-signal chip and board debugging has also been considered. There are many similarities in scan and boundary-scan techniques. The advantage, however of the latter, is that a fixed hardware and software protocol exists, which makes communication possible between the chips of different vendors. For details on boundary-scan techniques, refer to the official IEEE standard and the many papers that have appeared on the subject.

It is also possible to use boundary scan for testing analog circuits. To reduce the total overhead, however, it will be assumed here that the analog macros are part of a large mixed-signal system. A basic set-up of peripheral analog parts located with respect to the boundary-scan chain is shown in Fig. 12.14. The abbreviations $BS$, $PI$, and $PO$ denote boundary-scan, primary input and primary output, respectively.

As the analog parts reside at the peripheral boundary of the chip, the inputs are either well controlled by primary inputs or the outputs are well observed at the primary outputs of the analog part. Furthermore, it is assumed that A/D or D/A converters are available to interface with the digital parts. Hence, either

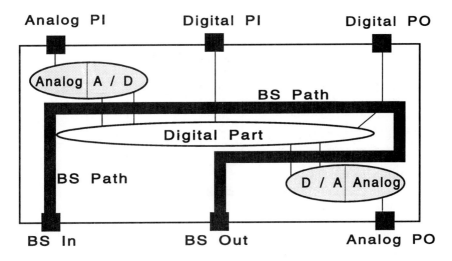

**FIGURE 12.14**
Testing analog peripheral circuits in a mixed analog/digital environment using boundary-scan techniques.

analog primary inputs and digitized analog internal outputs or digitized analog internal inputs and analog primary outputs are available. One has to keep in mind that the signals are sampled, and hence put restrictions on the stimuli and responses.

This method separates the analog tests from the digital ones. The analog macros are considered as 'external' devices. When operated in the test mode, the digitized outputs of the analog part can be connected via the boundary-scan shift register to external pins without the need to propagate through the relatively complex digital part. A basic set-up of an analog peripheral macro at the input is shown in Fig. 12.15.

The observability is increased. A similar case for the analog peripheral macros at the primary outputs is shown in Fig. 12.16. In this case, the digitized analog signals are provided at the input, and the outputs are directly measured at the outputs. It is obvious that the specifications of the data converters play a key role in the possibilities of these approaches. This approach also works for embedded macros. In the latter, the inputs as well as the outputs are controlled by the boundary-scan cells.

Boundary-scan can also be used in the macro test approach for control purposes where the analog blocks (e.g., op amp, oscillator) are embedded in the circuit. In this case, the embedded analog macros are preceded and followed by analog multiplexers and demultiplexers which are controlled by digital boundary-cell circuits. In this way, the boundary cells can control the analog demultiplexers providing isolation and access of analog macros.

In another approach, it has not been assumed that A/D or D/A converters are present near the analog parts. In this case, the data converters are incor-

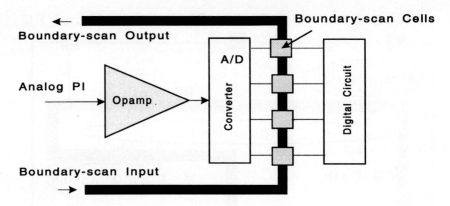

**FIGURE 12.15**

Analog peripheral macro at the primary input $(PI)$ and embedded digitized output connected to boundary-scan cells.

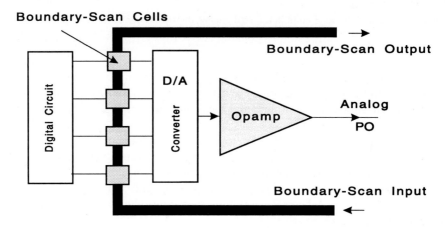

**FIGURE 12.16**

Analog peripheral macro at the primary output (PO) and embedded (digitized) analog input connected to boundary-scan cells.

porated in every boundary-scan cell, thus emulating something like an 'analog' boundary-scan cell. The effect is almost the same as previously discussed, and it can also serve embedded macros. It is clear that an AD/DA converter for a number of boundary-scan cells, depending on the number of bits of the converters, will require an unacceptable amount of overhead. However, this can be reduced if these converters are shared, or converters already available are used.

The analog boundary-scan cell can be used in the same way as the digital boundary-scan cell. As is clear from the previous approach that the analog boundary-scan cell suffers from the fact that it employs digital shift registers, which in turn make it necessary to employ A/D and D/A converters. It should be possible, however, to use the same technique as described in the previous section, where an analog shift register is employed, like a charge-coupled device or a bucket-brigade-type structure. Until now no examples have appeared in the literature.

It is anticipated that boundary-scan techniques, which are widely used in digital systems, will also have a significant influence on the design and test of analog and especially mixed-signal chips in the near future.

## 12.7  BUILT-IN SELF TEST

All previously discussed approaches have assumed that the proper analog test equipment was available at the input and output pins of the integrated circuit. This equipment could include stimuli like dc-voltages and ac-signals (sine, cosine, pulses, etc.) or noise sources. Furthermore, response evaluation equipment, like dc- and ac-multi-meters, counters or waveform analyzers are required.

A completely different approach to testing is to include the stimuli and response evaluation equipment on-chip. From a practical and economical point of view, the variation in different types of equipment has to be quite limited. Hence, the tests to be carried out and types of faults that can be detected will also be limited. There are, however, a lot of advantages to include Built-In Self Tests (BIST) in analog VLSI, e.g., in the area of maintenance.

In the digital design community, built-in self test is not uncommon any more. Essential parts here are the different digital shift registers which can access the required nodes, and in addition generate and evaluate digital test patterns. Sometimes ROMs or RAMs are also used to store test stimuli and responses. Analog self testing has been extremely uncommon until now, except for some local auto-calibration circuitry, which can be hardly designated as self-testing hardware. The large scale introduction of mixed analog and digital devices, however, is expected to change this rapidly.

An example of a chip in which the analog hardware is tested on-chip will be discussed. Self-test concepts similar to digital integrated circuits are used. The following hardware is required:

- an analog multiplexer
- a test-stimulus generator
- a test-response evaluator
- a test controller

*Analog multiplexer:* The analog multiplexer is used to isolate the analog inputs of the analog macro system from the chip input pins and to connect it to the analog test-stimulus generator.

*Test-stimulus generator:* The choice of the test-stimulus generator is very important, as it determines the test method used, and hence the area overhead and fault coverage. Usually, a functional test is used for analog circuits which can be carried out in the time domain or frequency domain. The time domain is attractive, because one only has to provide transients (e.g., a step function) at

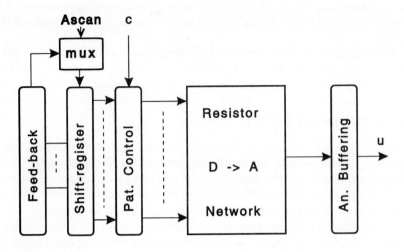

**FIGURE 12.17**
Basic block diagram of a digitally controlled analog test-stimulus generator.

the inputs. They are very simple to implement. The response analysis hardware, however, is very difficult to implement on-chip. A sine oscillator, or derivative, for testing in the frequency domain requires quite a bit of silicon area, and in addition, the test-response evaluation is not easy to implement. Therefore, use is made of changing discrete dc-voltages. It is implemented by a D/A conversion network, a digital feedback shift register, some combinational control logic and a buffering amplifier. A basic block diagram of the analog-stimulus generator is shown in Fig. 12.17.

Figure 12.18 illustrates a possible output signal that can be generated. The data and clock frequency of the Built-In Logic Block (BILBO) latches and the sampling rate determine the amplitude, sequence and phase of the signal.

*Test-response evaluator*: The response output signal of the macro is converted into a digital signal by an A/D converter, and the result is stored and evaluated by a conventional digital multi-input signature-analysis shift register (MISR). If more analog or digital macros have to be evaluated, a conventional BILBO is recommended for temporary storage of the data before signature analysis is carried out (Fig. 12.19). The expected response of the analog part can be obtained by simulation. It has to be noted in the latter case, that in contrast to digital circuits, not a single value of the analog response decides whether or not a circuit operates properly, but rather a range of permitted values due to allowed process deviations.

*Test controller*: The digital test controller, which can be activated externally or internally by power-on, controls the analog multiplexers for isolation, starts the test-stimulus and test-response hardware, initializes all registers involved and takes care of synchronization.

An overall view of the whole integrated system is shown in Fig. 12.19. The

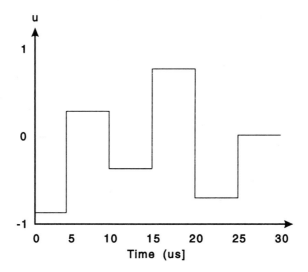

**FIGURE 12.18**
Part of the generated pseudo-randomly distributed amplitudes output signal of the test-stimulus device.

signal labeled '$u$' denotes the signal generated by the analog-stimulus generator, which is controlled by BILBO2 connected as a test-pattern generator. The responses of the analog macro labeled '2' can be externally measured. The other macro labeled '1' receives the signal $u$, after its inputs are disconnected by the analog multiplexer and its response is converted into a digital signal. It is then subsequently verified by BILBO1 which in this case is connected as a Multiple-Input Shift Register. In more complex cases, they can operate together with the separate register on top. The block labeled 'disconnect' can be used to disconnect remote loads for testing purposes.

An important issue of this self-testing technique is the quality of the test performed. In order to quantify this quality in this actual design, the fault coverage was calculated based on the assumption that only catastrophic errors are involved in the analog part. It was calculated by means of fault simulation that a fault coverage of more than 90% has been obtained, and also a large portion of soft deviation errors covered.

Another recently developed approach toward analog self-test claims to offer on-chip static and dynamic stimuli and measurement capability using a macro library approach. Among others, this library will offer A/D and D/A converters, oscillators, and voltage as well as current comparators. Even pseudo-noise on-chip generators have been proposed. It is claimed in this approach that deviation faults can also be detected.

An interesting approach toward the automatic testing of switched-capacitor circuits has recently been reported. In this case, a number of cascaded switched-capacitor biquadratic sections tested concurrently, by emulating the respective

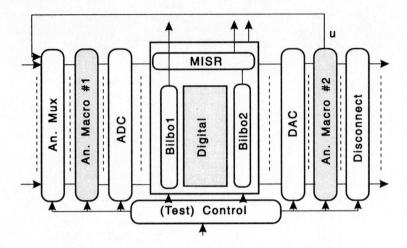

## FIGURE 12.19
Overall block diagram of a self testing, mixed analog/digital chip including test generation and test-evaluation hardware.

sections by means of a programmable biquad. The essential data of these emulated sections for this programmable biquad are stored digitally in the control section of the chip. By means of a voter circuit, essentially a flexible comparator with certain tolerance margins, the outputs of the actual sections are compared with the outputs of the corresponding emulated section. Next, conclusions are provided with regard to the faulty or fault-free behavior of the tested sections. The correct routing of the internal inputs and outputs of the actual sections is carried out by analog multiplexers activated by the control section.

One disadvantage of the presented approaches is the fact that only a limited functional test is carried out, and no data or statement is provided with regard to parametric behavior. Furthermore, very high-speed or high-precision analog parts will cause test problems because of phase shifts and the limited speed and accuracy of the A/D and D/A converters used. The suggested approach only makes sense in terms of test hardware overhead if the analog macros are part of a very large mixed-signal chip, thereby using the already available AD/DA converters and the digital self-test (BILBO) hardware extensively.

The other advantages and disadvantages of analog built-in self-test are similar to the digital case. Among the advantages are the reduction in test-pins, the short low parasitic input/output test lines, the possibility to test at full functional speed, reduction or omission of external ATE, built-in test knowledge and the relative ease of production testing and maintainability. Disadvantages include limited tests, additional design time, and the sometimes dramatic increase in silicon area.

In the future it can be expected that a Digital Signal Processing-like (DSP) approach of self-testing analog and mixed-signal systems will introduce the biggest change on-chip. It is extremely flexible in terms of stimuli generation and evaluation methods, and the test program could be stored digitally. The digital processing capability and D/A and A/D converters will be incorporated as part of the system in the future, thereby reducing the self-testing hardware and software overhead.

## 12.8   ANALOG TEST BUSSES

One of the remaining major approaches for improving testability of an electronic system or chip is to make use of a bus, usually referred to as a testability or test bus (T-Bus). The concept of using test busses in order to reduce the number of external test input and output pins is well-known in the digital world.

A standard has been established for testability busses, the IEEE P1149, which describes the protocol and other rules agreed upon at the various levels of hierarchy. The advantage of a testability bus is that it can support many different styles of testability, such as ad hoc, scan, and built-in self-test. Furthermore, it can reduce the test development time and increase the accuracy of diagnosis. It also facilitates the uniform test and diagnostic approach of analog systems.

Originally, the T-bus was a combination of an analog/digital serial/parallel bus, that allowed real-time data input and output to any addressable control and observability point. Unfortunately, the analog part of the standard, the P1149.4 Real-Time Analog Subset (RTAS) testability bus was abandoned because of the changes in the overall P1149 structure. Currently, a new mixed signal test bus standard is being developed because of the rapid commercial introduction of mixed-signal systems and their many test-related problems. It is anticipated that this test bus will also provide a means for controllability and observability for embedded analog macros.

It is not clear yet whether the bus should be a new standard or an addendum to the existing P1149 standards, or whether serial or addressable test busses should be considered. The same also holds for which characteristics should be specified (functions, electrical, timing, mechanical, data format) or which levels of integration (devices, boards, systems) should be covered.

Although not yet a standard, the general use of analog testability busses can provide many interesting features on a chip. One of the many possible realizations of a testability bus on a chip is shown in Fig. 12.20, and was originally suggested for analog printed-circuit boards. However, the components on the printed-circuit board can be considered as macros, while the board can be regarded as the chip.

In the example, one requires analog multiplexers and demultiplexers and the testability bus. The testability bus includes the signals *Ain*, *Ain.En*, *Aout*, *Aout.En* and *Address*. *Ain* denotes the analog input pin, and *Ain.En* the controlling enable signal pin. The address $N$ of the test point to be controlled is provided by the address bus lines (*Address*), and controls the demultiplexer. Hence, a real- time analog signal can be connected to a controllable test point if the demultiplexer is enabled and the proper address is provided.

In a similar way, a real-time analog output signal, *Aout*, can be observed at the output pin. In turn, the analog multiplexer is enabled by the signal *Aout.En*

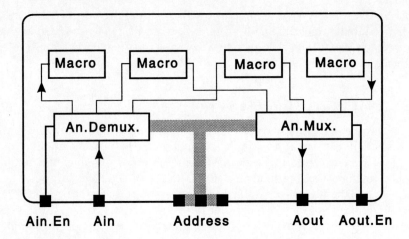

**FIGURE 12.20**
Setup of the architecture of a macro-based chip employing an analog bus. Selection of inputs and outputs of macros is carried out by means of addresses.

and the address $M$ of the test point to be observed and is controlled by the address lines *Address*. It is obvious that the multiplexers and demultiplexers should in no way obstruct or distort (e.g., limited bandwidth, gain, or noise) the passage of real-time signals. Also care should be taken to provide the correct impedance to apply signals and propagate the results to the output pin.

Apart from any standard or protocol, it is also possible to use a combination of (parallel) test lines and hence, construct a custom-like testability or test bus. If this approach is combined with the previously discussed macro-test approach, a powerful and structured testability scenario can be developed. An example of this idea is shown in Fig. 12.21. First, it is assumed that the entire system can be divided into a number of independently testable analog (and/or digital) macros. The suggested system can now be divided into three parts:

1.  A P-wire test bus going along all macros (P = 1, 2, 3, ....)
2.  Flip-flops (SR) and some logic (L), forming a conventional (scan-like) shift register and control logic
3.  A number of analog switches controlled by the flip-flops (SR) and associated control logic (L)

The proper control of the analog switches is obtained by loading the proper digital data in the shift registers. It is then possible to independently control and observe the inputs and outputs of each macro. Actually, the provision of a serial shift register reduces the number of address lines required in the previous approach. The price that has to be paid is less rapid access to and from test points. If activated, the inputs and outputs (or internal nodes) of macros are connected to the P-wire test bus. The advantage of this approach is that more

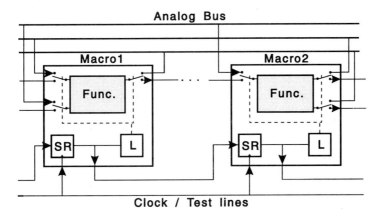

**FIGURE 12.21**
Example of a macro-based system using a three-wire analog bus and two control/test lines.

than one input signal can be controlled and more than one output can be monitored simultaneously. Although Fig. 12.21 shows a disconnection of a macro with its neighbors, the macro can also be loaded by the succeeding macro by not disconnecting this macro and hence showing actual loading conditions. Furthermore, the number of test lines can be reduced by adding more flip-flops, thereby observing a number of outputs sequentially. The number of wires $(P)$ in the test bus is up to the designer: it is a trade-off between area overhead and speed. In the example in Fig. 12.21, the parameter $P$ was chosen to be three. It is also possible to give an estimation of the area overhead $AO$ of this particular approach. It can be formulated as:

$$AO(\%) = (d/\sqrt{A})\sqrt{n}(L+2) \qquad (12.2)$$

where $d$ denotes the layout pitch of the test and control lines, $A$ the total chip area, $n$ the number of parameters to be measured, and $P$ the number of test lines. The number 2 in Eq. (12.2) results from the fact that two control lines are required, the test-data line and the clock line.

Industrial ICs have been designed using the previous approach. It is clear that careful system partitioning into independently testable macros is necessary. In addition, a sensible trade-off has to be made with regard to the number of test lines. An example has been included in the exercise section at the end of this chapter.

There are many other possibilities available with regard to test busses. One can make a distinction between stimuli and response test lines. Finally, it is likely that the test bus approach may be merged with the previously discussed boundary-scan approach.

## 12.9   DESIGN FOR ELECTRON-BEAM TESTABILITY

In the previous sections, several possibilities have been given to enhance the observability of nodes or parameters by means of inserting additional hardware providing electrical contact between the items to be measured and a conventional analog test system.

A somewhat different approach of measuring this analog data is by using an electron-beam system to measure internal upper-layer metal nodes on the chip. In this approach, the chip is placed in a vacuum and a sampling e-beam of primary electrons is directed to the requested node on the chip. The energies of the returning secondary electrons can be measured by a spectrometer and this provides information about the analog voltage level at that node. This type of measurement which is required in analog testing is referred to as 'waveform measurements' in order to distinguish it from many other methods of e-beam testing.

Although very well known and used in the digital world for fault diagnosis in digital integrated circuits, electron-beam machines are rarely used for pure analog purposes until now. However, the increased use of e-beam systems to carry out failure analysis on large (4 and 16M) dynamic memories where one is also interested in the analog properties of digital signals, e.g., the analog value of a sense amplifier output signal in volts and timing in general, is a first step in this direction. It is expected to increase further with the wide-spread introduction of mixed-signal chips. An example where SPICE analog circuit simulations and e-beam measurements are compared in a very high density circuit is shown in Fig. 12.22. Currently these kinds of measurements require a voltage resolution of around 15 mV, but this is expected to go down in the near future as the minimum design dimensions continue to shrink even more ( $< 0.2~\mu$m).

One of the reasons why it is not yet widely used is the difficulty in measuring absolute analog voltages accurately. In addition, an ac-signal is often superimposed on a dc-voltage. The latter, however, cannot be measured directly if the node is not located at the top surface of the chip. This is obvious because the measurement in this buried-layer case is actually a capacitive measurement requiring ac-signals. However, if one takes care during the design by bringing the nodes to be measured to the upper metal surface using vias, like in Fig. 12.23, one is also able to measure dc-voltages. Currently, the capabilities of most advanced electron-beam test systems are:

- Dynamic range: +/- 25 V
- Voltage resolution: 5 mV
- Voltage accuracy : < 3%
- Frequencies in excess of several hundreds of MHz
- E-beam spot diameter less than 0.1 $\mu$m

As the above data shows, the voltage resolution is at least an order of magnitude

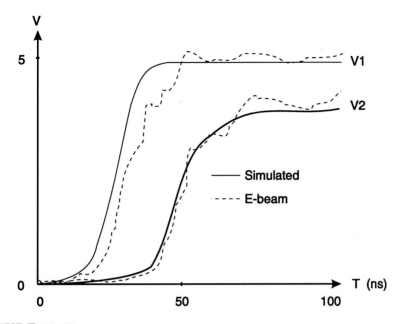

**FIGURE 12.22**

An example of comparisons between SPICE circuit simulations and measured signals from the electron-beam instrument using the waveform measurement method.

worse than present conventional analog testers. Although theoretically 0.08 mV resolution should be feasible, there are some physical phenomena and practical reasons which will limit it to 0.5 mV if sufficient testing time is allowed. One of the limiting factors is the introduction of noise by the whole measurement system. The resolution can to some extent be traded off with the measurement time required by averaging the measurements. Unfortunately, there is a quadratic relationship between the resolution and the required measuring time.

A situation which often occurs, especially in multi-layered (e.g., four interconnecting layers) ICs is the fact that a wire or node is obscured or inaccessible because one or more layers are on top of it, or it is buried deep in the chip. In the first case, one is unable to measure a voltage at that node. In the second case, large measurement inaccuracies can occur. Hence, especially in analog circuits, one wants to bring crucial measurement nodes, e.g., the inputs and outputs of an analog macro, to the top metal layer for accurate measurements. A probe test-point area of four times the electron-beam diameter can be safely used, e.g., $3 \times 3$ $\mu$m. For the sake of cross-talk and other spurious effects, one should surround the e-beam probing point with a grounded guard line if area overhead is of no major concern. In addition, no passivation layer should be deposited on these probing points as it will decrease the accuracy of the measurements.

By using the testability-analysis programs previously described, one is able to determine crucial observability points in advance during the design, hence

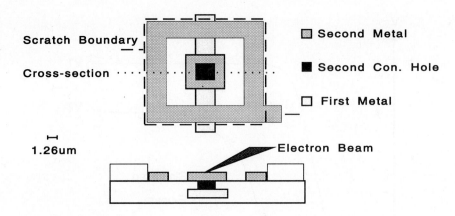

**FIGURE 12.23**
Electron-beam access of a buried layer through an upper-level metal layer using a via. The grounded guard ring is optional.

reducing the total number of electron-beam probe points. Next one can insert e-beam testing points automatically at these locations, as has already been the case for inserting scan flip-flops in the digital world. This relatively new approach toward testable design is usually referred to as design for e-beam testability (DFEBT). An example of the insertion of e-beam testing points in order to get contact with poly-silicon lines in an integrated circuit is shown in Fig. 12.24.

Among the advantages of using an e-beam approach are very low capacitive loading of the node, and this has little impact on the design. Finally, one can obtain a very high spatial resolution ( $< 0.1 \ \mu$m) and it is non-destructive. The additional probing point only adds a well-defined small capacitance to the netlist and can be inserted automatically. If the design data of a chip is available, the navigation of the e-beam to the requested nodes is very simple. Currently, software efforts are being made to carry out semi-automatic fault diagnosis in mixed-signal chips. The disadvantages are at this moment rather low voltage resolution and accuracy, the relative long measurement times in comparison with conventional testing approaches, and the investments to be made in an integrated electron-beam system. Still it can be considered as a new and potentially interesting method to improve the testability of analog as well as mixed-signal chips.

## 12.10   CONCLUSION

Design for testability of analog (VLSI) circuits has only recently gained widespread interest. Most concepts in this relatively new area basically originate from the digital world, where DFT has been around for about a decade.

However, the faults involved in analog circuits are much more complex than their digital counterparts, especially the deviation faults. Hence, definitions of faults are different, fault simulations are extremely time consuming, and the

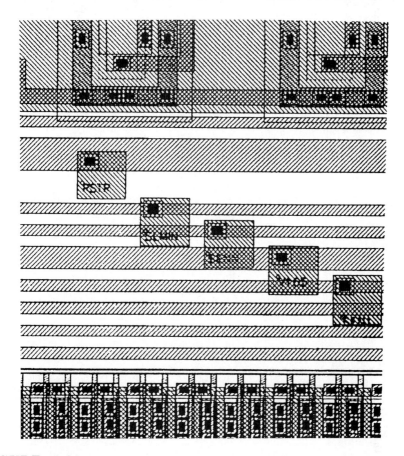

**FIGURE 12.24**
Example of e-beam test points in a complex chip in order to measure voltages at the buried poly-silicon lines.

measures of testability and fault coverage are fuzzier. Known concepts in digital design for testability like scan techniques, including boundary-scan, test busses and self-test, have been modified to fit analog testing needs. They usually support only a functional test. Clearly analog DFT is currently still an area in its infancy and much more research is required. Moreover, there is still a strong opposition from analog designers to apply DFT, and it is expected to take quite some time to change this attitude as has been the case in the past with digital designers.

It is expected however, that the steady increase in (BICMOS) mixed analog and digital ASICs will boost up the number and quality of analog DFT methods and analog designers will be forced to shift test problems to an early stage of design because time-to-market will require it. For further reading on this topic the reader is referred to [61].

## PROBLEMS

**12.1.** Identify several important facts which make DFT and testing for analog and mixed-signal circuits much more difficult than their digital counterparts.

**12.2.** Indicate why fault-simulation techniques and a fault-dictionary approach in analog circuits have only a limited practical use.

**12.3.** Consider the RCL network in Fig. 12.4 and the testability of R in Fig. 12.5. Argue at which radian frequencies useful information can be measured with regard to the resistor R, and why test point 2 is favored over test point 1.

**12.4.** Assume a serially connected string of analog macros (e.g., multipliers) in a chip design where testability analysis shows that the major system parameters are inseparable. Explain the problem in terms of testing and provide at least two methods to circumvent this problem. Also indicate the advantages and disadvantages of the methods used.

**12.5.** Most analog and certainly digital DFT circuits for observation are based on observing voltages only. In the analog and hence mixed-signal case however, it is sometimes required to measure internal currents. Indicate how it is possible to observe internal current values.

**12.6.** Your mixed-signal ASIC is to be used with other similar and digital components on a printed-circuit board. Indicate how you would provide testability of your analog and digital design parts in this environment.

**12.7.** You have designed a pure analog ASIC and are considering to incorporate self-test. Which approach would you choose?

**12.8.** The Philips TDA 8466 is an integrated PAL/NTSC decoder. It identifies and demodulates PAL/NTSC signals. Its I2C interface (and associated DACs) control the mode of the decoder and several basic functions like contrast. It is manufactured in a BICMOS process and has 40 pins. The basic set up is shown in Fig. P12.8(a). It consists of a decoder, I2C bus, DACs, and an interface, and finally an RGB processor. A CCD is used externally to act as a delay line. More detailed block mixed-signal schemes of the decoder and control part are shown in the Figs. P11.8(b) and P11.8(c), respectively. The schemes have been considerably simplified in terms of the number of functional blocks as well as interconnection lines.

In the decoder (Fig. P12.8(b)), the chroma signal is amplified and supplied to several detectors, e.g., NTSC and PAL. The Hue control is only active for NTSC. The burst detector controls the oscillator section. The automatic system manager (ASM) scans the systems. The sand-castle detector and logic provides timing information.

The RGB processor (Fig. P12.8(c)) has luminance and color, and color differences as input. After amplification, the luminance signal is applied to the RGB matrix, together with the color differences. Dependent on the system (PAL or NTSC), different calculations are carried out. Next these results are supplied to a fast-switch circuit on which the external R, G, B signals can be chosen. Finally, R, G, and B are amplified and controlled on a number of parameters, like contrast and brightness.

Provide a design-for-testability approach for this complex mixed-signal chip and provide a rough block scheme for a testable version of the system and argue your choices.

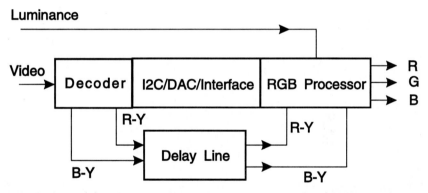

**FIGURE P12.8(a)** Basic block diagram of the integrated PAL/NTSC decoder TDA 8466.

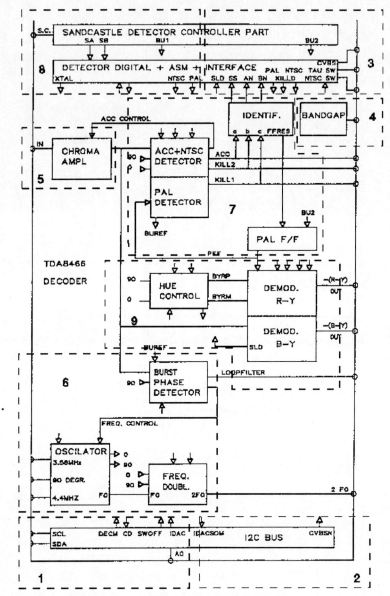

**FIGURE P12.8(b)** The decoder part of the TDA 8466. The new macro partitioning is indicated by the blocks with the dashed lines.

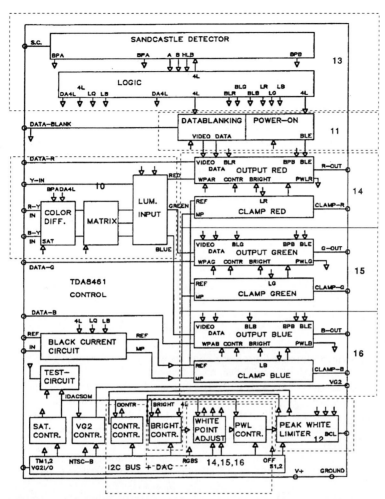

**FIGURE P12.8(c)** The RGB processor part of the TDA 8466. The new macro partitioning is indicated by the blocks with the dashed lines.

## REFERENCES

[1] R.R. Fritzemeier et al., "Fundamentals of testability," *IEEE Trans. on Industrial Electronics,* vol. 36, no. 32, pp. 117-128, May 1989.

[2] F.P.M. Beenker, "Macro-testing: unifying IC and board test," *IEEE Design and Test of Computers,* vol. 3, no.6, pp. 26-32, 1986.

[3] P.P. Fasang, "Analog/digital ASIC design for testability," *IEEE Trans. on Industrial Electronics,* vol. 36, no. 2, pp. 219-226, May 1989.

[4] K.D. Wagner and T.W. Williams, "Design for testability of mixed-signal integrated circuits," *International Test Conference,* 1988, pp. 823-828.

[5] J. Bond, "Mixed-signal malaise," *Test & Measurement World,* December 1991, pp. 43-48.

[6]  A. Jongepier et al., "Macro test for analog circuits: a feasibility study based on the TDA 8466," Philips Research Laboratories, Eindhoven, The Netherlands, 1988.

[7]  M. Syrzycki, "Modeling of spot defects in MOS transistors," *International Test Conference*, 1987, pp. 148-157.

[8]  L. Milor and V. Visvanathan, "Detection of catastrophic faults in analog integrated circuits," *IEEE Transactions on Computer-aided Design*, vol. 8, no. 2, pp. 114-129, February 1989.

[9]  A. Dorey and J.B. Hibbert, "Simplified test strategies for analog ICs," *European Test Conference*, 1991, p. 494.

[10]  M. Soma, "Probabilistic measures of fault equivalence in mixed-signal systems," *IEEE VLSI Test Symposium*, 1991, pp. 67-70.

[11]  J.E. Jagodnik and M.V. Wolfson, "Systematic fault simulation in an analog circuit simulator," *IEEE Transactions on Circuits and Systems*, vol. CAS-26, no. 7, pp. 549-554, July 1979.

[12]  M. Soma, "An experimental approach to analog fault models," *IEEE Custom Integrated Circuits Conference*, 1991, pp. 13.6.1-4.

[13]  A. Meixner and W. Maly, "Fault modeling for the testing of mixed integrated circuits," *International Test Conference*, pp. 564-572.

[14]  "Circuit analysis using Saber native mixed-mode simulation," *Application notes,* Analogy Inc, Beaverton (OR), USA, 1991.

[15]  J.W. Bandler and A.E. Salama, "Fault diagnosis of analog circuits," *Proceedings of the IEEE*, vol. 73, no. 8, pp. 1279-1320, August 1985.

[16]  P. Duhamel and J.C. Rault, "Automatic test generation techniques for analog circuits and systems: a review," *IEEE Trans. Circuits Syst.*, vol. CAS-26, no. 7, pp. 411-439, July 1979.

[17]  L.H. Goldstein, "Controllability/observability analysis of digital circuits," *IEEE Trans. Circuits Syst.*, vol. CAS-26, pp.685-693, no. 9, September 1979.

[18]  N. Sen and R. Saeks, "A measure of testability and its application to test point selection theory," *20th Midwest Symp. Circuits Syst.*, Lubbock, August 1977, pp. 576-583.

[19]  R.W. Priester and J.B. Clary, "New measures of testability and test complexity for linear analog failure analysis," *IEEE Trans. Circuits Syst.*, vol. CAS-28, no. 11, pp. 1088-1092, November 1981.

[20]  G. Iuculano et al., "Multi-frequency measurement of testability with application to large linear analog systems," *IEEE Trans. Circuits Syst.*, vol. CAS-33, no.6, pp. 644-648, June 1986.

[21]  G.N. Stenbakken et al., "Ambiguity groups and testability," *IEEE Trans. Instrumentation and Measurement*, vol. 38, no. 5, pp. 941-947, October 1989.

[22]  G.J. Hemink et al., "TASTE: A tool for analog system testability evaluation," *International Test Conference*, 1988, pp. 829-838.

[23]  G.J.Hemink et al., "Testability analysis of analog systems," *IEEE Trans. on computer-aided design*, vol. 9, no. 6, pp. 829-838, June 1990.

[24]  J. Torino, *Design to test*, A. van Nostrand Reinhold, New York, ISBN 0-442-00170-3, 1990.

[25]  M. Williams and J. Angel, "Enhancing testability of large-scale integrated circuits via test-points and additional logic," *IEEE Trans. on Computers*, vol. 22, no.1, pp.46-60, January 1973.

[26]  P. Fasang, "Analog/digital ASIC design-for-testability," *IEEE Trans. Industrial Electronics*, vol. 36, no.2., pp. 219-226, May 1989.

[27]  P. Fasang et al., "Design for testability for mixed analog/digital ASICs," *Proc. IEEE Custom Integrated Circuits Conference*, 1988, pp. 16.5.1-4.

[28]  J. Bond, "E-beams test, ion beams repair," *Test & Measurement World,* May 1991, pp. 59-64.

[29] K. Wagner and T. Williams, "Design for testability of analog/digital networks," *IEEE Transactions on Industrial Electronics,* vol. 36, no. 2, p. 227-230, May 1989.

[30] K. Wagner and T. Williams, "Design for testability of mixed-signal integrated circuits," *International Test Conference,* 1988, pp. 823-828.

[31] C.L. Wey, "Built-in self test (BIST) structure for analog circuit fault diagnosis," *IEEE Transactions on Instrumentation and Measurement,* vol. 39, no. 3, pp. 517-521, June 1990.

[32] C.L. Wey et al.,"Built-in self test for analog circuits," *Proc. 31st Midwest Symp. Circuits Syst.,* August 1988, pp. 862-865.

[33] C.L. Wey et al.,"Built-in self test (BIST) design of large-scale analog circuit networks," *Proc. IEEE Intl. Symp. Circuits Syst.,* May 1989, pp. 2048-2051.

[34] G. Schafer et al.,"Block-oriented test-strategy for analog circuits," *Proc. of European Solid-State Circuits Conference,* 1991, pp. 217-220.

[35] E. Klinkenberg, *Investigation on the applicability of BCCDs in analog testing,* University of Twente, The Netherlands, no. 060.8567, November 1991.

[36] M. Soma, "A design-for test methodology for active analog filters," *International Test Conference,* 1990, pp. 183-192.

[37] M. Soma, "Testable design of switched-capacitor circuits," *ISSCC Workshop on switched-capacitor design-for-test,* February 1992.

[38] C. Maunder and F. Beenker, "Boundary-scan, a framework for structured design-for-test," *International Test Conference,* 1987, pp. 714-724.

[39] *Boundary-Scan Architecture Standard Proposal,* version 2.0, JTAG Technical Committee, March 1988.

[40] P.P. Fasang, "Boundary-scan and its application to analog-digital ASIC testing in a board/system environment," *IEEE Custom Integrated Circuits Conference,* 1989, pp. 22.4.1-3.

[41] H. Vefling, "Test of analog components in a digital environment," *European Test Conference,* 1991, p. 522.

[42] M. Jarwala, "A framework for design for testability of mixed analog/digital circuits," *IEEE Custom Integrated Circuits Conference,* 1991, pp. 13.5.1-13.5.4.

[43] M.J. Ohletz, *Self testing monolithic analog-digital circuits,* Ph.D. thesis, University of Hannover, Germany, 1989.

[44] M.J. Ohletz, "Hybrid built-in self test (HBIST) for mixed analog/digital integrated circuits," *European Test Conference,* Munich , April 1991, pp. 307-316.

[45] R. Colbey, "Developments for built-in self-test of mixed ASICs," *European Test Conference,* Munich, April 1991, p. 512.

[46] J.L. Huertas, "Concurrent testing of analog filters using a programmable biquad," *Proc. Intl. Symp. Circuits Syst.,* April 1992, pp. 423-426.

[47] M. Mahoney, "DSP-based testing of analog and mixed-signal circuits," *IEEE Computer Society Press,* ISBN 0-8186-07585-8, 1987.

[48] P1149/D3 (IEEE) commission, *T-bus specification,* Chapter 1, September 1987.

[49] IEEE P1149.4, *Real-time analog testability bus protocol,* Draft.

[50] P1149 (IEEE) commission, *Mixed-signal test bus,* September 1991.

[51] J.Torino, *Design to Test,* A. van Nostrand Reinhold, New York, ISBN 0-442-00170-3, 1990

[52] A. Jongepier, "Testing analog circuits," *DFT workshop,* Garderen, The Netherlands, 1988.

[53] A. Jongepier et al., *Macro test for analog circuits,* RNR-40E/88-VI-223, Philips Research Labs & Philips Components, The Netherlands, June 1988.

[54] R.J. van Rijsinge, *Analog Industrial Test Development Aid,* University of Twente, no. 060.7490, September 1989, pp.16-17.

[55] M. Jarwala and S. Tsai, "A framework for design for testability of mixed analog/digital circuits," *IEEE Custom Integrated Circuits Conference,* 1991, pp. 13.5.1-13.5.4.

[56] E. Menzel and E. Kubalek, "Fundamentals of electron-beam testing of integrated circuits," *VLSI Testing and Validation Techniques,* IEEE Society Press, ISBN 0444879471, 1985.

[57] S. Gorlich et al., "Integration of CAD, CAT and electron-beam testing for IC internal logic verification," *International Test Conference,* September 1987, pp. 566-574.

[58] J. Kolzer et al., "Chip verification of 4-bit DRAMs by e-beam testing," *Microelectronic Engineering,* vol. 12, 1990, pp. 27-36.

[59] K.D. Herrmann and E. Kubalek, "Design for e-beam testability: a demand for e-beam testing of future device generations ?," *Microelectronic Engineering,* vol. 7, 1987, pp. 405-415.

[60] W.T. Lee, "Engineering a device for e-beam probing," *IEEE Design and Test of Computers,* pp. 36-48, June 1989.

[61] J.L. Huertas, "Test and design for testability of analog and mixed-signal integrated circuits: Theoretical basis and pragmatical approaches," *Chapter 2 in Circuit Theory and Design: Selected Topics in Circuits and Systems,* H. Dedieu (Editor), Elsevier, 1993.

# CHAPTER
# 13

## ANALOG VLSI INTERCONNECTS

## 13.1  INTRODUCTION

As VLSI processes evolve and improve, a well-known empirical relationship has emerged: as stated by Gordon Moore, the number of devices on state-of-the-art integrated circuits doubles approximately every two years. Moore's Law , as this observation has come to be known, has been representative of the industry's improvements over the past 20 years, and is likely to continue for at least one more decade, until fundamental limits in device physics are reached [4]. These improvements have taken three major forms: (i) minimum device feature sizes have decreased due to improved lithography, deposition, patterning, and etching tolerances, (ii) reduced relative spacing requirements between adjacent devices, due to improved device and isolation structures (e.g., trench capacitors and high-resistivity wafer substrates), (iii) the area of reliably manufacturable dice has increased, due to availability of large wafers with reduced defect densities and improved optical mask alignment technologies.

As this trend toward smaller and faster devices continues, the *system* performance of VLSI chips is increasingly determined by the electrical characteristics of the wires connecting these devices, as well as the interconnects between chips. These performance metrics include speed, power consumption, and power-delay product. Furthermore, wiring area often constrains the size of highly-connected analog circuit networks.

In this chapter, we will describe the various types of interconnect available in a CMOS process, explore their characteristics, and examine the constraints

585

that arise from their application. Future trends are developed in terms of a simple scaling model for MOS devices and processes. Finally, implications for large-scale analog systems are examined, with the development of a theoretical wiring model, and the examination of a system case-study.

## 13.2 PHYSICS OF INTERCONNECTS IN VLSI

At low-signal switching frequencies, at which the signal wavelength is long compared to the physical dimensions of the interconnect wires, the electrical behavior of the interconnect is described by its dc-resistance and capacitance. These quantities are in turn derived from the geometry of the interconnect and the materials involved.

### 13.2.1 Resistance of Interconnects

Conductivity is the process of movement of charged species (typically electrons and holes) under an applied electric field. It is convenient to parameterize this movement by the conductivity of the material, which relates the current density $J$ to the applied electric field $E$ by Ohm's Law:

$$\mathbf{J} = \sigma \mathbf{E} \tag{13.1}$$

Materials are commonly approximated as isotropic, so that $\sigma$ is a scalar. In a uniform conductor the macroscopic form of Ohm's law is easily derived, assuming uniform current density:

$$V = IR \tag{13.2}$$

Where

$$V = \int E dy = EL \tag{13.3}$$

$$I = \int \int J dx dz = JWT \tag{13.4}$$

Hence,

$$R = \frac{L}{\sigma W T} \tag{13.5}$$

In cases where $\sigma$ is not uniform (e.g., due to semiconductor doping gradients), it is still appropriate to use a simple expression for $R$ in cases where $L$ is long (i.e., the electric field $E$ is constant). Here:

$$R = \frac{L}{\int \int \sigma dx dz} \tag{13.6}$$

Typically, $\sigma$ is a function of $z$ only, so that:

$$R = \frac{L}{W \int \sigma dz} \tag{13.7}$$

In either case, it is common to measure resistance of a conductor in units of "ohms per square," reflecting the fact that only the ratio of $L$ to $W$ determines the total wire resistance (for a uniform conductor).

$$R/\Box = \frac{1}{\int \sigma dz} = \frac{1}{\sigma T} \tag{13.8}$$

In the general case, the symmetry of the conductor cannot be used to integrate Ohm's law, and a macroscopic resistance $R$ must be computed differently. Most commonly, Gauss' Law is applied:

$$\nabla \cdot \mathbf{E} = \frac{\rho}{\epsilon} \tag{13.9}$$

with the observation that, in a conductor in steady state, the charge density, $\rho$, must be identically zero, and that the electric field is derivable from the vector potential $V$ (the voltage in the conductor) by $\mathbf{E} = -\nabla V$. Hence,

$$\nabla^2 V = 0 \tag{13.10}$$

This form, known as Laplace's equation, must be integrated with the appropriate boundary conditions to find the voltage within the conductor. From that, we can also determine the current flowing through the conductor by integrating Ohm's law along any cross-section.

Several common geometries are illustrated in Fig. 13.1, along with their equivalent resistances. These values, normalized to the number of "squares" of equivalent conductor, are useful for quick calculations of the resistance of a particular geometry. It is worth emphasizing, however, that the photolithographic patterning accuracy of VLSI interconnect has a large effect on many of these structures; consequently, the best way to achieve precise matching between different components is to faithfully replicate the exact geometry in question.

## 13.2.2   Capacitance

The performance of large VLSI circuits is often determined by the capacitance of the interconnect. In digital systems, the switching speed is directly determined by the load capacitance seen by a given driver [7], and the power dissipation of the circuit can be reasonably estimated by:

$$P = \frac{1}{2}\alpha C V_{DD}^2 f \tag{13.11}$$

where $C$ is the total signal capacitance on the integrated circuit, $f$ is the clock (or switching) frequency, and $V_{DD}$ is the power-supply voltage. The factor $\alpha$ models the average number of nodes that switch on a given clock transition. Even if circuit techniques such as exponentially sized drivers are employed to reduce the signal propagation time, this power dissipation budget remains unchanged, and reflects the fundamental physical limitations of fully restored digital logic.

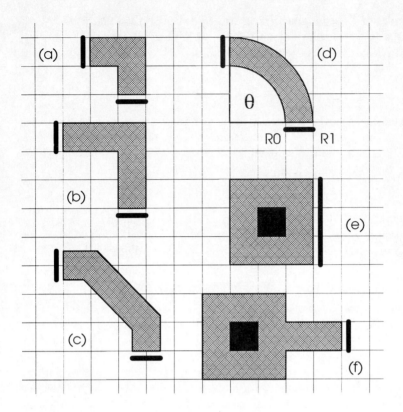

**FIGURE 13.1**

Equivalent resistance of various structures, measured in squares. The resistance is measured between equipotential surfaces denoted by the heavy lines. Part (a) measures $2.54$ squares, part (b) measures $4.55$ squares (indicating that the equipotential approximation used to analyze (a) is valid). Part (c) applies to mitered wires, and measures $4.81$ squares. Part (d) is for circular wire segments (of large radius), and yields a resistance for the arc of $R = \rho\Theta/ln(R_1/R_0)$; for a $90$ degree arc corresponding to the orthogonal case of part (a), the resistance of the arc is $2.27$ squares. Part (e) is a contact cut, with the resistance measured between the inner dark square and the heavy line on the right; this structure comprises $0.41$ squares. Part (f) is a common contact structure to a narrow wire, and corresponds to $2.64$ squares of resistance.

In analog circuits, wiring and node capacitances must be considered for a second reason: parasitic capacitances are usually critical in determining the functionality of an analog component. Modeling such parasitics is essential at the circuit simulation level of design. In this section, we will concentrate on the geometry of diffused and metal interconnect, and develop expressions for their capacitance.

### 13.2.3 Capacitance of Metalized Interconnect

In a good electrical conductor, electrical charge redistributes itself to make the electrical field inside the conductor identically zero (otherwise, current would flow). In practice, this means the charge is always located on the surface of the conductor.

The interconnect capacitance to dielectric-insulated layers in a VLSI process is determined by three components: a parallel-plate component, a fringing-field component, and a wire-to-wire component. The parallel-plate component is generally a capacitance to substrate, and is proportional to the area of the interconnection wire:

$$C_{pp} = \epsilon_{or} \frac{W \cdot l}{t_{or}} \tag{13.12}$$

where $t_{or}$ is the thickness of the field oxide, and $\epsilon_{or} = \epsilon_r \epsilon_o$ (where $\epsilon_r$ is the relative permittivity of the dielectric (3.9 for $SiO_2$) and $\epsilon_o$ is the permittivity of free space: 8.85 x $10^{-12} F/m$). For a typical $2\mu$ CMOS process (with $3\mu$ metal widths over $0.5\mu$ field oxide), this capacitance corresponds to about $2pF$ for a minimum-size wire running the width of a $10mm$ integrated circuit.

Unfortunately, the parallel-plate capacitance is only part of the story. At reduced linewidths, fringing capacitances become more significant as the thickness of the interconnect metallization can no longer be ignored. An excellent approximation to the parallel-plate and fringing capacitance of a metal wire (of width $W$ and thickness $H$) can be obtained by considering the conductor as two components: a parallel-plate capacitor of width $W - H/2$, and a cylindrical wire of diameter $H$ [17]. The resulting capacitance per unit length is then:

$$\frac{C}{l} = \epsilon_{or} \left\{ \frac{W}{t_{or}} - \frac{H}{2t_{or}} + \frac{2\pi}{ln\left[1 + \frac{2t_{oz}}{H}1 + \sqrt{1 + \frac{H}{t_{oz}}}\right]} \right\} \tag{13.13}$$

This expression can be plotted as a function of $W/t_{or}$ for various values of $H/t_{or}$. This graph is shown in Fig. 13.2.

The third component of interconnect capacitance is the wire-to-wire capacitance between adjacent metalization lines. This capacitance has the additional disadvantage of contributing crosstalk between neighboring signal lines, that can cause significant performance degradation if not carefully considered and controlled. For example, many highly-connected analog neural network architectures [12] employ differential amplifiers with complementary outputs in which both the inputs and the outputs to a single amplifier run in parallel for the entire width of the die; the effect of the additional Miller capacitance must be carefully considered in the case of negative feedback (as, of course, must be the stability of the circuit in the case of positive feedback).

Although the precise computation of wire-to-wire capacitance requires two- and three-dimensional modeling of the geometry [10, 3], the qualitative effect is predicted rather well by the preceding discussion of fringing fields. In fact, a useful working approximation is that in the case where the metal thickness is

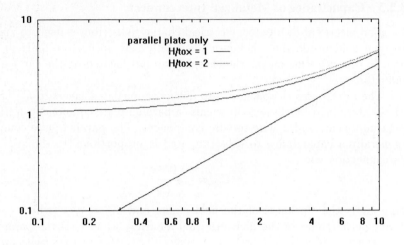

**FIGURE 13.2**

Parallel-plate and fringing capacitance of a single wire. The abscissa is calibrated in $pF/cm$; the ordinate is the ratio of the wire width to the oxide thickness. The straight line shows the contribution due to the parallel plate approximation alone. Clearly, as the width of the interconnect approaches the thickness of the oxide dielectric, the total capacitance is limited by the fringing capacitance, which is approximately constant at $1pF/cm$.

roughly equal to the dielectric thickness, and the wire width is twice this quantity and equal to the minimum spacing between adjacent wires (i.e., wire spacing $\approx W \approx 2H \approx 2t_{ox}$), the total capacitance of a given node is effectively doubled. Such geometries are typical in modern CMOS processes. A more detailed model for the geometry shown in Fig. 13.3 is developed in [10], and leads to the approximation:

$$\frac{C}{l} = \epsilon_{ox} \left[ \frac{115W}{t_{ox}} + 2.8 \left( \frac{H}{t_{ox}} \right)^{0222} \right]$$
$$+ \epsilon_{ox} \left[ \frac{006W}{t_{ox}} + \frac{166H}{t_{ox}} - 0.14 \left( \frac{H}{t_{ox}} \right)^{0222} \right] \left( \frac{t_{ox}}{S} \right)^{134} \qquad (13.14)$$

where $W$ is the width of the metalization, $H$ is the thickness of the metal wire, and $S$ is the separation of the adjacent lines.

From this discussion and from Fig. 13.2, it can be seen that fringing capacitances and wire-to-wire capacitances can dominate the parallel-plate capacitance, and must be carefully considered in critical applications.

### 13.2.4 Capacitance of Diffused Semiconductor Interconnects

Although the geometry and physical origin of the charge separation is different for diffused wires relative to dielectric-insulated metalization interconnect, the basic underlying cause of capacitance is unchanged: capacitance measures the

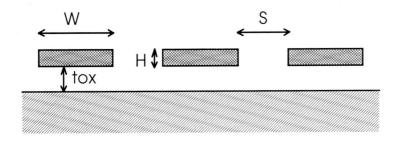

**FIGURE 13.3**
Capacitance between central wire and ground plane and adjacent wires.

energy required to place additional electrical charge on a given node. In a reverse-biased semiconductor junction, this energy is required to increase the width of the depletion region.

The incremental capacitance of a semiconductor junction has exactly the same form as the capacitance of a parallel-plate structure:

$$C(V) = \frac{dQ}{dV} = \frac{\epsilon A}{t} \tag{13.15}$$

where $\epsilon = \epsilon_0 \epsilon_r$ (and the relative permittivity of silicon is $\epsilon_r = 11.8$). The voltage dependence of the junction capacitance is captured in the thickness of the depletion layer, $t$. For a step junction (one in which the doping transition is sharp between the p-region (with $N_A$ acceptors per unit volume) and the n-region (with $N_D$ donor atoms per unit volume)), the thickness of the depletion layer is:

$$t = \sqrt{\frac{2\epsilon(V_0 - V)}{q}\left(\frac{N_A + N_D}{N_A N_D}\right)} \tag{13.16}$$

where $V_0$ is the junction's built-in potential (typically around 0.7 Volts) and $V$ is the bias applied to the junction (and is negative for the typical case of diffused interconnect that is intended to be electrically isolated from the semiconductor substrate).

A more realistic case, however, is for a graded junction, where the doping density varies approximately linearly about the junction. If we model the doping profile by $N_D - N_A = Gx$, where $x$ measures the distance from the junction, and if we assume that the depletion region is entirely contained within this linearly doped region, the thickness of the depletion region can be shown to be:

$$t = \left[\frac{12\epsilon(V_0 - V)}{qG}\right]^{\frac{1}{3}} \tag{13.17}$$

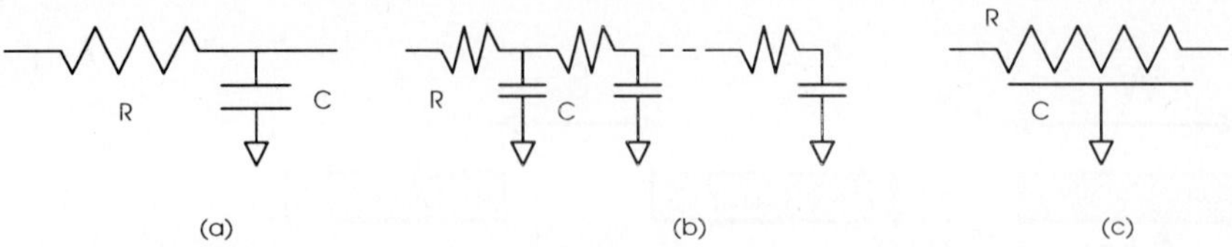

(a)  (b)  (c)

**FIGURE 13.4**

Modeling interconnect with distributed resistance and capacitance. (a) shows a lumped model; (b) is a discrete model incorporating $N$ stages, each with lumped resistance $R$ and capacitance $C$; (c) is a fully distributed model, where every part of the interconnection contributes some capacitance and some resistance.

### 13.2.5 Propagation Delay of Interconnect

At low signal frequencies (i.e., signals for which the time-of-flight propagation delay down the wire is much shorter than the period of the signal frequencies of interest), the propagation delay is determined by the distributed resistance and capacitance of the wire. At higher frequencies, transmission-line analysis must be used, which considers the inductance of the interconnect, and incorporates detailed information of the geometry to calculate reflections.

At low frequencies, interconnect may be modeled either as lumped loads, or as distributed loads. The selection is usually based on required accuracy and ease of calculation. In the lumped case (Fig. 13.4), an interconnect is modeled by a single resistor and a single capacitor The response of a lumped $RC$ network to a step input is:

$$V(t) = 1 - e^{-tRC} \qquad (13.18)$$

This form is convenient for rough calculations, as propagation delays and signal rise times can be computed directly. Both of these quantities measure the response of the network to a step input: propagation delay is usually taken to be the time required for the output signal to reach 50% of the input step height (and is approximately $0.7RC$ for the lumped network), while rise time is typically measured as the time required for the output to go from 10% to 90% of the input signal (the rise time of the lumped $RC$ network is $0.9RC$).

This lumped model, and several other lumped models (discussed in [11]) are also convenient because they allow estimates for wiring delays [6]. Basically, these models predict that the propagation delay from one node to another is the sum of the propagation delays between all the intermediate nodes and the output node; the propagation delay of a particular node is the product of its output resistance and the total capacitance it must drive downstream. Fig. 13.5 shows an example.

Thus, it can be seen that long $RC$ chains (of the type in Fig. 13.4(b)) will have extremely long propagation delays. In fact, the model predicts that an n-stage $RC$ line will have a 50% propagation delay of:

$$t_{50\%} = \frac{RCn(n+1)}{2} \qquad (13.19)$$

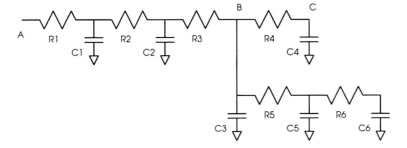

**FIGURE 13.5**

Propagation delay approximation down a $RC$ tree. The propagation delay from $A$ to $B$ is approximately $AB = R1(C1 + C2 + C3 + C4 + C5 + C6) + R2(C2 + C3 + C4 + C5 + C6) + R3(C3 + C4 + C5 + C6)$, and the delay from $B$ to $C$ is approximately $BC = R4 \cdot C4$; the total delay from $A$ to $C$ is then $AB + BC$.

For large $n$, this propagation delay is proportional to $n^2$. For long enough interconnect, this delay far exceeds the propagation delay of buffer stages, and system performance can be substantially improved by adding such buffers to break up the long wire delay. If a logic buffer has delay $\delta$, the optimum segmentation is:

$$\text{RC stages between buffers} = \sqrt{\frac{2\delta}{RC}} \tag{13.20}$$

$$\text{Total delay (wire + buffers)} = n\left[\sqrt{2\delta RC} + \frac{RC}{2}\right] \tag{13.21}$$

The frequency response of a fully-distributed $RC$ delay line (Fig. 13.4(c)) is:

$$H(s) = \frac{1}{s \cosh \sqrt{sRC}} \tag{13.22}$$

where $R$ and $C$ are the total resistance and capacitance, and are uniformly distributed over the length of the interconnect. The time-domain response for a unit step input is shown in Fig. 13.6. The 10% to 90% rise time is approximately $0.9RC$, and the 50% propagation delay is approximately $0.4RC$.

Thus, it can be seen that the simple lumped approximation treated above is very pessimistic; a correction commonly made for the RC-tree propagation delay computation above is that a distributed $RC$ line only contributes one-half of its capacitance when driven through its own resistance. For example, the delay computation of Fig. 13.5 would become:

$$\Delta_{AB} = R_1\left(\tfrac{C_1}{2} + C_2 + C_3 + C_4 + C_5 + C_6\right) + R_2\left(\tfrac{C_2}{2} + C_3 + C_4 + C_5 + C_6\right)$$
$$+ R_3(C_3 + C_4 + C_5 + C_6) \tag{13.23}$$

$$\Delta_{BC} = \frac{R_4 C_4}{2} \tag{13.24}$$

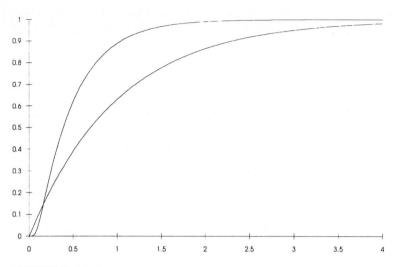

**FIGURE 13.6**
Comparison of response to unit step input for both distributed $RC$ lines and lumped $RC$ load.

where

$$\text{total delay} = \Delta_{AB} + \Delta_{BC} \tag{13.25}$$

assuming that $C1$, $C2$, and $C4$ became distributed $RC$ wires, while $R3$ and $C3$ remained lumped loads.

## 13.3   SCALING OF INTERCONNECTS

Interconnection delay has already become a performance-limiting factor in present-day CMOS technologies. In advanced processes (submicron CMOS, BiCMOS, GaAs, etc.) improved transistor performance makes the effect of interconnect parasitics felt even more strongly. In this section, we examine a simple model for predicting system performance as process feature sizes decrease due to improved lithography and patterning. The model follows the approach taken by [7]
and [1].

Two principal factors contribute to the increase in single-chip system complexity: (i) improvements in patterning, and new doping profiles and inter-device isolation techniques permit smaller devices with smaller spacing between devices, and (ii) improvements in the quality of raw wafers and in the yield of the patterning operations permit larger die to be cost-effectively manufactured. We model each of these improvements by a single scale factor: all device fabrication dimensions are reduced by a factor $S$ (e.g., minimum lengths and widths, oxide and patterned layer thicknesses, and minimum patterned layer separations), while the practical die size increases by a factor $D$.

**TABLE 13.1**
Scaling of local and global interconnect electrical parameters.

|  | Local Interconnect | Global Interconnect |
|---|---|---|
| Feature size (H, W, Sep) | $1/S$ | $1/S$ |
| Wire length | $1/S$ | $D$ |
| Wire resistance | $S$ | $S^2 D$ |
| Wire capacitance | $1/S$ | $D$ |
| Wire RC delay | 1 | $S^2 D^2$ |

We are then concerned with the electrical properties of devices and interconnect. In particular, we consider separately two categories of interconnect: local interconnect comprised of wires that scale according to $S$ only, and global interconnect whose length varies proportionately to $D$, Tab. 1.

Because MOS devices exhibit an on-resistance that does not depend on $S$ (if we also assume that the power supply voltage is reduced by a factor $S$), it is obvious that the performance of future VLSI systems will be determined by interconnect delays of long wires. Furthermore, the length that a wire must reach before it is considered a "long" wire is scaled as $1/S$; thus, any wire that is not strictly local (i.e., has a constant length, measured in units of minimum feature size) will contribute larger delays as processes improve! This problem is compounded by the fact that individual MOS devices are increasing in speed by a factor $S$, and could ideally be operated at correspondingly higher system clock frequencies.

It is therefore imperative that VLSI designers consider interconnect delay at the architecture level of system design. High-performance systems of the future will feature locally connected components, with system design effort expended on pipelining (and other techniques that trade off communication bandwidth for processing latency), and circuit design effort expended on repeaters (to reduce the load that must be driven by a single stage) and exponential horns (to match circuits' drive strengths to interconnect loads).

## 13.4   A MODEL FOR ESTIMATING CIRCUIT WIRING DENSITY

The examination of the role of interconnect at the system level begins with the quantization of silicon resources dedicated to interconnect. The need to consider complexity of wiring as an essential component of any algorithm has been a key contribution to computer science from VLSI design. Whereas in digital VLSI, wiring considerations give rise to an area–time tradeoff, in analog VLSI (in the absence of time multiplexing) wiring complexity determines the scalability of an architecture or algorithm. This constraint is amply illustrated by the full interconnect of the Hopfield associative memory [12]. A fully-connected architecture requires $O(N^2)$ area, where $N$ is the number of neurons in the collective system.

Much of VLSI complexity theory has concentrated on efficient embeddings of particular computational graphs, under particular models for interconnect [15, 16]. The basic procedure is to select a class of graphs, and derive an information-theoretic lower bound for a separator (or partitioning) operator. This separator is then combined with a simple synchronous model for data communication to determine a minimum $AT^2$ constraint on the computation.

In this section, we take a somewhat different approach: we ignore the time cost of communication (we assume that our analog networks use dedicated wires for all signals), and consider only the connectivity of graphs that can be generated when we define a set of layout parameters that specify acceptable component placements. This procedure will define a necessary (but not sufficient) constraint on the connectivity of a circuit graph, in contrast to traditional approaches, which generate the worst-case solution for a class of graphs, not the best-case for a specific graph (i.e., solutions which are "not universally optimal, but existentially optimal" [2]).

Under a connectivity model based on the intrinsic dimensionality of the interconnect technology, we will derive a wire partition function to bound the number of wires that can cross a closed perimeter. We will show how this function is consistent with a common empirical wiring relation known as Rent's rule. Finally, we will consider acceptable distributions of wire lengths within regular layouts, and derive a physical constraint on the average wire length predicted by the model.

### 13.4.1   Rent's Rule

A circuit may be considered as a graph, the vertices of which represent elements, linked by arcs corresponding to wires. Rent's rule [5] is an empirical relationship that relates the number of wires crossing an arbitrary boundary containing a group of elements:

$$P = P_0 N^b \tag{13.26}$$

where $P$ is the number of wires that need to cross the border, $N$ is the number of elements (each of which has $P_0$ pins) contained within the border, and $b$ is an assumed constant. Analysis of several large designs has indicated that typical values lie in the range $b \in [0.5, 0.7]$ [5].

### 13.4.2   An Operational Definition of Optimal Design

In this section, we introduce an aesthetic evaluation metric by which we can measure the quality of a design's implementation. Unlike the information-theoretic lower bounds developed by classical VLSI complexity theory [15, 16], we are less interested in bounds on classes of graphs than in considering the optimality of the layout of a specific instance within the class. In particular, we are interested in specifying minimum acceptance levels, below which a layout is considered too

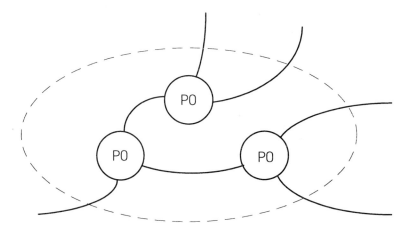

**FIGURE 13.7**
Estimating the number of wires crossing the perimeter.

inefficient, and hence not economically viable. Finally, we will show how our constraint conforms to an empirical wiring relation known as Rent's rule.

We begin with a key observation regarding manufactured systems. We consider such systems to be examples of "optimal systems"—they have successfully satisfied the constraints imposed by the end-application (e.g., cost, size, power, weight, reliability, manufacturability).

The observation we make is the following:
**Observation 1:** In any *system* consisting of a collection of *components*, these components are distributed spatially in an approximately uniform manner.

The phrasing of this observation is intentionally general; we intend for the results of this section to be applicable to many different systems, ranging from integrated circuits to printed-circuit boards to entire racks and boxes. In this section, we combine this observation with the physical dimensionality of interconnect wiring to obtain an upper bound partition function. This function bounds the number of wires crossing a manifold containing a number of components of the system in question (see Fig. 13.7). We will derive a particular function for systems that are intrinsically two-dimensional (e.g., integrated circuits and printed-circuit boards) and repeat the derivation for three-dimensional systems (e.g., computer back-planes and other aggressive interconnect technologies).

We begin by quantifying Observation 1: In a two-dimensional space, assume the density of components is uniform and equal to $\rho$ components per unit area. Furthermore, assume that a fraction $\gamma$ of the total area of the system is consumed by components. If we model the components as identical and square, of dimension $x \cdot x$, we have

$$x^2 = \frac{\gamma}{\rho} \tag{13.27}$$

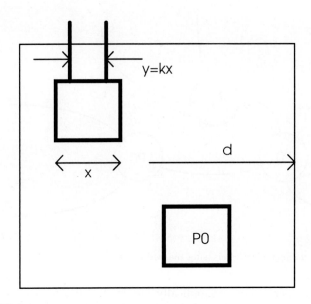

**FIGURE 13.8**
A model for circuit connectivity. The area in question measures $d$ by $d$, and has perimeter $8d$.

We now apply this result to the system in Fig. 13.8: a square area of dimensions $2d$ by $2d$. We wish to compute the number of wiring tracks crossing the perimeter as a function of the number of components within the boundary. We begin by parameterizing the space occupied by these wiring tracks (i.e., the wiring pitch) in terms of the single dimension of the model: $x$, the size of the components.

Let the average wire spacing $y = kx$. In the two-dimensional model, $k$ is a scalar, because $x$ and $y$ are simple linear dimensions.

The component count within the boundary is then

$$N = 4d^3\rho \qquad (13.28)$$

The maximum pin count per element is limited by the number of wiring channels that can abut the perimeter of a single component:

$$P_0 = \frac{4x}{y} = \frac{4}{k} \qquad (13.29)$$

### 13.4.3  A Model for Circuit Connectivity

Estimating the number of wiring channels that cross the perimeter of the region is somewhat more subtle. The analysis is complicated by the need to model the routing capacity through the perimeter $\Gamma$ in regions where $\Gamma$ lies within a component. We make the (rather pessimistic) assumption that no routing can occupy the same physical location as the components. This restriction is appropriate for integrated circuits, where programmable logic devices (PLDs), gate arrays, and standard-cell designs tend to have pre-allocated wiring channels between rows of cells, with severely limited (if any) routing capability through these cells. In particular, this restriction is appropriate for the architecture presented in the following section. It is less appropriate for technologies such as multi-layer PCBs; however, as we shall see, this approximation only affects certain multiplicative constants—it does not affect the general form of our results.

For randomly placed components, the probability that any given point on the perimeter lies within a component is $\gamma$; consequently, the amount of perimeter available to wiring tracks is $8d(1 - \gamma)$. The maximum number of wiring tracks is then

$$P = \frac{8d(1 - \gamma)}{kx} \tag{13.30}$$

Combining the previous three equations, and recalling $x^2 = \gamma/\rho$, we obtain the partition function

$$P = \frac{1 - \gamma}{\gamma^{12}} P_0 N^{12} \tag{13.31}$$

As a self-consistency check, we can compare this expression with the conventional statement of Rent's rule: $P = P_0 N^b$. We obtain equivalence between the two forms (at $b = 1/2$) for $\gamma = 0.38$. This result corresponds to an area utilization of approximately 40%, a typical value for semi-automated VLSI designs (gate arrays, standard-cells, etc.).

We can repeat this analysis for three dimensional systems (in which wires now have finite cross-sectional area). If we set $y = kx^3$, and observe that $x^3 = \gamma/\rho$, we obtain:

$$P = \frac{1 - \gamma}{\gamma^{23}} P_0 N^{23} \tag{13.32}$$

In summary, this analysis yields wiring bounds that are consistent with Rent's rule in form, and suggest that the typically observed values for Rent's rule exponents (typically in the range $0.5, 0.7$ [5]) are due to dimensional constraints of the implementation technology.

## 13.4.4 Contact Density

An interesting consequence to the partition function is obtained by considering the number of contacts that must be placed in order to satisfy the above wiring limit. Clearly, not all of the wires that originate within the perimeter of the region in Fig. 13.8 can propagate through the boundary. Each of the other wires must terminate at a contact with at least one other internal wire. For the purposes of this analysis, we assume, for simplicity, that precisely two wires meet at a contact (this assumption does not affect the form of the final result).

The total number of wires originating within a region measuring $L X L$ (or, $d = L/2$) is

$$P_{total} = P_0 L^2 \rho \qquad (13.33)$$

The maximum number that can pass through the region's perimeter is given by

$$P = \frac{4L(1 - \gamma)}{kx} \qquad (13.34)$$

The number of required contacts is then

$$P_c = \frac{1}{2}\left[ P_0 L^2 \rho - \frac{4L(1 - \gamma)}{kx} \right] \qquad (13.35)$$

and the average area per contact is (in the large $L$ limit)

$$A_c = \frac{L^2}{P_c} = \frac{2}{\rho P_0} \qquad (13.36)$$

We can interpret this expression by observing that $1/\rho$ is the average area per component, in which $P_0$ pins originate. Each pin must terminate at a contact—the factor of 2 indicates that each contact has two wires connected to it.

If we compute the area density of contacts as a function of $d$, we obtain

$$\frac{\partial P_c}{\partial A} = \frac{P_0}{x}\left[ \frac{\gamma}{2x} + \frac{\gamma - 1}{8d} \right] \qquad (13.37)$$

Thus, the area density of contacts as a function of position $d$ tends to a uniform limit. This result is consistent with our original assumption that components are uniformly distributed throughout the region.

### 13.4.5   Limits on Wire Length

Although these results place upper bounds on the number of available wiring channels, they tell us nothing about the macroscopic properties of the interconnect. In particular, it would be useful to consider the distribution of wire lengths predicted by the model, and to verify that this distribution satisfies the dimensional constraints of the implementation technology.

Before we can compute the average wire length, we must consider the rate at which wires from a given cell terminate as a function of distance. Consider the two-dimensional case illustrated in Fig. 13.9; a circular partition is drawn a distance r from the cell in question. Applying the general form of Rent's rule, we have

$$P = P_0 N^b \qquad (13.38)$$

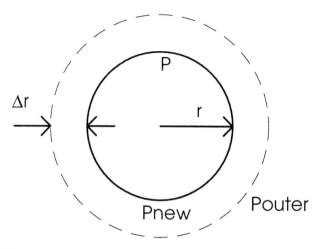

**FIGURE 13.9**
Model for predicting distribution of wire lengths. The system consists of a regular array of identical cells of area $\pi r_c^2$ each. $P$ wires cross the perimeter at $r$, while $P_{outer}$ cross the perimeter at $r + \Delta r$. The number of new pins appearing in the annulus is $P_{new}$.

$$P = P_0 \gamma^b \left[ \frac{r}{r_0} \right]^{2b} \tag{13.39}$$

It can be shown [13] that the probability of a wire originating in the center of the array has not yet contacted another pin after traveling a distance $r$ is:

$$f(r) = 1 - A_0 \left[ \frac{r}{r_0} \right]^b exp \left[ -\frac{\gamma^{1-b}}{2-2b} \left[ \frac{r}{r_0} \right]^{2-2b} \right] \tag{13.40}$$

where

$$A_0 = \left[ \frac{r_c}{r_0} \right]^{-b} exp \left[ \frac{\gamma^{1-b}}{2-2b} \left[ \frac{r_c}{r_0} \right]^{2-2b} \right] \tag{13.41}$$

Armed with the function $f(r)$, we can now compute the probability density function $p_f(r)$ that specifies the probability that a wire terminates between a distance $r$ and $r + \Delta r$:

$$p_f(r) = \frac{\partial f(r)}{\partial r} \tag{13.42}$$

The average length of wires connected to a component is then

$$\langle w \rangle = \int_{r=r_c}^{\infty} r p_f(r) dr \tag{13.43}$$

If the limits of integration are changed from $r = r_c$ to $r = 0$, a closed form

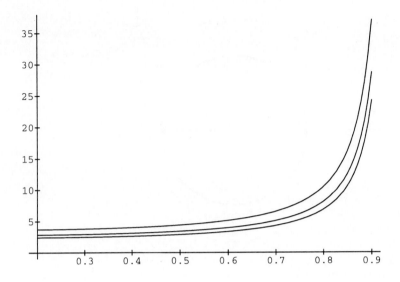

**FIGURE 13.10**

Average wire length predicted by model. The wire length is in units of $r_0$, and is plotted as a function of the Rent's rule coefficient $b$, for $\gamma = 0.3, 0.5$, and $0.7$.

solution exists. The error introduced by this approximation is less than 5% for Rent's rule exponents $b > 0.3$:

$$
\langle w \rangle \approx \frac{A_0}{2-2b} \left\{ \frac{1-b}{r_0^{2-2b}} \left[ \frac{1-b}{(2-2b)r_0^{2-2b}} \right]^{-\frac{3-b}{2-2b}} \Gamma \left[ \frac{3-b}{2-2b} \right] \right.
$$

$$
\left. -b \left[ \frac{1-b}{(2-2b)r_0^{2-2b}} \right]^{-\frac{b+1}{2-2b}} \Gamma \left[ \frac{b+1}{2-2b} \right] \right\}
\tag{13.44}
$$

The average wire length predicted by the model is shown in Fig. 13.10.

## 13.4.6   Area Limits on Average Wire Length

For a regularly tiled array of cells, the area of wire connected to each cell must be less than the area of the cell itself. This result is obtained by "folding back" all wires from a particular cell onto the cell itself, and can be expressed as

$$
\sum_{L=0}^{L_{max}} W L n(L) = A
\tag{13.45}
$$

where $W$ is the wire pitch (the minimum center-to-center distance between two wires), $L$ is the length of a given wire, and $n(L)$ is the number of wires of that length [8]. In the continuous limit, the summation is replaced by an integral using the probability distribution function computed above. If we attribute each wire equally to the two cells it connects, and we substitute $A = \pi r_c^2$, we obtain

$$\frac{W P_0 \langle w \rangle}{2} \leq \pi r_c^2 \tag{13.46}$$

giving

$$\langle w \rangle \leq \frac{2\pi r_c^2}{\gamma W P_0} \tag{13.47}$$

### 13.4.7  Summary

In this section, we have developed a simple wiring model based on an analysis of the intrinsic dimensionality of the interconnection medium. The essential assumption we made is that components are uniformly distributed throughout the spatial extent of the system. This postulate leads directly to the result that a power law relates available wiring density to component count. We have presented this result as a specific case of Rent's rule, a general empirical wire partition relation.

Based on Rent's rule, we have derived a closed-form approximation to the average wire length for cells in regular arrays. Finally, we have reintroduced physical wiring constraints to obtain yet another upper bound on the connectivity of cells.

### 13.5  A CONFIGURABLE ARCHITECTURE FOR PROTOTYPING ANALOG CIRCUITS

Even with fast-turnaround fabrication, a considerable fraction of prototyping lead-time is consumed by the VLSI fabrication step. This delay is particularly expensive when we consider prototyping subcircuits, or exploring new circuit structures and logic forms as a prelude to a large design project. The fabrication step is beyond the control of the designer, raising technical issues of yield and parametric behavior, as well as the logistic issues of cost, time, and the need to interface with an external vendor.

This section presents the design of a VLSI prototyping platform, in which the PROTOCHIP, a field-reconfigurable integrated circuit, is used to physically wire up the circuit to be tested. Field programmable analog arrays, discussed in Chapters 3 and 15, achieve similar goals. The PROTOCHIP is particularly intended for synthesis and testing of analog neural-network architectures [8], which generally require large numbers of active elements, but usually with relaxed precision requirements. These demands make neural networks a poor match for conventional software simulators. In addition to the obvious advantages of a hardware accelerator and simulator for fast prototyping, this system provides the first opportunity to attain a truly integrated design–fabricate–test environment.

The basic structure of the PROTOCHIP is a set of circuit elements arranged within a hierarchy of interconnect switches, facilitating the synthesis of the desired circuit. This programming process is controlled in the field by a workstation. Because this chip can effectively be considered an extension of a design-tool suite, similar performance criteria are applicable:

1. **Universality.** Given that the circuits to be prototyped generally are designed after the PROTOCHIP has been fabricated, can the user be reasonably assured the system is capable of testing a usefully wide range of circuits?
2. **Functionality and performance.** Do the prototyped circuits function correctly? How useful is the tool as a simulator?
3. **Programmability.** How directly can the PROTOCHIP be configured into the desired circuit for testing?
4. **Scalability.** Can the prototyping-chip architecture exploit the additional capacity of improving VLSI processes? In particular, can we exploit hierarchical architectures and modular programming algorithms, in order to remain efficient as transistor densities and counts increase.
5. **Flexibility.** Can the design of the PROTOCHIP itself be usefully adapted for synthesizing particular classes of circuit networks? The system can be custom-tailored at two levels: the basic circuit elements and interconnect structure are bound at fabrication time, and the actual PROTOCHIP is configured electrically in the field at runtime.

This section concentrates on three aspects of the PROTOCHIP's design. First, an analysis of a general model for circuit connectivity is used to explore the requirements on the interconnect matrix. The case for a hierarchical architecture is presented, based on a desire to minimize these wiring resources, as are some general observations regarding the organization of large systems. Second, interconnection switch technologies are discussed. Third, the structure and design of the fabricated PROTOCHIP is discussed, and test results are presented.

## 13.5.1 Connectivity of Circuits

The general form of the prototyping chip is a collection of functional elements interconnected by a programmable switch matrix. This matrix is responsible for both the great flexibility of the system and also is that system's most severe constraint. Because the design of the switch matrix is bound when the PROTOCHIP is fabricated, it needs to accommodate unanticipated circuits, while not incurring a prohibitive area penalty.

To this end, it is highly instructive to examine some general interconnect properties of circuits. This section uses Rent's rule as a connectivity model to investigate various wiring strategies. The ultimate objective is the development of a hierarchical interconnect matrix.

It is important to distinguish between topology and structure; the former deals exclusively with connectivity, the latter adds placement and layout constraints. The high cost of wiring in VLSI circuits makes consideration of structure imperative, as do several objectives for particular networks (e.g., an imaging array should be regularly placed).

The least restrictive model of circuit connectivity requires every element to be connectable to any other. Defining N to be the number of elements permitted on the chip, and $P_L$ to be the number of ports (pins) on each of these leaf elements, the required number of switches is

$$N_s = P_L^2 N^2 \qquad (13.48)$$

Of course, the requirement that any pin connect arbitrarily to any other pin (and possibly to any number of other pins) requires the full generality of a crossbar switch, with its associated quadratic overhead. We define a useful measure of circuit-wiring complexity to be

$$\beta = \frac{N_s}{P_L^2 N} \qquad (13.49)$$

which can be thought of as the number of other elements to which a particular element must fan out. For the crossbar,

$$\beta_{crossbar} = N \qquad (13.50)$$

Now consider an interconnect chip with $N$ elements. If the interconnect is flat (nonhierarchical), applying Rent's rule once gives the following number of outputs:

$$P_{out} = P_L N^b \qquad (13.51)$$

The interconnection matrix can then be realized by a crossbar with $P_L N$ inputs and $P_{out}$ outputs. The total number of switches is then

$$N_s = P_L^2 N^{b+1} \rightarrow \beta = N^b \qquad (13.52)$$

which, for large $N$, represents a sizable improvement over the full-crossbar interconnect.

Another approach to connecting the $N$ elements is incremental in nature: Suppose $k$ elements are connected simultaneously with a crossbar switch; then, $k$ of these meta-elements are connected. Repeating this procedure gives rise to the k-ary tree in Fig. 13.11, of height $L = log_k N$.

The number of switches required to implement the $i$th level of interconnect is then given by

$$
\begin{aligned}
N_{Si} = &\ (\text{inputs to each matrix at level } i \cdot (\text{output from level } i) \\
&\ (\text{number of } i\text{th level matrices}) \\
&\ \text{Level 1}: N_{S1} = (kP_L)(P_L k^b)(\tfrac{N}{k}) \\
&\ \text{Level 2}: N_{S2} = (k(P_L k^b))((P_L k^b)k^b)(\tfrac{N}{k^2}) \\
&\ \text{Level i}: N_{Si} == k P_{i-1}^2 k^b (\tfrac{N}{k^i})
\end{aligned}
\qquad (13.53)
$$

where

$$P_i = \text{outputs of matrix at level } i = \begin{cases} P_L k^b & \text{if } i = 1 \\ k^b P_{i-1} & \text{otherwise} \end{cases} \qquad (13.54)$$

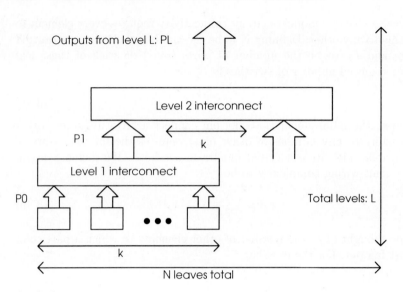

**FIGURE 13.11**

Hierarchical interconnect parameters. The total number of levels is $L = \lceil log_k N \rceil$; the outputs from the leaves are $P0 = P_L$, the outputs from the level 1 interconnect are $P1 = P_L k^b$, and the outputs from level L are $PL = P_L N^b$.

The solution to this recurrence is $P_i = P_L k^{bi}$, which gives the number of switches at level $i$ to be

$$N_{Si} = P_L^2 N k^{(2b-1)i} k^{1-b} \tag{13.55}$$

The total number of switches is the sum over all levels in the hierarchy:

$$N_s = \sum_{i=1}^{log_k N} N_{Si} = P_L^2 \frac{k^b}{k^{2b-1} - 1} (N^{2b} - N) \tag{13.56}$$

Giving

$$\beta = \frac{k^b}{k^{2b-1} - 1} (N^{2b-1} - N) \tag{13.57}$$

Now $2b - 1 < b$ for all $0 < b < 1$, and, asymptotically, this expression gives us fewer switches than the case of a single stage of interconnect. In addition, an interesting case results from $b = 0.5$, which yields the greatest savings

$$\beta = \frac{\sqrt{k}}{\ln k} \ln N \tag{13.58}$$

The number of switches for the hierarchical interconnect compared to the flat Rent's rule interconnect is plotted in Fig. 13.12 for $k = 2$.

We can obtain further gains by interleaving processing elements with the interconnect hierarchy. In general, for a tree with branching factor $k$, the addition

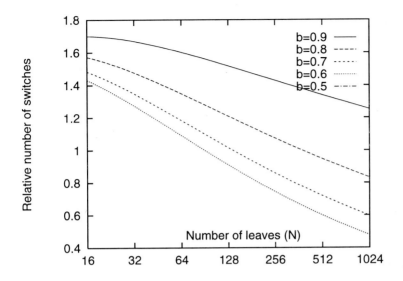

**FIGURE 13.12**
Number of switches in hierarchical interconnect relative to flat Rent's rule interconnect (plotted for $k = 2$).

of one fixed fanout element at each interconnect node yields

$$\beta = \beta_{hier} \frac{k-1}{k} \qquad (13.59)$$

where $\beta_{hier}$ is the quantity computed previously (and N is the number of leaves at the bottom of the tree).

## 13.5.2    Hierarchy and Abstraction in System Design

In addition to assisting in the management of the basic issues of wiring complexity and length, the hierarchy in the PROTOCHIP can be used to take advantage of system-level organization in the network to be embedded. In particular, the top-down design style expounded in structured design methodologies [7] can be exploited; we can simplify greatly the mapping from design to silicon by using for interconnect the same hierarchical specification present in the functional description of the design. In short, most circuits that we wish to prototype are hierarchical, and therefore the general PROTOCHIP architecture uses hierarchy as well. Furthermore, such circuits generally are easily scalable to larger systems, as is the PROTOCHIP architecture itself. Several other issues concerning hierarchy in system synthesis are discussed in the literature [9].

A hierarchical design has three other advantages. First, the electrical properties of the resulting circuits are improved, due to (1) lower capacitances, because the wires are shorter and there are fewer switches, and (2) the ability (and space) to place buffer stages within the interconnect itself. This possibility of integrating processing with interconnect also is appealing for networks that are inherently specified recursively (as are, e.g., many neural-network architectures). Second, the partitioning of a circuit into a hierarchy permits (1) nonuniform hierarchies to be tailored a priori to specific architectures (e.g., neural-networks, processing surfaces, systolic arrays), (2) high-density special cells to be incorporated, and (3) subtle embeddings of subcircuits to be stored in libraries and easily integrated into subsequent designs. Third, hierarchy simplifies the embedding problem by providing a convenient interface with CAD tools, most of which encourage or enforce a hierarchical style. In addition, most interchange formats (CIF, NTK, EDIF, etc.) intrinsically support hierarchy.

### 13.5.3 Embedding a Circuit Graph

The ultimate workability of this prototyping system is determined by the ease with which the desired circuit graph can be embedded onto the interconnect matrix. Basically, there are four approaches to this problem:

1. Exploit the hierarchy present in the circuit specification as much as possible. When this approach results in an unroutable system, resort to one of the other approaches to solve the embedding of the problem spot.
2. Exploit structural information available from the design tool. For example, a schematic editor also identifies approximate relative placement of subcircuits; schematics usually are drawn with similar objectives (e.g., minimal global and random wiring) as required by the embedding algorithm.
3. Ignore any hierarchy in the specification, and attempt to synthesize a partition of the circuit graph that can be embedded directly.
4. Accept a user-provided interactive routing strategy.

These approaches are described in detail in [13], and the universality of the hierarchical interconnect is investigated (and algorithms for embedding circuit netlists on the interconnect tree are developed and tested). Also, various technologies available for implementation of the switches are explored, and the electrical performance and system implications of CMOS transmission-gate and nonvolatile laser-programmed switches are considered. Finally, electrical performance specifications are presented. An example is shown in Fig. 13.16, actual performance of an analog delay line used in a continuous-time cochlear filter.

### 13.5.4 Analog Leaf Cell

One approach to maximizing the versatility of the PROTOCHIP while maintaining a uniform hierarchy is to make the leaf cells themselves programmable. The analog leaf of Fig. 13.13 consists of two blocks of partially preconnected n-transistors and p-transistors. Each programming switch connects a horizontal wire, which is connected to at least one transistor gate, to a vertical wire, which

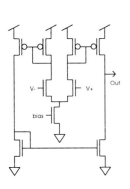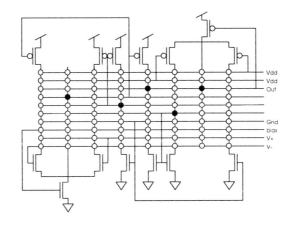

## FIGURE 13.13

Programmable analog leaf cell, configured as a wide-range transconductance amplifier. Each circle in the $10 \times 8$ matrix is a single programmable switch, which, when closed, connects the vertical and horizontal wires at that point. Applying the appropriate gate bias voltage disables the transistors that are not utilized.

is connected to at least one of a p-drain and an n-drain. Commonly used analog building blocks, such as current mirrors and differential pairs, are prewired in an attempt to reduce the complexity and size of the programming matrix. If a full crossbar were used to interconnect seven three-terminal p-transistors to an equal number of n-devices, $(3 \cdot 7)^2 = 441$ switches would be required; with the merged devices, only 80 switches are used. Yet the compact design is sufficient to construct a rich set of analog circuits, including wide-range transconductance amplifiers and horizontal resistors, as well as more conventional NAND/NOR and similar gates. In addition, a leaf cell is capable of configuring a fully independent floating three-terminal transistor.

### 13.5.5   Physical Placement of Leaves and Interconnects

The first version of the PROTOCHIP contained a binary tree of interconnect laid out as an H-tree (Fig. 13.14), in order to maximize the density of the switches and to simplify their programming. This structure permits the placement of the leaf cells on a regular grid. Address lines to set the switches are on a uniform mesh, with decoders located around the perimeter of the chip. We chose this approach over a hierarchical decoder because it is simpler. The mapping between hierarchical interconnect matrix coordinates and flat Cartesian coordinates is performed by the embedding compiler.

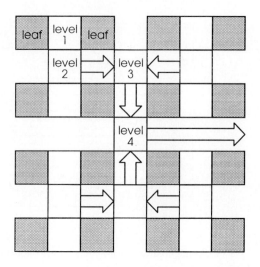

**FIGURE 13.14**

Binary H-tree hierarchical interconnect structure. The area of the silicon is divided between (1) leaf cells, (2) crossbar interconnect switches (shaded), (3) routing wires with optional buffer stages (arrows), and (4) space for processing elements interleaved in hierarchy.

### 13.5.6 Experiments with the PROTOCHIP

An example of the operation of the hierarchical interconnect PROTOCHIP is demonstrated by configuring a well-understood system: the follower–integrator analog delay line described in Chapter 8 of [8]. In theory, a delay line consisting of cascaded first-order sections (see Fig. 13.15) will exhibit a transfer function

$$\frac{V_n}{V_{in}} = \frac{e^{-jn}}{1 + \frac{n}{2}(\omega\tau)^2} \tag{13.60}$$

where $V_n$ is the output at the nth tap, $\tau$ is the time constant of a single stage, and $\omega$ is the frequency component of the signal in question.

Hence, at the cutoff frequency $\omega_c$ (the frequency at which the output amplitude is half the input amplitude), the delay-bandwidth product of the delay line is

$$\omega_c\tau = \sqrt{\frac{2}{n}} \tag{13.61}$$

Consequently, we expect the total phase delay to be a linear function of the distance along the delay line, and the rise-time of a step edge ($t_r$ is approximately the reciprocal of the bandwidth $\omega_c$) to be

$$t_r^2 = \frac{\tau^2 n}{2} \tag{13.62}$$

Experimental data are plotted in Fig. 13.16 and Fig. 13.17, illustrating the correct electrical behavior of a follower-integrator delay line constructed using the PROTOCHIP.

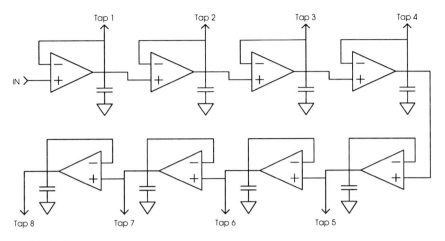

**FIGURE 13.15**
A follower–integrator delay line constructed of first-order sections.

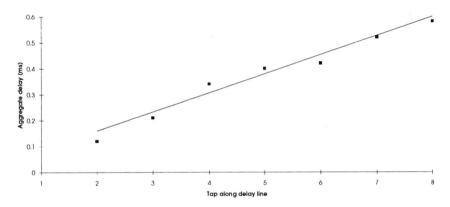

**FIGURE 13.16**
Waveform delay as a function of distance down the delay line.

### 13.5.7 Summary

The test PROTOCHIP has demonstrated the workability of a dynamically configurable array for prototyping analog circuits. The system described in this section is easy to use, provides fast turnaround (from schematic to ready-to-test chip in under 1 minute), and good performance and simulation accuracy, and is able to handle networks of large size.

The ability to specify and instrument a circuit interactively and to perform a simulation is a great convenience to the user. We feel that an integrated electrical prototyping tool would be a valuable adjunct to a VLSI design tool set, and that the PROTOCHIP is a first step toward a truly unified design environment.

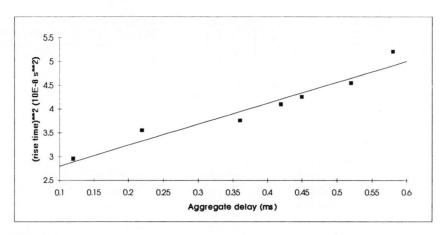

## FIGURE 13.17
Waveform (rise time)$^2$ as a function of delay for the follower–integrator delay line.

## PROBLEMS

**13.1.** Find the resistance between the inside and outside edges of the conductive ring shown in Fig. P13.1.

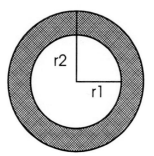

**FIGURE P13.1** A ring of conductive material (assume a sheet resistance $\rho$) has ideal circular contacts at radii $r1$ and $r2$. Find the resistancebetween these contacts.

**13.2.** Instead of the ring of Prob. 13.1, we now have a square region. Using numerical techniques, find the resistance between the inner and outer square (see Fig. P13.2).

**13.3.** Finding the resistance per square of diffused interconnect is complicated by the need to account for the change of doping, and hence resistivity, with distance into the substrate. In this problem, we present a simple model, and encourage the reader to work the problem through in closed form. In practice, such calculations would be performed numerically by process simulation tools.

(*a*) Find the resistance per square of a diffused wire, with a surface doping concentration of $5 \cdot 10^{16} \, cm^{-3}$ phosphorus atoms, diffused at 1100C for 1 hour. Hint 1: Model the conductivity of the silicon by $\sigma = \frac{N}{4 \cdot 10^{15}} (\Omega \cdot cm)^{-1}$. Hint 2: Model the doping profile by $N(z) = N_0 \, erfc(\frac{z}{2\sqrt{Dt}})$, where $D$ is the

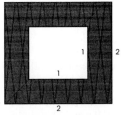

**FIGURE P13.2** Resistance of a rectangular region.

diffusion coefficient, and $t$ is the time over which the dopants were diffused. At $1100C°$, the diffusion coefficient of phosphorus is $3 \cdot 10^{-13}\,cm^2/sec$.

($b$) Find the resistance per square of a diffused wire, in the case of an unknown diffusion profile. What quantity must be given in order to solve this problem. Identify the limitations of this approach.

**13.4.** Explain how to find capacitance by using Laplace's equation to solve electric field distribution, and Gauss' law to compute the charge on a capacitor. Use this technique to find the capacitance per unit length of the coaxial cable shown in Fig. P13.4.

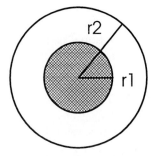

**FIGURE P13.4** The coaxial cable has a dielectric with relative permittivity 1.

**13.5.** Use the result of Prob. 13.3 to determine the capacitance of the structure in Fig. P13.5 numerically. (Hint: use the symmetry of the problem to bound the region that needs to be analyzed).

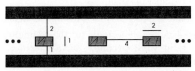

**FIGURE P13.5** Cross-section view of stripline. The central conductors are all at the same potential, as are the top and bottom surfaces. Find the capacitance between one of the central conductors and the top and bottom surfaces.

**13.6.** Use a circuit simulator (such as SPICE) to model a distributed RC line, and compare the results of a lumped model. For simplicity, assume a $1\Omega$ resistance and a 1F capacitance. In each case, measure the rise time and propagation delay.

**13.7.** Use SPICE to check the approximate calculation of propagation delay for RC trees (Fig. P13.5). Do the simulation for both lumped models and distributed RCs. Compare your results to those predicted by the model.

**13.8.** Use SPICE to compute the frequency domain transfer function for the lumped and distributed RC lines, and compare them to theoretical predictions.

## REFERENCES

[1] H.B. Bakoglu, *Circuits, Interconnect, and Packaging for VLSI*, Addison-Wesley, Reading, MA, 1990.

[2] S.N. Bhatt and F.T. Leighton, "A framework for solving VLSI graph layout problems," *Journal of Computer and System Sciences*, 28:300-343.

[3] W.H. Chang, "Analytical IC metal-line capacitance formulas," *IEEE Transactions on Microwave Theory Technology*, vol. MTT-24, pp. 608-611, September 1976.

[4] B. Hoeneisen and C.A. Mead, "Fundamental limitations on micro-electronics - 1. MOS Technology," *Solid-State Electronics*, vol. 15, 1972, pp. 819-829.

[5] B. Landman and R. Russo, "On a pin vs. block relationship for partitions of logic graphs," *IEEE Transactions on Computers*, pp. 1469-1479.

[6] T.-M. Lin, "A Hierarchical Timing Simulation Model for Digital Integrated Circuits and Systems," Ph. D. thesis, California Institute of Technology, 1984. 5133:TR:84.

[7] C.A. Mead and L. Conway, *Introduction to VLSI Systems*, Addison-Wesley, Reading, MA, 1980.

[8] C.A. Mead, *Analog VLSI and Neural Systems*, Addison-Wesley, 1989.

[9] S.M. Rubin, *Computer Aids for VLSI Design*, Addison-Wesley, 1987.

[10] T. Sakurai and K. Tamaru, "Simple formulas for two- and three-dimensional capacitances," *IEEE Transactions on Electron Devices*, vol. ED-30, pp. 183-185, February 1983.

[11] T. Sakurai, "Approximation of wiring delay in MOSFET LSI," *IEEE J. of Solid-State Circuits*, vol. SC-18, pp. 418-426, August 1983.

[12] M.A. Sivilotti, M.R. Emerling, C.A. Mead, *VLSI architectures for implementation of neural networks*, In: Neural Networks for Computing (Snowbird 1987), J. Denker, ed.

[13] M.A. Sivilotti, *Wiring Considerations in Analog VLSI Systems, with Application to Field-Programmable Networks*, Ph.D. thesis, California Institute of Technology, 1991.

[14] S.M. Sze, *Physics of Semiconductor Devices*, John Wiley & Sons, New York, 1981.

[15] C.D. Thompson, *A Complexity Theory for VLSI*, Ph.D. thesis, Carnegie-Mellon University, Technical Report CMU-CS-80-140, 1980.

[16] J.D. Ullman, *Computational Aspects of VLSI*, Computer Science Press.

[17] C.P. Yuan and T.N. Trick, "A simple formula for the estimation of the capacitance of two-dimensional interconnects in VLSI circuits," *IEEE Electron Device Letters*, vol. EDL-3, pp. 391-393, December 1982.

# CHAPTER
# 14

## STATISTICAL MODELING AND SIMULATION

## 14.1  INTRODUCTION

The development of precision analog circuits requires both a thorough understanding of basic circuit design techniques and a knowledge of the way in which
circuit performance is affected by transistor nonidealities. A key task of the
analog designer is to design around device imperfections and variations in the
fabrication process. Some nonidealities, such as mobility degradation in the MOS
transistor, can easily be included in simulation models. With the aid of circuit
simulators, such as SPICE [1], the effect of such nonideal transistor behavior can
be included in the design phase.

The effect of variations in the fabrication process is not easily characterized.
For instance, from wafer to wafer, the value of MOSFET threshold voltage may
vary by as much as $150mV$. Similarly two transistors of the same size on the
same wafer can have threshold voltages which differ by tens of millivolts. These
variations are random in nature, and must be described by statistical methods.

In this chapter, some basic statistical tools are presented which can aid
in the design of robust analog circuits, i.e., they are insensitive to variations in
the fabrication process. A few approaches to statistical modeling of the MOS
transistor are also included. An accurate, yet general, statistical device model
is required if statistical simulation is to eventually become a standard practice
among analog circuit designers. Finally, the statistical variation of some basic

615

analog circuit blocks will be calculated both analytically and through simulation as an example of the way statistical circuit analysis techniques can aid the circuit designer.

## 14.2 REVIEW OF STATISTICAL CONCEPTS

Before discussing statistical modeling and simulation, a few key ideas from statistics must be introduced. In this section, we first review some basic concepts involving probability density functions, focusing primarily on the normal distribution [2, 3]. Then, we derive the relationship between the variance of a function of random variables and the variances of the individual random variables. This technique, known as the transmission of moments [4], provides a very useful conceptual link between the performance variation of an integrated circuit and the underlying transistor parameter variations.

### 14.2.1 Random Variables and Probability Density Functions

All random variables (such as the threshold voltage, $V_T$ or current factor, $\beta$, of a randomly selected MOS transistor) can be characterized by a probability density function, or pdf. For a continuous random variable $X$, the pdf, $\phi(x)$, is defined by

$$P(x_1 < X < x_2) = \int_{x_1}^{x_2} \phi(x)dx \tag{14.1}$$

where $P(x_1 < X < x_2)$ is the probability that the value of the random variable $X$ lies somewhere between the values $x_1$ and $x_2$. Since $X$ is a real number, it is clear that

$$P(-\infty < X < \infty) = 1 = \int_{-\infty}^{\infty} \phi(x)dx \tag{14.2}$$

which implies that there is a probability of 1 (absolute certainty) that $X$ will take on some real number value.

For any function $f$ of the random variable $X$, we can define a quantity $E[f(X)]$ known as the expected value of $f(X)$

$$E[f(X)] = \int_{-\infty}^{\infty} f(x)\phi(x)dx \tag{14.3}$$

It follows directly from Eq. (14.3) that for any constant $a$, $E[af(X)] = aE[f(X)]$. In calculating the expected value of $f$, the values of $f(x)$ are weighted by their corresponding probability of occurrence, $\phi(x)$, and added together. This process results in an "average" value of $f(x)$. In other words, if $n$ occurrences of the random variable $X$ are measured (assuming each occurrence is independent of the others) and a value of $f(x)$ is associated with each occurrence, then the average of $f$ over the $n$ occurrences should converge to $E[f(X)]$ for large values of $n$.

We can use the concept of expected value to introduce two important quantities: the mean and variance of a random variable $X$. The mean of $X$ is defined as

$$Mean(X) = E[X] = \int_{-\infty}^{\infty} x\phi(x)dx \qquad (14.4)$$

From the discussion of expected values, it is clear that the mean of $X$ is the average or "most likely" value of the random variable $X$ (the maximum value of $\phi(x)$ does not necessarily occur at the mean). The variance of the random variable $X$ is defined as the expected value of the square of the difference of $X$ from the mean, that is

$$Var[X] = E[(X - E[X])^2] \qquad (14.5)$$

It is easy to show that $Var[X]$ can also be written

$$Var[X] = E[X^2] - (E[X])^2 \qquad (14.6)$$

The variance of the random variable $X$ is a measure of the *spread* of the pdf of $X$. Another commonly used measure of *spread* is the standard deviation, which is simply the square root of the variance.

## 14.2.2 The Normal Distribution

Most process and device parameter variations tend to be characterized by a special type of pdf known as a normal (or gaussian) distribution [5]. Due to the unique properties of the normal distribution, the approximation of normally distributed device parameters eases the tasks of statistical device modeling and circuit analysis.

The pdf of the normal distribution with parameters $\mu$ and $\sigma$ is given by

$$\phi_N(x) = \frac{1}{\sqrt{2\pi}\sigma} \exp[-\frac{1}{2}(\frac{x-\mu}{\sigma})^2] \qquad (14.7)$$

It is a simple exercise to show that if $X$ is a normally distributed random variable, then

$$E[X] = \mu, \qquad Var[X] = \sigma^2 \qquad (14.8)$$

Thus, the normal pdf, $\phi_N(x)$, is fully characterized by its mean, $\mu$, and its standard deviation, $\sigma$.

As discussed earlier, device parameters such as the threshold voltage, $V_T$, the current factor, $\beta$, or the body effect parameter, $\gamma$, of a randomly selected MOS transistor can be characterized by the normal distribution. However, since the variations in the transistor parameters are all caused by the same set of process variations, the transistor parameters variations are not independent of each other. For example, since both the threshold voltage and the body effect parameter of an NMOS transistor are increasing functions of doping density, a random variation in the substrate doping will tend to affect both $V_T$ and $\gamma$ in the same way. Thus, instead of being characterized by a set of independent, single-variable density functions, the entire set of transistor parameters is properly characterized by a joint pdf.

If $X$ and $Y$ are random variables, they can be characterized by a joint pdf, $\phi(x, y)$, such that [6, 7]

$$P[(x_1 < X < x_2) \bigcap (y_1 < Y < y_2)] = \int_{x_1}^{x_2} \int_{y_1}^{y_2} \phi(x, y) dy dx \qquad (14.9)$$

If $X$ and $Y$ are jointly normally distributed, their pdf, $\phi_N(x, y)$, is given by

$$\phi_N(x, y) = \frac{1}{2\pi\sigma_r\sigma_y\sqrt{1 - \rho_{xy}^2}} \exp\{\frac{-1}{2(1 - \rho_{xy}^2)}[\frac{(x - \mu_x)^2}{\sigma_x^2}$$

$$+ \frac{(y - \mu_y)^2}{\sigma_y^2} - \frac{2\rho_{xy}(x - \mu_x)(y - \mu_y)}{\sigma_x\sigma_y}]\} \qquad (14.10)$$

The joint pdf, $\phi_N(x, y)$, is completely characterized by the parameters $\mu_x$, $\mu_y$, $\sigma_x$, $\sigma_y$, and $\rho_{xy}$. The first four of these are the means and standard deviations of the variables $X$ and $Y$, while the parameter $\rho_{xy}$ is known as the correlation between $X$ and $Y$, and is defined as [2, 3]

$$\rho_{xy} = E[\frac{(x - \mu_x)}{\sigma_x} \frac{(y - \mu_y)}{\sigma_y}] \qquad (14.11)$$

It can be shown that the correlation between two random variables is always between -1 and 1. The correlation is a measure of how closely related are the random variables $X$ and $Y$. That is, if $\rho_{xy}$ is close to 1, then $X$ and $Y$ tend to be high or low at the same time. Similarly, if $\rho_{xy} = -1$, $Y$ tends to be low (high) whenever $X$ is high (low). Finally, if $\rho_{xy} = 0$, $X$ and $Y$ are independent.

For a set of N interrelated random variables, $X_1, \ldots, X_N$, we can define a joint pdf for all N variables. If the $X_i$ are jointly normally distributed, the joint pdf is characterized by the set of N means, $\mu_{r_i}$, the set of N standard deviations, $\sigma_{r_i}$, and the set of N(N-1)/2 correlations, $\rho_{r_i r_j}$ with $i \neq j$.

## 14.2.3   Functions of Random Variables; The Transmission of Moments Formula

For statistical modeling and analysis, one of the most important properties of the normal distribution is the variance of a linear function of normally distributed variables. If $Y$ is the linear combination of N independent, normal random variables

$$Y = a_1 X_1 + a_2 X_2 + \ldots + a_N X_N \qquad (14.12)$$

then, the variance of $Y$ is

$$Var[Y] = \sum_{i=1}^{N} a_i^2 \sigma_{r_i}^2 \qquad (14.13)$$

If the $X_i$ is not independent, then the variance of $Y$ also depends on the correlation parameters, $\rho_{r_i r_j}$,

$$Var[Y] = \sum_{i=1}^{N} a_i^2 \sigma_{r_i}^2 + 2 \sum_{i=1 j=i+1}^{N} a_i a_j \rho_{r_i r_j} \sigma_{r_i} \sigma_{r_j} \qquad (14.14)$$

Next, suppose $Y = f(X_1, \ldots, X_N)$ is a nonlinear function of the $X_i$'s. We can expand $Y$ in a first-order Taylor series

$$Y = f(x_1, \ldots, x_N) \approx f(\mu_{r_1}, \ldots, \mu_{r_N}) + \sum_{i=1}^{N} \frac{\partial f}{\partial x_i} \Delta x_i \qquad (14.15)$$

where the partial derivatives $\frac{f}{r_i}$ are evaluated at $\mu_{r_i}$; the mean values of the $X_i$'s. Equation (14.15) approximates the nonlinear function $f$ by a linear function. Equation (14.14) can then be used to approximate the variance of $Y$ as [4]

$$\sigma_y^2 = \sum_{i=1}^{N} \left(\frac{\partial f}{\partial x_i}\right)^2 \sigma_{r_i}^2 + 2 \sum_{i=1 j=i+1}^{N} \frac{\partial f}{\partial x_i} \frac{\partial f}{\partial x_j} \rho_{r_i r_j} \sigma_{r_i} \sigma_{r_j} \qquad (14.16)$$

This is known as the transmission of moments formula. Since its derivation depended on the representation of the nonlinear function $f$ by a first-order Taylor series, the use of Eq. (14.16) can lead to significant errors if the variances of the $X_i$'s are large.

The assumption of normally distributed device characteristics will be used throughout this chapter. With this assumption of statistical modeling or analysis of analog circuits requires arranging the governing device equations in such a way that Eq. (14.14) or Eq. (14.16) can be applied. For example, we could calculate the variance of the drain current of a MOSFET as a function of the variances of such transistor parameters as the threshold voltage or the current factor.

### 14.2.4   Monte Carlo Simulation

The transmission of moments formula in Eq. (14.16) is a straightforward way to calculate the variance of an output as a function of the means and variances of the input variables. Theoretically, it is possible to relate the outputs of any integrated circuit to the device parameters of the transistors from which it is formed. However, on a practical level, this is difficult, if not impossible, for most integrated circuits. For example, it would be extremely difficult to find a closed-form function which relates the open-loop gain of an operational amplifier to the BSIM parameters of its transistors.

A technique known as Monte Carlo simulation [4] can get around this difficulty. We will first illustrate Monte Carlo simulation with a function of several independent random variables and then explain its use with a circuit simulator to investigate random variations in integrated circuits.

Suppose we want to find the mean and variance of a normal random variable with mean $\mu_{X_1}$ and variance $\sigma_{X_1}$. Using an ideal random number generator, we could *generate* a set of a large number of samples of the variable $X_1$, i.e., $\{X_{11}, X_{12}, \ldots, X_{1M}\}$. For large M (1000 or more), the sample mean and variance of this finite population will be close to the actual mean and variance of the variable $X_1$. If we evaluate $f$ at each of the $X_i$ sample points, that is, $f_i = f(X_i)$, we would find that the sample mean and variance of the set $\{f_1, \ldots, f_M\}$ are close to the actual mean and variance of the random function $f(X_i)$.

This technique could easily be generalized to a function of several random variables $f(X_1, \ldots, X_N)$. We can use the random generator to obtain M random samples of the parameters, $[(X_{11}, X_{21}, \ldots, X_{N1}), \ldots, (X_{1M}, X_{2M}, \ldots, X_{NM})]$. (Note that each sample contains one value of each of the random variables $(X_1, X_2, \ldots, X_N)$). As before, we would evaluate the function $f$ at each sample point

$$f_1 = f(X_{11}, X_{21}, \ldots, X_{N1})$$

$$f_2 = f(X_{12}, X_{22}, \ldots, X_{N2})$$

$$\vdots$$

$$f_M = f(X_{1M}, X_{2M}, \ldots, X_{NM}) \tag{14.17}$$

and calculate the sample mean and variance of $\{f_1, f_2, \ldots, f_M\}$. For large M, we would find that this sample mean and variance closely approximate the actual mean and variance of $f$.

The real value of Monte Carlo simulation lies in the fact that it does not require a closed-form function, $f$, relating the output to the set of input parameters $(X_1, X_2, \ldots, X_N)$. For example, if the variables $X_i$ are transistor parameters (for example $V_T$, $\beta$, $\gamma$, etc.), Monte Carlo sampling could be used to generate random values of each parameter with its correct mean and variance. Then, the value of any output function (such as open-loop gain or input offset voltage of an op amp) can be *evaluated* for each parameter sample using a circuit simulator such as SPICE or APLAC [8]. The sample mean and variance of this output set will then be very close to the actual mean and variance of the output function.

In order to perform accurate Monte Carlo simulations of integrated circuits, we must insure that the random values for the transistor parameters produced by the random number generator follow the correct probability distribution. That is, for large numbers of samples (greater than 1000), the sample mean and variance of each generated parameter should be very close to the actual mean and variance of the parameter. In addition, the sample correlations between generated parameters should be close to the actual parameter correlations.

Many techniques exist for generating independent, normally distributed random numbers with a given mean and variance. This subject is beyond the scope of this work and will not be discussed in any further detail. However, the next section will discuss a technique known as Principal Component Analysis (PCA) which can preserve the correlations between a set of normally distributed random parameters. PCA will enable us to generate sets of correlated transistor parameters and thus perform Monte Carlo simulations of integrated circuits.

## 14.3 CORRELATIONS AND PRINCIPAL COMPONENT ANALYSIS

As discussed in a previous section, random variations in MOS process parameters, such as oxide thickness or doping density, affect more than one transistor parameter [9]. Thus, the device parameters of an MOS transistor, such as threshold voltage or mobility, are correlated. An accurate statistical device model must describe the correlations between parameters, as well as the parameter means and variances.

To illustrate the need to account for parameter correlations within a single transistor, a scatter plot of two model parameters, the flat band voltage (VFB) and the body effect coefficient (BE) is presented in Fig. 14.1, while a similar plot of VFB and the linear region mobility (MUS) is shown in Fig. 14.2. Each point on both plots corresponds to the *measured* values of VFB and BE or VFB and MUZ for one transistor from a random sample of similar transistors on the same die. It should be noted that the assumption that these three parameters are normally distributed appears to be valid.

The scatter plot of Fig. 14.1 indicates a very high correlation ($\rho = -0.897$) between measured values of VFB and BE. This correlation is most likely due to each parameter's dependence on the same underlying process parameters, i.e., the substrate doping and the gate oxide thickness. Clearly, the correlation between VFB and BE makes certain combinations of the two parameters far more probable than others. For instance, the values (BE, VFB) = (1.03, -0.804), (1.03, -0.796), (1.038, -0.804), and (1.038, -0.796) are all equidistant from the mean point, (BE, VFB) = (1.034, -0.80). Thus, if VFB and BE were independent, all four points would have equal probability of occurrence. However, from the scatter plot, the first and last points are clearly much less probable combinations. The scatter plot of Fig. 14.2 shows a very low correlation ($\rho = 0.289$) between measured values of VFB and MUZ. Clearly, the joint distribution between two nearly uncorrelated parameters is much different from the joint distribution of two highly correlated parameters.

### 14.3.1 Principal Component Analysis

If $X$ and $Y$ are unit-normal random numbers (that is, have a normal distribution with mean = 0 and variance = 1), with correlation $\rho$, it is possible to write $X$ and $Y$ as linear combinations of two independent unit-normal random numbers, $C_1$ and $C_2$. That is,

$$X = \sqrt{\frac{1+\rho}{2}}C_1 + \sqrt{\frac{1-\rho}{2}}C_2 \tag{14.18}$$

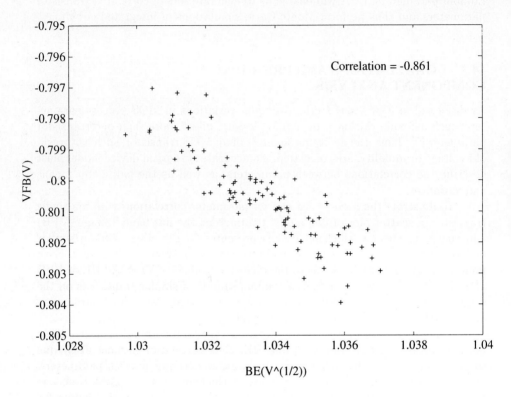

**FIGURE 14.1**
NMOS flat band voltage, VFB, and body effect coefficient, BE.

and,

$$Y = \sqrt{\frac{1 + \rho}{2}} C_1 - \sqrt{\frac{1 - \rho}{2}} C_2 \tag{14.19}$$

Since $C_1$ and $C_2$ are unit-normal random numbers, $E[C_1^2] = E[C_2^2] = 1.0$. In addition, $C_1$ and $C_2$ are independent, so that $E[C_1 C_2]$ is zero.

Using this information, the variance of $X$ in Eq. (14.18) is

$$\sigma_X^2 = E[(X - \mu_X)^2] = E[X^2]$$

$$= E[(\frac{1 + \rho}{2}) C_1^2 (\frac{1 - \rho}{2}) C_2^2 + \sqrt{1 - \rho^2} C_1 C_2]$$

$$= (\frac{1 + \rho}{2}) E[C_1^2] + (\frac{1 - \rho}{2}) E[C_2^2] = \frac{1 + \rho + 1 - \rho}{2} = 1 \tag{14.20}$$

Similarly, the variance of $Y$ in Eq. (14.19) is $\sigma_Y^2 = 1$. In addition, the correlation between $X$ and $Y$ in Eqs. (14.18) and (14.19) is

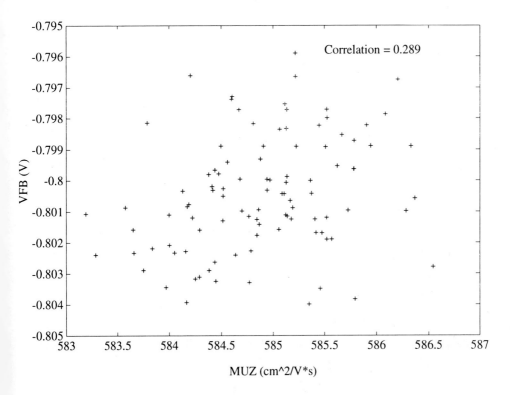

**FIGURE 14.2**
NMOS flat band voltage and linear region mobility.

$$E[\frac{(X - \mu_X)}{\sigma_X} \frac{(Y - \mu_Y)}{\sigma_Y}] = E[XY] = E[(\frac{1 + \rho}{2})C_1^2 - (\frac{1 - \rho}{2})C_2^2]$$

$$= (\frac{1 + \rho}{2}) - (\frac{1 - \rho}{2}) = \rho \qquad (14.21)$$

Thus, the linear combinations in Eqs. (14.18) and (14.19) preserve the correlation between the random variables $X$ and $Y$, as well as their means and variances.

The technique shown above can be generalized to the case of $N$ correlated unit-normal random variables $(X_1 \ldots X_N)$ using a technique known as Principal Component Analysis (PCA) [6, 10, 11, 12, 13]. PCA provides a linear transformation between a set of $N$ independent unit-normal random numbers $(C_1 \ldots C_N)$ called principal components and the original correlated unit-normal random numbers $\{X_i\}$.

Let $\rho$ be the $N \times N$ symmetric matrix whose components are defined as follows

$$\rho_{ij} = 1 \qquad i = 1, \ldots, N$$

$$\rho_{ij} = E[X_i X_j] \qquad i \neq j \tag{14.22}$$

In other words, the $ij^{th}$ component of $\rho$ is the correlation between the variables $X_i$ and $X_j$. This is known as the correlation matrix for the $X_i$ variables.

Principal component analysis states that the linear transformation

$$\mathbf{X} = \mathbf{U}\mathbf{\Lambda}^{12}\mathbf{C} \tag{14.23}$$

maps the vector of independent unit-normal variables (principal components), $\mathbf{C} = [C_1, \ldots, C_N]^T$ into the vector of correlated random numbers, $\mathbf{X} = [X_1, \ldots, X_N]^T$. $\mathbf{\Lambda}$ is a diagonal matrix containing the eigenvalues of the correlation matrix, and $\mathbf{U}$ is the matrix of normalized (that is $U_i^T U_i = 1$) eigenvectors corresponding to the elements in $\mathbf{\Lambda}$.

Equation (14.23) relates the variables $X_i$ to the principal components $C_i$ by a set of linear equations

$$X_1 = A_{11}C_1 + A_{12}C_2 + \ldots + A_{1N}C_N$$

$$\vdots$$

$$X_N = A_{N1}C_1 + A_{12}C_2 + \ldots + A_{NN}C_N \tag{14.24}$$

Often, the coefficients of all but the first M principal components in Eq. (14.24) are very small. Thus, we can neglect these $N - M$ insignificant principal components and approximate the $X_i$ as

$$X_1 = A_{11}C_1 + A_{12}C_2 + \ldots + A_{1M}C_M$$

$$\vdots$$

$$X_N = A_{N1}C_1 + A_{N2}C_2 + \ldots + A_{NM}C_M \tag{14.25}$$

### 14.3.2   Correlations Between MOS Transistors

As mentioned before, the parameters of a randomly selected MOS transistor form a set of correlated normal random numbers. Thus, we can use PCA to relate the correlated transistor parameters to a set of independent unit-normal principal components. However, since the PCA expression of Eq. (14.23) works with unit-normal variables, we must first form a set of unit-normal parameters from the transistor parameters.

Clearly, for each transistor parameter, $P$, with mean, $\mu_P$, and standard deviation, $\sigma_P$, the transformation

$$P' = \frac{P - \mu_P}{\sigma_P} \tag{14.26}$$

produces a unit-normal random number. The set of unit-normal random numbers generated by Eq. (14.26) has the same correlation matrix as the original transistor parameters.

**Example 14.1.** Suppose that the correlation matrix for the BSIM parameters VFB, K1, K2, and MUZ extracted from a set of NMOS transistors is shown in Tab. 14.1. After calculating the eigenvalues and eigenvectors of the correlation matrix, Eq. (14.23) gives the following set of equations relating the BSIM parameters normalized by Eq. (14.26) to a set of independent unit-normal principal components $\{C_1, C_2, C_3, C_4\}$

$$VFB' = -0.903C_1 + 0.236C_2 + 0.361C_3$$

$$K1' = 1.00C_1$$

$$K2' = 0.875C_1 + 0.340C_2 + 0.345C_3$$

$$MUZ' = -0.102C_1 + 0.973C_2 - 0.208C_3 \qquad (14.27)$$

The coefficients of the fourth principal component in the above equations were so small that $C_4$ can be neglected.

PCA accounts for parameter correlations through a set of linear equations; any nonlinear relationships between the original variables will be approximated as linear in the PCA analysis. The linear approximation of nonlinear relationships between MOSFET model parameters should be sufficient for our purposes, although some error will be incurred for large relative parameter variations.

## 14.4  STATISTICAL DEVICE MODELING

Random variations in integrated circuit processes result in random variations in transistor parameters [14]. For example, fabrication equipment tends to drift over time, resulting in a variation in the average values of transistor parameters from one wafer lot to the next. Parameter values on a given wafer may also depend on its location in a multi-wafer fabrication machine. Furthermore, gradients in temperature and implantation density and angle cause the parameters of an individual die to be dependent on relative placement of the die on a wafer.

For the purposes of statistical circuit modeling, the lot-to-lot, wafer-to-wafer, and die-to-die variations of transistor parameters described above can be lumped together into a single term known as interdie parameter variation [5]. If we assume that transistor parameters are normally distributed, the interdie variation of any parameter, P, is completely characterized by its variance, $\sigma_{inter}^2(P)$, and the mean value of the parameter for the entire process, $\mu_{process}(P)$.

**TABLE 14.1**
Correlation matrix for PCA calculation example.

|     | VFB | K1 | K2 | MUZ |
|-----|-----|-----|-----|------|
| VFB | 1.0 | -0.9011 | -0.5842 | 0.2467 |
| K1 | -0.9011 | 1.0 | 0.8837 | -0.0884 |
| K2 | -0.5842 | 0.8837 | 1.0 | 0.1693 |
| MUZ | 0.2467 | -0.0884 | 0.1693 | 1.0 |

In addition to interdie parameter variation, random process variations can cause parameter mismatch between two same-size transistors on the same die. This phenomenon, known as intradie parameter variation, results from a variety of sources such as local variation of oxide trapped charge or variation of gate oxide thickness.

From a modeling perspective, interdie variability can be modeled easily. Since it affects all transistors in a given circuit equally, it can be represented as an equal deviation in the parameter value of every transistor in the circuit. Transistor mismatch is much more difficult to model than interdie parameter variation, since the degree of mismatch between two devices in a circuit is dependent on both the size and relative location of the devices.

Many previous statistical simulation techniques and tools contend only with interdie variances [15, 16, 9, 17]. Others model intradie transistor mismatch through a few lumped statistics such as matched and die-level parameter variances [18]. These models do not account for the effects of circuit layout or device size on parameter mismatch variances.

For a standard digital logic gate, both power dissipation and delay time are dependent on raw transistor current values; small mismatches of transistor characteristics between two transistors on the same chip have negligible effect on the gate operation. Thus, interdie parameter variation has a much larger effect on the statistical distribution of digital circuit performance than parameter mismatch.

In analog circuits, both interdie and intradie variability contribute greatly to variances in circuit performance. While variations in many circuit specifications, such as the gain-bandwidth, the slew rate, and the phase margin, are caused mostly by interdie process variation, variations in other circuit outputs are more dependent on device mismatch. One key circuit specification which is wholly dependent on mismatch is the input offset voltage [19]. It is extremely important for realizing high precision data converters, voltage references, and comparators. Therefore, a statistical model which includes both interdie variability and device mismatch is necessary for accurate statistical simulations of analog circuits.

### 14.4.1 Models for Interdie Variation

The variation of most circuit performance criteria is primarily due to interdie process variations. The statistical uncertainty in signal delay and power dissipation of a logic gate and the gain-bandwidth and phase margin of an operational amplifier are most strongly affected by die-to-die variations of device characteristics. For these performance criteria, simple statistical parameter models which only account for interdie parameter variances can be employed to describe circuit variability.

As mentioned before, interdie parameter variation can be represented as a random perturbation in the device parameters of every transistor on a given die. Since interdie variation does not introduce any mismatch in transistor parameters, this perturbation has the exact same value for all transistors on the die. For example, if the threshold voltage of a given process has a mean value $\mu_{process}(V_T)$ and a variance of $\sigma^2_{inter}(V_T)$ caused by interdie parameter variation, then the threshold voltage of every transistor on a randomly-selected die fabricated in the process can be written

$$V_T = \mu_{process}(V_T) + \sigma_{inter}(V_T)R_{inter}(V_T) \tag{14.28}$$

Similarly, the current factor of every transistor can be written

$$\beta = \mu_{process}(\beta) + \sigma_{inter}(\beta)R_{inter}(\beta) \tag{14.29}$$

where $R_{inter}(V_T)$ and $R_{inter}(\beta)$ are unit-normal random numbers. It is important to note that the random numbers $R_{inter}(V_T)$ and $R_{inter}(\beta)$ have the same correlation coefficient as $V_T$ and $\beta$.

One simple method to perform interdie variance simulations on an integrated circuit is to choose a few physically meaningful statistically independent (uncorrelated) principal factors, such as the channel length, channel width, flat-band voltage, and gate oxide thickness [9], to describe the MOSFET. Since these parameters are independent, they can be generated independently in a Monte Carlo sampling scheme (see next section) without concern for preservation of parameter correlations. Unfortunately, determining a complete set of principal factors which can accurately describe the performance variation of every circuit is difficult if not impossible.

A second method to model interdie process variations is through Monte Carlo simulations with independent process parameters such as diffusion sheet resistivities, diffusion profiles, and gate oxide thickness [14]. These process parameters can be experimentally related to the device model parameters such as $V_T$ and $\beta$ to allow circuit simulations. Once this link is determined, there is no need for statistical techniques such as PCA since the process parameters generated by Monte Carlo sampling are independent. Unfortunately, finding accurate relationships between process parameters and device parameters is not an easy task. If the equations relating the device parameters to the process parameters are not accurate, then circuit simulations based on the Monte Carlo sampled parameters will not accurately represent the statistical variation of the circuit performance.

As we shall see in an upcoming section, Principal Component Analysis gives us a way to use Monte Carlo sampling techniques directly on correlated device parameters such as $V_T$, $\beta$, and $\gamma$. Working directly with device parameters avoids the need to find the relationship between process and device parameters, and allows for accurate circuit simulations.

### 14.4.2  Models for Intradie Variations

Many important performance criteria for analog circuits, including the input offset voltage and harmonic distortion of an amplifier, depend strongly on device parameter matching. In order to include random mismatch between devices within a circuit, the statistical model must produce a different set of model parameters for each transistor in a circuit.

Empirical evidence has consistently shown that the variance of the parameter mismatch between two transistors *increases* with decreasing gate area and with increasing separation distance between devices. In other words, small, widely spaced devices have a statistically larger mismatch than large, closely spaced devices. The work of Pelgrom and others [20, 21, 22] showed that the variance of the mismatch between transistors 1 and 2 of a general model parameter, P, can be represented by

$$\sigma^2(P_1 - P_2) = \frac{a_P}{2W_1 L_1} + \frac{a_P}{2W_2 L_2} + s_p^2 D_{12}^2 \qquad (14.30)$$

where $D_{12}$ is the distance between transistors 1 and 2, $W_1 L_1$ and $W_2 L_2$ are the gate areas of transistors 1 and 2, and $a_P$ and $s_P$ are process dependent fitting constants.

Three properties of the parameter mismatch variance model given in Eq. (14.30) make it ideal for use in a statistical modeling scheme. First, the variance model was derived for a general parameter and, therefore, should be valid for any set of MOSFET parameters including level 2 ($V_T$, $K_P$, $\gamma$, $\lambda$, etc.) and BSIM (VFB, K1, K2, MUZ, etc.). Second, the mismatch variance model includes two of the three circuit parameters which have the greatest effect on the variability of analog circuits: device size and circuit layout. The effect of biasing on circuit variability is preserved by the circuit simulator as will be shown in a later section. Third, the three terms in the mismatch variance model are independent and are assumed to be normally distributed; therefore, they can be accounted for separately when calculating transistor parameter sets in a Monte Carlo scheme.

### 14.4.3  A Combined Statistical Device Model

The SMOS (Statistical MOS) model [13, 23, 24, 25, 26, 27] combines interdie and intradie statistical device models in a manner compatible with existing circuit simulators. Using this model, parameters in standard device models, such as SPICE level 2 or BSIM, can be calculated for each transistor in way which accounts for interdie and intradie parameter variations while preserving correlations between parameters within the same device. The SMOS model for a general parameter, P (which can represent parameters such as $V_T$, $\beta$, $V_{FB}$, etc.), of the M'th transistor of a circuit can be written as

$$P_M = \mu_{inter}(P) + \sqrt{\frac{a_P}{2W_M L_M}} R_P(M) + s_P(X_M R_{PX} + Y_M R_{PY}) \qquad (14.31)$$

where

$$\mu_{inter}(P) = \mu_{process}(P) + \sigma_{inter}(P) R_{Pinter} \qquad (14.32)$$

The $\mu_{inter}(P)$ term models the interdie variation of the parameter P about the process mean, $\mu_{process}(P)$. This term is included in the parameter mean model since for a given circuit simulation every transistor in the circuit has the same $\mu_{inter}(P)$ value. The remaining terms in Eq. (14.31) model the intradie variation of the values of P about the chip mean, $\mu_{inter}(P)$ for each transistor in the circuit. The distance dependence in Eq. (14.30) is included through $X_M$ and $Y_M$, the coordinates of the centroid of the gate of transistor M with respect to the centroid of the circuit layout. The $R_P$ variables are unit normal random numbers which preserve parameter correlations. For example, if we use a level 2 MOS model, the value of $R_{V_T}$ for each transistor should be correlated with the value of $R$ for the same transistor. These random numbers can be generated using PCA, given a knowledge of the parameter correlation matrix for the set of parameters being modeled from a specific process line.

**Example 14.2. Parameter Calculation using the SMOS Model.** A circuit layout, shown in Fig. 14.3 provides information concerning the gate area and centroid position of two transistors, represented by rectangles. Note that the coordinates of the centroid of each transistor are given in $\mu m$ relative to an arbitrary origin. Table 14.2 supplies the parameter statistics required by the SMOS model. Four BSIM [28] parameters, VFB (flat-band voltage), K1 (body effect coefficient), K2 (charge sharing coefficient), and MUZ (linear region mobility) will be calculated to preserve the statistics from Tab. 14.2 and the parameter correlations in Tab. 14.1.

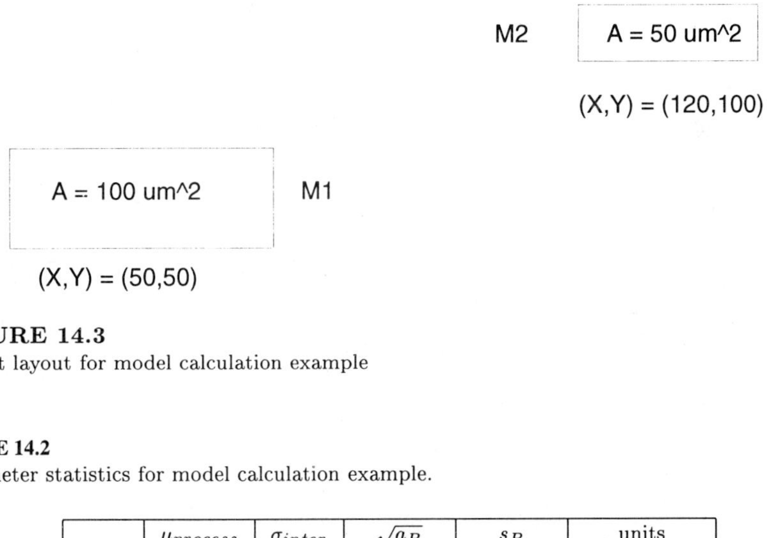

M2    A = 50 um^2

(X,Y) = (120,100)

A = 100 um^2    M1

(X,Y) = (50,50)

**FIGURE 14.3**
Circuit layout for model calculation example

**TABLE 14.2**
Parameter statistics for model calculation example.

|  | $\mu_{process}$ | $\sigma_{inter}$ | $\sqrt{a_P}$ | $s_P$ | units |
|---|---|---|---|---|---|
| VFB | -0.7923 | 0.08 | 0.0623 | 7.33e-05 | V |
| K1 | 1.0631 | 0.11 | 0.0714 | 4.53e-05 | $V^{1/2}$ |
| K2 | 0.1297 | 0.01 | 0.0178 | 7.00e-06 | none |
| MUZ | 611.9 | 60.0 | 27.9 | 0.0399 | $cm^2/(V*s)$ |
|  | unit | unit | unit*um | unit/um |  |

Inserting the layout information and parameter statistics into Eq. (14.31) results in a set of parameter equations for M1 and M2

$M1:$

$$VFB_1 = -0.7923 + 0.08VFB'_{inter} + 4.41 * 10^{-3}VFB'(1)$$
$$+3.67 * 10^{-3}VFB'_X + 3.67 * 10^{-3}VFB'_Y$$

$$K1_1 = 1.063 + 0.11K1'_{inter} + 5.05 * 10^{-3}K1'(1)$$
$$+2.27 * 10^{-3}K1'_X + 2.27 * 10^{-3}K1'_Y$$

$$K2_1 = 0.1297 + 0.01K2'_{inter} + 1.26 * 10^{-3}K2'(1)$$
$$+3.50 * 10^{-4}K2'_X + 3.50 * 10^{-4}K2'_Y$$

$$MUZ_1 = 611.9 + 60.0MUZ'_{inter} + 1.97MUZ'(1)$$
$$+2.00MUZ'_X + 2.00MUZ'_Y$$

$M2:$

$$VFB_2 = -0.7923 + 0.08VFB'_{inter} + 6.23 * 10^{-3}VFB'(2)$$
$$+8.80 * 10^{-3}VFB'_X + 7.33 * 10^{-3}VFB'_Y$$

$$K1_2 = 1.063 + 0.11K1'_{inter} + 7.14 * 10^{-3}K1'(2)$$
$$+5.44 * 10^{-3}K1'_X + 4.54 * 10^{-3}K1'_Y$$

$$K2_2 = 0.1297 + 0.01K2'_{inter} + 1.78 * 10^{-3}K2'(2)$$
$$+8.40 * 10^{-4}K2'_X + 7.00 * 10^{-4}K2'_Y$$

$$MUZ_2 = 611.9 + 60.0MUZ'_{inter} + 2.97MUZ'(2) +$$
$$4.80MUZ'_X + 4.00MUZ'_Y \tag{14.33}$$

To preserve parameter correlations, the primed variables in Eq. (14.33) are generated using the PCA equations shown in Eq. (14.27). Thus, inserting the PCA equations into Eq. (14.33) results in an equation relating the model parameters to the independent, unit-normal principal components $C_1, C_2, C_3$. For this example, five sets of principal components (15 total random numbers) are required to calculate a model deck for these two transistors. Each additional transistor will require one more set of computer-generated principal components.

## 14.5  STATISTICAL CIRCUIT SIMULATION

Once a statistical model which meets the needs of the user has been developed, a method to employ this model in statistical circuit simulations must be chosen. Three statistical simulation techniques will be considered in this section: worst-case analysis, Monte Carlo simulation, and response surface methodology (RSM). Worst-case analysis is both a statistical modeling and simulation technique, while Monte Carlo simulation and RSM require an accurate statistical device model in order to produce accurate circuit output distributions. The advantages and limitations of each technique will be discussed for the case of analog MOS integrated circuits.

### 14.5.1  Worst-Case Analysis

Worst case analysis is the simplest and most commonly used technique to perform statistical circuit analysis [4, 29, 30]. To determine the worst-case circuit performance, circuit simulations are performed at points in the parameter space which may be termed process corners. The process corners are defined as the $\pm 3\sigma$ (that is, three standard deviations above and below the mean) values of a representative device characteristic. Often, the drain saturation current ($I_{Dsat}$) at a given bias, e.g., $V_{GS} = V_{DS} = 5V$, is chosen to be the representative characteristic because of its relation to the propagation delay through a digital gate. Model decks are then found to represent the $+3\sigma$ values (fast-p, fast-n) and $-3\sigma$ values (slow-p, slow-n) of $I_{Dsat}$. Therefore, to simulate the worst-case propagation delay of a digital gate, four circuit simulations are required: fast-p/fast-n, fast-p/slow-n, slow-p/fast-n, and slow-p/slow-n.

Calculation of statistically significant worst-case model decks for $I_{Dsat}$ requires a data-base of extracted model parameters including the value of $I_{Dsat}$ for each transistor. PCA can be performed on the extracted model parameters and $I_{Dsat}$ to arrive at a set of equations relating the normalized parameter set to independent, unit-normal principal components. For example, if only two model parameters, $V_T$ and $\beta$, are considered, the set of PCA equations might look like:

$$I'_{Dsat} = 0.9C_1 - 0.435C_2$$

$$V'_T = -0.6C_1 + 0.8C_2$$

$$\beta' = 0.707C_1 - 0.707C_2 \tag{14.34}$$

Then, a pair of values $\{C_1, C_2\}$ is chosen which gives $I_{Dsat} = +3$ in Eq. (14.34). The values of $V'_T$ and $\beta'$ obtained from Eq. (14.34) for this $\{C_1, C_2\}$ pair constitute a 'fast' model deck; that is, a transistor parameter set which gives a value of $I_{Dsat}$ which is $3\sigma$ above $\mu(I_{Dsat})$. Similarly, a pair of $\{C_1, C_2\}$ can be chosen to give a 'slow' parameter deck.

PCA assumes a linear relationship between model parameters. Since this assumption may not be valid over the entire parameter space, the set of minimum valued $\{C_1, C_2\}$ which give $I_{Dsat} = \pm 3\sigma$ should be chosen. In this case, a fast model deck could be calculated by using $C_1 = 2.7$ and $C_2 = -1.305$, resulting in $V_T' = -2.66$ and $\beta' = 2.83$. These numbers are denormalized using the inverse of Eq. (14.26) to complete the calculation of a fast model deck.

While worst-case analysis is relatively simple, it has obvious limitations for the analog circuit designer. The worst-case model decks are useful only for the device characteristic for which they were derived (in this case, $I_{Dsat}$). Specifications for analog circuits are based on many performance criteria, which can not be modeled by variations in a single device characteristic such as saturation drain current. While it is possible to calculate worst-case model decks for any device characteristic, the multitude of model decks required to account for every analog performance criterion makes this technique unsuitable for analog circuit simulation.

### 14.5.2 Monte Carlo Simulation

Monte Carlo simulation is a "technique whereby a random (statistical) exploration of the component space is performed" [4]. For statistical circuit simulation, the input component space of the simulation can either be the space of transistor model parameters or a set of process parameters [14], depending on which statistical model is chosen. A statistical device model is used to calculate sets of device model parameters which preserve both the distribution and correlation between each model parameter. If a large number of these model decks are entered into a circuit simulator, the result is a circuit output distribution representing the variation of circuit performance for that process.

Monte Carlo simulations are limited by the accuracy of the statistical model and the CPU time required to perform the many circuit simulations. The accuracy of Monte Carlo simulation is wholly dependent on the ability of the statistical model to represent the process variation. For instance, statistical simulation of the dc-offset of an amplifier is not possible unless the statistical device model includes the effect of device mismatch. In the past, the CPU time required to perform 1000 (or 10,000) circuit simulations made Monte Carlo simulation prohibitive. However, advances in computer technology have now made these simulations a viable option. Thus despite its limitations, the Monte Carlo technique remains the most general and dependable method to perform statistical simulations.

**Example 14.3. Monte Carlo Simulation of a Voltage Reference Circuit.**

To illustrate the way that Monte Carlo simulation can be used to estimate circuit performance variation, the voltage reference circuit shown in Fig. 14.4 will be simulated using the following model for the drain current in saturation

$$I_D = (\beta/2)(V_{GS} - V_T)^2(1 + \lambda V_{DS}) \tag{14.35}$$

For this simple device model, we can perform circuit simulation by equating the PMOS and NMOS drain currents and solving iteratively for $V_{ref}$. Since $V_{ref}$

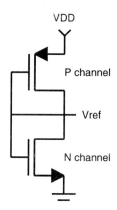

**FIGURE 14.4**
Voltage reference circuit

**TABLE 14.3**
Correlation matrix for Monte Carlo example.

|        | $V_{Tp}$ | $V_{Tn}$ | $\beta_p$ | $\beta_n$ | $\lambda_p$ | $\lambda_n$ |
|--------|--------|--------|--------|--------|--------|--------|
| $V_{Tp}$   | 1      | -0.246 | -0.248 | -0.132 | 0.036  | 0.054  |
| $V_{Tn}$   | -.246  | 1      | .237   | .039   | .033   | -.146  |
| $\beta_p$      | -.248  | .237   | 1      | .763   | .474   | .446   |
| $\beta_n$      | -.132  | .039   | .763   | 1      | .235   | .763   |
| $\lambda_p$      | .036   | .033   | .474   | .235   | 1      | .318   |
| $\lambda_n$      | .054   | -.146  | .446   | .763   | .318   | 1      |

is dependent on the "raw" value of the device parameters and not on parameter mismatch, variation in $V_{ref}$ is mainly due to interdie parameter variations. Therefore, a statistical model which accounts for only interdie variations is sufficient for this circuit. Tables 14.3 and 14.4 contain the parameter statistics required to calculate significant model decks for both the n-channel and p-channel transistors.

PCA can be performed on the correlation matrix, (Tab. 14.3) leading to a set of equations for the normalized parameter values

$$V'_{Tp} = -0.1943C_1 + 0.7261C_2 + 0.3857C_3 + 0.5148C_4 - 0.1453C_5$$

$$V'_{Tn} = 0.1242C_1 - 0.7819C_2 + 0.2870C_3 + 0.5079C_4 + 0.1810C_5$$

$$\beta'_p = 0.8789C_1 - 0.2256C_2 + 0.0890C_3 - 0.0075C_4 - 0.3829C_5$$

$$\beta'_n = 0.9081C_1 + 0.0763C_2 - 0.2961C_3 + 0.1734C_4 - 0.1030C_5$$

$$\lambda'_p = 0.5566C_1 + 0.1026C_2 + 0.7228C_3 - 0.3689C_4 + 0.1299C_5$$

$$\lambda'_n = 0.7833C_1 + 0.3960C_2 - 0.2199C_3 + 0.1167C_4 + 0.3919C_5 \qquad (14.36)$$

**TABLE 14.4**
Parameter statistics for Monte Carlo simulation example.

|            | mean   | stdev | units       |
|------------|--------|-------|-------------|
| $V_{Tp}$   | -0.735 | 0.05  | $V$         |
| $V_{Tn}$   | 0.8    | 0.05  | $V$         |
| $\beta_p$  | 68     | 7.5   | $\mu A/V^2$ |
| $\beta_n$  | 75     | 7.5   | $\mu A/V^2$ |
| $\lambda_p$| 0.02   | 0.002 | $1/V$       |
| $\lambda_n$| 0.01   | 0.001 | $1/V$       |

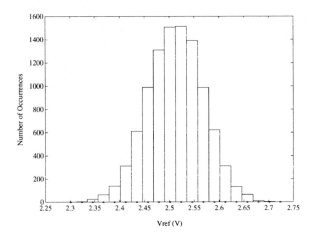

**FIGURE 14.5**
Output distribution of $V_{ref}$ from Monte Carlo simulations.

The relationship between the normalized transistor parameters in Eq. (14.36) and the actual transistor parameters is given by

$$V_{Tp} = mean(V_{Tp}) + sigma(V_{Tp})V_{Tp}'$$

$$V_{Tn} = mean(V_{Tn}) + sigma(V_{Tn})V_{Tn}'$$

$$\beta_p = mean(\beta_p) + sigma(\beta_p)\beta_p'$$

$$\beta_n = mean(\beta_n) + sigma(\beta_n)\beta_n'$$

$$\lambda_p = mean(\lambda_p) + sigma(\lambda_p)\lambda_p'$$

$$\lambda_n = mean(\lambda_n) + sigma(\lambda_n)\lambda_n' \tag{14.37}$$

For each simulation, random numbers are generated from a unit-normal distribution for each sampling variable, $C_1 \ldots C_5$. Values for these sampling variables are then entered into Eqs. (14.36) and (14.37) to complete calculation of device models for a single simulation. After 10,000 simulations, the distribution of

$V_{ref}$ values, shown in Fig. 14.5, has a mean value of mean$(V_{ref}) = 2.5014$ V and a standard deviation of sigma$(V_{ref}) = 0.0556$ V. Note that the sampling variables are linked to the circuit output through the device model and simulation program. Thus, the circuit simulator is inside the Monte Carlo loop, meaning that 10,000 simulations using this statistical technique will take 10,000 times more CPU time as a single circuit simulation.

### 14.5.3  Response Surface Methodology

To combat the large CPU time required to perform Monte Carlo simulations, response surface methodology (RSM) [31, 32, 33] can be employed to simplify the link between circuit output and the independent sampling variables. The goal of RSM is to lessen the CPU time by reducing the amount of time required for a single circuit simulation. This is done by taking the circuit simulator out of the Monte Carlo loop and replacing it with a polynomial equation relating the circuit output to the sampling variables. However, before this can be achieved, a simple link between each circuit output and the independent sampling variables $\{C_1, \ldots, C_5\}$ must be determined. Experimental design, effects analysis, regression and analysis of variance techniques are tools typically used to find this link. While a complete study of experimental design and analysis is beyond the scope of this chapter, a working knowledge of these techniques will be presented.

**Experimental Design**  The first step to finding a simple polynomial model for a given circuit output is to perform a screening, or 2-level, experiment on the sampling variables. The goal of the screening experiment is to determine which sampling variables do not contribute significantly to the circuit output variability and eliminate these variables from subsequent experiments. To perform this screening experiment, high and low levels, typically $\pm 3\sigma$ values, are chosen for each sampling variable and entered into an experimental matrix. An example of a screening experimental matrix is the Plackett-Burman design shown in Tab. 14.5. The first seven rows of this Plackett-Burman design are created using a generating vector, in this case $(+ + + - + - -)$. This generating vector is placed in the $C_1$ column. The next column is generated by making the last value in the $C_1$ column the first value in $C_2$, and shifting the rest of $C_1$ down one row. This procedure is repeated until the number of columns equals the number of sampling variables. The final row contains all "-" levels. This generating vector is sufficient for up to 7 sampling variables. If more variables need be screened, a different generating vector, e.g., $(+ + - + + + - - - + -)$ can be chosen.

To analyze a screening experiment, an effects table is created to compare the significance of each linear term. An effect for an input variable, $X_1$, is calculated by subtracting the experimental results associated with the "-" value of $X_1$ from the results for which $X_1$ has a "+" value. The sum of squares for $X_1$ can be calculated using

$$SS(X_1) = N/4 * [avg(X_1 = +1) - avg(X_1 = -1)]^2 \qquad (14.38)$$

**TABLE 14.5**
Placket Burman Matrix for seven input variables.

| Run | x1 | x2 | x3 | x4 | x5 | x6 | x7 |
|-----|----|----|----|----|----|----|----|
| 1 | + | - | - | + | - | + | + |
| 2 | + | + | - | - | + | - | + |
| 3 | + | + | + | - | - | + | - |
| 4 | - | + | + | + | - | - | + |
| 5 | + | - | + | + | + | - | - |
| 6 | - | + | - | + | + | + | - |
| 7 | - | - | + | - | + | + | + |
| 8 | - | - | - | - | - | - | - |

where $N$ is the number of experiments and $[avg(X_1 = +1) - avg(X_1 = -1)]$ is the effect of $X_1$. The SS value is an indication of the relative importance of each input parameter. A comparison of these values determines which parameters contribute significantly to the output variability and which variables can be discarded. Note that all quadratic and many interaction dependencies can not be separated from the linear terms. However, this confounding of terms generally is not a problem provided care is taken when screening input variables.

A modeling experiment is then performed on the significant sampling variables. Modeling experiments contain three or more levels for each input variable to allow least-squares fitting of the experimental results to a quadratic model. A sample modeling experiment, the Box-Behnken design, is shown in Tab. 14.6 for four input variables where +1, -1, and 0 correspond to the high, low input value, and intermediate input values, respectively. An experiment is designed at each possible combination of "+1" and "-1." A Box-Behnken design can be generated for any number of input variables by including four experimental runs for every unique pair of input variables.

The final model equation for the circuit output is the found by using least-squares multiple regression. To find a least-squares fit to the equation

$$y = b_0 + b_1 X_1 + b_2 X_2 + \dots \tag{14.39}$$

a set of equations known as the normal equations must be solved

$$sum(y) = b_0 N + b_1 sum(X_1) + b_2 sum(X_2) + \dots$$

$$sum(yX_1) = b_0 sum(X_1) + b_1 sum(X_1^2) + b_2 sum(X_1 X_2) + \dots$$

$$sum(yX_2) = b_0 sum(X_2) + b_1 sum(X_1 X_2) + b_2 sum(X_2^2) + \dots$$

$$\vdots \tag{14.40}$$

Quadratic or interaction terms can also be included in the regression equation. The normal equations can be solved easily by any matrix solver program. The

**TABLE 14.6**
Box-Behnken Matrix for four input parameters.

| Run | $X_1$ | $X_2$ | $X_3$ | $X_4$ |
|-----|-----|-----|-----|-----|
| 1-4 | +-1 | +-1 | 0 | 0 |
| 5-8 | +-1 | 0 | +-1 | 0 |
| 9-12 | +-1 | 0 | 0 | +-1 |
| 13-16 | 0 | +-1 | +-1 | 0 |
| 17-20 | 0 | +-1 | 0 | +-1 |
| 21-24 | 0 | 0 | +-1 | +-1 |
| 25 | 0 | 0 | 0 | 0 |

quality of the model equation can be represented by an $R^2$ value.

$$R^2 = sum[(\hat{y} - \overline{y})^2]/sum[(y - \overline{y})^2] \tag{14.41}$$

where $y$ is the measured data, $\overline{y}$ is the sample mean of the measured data, and $\hat{y}$ is the model prediction for the data. The $R^2$ value measures the percentage of system variability described by the model. Ideally, this value would be very close to 1.

Once this quadratic model, or response surface, is determined, statistical simulations can be performed directly on this equation, bypassing the circuit simulator. Unlike standard Monte Carlo simulation, thousands of these simulations can be completed in less than one second of CPU time. This entire procedure, including the experimental design and simulations, regression, and statistical simulations, can be automated to reduce the total time from problem definition to statistical simulation results. However, the entire procedure must be repeated for each desired circuit output since a different quadratic model, composed of different sampling variables, will be required to represent each output.

Like standard Monte Carlo simulations, the RSM technique for statistical simulation is limited by the accuracy of its statistical model. Additional inaccuracies will also be introduced by the quadratic modeling, since it is doubtful that every circuit output can be described as well by a quadratic model as it can through use of the device model. However, for most applications, RSM remains a viable alternative or supplement to standard Monte Carlo simulations.

**Example 14.4. Statistical Simulation of the Voltage Reference Circuit using RSM.**
To illustrate the differences between full Monte Carlo simulation and RSM, the voltage reference circuit of Example 13.3 will now be analyzed using RSM. For this example, the parameter statistics provided in Tabs. 14.3 and 14.4 along with the statistical model given by Eqs. (14.36) and (14.37) are used for each circuit simulation.

The first step toward finding a polynomial model for $V_{ref}$ in terms of the independent principal components, $C_1 \ldots C_5$, is to perform a screening experiment.

Simulations are performed according to the Plackett-Burman experimental design defined in Tab. 14.7. The simulation results are summarized by the effects table shown in Tab. 14.8. The sum of squares terms, calculated using Eq. (14.38), indicate that 95.8% of the simulation variability is the result of three input variables, $C_2, C_3, C_4$. These three input variables are deemed significant, and will be carried on to the modeling experiment.

**TABLE 14.7**
Plackett-Burman Matrix and simulation results for screening experiment.

| Run | $C_1$ | $C_2$ | $C_3$ | $C_4$ | $C_5$ | $V_{ref}$ |
|-----|-------|-------|-------|-------|-------|-----------|
| 1 | +3 | -3 | -3 | +3 | -3 | 2.6248 |
| 2 | +3 | +3 | -3 | -3 | +3 | 2.3466 |
| 3 | +3 | +3 | +3 | -3 | -3 | 2.4543 |
| 4 | -3 | +3 | +3 | +3 | -3 | 2.359 |
| 5 | +3 | -3 | +3 | +3 | +3 | 2.6846 |
| 6 | -3 | +3 | -3 | +3 | +3 | 2.1398 |
| 7 | -3 | -3 | +3 | -3 | +3 | 2.7235 |
| 8 | -3 | -3 | -3 | -3 | -3 | 2.6295 |

**TABLE 14.8**
Effects table for screening experiment.

| Input | Effect | SS | % |
|-------|--------|------|------|
| $C_1$ | 0.2585 | 0.1336 | 2.9 |
| $C_2$ | -1.3627 | 3.7139 | 80.6 |
| $C_3$ | 0.4807 | 0.4621 | 10.0 |
| $C_4$ | -0.3457 | 0.2390 | 5.2 |
| $C_5$ | -0.1731 | 0.0599 | 1.3 |

A Box-Behnken modeling experiment and simulation results are shown in Tab. 14.9. From this data, a quadratic model for $V_{ref}$ is determined using the normal regression equations in Eq. (14.40) and a matrix solving program.

$$V_{ref} = 2.5005 - 0.0514 * C_2 + 0.0165 * C_3 - 0.0084 * C_4 +$$
$$116.7E - 6 * C_2^2 - 19.4E - 6 * C_3^2 + 191.7E - 6 * C_4^2 +$$
$$169.4E - 6 * C_2C_3 + 230.6E - 6 * C_2C_4 - 400.0E - 6 * C_3C_4 \qquad (14.42)$$

The $R^2$ value for this model is 0.9999, indicating excellent model fit to the data.

At this point, Monte Carlo simulations of Eq. (14.42) itself can be performed, resulting in mean($V_{ref}$)=2.5029 V, sigma($V_{ref}$)=0.0545 V, and a distribution of $V_{ref}$ very similar to Fig. 14.5. However, Monte Carlo simulations are not

**TABLE 14.9**
Box-Behnken Matrix and simulation results for modeling experiment.

| Run | $C_2$ | $C_3$ | $C_4$ | $V_{ref}$ |
|-----|-------|-------|-------|-----------|
| 1 | +3 | +3 | 0 | 2.396 |
| 2 | +3 | -3 | 0 | 2.2941 |
| 3 | -3 | +3 | 0 | 2.7014 |
| 4 | -3 | -3 | 0 | 2.6056 |
| 5 | +3 | 0 | +3 | 2.324 |
| 6 | +3 | 0 | -3 | 2.37 |
| 7 | -3 | 0 | +3 | 2.6282 |
| 8 | -3 | 0 | -3 | 2.6825 |
| 9 | 0 | +3 | +3 | 2.5227 |
| 10 | 0 | +3 | -3 | 2.5804 |
| 11 | 0 | -3 | +3 | 2.4309 |
| 12 | 0 | -3 | -3 | 2.4742 |
| 13 | 0 | 0 | 0 | 2.5005 |

necessary, since the output variance can be calculated directly from Eq. (14.42) using the transmission of moments formula, Eq. (14.16). These calculations result in mean($V_{ref}$)=2.5006 V and sigma($V_{ref}$)=0.0546 V, very close to the values obtained through Monte Carlo simulations of Eq. (14.42).

An added benefit of a model equation similar to Eq. (14.42) is the ability to create contour plots showing the output response surface. Inserting an inverted version of Eq. (14.36) into the output model equation, a relation between $V_{ref}$ and the normalized model parameters can be calculated. For instance, an equation relating $V_{ref}$ to two normalized parameters, $V'_{Tp}$ and $V'_{Tn}$, with all other parameters at their mean values is

$$V_{ref} = 2.5005 + 0.02525 * V'_{Tp} + 0.02981 * V'_{Tn} + 118.8E - 06 * V'_{Tp}V'_{Tn}$$

$$+124.2E - 06 * V'^2_{Tp} - 77.8E - 06 * V'^2_{Tn} \qquad (14.43)$$

From this relation, the contour plot shown in Fig. 14.6 can be drawn. However, we must remember that the normalized model parameters are not, in general, independent. Using this technique to investigate response surfaces, correlations between model parameters are ignored. In this case, Fig. 14.6 does accurately represent the response surface of $V_{ref}$, since $V_{Tp}$ and $V_{Tn}$ have negligible correlations to themselves and the other model parameters.

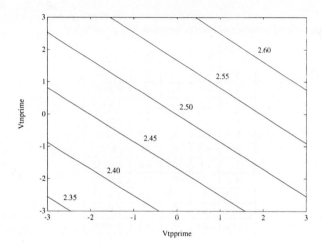

**FIGURE 14.6**
Contour plot of output response surface

## 14.6 EXAMPLES OF STATISTICAL CIRCUIT SIMULATIONS

In this section, we will examine the effects of process variations on the current mismatch of a pair of NMOS transistors. Then, we will use this result to examine the effects of mismatch on the performance of an NMOS current mirror and an NMOS differential pair with a PMOS current mirror load. These two circuits are fundamental building blocks in most CMOS analog integrated circuits. Thus, a knowledge of the effects of mismatch and interdie process variations in these two circuits can provide insight into the effects of process variation on larger circuits.

Both the current mirror and the differential pair are simple enough that their current and voltage characteristics can be solved analytically using a simple first-order model for the MOSFET drain current in saturation [34].

$$I_D = \frac{\beta}{2}(V_{GS} - V_T)^2(1 + \lambda V_{DS}) \tag{14.44}$$

Then, the transmission of moments formula discussed earlier can be used to relate the circuit output variances to the transistor parameter variations.

For each circuit, we will choose one or more standard performance criteria (such as output current, voltage gain, or offset voltage) and obtain analytical solutions for the effects of parameter variations on these criteria. Then, we will present the results of statistical simulations (1000 Monte Carlo simulations for each solution) of the circuit using the SMOS model in the APLAC circuit analysis program described in an earlier work [13]. Note that much of the analyses and simulations of this section appeared previously in [27].

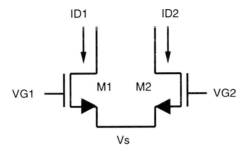

## FIGURE 14.7
General mismatched transistor pair.

### 14.6.1   General Transistor Pair

Figure 14.7 shows a general NMOS transistor pair with mismatched parameters and mismatched gate and drain biases.

Using the simple MOS drain current formula of Eq. (14.44), the current through each transistor is

$$I_{D1} = \frac{1}{2}(\beta + \frac{\Delta\beta}{2})(V_{GS} + \frac{\Delta V_{GS}}{2} - V_T - \frac{\Delta V_T}{2})^2$$

$$(1 + (\lambda + \frac{\Delta\lambda}{2})(V_{DS} + \frac{\Delta V_{DS}}{2})) \tag{14.45}$$

$$I_{D2} = \frac{1}{2}(\beta - \frac{\Delta\beta}{2})(V_{GS} - \frac{\Delta V_{GS}}{2} - V_T + \frac{\Delta V_T}{2})^2$$

$$(1 + (\lambda - \frac{\Delta\lambda}{2})(V_{DS} - \frac{\Delta V_{DS}}{2})) \tag{14.46}$$

Subtracting Eq. (14.45) from Eq. (14.46) and neglecting second-order terms such as $\Delta\beta\Delta V_T$ or $\Delta V_T^2$, and assuming small gate and drain bias mismatches, we find that

$$\frac{\Delta I_D}{I_D} = \frac{\Delta\beta}{\beta} - 2\frac{\Delta V_T}{V_{GS} - V_T} + V_{DS}\frac{\Delta\lambda}{1 + \lambda V_{DS}}$$

$$+2\frac{\Delta V_{GS}}{V_{GS} - V_T} + \lambda\frac{\Delta V_{DS}}{1 + \lambda V_{DS}} \tag{14.47}$$

where $I_{DS} = \frac{\beta}{2}(V_{GS} - V_T)^2(1 + \lambda V_{DS})$ is the drain current through each transistor with all mismatch terms equal to zero.

Assuming a simple MOS model for the drain current of a PMOS transistor

$$I_D = \frac{\beta}{2}(V_{GS} - V_T)^2(1 + \lambda V_{SD}) \tag{14.48}$$

with a positive drain current defined as flowing from source to drain, the relative drain current mismatch for the PMOS mismatched pair of Fig. 14.8 is

$$\frac{\Delta I_D}{I_D} = \frac{\Delta \beta}{\beta} - 2\frac{\Delta V_T}{V_{GS} - V_T} + V_{SD}\frac{\Delta \lambda}{1 + \lambda V_{SD}}$$

$$+2\frac{\Delta V_{GS}}{V_{GS} - V_T} + \lambda\frac{\Delta V_{SD}}{1 + \lambda V_{SD}} \tag{14.49}$$

This is the same as the mismatch equation for the NMOS pair with $V_{DS}$ replaced by $V_{SD}$ and $V_{GS}$ replaced by $V_{SG}$.

### 14.6.2   NMOS Current Mirror

Figure 14.9 shows a standard NMOS current mirror fed by an ideal source carrying current $I_{SS}$. We would like to determine the drain current mismatch

$$\Delta I = I_{SS} - I_{OUT} \tag{14.50}$$

as a function of the nominal values of the transistor parameters $\beta$ and $V_T$ and $\lambda$, and the parameter mismatches $\Delta V_T = V_{T1} - V_{T2}$, $\Delta \beta = \beta_1 - \beta_2$, and $\Delta \lambda = \lambda_1 - \lambda_2$.

We will define $\beta$, $\Delta \beta$, $V_T$, $\Delta V_T$, $\lambda$ and $\Delta \lambda$ such that

$$V_{T1} = V_T + \frac{\Delta V_T}{2}$$

$$V_{T2} = V_T - \frac{\Delta V_T}{2}$$

$$\beta_1 = \beta + \frac{\Delta \beta}{2}$$

$$\beta_2 = \beta - \frac{\Delta \beta}{2}$$

$$\lambda_1 = \lambda + \frac{\Delta \lambda}{2}$$

$$\lambda_2 = \lambda - \frac{\Delta \lambda}{2} \tag{14.51}$$

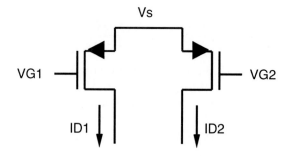

**FIGURE 14.8**
PMOS mismatched transistor pair.

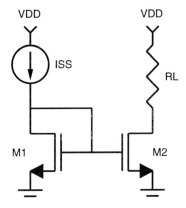

**FIGURE 14.9**
NMOS current mirror

Since $\Delta V_T$ or $\Delta \beta$ can take on either positive or negative values, there is no loss of generality in the definitions of Eq. (14.51).

Since the gates of M1 and M2 in Fig. 14.9 are connected, it is clear that $\Delta V_{GS} = 0$. If $R_L$ is chosen so that the nominal value of $I_{OUT}$ is equal to $I_{SS}$, then we also have $\Delta V_{DS} = 0$. Using Eq. (14.47) subject to these constraints, the relative current mismatch of the current mirror is

$$\frac{\Delta I}{I_{SS}} = \frac{\Delta \beta}{\beta} - 2\frac{\Delta V_T}{V_{GS} - V_T} + V_{DS}\frac{\Delta \lambda}{1 + \lambda V_{DS}} \qquad (14.52)$$

where $V_{GS}$ and $V_{DS}$ are the nominal values with no parameter mismatch.

We can use the transmission of moments formula (Eq. (14.16)) to calculate the variance of the current mismatch in Eq. (14.52)

$$\frac{\sigma^2(\Delta I)}{I_{SS}^2} = \frac{\sigma^2(\Delta \beta)}{\beta^2} + 4\frac{\sigma^2(\Delta V_T)}{(V_{GS} - V_T)^2} + V_{DS}^2\frac{\sigma^2(\Delta \lambda)}{(1 + \lambda V_{DS})^2} \qquad (14.53)$$

The discussion of Pelgrom's model for intradie parameter variation showed that $\sigma^2(\Delta \beta)$, $\sigma^2(\Delta V_T)$, and $\sigma^2(\Delta \lambda)$ decrease as the gate areas of M1 and M2 are increased, and as the separation distance between the transistors decreases. Thus, Eq. (14.53) shows that the drain current mismatch can be reduced by making the transistor areas large and placing the transistors close together.

It is also clear that the threshold voltage variation has a more pronounced effect on the current mismatch for small values of $V_{GS} - V_T$, that is, for small values of $I_{SS}$, while the drain conductance variation has a more pronounced effect for large values of $V_{DS}$, that is, large values of $I_{SS}$.

Current mismatch in both NMOS and PMOS current mirrors was simulated with the APLAC implementation of the SMOS algorithm. For all simulations, the load resistance was set to $R_L = 10.0K\Omega$, $V_{DD}$ was set to $5.0V$, and the transistor $W/L = 1.5$.

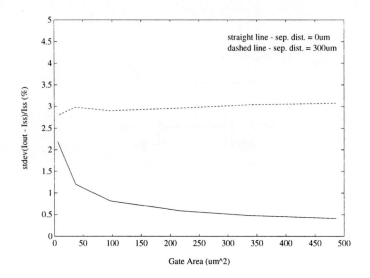

**FIGURE 14.10**
Relative standard deviation of NMOS current mirror mismatch for $I_{SS} = 50$ uA.

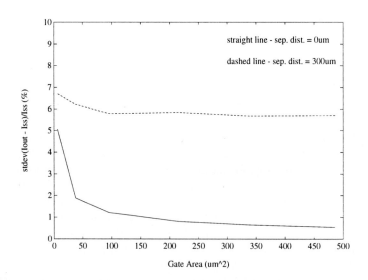

**FIGURE 14.11**
Relative standard deviation of NMOS current mirror mismatch for $I_{SS} = 10$ uA.

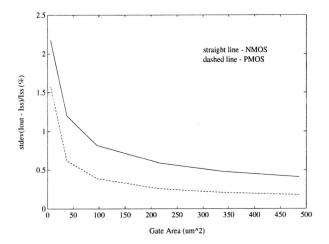

**FIGURE 14.12**
Relative standard deviation of NMOS and PMOS current mirror mismatch.

Figures 14.10 and 14.11 indicate the effects of both transistor separation distance and bias current on the relative variation of current mismatch in an MOS current mirror. As was discussed earlier in this section, increased separation distance between the mirror transistors can significantly increase the current mismatch variation. In addition, it is also clear that the the current mismatch variation increases with decreasing bias current.

Figure 14.12 compares the current mismatch variation of NMOS and PMOS current mirrors, for the case of $(W/L)_n = (W/L)_p = 1.5$ and $I_{SS} = 50\mu A$. Because PMOS mobility is lower (by a factor of between 2.0 and 3.0 for this process) than NMOS mobility, biasing equal sized current mirrors of each type with the same $I_{SS}$ results in a larger value of $|V_{GS} - V_T|$ for the PMOS mirror. From Eq. (14.53), it is clear that a larger value of $V_{GS} - V_T$ results in reduced current mismatch variance. Thus, the PMOS current mismatch in Fig. 14.12 is smaller than the NMOS current mismatch.

### 14.6.3  CMOS Differential Pair with Current Mirror Load

**Interdie Variation of Open Loop Gain**  The open circuit voltage gain of the perfectly matched differential amplifier is given by

$$A_V = \frac{g_{mn}}{g_{dsn} + g_{dsp}} = \frac{v_{out}}{v_{id}} \qquad (14.54)$$

where $v_{id} = v_{g1} - v_{g2}$, $g_{mn}$ is the transconductance of the NMOS transistors M1 and M2, and $g_{dsn}$ and $g_{dsp}$ are the drain conductance of M2 and M4, respectively.

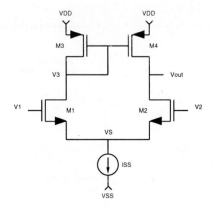

**FIGURE 14.13**
CMOS differential pair.

Substituting for $g_{mn}$, $g_{dsn}$, and $g_{dsp}$ in terms of the transistor parameters and the bias current $I_{SS}$, we have

$$A_V = \left(\frac{2}{\lambda_n + \lambda_p}\right)\sqrt{\frac{\beta_n}{I_{SS}}} \tag{14.55}$$

The derivations of Eqs. (14.54) and (14.55) will not be repeated here.

We can apply the transmission of moments formula to relate the variance of the voltage gain to the interdie variances of the transistor parameters. That is, neglecting correlations

$$\sigma^2(A_V) = \left(\frac{\partial A_V}{\partial \beta_n}\right)^2 \sigma^2(\beta_n) + \left(\frac{\partial A_V}{\partial \lambda_n}\right)^2 \sigma^2(\lambda_n) + \left(\frac{\partial A_V}{\partial \lambda_p}\right)^2 \sigma^2(\lambda_p) \tag{14.56}$$

Taking the required partial derivatives of the voltage gain from Eq. (14.55), we see that

$$\frac{\sigma^2(A_V)}{A_V^2} = \frac{1}{4}\frac{\sigma^2(\beta_n)}{\beta_n^2} + \frac{\sigma^2(\lambda_n) + \sigma^2(\lambda_p)}{(\lambda_n + \lambda_p)^2} \tag{14.57}$$

Three important points are evident from Eq. (14.57). First, it is clear that variations in NMOS or PMOS threshold voltages or PMOS mobility have no effect on the voltage gain. Second, the relative variance of the voltage gain (i.e., $\sigma^2(A_V)/A_V^2$) is independent of the bias current $I_{SS}$. Finally, since the ratio $\sigma^2(\beta_n)/\beta_n^2$ is independent of $(W/L)_n$, $\sigma^2(A_V)/A_V^2$ is also independent of $(W/L)_n$.

Table 14.10 shows the results of statistical simulations of the differential amplifier. Note that the amplifier was simulated with $V_{DD} = -V_{SS} = 5.0V$. In all cases, $W/L$ of the PMOS current mirror load was adjusted so that the dc-value of $V_{out}$ is approximately 0 V for $V_1 = V_2 = 0$. It is clear that the percent variation of the voltage gain, $\sigma(A_V)/A_V$, is indeed independent of both $I_{SS}$ and $\beta_n$.

**TABLE 14.10**
Interdie variation of open-loop gain of differential amplifier.

| NMOS W/L | PMOS W/L | $I_{SS}$ | mean($A_V$) | $\sigma(A_V)/mean(A_V)$ |
|---|---|---|---|---|
| | | $(\mu A)$ | $(V/V)$ | % |
| 50/4 | 2/10 | 50.0 | 49.4 | 12.1 |
| 25/4 | 2/10 | 50.0 | 34.8 | 11.8 |
| 100/4 | 2/10 | 50.0 | 69.8 | 13.0 |
| 50/4 | 4/10 | 100.0 | 32.9 | 11.4 |
| 25/4 | 4/10 | 100.0 | 24.1 | 12.3 |
| 100/4 | 4/10 | 100.0 | 46.47 | 11.9 |

**Effect of Mismatch on Offset Voltage**  Next, we will apply the current mismatch formula for the general transistor pair to both transistor pairs in the differential amplifier in order to relate the input offset voltage to the parameter mismatches between the transistors in each pair.

A straightforward application of Eq. (14.47) to the NMOS source coupled pair in Fig. 14.13 gives the input voltage difference needed to make $V_{OUT} = V_3$ (that is $\Delta V_{DSn} = 0$)

$$V_{OS} = \Delta V_{GSn} = \Delta V_{Tn} - \frac{V_{GSn} - V_{Tn}}{2\beta_n}\Delta\beta_n -$$

$$\frac{V_{GSn} - V_{Tn}}{2(1 + \lambda_n V_{DSn})}V_{DSn}\Delta\lambda_n - \frac{V_{GSn} - V_{Tn}}{I_{SS}}\Delta I_D \qquad (14.58)$$

Then, we can apply Eq. (14.49) to the PMOS current mirror to calculate $\Delta I_D/I_D$

$$\frac{\Delta I_D}{I_D} = \frac{\Delta I_D}{I_{SS}/2} = \frac{\Delta\beta_p}{\beta_p} - 2\frac{\Delta V_{Tp}}{V_{GSp} - V_{Tp}} + V_{SDp}\frac{\Delta\lambda_p}{1 + \lambda_p V_{SDp}} \qquad (14.59)$$

where we have utilized the fact that $\Delta V_{GSp} = 0$ for the current mirror and $\Delta V_{SDp} = 0$ since $V_{OUT} = V_3$. Since $\Delta I_D$ is the same for the NMOS pair as it is for the current mirror, we can substitute the expression for $\Delta I_D/I_{SS}$ from Eq. (14.59) into Eq. (14.58). This gives the expression we have been looking for

$$V_{OS} = \Delta V_{Tn} - \frac{V_{GSn} - V_{Tn}}{V_{GSp} - V_{Tp}}\Delta V_{Tp} - \frac{(V_{GSn} - V_{Tn})}{2}\left(\frac{\Delta\beta_n}{\beta_n} - \frac{\Delta\beta_p}{\beta_p}\right.$$

$$\left. + \frac{V_{DSn}}{1 + \lambda_n V_{DSn}}\Delta\lambda_n - \frac{V_{SDp}}{1 + \lambda_p V_{SDp}}\Delta\lambda_p\right) \qquad (14.60)$$

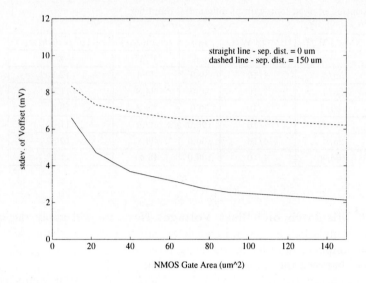

**FIGURE 14.14**
Standard deviation of input offset voltage versus gate area of NMOS input transistors.

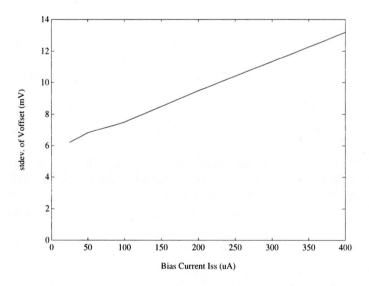

**FIGURE 14.15**
Standard deviation of input offset voltage versus bias current $I_{SS}$.

As before, the transmission of moments formula, Eq. (14.16), can be applied to Eq. (14.60) to relate the variance of the input offset voltage, $\sigma^2(V_{OS})$, to the variances of the transistor parameters.

$$
\begin{aligned}
\sigma^2(V_{OS}) = {}& \sigma^2(\Delta V_{Tn}) + \left(\frac{V_{GSn} - V_{Tn}}{V_{GSp} - V_{Tp}}\right)^2 \sigma^2(\Delta V_{Tp}) \\
& + \left(\frac{V_{GSn} - V_{Tn}}{2}\right)^2 \left[\frac{\sigma^2(\Delta\beta_n)}{\beta_n^2} + \frac{\sigma^2(\Delta\beta_p)}{\beta_p^2}\right. \\
& + \left(\frac{V_{DSn}}{1 + \lambda_n V_{DSn}}\right)^2 \sigma^2(\Delta\lambda_n) \\
& \left. + \left(\frac{V_{SDp}}{1 + \lambda_n V_{SDp}}\right)^2 \sigma^2(\Delta\lambda_p)\right]
\end{aligned}
\tag{14.61}
$$

Three important observations can be made from Eqs. (14.60) and (14.61). First, for most differential amplifiers, $(W/L)_n \gg (W/L)_p$, and thus $|V_{GSp} - V_{Tp}| \gg V_{GSn} - V_{Tn}$. Thus, the threshold voltage mismatch of the input pair transistors has a much greater effect on the offset voltage than the threshold mismatch of the current mirror transistors. Second, to provide adequate open-loop gain, the bias current $I_{SS}$ is generally kept small and the $W/L$ of the input transistors is generally large. Thus, $V_{GSn} - V_{Tn} < V_{Tn}$. Assuming that for a given size transistor $\sigma(V_{Tn})/V_{Tn} \approx \sigma(\beta_n)/\beta_n$, it is clear from Eq. (14.61) that for small bias currents the offset voltage variance is much more a function of the threshold voltage mismatch of the input pair, $\Delta V_{Tn}$, than it is of any other parameter mismatch. Finally, as the bias current increases, holding the device areas constant so that parameter mismatch variance stays the same, the offset voltage variance should increase.

The APLAC implementation of the SMOS mismatch model was used to perform Monte Carlo simulations of the input offset voltage of the differential amplifier. The results are shown in Figs. 14.14 and 14.15.

Figure 14.14 shows the simulated results of the standard deviation of the input offset voltage versus input transistor gate area. The figure has results both for the case of no separation distance between the paired transistors and for the case of a $150\mu m$ separation distance. The bias current $I_{SS}$ was set to $50\mu A$ and the PMOS load transistors sized $W = 8\mu m$ and $L = 40\mu m$ for all simulations shown. It is clear from this figure that to reduce the input offset voltage, the input transistors should have large gate areas and should be laid out with a common centroid geometry to eliminate the effects of spacing-dependent parameter mismatch (see appendix on layout techniques).

Figure 14.15 indicates the effect of bias current on the offset voltage variance. All the simulations in this figure were performed with the NMOS input pair sized at $W = 5\mu m$ and $L = 2\mu m$, and the area of the PMOS current mirror maintained at $320\mu m^2$ (the width-to-length ratio of the PMOS transistors was adjusted so that the nominal output voltage was 0 V). With the device areas constant, the transistor parameter variance was the same for all simulations in this figure. It is clear that, as discussed earlier, increasing the bias current degrades the input offset voltage.

## 14.7   CONCLUSION

As feature sizes in MOS processes move into the submicron range, and power supply voltages are reduced, the effect of both device mismatch and interdie process variations on the performance and reliability of analog integrated circuits is magnified. In order to fully utilize the capabilities of a given process, a circuit designer needs to have both a complete knowledge of the statistical distributions of transistor parameters produced by the process and a way to determine accurately what effect variations in these parameters have on circuit performance.

Models such as the SMOS model [13, 24, 25] presented in this work and techniques such as Monte Carlo simulation, response surface methodology, and worst-case analysis can be used to translate parameter variation into circuit performance variation. Statistical design methodologies such as these will assist circuit desginers in bringing high performance and high yield to the next generation of analog MOS integrated circuits.

## PROBLEMS

**14.1.** Show that the variance of a random variable $X$, given by Eq. (14.5) can be written as

$$Var[X] = E[X^2] - (E[X])^2 \qquad (14.62)$$

**14.2.** Using the expression for the pdf of the normal distribution Eq. (14.7) of the text, show that for a normally distributed random variable $X$,

$$mean[X] = E[X] = \mu \quad and \quad Var[X] = \sigma^2 \qquad (14.63)$$

*Hint:* use the fact that

$$\int_{-\infty}^{\infty} e^{-z^2/2} dz = \sqrt{2\pi} \qquad (14.64)$$

**14.3.** Let $\mathbf{X} = [X_1, \ldots, X_N]^T$ be a vector of correlated unit-normal random numbers with the correlation between $X_i$ and $X_j$ given by $\rho_{ij} = E[X_i X_j]$. Show that if $X_i$ and $X_j$ are written as linear combinations of $N$ independent principal components

$$X_i = A_{i1}C_1 + A_{i2}C_2 + \ldots + A_{iN}C_N$$
$$X_j = A_{j1}C_1 + A_{j2}C_2 + \ldots + A_{jN}C_N$$

then, $\rho_{ij}$ can be written

$$\rho_{ij} = \sum_{k=1}^{N} A_{ik}A_{jk} \qquad (14.65)$$

**14.4.** Find $\Delta I$ for the circuit in Fig. P14.4 in terms of $\Delta\beta$, $\Delta V_T$, and $\Delta R$ using the following equation for MOSFET drain current

$$I_D = \frac{\beta}{2}(V_{GS} - V_T)^2 \qquad (14.66)$$

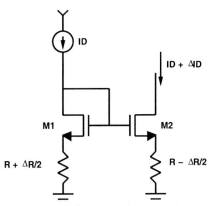

**FIGURE P14.4** Current mirror with source degeneration.

**14.5.** Find the offset voltage (that is, the value of $V_{OS} = V_{G1} - V_{G2}$ which makes $V_{O1} = V_{O2}$) of the resistively-loaded differential pair in Fig. P14.5 in terms of the mismatches in $V_T$, $\beta$, $\lambda$, and $R$.

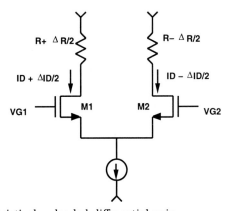

**FIGURE P14.5** Resistively - loaded differential pair.

**14.6.** Figure P14.6 shows an NMOS current mirror with the $W/L$ ratios of both transistors shown. Assuming a simple square law MOS model with the following parameter values $\mu C_{ox} = 30\mu A/V^2$, $V_T = 0.8V$, and $L_1 = L_2 = L$, and assuming that all transistor parameters are matched, *except for $V_T$*, answer the following:

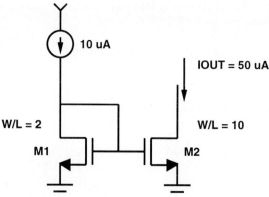

**FIGURE P14.6** NMOS current mirror.

 (a) Neglecting distance dependent mismatch, find an expression for $\sigma^2(I_{OUT})$ in terms of the Pelgrom model fitting constant $a_{V_T}$ and the transistor length $L$.

 (b) Given a value of $a_{V_T} = 3.6 * 10^{-5} V^2 \mu m^2$, find the minimum $L$ so that $\frac{\sigma(I_{OUT})}{I_{OUT}} < 0.001$.

**14.7.** This exercise illustrates the need to account for transistor parameter correlations when analyzing the performance variance of transistor circuits.

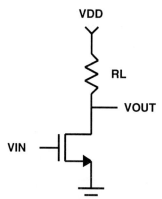

**FIGURE P14.7** Resistively-loaded common source amplifier.

 (a) Consider the resistively-loaded common-source amplifier shown in Fig. P14.7. Assuming the simplified MOS saturation drain current model $I_D = \frac{1}{2}\beta(V_{GS} - V_T)^2$ (neglect channel length modulation), and assuming that the transistor parameters $\beta$ and $V_T$ have a correlation coefficient of $\rho_{\beta V_T}$, use the transmission of moments formula to relate the variance of the small signal gain $A_V = v_{OUT}/v_{GS}$ to the variances of $\beta$, $V_T$, and $R_L$.

 (b) Calculate $\frac{\sigma(A_V)}{A_V}$ assuming $\frac{\sigma(V_T)}{(V_{GS} - V_T)} = \frac{\sigma(\beta)}{\beta} = \frac{\sigma(R_L)}{R_L} = 0.05$, and $\rho = 0$, that is, $V_T$ and $\beta$ are independent.

(c) Repeat part (b) with $\rho = -0.9, -0.5, 0.5$, and $0.9$. Comparing these results with the result of (b), how important is the correlation between $\beta$ and $V_T$?

**14.8.** For the current mirror shown in Fig. P14.8, assume that all parameters are matched except for the threshold voltage, and assume that the nominal $V_T = 0.75V$.

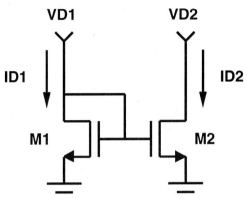

**FIGURE P14.8** Current mirror.

(a) Calculate the percentage of drain current mismatch $\frac{\Delta I_D}{I_D}$ if $V_{D1} = V_{D2} = 0.95V$ and $\Delta V_T = 2mV$.

(b) Recalculate $\frac{\Delta I_D}{I_D}$ for $V_{D1} = 0.95V$ and $V_{D2} = 3.95V$. Assume $\lambda = 0.005V^{-1}$. What does this result say about the importance of equalizing the load seen by transistors in a current mirror?

**14.9.** Figure P14.9 shows a cascode current mirror. Assume $\beta$ and $\lambda$ are equal for all four transistors and neglect the body effect. Using Eq. (14.52) of the text, develop an expression for $\frac{\Delta I_D}{I_D}$ where $\Delta I_D = I_{D1} - I_{D2}$ as a function of the $V_T$ mismatches between transistors $M3$ and $M4$ and between $M1$ and $M2$, and the drain voltage mismatch $V_{D3} - V_{D4}$. Calculate $\frac{\Delta I_D}{I_D}$ for the case of $\Delta V_T(M1 - M2) = 2mV$, $V_{GS} - V_T = 0.2V$, and $V_{D1} - V_{D2} = 3V$. Compare this with the results of Prob. 14.8. Does the drain voltage mismatch play as large a role in the cascode current mirror as it does in the standard current mirror?

**14.10.** Consider the circuit shown in Fig. P14.10. Assume that at time $t = o^-$ $C_L$ is charged to $V_{DD}$. A step input of height $V_{DD}/2$ is applied to $V_{IN}$ at time $t = 0^+$ to discharge $C_L$.

(a) Using the simple MOS drain current model (neglect $\lambda$), develop an expression for $t_d$, the time required to discharge $C_L$. To simplify the expression, assume that $M1$ is in saturation throughout the discharging process.

(b) Using the method of moments, write an expression for the percent variation of $t_d$, $\sigma(t_d)/t_d$, as a function of the variances of $V_T$, $C_{OX}$, $L_{EFF}$, and $C_L$. Assume all parameters are statistically independent.

(c) Substitute $V_{DD} = 5V$, $V_T = 0.75V$, $L_{EFF} = 1.5\mu m$, $\sigma(V_T) = 10mV$, $\sigma(L_{EFF}) = 0.05\mu m$, and $\sigma(C_{OX})/C_{OX} = \sigma(C_L)/C_L = 0.033$ into the

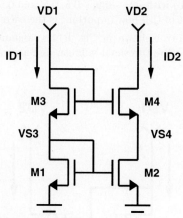

**FIGURE P14.9** Cascode current mirror.

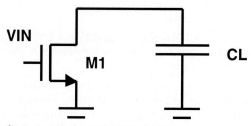

**FIGURE P14.10** Capacitively-loaded NMOS transistor.

expression developed in part (b). Does the variation of $V_T$ have a significant effect on the variation of $t_d$?

(d) Suppose that $C_L$ is dominated by the gate capacitance of another transistor (this would be the case, for example, if a digital logic gate is driving a nearby gate), and thus is strongly correlated with $C_{OX}$. Would you expect more or less variation in $t_d$ than for the case of a completely independent $C_L$?

# REFERENCES

[1] P. Antognetti and G. Massobrio, *Semiconductor Device Modeling with SPICE*, McGraw-Hill, 1988.

[2] P. Z. Peebles, *Probability, Random Variables, and Random Signal Principles*, McGraw-Hill, 1987.

[3] R. J. Larsen and M. L. Marx, *An Introduction to Mathematical Statistics and Its Applications*, Prentice-Hall, 1986.

[4] R. Spence and R. S. Soin, *Tolerance Design of Electronic Circuits*, Addison Wesley Publishing Co., 1988.

[5] P. J. Rankin, "Statistical modeling for integrated circuits," *IEEE Proceedings G, Electronic Circuits and Systems*, vol. 129, pp. 186–191, August 1982.

[6] D. F. Morrison, *Multivariate Statistical Methods*, McGraw-Hill, 1976.

[7] T. W. Anderson, *An Introduction to Multivariate Statistical Analysis*, McGraw-Hill, 1958.

[8] Helsinki University of Technology, Circuit Theory Laboratory and Nokia Research Center, *APLAC – An Object-Oriented Analog Circuit Simulator and Design Tool, 6.0 User's Manual and Reference Manual*, October 1991.

[9] P. Yang, D. E. Hocevar, P. F. Cox, C. Machala, and P. K. Chatterjee, "An integrated and efficient approach for MOS VLSI statistical circuit design," *IEEE Transactions on Computer-Aided Design*, vol. CAD-5, pp. 5–14, January 1986.

[10] E. E. Cureton and R. B. D'Agostino, *Factor Analysis: An Applied Approach*, Lawrence Erlbaum Associates, 1983.

[11] B. Flury, *Common Principal Components and Related Multivariate Models*, John Wiley and Sons, 1988.

[12] A. A. Afifi and S. P. Azen, *Statistical Analysis: A Computer-Oriented Approach*, Academic Press, 1972.

[13] C. Michael, *Statistical Modeling for Computer-Aided Design of Analog MOS Integrated Circuits*, Ph.D. thesis, The Ohio State University, 1991.

[14] S. R. Nassif, A. J. Strojwas, and S. W. Director, "FABRICS II: A statistically based IC fabrication process simulator," *IEEE Transactions on Computer-Aided Design*, vol. CAD-3, pp. 40–46, January 1984.

[15] N. Herr and J. J. Barnes, "Statistical circuit simulation modeling of CMOS VLSI," *IEEE Transactions on Computer-Aided Design*, vol. CAD-5, pp. 15–22, January 1986.

[16] P. Cox, P. Yang, S. S. M. Shetti, and P. Chatterjee, "Statistical modeling for efficient parametric yield estimation of MOS VLSI circuits," *IEEE J. of Solid-State Circuits*, vol. SC-20, pp. 391–398, February 1985.

[17] T. K. Yu, S. M. Kang, I. N. Hajj, and T. N. Trick, "Statistical performance modeling and parametric yield estimation of MOS VLSI," *IEEE Transactions on Computer-Aided Design*, vol. CAD-6, pp. 1013–1022, November 1987.

[18] S. Inohira, T. Shinmi, M. Nagata, T. Toyabe, and K. Iida, "A statistical model including parameter matching for analog integrated circuit simulation," *IEEE Transactions on Electron Devices*, vol. ED-6, pp. 2177–2184, October 1985.

[19] P. E. Allen and D. R. Holberg, *CMOS Analog Circuit Design*, Holt, Rinehart & Winston Inc., 1987.

[20] M. J. M. Pelgrom, A. C. J. Duinmaiger, and A. P. G. Welbers, "Matching properties of MOS transistors for precision analog design," *IEEE J. of Solid-State Circuits*, vol. SC-24, pp. 1433–1439, October 1989.

[21] J. B. Shyu, G. C. Temes, and F. Krummenacher, "Random error effects in matched MOS capacitors and current sources," *IEEE J. of Solid-State Circuits*, vol. SC-19, pp. 948–955, December 1984.

[22] K. R. Lakshmikumar, R. A. Hadaway, and M. A. Copeland, "Characterization and modeling of mismatch in MOS transistors for precision analog design," *IEEE J. of Solid-State Circuits*, vol. SC-21, pp. 1057–1066, December 1986.

[23] C. Michael and M. Ismail, "Simulation of mismatch induced variance in short-channel analog CMOS circuits," in *Proceedings of the 1990 SRC Techcon*, pp. 351–354, 1990.

[24] C. Michael and M. Ismail, "Statistical modeling of device mismatch for analog MOS integrated circuits," *IEEE J. of Solid-State Circuits*, vol. SC-27, pp. 154–166, February 1992.

[25] C. Michael, C. Abel, and M. Ismail, "SMOS: A CAD-compatible statistical model for analog MOS integrated circuit simulation." International Journal on Circuit Theory and Applications, vol. 20, pp. 327–347, March 1992.

[26] C. Michael and M. Ismail, "Statistical modeling for computer-aided design of MOS VLSI circuits," Kluwer Academic Publishers, 1993.

[27] C. J. Abel, "A complete statistical MOS model and its application to analog and digital integrated circuits," Master's thesis, The Ohio State University, 1992.

[28] B. J. Sheu, D. L. Scharfetter, P. K. Ko, and M. C. Jeng, "BSIM: Berkeley short-channel IGFET model for MOS transistors," *IEEE J. of Solid-State Circuits*, vol. SC-22, pp. 558–566, August 1987.

[29] S. R. Nassif, A. J. Strojwas, and S. W. Director, "A methodology for worst-case analysis of integrated circuits," *IEEE Transactions on Computer-Aided Design*, vol. CAD-5, pp. 104–113, January 1986.

[30] E. D. Boskin and C. J. Spanos, "Worst-case device characterization for statistical circuit design," in *Proceedings of the 1990 SRC Techcon*, pp. 165–168, 1990.

[31] G. E. P. Box, W. G. Hunter, and J. S. Hunter, *Statistics for Experimenters*, John Wiley and Sons, 1978.

[32] A. Khuri, *Response Surfaces: Designs and Analyses*, Marcel Dekker, 1987.

[33] G. E. P. Box, *Empirical Model Building and Response Surfaces*, John Wiley and Sons, 1987.

[34] H. Schichman and D. A. Hodges, "Modeling and simulation of insulated-gate field-effect transistor switching circuits," *IEEE J. of Solid-State Circuits*, vol. SC-3, pp. 285–289, September 1968.

# CHAPTER
# 15

# ANALOG COMPUTER-AIDED DESIGN

## 15.1  INTRODUCTION

In this chapter we explore how Computer-Aided Design (CAD) tools can aid in the design of analog signal processing ICs. First, we review the design process arbitrarily dividing the task into system-level design and cell-level design. Then we examine the existing and emerging approaches for carrying out analog IC design at both of these levels.

For both the company attempting to get its new product to market or the student attempting to get their IC design finished by a deadline, the biggest challenge is to rapidly, efficiently, and correctly design a mixture of analog and digital circuits for a specific application. This type of IC is often referred to as a mixed-signal application specific IC (ASIC). The ever increasing number of transistors that can be placed on a single IC, coupled with a desire to decrease manufacturing cost, leads inevitably in the direction of reducing complete systems to one or a few ASICs rather than a collection of standard parts on an application specific printed circuit board. Naturally this tendency leads to a great increase in the number of ASIC designs done each year.

Because existing analog design methodologies cannot keep pace with the increasing number of ASIC designs, there is a pressing need to develop analog CAD tools that increase the productivity of the analog designer. Already, there exist many powerful CAD tools such as logic synthesis and standard-cell place-and-route which aid in the rapid assembly of complex digital ASICs. With these tools it is not uncommon for large ($\geq 100,000$ transistors) digital ASICs to go from algorithm to completed IC masks in only a few weeks. In contrast, it is not unusual for an analog designer to spend several *months* on the design of a small ($\leq 30$ transistors) analog cell. In part, the difference stems from the more complex ways in which analog circuits use transistors and in part from the lack of sophisticated CAD tools to aid in the analog design process.

Our goal in this chapter is to expose the reader to some of the CAD tools that can aid in the analog design process and to provide an overview of the important problems that must be addressed. Where appropriate, we will also provide a brief description of the underlying algorithms and approaches upon which these tools are based. Note, this chapter is not meant to be a comprehensive survey of all CAD tools for analog design. Instead we select a few examples in order to illustrate both the problems that must be faced in the analog design process and some approaches to their solution.

In order to narrow the scope of our explorations, in this chapter we will focus only on CAD tools rated to the design process-converting performance specifications for an analog signal processing system into mask geometry. We call this process of creating a complete circuit schematic and mask geometry from performance specifications *synthesis*. Note, there are many other aspects of CAD for analog circuit design that will not be covered; e.g., simulation (see Chapter 14).

The analog IC design task can be decomposed into two components: system level design and cell level design. And, at both the system-level and the cell-level we have two quite dissimilar tasks: the circuit design task (specifying the "performance" of the primitive objects) and the physical design task (generating mask geometry). In this chapter we will first consider system-level circuit design and then cell-level circuit design. This will be followed by a discussion of the system-level layout problem, and we will conclude by addressing the cell-level layout problem. Layout of primitive analog cells and other important physical details are discussed in Chapter 16. In all of these sections we will describe existing approaches to the task and explore how CAD tools might be used.

Before we begin a discussion of the analog IC design process, however, we will briefly discuss the fact that there are many different genres of analog IC designs. For example, there are high volume analog building block ICs (e.g., a 741 op amp or a data converter), medium/low volume mixed-signal ASICs (e.g., local area network interface IC ), and one-of-a-kind university/experimental mixed analog/digital IC prototypes (e.g., the silicon retina [1]). Each of these different genres of design place very different demands on the analog designer and hence on the CAD tools that help the designer.

Successful analog building block ICs are usually manufactured in very large quantity. For these high volume analog parts production cost and performance are the most important considerations. In general, development costs will be unimportant because of the vast quantity of parts that will be sold. Therefore, designers of high volume parts devote tremendous effort to optimizing performance, minimizing cell area, and maximizing IC yield. Full custom manual design is typical for this type of product. It is likely that CAD tools that automatically synthesize analog circuits from specifications will not be applicable in this case. However, CAD tools that support the designer may still be of value; for example, synthesis tools might aid the expert designer in quickly exploring many alternative circuit topologies in order to choose one for further hand refinement.

The volume in which mixed-signal ASIC designs are manufactured can vary from tiny (e.g., a medical imaging signal processor ) to extremely large (e.g., a telephone subscriber line interface IC). However, in most cases the profit which can be made from an ASIC design is a strong function of the time from concept to completed IC (time to market). For example, a product can go from profit to loss because of a six-month delay in the design process. Hence rapid design is the key in the ASIC market. Various performance specifications must be met, but typically there is little incentive to optimize the performance beyond the requirements dictated by the specific application, unlike building block analog ICs. Because of the strong desire to minimize time to market, CAD tools for synthesis can potentially have great impact on ASIC design by speeding up large portions of the design process. In this case, the decrease in total design time (and hence time to market) is worth the penalty in power consumption and die area that typically occurs when using automated synthesis CAD tools. It is interesting to note, however, that for successful large volume ASIC parts, it is often the case that CAD tools are used in an initial design that fills orders early in the product life cycle, and that a full custom manual design is used later in the product's life cycle to achieve lower production costs.

For the prototype analog IC, the number of circuits that will be manufactured is generally quite small. Design time and costs must be minimized since they cannot be compensated by IC sales. On the academic side, we wish to decrease the design time in order to allow students to attempt larger more interesting system designs. Often for these designs, minimizing silicon area is not a priority because total project cost is dominated by design cost rather than manufacturing cost. All that is necessary is to make sure the analog circuit performs the desired function. This is clearly a case where CAD tools for synthesis can have great impact. Since the focus of this chapter is on CAD tools, we will concentrate on methods used in the design of ASICs and prototype analog ICs.

## 15.2   AUTOMATING ANALOG CIRCUIT DESIGN

In this section we address the task of converting performance specifications into a complete circuit diagram including device sizes and passive element values. First we will explore the structure of the analog design problem and discuss hierarchical decomposition. Then we will examine the characteristics of analog design at two distinct levels: the system level and the cell level.

## 15.2.1 Divide and Conquer -- Hierarchical Decomposition

Consider the design of a complex analog signal processing function. We find that the designer typically breaks down the design task into several parallel sub-design tasks by exploiting the hierarchical nature of analog circuits. Unlike digital circuits, which have an extremely well defined hierarchical decomposition (e.g., the transistor level, the gate level, and the register transfer level), there are no precise standards for how analog systems can be decomposed into lower complexity blocks. In digital system design, these well defined layers of hierarchy are very important. Individual design steps need only translate from one level to the next lower level (e.g., from the register transfer level to the gate level) which is often a much simpler task than going from the specifications for the digital system directly to transistors.

When several decompositions exist for a given analog system, we need to be able to compare their "quality." If it is possible for all of the sub-design tasks to proceed without any information about the results of other sub-design tasks, then the hierarchical decomposition is perfect and all of the sub-design tasks can proceed in parallel. Unfortunately this is only rarely achieved. However, we can use the amount of information about other sub-design tasks that is required in each sub-design task as a measure of the quality of a given hierarchical decomposition. In this case the less information that must flow between sub-design tasks, the better the hierarchical decomposition. For example, the simple Miller-compensated unbuffered op amp can be decomposed in many possible ways; e.g., into a transconductance amplifier, current mirrors, and a differential pair or into a first-stage amplifier and a second-stage amplifier. Following up both of these possible hierarchical decompositions, it quickly becomes clear that dividing the op amp into two amplifier stages was a poor choice because almost no decisions in the first stage can be made without knowledge of the second-stage decisions and vice versa. On the other hand, each of the sub-design tasks in the first hierarchical decomposition can proceed nearly independent of the others. Figure 15.1 illustrates a simple hierarchical decomposition of the steps involved in the analog system design process. For simplicity, we have illustrated only two levels of hierarchy – the system level and the cell level.

## 15.2.2 System-Level Analog Design

Let us take an example of a more complex analog system: a pipelined analog-to-digital (A/D) converter (Fig. 15.2). First, the pipelined A/D converter is broken down into a number of individual stages. The specifications for each stage must be chosen such that when the stages are assembled, the overall A/D converter meets its performance specifications. In this example each stage contains a comparator and a switched capacitor amplifier that. Finally, the amplifier is broken down into a high performance op amp and assorted capacitors and switches. The comparator, the op amp, and the switches and capacitors are then treated as fundamental cells and are not decomposed further.

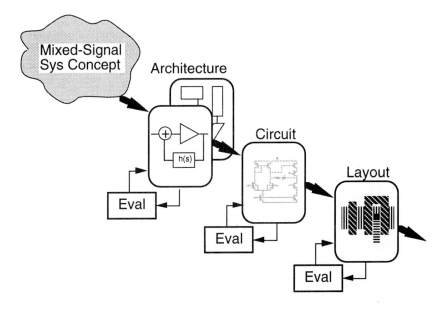

**FIGURE 15.1**
First-order decomposition of Analog System design process.

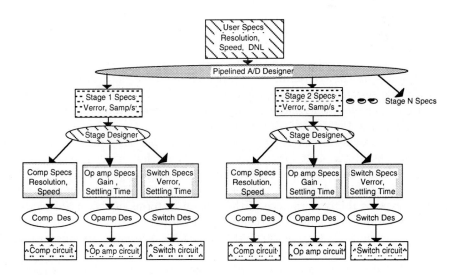

**FIGURE 15.2**
Possible hierarchical decomposition of a pipelined A/D converter.

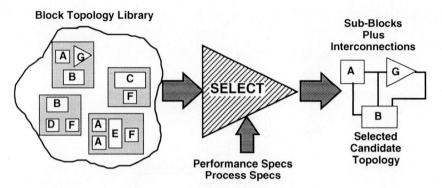

**FIGURE 15.3**
Selecting a candidate topology for a particular block.

Let us walk through this design example following the steps a typical analog designer might take. First, the architecture for the pipelined A/D converter is chosen. Some of these choices include: how many stages will be used, how many bits will be resolved in each stage, and what kind of digital error correction will be employed. The general term for this process is *topology selection* (Fig. 15.3). The designer starts by selecting a topology of blocks that will perform the desired overall function. What guides the designer's topology choice? In some cases, approximate analytical expressions can be written that roughly predict the performance, die area, and power requirements of each topology. In this case it is possible to directly compare various topologies and choose one that meets the required performance with the minimum area or power. Unfortunately, such analytical expressions may not be available and the designer may have to extrapolate from experience with similar system designs. And, if sophisticated CAD tools are available to help speed the design process, another possibility is to completely design several different topologies and see which one is best.

Having selected a topology, the next step is to choose performance specifications for each block that makes up that topology such that if every block meets its specifications, then the overall system will achieve the required performance. We call this process of converting specifications at one level into specifications for the blocks at the next lower level in the hierarchy *refinement* (Fig. 15.4). There are many approaches to refinement. A common approach for the human designer is to either derive, or to look up in a textbook, approximate formulae that express the performance of the system in terms of the performance of its blocks. Note, these equations themselves do not constitute a design method, but only a way of evaluating the systems performance given the performance of every block in the topology. A common design approach is to simplify these equations and to add reasonable heuristic assumptions (e.g., after meeting the performance specifications, die area and power consumption should always be minimized) until a set of specifications for the blocks can be derived. First, because we simplified the original equations in order to solve them, we should plug

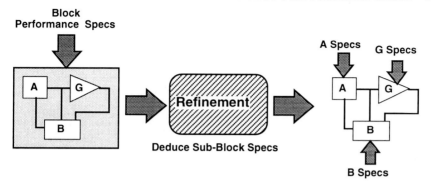

**FIGURE 15.4**
Choosing specifications for each block in a topology.

the set of specifications we derived for each of the blocks into the unsimplified equations in order to verify that they do indeed satisfy the required system performance specifications. If not, we can make small adjustments until they do – hopefully we are close to the correct answer. Another alternative at this point is to employ some form of simulation to predict the system performance given the performance of every block in the topology. For some systems (e.g., 1-bit oversampled data converters), analytical equations are so difficult that the designer may have to start with a rough guess for the blocks and iterate using simulation right from the start.

Having carried out the process of selection and refinement until we have determined specifications for all of the cell-level blocks, we next design and assemble them, and we are done. Unfortunately, this is rarely the case. Since we assigned performance specifications to cells without a detailed knowledge about how those cells would be implemented, we may discover that one or more of the cells simply cannot be designed to meet its performance requirements. The human designer relies heavily on experience in order to avoid requiring unreasonable performance from a cell. However, novice designers, and experienced designers working in a new technology, often find that many iterations through this design process are necessary. We call this process of verifying the performance and designability of lower level blocks, and then changing prior decisions to move toward a more realizable design, *backtracking*. The philosophy for backtracking suggested in Fig. 15.5 is to always make a change in the most recent heuristic or topology selection decision because it will affect the fewest number of other cells.

## 15.2.3   Cell-Level Analog Design

Many of the early successes in the development of CAD tools for digital ASICs were based on the use of hand designed and optimized libraries of cells which performed basic digital functions – e.g., a 2-input NAND gate or a D-latch. Similarly, many of the early successes in the development of analog synthesis

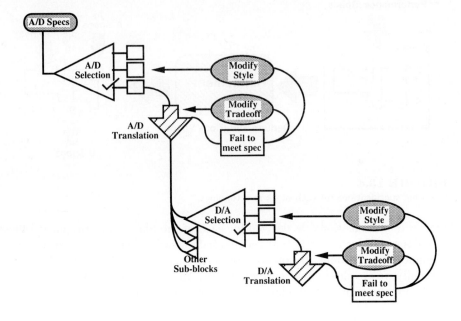

**FIGURE 15.5**
Backtracking to handle cell design failures.

tools were also based on hand-designed and optimized libraries of cells which performed basic analog functions; e.g., an op amp or a comparator. Early examples using this style included the automatic synthesis of data converters [2] and switched capacitor filters [3].

Although there are several noteworthy applications for which the cell library approach has worked, it has generally failed to handle most analog system design problems. The reason becomes clear if we consider the number and kinds of performance specifications that characterize a cell. For example, compare a NAND gate and an op amp. The NAND gate is characterized by its truth table, some specification on propagation delay, output levels, input levels, and load driving capabilities. However, for a given logic family, only the propagation delay and the load driving capability are likely to change across a wide range of designs, and most requirements are probably satisfied by a few values of each. On the other hand, the number of performance specifications that characterize an op amp are numerous ($\geq$ 20). A few of the most common ones include: open-loop gain, unity-gain bandwidth, input common-mode range, output swing range, common-mode rejection ratio, slew rate, etc. Moreover, different designs for different applications might require very different combinations of these op amp performance specifications. Hence, a "complete" analog cell library, one which has the right cell to meet all system design requirements, is impossible to create and maintain. A library with cells having two values, high and low, for each of 20 performance specifications would contain over 1,000,000 cells! This need to handle many performance specifications for analog cells is one of the primary motivations to develop more general cell-level synthesis tools. The above argument may over estimate the problem: for example, a cell with more slew rate than necessary will still work, though it may consume more power than a cell that just meets the slew rate specification. However, most ASIC design companies have found that while analog cell libraries can be useful as a starting point, most ASIC designs require modification of one or more analog cells in order to meet the overall system performance requirements.

In many ways the steps involved in automating cell-level analog circuit design are not different than those at the system level – we must still perform topology selection and refinement. Topology selection proceeds much as for system-level design; however, refinement can be quite different. The primary difference is that instead of mapping the cell's performance specifications into specifications for unknown sub-blocks, we are mapping them into devices sizes. The performance of a device as a function of its sizes can be determined by using device model equations. Therefore, at the cell level it is possible to predict the cell's performance given a choice for all of the device sizes and operating points. This ability to numerically predict cell performance encourages a numerical optimization style for refinement at the cell level.

Most methods of cell-level analog circuit design ultimately rely on some form of numerical optimization in order to determine the device sizes and operating points. Figure. 15.6 illustrates the most general form of optimization-based analog circuit design. The two distinguishing characteristics of an automated analog design system are how performance is evaluated and the optimization algorithm which determines the new values of the design variables (i.e., the device sizes and bias currents).

In this section we will begin by describing analog cell-design techniques which are based on using circuit simulation to predict cell performance. We will then describe "equation-based" performance prediction which is the basis for most of the automated analog cell design work to date. Next, we will point out the difficulties imposed by the equation-based approach and briefly describe one of the current research directions attempting to overcome these barriers. An example of the automatic design of a high-performance op amp will be presented. Finally, we will discuss the importance of manufacturing variations and operating range on the design of analog cells.

**Simulation-Based Analog Cell Design:** The most common method for predicting the performance of analog cells numerically is a circuit simulator. In general, a circuit simulator must solve a set of nonlinear equations (given by the device models) for a set of dc-node voltages that satisfy KCL – we call this a "dc-solve." The simulator can then determine various ac-transfer functions of the small-signal circuit linearized about this dc-operating point – we call this "ac-analysis." Or, the simulator can numerically integrate the response of the circuit to a transient input using the dc-operating point as an initial condition

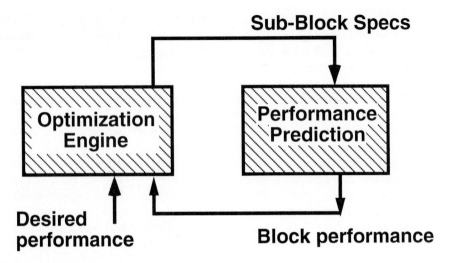

**FIGURE 15.6**
General use of optimization based on performance prediction for refinement to device sizes.

– we call this "transient analysis." For more information about simulation, refer to one of the standard texts in this area; e.g., [4].

Early attempts at synthesizing analog circuits typically used some form of circuit simulation in order to predict circuit performance and some form of "downhill" algorithm for numerical optimization [5-8]. Unfortunately, due to the large computational cost of circuit simulation for even small analog circuits, only a limited number of points (distinct sets of sizes and bias currents) in the design space can typically be evaluated. For example, approximately 4 seconds of computer time is required to simulate the basic dc- and ac-performance measures for the op amp that will be presented as an example later in this chapter. Thus, within an hour, less than 1,000 combinations of the design variables can be evaluated. If we are only tuning the size of one device while keeping the sizes of all other devices fixed, then this is enough combinations to obtain a good design. However, in order to obtain good designs when choosing widths and lengths for many transistors, we must evaluate many more designs. This problem is further aggravated by methods that directly compute derivatives because determining the partial derivative, with respect to each design variable then requires additional simulations. Because of its excellent accuracy, simulation-based optimization is the method of choice when only a few design variables are being determined, and the rest of the design variables are fixed. In addition, many commercial CAD tools offer some form of simulation-based optimization; e.g., [9].

**Equation-Based Analog Cell Design:** If CAD tools are to aid the designer in the process of simultaneously choosing sizes for all the devices in an analog cell, some more efficient means for evaluating performance is necessary. A much faster approach to predicting the performance of an analog cell is to mimic in the computer the process by which the expert human circuit designer predicts performance; a.k.a., cocktail napkin analysis. As the story goes, any circuit worth designing can be analyzed over drinks using equations scratched onto a cocktail napkin. The point of this bit of circuit designer lore is that it is often possible for the expert designer to greatly simplify a circuit keeping only the second-order effects that are important in the analysis at hand. In fact, it may well be this ability to identify which terms in a complex analytical expression can be ignored and which must be kept that separates the novice designer from the expert designer. There have been numerous approaches to solving the cell-level analog circuit design problem that have employed some form of analytical equations to predict the approximate behavior of both analog circuits and devices [10-15]. This method has even been employed in a commercial system for automating analog cell design, IDAC[10]. The computational cost of evaluating these approximate analytical equations is extremely low, making it possible to explore an extremely large number of points in the design space. Methods based on using analytical equations (equation-based methods) to evaluate performance have been able to successfully synthesize analog circuits without requiring any starting point information.

There are two primary drawbacks to equation-based optimization methods: inaccurate device models and the difficult and time-consuming nature of the process of acquiring the equations from expert analog designers. Most equation-based optimization methods employ device model equations that are much simpler than the device models used in circuit simulation. This is typically done so that the dc-biasing equations can be solved more easily. Unfortunately, as device dimensions continue to scale downward, analytically tractable device model equations for small devices become ever more difficult to develop and verify, increasing the likelihood that synthesized designs will not meet the required performance specifications. Therefore, it would be desirable to make use of the same device model equations during automated circuit design as those used in circuit simulation.

The second barrier to the wide-spread acceptance of equation-based optimization methods for cell-level analog circuit design is the difficulty of creating the equations. It may take an expert analog circuit designer a considerable amount of effort to determine just which terms in a complex expression for the performance of an analog circuit can be ignored and which are important. And, it is often the case that these determinations are a function of the performance specifications themselves – what is negligible in one application might be very important in a different application of the same analog cell. One approach to automating the process of creating these equations is to derive them symbolically, which is termed symbolic simulation [15]. Unfortunately, to date, only linear or nearly-linear systems of equations can be solved symbolically and the growth in the number of terms in the symbolic expressions with the number of elements in the circuit tends to be exponential. However, pruning strategies may allow approximate symbolic expressions of tractable size to be derived.

The large investment required to develop a set of equations that predict the performance of an analog cell as a function of the device sizes has limited the applications of equation-based methods primarily to analog cells which are extremely common; e.g., an op amp. In this case the large time invested in creating the automatic design tool can be amortized over many automatically generated designs. However, the direction of research in this area is toward decreasing the number of equations which must be provided by the expert designer and to developing user interfaces which make it simpler to present these equations in computer readable form.

**A Hybrid Approach to Automated Analog Cell Design:**   As an example of one direction research is moving in the area of analog design, we present a hybrid approach which attempts to combine the good features of simulation-based methods with the good features of equation-based methods. The hybrid approach is embodied in a tool called ASTRX/OBLX [16]. This particular hybrid approach derives from the equation-based method. However, it incorporates modifications to (1) require only simulator device models, and (2) to decrease the number of equations that the expert designer must provide, particularly the ones that are difficult to derive.

In our experience, deriving highly complex equations to predict linear circuit performance (e.g., an equation to predict the phase margin of an op amp) by hand can be both tedious and error prone. Although the usual simplifications found in textbooks can be applied to determine pole locations to first order, the errors here become very significant in high performance designs. The use of symbolic methods or efficient numerical methods to handle linearized transfer functions of all types greatly frees the expert analog designer, when creating an automatic design tool for a new circuit topology, from the need to derive expressions for small-signal transfer functions. This is particularly advantageous for high performance analog designs in which a number of poles and zeros may all play a complicated role in determining performance.

Therefore, ASTRX/OBLX incorporates an extremely efficient method for solving small-signal circuits numerically eliminating the need for the designer to provide any small-signal transfer function equations. For example, instead of asking the expert designer to create an approximate equation to predict the phase margin of an op amp, we can simply solve for the phase margin numerically. However, we must remember that if we carry numerical solutions too far, we will simply end up with a circuit simulator. Therefore, the goal is to decrease the number of equations that must be provided by the expert analog circuit designer as much as possible without increasing the computer time required to the order of that required by circuit simulation. One of the keys to achieving this goal is to use very efficient numerical methods for determining the transfer functions of linear systems, in this case Asymptotic Waveform Evaluation (AWE) [17].

Figure 15.7 illustrates how the performance of an analog cell is determined in ASTX/OBLX. It starts by assuming that the transistor sizes and the circuit's node voltages (Vs) are given. There are two approaches to determining the node

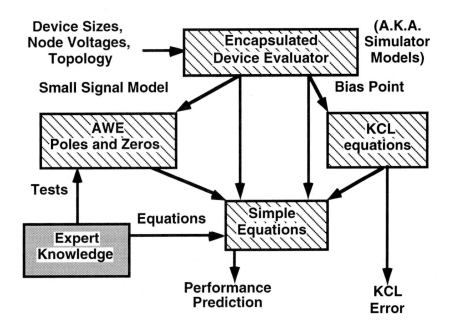

**FIGURE 15.7**
Block diagram showing how analog circuit performance is predicted.

voltages. We can either take the specified device sizes and solve the circuit for dc-equilibria, or we can add the set of node voltages to the set of design variables and the KCL equations as constraints for the optimizer. One drawback of doing a dc-solve each time the device sizes are changed is that solving the dc-problem is one of the more difficult aspects of circuit simulation. Therefore, if we were to follow this approach, we might as well return to using a circuit simulator to predict performance.

Since we know that the optimizer is going to do a long sequence of performance evaluations, we may wish to relax the accuracy of the dc-solutions early in the design process. We can make the node voltages unknowns that are determined by the optimizer. In essence, the optimizer need only arrive at node voltages that correctly solve the KCL equations at the end of the optimization process when a solution is found. Early in the design process, the sum of the currents from all devices attached to a given node may not equal zero (a KCL violation) because the optimizer does not determine the precise dc-equilibrium node voltages for each set of device sizes.

From the device sizes and operating points, all device currents and small-signal parameters can be determined by employing device models identical to those used in circuit simulation. From this we can construct small-signal circuits which can be evaluated numerically or by invoking expert designer provided equations. Nonlinear behavior such as slew rate is estimated using the expert designer's equations, values of currents, and device parameters. The expert de-

signer can provide equations that predict any performance of interest as long
as they are expressed in terms of the small-signal parameters, the dc-bias information that is determined from device evaluations, and the transfer function
predicted by numerical analysis of the small-signal circuit. Note, the cell can
be placed in any number of different test circuits (as specified by the expert
designer) in order to characterize its performance. In addition, different performance measures may be evaluated at different dc-operating points. Evaluating
performance measures at different dc-operating points requires starting at the
top of Fig. 15.7 by calling the device evaluator with the same device sizes but
different sets of node voltages.

**Choice of Optimization Algorithm:** So far we have only discussed approaches to evaluating the performance of the analog cell. There are many possible numerical optimization methods that can be used to solve the analog cell
design problem. Each approach has its own advantages and disadvantages. For
an overview, see [18]. One common approach is to use an unconstrained nonlinear optimizer; e.g., [12]. There are two main difficulties with the unconstrained
optimization approach. First, there is no easy way to specify the KCL requirement – an unconstrained optimizer will always trade slightly better performance
for not satisfying KCL. Second, when multiple performance specifications must
be met, unconstrained optimizers require that a weight indicating the relative
importance be assigned to each specification so that they can be added together
to form a scalar cost function. It may be difficult for the designer to set these
weights. Use of nonlinear constrained optimization techniques avoids both of
these problems [19, 20]. Forcing KCL violating currents to equal zero can be
included as a constraint along with other performance constraints and device
operating region constraints.

A problem with nonlinear optimization in general is that there may be
multiple local minima. Most nonlinear optimization algorithms will tend to find
the local minima nearest to the starting point and may not be able to find a
much better solution, even if one exists. A numerical optimization technique that
searches for globally optimal solutions and is relatively independent of starting
point is simulated annealing. For an overview of simulated annealing see [21].
Because of the nature of the analog design problem formulation, the annealer's
cost function and move generation mechanism must be modified [16]. Specifically,
we have a mixture of continuous (e.g., node voltages) and discrete variables (e.g.,
device sizes) which must be optimized. And in order to force simulated annealing
to act as a constrained optimizer, the weights for the various terms in the cost
function must be modulated dynamically during the annealing process.

**Example 15.1.    Op Amp Design.**

For our design example, let us consider the case where we are synthesizing
an op amp circuit with no initial idea about what the device sizes should be.
Therefore, we will use simulated annealing as an optimization strategy in order
to avoid the local minima problem. Consider the design of the high slew rate

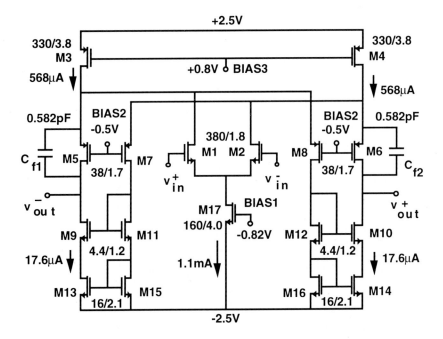

**FIGURE 15.8**
Schematic diagram of fully-differential high slew-rate op amp topology.

fully-differential op amp topology [22, 23] shown in Fig. 15.8.

For simplicity, let us imagine that the performance specifications that we are interested in are the following: (1) dc-gain, (2) unity-gain frequency (UGF), (3) phase margin, (4) slew rate, (5) output swing range, and (6) input common-mode range. Further, we need to be given operating conditions: e.g., that the circuit will be driving a purely capacitive load, $C_L$, and what the power supplies are $(+/-2.5V)$. The op amp is then connected as an inverting amplifier with the noninverting input tied to ground. The KCL violating currents are calculated using this circuit.

Numerical evaluation of the linearized small-signal circuit for the inverting amplifier can then be used to predict performance measures (1)-(3). However, because they involve large signal or transient analyses, equations predicting performance measures (4)-(6) must be provided by the expert designer as shown below.

$$SlewRate = \frac{I_{DS3}+I_{DS4}}{4(C_L+C_{DB6}+C_{DB10}+C_F)}$$

$$V_{OL} = V_{DS13/14} + V_{DS-SAT9/10}$$

$$V_{OH} = V_{DD} - V_{DS3/4} - V_{DS-SAT5/6}$$

$$V_{IL} = V_{GS1/2} + V_{DS-SAT17}$$

**TABLE 15.1**
Four folded-cascode designs with varying UGF. The CPU hours are on a 20MIP workstation.
These designs are in a $1.2\mu$ CMOS process.

| | Spec | 1 | 2 | 3 | 4 |
|---|---|---|---|---|---|
| | | Pred /Sim | Pred /Sim | Pred /Sim | Pred /Sim |
| $V_{DD}$ / $V_{SS}$ | +2.5V/-2.5V | | | | |
| $C_{Load}$ | 1.0 pF | | | | |
| Gain(dB) | 83 | 86/87 | 84/86 | 84/84 | 84/80 |
| UGF (MHz) | > 90 | 99/92 | 120/121 | 145/146 | 237/239 |
| $\phi$ Marg. (deg) | 45 | 91/91 | 90/90 | 88/88 | 86/86 |
| Slew Rate ($V/\mu s$) | 150 | 150/145 | 150/145 | 150/145 | 150/145 |
| Out Swing (V) | 2.5 | 2.6/2.5 | 2.6/2.5 | 2.6/2.5 | 2.6/2.5 |
| In C-M Range (V) | 1.0 | 1.7/1.7 | 1.7/1.7 | 1.7/1.7 | 1.7/1.7 |
| Area ($\mu m^2$) | - | 5500 | 7000 | 6600 | 9100 |
| Power ($mW$) | - | 1.27 | 1.87 | 2.30 | 5.66 |
| Perf. Eval ($10^3$) | - | 60 | 60 | 63 | 72 |
| CPU (Hours) | - | 1.0 | 1.0 | 1.0 | 1.2 |

$$V_{IH} = V_{DD} + V_{GS1/2} - V_{DS3/4} - V_{DS-SAT1/2}$$

One final equation, which is needed to guarantee that $I_{DS3/4}$ are actually available during a slewing transient, makes sure that $M3/4$ stay in the saturated region of operation. That equation is

$$BIAS2 - V_{GS5/6/7/8}(I_D = I_{DS3}) < V_{DD} - V_{DS-SAT3/4}$$

Note, the expert designer must also provide simple equations that relate the unity-gain frequency and the phase margin to the poles and zeros determined by AWE. However, this relationship is the same for all op amps and can be specified once; e.g., a general procedure for computing phase margin from a set of poles and zeros. Note that all of the equations which must be provided by the expert human designer are quite simple and straightforward to derive.

Table 15.1 summarizes the performance of several fully-differential op amps that have varying unity-gain frequencies and areas. The table also emphasizes one of the important points about simulated annealing. Although it is able to escape local minima, it is not guaranteed to find a global optima. The second design is slightly inferior to the other three in terms of the area required for the performance obtained. It is standard practice to synthesize a number of circuits to the same specifications and then to pick the best ones. The excellent correspondence between the predicted and simulated values of unity-gain frequency and phase margin are a testimony to the accuracy of the device models and the ability of AWE to accurately characterize the small-signal transfer function.

**Manufacturing Variations and Operating Range:** Two important aspects of the design process, whether manual design or optimization-based design, are the impact of variations in the manufacturing process and the impact of operating range specifications.

The integrated circuit fabrication process exhibits inherent fluctuations in parameters, which appear as correlated variations in the device model parameters from the perspective of the analog design tool. There are many possible approaches to managing these statistical variations in the device model parameters. One common approach is to use Monte Carlo analysis to predict the distribution of performance parameters as a function of device model parameters. By applying statistical techniques, the correlations between variations in model parameters can be measured and a statistically independent set of disturbance variables can be determined. Typically, most of the variation in the model parameters can be explained with only a few variables (typically between 5 and 11 independent variables are needed for a CMOS process). In addition to these independent variables which model global manufacturing process variations, an independent variable with an appropriate statistical distribution is also needed for each pair of matching elements. In the face of large global process variations (20% variations are typical) analog designers achieve high precision by designing circuits whose performance relies on the matching between two components, not on their absolute value. If performance estimation is to accurately predict the performance of manufactured circuits, then it must also take into account the small mismatches which occur between nominally identical components. Note, as will be discussed in the next section, the physical layout of these components will have a significant impact on the variance of the mismatch between them.

Monte Carlo analysis is performed by picking a large number (e.g., 300) of sets of these independent variables according to their probability distributions as determined by the actual fabrication process, and then computing the performance for the analog cell for the device models determined by each set of independent variables. The fraction of the designs which satisfy all of the performance requirements is called the "yield." An simple example of this type of yield analysis is shown in Fig. 15.9.

Unfortunately, computing the yield for the analog cell at each step of the design (optimization) process may be computationally impractical, even when using equation-based methods. Therefore, several alternative approaches with decreased computational cost have been developed. One very common approach is to use a small number of sets of the independent process variables that are carefully chosen to be "representative" of the variations in the overall fabrication process. For example, sixteen sets of variables can give a representative indication of the behavior of the circuit with respect to global fabrication disturbances, while requiring much less computational effort than full Monte Carlo analysis. If the analog cell meets its performance specifications at all sixteen points then it can be expected to have good yield during the actual manufacturing process. Of course, choosing the sixteen points can be difficult. And, new circuit topologies can exhibit heretofore unknown sensitivities to the correlations between model parameter variations. This technique represents an excellent compromise between accuracy and computational effort for use during the design process. However,

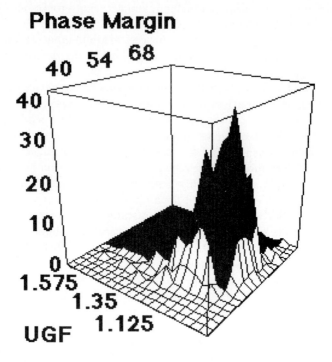

**FIGURE 15.9**
Variation in phase margin and unity-gain frequency of a two-stage Miller compensated op amp with manufacturing variations. The black region corresponds to designs which exceed both the phase margin specification (45 degrees) and the unity-gain frequency specification (1 MHz). The white regions in front and on the left are designs which do not meet the performance specifications.

the synthesized analog cell design should be subjected to a complete Monte Carlo analysis before it is manufactured.

Another common approach has been called $N\sigma$ design, where an $N$ of 3 is common. In this approach, we assume that the worst-case performance occurs as we move further from the nominal parameter values; i.e., the worst-case performance will occur when the independent variables take on values of $+/-N\sigma$. Assuming that there are $M$ independent variables, one approach would be to examine all possible combinations of the $M$ independent variables being at their $+/-N\sigma$ values. This would be similar to the digital design philosophy of fast-fast, fast-slow, slow-fast, and slow-slow for the speed of NMOS and PMOS devices in a CMOS circuit design. In order to ensure good yield, the circuit should be designed to meet its performance specifications at each of these "corners" of the process space. However, if there are $M$ independent variables, then circuit performance must be evaluated at $2^M$ possible corners plus the nominal point. Although this is acceptable for $M = 2$, as in the digital case, as $M$ increases the computational time required rapidly exceeds that needed to perform Monte

Carlo analysis.

The exponential increase in the number of performance evaluations with increasing $M$ can be avoided by changing only one process variable to its $+/-N\sigma$ values while keeping all of the other independent disturbance variables constant. In this case, the circuit performance must be evaluated at only $2M$ possible corners plus the nominal point. In order to ensure good yield, the circuit should be designed to meet its performance specifications at each of these corners of the process space. This strategy is often applied by the expert human designer. For example, when biasing a transistor, the human designer will make sure that the transistor remains in the desired operating region even if $V_T$ of the transistor varies from its nominal to its $+/-N\sigma$ values. This method is also particularly easy to apply to equation-based analog cell design because any time a model parameter appears in an equation, we can just plug in its worst-case value from all of the possible combinations of independent variables. For example, if a $g_m$ appears in the numerator of an expression which has a specified minimum value, then we can plug in the minimum $g_m$ which occurs for all possible corners.

Another type of performance constraints, which are conceptually quite different from manufacturing variations, are operating range constraints. For example, a particular op amp should meet its minimum gain specification over a range of power supply voltages and temperatures. These specifications of operating range can actually be regarded as an infinite set of constraints on the gain at every allowed combination of power supply voltage and temperature. Ignoring operating range constraints when synthesizing analog circuits can result in a design which meets specifications at the nominal operating point, but fails to meet specifications somewhere in the allowed operating range. For example, Fig. 15.10 illustrates the variation in phase margin of an op amp of the type described above that was synthesized by ASTRX/OBLX using only nominal operating point performance measurements. For reference, the nearly flat line on the figure shows the behavior of a manual design of this same op amp topology [23]. It is dramatically obvious from these curves that analog cell design cannot be based on only the nominal operating point, but must take into account the full operating range specifications.

The simplest approach to coping with operating range specifications assumes that the variation in any performance measure with any operating range variable is monotonic. That is, the minimum gain will either occur at the maximum allowed temperature or the minimum allowed temperature. Similarly, the minimum gain will either occur at the maximum power supply voltage or the minimum. Further, it is often the case that for many variables the end of the range which will yield the worst value for a particular performance measure can be identified in advance by the expert designer. In this case, instead of evaluating performance at the nominal operating point we evaluate each performance measure at the worst combination of operating range variables for that measure. Note, different performance measures will require evaluation at different values of the operating range variables. Unfortunately, there are some performance measures and some operating range variables which are not related monotonically.

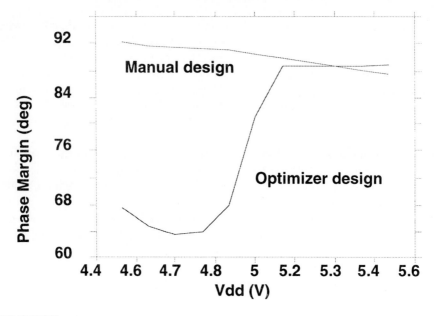

**FIGURE 15.10**

Variation in phase margin with power supply voltage for a manually designed op amp and one designed using only nominal process parameters and operating points.

However, methods for coping with this situation are beyond the scope of this chapter.

In conclusion, automating analog cell design is an area which is currently undergoing intense research activity. Analog CAD tools do exist that can aid the designer in tuning a design (starting from a good design and moving to one which is better) or exploring for the best value of a small number of design variables; e.g., [9]. And, analog CAD tools that can aid in the process of determining initial device sizes for an entire analog cell using equation-based methods have also been developed but still require a formidable effort in order to enter a new analog cell topology; e.g., [10]. Finally, new approaches are currently being developed which will lead to analog CAD tools that (1) will determine device sizes for an entire analog cell, (2) will not require an onerous amount of expert designer effort to develop a new circuit topology, and (3) will take manufacturing variations and range variable constraints into account.

## 15.3   AUTOMATING ANALOG LAYOUT

In this section we will discuss how mask geometry for an analog IC is generated, both at the system level and at the cell level. Before discussing approaches to generating analog layouts automatically, it is important to review some of the properties that distinguish analog IC layout from digital IC layout.

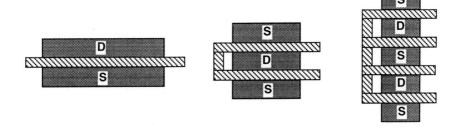

**FIGURE 15.11**
Examples of possible layout variants for a single MOS device.

## 15.3.1   Analog Layout Challenges

In general, the wires required to connect together the elements of a circuit will introduce additional parasitics which degrade performance. We can arbitrarily divide the degradations introduced into analog circuits and systems by the IC layout process into four categories: *device layout parasitics, wiring parasitics, signal couplings,* and *matching deficiencies.* We next describe each of these in more detail, and also describe approaches to minimizing their impact.

   *Device layout parasitics* are parasitic elements added to the nominal device or element by the process of generating the specific layout geometry. For example, for an MOS device with a large width, there are many possible layout variants (Fig. 15.11) each corresponding to a different number of gate folds and hence a different aspect ratio. During the layout process the human layout designer, or the automatic layout system, needs to be able to switch between several of these layout variants in order to best fit the device into the available space. Preselecting a single geometry for each device may significantly increase the total die area required by the cell. Unfortunately, each of these layout variants has a different size of parasitic reverse biased diode from the source and drain to the well or substrate; and, the parasitic capacitance of these reverse biased diodes often has a significant impact on the performance of the cell. Therefore, the transistor sizing process, which requires reasonably accurate knowledge of device parasitics, and the layout process, during which specific shape variants with different parasitic capacitances are being selected, are intimately coupled and must proceed simultaneously in order to fully optimize performance and die area.

   One strategy which helps to decouple the impact on performance of changing variants is to require at least 1 fold for MOS devices larger than some minimum width. There is almost a factor of two decrease in the capacitance between the drain and the well or substrate when going from no folds in the gate to one fold in the gate. However, the difference between 1 fold and 3 folds is only a few percent (only a difference in the total perimeter of the drain). Therefore, a common strategy during design is to assume that devices larger than some minimum width will always be folded at least once and will have an even number of gate stripes. This is a relatively slight restriction on the layout process and generally does not significantly increase the die area of the cell over the case where no limitation is imposed on the number of folds. During the transistor sizing process, we can compute approximate device parasitics by assuming that the number of folds is fixed, or even that it is adjusted to yield a device whose overall aspect ratio is closest to square. This is an example of the kind of strategies which can be adopted to minimize the coupling between the circuit design and the impact of layout parasitics.

Parasitics for passive elements may also be very important, and are often neglected by layout-to-circuit extraction programs. Common examples include the bottom plate parasitic capacitance for a precision capacitor, the parasitic capacitance to the substrate or well associated with a resistor, and the parasitic diode that occurs when a floating well is used.

Another extremely important impact of layout on device parasitics is the possibility of separate devices sharing pieces of geometry. For example, the sum of the drain-to-substrate capacitance and the source-to-substrate capacitance of a cascoded current mirror (Fig. 15.12) can be dramatically decreased by sharing the drain diffusion of the current mirror transistor with the source diffusion of the cascode transistor. In essence, this is a generalization to multiple devices of the type of geometry sharing already introduced in the folding of a single device. From the perspective of source and drain diffusions, every diffusion stripe has two edges, and if it is electrically possible to use both as terminals of a device, then the total diffusion parasitic capacitance will be reduced. Other advantages of merging diffusion regions are that the total area of the cell is reduced and that the complexity of the routing task is decreased (one wire is removed from the list of wires that must be routed). Note, it is straightforward to identify all possible diffusion regions which can be merged given only the netlist: any electrically connected diffusions of the same type can be merged.

However, it is not the case that an equal amount of parasitic capacitance is equally harmful on every node in the circuit. For example, diffusion capacitance on the source of the current source transistor has no effect on the circuit (it is shorted by the connection to a power supply) while diffusion capacitance on the source of the cascode transistor does impact the circuit performance. Either the human or the CAD tool performing the layout needs to know the relative importance of parasitic capacitance at every node in the circuit.

Fortunately, for small circuits it is possible to use sensitivity analysis based on circuit simulations to automatically compute a set of upper bounds on all of the device capacitances that, if satisfied, will guarantee that the circuit will meet its performance requirements [24]. These limits on parasitic capacitances can be used to guide the selection of diffusion mergings.

Note, the value of diffusion merging is primarily of importance for MOS devices. In the case of bipolar devices one can sometimes merge base regions, often merge collector regions, and almost always merge isolation diffusions. However, because of the complex three-dimensional nature of current flow in a bipolar

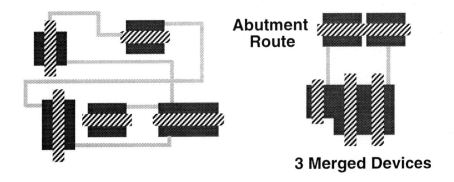

**Abutment Route**

**3 Merged Devices**

**FIGURE 15.12**
Examples of possible device mergings. The group of devices on the left is not merged. The group of devices on the right implements the same circuit, but employs merging and routing by abutment.

transistor, as opposed to the approximately two-dimensional current flow in an MOS transistor, it is frequently the case that experimental measurements are used to characterize specific device geometries electrically, and that only those geometries that have been characterized are typically used in the design of ICs.

*Wiring parasitics* are the result of connecting devices and passive elements together with wires which have series resistance and parallel capacitance to the substrate, well, or other region over which they are routed. Both the parasitic series resistances and the parasitic capacitances typically degrade the performance of an analog circuit. Therefore our goal in creating the layout is to minimize the length of the wires which simultaneously minimizes both the series resistance and the parallel capacitance.

Series resistance is fundamentally different from parallel capacitance. The impact on performance of wiring resistance depends not only on the total length of wire on each layer but even on the topology by which the terminals of a multi-point net are connected (Fig. 15.13). This is because wire resistance allows the current drawn by one module on a single net to influence the voltage seen at other modules to the extent that they share resistance. One common way to distribute signals such as reference voltages which can be connected to many modules and which are very sensitive to small errors in voltage is to use a star connection (Fig. 15.13). The advantage of a star connection is that no module shares any wiring resistance with any other module. The performance bounds on variation in the distributed voltage and knowledge about the current drawn by a module are all that are needed in order to determine an upper limit on the series resistance from that module to the star (the voltage source point). Note, the star connection approach has the serious disadvantage that it can require dramatically more die area because it requires substantially longer wires than would be used in a standard routing method. On the other hand, if the allowable voltage deviations and the range of input currents are known for every module,

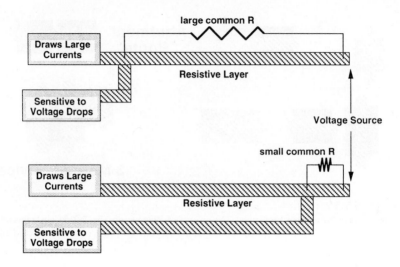

**FIGURE 15.13**
Examples of impact of series resistance on different net architectures.

then various segments of the route can be automatically sized in a manner which results in acceptable deviations in voltage throughout the entire net. This is particularly true for voltage references and power supply nets. Because of the tedious nature of these calculations they are best done by an automatic analog CAD tool [25].

Just as for device parasitics, it is not the case that an equal amount of parasitic capacitance is equally harmful on every node in the circuit nor that an equal amount of series resistance is equally harmful in every wire. Either the human or the CAD tool performing the layout needs to know the relative importance of the parasitic capacitance at every node and the parasitic resistance of every connection in the circuit. Fortunately, it is possible to use sensitivity analysis based on circuit simulations to automatically compute a set of upper bounds on the parasitic capacitances between every pair of nodes in the circuit that, if satisfied, will guarantee that the circuit will meet its performance requirements [24]. These limits on parasitic capacitances can be used to guide both placement and routing in order to meet the performance requirements. Determining the parasitic resistance bounds is made more complex because the actual targets depend on the precise manner in which the a network is decomposed. After a network is decomposed into a set of two-point wire segments, an upper bound can be placed on the resistance of each two-point wire segment. Unfortunately, it is not until we are in the middle of the layout process that we can tell which decomposition of a network will fit best into the physical space available. Although it would be technically possible to choose a single fixed decomposition into two-point wire segments for every net before beginning physical placement and routing, in most cases this would result in a substantial increase in the die area required by the cell. A better approach is to specify bounds on the deviations in the voltages at each module in the net, as well as ranges of current demand. Using this information, the layout tool can determine how well any particular net decomposition satisfies the specified voltage deviations.

*Signal couplings* allow signals on one net to cause a disturbance on another net. The size of disturbance which can be tolerated on an analog signal is typically much smaller than the noise margin for a digital signal. For example, an analog circuit with a full-scale signal range of 2 volts and a performance specification that requires a signal-to-noise ratio (SNR) of 60dB, or alternatively an accuracy of 0.1%, can only tolerate a 2 mV disturbance induced by other signals.

Where do these disturbances come from? There are many possible sources. The primary source of these disturbances are capacitive couplings between signal lines or between signal lines and devices. For example, clock lines, other digital signal lines, or even high swing analog signals that cross over, or even run near lines carrying sensitive analog signals can induce disturbances. Unfortunately, accurate prediction of the precise capacitance between two wires would require a full 3-D fields analysis; however, reasonable ways to quickly estimate the capacitance can be used. For example, when two wires cross, or a wire crosses over a device, we can assume a perfect parallel plate assumption and simply look up the capacitance per unit area between layer 1 and layer 2 in a table for that process and multiply by the overlapping areas. Of course this is likely to underestimate the actual capacitance, but it is usually a good first approximation. When two wires run adjacent to each other they also have a mutual capacitance which couples signals between them. Unfortunately, many extractors compute capacitance only based on vertical overlap. A simple way to estimate the capacitance due to a parallel run of two wires is build a table for the target fabrication technology of the 2-D capacitance between wires every possible pair of layers in isolation at various separations. Then to compute the actual capacitance simply look up the correct pair of layers and interpolate to the actual separation to get a value for capacitance per unit length, and then to multiply by the actual length in the layout.

Remember that this is only a simple first approximation as the actual fields will be much more complicated due to the existence of electrically conductive material on other layers in the vicinity and shapes such as corners which are not well represented by a 2-D approximation. One final comment on capacitive couplings. One obvious approach to avoiding capacitive couplings is to simply place the objects being connected such that the sensitive wire can be kept far away from all other signals which would disturb it. Unfortunately this is not always possible; in fact, some circuits have topologically unavoidable crossings build into them. An important option to keep in mind when two signals must cross topologically, but cannot couple capacitively, is to insert a shield layer [26]. For example, in a fabrication process with two metal layers, first-level metal can be used as a shield while the two signals cross on second-level metal and poly (Fig. 15.14).

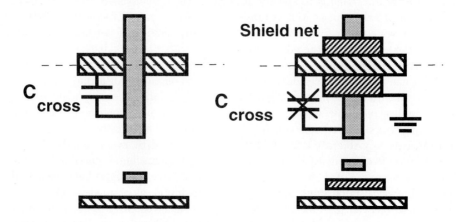

**FIGURE 15.14**
Layout for crossing structures. The picture on the left illustrates the case where a two-wires cross and a coupling capacitance occurs. On the right, the capacitive crosstalk is blocked by a third shielding layer.

The lines that bring power to an analog cell can also couple signals from other portions of the chip onto signal nodes within the cell. Of particular importance is the coupling of signals from CMOS digital logic, which tends to generate current spikes at switching times, onto sensitive analog signal nodes. Note, logic families which draw constant current, e.g., emitter-coupled logic (ECL), do not generate nearly as large of transients on the power supply net. A standard approach to avoid coupling between digital and analog circuits is to have separate analog and digital power supplies which are only merged off chip. In this manner, we can avoid coupling digital switching noise into the analog power supplies. Unfortunately, there is another method for coupling signals on-chip – through the substrate.

Signals on one part of a chip can induce currents into the substrate that result in a disturbance at sensitive analog nodes. This is particularly a problem because typical design practice for digital CMOS cell libraries is to connect the substrate contact directly to the appropriate digital power supply in every cell - e.g., in an N-Well process the P-type substrate would be tied to ground. Any noise introduced onto the digital ground line by the dynamic currents typical of switching CMOS logic are therefore coupled directly into the substrate by the substrate contacts and then spread outward through the resistive substrate toward the analog portion of the IC. An obvious solution is to separately wire the substrate contacts in every digital cell, but this solution is rarely adopted because of the area increase that the extra wiring would cause in digital cell libraries. Remember that a typical mixed-signal ASIC might be 90% digital and only 10% analog.

In a bulk CMOS process, we can improve the situation by isolating the analog cells from the digital cells and by placing a wide guard ring of substrate contact tied to a quiet voltage reference around the entire analog region. Much of the currents induced in the substrate by the digital signals will then be collected by the guard ring and will not enter the analog portion of the die. In a bulk CMOS process, a large number of substrate contacts in the analog cells, also tied to a quiet voltage reference, will further attenuate the digital signals transmitted through the substrate by resistive division.

Interestingly, when the CMOS devices are grown in an epitaxial layer on top of a low resistance substrate (epi CMOS), additional substrate contacts may only worsen the coupling of digital switching noise into the analog circuits [27]. This is the case when the low resistance substrate cannot be attached to the die with a low impedance (especially at high frequencies where digital switching noise is worse) substrate connection because then the substrate acts as a low impedance source for digital switching noise throughout the entire IC. In this case adding more substrate contacts in the analog cells only couples more digital switching noise onto the analog power supply line. Note, when a low impedance connection can be made to a low impedance substrate, then we have the desirable situation in which a low impedance ground plane is running under the entire IC.

One last source of coupling which is often overlooked is due to the bond wires that connect the die to the package. Mutual inductance coupling between adjacent bond wires can couple signals. Fortunately, we can avoid this type of coupling by separating sensitive analog signal bonding pads from bonding pads carrying high swing high frequency signals.

*Matching deficiencies* cause a systematic mismatch between elements which the designer assumed were perfectly matched. Note, in this case we are specifically discussing systematic mismatches – ones which recur in every die. There are of course random variations which are introduced by the fabrication process as well. Note, matching is of particular importance in the design of analog circuits. In part, this is because there are quite wide variations in device parameters; therefore, in order to achieve the high precision that is required, analog circuits are often designed so that their operational precision is based on ratios of matched devices rather than on the absolute value of any device parameter. Therefore, the layout process must be careful to preserve the matching property of devices and passive elements. The most common ways in which layout influences matching are through device size, device orientation, and device proximity.

Any time precise integer ratios are required between devices, the most accurate method of achieving an N:1 ratio is by laying out N+1 identical devices and wiring N of them in parallel. This is because device areas that are defined on a mask are all offset by a small, but variable, amount of shrinkage or bloating. For simplicity we will refer to this as shrinkage and consider bloating to be a negative shrinkage. Simply making an N:1 ratio in mask area alone would be in error due to the mask shrinkage. For example, to achieve a precise capacitor ratio of 27.3:1 we would lay out 28 identical unit size capacitors plus one capacitor with a nominal value of 0.3 times that of the unit size capacitors. 27 of the capacitors and the 0.3X capacitor would be wired in parallel resulting in a precise ratio of 27.3:1. Variations in the shrinkage will have no impact on the 27:1 ratio, but will only affect the 0.3X capacitor. In this example, a 10% variation in the 0.3X capacitor due to variations in the shrinkage will only result in roughly a 0.1% error in the overall ratio – which is roughly on the order of the random component of variations in capacitor matching.

A more subtle aspect of shrinkage is that it can be anisotropic. For example, ion implants are applied slightly off of normal axis of the wafer to control their penetration into the crystal lattice. However, this can result in the shrinkage on the left side of a gate stripe being greater than the shrinkage on the right side. An example of matching which requires great care in layout is the creation of a half-size dummy switch to cancel the clock feedthrough and charge injection that results from turning off an MOS switch. Since a precise 2:1 ratio is required three identical devices should be used. On the dummy switch, both the source and drain are tied to the charge summing node thus the clock feedthrough comes from both a diffusion to the left of a gate and one to the right of a gate (Fig. 15.15). The best way to wire the other two devices together to act as a switch is therefore to cross-couple them connecting one diffusion on the left of a gate and one diffusion on the right of a gate to the charge summing junction. Many matched elements exhibit this type of sensitivity to anisotropic shrinkage and care must always be taken to assure that the resulting structures match independent of this anisotropy.

One final difficulty about shrinkage is that it is sometimes dependent upon other mask geometry in the vicinity of an edge. For example, in an array of precisely matched resistors each resistor has a neighboring resistor to its right and its left except for the two end resistors. This physical arrangement can result in the end resistors shrinking either more or less than all of the other resistors resulting in a systematic mismatch. The normal solution to this type of problem is to ensure that the local environment of all matching components is identical. For the example of an array of resistors, we add a dummy resistor which is not used electrically at both ends of the array so that the environment of the electrical end resistors is the same as all of the other resistors. This same principle applies to arrays of capacitors or transistors. The degree of sensitivity is strongly dependent on the layers in question and on the distance to the other geometric features.

Matching of components is also degraded by placing them far apart (see Chapter 14). In any IC fabrication process, there are slow spatial variations in many of the fundamental device properties; e.g., oxide thickness, doping densities, etc. Therefore as two devices intended to match are placed closer together, the expected value of their mismatch will decrease. Another coupling mediated by the substrate that also appears as a slow spatial variation is thermal coupling. Power generated in one part of an IC causes a thermal gradient across the substrate and changes device parameters (e.g., $V_{BE}$) of matched devices by different amounts – effectively creating a slow variation in device parameters with position.

The simplest method for coping with spatial variations is to keep matching components close together. For illustration, consider two matched equal value

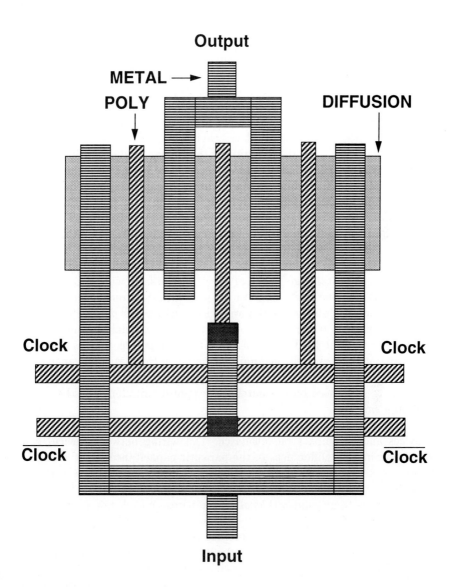

**FIGURE 15.15**
A method for laying out a switch transistor and a half-size dummy transistor to be used to cancel the charge injection from the switch. This layout structure is immune to asymmetries between source and drain diffusions.

capacitors, $C1$ and $C2$. Assuming that the two capacitors together fit within a square, then the center-to-center distance between the capacitors is determined purely by the value of the capacitor (i.e., the area of the capacitor), and large value capacitors will have large center-to-center spacing (Fig. 15.16(a)). Slow variations in capacitor oxide will therefore significantly affect the matching of the two capacitors. One technique for improving the matching of capacitors is called the "common centroid" layout style. Just as the name implies, both capacitors share the same "center" of mass. How can we do this with our two-capacitor example? As shown in Fig. 15.16(b), we can divide each capacitor in half and then place the halves in diagonally opposite corners. Both capacitors will have an effective center in the same place, therefore, any linear gradient in oxide thickness, no matter what the direction of that linear gradient, will not affect the matching of the two capacitors. The common centroid method can also be applied to transistors or any other matching elements. Note however, matching is still limited by random mismatch and by slow variations with quadratic and higher order spatial moments. Figure 15.16(c) illustrates that we can go to even finer divisions, in this case dividing each of the original capacitors up into eight pieces, which will diminish the sensitivity of the mismatch to quadratic spatial variations in process parameters. However, because further dividing the original capacitors also increases the total edge length of the capacitors, we will eventually discover that the mismatch due to random lithographic variations on edges increases more than the mismatch due to spatial variations in process parameters decreases. Beyond this point, further division of the device into smaller elements actually increases the mismatch. Because of these two competing sources of mismatch, there is an optimum size for the "unit" elements of the matching capacitors. In a modern CMOS process using a polysilicon-oxide-polysilicon capacitor structure, the optimum size based on mismatch for a square unit element is between $20 - 30\mu m$ on a side.

All of the above discussions of mismatch have focused on the placement of devices and its impact on mismatch. The wiring between devices can also result in mismatch. Differences in the parasitic capacitance in the wiring of two precisely matched capacitors can easily destroy the performance of the overall circuit. In general, the only way to guarantee that two parts of a symmetric circuit match is to guarantee that all of the wires on each side contain the same length on each wiring layer and have the same exact crossings with other wires or devices. For circuits in which there are two nearly separate paths with each device in one path having a matching device in the other path, the easiest way to do this is to simply lay out half of the circuit and then generate its mirror image for the other half of a circuit. Note, simultaneously requiring matched devices, which must have the same orientation of gate stripe, and matched wiring, which requires mirror symmetry in the wiring and hence in the device contacts, forces the gate stripes of MOS devices to be perpendicular to the mirror symmetry line. Also, many circuits which are nearly symmetric have a pair of nets which cross at the mirror symmetry line; e.g., fully-differential regenerative comparators. A standard solution is to create a crossing cell which straddles the mirror symmetry

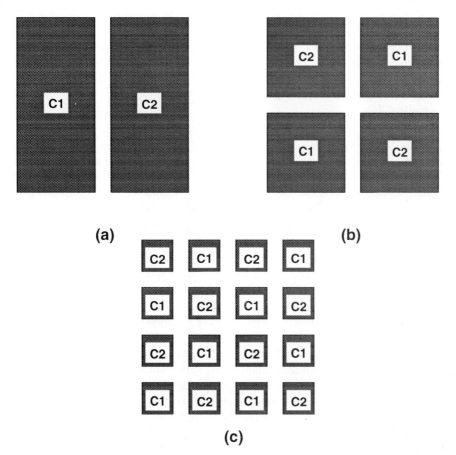

**FIGURE 15.16**
Common centroid layout method for laying out matching elements reduces the mismatch caused by gradients in process parameter values. (a) is a pair of matching capacitors. (b) illustrates a common centroid layout using 2 unit elements for each capacitor, and (c) using 8 unit elements for each capacitor.

line and has two terminals on each side. Internally, a crossing is generated with extra material and layer changes added to achieve the best possible matching between the two nets. Circuits which have transistors which do not clearly belong to one path or the other can often be modified to make them fully symmetric. For example, the tail current source transistor of a fully-differential op amp can be split into two transistors, each of half the width of the original, wired in parallel.

We have now completed a brief overview of some of the important issues that must be addressed in the placement and routing of analog circuits. In the next two sections, we will look at some examples of the kinds of automation approaches that can be applied to analog IC physical design.

## 15.3.2 Automating System-Level Analog Layout

In this subsection we will explore two methods that have been employed in order to assemble mixed-signal IC systems. The first method assumes that all of the cells have been created with equal height, and that they can be placed in rows. The second method assumes that the cells are rectangular, but of varying heights and widths.

**Row-Based System Layout:** Figure 15.17 illustrates how row-based standard cells can be used in a mixed analog/digital system design. Algorithms for determining the order of the cells within each row, and which cells should be placed in which rows have been adapted from solutions to the corresponding digital standard cell task [28]. For a digital design, the quality of a proposed ordering is determined primarily by the total wire length required. However, in the case of an analog design avoiding coupling between critical nets may be far more important than minimizing wire length. Because the number of cells in an analog system may be rather small, the placement of the cells may also be done manually.

An early attempt at addressing this problem divided all nets into two classes: analog nets and digital nets. In order to prevent corruption of sensitive analog signals by digital signals, the channels are segregated into only analog signals or only digital signals [29]. The cells must have all have their digital signal connections on one end and their analog signal connections on the other end (Fig. 15.17). The main drawback to this strategy is that critical couplings can occur between more than just two classes of nets. For example, the capacitance between the analog output of an oscillator and the input to a phase detector may also need to be kept small. In general, we would like to be able to specify bounds on parasitics between every pair of nets. However, computation of sensitivities for entire systems may prove difficult. In this case, we would at least like to specify a number of different classes of signals, and to be able to indicate which classes should avoid coupling with which other classes.

**Slicing Structure-Based System Layout:** In many cases analog cells have many different heights and widths and it is difficult or wasteful of area to force a row-based design style. In this case, we can adopt a macrocell place and route strategy. The system-level macrocell style is characterized by two phases. The first phase is placement: each cell is oriented and placed on a plane. The second phase is routing: all of the terminals of the devices are connected together as required by the netlist. Although we could place the cells in any arbitrary pattern, existing CAD tools have generally chosen a slicing structure; e.g., [30] and [31]. A slicing structure is one in which the cells can be separated by a sequence of horizontal or vertical cuts. Slicing structures have the advantage that it is always possible to expand the space between any pair of cells if more wires need to be inserted. Again, algorithms for determining the best placement of cells in the slicing structure have generally been adapted from solutions to the

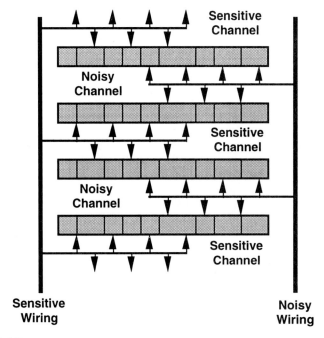

**FIGURE 15.17**
Typical row-based mixed analog/digital floorplan.

corresponding digital cell placement problem; and are primarily concerned with minimizing the total wire length [30].

Once the blocks are placed, the regions between them can be defined as routing channels and expanded as dictated by the number and width of the wires that traverse that channel. Figure 15.18 illustrates how more complex mixed analog and digital cells of varying heights and widths can be placed in a slicing structure. It is still extremely important for the routing tools to be aware of and to prevent corruption of sensitive analog signals by digital signals or noisy analog signals [31]. We can divide the routing problem into two parts: a global signal route planning and a detailed routing. The global route planner determines which intermediate channels will be used by a signal that must get from one channel to another channel and the detailed router inserts layout geometries to connect all of the signals that are internal to that channel. Note, the global signal route planning is primarily a topological problem, and its solution does not generate any mask geometry, but only a path for each net and a pin ordering at the end of each channel. The global route planner must take into account the sensitivity of every net to every other net. Of particular importance are decisions which result in, or avoid, a crossing between nets that must avoid each other. Then within each channel region, a detailed router generates mask geometry in order to maintain the desired signal avoidance properties. This overall approach is much more flexible than the alternating analog and digital channels idea presented earlier

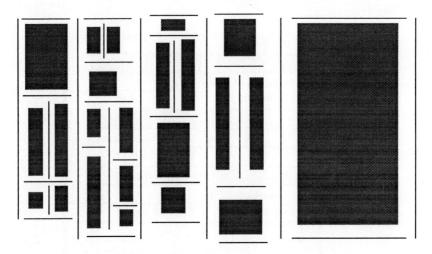

**FIGURE 15.18**

Typical macrocell mixed analog/digital floorplan. In this example there are 23 cells, approximately 10 analog and 13 digital. The lines between the cells indicate the slicing planes, which are also the channels for routing.

in this section because it can accommodate many possible kinds of avoidance classes. For example, some nets might only be sensitive during one clock phase and could be routed over signals that are noisy only during a different clock phase.

### 15.3.3 Automating Cell-Level Analog Layout

There are a wide variety of approaches for laying out analog cells. Currently, the most common approach for the layout of analog cells is fully custom manual layout. However, this is a very time consuming approach. In this subsection we will overview several of the approaches that have been adopted to decrease the time required to perform analog cell-level layout. Although none of these approaches can match the density of full custom manual layouts, they range widely in density. We begin by describing two approaches that are actually both cell-level and system-level approaches because they provide no clear division between cell and system layout: mask programmed and electrically programmed arrays. We then discuss two approaches that are closer to custom cell layout both in density and because they generate layouts for the cells which must then be placed and wired together at the system level: fixed floorplan module generators and macrocell place and route systems.

**Mask Programmed Design:** The idea of wiring together predefined cells from a library at the system level can be extended to the cell level by taking a set of predefined transistors instead of a set of predefined cells. Since these transistors can be interconnected and biased arbitrarily, a wide range of cell-level circuits can be represented in this manner. Usually, resistors with many taps are included with the transistors. By choosing which taps to connect, the value of the resistor can be varied. Figure 15.19 is an example of how a bipolar transistor array might be structured. In some cases all of the processing steps associated with the transistors and resistors are performed ahead of time. Then, for a particular design, the metal interconnect layers are placed in order to implement the desired cell and system functions. Specialized CAD tools have been developed to help automate much of the task of selecting which transistors physically correspond with which transistors in the schematic ("placement") and generating the mask geometry for the interconnections [32].

The density achieved by this approach can be quite low – often many of the transistors are left unused because the number of each type of transistor that will be needed varies from one design to the next. And, because the density of transistors is substantially lower than that which could be achieve in a fully custom design, the total length of wires that connect the devices is longer, and hence the parasitic capacitance of the interconnect is higher. Therefore, the maximum possible speed for this style of device layout is somewhat lower than that possible with full custom.

On the other hand, the time from completed design until the prototype chip is ready for testing can be very short because fewer masks need to be generated and fewer processing steps need to be performed. When time to market is critical, the decrease in fabrication time for a prototype may more than make up for the relatively poor utilization of silicon die area. In addition, this approach allows the cost of all of the other masks for a given transistor array to be shared across many prototypes.

The transistor array methodology has been widely accepted, especially for prototyping, for the design of bipolar ICs. However, its use with MOS transistors has been very limited. Just as there are too many performance specifications for analog cell libraries to be practical, there are too many "performance specifications" for an MOS transistor (width's and length's) for MOS transistor arrays to be practical. Although bipolar transistors come in various emitter sizes and the number of base and collector contacts may vary, as long as the collector current is below a given limit, a minimum size transistor can be used in many cases.

**Electrically Programmed Design:** A recent innovation in the implementation of analog circuits is the use of an array of MOS transistors and passive components coupled with interconnect and switches that can be electrically programmed [33, 34]. This approach is patterned after the very successful approach taken for digital circuits in the form of field programmable gate arrays (FPGAs). There are many possible choices for the primitive cells in a field programmable analog array (FPAA) ranging from isolated transistors (i.e., a field programmable transistor array ) to complex cells (i.e., a field programmable standard cell array ). The tradeoff which must be made is between compromising performance and task specificity. When smaller blocks of elements are being assembled to perform a given function, more switches are required – less of the routing is pre-

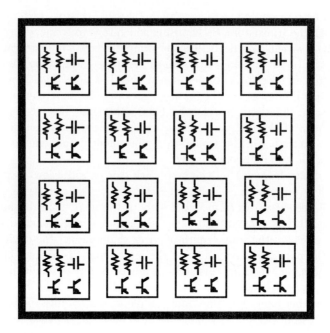

**FIGURE 15.19**
Example of the floorplan of BJT transistor array.


defined. However, because the switches introduce significant series resistance and parasitic capacitance, the more switches required to implement a given function the lower the performance (especially bandwidth) that can be achieved. The decrease in bandwidth over a full custom implementation can be an order of magnitude or more! However, in many ASIC applications the bandwidth of the analog interface circuitry is extremely low (e.g., monitoring the temperature in a barrel of wine), hence the bandwidth restrictions imposed by the switches may be unimportant for some applications.

We can decrease the number of switches, and the associated loss of the bandwidth, by interconnecting blocks that contain more elements than a single transistor. On the other hand, while an array of transistors and passive elements can be configured to implement almost any analog function (assuming there are enough elements), if we start with an array of switched capacitor biquadratic filter segments, we may only be able to build switched capacitor filters. Because of this tradeoff, it should be no surprise that a majority of FPAA designs to date have tended to choose some point in between these two extremes for the blocks that will be assembled. A typical choice is to have blocks which consist of several (2-5) transistors preconnected in ways which are likely to appear in analog circuits [33].

The attraction of the FPAA idea is its rapid turnaround time. Instead of weeks or months, programming the FPAA may only take minutes. This makes it particularly attractive for very low volume applications and for prototyping. The design of the circuit may be complicated by the additional constraints imposed by the number and characteristics of each analog block. And, although not apparent at first, there is still a layout problem. In this case it corresponds simply to choosing which of the blocks of a given type are associated with a block of that type on the schematic; i.e., which way are the switches that control the configuration set. This selection may be complicated by the pattern of allowed interconnect. In order to lower the parasitics associated with switches that are off, FPAAs generally have several layers of interconnects. There may be local switches allowing arbitrary connection of a group of blocks, but only a limited number of switches between that group of blocks and a more global interconnect bus. Therefore the layout process must manage resources in such a manner that all of the required interconnections can be implemented. In addition, because the parasitics due to the switches (and hence the bandwidth limitations imposed) depend on the particular mapping between blocks on the schematic and blocks on the FPAA, it may be necessary to not only find one mapping, but to find one that also meets the required performance constraints.

Note, all of these features did not come without a price. The percentage of the silicon die area that is actually used to implement the desired analog circuit is only a small fraction of the FPAA die area (often below 20%). First, not all of the blocks may be usable in a particular design, just as the case for transistor arrays but the problem is worse here because there is generally a larger variety of blocks. Second, the area required by the switches and the memory associated with the switches may be much larger than the area occupied by the blocks themselves. Again this simply suggests that FPAAs will only be attractive for very low volume applications and for prototyping.

**Fixed Topology Module Generators:**  In order to achieve densities closer to full custom layout, we must generate just the devices that are needed in a particular cell, unlike the above two approaches which constructed cells from a predefined array of elements. A common approach is to have a predefined topological relationship between the elements of a cell that is specified by an expert designer. When the transistor sizes are changed in order to meet new performance goals, the devices are expanded and the wiring is lengthened in order to maintain the topological relationship between the elements. This approach has been applied to create "module generators" for common analog cells; e.g., op amps [35]. One very simple way to implement this approach is to have a fixed height cell in which each horizontal region is dedicated to one MOS device (e.g., Fig. 15.20). As the width of the MOS device changes, the number of gate stripes can be increased or decreased. In the style shown in Fig. 15.20 the wires underneath the transistor are used to connect other transistors and components together. When a new gate stripe is added, all of the wiring that crosses beneath the transistor is stretched. Certain noncritical passive elements, such as the compensation capacitor in a two-stage op amp, can also be used to fill in open space.

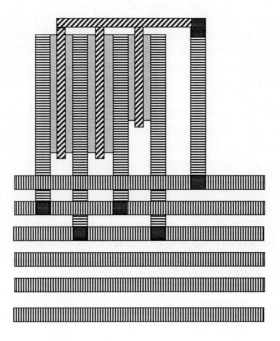

**FIGURE 15.20**
Example of an expandable transistor for use in a simple automatic module generator.

This approach tends to work best when all of the devices are large. If many devices are small and have only single gate stripes that do not span the full height of the cell, then area may be wasted. However, the above module generation method can be generalized so that each transistor does not have to span the full height of the cell [13], though it is generally the case that a specific topological arrangement of devices will work well only over a limited range of device sizes. Therefore, it is common to create multiple topological arrangements for a single analog cell and to choose the one that best fits the specified device sizes.

**Macrocell Place and Route:**  A more general approach to analog cell layout is the macrocell place and route style [30, 36, 37]. At the cell level, the macro-cell style is characterized by three phases. First, several variants (with different numbers of folds in the gate) of each device are generated. Second, a particular variant of each device is selected and placed. The placement may be either arbitrary [37] or in a slicing structure [30]. The third and final phase is routing, connecting all of the terminals of the devices together as required by the netlist.

One of the major attractions of the macrocell place and route style, compared to the previous fixed topology module generation style, is that it does not require the human designer to provide any specific knowledge about how to lay out the elements. Only information about electrical requirements between nets

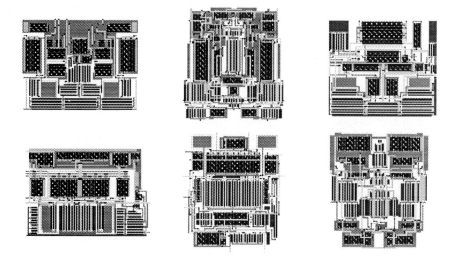

**FIGURE 15.21**

Comparison of hand layouts with computer generated layouts. The two layouts in the middle were automatically generated. The two on the left and the two on the right were manually generated.

(e.g., required symmetries and coupling avoidance specifications) needs to be provided. In addition, since it treats all objects as cells, merging of devices with previously generated subcell layouts is easy.

At the cell level, especially in the case of CMOS layouts, we find great variations in the size of the cells being placed; e.g., from minimum size transistors to very large transistors. In this case, requiring that the cell placement take the form of a slicing structure can result in a substantial waste of die area. However, when applying the macrocell place and route approach to nonslicing structures, it is not always possible to introduce additional space between cells for wiring without significantly distorting the layout. Because the placement process is done before the routing process begins, the placer does not know where the wires need to go and cannot leave room for them. The most common way to cope with this situation is to set aside a "halo" of blank space around every cell. The thickness of these halos can be adjusted using heuristics based on the number of connections to that cell and on some estimate of the congestion of wiring traffic in the region of a cell. If a placement cannot be routed, we can increase the thickness of all of these halos and then try to place and route the circuit again.

Research is currently focusing on two approaches to improve this situation. The first approach is to leave large enough halos so that the routing can always be completed, and then to follow the routing process with a special compaction process that takes the analog performance constraints into account. The second approach is to perform the placement and the routing simultaneously [38]. In this case wires and cells compete for space simultaneously.

Figure 15.21 compares manually generated layouts with layouts generated automatically by [37]. The two layouts in the middle were generated automatically. The circuit is a high-speed, fully-differential comparator. The automatic layout tool was given information that specified the required net symmetries, device matches, net classes, and class avoidances. The die area of all six layouts is comparable. The performance, in terms of decision speed and systematic offset, of the computer generated layouts was actually slightly better than that of any of the manual layouts. In part, this is because the computer generated truly symmetric nets where required, but the humans performing layout tended to make small mistakes in forcing two nets to match.

## 15.4 CONCLUSION

In this chapter we have given an overview of many of the important constraints that must be respected in order to generate operational analog signal processing integrated circuits. We explicitly looked at the structure of the analog IC design process at both the system and cell levels. Further, we have explored how computer-aided design tools can help in the synthesis of analog signal processing ICs.

## REFERENCES

[1] M. A. Sivilotti, M. A. Mahowald, and C. A. Mead, "Real-Time Visual Computations Using Analog CMOS Processing Arrays," in *1987 Stanford Conference on Very Large Scale Integration*, pp. 295–312, MIT Press, March 1987.

[2] P. E. Allen and P. R. Barton, "A Silicon Compiler for Successive Approximation A/D and D/A Converters," in *Proc. IEEE Custom Integrated Circuiuts Conf.*, 1986.

[3] P. E. Allen and E. R. Macaluso, "AIDE2: An Automated Analog IC Design System," in *Proc. IEEE Custom Integrated Circuiuts Conf.*, 1985.

[4] J. Vlach and K. Singhal, *Computer methods for circuit analysis and design*. New York: Van Nordstrand Reinhold, 1983.

[5] G. Hachtel and R. Rohrer, "Techniques for the Optimal Design and Synthesis of Switching Circuits," *Proc. IEEE*, vol. 55, November 1967.

[6] W. T. Nye, et. al., "DELIGHT.SPICE: An Optimization-Based System for the Design of Integrated Circuits," *IEEE Transactions On Computer-Aided Design*, vol. 7, pp. 501–519, April 1988.

[7] J.-M. Shyu and A. Sangiovanni-Vincentelli, "ECSTASY:A New Environment for IC Design Optimization," in *Proceedings of ICCAD*, pp. 484–487, 1988.

[8] Robert K. Brayton, et. al., "A Survey of Optimization Techniques for Integrated-Circuit Design," *Proceedings of the IEEE*, vol. 69, pp. 1334–1362, October 1981.

[9] Meta-Software, *HSPICE User's Manual*. Campbell, CA: Meta-Software Inc., 1989.

[10] M. G. R. Degrauwe, et. al., "IDAC: An Interactive Design Tool for Analog CMOS Circuits," *IEEE J. of Solid-State Circuits*, vol. SC-22, pp. 1106–1116, December 1987.

[11] R. Harjani, R. A. Rutenbar, and L. R. Carley, "OASYS: A Framework for Analog Circuit Synthesis," *IEEE Transactions on Computer-Aided Design*, vol. 8, pp. 1247–1266, December 1989.

[12] H. Y. Koh, C. H. Sequin, and P. R. Gray, "OPASYN: A Compiler for CMOS Operational Amplifiers," *IEEE Transactions on Computer-Aided Design*, vol. 9, pp. 113–125, February 1990.

[13] E. Berkcan, M. d'Abreu, and W. Laughton, "Analog Compilation Based on Successive Decompositions," in *Proceedings of 25th Design Automation Conference*, pp. 369–375, ACM/IEEE, 1988.

[14] G. G. E. Gielen, H. C. C. Walscharts, and W. M. C. Sansen, "ISAAC: A Symbolic Simulator for Analog Integrated Circuits," *IEEE J. of Solid-State Circuits*, vol. 24, pp. 1587–1597, December 1989.

[15] G. G. E. Gielen, H. C. C. Walscharts, and W. M. C. Sansen, "Analog circuit design optimization based on symbolic simulation and simulated annealing," *IEEE J. of Solid-State Circuits*, vol. 25, pp. 707–714, June 1990.

[16] E. S. Ochotta, R. A. Rutenbar, and L. R. Carley, "Equation-Free Synthesis of High-Performance Linear Analog Circuits," in *Proc. Brown/MIT Conference on Advanced Research in VLSI and Parallel Systems*, March 1992.

[17] L. Pillage and R. Rohrer, "Asymptotic waveform evaluation for timing analysis," *IEEE Transactions on Circuits and Systems*, vol. CAD-9, April 1990.

[18] R. Fletcher, *Practical Methods of Optimization, 2nd Edition*. New York: Wiley, 1987.

[19] P. C. Maulik and L. R. Carley, "Automating analog circuit design using constrained optimization techniques," in *Proceedings of ICCAD*, November 1991.

[20] P. C. Maulik, L. R. Carley, and D. J. Allstot, "Sizing of Cell-Level Analog Circuits Using Constrained Optimization," *IEEE J. of Solid-State Circuits*, vol. SC-28, March 1993.

[21] R. A. Rutenbar, "Simulated Annealing Algorithms: an Overview," *IEEE Circuits and Devices Magazine*, vol. 5, January 1989.

[22] K. Nakamura and L. R. Carley, "A current-based positive-feedback technique for efficient cascode bootstrapping," in *Proc. VLSI Circuits Symposium*, June 1991.

[23] K. Nakamura and L. R. Carley, "An Enhanced Fully-Differential Folded-Cascode Op Amp," *IEEE J. of Solid-State Circuits*, vol. 27, April 1992.

[24] U. Chowdhury and A. Sangiovanni-Vincentelli, "Constraint Generation For Routing Analog Circuits," in *Proceedings of Design Automation Conference*, pp. 561–566, ACM/IEEE, 1990.

[25] B. R. Stanisic, R. A. Rutenbar, and L. R. Carley, "Power Distribution Synthesis for Analog and Mixed-Signal ASICs in RAIL," in *Proceedings of Custom Integrated Circuits Conference*, IEEE, May 1993.

[26] U. Chowdhury and A. Sangiovanni-Vincentelli, "Constraint Based Channel Routing for Analog and Analog/Digital Circuits," in *Proceedings of ICCAD*, pp. 198–201, November 1990.

[27] T. J. Schmerbeck, R. A. Richetta, and L. D. Smith, "A 27MHz Mixed Analog/Digital Magnetic Recording Channel DSP Using Partial Response Signalling with Maximum Likelihood Detection," in *Proceedings of International Solid-State Circuits Conference (ISSCC)*, pp. 136–137, 1991.

[28] C. Sechen, "Chip-planning, placement and global routing of macro/custom cell integrated circuits using simulated annealing," in *Proc. 25th ACM/IEEE Design Automation Conf.*, pp. 73–80, June 1988.

[29] A. E. Dunlop, G. F. Gross, D. D. Kimble, M. Y. Luong, K. J. Stern, and E. J. Swanson, "Features in LTX2 for Analog Layout," in *Proceedings of ISCAS*, IEEE, 1985.

[30] J. Rijmenants, et. al., "ILAC: An Automates Layout Tool for Analog CMOS Circuits," in *Proc. IEEE Custom IC Conf.*, May 1988.

[31] S. Mitra, S. K. Nag, R. A. Rutenbar, and L. R. Carley, "System-level Routing of Mixed-Signal ASICs in WREN," in *Proceedings of ICCAD*, November 1992.

[32] J. Trnka, R. Hedman, G. Koehler, and K. Ladin, "A device level auto place and wire methodology for analog and digital masterslices," in *Proceedings of the 1988 International Solid-State Circuit Conference*, pp. 260–261, February 1988.

[33] E. K. F. Lee and P. G. Gulak, "A CMOS Field-Programmable Analog Array," *IEEE J. of Solid-State Circuits*, vol. 26, pp. 1860–1867, December 1991.

[34] M. Ismail and S. Bibyk, "CAD Latches Onto New Techniques for Analog ICs," *IEEE Circuits and Devices Magazine*, vol. 7, pp. 11–17, September 1991.

[35] J. Kuhn, "Analog Module Generators for Silicon Compilation," *VLSI Systems Design*, May, 1987.

[36] D. J. Garrod, R. A. Rutenbar, and L. R. Carley, "Automatic Layout of Custom Analog Cells in ANAGRAM," in *Proc. 1988 IEEE Int'l Conf. on CAD*, November 1988.

[37] J. M. Cohn, D. J. Garrod, R. A. Rutenbar, and L. R. Carley, "KOAN/ANAGRAM II: New Tools for Device-Level Analog Placement and Routing," *IEEE J. of Solid-State Circuits*, vol. 26, pp. 330–342, March 1991.

[38] J. M. Cohn, D. J. Garrod, R. A. Rutenbar, and L. R. Carley, "Techniques for Simultaneous Placement and Routing of Custom Analog Cells in KOAN/ANAGRAM II," in *Proceedings of ICCAD*, pp. 394–397, November 1991.

# CHAPTER
# 16

# ANALOG AND MIXED ANALOG-DIGITAL LAYOUT

## 16.1 INTRODUCTION

In recent years considerable interest has developed in mixed analog-digital circuit design. By combining analog signal conditioning with complex digital processing on the same substrate, an entire system is realized on a single chip. However, mixing together analog and digital circuits creates specific technological and design methodological problems which have been discussed in prior chapters. One final issue to be considered is the layout. It is the final step of the design process where the electrical description of a circuit is transformed into its physical representation. The layout is then used for realizing the masks required in the chip fabrication.

Before going into specific details, it is worthwhile remembering that the layout of a digital circuit is completely different from the layout of an analog circuit. A digital circuit is made by interconnecting simple blocks. The layout of the basic blocks is performed manually or with automatic tools to build up a library of basic cells. The major task is then to place the cells and route between them. The designer performs these two steps with the help of powerful CAD tools, especially when very large networks must be created. The layout of a digital circuit is generated to minimize area and signal delay. Generally speaking, this involves looking at the interconnections among blocks and not so much at the individual transistors.

By contrast, the layout of an analog circuit contains networks with limited complexity. A given analog integrated circuit uses the same cells a limited number of times and different designs very rarely reuse specific cells. Therefore, the layout of analog circuits mainly involves optimizing transistor layouts with much less concern for interconnections. Important design criteria influenced by the layout are accuracy and noise immunity. Using these criteria, a number of hints and practical recommendations may be derived.

When analog and digital circuitry are integrated onto the same chip, additional problems arise. The noise generated by the digital circuitry may couple into the analog circuitry and corrupt the overall analog circuit performance. Therefore, the control of the noisy interaction is vital in mixed circuits. This can be done by careful circuit design (for example, by achieving good PSRR), but mostly it is more a matter of scrupulous layout.

This chapter presents layout techniques of basic elements including transistors, resistors, and capacitors. Then, we discuss the layout of leaf cells (mainly operational amplifiers) and finally we examine strategies for mixed analog-digital circuit layout. A number of layout examples illustrate all of these points.

## 16.2 CMOS TRANSISTOR LAYOUT

The heart of the layout of an MOS transistor is simply the crossing of two rectangles: one is polysilicon and the other is diffusion (Fig. 16.1). The polysilicon defines two separate areas in the diffusion rectangle which are the source and the drain terminals. Depending on the technology used, it is necessary to realize the diffusion inside or outside a well to obtain n-channel or p-channel transistors.

The two rectangles do not complete the layout of the MOS transistor; it is necessary to design a number of other patterns. In particular, they are the electrical connection of the source, drain, and gate to the rest of the circuit. Metal lines normally make these connections; less frequently they are made by polysilicon or diffusion. Figure 16.2 shows a typical layout of a minimum area MOS transistor. A set of design rules defines the size of the contacts as well as their overlap with the structures below or the overlap of metal with the contacts themselves. The design rules guarantee the best tradeoff between fabrication success and layout compactness. They also fix the minimum width of lines, minimum distance between them and so on.

For analog applications, the aspect ratio $(W/L)$ of transistors is fairly high; therefore, it is necessary to design wide structures. In this case, it is important to remember that the diffusion used to realize the source and drain terminals has a non-negligible specific resistance (around $100\Omega/\square$). A few squares may result in an unacceptable drain resistance. Figure 16.3(a) shows an example of a poor transistor layout. The source and drain contacts are made at each end of the diffusion. The structure corresponds to the equivalent circuit shown in Fig. 16.3(b)

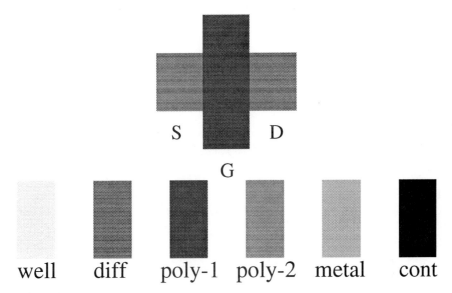

**FIGURE 16.1**
(a) Basic layout of an MOS transistor and (b) legend of layers.

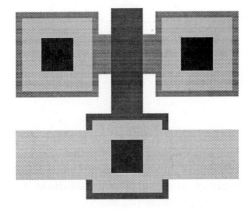

**FIGURE 16.2**
Layout of an MOS transistor.

where elemental transistors and stray resistances are used. Figure 16.3(c) shows the correct layout of the same transistor; a number of contacts short the distributed resistance of the two diffused areas. The contacts are placed at the minimum distance permitted by the design rules. The use of many contacts across the width of the transistor instead of only one improves the reliability.

When the aspect ratio of a transistor is very large, the resulting layout becomes unmanageable. In this case, it is a good idea to split a wide transistor into the parallel connection of a number of elements, $n$, as seen in Fig. 16.4 [1]. The electrical performance of the parallel structure is equivalent to a single transistor

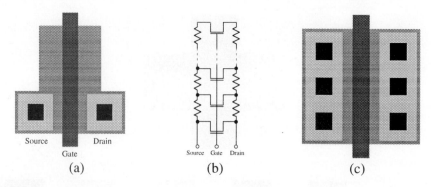

Source     Drain
Gate
(a)

Source   Gate   Drain
(b)

(c)

**FIGURE 16.3**

Layout of an "analog" MOS transistor with large aspect ratio (a) poor layout, (b) equivalent circuit of (a) and (c) correct layout.

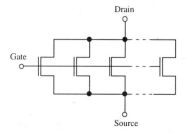

Drain

Gate

Source

**FIGURE 16.4**

Equivalent circuit of a wide transistor split into $n$ parallel parts.

whose width is equal to the total width of the parallel elements. The transistor layout in Fig. 16.5 with $n = 4$ has a shape that better meets the requirements. Additionally, this layout has the advantage that the parasitic capacitance associated with the reversed-biased diffusion-substrate diode is reduced. For a single transistor, the parasitic capacitances $C_{sb}$ and $C_{db}$ are proportional to the width, $W$, of the transistor. For split transistors, $C_{sb}$ and $C_{db}$ are reduced by a factor of $\frac{n+1}{2n}$ if $n$ is odd, however, if $n$ is even $C_{sb}$ is reduced by $\frac{1}{2}$ while $C_{db}$ is reduced by $\frac{n+2}{2n}$ [2].

This parasitic reduction is quite important for high speed applications. The practice of splitting a transistor into the parallel connection of a given number of parts is also useful for improving the matching between elements. A major source of transistor mismatches is due to gradients that exist in the fabrication process. To minimize the gradient effect, two transistors that must be matched to each other should be very close. This may be difficult for wide transistors. However, the use of interdigitized arrangements allow us to place matched transistors close to each other and a manageable shape in the layout is obtained [3]. Figure 16.6 shows three possible layout designs of two matched transistors as required by a

**FIGURE 16.5**
Layout of a split transistor $(n = 4)$.

differential pair showing normal (Fig. 16.6(a)), interdigitized (Fig. 16.6(b)), and with common centroid symmetry (Fig. 16.6(c)).

Recall that different orientations of transistors cause mismatch. This is certainly true for orthogonal elements (Fig. 16.7(a)), but it is also the case for mirrored elements where one transistor has the drain on the right side and the other transistor has its drain on the left side (Fig. 16.7(b)). To be sure that corresponding elements match, it is useful to split them into the parallel connection of an even number of equal parts; half of them will have the drain at the right and half at the left side. Finally, another source of transistor mismatch is produced by the boundary dependent etching of polysilicon gates [4]. Figure 16.8 helps to illustrate this effect. The etching of unwanted patterns defines the gate of an MOS transistor. However, the etching, even for a nonisotropic process, continues to cut under the protection (undercut). This effect depends on the boundary and the undercut is more efficient for free space around the pattern to be created. Therefore, the length of the gates $G2$ and $G3$ is reduced less than that of the gates of the terminal elements $G1$ and $G4$. This effect can be significant for precise applications but it can be compensated for by using dummy elements (Fig. 16.9).

## 16.3 RESISTOR LAYOUT

Resistors are important elements in CMOS analog integrated circuits. They are used in basic blocks such as op amps and comparators, but much more frequently in analog systems like data converters and filters.

An integrated resistor is fabricated by using one of the highly resistive layers available in a CMOS technology: diffusion ($p^+$, $n^+$ or well) or polysilicon [5]. All these layers have a given specific resistance, $R_q$, which defines the resistance of a square of the layer as measured between two opposite sides of the square. In highly doped diffusion sheets or in polysilicon sheets, this parameter is typically a few tens of $\Omega/\square$, while in wells, it is on the order of $k\Omega/\square$. The absolute accuracy of integrated resistors is less than 30% and they exhibit poor temperature and voltage coefficients. Matching between resistors is much better and can be kept,

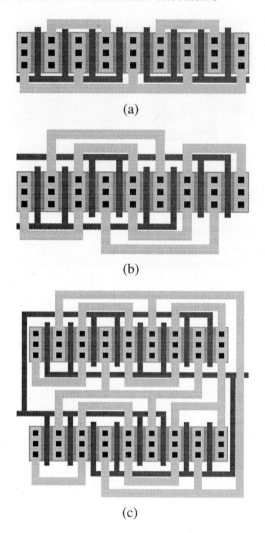

(a)

(b)

(c)

**FIGURE 16.6**
Layout of a differential pair showing (a) normal, (b) interdigitized, and (c) common centroid.

with careful layout, below 0.1%. The typical structure of an integrated resistor is shown in Fig. 16.10, where a piece of resistive layer is connected to the metal terminals by two ohmic contacts. If the sheet resistance of the strip is $R_q$, its total resistance is expressed by:

$$R = 2R_{cont} + (W/L)R_q \tag{16.1}$$

where $W$ and $L$ are effective width and length and $R_{cont}$ is the localized resistance of the contact between the ends of the resistor and the metal connections.

The relatively small specific resistance of the layers necessitates the use of structures with considerably large aspect ratios $(W/L)$. For this, it is a normal

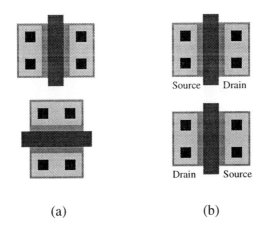

(a)                                    (b)

**FIGURE 16.7**
Examples of poor transistor layouts where the transistors have different orientations.

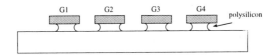

**FIGURE 16.8**
Boundary dependent etching.

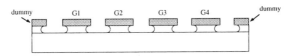

**FIGURE 16.9**
Compensation of boundary dependent etching with dummy elements.

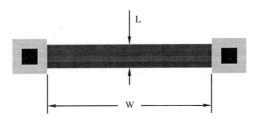

**FIGURE 16.10**
Layout of an integrated resistor.

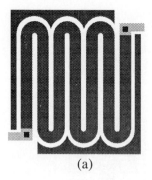
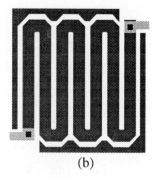

(a)                                          (b)

**FIGURE 16.11**
Resistor layout with rounded corners and dummy strips (a) and layout with $45°$ corners and dummy strips (b).

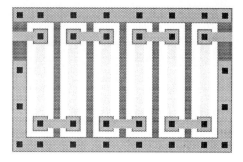

**FIGURE 16.12**
Layout of a resistor made by well diffusion.

practice to utilize a serpentine layout. One of the problems to overcome is to correctly count the squares located at the corners of the serpentine. As a rough estimate, they are computed as half of a square. Another problem is that there is more active etching at the edges of the corner which makes them rounded and the shape is poorly controlled. For precise applications, it is a good practice to design rounded (Fig. 16.11(a)) or $45°$ corners (Fig. 16.11(b)). With this solution, better shape control is achieved. Resistors with very high value must be made with well layers. However, these resistors may experience the noisy interaction of the body of the resistor with the substrate and insulation between parts of the same resistor. Both these problems can be solved using the layout shown in Fig. 16.12. A substrate bias surrounding the wells constitutes the resistor and ensures better insulation thus avoiding lateral stray resistance and at the same time establishing a guard ring to prevent noise injection from the substrate.

Boundary dependent over-etching is, as already discussed for MOS transistors, a major source of errors. Dummy strips placed around the resistor, as shown in Fig. 16.11, define the same boundary and help to improve resistor value control. It is worth remembering that the effective width of the strip will be reduced

**FIGURE 16.13**
Layout of two matched interdigitized resistors.

with respect to the designed value; however, with dummy strips this reduction is kept constant over the entire arrangement.

The absolute value of resistors also depends on the edge terminations and the contacts between the resistor strips and the metal lines [6]. These two sources of errors must be minimized with a careful layout. In particular, the edge termination must provide the best transition between the resistive strip and the contact region. If the laminar current flow is perturbed, a localized resistance as large as one square or more of resistive material could result. Moreover, the contact between the strip and the metal must be ohmic with a low resistive value. To get an ohmic contact, it is necessary to use highly doped silicon; in the well a $p^+$ or a $n^+$ diffusion should be interposed between well and metal contacts to avoid rectifying effects. Depending on the technology, the value of the contact resistance, $R_{cont}$, can range from 5 to $30\Omega$. Specific problems arise when two resistors must be matched with each other. The main sources of mismatch stem from process gradients. Therefore, the layout should minimize the distance between the elements to be matched. For long (serpentine) strips, this is not trivial. However, an interdigitized arrangement, as shown in Fig. 16.13, helps in obtaining satisfactory matching (better than 0.5%), since the centroids of the two resistors are kept close to each other [7].

## 16.4   CAPACITOR LAYOUT

MOS integrated technology allows a capacitor to be made very naturally. The plates are made using low resistive layers like polysilicon and highly doped silicon with silicon oxide being an excellent material for realizing the dielectric. Parallel plate elements, like the ones shown in Fig. 16.14, are common structures for integrated capacitors. The thickness of the oxide is in the range of $20 - 70nm$ for modern technologies. Therefore, the specific capacitance, given by $\epsilon_0\epsilon_r/t_{ox}$, is in the range of $0.5 - 1.7fF/\mu m^2$. It follows that a square plate with $25\mu m$ on a side corresponds to a capacitance of $0.3pF - 1pF$ [8].

In reality, the value of an integrated capacitor is not equal to the product of the specific capacitance of the designed area of the plate. The etching that defines

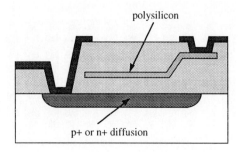

**FIGURE 16.14**
Capacitor in CMOS technology.

**FIGURE 16.15**
Layout of a capacitor.

the plate proceeds under the protective mask (undercut effect) and reduces the designed area:

$$A_{eff} = (W - 2x)(L - 2x) \approx WL - 2(W + L)x = A - Px \qquad (16.2)$$

where $A$ is the designed area and $P$ is the perimeter. Thus, the effective area is smaller than the designed area by an amount proportional to the perimeter of the plate. This result is important when matched capacitors must be designed. In order to also match the reduction effect due to the undercut, it is necessary to keep the area-perimeter ratio constant.

Figure 16.15 shows a capacitor layout realized with two polysilicon levels. The plate is square (to minimize the relative error in the length and width definition) with 45° corners. The bottom plate is larger than the top plate; therefore, the latter one defines the area of the capacitor. The contact on the top plate is obtained through a termination on the thick oxide region. Some technologies allow us to realize the contact on the top of polysilicon in the thin oxide region.

Because of fabrication inaccuracies, oxide thickness can be affected by errors [9]. When these errors correspond to a gradient, their first order effects can be cancelled by arranging matched elements with a common centroid symmetry [10]. Figures 16.16 and 16.17 present the layout of two equal capacitors which are split into eight equal parts connected in parallel. Thanks to the symmetry of the structure, a gradient either in the $x$ or $y$ direction does not introduce mismatch

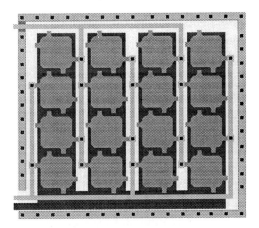

**FIGURE 16.16**
Layout of two matched capacitors with common centroid symmetry.

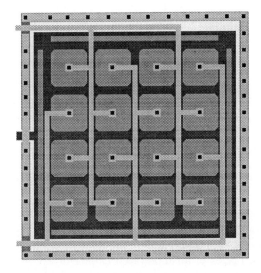

**FIGURE 16.17**
Layout of two matched capacitors with common centroid symmetry. Contacts on the top of thin oxide are permitted.

between capacitors. Moreover, in Fig. 16.17 making the contact on the top of polysilicon allows more compact structures to be designed. Finally, it should be noted that in Fig. 16.17 dummy strips were used to obtain the same boundary conditions.

Very often it is necessary to design capacitors with a given ratio where their absolute value is not important. The accuracy of the system depends only on the matching. In this case, it is crucial to control the area-perimeter ratio of the capacitor plates. When the capacitor ratios are integer multiples, it is a

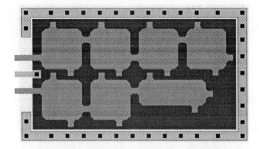

**FIGURE 16.18**
Layout of a non-integer multiple of a unit capacitor.

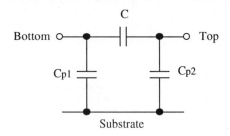

**FIGURE 16.19**
Equivalent circuit of an integrated capacitor.

common practice to realize the bigger capacitor as the parallel connection of a number of small capacitor elements. The area-perimeter ratio is automatically kept constant. For a noninteger capacitor ratio, a number of unit-sized capacitors and an additional capacitor between 1 and 2 are connected in parallel. The unit capacitor is normally square and the nonunit element is rectangular with an area-perimeter ratio equal to one of the unit capacitors. Figure 16.18 shows two capacitors with values of 4 and 3.2 and one common terminal.

Since the two plates of an integrated capacitor are very close to the substrate, a non-negligible parasitic capacitance exists. Therefore, the structure created is not a simple capacitor but the equivalent network shown in Fig. 16.19. The two parasitic capacitors between the two plates and the substrate determine the coupling between the substrate and the circuit. This coupling is dangerous because it can be the source of noise injection in critical points of the analog circuits. Therefore, it is a good practice to shield the plates of the capacitor from the substrate with a well biased at a "quiet" voltage (Fig. 16.20).

## 16.5 ANALOG CELL LAYOUT

Analog cells are the basic component of analog signal processors. They very often consist of both operational amplifiers and comparators. The complexity of analog

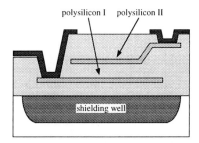

**FIGURE 16.20**
Cross-section of a poly-poly capacitor with a shielding well.

cells is limited, i.e., the number of transistors ranges from a few units to some tens of units. For blocks with limited complexity, the layout can be produced manually or with simple tool aids without a big risk of making errors. However, since the performance of these circuits depends on layout, their design is a very critical task that requires a certain amount of caution. It is important to follow all the suggestions that have been given previously. Some of these guidelines include:

- Use transistors with the same orientation (preferable all those in a cell).
- Minimize the source or the drain contact area by stacking transistors (for reducing the substrate parasitic capacitance).
- Respect the symmetries that exist in the electrical network as well as in the layout (to limit offset).
- Use low resistive paths (metal and not polysilicon) when a current needs to be carried (to avoid parasitic drop voltages).
- Shield critical nodes (to avoid undesired noise injection).

The first two recommendations find a practical application in the so-called full stacked technique [2]. The first step of this approach is the choice of the transistor aspect ratios in a way to make the layout more convenient. This practice is called layout-oriented design. In any analog cell, the size of a few critical transistors influences the performance. For the other transistors, the sizes are not very critical. Therefore, it is possible to change, within limits, the size of these non-determining transistors to achieve a good layout. Namely, it is possible to enlarge or to diminish the width of a transistor to stack it with other transistors of the cell in a full stacked arrangement. In this case, we achieve the benefits of reducing the parasitics. Moreover, it is easier to achieve matching and to respect the electrical symmetries.

Consider, for example, the two-stage transconductance operational amplifier (OTA) shown in Fig. 16.21. The transistors of the input differential pair should match, therefore their layout should be interdigitized (or possibly arranged in a common centroid fashion). The p-channel transistors are $M_3$, $M_4$,

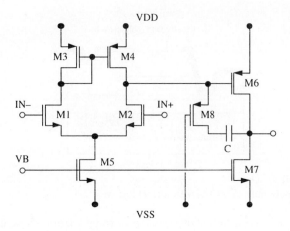

**FIGURE 16.21**
Two-stage OTA.

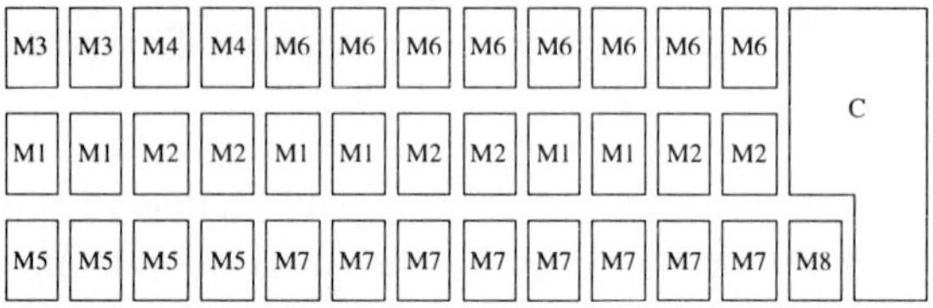

| M3 | M3 | M4 | M4 | M6 | M6 | M6 | M6 | M6 | M6 | M6 | M6 | |
|----|----|----|----|----|----|----|----|----|----|----|----|---|
| M1 | M1 | M2 | M2 | M1 | M1 | M2 | M2 | M1 | M1 | M2 | M2 | C |
| M5 | M5 | M5 | M5 | M7 | M7 | M7 | M7 | M7 | M7 | M7 | M7 | M8 |

**FIGURE 16.22**
Placement of transistor elements in a stacked fashion.

and $M_6$. Transistors $M_3$ and $M_4$ must match and $M_6$ must be wider by a given ratio to obtain a symmetrical slew rate. This last condition is not very important as it depends on the output capacitive load whose value is not completely controlled. Therefore, if the aspect ratio, $W/L$, of $M_6$ is to within four times that of $M_3$, the three transistors can be placed on the same stack as shown in the schematic topology of Fig. 16.22. Zero systematic offset together with this choice of $W/L$ of $M_6$ require that the aspect ratio of $M_7$ be equal to twice that of $M_5$. Therefore, it is possible to arrange $M_5$ and $M_7$ in the same stack. Thus, the layout of the cell is that of an interconnection of three stacks, two made by n-channel transistors and one by p-channel transistors.

Once the placement is defined, it is necessary to describe the routing. The routing for the circuit in Fig. 16.21 is shown in Fig. 16.23. Only the connections to the gates are made by polysilicon, all the other connections are made of first-level metal. If the technology provides two levels of metals, the second metal layer is available for higher level interconnections. A layout with the full stack technique is both compact and regular. The cell is rectangular, thus permitting easy use in an analog system. The $V_{DD}$ and $V_{SS}$ supply lines run parallel and cross the cell making a direct biasing connection possible when cells of the same

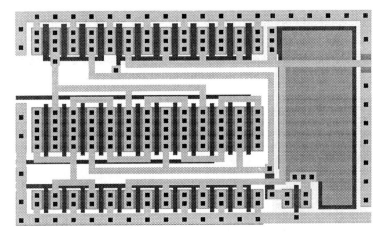

**FIGURE 16.23**
Layout of the OTA of Fig. 16.21.

| D | M3 | M4 | M4 | M3 | D | M6 | M6 | M6 | M6 | M6 | M6 | M6 | M6 | |
|---|----|----|----|----|---|----|----|----|----|----|----|----|----|---|
| M1 | M1 | M2 | M2 | M1 | M1 | M2 | M2 | M1 | M1 | M2 | M2 | C | | |
| D | M5 | M5 | M5 | M5 | M7 | M7 | M7 | M7 | M7 | M7 | M7 | M7 | D | M8 |

**FIGURE 16.24**
Use of dummy transistors in the placement of transistor elements (OTA of Fig. 16.21).

height are placed side by side. The pole splitting compensation is achieved by the transistor $M_8$ and the capacitor made by a poly1-poly2 structure at one side of the cell. To limit the coupling with the substrate, the layout uses a well under the capacitor.

As previously pointed out, the boundary dependent etching affects the width of transistors. Therefore, the width of the terminal elements of stacked structures is slightly smaller than the width of the internal elements and this may produce a systematic offset in the circuit. Dummy transistors allow us to overcome this problem. The biasing of the gate of the dummy elements must ensure that they are switched off, that is the p-channel gate is connected to $V_{DD}$ and the n-channel gate to $V_{SS}$. Figures 16.24 and 16.25 show the schematic topology and the layout of the two-stage transconductance amplifier of Fig. 16.21 with the addition of dummy transistors. It is worth noting that the use of dummy transistors not only helps in solving the problem of boundary dependent etching but also furnishes an additional degree of freedom. Dummy transistors placed in between elements of any given stack can provide insulation.

Figures 16.26 and 16.27 show additional examples of the full stacked layout of analog cells (a mirrored cascode OTA and a folded cascode OTA). The figures show the splitting of transistors into a given number of parts to achieve a full

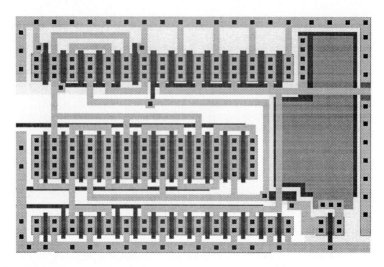

**FIGURE 16.25**
Layout of the solution given in Fig. 16.24.

stack arrangement and the required routing. The resulting layouts respect the symmetries, moreover, they are compact and regular.

Switches are frequently used in analog systems, which are made by one or a pair of complementary MOS transistors. Typically, switches are opened and closed at a relatively high frequency (hundreds of kHz or MHz) when they are used in switched capacitor circuits or data converters. They operate in a static fashion when used for achieving calibration or programming of analog functions. In the latter case, the layout of the transistors used is not so important. By contrast, in the former case, it is necessary to pay extreme attention since digital signals drive the transistors' gates while the sources and the drains are connected to analog nodes. This facilitates the noisy interaction of the digital and analog sections of the circuit.

This consideration would suggest the use of minimum area transistors whenever possible. This minimizes the parasitic capacitance between the substrate and the drain and source connections, therefore, limiting the coupling through the substrate of the analog and digital parts. Moreover, the crossing of any lines carrying analog signals with lines carrying the digital controls must be avoided. Unfortunately, in some cases this is not possible to achieve. For example, in the simple circuit in Fig. 16.28, given that the switch is constructed of complementary transistors, the crossing of the lines driving the p-channel transistors from the input node cannot be avoided. However, this configuration limits the coupling to when the switches are on. The layout of Figs. 16.29(a) and (b) shows additional suggestions for reducing noisy interaction [11]. A substrate bias separates the switches and the bus carrying the digital signals defines a protective guard ring. Moreover, the transistor that is made in the well faces the analog

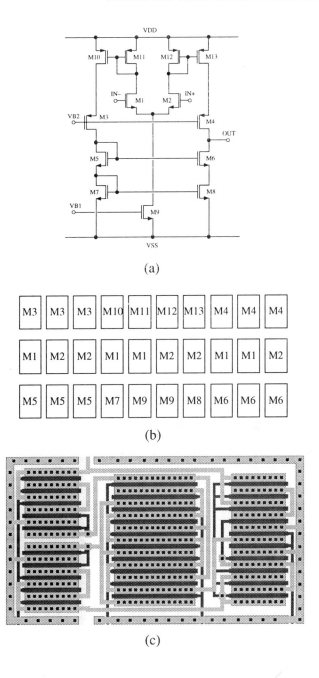

**FIGURE 16.26**
a) Mirrored cascode OTA, b) transistor placement, and c) layout (rotated 90 degrees).

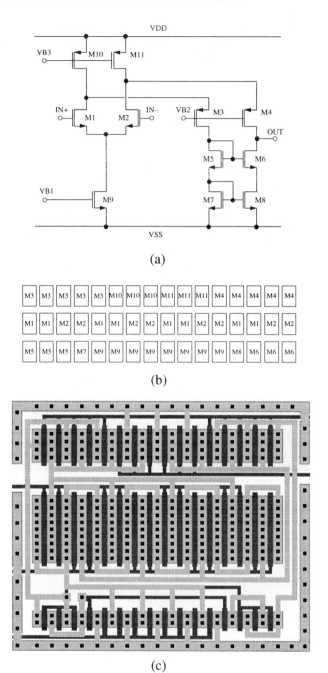

**FIGURE 16.27**
a) Folded cascode OTA, b) transistor placement, and c) layout.

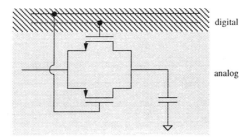

digital

analog

**FIGURE 16.28**
Schematic of a simple sample and hold.

side. This determines a sort of self-protective barrier (created by the well) which limits noise transfer.

### 16.5.1   Analog Floor Plan

Basic analog cells are connected to each other and possibly to passive components to implement a given processing function. An active filter, for example, needs to use a number of operational amplifiers usually equal to the order of the filter. Moreover, it uses capacitors and resistors (or transistors) for a continuous-time operation, or it uses switches for a switched capacitor (sampled data) operation. To facilitate the layout at a higher level, the analog cells should have a rectangular shape with the input and output terminals placed in their proper position. Moreover, the biasing should cross the cell so that biasing busses can be achieved more easily. Figure 16.30 shows the layout of the OTA of Fig. 16.26(c) modified according to these suggestions. Figure 16.31 shows a significant part of a typical analog floor plan. The blocks are placed side by side with the input and output terminals facing the same side. These terminals can be connected to a capacitor array, as shown in the floor plan of a single-ended switched capacitor filter of Fig. 16.32. The capacitor array is made of a parallel connection of unit-sized and some nonunit capacitors as previously discussed. A well under the array allows us to protect the circuit against the injection of noise from the substrate. A digital bus controls the switches, which are placed as far as possible from the operational amplifiers. In addition, a substrate bias and well separates the capacitor array from the switches.

Figure 16.33 shows the floor plan of a fully-differential switched-capacitor filter. The operational amplifiers occupy the middle of this layout. The capacitor arrays surround the two sides of the op amps while the switches are at the extreme ends. This forms a symmetrical arrangement which is required to obtain the electrical symmetry. Moreover, the bus used to carry the digital signals forms a c-like connection to drive the switches without crossing the analog signals.

The floor plan of data converters also follow these guidelines. However, since the number of operational amplifiers is in general limited, much more attention

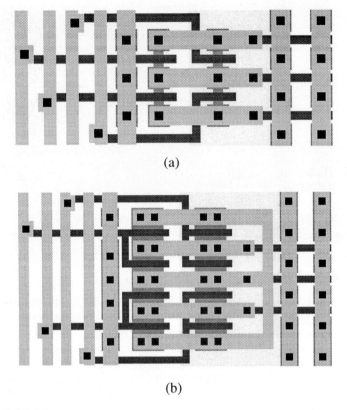

(a)

(b)

**FIGURE 16.29**
Layout of a toggle switch with a) minimum area transistors and b) wide transistors.

should be paid to the passive components and to the switches.

D/A converters make use of comparators, whose number depends on the specific architecture used. Algorithmic or successive approximation converters use only one comparator; whereas in contrast, flash architectures use a large number of comparators. Since a comparator is placed at the boundary between the analog and the digital part, the floor plan which includes comparators must follow the same procedures as used for circuits which have switches. That is, techniques which reduce the noisy interaction between the analog and digital sections.

The layout of an 8-bit two-step flash is shown in Fig. 16.34. The circuit is based on course and fine division of the reference voltage by a given resistive network [12]. Two banks of comparators sit on the two sides of the resistive network. A protection ring runs around the resistive network. However, after the evaluation of the most significant bits (MSBs), a fine division of the reference voltage must be selected, which requires the use of digital decoding signals. Therefore, they can corrupt the reference voltages used for the MSB and the LSB determination. A protective well under the poly resistors and a metal-1

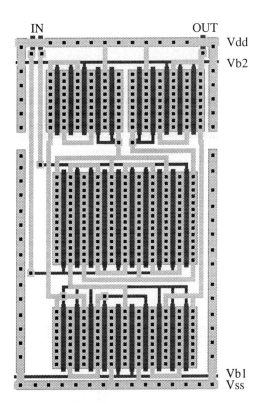

**FIGURE 16.30**
Layout of an OTA as a cell for side-to-side use.

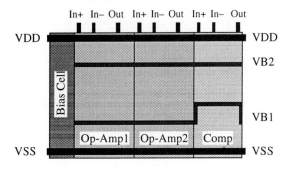

**FIGURE 16.31**
Floor plan including bias cell, op amps, and comparator.

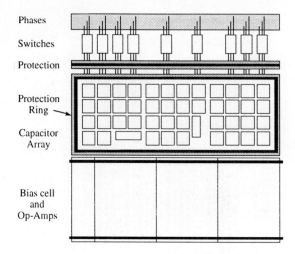

**FIGURE 16.32**
Typical floor plan of a single-ended SC filter.

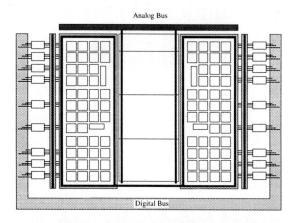

**FIGURE 16.33**
Typical floor plan of a fully-differential SC filter.

shield placed at the metal-2 logic crossings reduce this effect (Fig. 16.35).

## 16.6 MIXED ANALOG-DIGITAL LAYOUT

Modern technologies allow us to make very complex circuits which include both analog and digital functions. One of the major design problems with these categories of circuits is the noisy interaction of the digital and analog parts. The main sources of this important limitation are:

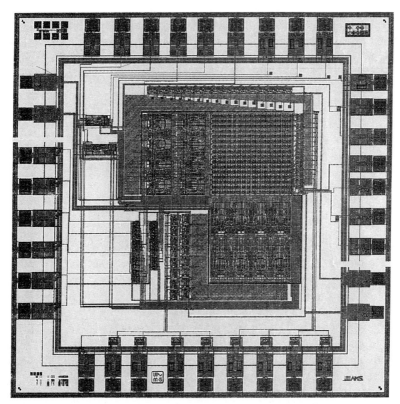

**FIGURE 16.34**
Layout of an 8-bit two-step flash ADC.

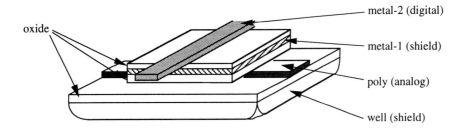

**FIGURE 16.35**
Double shielding of a poly resistor.

- capacitive coupling (direct or through the substrate),
- noise injection through the power supply lines.

We have already examined the first effect in a number of situations and we have also previously discussed adequate layout solutions to limit the capacitive coupling. Shielding with plates connected to a quiet voltage are the usual defense. The shielding layers can be either a well for defense against any noise coming from the substrate or metal to mask the noisy lines (or the noise coming from the plastic package).

The noise on the substrate is produced by two mechanisms: the capacitive coupling of logic nodes with the substrate, and the impact ionization effect responsible for substrate currents. A suitable control of parasitic capacitances of the substrate is enough to limit the first effect. Typically, drain contacts in wide transistors of logic gates are kept as small as possible. The area of the drain-to-substrate diodes is consequently small and coupling with the substrate is reduced. This approach also gives rise to a distributed resistance along the drain (Fig. 16.3(b)). Such a situation is not detrimental but rather creates a beneficial result, i.e., the speed of the transistor is reduced and the derivative of the output current in transition between logic states is also reduced. This, as we will discuss later, helps to control the voltage drop on bias connections.

In addition to these points, we should also discuss the effect of lateral coupling [3]. This occurs when two metal lines run parallel for a given path. This type of coupling occurs when an analog signal (for example a bias voltage) and a digital signal (or bus, for example, the phases of a switched capacitor filter) are run in close proximity. The usual means of reducing this problem is to place the lines carrying analog signals reasonably far away from the lines carrying digital signals. Moreover, a dummy line placed between the analog and digital lines biased at a "quiet" voltage achieves a horizontal shielding (Fig. 16.36). In addition, when possible, the use of different levels of metal allows us to reduce lateral coupling.

Another important source of noise are disturbances in the analog sections through the power supply lines. Designs with good power supply rejection ratio (PSRR) can mitigate this problem. Unfortunately, this is effective only at low frequencies. At high frequencies, due to unavoidable capacitive couplings, the PSRR is very poor and circuits are not capable of any self-defense against the noise coming through the supply lines. Therefore, the only solution is to control the noise generation at its source.

The noise which affects power supply lines is mainly due to a voltage drop across the supply connections. Depending on the layout, the connections from the external pin to the analog section of the chip is shown by the equivalent schematic of Fig. 16.37. The inductance, $L$, represents the inductive coupling between the chip and the padframe plus the inductance of the frame-pin connection. The resistor, $R_1$, represents the resistance of the common metal path and $R_2$ corresponds to the resistance of the analog path. Therefore, the voltage drop

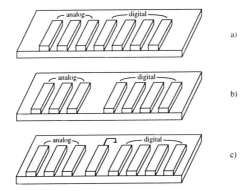

**FIGURE 16.36**
a) Mixed A-D bus, b) separation of analog and digital lines to reduce horizontal coupling, and c) dummy line for horizontal shielding.

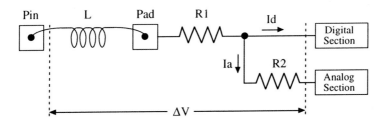

**FIGURE 16.37**
Equivalent circuit of the supply connection in mixed A-D circuit.

from the external pin and the analog section is given by:

$$\Delta V = R_1(I_a + I_d) + R_2 I_a + L\frac{d(I_a + I_d)}{dt} \tag{16.3}$$

In practical cases, the current $I_a$ of the analog section has a spectrum concentrated at low frequencies. In contrast, the spectrum of the current of the digital section, $I_d$, contains high frequency components as it is made of sharp pulses. They are almost synchronous with the digital clock and correspond to the charging and discharging of the output capacitance of logic gates.

The specific resistance of metal lines is approximately tens of $m\Omega$s per square, therefore, a typical value of the resistances $R_1$ and $R_2$ is a fraction of an ohm. The inductance of bonding connections contributes (as a rule of the thumb) 1 nH per millimeter of connections. Therefore, depending on the distance from the pad to the pin, $L$ can range from between a few units to tens of nHs.

The voltage drop expressed by Eq. (16.3) depends on the specific situation. However, when output drivers have to control large capacitive loads (typical values are around $100pF$), the last term in Eq. (16.3) is dominant. The peak

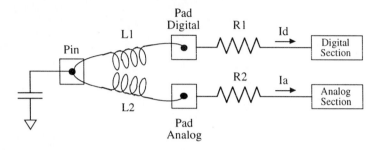

**FIGURE 16.38**
Filter separation of the analog and digital supplies with double bonding.

value of the derivative of the current can be several million amperes per second. Therefore, if the current in the analog or digital sections are in the range of mAs, the first of the two terms contribute mVs, while the third term is responsible for spikes whose peak values are tens or hundreds of mVs.

These spikes can be reduced by limiting the value of inductance or the time derivative of the digital current. The latter is achieved by careful design of the output buffers, the former by using pads with a very short connection to the pins and, possibly, by using multiple bondings.

It is also possible to diminish the noisy interaction of analog and digital power supplies by placing a proper filter in between. Figure 16.38 shows an example of this. The double bonding of the supply connections to the analog and the digital sections defines a suitable de-coupling filter together with the contribution of the external capacitive load. Figure 16.39 shows another solution which requires the use of an additional pin. Furthermore, the external parasitic inductance of the connection between the two pins improves the de-coupling operation. However, the solution in Fig. 16.39 cannot be used for both the supply terminals. The supply used to bias the substrate cannot adopt this method since any delay between the circuit and the substrate bias can start latch-up.

The design of floor plans of mixed analog-digital circuits is very important. The guidelines for this design step corresponds to the general philosophy already discussed for switched capacitor circuits. To avoid the noisy interaction of the analog and digital sections, a clear separation of the two parts should be used. Critical analog blocks (and the critical nodes of these parts) must be kept as far as possible from the digital part with an interposing shielding (horizontal and vertical) and guard rings to avoid any interaction and to catch noise and divert it to quiet points. The power supply connections must use a pad that is close to the frame to minimize parasitic inductance. Finally, the distribution of the power supplies to the digital and to the analog parts must be separated and then merged only at the pads.

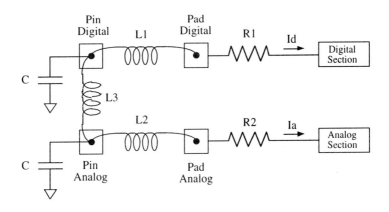

**FIGURE 16.39**
Filter separation of the analog and digital supplies by using two pins.

## 16.7   CONCLUSION

This chapter has considered the problem of the layout of analog and mixed analog-digital circuits. Layout techniques of analog components and of basic cells have been extensively discussed. A number of practical examples have shown the variety of problems that occur in purely analog and, much more frequently, in mixed A-D solutions. Together with the identification of the problems, a number of practical hints and suggestions for their solution have been given. Some of the given suggestions may appear trivial, but, if looked at closely, the performance of the circuits can be significantly improved.

## REFERENCES

[1] K. C. Hsieh, P. R. Gray, D. Senderowicz, D. G. Messerschmitt, "A low noise chopper stabilized differential switched capacitor filtering technique," *IEEE J. Solid-State Circuits,* vol. SC-16, pp. 708-715, December 1981.

[2] U. Gatti, F. Maloberti, V. Liberali, "Full stacked layout of analogue cells," *Proc. IEEE Intl. Symp. Circuits and Syst.,* 1989, pp. 1123-1126.

[3] P. O'Leary, *Analogue-Digital ASICs,* Editors R. S. Soin, F. Maloberti and J. Franca, Peter Peregrinus Ltd., 1991. Chapter 10, "Practical Aspects of Mixed Analogue and Digital Design."

[4] S. M. Sze, *VLSI Technology,* McGraw Hill, New York 1983.

[5] P. R. Gray and R. G. Meyer, *Analysis and Design of Analog Integrated Circuits - 2nd Ed.,* J. Wiley & Sons, New York 1984.

[6] A. B. Grebene, *Bipolar and MOS Analog Integrated Circuit Design,* J. Wiley & Sons, New York 1984.

[7] D. J. Allstot and W. C. Black, "Technological design considerations for monolithic MOS switched capacitor filtering systems," *Proc. of IEEE,* vol. 71, pp.967-986, August 1983.

[8] Austria Mikro Systeme, "AMS Design Rules CMOS Manual," 1992.

[9] J. L. McCreary, "Matching properties, voltage and temperature dependence of MOS capacitors," *IEEE J. Solid-State Circuits,* vol. SC-16, pp. 608-616, December 1981.

[10]  J. L. McCreary and, P. R. Gray, "All-MOS charge redistribution analog-to-digital conversion techniques," *IEEE J. Solid-State Circuits,* vol. SC-10, pp. 371- 379, 1975.

[11]  R. Gregorian and G. Temes, *Analog MOS Integrated Circuits,* J. Wiley & Sons, New York 1986.

[12]  M. J. M. Pelgrom, "A 10-b 50-MHz CMOS D/A converter with 75-W buffer," *IEEE J. Solid-State Circuits,* vol. SC-25, pp. 1347-1352, December 1990.

[13]  D.K. Su, J.J. Loinaz, S. Masui and B.A. Wooley, "Experimental results and modeling techniques for substrate noise in mixed-signal integrated circuits," *IEEE J. Solid-State Circuits,* vol. SC-28, pp. 420-430, April 1993.

# APPENDIX
# A

## SPICE MODEL PARAMETERS

This appendix contains model parameter sets for both 2 $\mu$ CMOS (level 2) and bipolar (Gummel-Poon) transistors which may be used to perform the simulations suggested in the homework problems, to reinforce the understanding of circuits discussed throughout the book, or to study nonideal and second-order effects that are often difficult to analyze by hand. It also relates the SPICE notation for each parameter to notation used throughout the book (see Tabs. A.1 and A.2).

```
*
* NMOS Model Parameters
*
.MODEL nmod NMOS LEVEL=2 LD=0.225112U TOX=405.000008E-10
+ NSUB=2.256420E+16 VTO=0.972134 KP=4.954000E-05 GAMMA=1.0151
+ PHI=0.6 UO=581 UEXP=0.217189 UCRIT=115146
+ DELTA=1.36044 VMAX=68535.3 XJ=0.250000U LAMBDA=2.734263E-02
+ NFS=2.859612E+12 NEFF=1 NSS=1.000000E+10 TPG=1.000000
+ RSH=27.280000 CGDO=2.879052E-10 CGSO=2.879052E-10
+ CGBO=3.840453E-10 CJ=4.108700E-04 MJ=0.465074
+ CJSW=4.837600E-10 MJSW=0.351006 PB=0.800000
* Weff = Wdrawn - $\delta W$
* The suggested $\delta W$ is 0.16 um
```

```
*
* PMOS Model Parameters
*
.MODEL pmod PMOS LEVEL=2 LD=0.177433U TOX=405.000008E-10
+ NSUB=3.956783E+15 VTO=-0.747971 KP=2.549000E-05 GAMMA=0.4251
+ PHI=0.6 UO=299 UEXP=0.193338 UCRIT=5462.67
+ DELTA=0.912857 VMAX=29720.9 XJ=0.250000U LAMBDA=5.812003E-
02
+ NFS=1.000000E+11 NEFF=1.001 NSS=1.000000E+10 TPG=-1.000000
+ RSH=107.400000 CGDO=2.269265E-10 CGSO=2.269265E-10
+ CGBO=3.471611E-10 CJ=1.893400E-04 MJ=0.439638
+ CJSW=2.264000E-10 MJSW=0.207285 PB=0.700000
* Weff = Wdrawn - $\delta W$
* The suggested $\delta W$ is 0.01 um


*
*Vertical NPN Model Parameters
*
.MODEL BN2B4 NPN
+ BF=160 IS=1.5-16 NF=1.0030 NE=1.3199 VAF=90 IKF=6.690E-02
+ ISE=2E-16 RE=11 RC=150.00 RB=300 RBM=4.96 ISC=2E-14 TF=40E-
12
+ NC=1.0382 +CJE=0.035E-12 MJE=0.5050 VJE=0.85 CJC=0.025E-12
+ MJC=0.4990 VJC=0.80 CJS=0.2E-12 MJS=0.2033 VJS=0.70


*
*Vertical PNP Model Parameters
*
.MODEL BPV2B4 PNP
+ BF=55 IS=1.5E-16 NF=1.0030 NE=1.3199 VAF=30 IKF=6.690E-02
+ ISE=2E-16 RE=15 RC=300 RB=400 RBM=4.96 ISC=2E-16 TF=100E-12
+ NC=1.0382 CJE=0.050E-12 MJE=0.5050 VJE=0.85 CJC=0.045E-12
+ MJC=0.4990 VJC=0.80 CJS=0.2E-12 MJS=0.2033 VJS=0.70


*
*Lateral PNP Model Parameters
*
.MODEL BPLAT PNP
+ BF=30 IS=2E-15 NF=1.05 NE=1.35 VAF=40 IKF=8.45E-02
+ ISE=2.5E-16 RE=20 RC=120 RB=200 RBM=15 ISC=2.5E-16 TF=30E-
9
+ NC=1.5 CJE=0.25E-12 MJE=0.45 VJE=0.80 CJC=2.0E-12
+ MJC=0.445 VJC=0.78 CJS=2.5E-12 MJS=0.15 VJS=0.68
```

**TABLE A.1**
**MOS Transistor Parameters**

| SPICE Symbol | Text Symbol | Description |
|---|---|---|
| VTO | $V_{T0}$ | Zero-Bias Threshold Voltage |
| KP | $\mu C_{OX}, K$ | Transconductance Parameter |
| GAMMA | $\gamma = \sqrt{2q\epsilon N_A}/C_{OX}$ | Body Effect Parameter |
| PHI | $2\phi_f$ | Surface Inversion Potential |
| LAMBDA | $\lambda$ | Channel-Length Modulation Parameter |
| CGSO | $C_{ovl}$ | G-S Overlap Capacitance |
| CGDO | $C_{ovl}$ | G-D Overlap Capacitance |
| CJ | $C_{j0}$ | S/D Junction Capacitance |
| MJ | $n$ | Junction Capacitance Exponent |
| CJSW | $C_{jsw0}$ | S/D Sidewall Capacitance |
| MJSW | $n$ | Sidewall Capacitance Exponent |
| PB | - | Junction Built-In Potential |
| TOX | $t_{OX}$ | Oxide Thickness |
| NSUB | $N_A, N_{sub}$ | Substrate Doping Density |
| NSS | - | Surface State Density |
| XJ | $X_j$ | S/D Junction Depth |
| LD | $L_D$ | S/D Lateral Diffusion |
| UO | $\mu$ | Surface Mobility |
| UEXP | - | Mobility Degradation Exponent |
| UCRIT | - | Mobility Degradation Critical Field |
| DELTA | - | Narrow Channel Effect on $V_T$ |
| VMAX | - | Maximum Carrier Drift Velocity |
| NFS | - | Fast Surface State Density |
| NEFF | - | Total Channel Charge Coefficient |
| RSH | - | S/D Diffusion Sheet Resistance |

**TABLE A.2**
**Bipolar Transistor Parameters**

| SPICE Symbol | Text Symbol | Description |
|---|---|---|
| IS | $I_S$ | Forward Saturation Current |
| BF | $\beta$ | Forward Current Gain |
| BR | - | Reverse Current Gain |
| VAF | $V_A$ | Early Voltage |
| RB | $r_b$ | Base Resistance |
| RE | $r_e$ | Emitter Resistance |
| RC | $r_c$ | Collector Resistance |
| TF | $\tau_f$ | Forward Transit Time |
| TR | - | Reverse Transit Time |
| CJE | $C_{je0}$ | B-E Junction Capacitance |
| VJE | - | B-E Built-In Potential |
| MJE | - | B-E Junction-Capacitance Exponent |
| CJC | $C_{jc0}$ | B-C Junction Capacitance |
| VJC | - | B-C Built-In Potential |
| MJC | - | B-C Junction-Capacitance Exponent |
| CJS | $C_{jcs0}$ | C-S Junction Capacitance |
| VJS | - | C-S Built-In Potential |
| MJS | - | C-S Junction-Capacitance Exponent |
| NF | - | Forward Current Emission Coefficient |
| NE | - | B-E Leakage Emission Coefficient |
| IKF | - | Corner for Forward Beta High Current Roll-Off |
| ISE | - | B-E Leakage Saturation Current |
| ISC | - | B-C Leakage Saturation Current |
| NC | - | B-C Leakage Emission Coefficient |

# Index